Metabolic Encephalopathy

David W. McCandless
Editor

Metabolic Encephalopathy

 Springer

Editor
David W. McCandless
The Chicago Medical School
Rosalind Franklin University of Medicine and Science
Department of Cell Biology & Anatomy
3333 Green Bay Road
North Chicago, IL 60064
USA
david.mccandless@rosalindfranklin.edu

ISBN: 978-1-4419-2712-5   e-ISBN: 978-0-387-79112-8
DOI: 10.1007/978-0-387-79112-8

This volume is dedicated to Jason, Michael, and James

Preface

The last several years have seen a burgeoning of technical advances in areas of medicine such as imaging and diagnosis, and this trend is sure to continue. These developments have special significance in areas such as internal medicine, neurology, and radiology. The further realization that early diagnosis of disorders which have a biochemical (metabolic) basis has the potential for rapid reversal/improvement, if not outright cure, argues strongly for increased awareness.

One example of reversible metabolic encephalopathy is that seen in thiamine deficiency. Both animal models and humans with "pure" thiamine deficiency develop highly specific neurological symptoms. These symptoms can be reversed completely, or dramatically improved, often within hours, by the administration of thiamine. These results are instrumental in leading to the concept of "metabolic encephalopathy", a disorder without structural brain changes. From a historical standpoint, there are increasing numbers of such disorders, which are amenable to successful treatment. Sadly, for example, in the case of kernicterus, managed health care has led to an increase in the number of cases due to the early release from hospitals of newborn infants after birth, even before the onset of jaundice.

The area of metabolic encephalopathies is unique in that animal models of disease closely mimic the symptoms seen in human disorders, allowing excellent correlative studies. Seizures are an example of this feature. Many animal models of experimentally induced seizures are available for study in mice and rats, and the neurochemical alterations before and after anticonvulsive therapy can be carefully studied. These more or less direct comparisons permit a more rapid application of results from animal studies to humans.

Given the above, the format for this book is that most chapter contributors have been asked to consider both animal and human studies and to integrate them into statements of mechanisms of biochemical alterations and treatments. The authors have written chapters dealing with the most commonly seen metabolic encephalopathies, and those in which the diagnosis and treatment have advanced and benefited from technological studies. In many cases, it will be seen that there is an underlying alteration in energy metabolism in specific brain regions, which can be documented in animal studies. As the deficit is reversed, the energy metabolites revert toward normal, and the symptoms lessen. Imaging studies in humans support these animal findings, and the rapidity with which this happens argues strongly for a metabolic lesion, not a structural one. Again, if treated early, structural changes can be minimized, or eliminated from consideration.

These results emphasize another very interesting aspect of metabolic encephalopathies: the diverse anatomical localization of cerebral effects. For example, bilirubin is highly selective in its localization, whereas thiamine deficiency has a completely different localization and is also highly specific. The reasons for this cerebral specificity are largely unclear, but some studies show vastly different metabolism between different brain regions. The cerebellum, for example, is biochemically different from adjacent areas.

In summary, this book on metabolic encephalopathies is meant to combine and correlate animal and human studies. It is hoped that increased awareness of the importance of early diagnosis and treatment of these disorders may result in a lowering of the incidence of structural changes, and morbidity. These disorders hold a special fascination for both basic scientists and clinical investigators because they are accessible, treatable, and there exist good animal models for study. Therefore, this book pulls together basic and clinical neuroscience issues in the treatment of specific metabolic encephalopathies.

This book would not have been possible without the participation and contributions of the many contributors, and I am grateful for their efforts. The editor wishes to acknowledge the expert secretarial and organizational skills of Mrs. Cristina I. Gonzalez and Mrs. Vilmary Friederichs, who have facilitated the production of this book.

We also wish to thank Ms. Ann Avouris of Springer Science and Business Media for her dedicated help in bringing this volume to completion.

Contents

Contributors

Allen I. Arieff
Department of Medicine, University of California School of Medicine,
San Francisco California, and Cedars-Sinai Medical Center, Los Angeles,
California, USA

Roland N. Auer
Departments of Pathology and Clinical Neuroscience, Faculty of Medicine,
University of Calgary, Calgary, Alberta, Canada

Ivan J. Boyer
DABT, Noblis (Mitretek), Falls Church, Virginia, USA

Roger F. Butterworth
Neuroscience Research Unit, Hopital Saint-Luc, University of Montreal,
Montreal, Quebec, Canada

Kate Chandler
Department of Veterinary Clinical Sciences, Royal Veterinary College,
Hatfield, UK

Pi-Shan Chang
Department of Clinical and Experimental Epilepsy, Institute of Neurology,
University College London, London, UK

John M. DeSesso
Center for Science and Technology, Noblis, Falls Church, Virginia; and
Department of Biochemistry and Molecular and Cellular Biology, Georgetown
University Medical Center, Washington, D.C., USA

Joseph DiCarlo
Pediatric Critical Care Medicine, Children's Hospital Los Angeles, Los Angeles,
CA 90027

Peter R. Dodd
School of Molecular and Microbial Sciences, University of Queensland,
Brisbane, Australia

Peter Ferenci
Department of Medicine III, Gastroenterology and Hepatology, Medical
University of Vienna, Vienna, Austria

Florian M. Gebhardt
School of Molecular and Microbial Sciences, University of Queensland,
Brisbane, Australia

Gary E. Gibson
Weill Medical College of Cornell University, Burke Medical Research Institute,
White Plains, New York, USA

Monisha Goyal
Department of Neurology, School of Medicine, Case Western Reserve University,
Cleveland, Ohio, USA

Brian H. Harvey
Division of Pharmacology, School of Pharmacy, North-West University,
Potchefstroom, South Africa

George I. Henderson
Department of Medicine, Division of Gastroenterology and Nutrition, University
of Texas Health Science Center, San Antonio, Texas, USA

Saravanan Karuppagounder
Weill Medical College of Cornell University, Burke Medical Research Institute,
White Plains, New York, USA

Smruta Koppaka
Department ofAnatomy, Case Western Reserve University, School of Medicine,
Cleveland, Ohio, USA

Maryam R. Kashi
Department of Medicine, Division of Gastroenterology and Nutrition, University
of Texas Health Science Center, San Antonio, Texas, USA

Joseph C. LaManna
Department of Anatomy, Case Western Reserve University, School of Medicine,
Cleveland, Ohio, USA

W. David Lust
Laboratory of Experimental Neurological Surgery, Department of Neurological
Surgery, Case Western Reserve University, Cleveland, Ohio, USA

David W. McCandless
Department of Cell Biology and Anatomy, The Chicago Medical School,
Rosalind Franklin University of Science and Medicine, North Chicago, Illinois,
USA

Jeffrey W. McCandless
NASA Ames Research Center, Moffett Field, California, USA

Michelle Puchowicz
Department of Anatomy, Case Western Reserve University, School of Medicine, Cleveland, Ohio, USA

Svetlana Pundik
Research and Neurology Service, Cleveland VA Medical Center, The Neurological Institute, University Hospitals of Cleveland, Case Medical Center, Case Western Reserve University, School of Medicine, Cleveland, Ohio, USA

Kottil W. Rammohan
Department of Neurology, The Ohio State University, Columbus, Ohio, USA

John W. Rumsey
Department of Biomolecular Sciences, Research Pavilion, University of Central Florida, Orlando, Florida, USA

Vivienne Ann Russell
Department of Human Biology, Faculty of Health Sciences, University of Cape Town, Observatory, South Africa

Steven Schenker
Department of Medicine, Division of Gastroenterology and Nutrition, University of Texas Health Science Center, San Antonio, Texas, USA

Kleopatra H. Schulpis
Inborn Errors of Metabolism Department, Institute of Child Health, Research Centre, "Aghia Sophia" Childrens Hospital, Athens, Greece

Jennifer Stewart
Department of Medicine, Division of Gastroenterology and Nutrition, University of Texas Health Science Center, San Antonio, Texas, USA

Jose I. Suarez
Vascular Neurology and Neurocritical Care, Department of Neurology and Neurosurgery, Baylor College of Medicine, Houston, Texas, USA

Stylianos Tsakiris
Department of Physiology, Medical School, University of Athens, PO Box 65257, Athens, Greece

Matthew Walker
Department of Clinical and Experimental Epliepsy, Institute of Neurology, University College London, London, UK

Karin Weissenborn
Department of Neurology, Medical School of Hannover, Hannover, Germany

Richard C. Wiggins
National Health and Environmental Effects Research Laboratory, Office of Research And Development, U.S. Environmental Protection Agency, Research Triangle, North Carolina, USA

Jennifer Zechel
Laboratory of Experimental Neurological Surgery, Department of Neurological Surgery, Case Western Reserve University, Cleveland, Ohio, USA

Chapter 1
Functional Anatomy of the Brain

John M. DeSesso

Introduction

The central nervous system comprises the brain and spinal cord which provide sensation, control of movement, emotion, aesthetics, reason and self-awareness. The tissue that makes up the central nervous system is highly differentiated and exceedingly ordered, yet plastic. The central nervous system is well protected throughout by a fluid-filled, tri-layered, connective tissue covering (the meninges) and various osseous claddings. The cranium provides a rigid armor for the brain whereas the vertebral column constitutes the flexible protection of the spinal cord. Because the focus of this volume is the study of various disease states that affect the functions of the brain, it is important to understand the normal relationship of the brain to its surrounding structures including its bony case and its connective tissue coverings, its blood supply, and its internal organization, as well as how the perturbation of the relationships among these structures can impact brain functions. It is the purpose of this chapter to present an overview of this information. More detailed anatomical information is readily available in textbooks of gross anatomy and neuroscience.

Protective Structures

The brain resides within the cranial cavity. The bony roof and sides of the cranial vault make up the calvaria, which is composed of frontal, temporal and parietal bones and a small portion of the occipital bone. The floor of the cranial vault is divided into three depressions or fossae: the anterior fossa extends from the region superior to the orbits and nasal cavity caudally as far as the posterior margin of the lesser wing of the sphenoid; the middle fossa occupies the region between the lesser wing of the sphenoid and the anterior border of the petrous portion of the temporal bone; and the posterior fossa, is underlain by the remainder of the temporal bones and the occipital bone.

Like all bones, those that make up the cranium are invested by a tissue lining, the periosteum. On the interior of the cranial vault, the periosteum is a specialized,

D.W. McCandless (ed.) *Metabolic Encephalopathy*,
doi: 10.1007/978-0-387-79112-8_1, © Springer Science + Business Media, LLC 2009

thickened tissue (dura mater) which is the outermost of the meninges that help to protect the brain. Within the cranial vault, the dura mater reflects off the walls of the cranium as horizontal or vertical septa that help to support or restrict the movement of the brain within the cranial vault. Each septum has a free margin. A vertical reflection is a falx; a horizontal reflection is a tentorium. The falces restrict the brain's lateral movement, as when one turns one's head too quickly. The falx located between the cerebral hemispheres is the falx cerebri; the smaller one between the cerebellar hemispheres is the falx cerebelli (Fig. 1.1). The tentoria support some regions of the brain and prevent compression of the structures below them. The most important of these is the tentorium cerebelli, which supports the occipital lobe where it overlies the cerebellum. The free margin of the tentorium

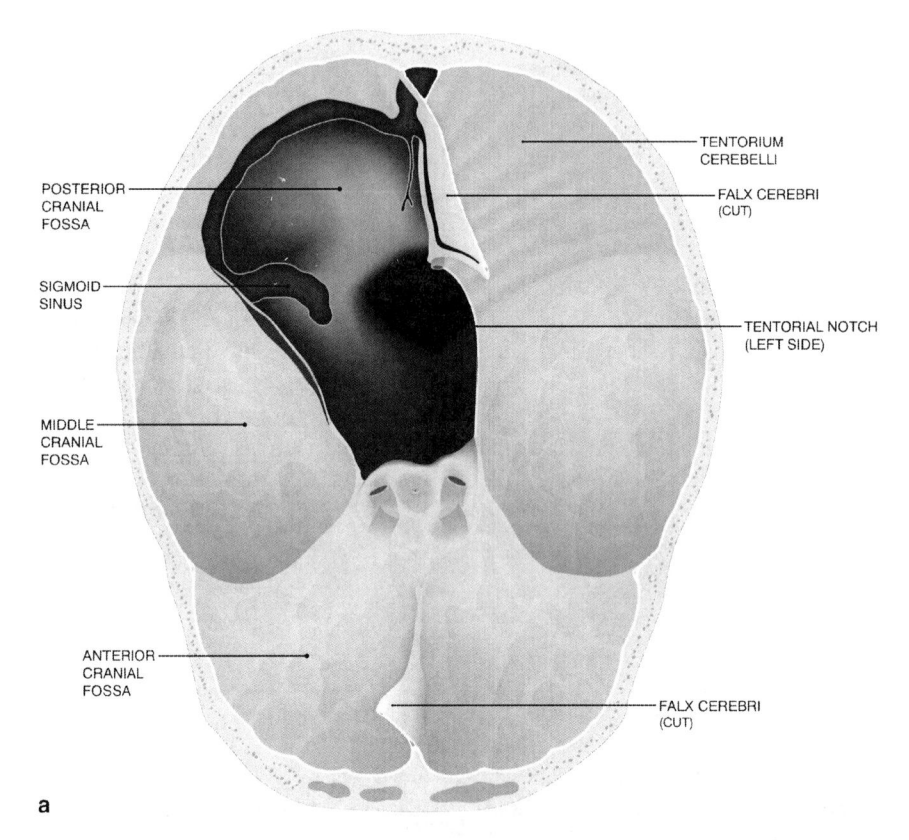

a

Fig. 1.1 (**a**) Interior of the cranial vault illustrating the major supporting and protective structures, including the cranial fossae. Reflections of dura mater off the calvaria form important structures that help support the weight and restrict the motion of the brain within the cranial vault. (**b**) Vertical reflections are falces and include the large falx cerebri between the two cerebral hemispheres and the smaller falx cerebelli (not shown) that separates the cerebellar hemispheres. The most significant horizontal reflection is the tentorium cerebelli which supports the occipital lobe of the brain and prevents it from crushing the underlying cerebellum, which resides in the posterior cranial fossa (redrawn after Drake et al., 2005) (*See also* Color Insert)

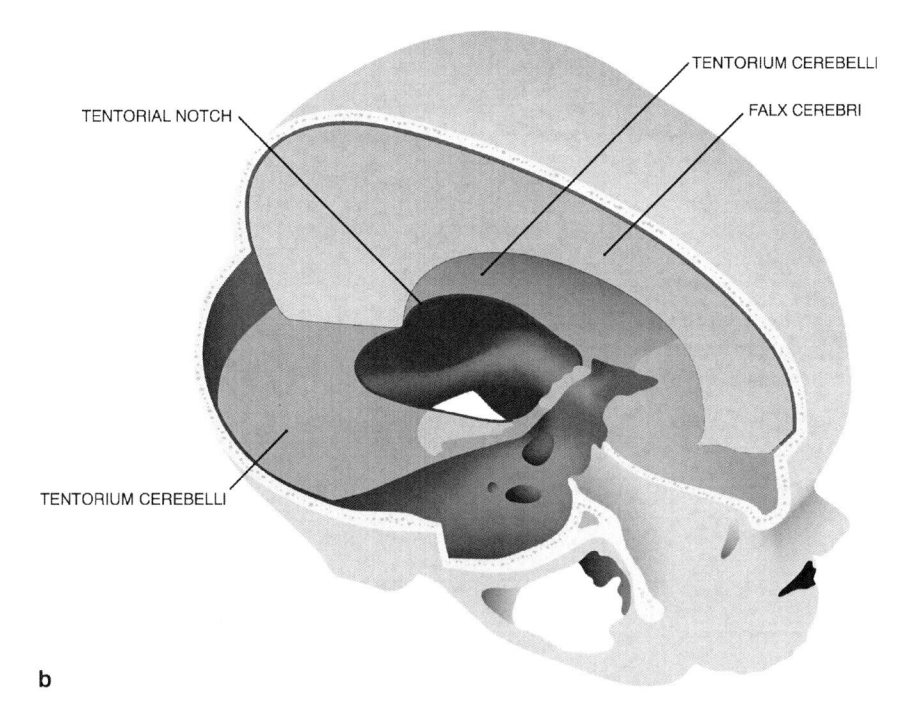

TENTORIUM CEREBELLI

FALX CEREBRI

TENTORIAL NOTCH

TENTORIUM CEREBELLI

b

Fig. 1.1 (continued)

cerebelli is a U-shaped opening (the tentorial notch) that allows the brainstem to connect to the rostral regions of the brain, including the thalamus and cerebral hemispheres.

Organization of the Central Nervous System

The divisions of the cranial central nervous system include the cerebral hemispheres, the diencephalon (thalamus and hypothalamus), the brainstem (midbrain, pons and medulla oblongata) and the cerebellum (Fig. 1.2). Each cerebral hemisphere occupies one half of the cranial vault and can be subdivided into four lobes (frontal, parietal, temporal, occipital), the insula and the limbic lobe. The first four lobes are named for the cranial bones that overlie them. With respect to the floor of the cranial cavity, the frontal lobes lie in the anterior cranial fossa; the brainstem and cerebellum occupy the posterior cranial fossa; the remaining structures are found either in the middle fossa or within the portion of the cranial vault above the tentorium cerebelli. The insula is covered by the temporal lobe and is not observable unless the temporal lobe is retracted. The limbic system is a continuous interior

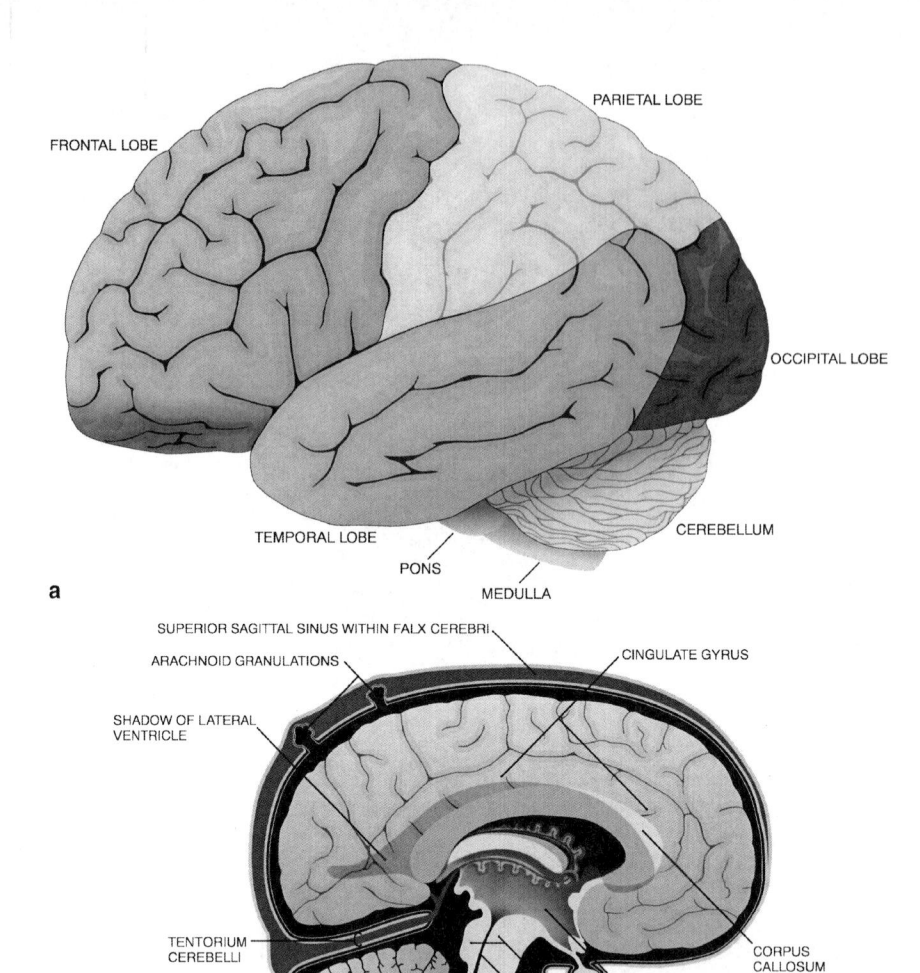

a

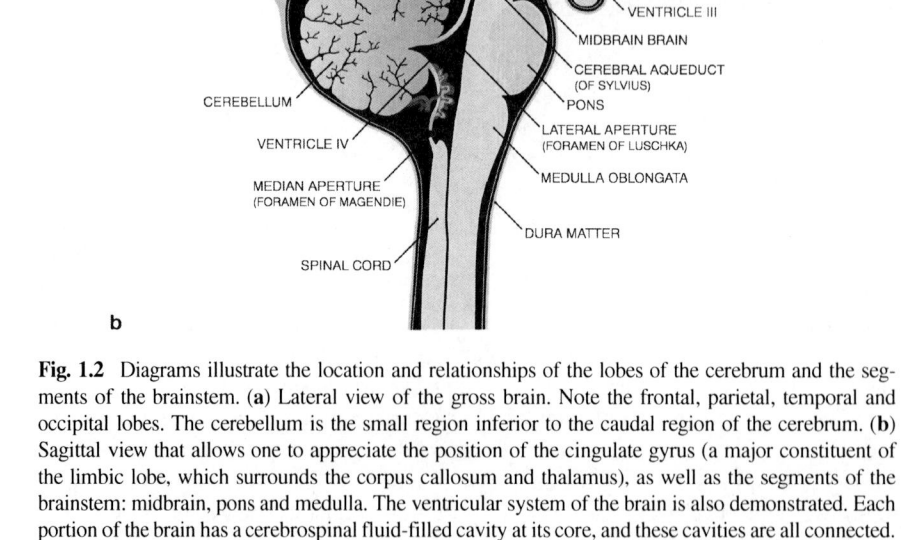

b

Fig. 1.2 Diagrams illustrate the location and relationships of the lobes of the cerebrum and the segments of the brainstem. (**a**) Lateral view of the gross brain. Note the frontal, parietal, temporal and occipital lobes. The cerebellum is the small region inferior to the caudal region of the cerebrum. (**b**) Sagittal view that allows one to appreciate the position of the cingulate gyrus (a major constituent of the limbic lobe, which surrounds the corpus callosum and thalamus), as well as the segments of the brainstem: midbrain, pons and medulla. The ventricular system of the brain is also demonstrated. Each portion of the brain has a cerebrospinal fluid-filled cavity at its core, and these cavities are all connected. Cerebrospinal fluid is produced by the choroid plexus within the ventricles and it escapes the ventricular system through foramina in the roof over the fourth ventricle at the pontomedullary junction (*See also* Color Insert)

structure that surrounds the rostral portions of the brainstem and diencephalon near the midline and is made up of portions of the frontal, parietal and temporal lobes.

The lobes of the cerebrum surround a stalk of nervous tissue that connects the spinal cord with the upper neural centers of the cerebrum. These midline structures include the thalamus and the brainstem. The thalamus protrudes into the middle cranial fossa, above the level of the clinoid processes of the sphenoid bone. It serves as a reciprocal gateway between the cerebral cortex and brainstem that conveys extensive sensorimotor and autonomic information.

Caudal to the thalamus are the midline structures of the brainstem: the midbrain, pons, and medulla oblongata. Dorsal to the pons, but inferior to the tentorium cerebelli, is the cerebellum which is the prominent structure in the inferior cranial fossa. The brainstem as a whole is concerned with somatosensory information from the neck and head as well as the specialized senses of taste, audition and balance. It acts as a conduit for ascending and descending pathways of motor and sensory information between the cortical regions of the brain and the body. In addition, the brainstem is responsible for mediating levels of consciousness and arousal.

The innermost region of the brain consists of a series of connected cerebrospinal fluid-filled cavities, the ventricles (Fig. 1.2). Each cerebral hemisphere contains a lateral ventricle; the midline cavity associated with the thalamus is the third ventricle, which communicates with the lateral ventricles by means of the foramina of Munro. The continuation of the ventricular system caudally through the midbrain is by means of a narrow cerebral aqueduct (of Sylvius), which empties into the fourth ventricle, lying between the ventrally placed pons and medulla oblongata and the dorsal cerebellum. The ventricular system continues into the spinal cord as the central canal. The liquid within the ventricular system is the cerebrospinal fluid, which is made by the highly vascular choroid plexus found within the four ventricles, and which fills the space between the arachnoid and pia maters that surround the central nervous system. The communication to this subarachnoid space occurs through a set of openings located in the roof of the fourth ventricle: the two laterally placed foramina of Luschka and the midline foramen of Magendie. Cerebrospinal fluid flows from its origin at the intraventricular choroid plexus, caudally towards the fourth ventricle where some of it exits the ventricular system to fill the subarachnoid space surrounding the brain and spinal cord. The fluid that invests the brain comes into contact with specialized tissue associated with the venous drainage of the superior aspects of the cerebrum (arachnoid granulations) where, under normal conditions, it enters the venous system, thereby preventing overfilling and distension of the ventricular and subarachnoid systems.

The central nervous system is a tubular structure that is composed of a relatively thick, but highly organized, layer of neuron cell bodies with their attendant cellular processes and numerous supporting glial cells. Populations of neuronal cell bodies are collocated in regions of the central nervous system that look beige (gray matter) when they are observed in the fresh condition. Areas of the central nervous system that contains large amounts of myelinated axons are vanilla-colored or light pink in the fresh condition and are termed "white matter." In the spinal cord, brain stem and thalamus, white matter is found on the outer surface and gray matter is located deep in the walls of the neural tube. In the cerebrum and cerebellum, gray matter is located on the surface and the white matter is deep. There are important substructures associated with both the white and gray matters. Bundles of axon fibres that traverse from one region of the central nervous system to another are tracts (also variously called fasciculi, lemnisci, radiations, or commissures — when they interconnect the hemispheres), which are often named for the areas of the central nervous system that they connect. Discrete clusters of neuronal cell bodies are nuclei, most of which have distinct names. For the most part, the neurons in a given nucleus share the same modalities and produce the same neurotransmitter substance.

When viewed in cross-section, the brainstem can be divided into geographically distinct regions, which are identified by means of the relationship of the walls (including both white and gray matter), to the lumen (Fig. 1.3). Thus, the lumen forms the center of the tube and the entire brainstem wall that is dorsal (or superior) to the lumen forms the tectum (from the Latin word for roof). In humans, the tectum remains as a distinct structure only in the midbrain; in the pons and medulla, the large cerebellum arises from the region that would have been the tectum. The territory of the wall that surrounds the rest of the lumen is termed the tegmentum (from the Latin word for covering). The tegmentum does not include all portions of the wall inferior to the lumen, as there are segments of the brainstem (e.g., basilar pons and cerebral peduncles) that have well-developed areas with specific functions and are considered to be distinct from the tegmentum. Throughout the brainstem is a region of gray matter that forms a neuronal mass extending from the rostral spinal cord throughout the brainstem, and into the thalamus and hypothalamus. This ill-defined structure is the reticular formation: a collection of large and intermediate-sized neurons that are loosely arranged into nuclei which form columns that run

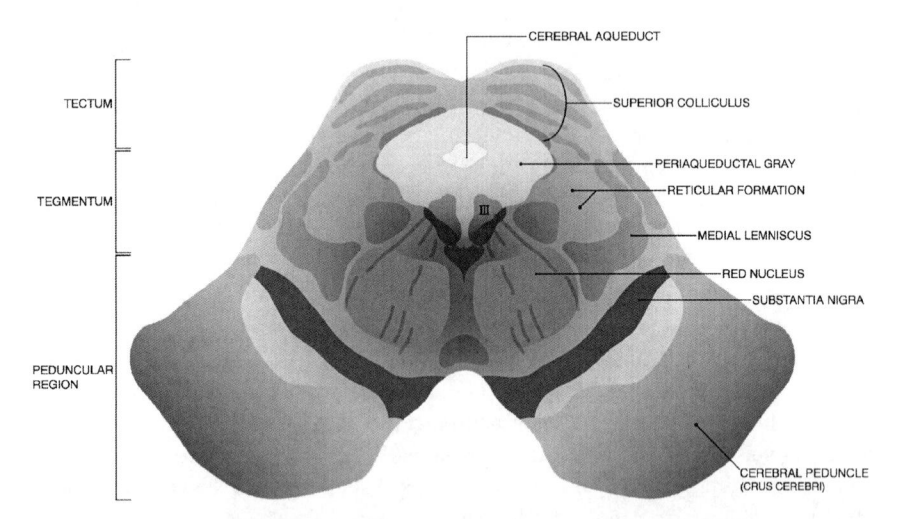

Fig. 1.3 A cross-section through the midbrain that illustrates the tectum, tegmentum and peduncular regions. Within the wall of the midbrain, note the positions of the periaqueductal gray matter, red nucleus, medial lemniscus, substantia nigra, and territory occupied by the reticular formation. (*III* nucleus of cranial nerve III, the oculomotor nerve) (*See also* Color Insert)

parallel to the long axis of the brainstem. The columns are located in the midline (raphe nuclei), just lateral to the midline (the paramedian reticular nuclei), and farther laterally (the lateral reticular nuclei).

The midbrain is of particular importance with regard to auditory and visual reflexes, regulation of arousal, and as a conduit between higher and lower centers of the central nervous system. Consequently, the anatomy of a cross-section of this relatively uncomplicated part of the brainstem will be discussed (Fig. 1.4). The tectum

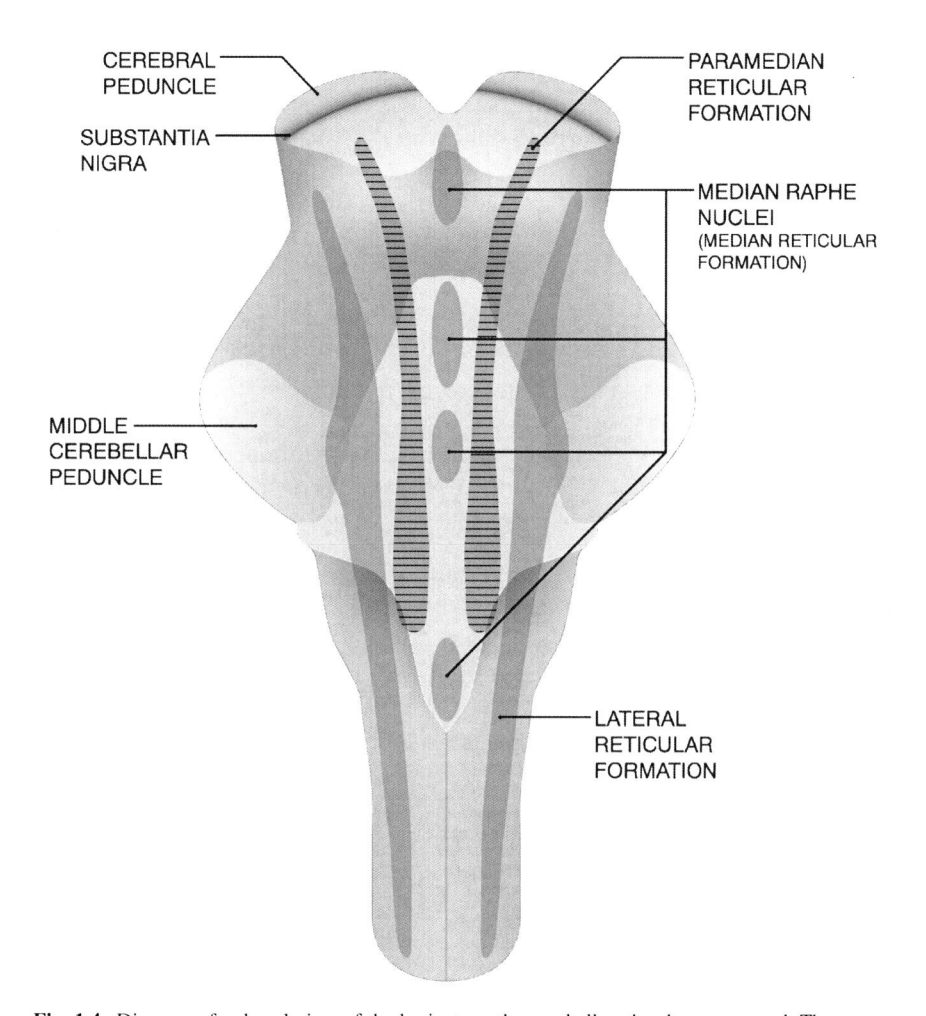

Fig. 1.4 Diagram of a dorsal view of the brainstem; the cerebellum has been removed. The extent of the reticular formation within the brainstem is illustrated. The reticular formation is a polysynaptic network that consists of three regions: a series of midline raphe nuclei (the median reticular formation, which is the site of origin of the major serotonergic pathways in the nervous system); this is flanked bilaterally by the paramedian reticular formation (an efferent system of magnocellular neurons with ascending and descending projections); and farthest from the midline, the lateral reticular formation, consisting of parvocellular neurons that project transversely (*See also* Color Insert)

comprises two pairs of grossly observable elevations (colliculi). The superior colliculi mediate visual reflexes and coordination of head movements and eyes towards a visual stimulus, including those associated with the saccadic movements involved in reading. The inferior colliculi coordinate analogous reflex movements of the head and ears associated with auditory stimuli. The periaqueductal gray matter, which is a descending pathway important in modulating pain, surrounds the lumen of the cerebral aqueduct. The important structures in the tegmentum include the nucleus of the oculomotor nerve (CN III) near the midline; the prominent, orb-shaped red nucleus, which is part of the extrapyramidal motor pathway that controls large muscles of the arm and shoulder; the slit-like substantia nigra which is an essential part of the dopaminergic system (involved in reward and addiction) and whose degeneration underlies Parkinson's disease; the medial lemniscus, which carries proprioceptive and touch information from the gracile and cuneate nuclei to centers in the thalamus; and the lateral reticular formation which plays a pivotal role in stimulating and maintaining arousal of the upper centers of the central nervous system.

Vascular Supply

All arterial blood supply to the brain and brainstem traverses branches of either the internal carotid or vertebral arteries (Fig. 1.5). These arteries, in turn, receive blood from major branches of the arch of the aorta: the internal carotid is a major division of the common carotid artery while the vertebral is derived from the subclavian artery. The blood supply to the brainstem, cerebellum, occipital lobe and the inferior aspect of the temporal lobe is derived from branches of the vertebral system. The frontal, parietal, upper 75% of the temporal lobes and the insular cortex receive their blood supply from the middle and anterior arteries, both of which are branches of the internal carotid system. Although the vertebral and carotid systems supply distinct areas of the brain and brainstem, the two systems are structurally joined by means of a multi-sided system of interconnected vessels (the circle of Willis) located at the base of the brain where they surround the stalk of the pituitary gland, the optic chiasm and optic tracts, and the hypothalamus. The basilar artery (derived from the fused vertebral arteries) terminates as the posterior cerebral arteries. The internal carotid arteries contribute the anterior and middle cerebral arteries and the posterior communicating arteries. The anastomosis is completed by the short anterior communicating artery between the two anterior cerebrals and the paired posterior communicating arteries between the posterior cerebral arteries and the middle cerebral artery. The latter arteries connect the vertebral and carotid blood supplies. Interestingly, the diameters of the arteries vary considerably, especially in the case of the posterior communicating arteries, which frequently may be extremely small on one side or even absent. Consequently, the anastomosis is often only a potential channel and tracer studies in adults have shown that the two blood streams (vertebral and internal carotid) do not mix.

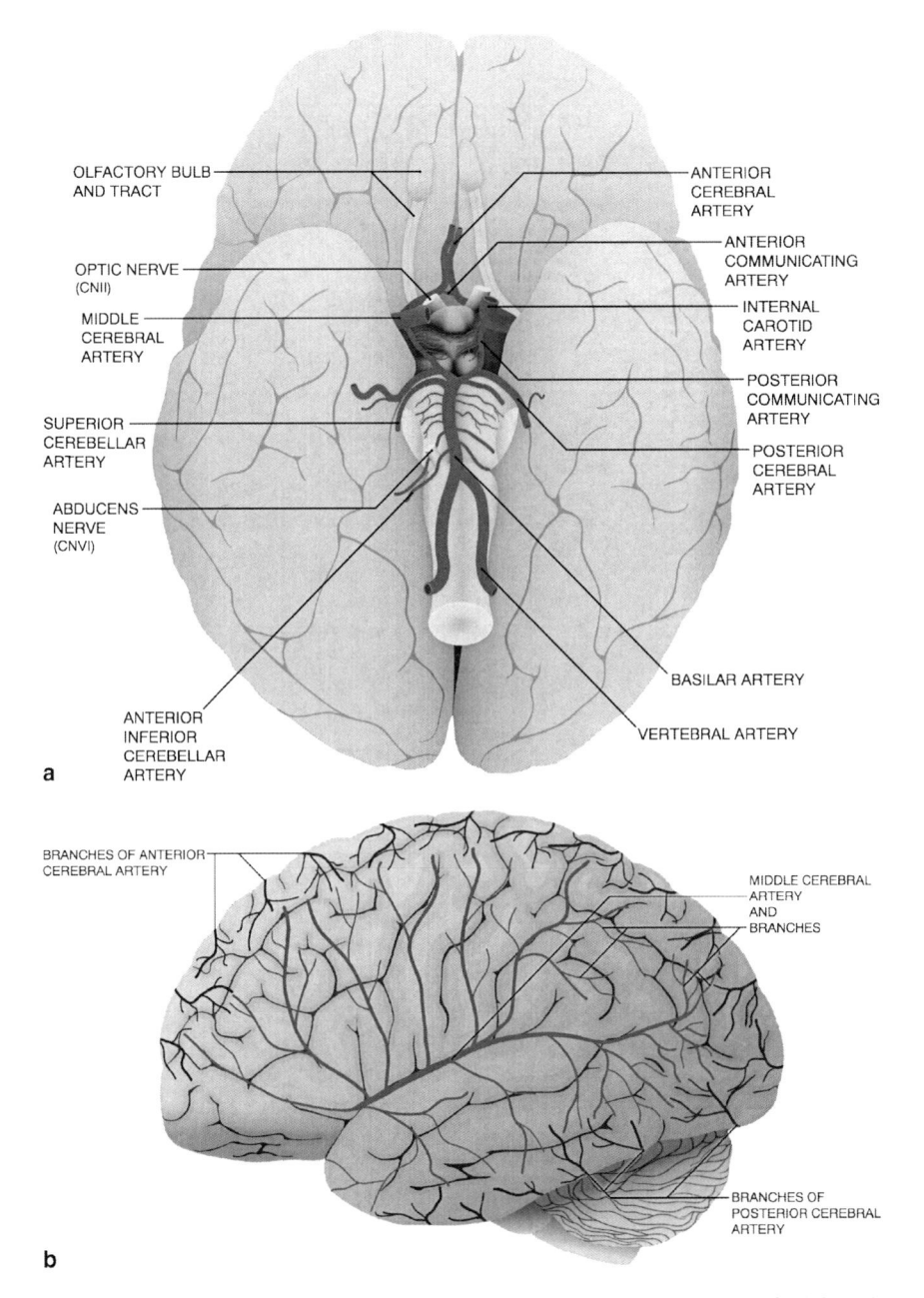

Fig. 1.5 Blood supply to the brain. (**a**) Vessels that contribute to the arterial circle of Willis at the base of the brain. Note contributions from the vertebral and internal carotid systems. Throughout its length, each vessel that participates in the arterial circle gives off numerous small, unnamed branches that penetrate the brainstem. (**b**) Distribution of blood supply to the lateral surface of the cerebrum is illustrated. The middle cerebral artery is the prominent vessel, the anterior cerebral artery vascularizes the territory on either side of the falx cerebri and a narrow strip of superior surface of the cerebrum. (**c**) Sagittal section that depicts the distribution of blood flow to the cerebrum. Note that the anterior and middle cerebral arteries carry blood from the internal carotid arteries, whereas the blood to the posterior cerebral arteries comes from the vertebral/basilar artery system (*See also* Color Insert)

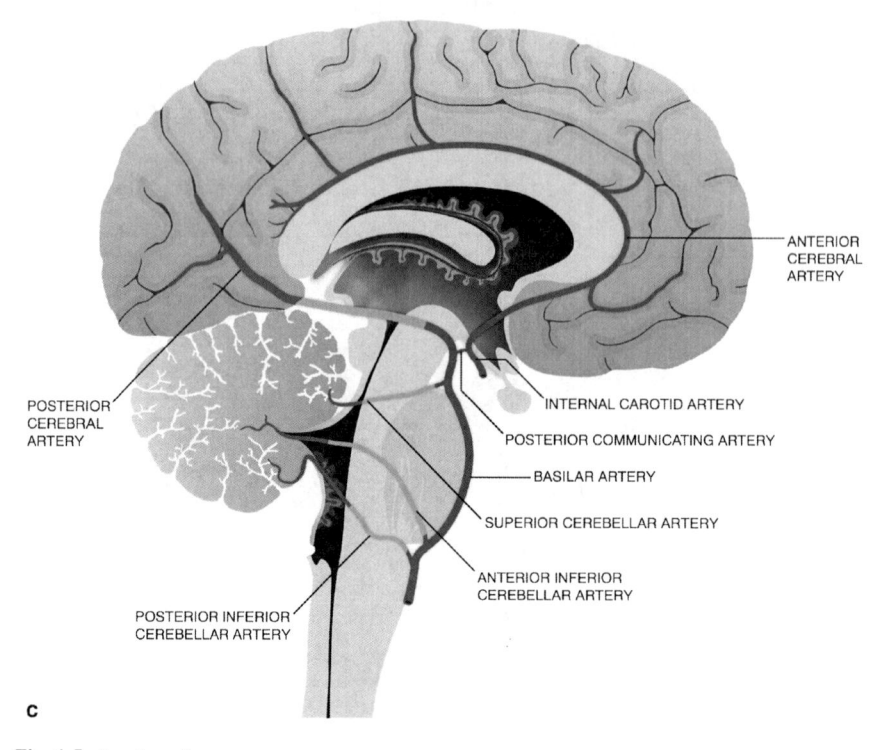

c

Fig. 1.5 (continued)

With regard to tissue blood supply, the organization of the brain and brainstem differs from that of the rest of the body in that there are no anastomoses within the nervous tissue. Each arterial branch is a functional end artery; if it were to be occluded, the territory of the brain that it supplied would become hypoxic and ischemic. Because the blood supply to the brain and brainstem is critical for normal cognitive function, a more complete description of the arterial supply follows.

The vertebral arteries branch from the subclavian arteries in the root of the neck and ascend within the foramina of the transverse processes of six of the cervical vertebrae (C6 – C1). Upon exiting the transverse foramina of C1, the vertebral arteries enter the skull through foramen magnum and approach each other in the midline where they fuse to form the basilar artery at approximately the level of the pontomedullary junction. The basilar artery travels rostrally in a groove on the base of the pons until it terminates as the superior cerebellar and posterior cerebral arteries. The basilar distributes blood via numerous small vessels that enter the pons to supply the pontine nuclei, corticospinal tract, and the pontine portion of the reticular formation. At its termination, the basilar artery gives off the superior cerebellar and posterior cerebral arteries each of which (plus the posterior communicating arteries) gives off numerous small arteries that penetrate the posterior perforated substance to supply the midbrain. Thus, geographically distinct regions of the midbrain receive blood supply from the posterior cerebral, posterior communicating, and superior cerebellar arteries. Each of these arteries gives off numerous small branches throughout their extents; these branches penetrate the nervous tissue to supply the various regions of the brainstem. The approximate areas of vascular distribution are depicted in the diagrammatic representation of a cross-section of the midbrain in Fig. 1.6. The bulk of the lateral midbrain reticular formation is supplied by the superior cerebellar artery; whereas the colliculi, periaqueductal gray matter, and raphe nuclei receive blood from the posterior cerebral artery; and the majority of the cerebral peduncles are vascularized by branches from the posterior communicating arteries.

Functions of Brain Regions

The states of consciousness, cognition and self-awareness are the result of highly complex and integrated functions of the brain. While it may be simplistic to segregate the functions of the brain into discrete activities that are carried out only in specific lobes of the brain, it is clear that regions of the brain that perform similar or related functions are often situated in anatomical proximity to each other.

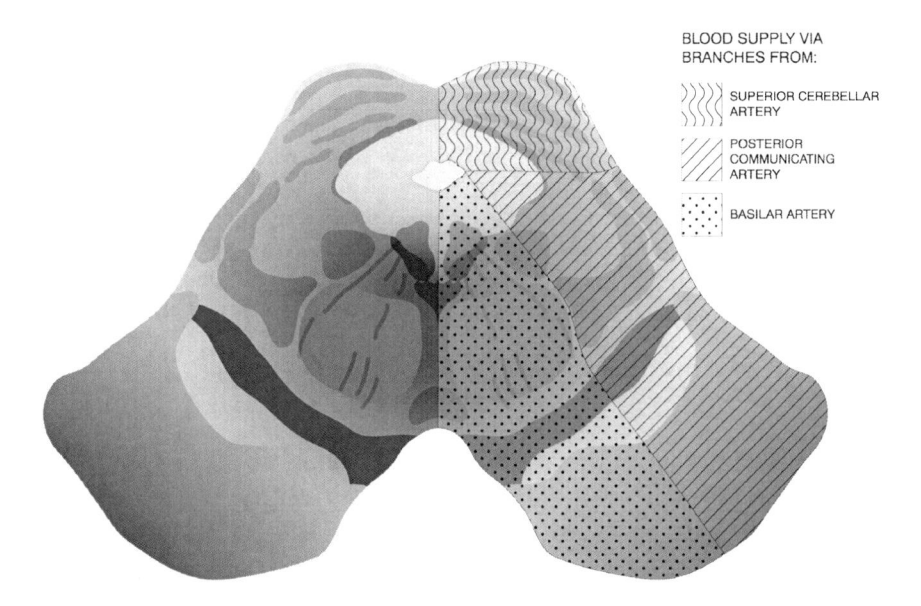

BLOOD SUPPLY VIA
BRANCHES FROM:

SUPERIOR CEREBELLAR ARTERY

POSTERIOR COMMUNICATING ARTERY

BASILAR ARTERY

Fig. 1.6 Cross-section illustrating the distribution of blood to the walls of the midbrain. The tectum is supplied by branches of the superior cerebellar artery. The medial aspects of the peduncular and tegmental regions are vascularized by branches of the basilar artery. The lateral peduncular and tegmental regions are supplied by branches of the posterior communicating artery. Note that these small arteries, which enter the walls of the central nervous system from the periphery, are functional end arteries and do not anastomose with adjacent arteries (*See also* Color Insert)

Provided that one remains cognizant of the simplification and that the borders of the lobes of the brain were arbitrarily determined by early anatomists, it is possible to observe that motivation and motor functions emanate from the frontal lobe; sensory information is interpreted and integrated in the "association cortex" of the parietal lobe; visual input is decoded and interpreted in the occipital cortex; hearing and declarative memory functions are centered, in large part, within the temporal lobe; and the insular cortex and limbic lobe are central to emotion. Highly complex functions occur at the intersections among these areas. For instance, the precentral gyrus of the frontal lobe (motor cortex) lies parallel to the postcentral gyrus of the parietal lobe (sensory cortex) and the somatotopic organization of these gyri is nearly identical, allowing for the rapid coordination of sensory and voluntary motor functions for given areas of the body. Similarly, functions relating to receptive communication reside in the area where the parietal and temporal lobes abut one another in the dominant hemisphere (Wernicke's area). Localized damage to specific portions of these areas, regardless of cause, will result in the loss of specific capabilities. As an example, damage to the temporal lobe could result in temporal lobe epilepsy.

The activities that are managed by the various areas of the brainstem are vital for normal functioning of the body and survival, but their control does not reach to the level of consciousness. These brainstem areas include centers in the tegmental pons and medulla that control the performance of cardiovascular, respiratory, and metabolic functions as well as interconnections between the brainstem and the cerebellum and thalami.

The reticular formation is an important structure located deep within the brainstem, which extends rostrally from the upper medulla through the pons and midbrain to the thalamus and forebrain. In its caudal (medullary and pontine) region, the reticular formation contributes to regulation of autonomic functions and to horizontal, conjugate eye movements. The ascending projections of the reticular formation, which traverse the midbrain, contribute to the ascending reticular activating system, which is responsible for controlling the state of arousal of forebrain structures and, thus, one's level of consciousness. This region is also associated with control of the waking and sleep cycle.

Physically Induced Alterations in Consciousness

Given the anatomical and functional complexity of the central nervous system, it is not surprising that a variety of potential problems may afflict individuals with reduced consciousness. These include such possibilities as a generalized reduction in the functions of much of the cerebral cortex due to generalized encephalitis or reduced physiologic functions of the brain due to intoxication or metabolic disease. More geographically restricted pathologies can be caused by vascular disruption or physical compression of portions of the brain. The subject of most of this book will relate to disease states, metabolic disorders and exogenous intoxications (both from drugs and environmental chemicals). For the purposes of the present discussion, consideration is restricted to the anatomical causes of reduced consciousness.

It is important to note that statements about an individual's general level of consciousness appear in common usage (e.g., drowsy, lethargic, alert); however, the actual clinical assessment of one's state of consciousness is a complicated exercise that is beyond the scope of this chapter (see detailed discussions in Weisberg et al., 2004; Plum and Posner, 1972). Nevertheless, it is easy to comprehend the notion that with the progression of a given pathological condition that impacts a region of the brain that is involved with consciousness, one's ability to respond to environmental stimuli will decrease. Thus, there are grades of impaired consciousness that are associated with various pathological conditions.

With respect to the state of consciousness, the most important configuration is the ascending reticular activating system, the projections of which traverse the midbrain. In general, anatomical causes for reduced levels of consciousness that relate to the central nervous system stem primarily from distortion of the anatomical integrity of important brain structures or, in some cases, compromise of the vascular supply. The former types of lesions can be subdivided into those caused by displacement of structures caused by space filling lesions (e.g., hydrocephaly, tumors, abscesses, hematomata and edema). As any of these lesions increases in size, the intracranial volume available for normal brain tissue is reduced. Eventually, compression of the brainstem results in distortion of the reticular formation with consequent dysfunction that will affect consciousness. Dislodgements of anatomical structures, such as a herniation of the midbrain through the tentorial notch, could also result in the squeezing of the midbrain such that the lateral reticular formation is compromised with consequent impact on arousal. Although dislodgements and displacements result from very different causes, both can ultimately compress internal structures of the brain, triggering symptomatically identical dysfunction.

Conclusion

The brain and brainstem are exceptionally complex anatomically, histologically, physiologically, and pharmacologically. Some cognitive functions occur in particular geographic regions with input via axonal projections from other areas of the central nervous system. It is not surprising that myriad perturbations (e.g., vascular, metabolic, inflammatory) can impact their function. Among them is a subset of space-filling lesions that could physically distort the anatomy of the regions that regulate the state of arousal in the cerebrum. This chapter has reviewed the anatomy of the brain and brainstem, with particular attention to the midbrain and the ascending pathways of the reticular formation to call attention to the possibility that impaired consciousness can emanate from pathologies stemming from the distortion of the normal anatomical relationships in the brainstem, and that the outcomes of these pathologies are often identical to those caused by disease states.

Acknowledgments The author gratefully recognizes the artistic talent and patience of Mr. Ken Arevalo, who painstakingly crafted the illustrations for this manuscript.Preparation of this manuscript was funded in part by the Noblis Sponsored Research Program.

References Cited

Plum, F. and J. B. Posner (1972), *The Diagnosis of Stupor and Coma, 2nd Edition*, F. A. Davis Company, Philadelphia

Weisberg, L. A., C. A. Garcia, and Strub R. L. (2004), *Essentials of Clinical Neurology, 3rd Edition*, Mosby, St. Louis, 786 p

General References

Arslan, O. (2001), *Neuroanatomical Basis of Clinical Neurology*, Parthenon Publishing, New York

Drake, R. L., Vogl, and W. Mitchell A. W. M. (2005), *Gray's Anatomy for Students*, Elsevier, Philadelphia

Kandel, E. R., J. H. Schwartz, and T. M. Jessell (1991), *Principles of Neural Science, 3rd Edition*, Appleton & Lange, Norwalk, CT

Lockhart, R. D., G. F. Hamilton, and F. W. Fyfe (1969), *Anatomy of the Human Body, 2nd Edition*, Lippincott, London

Moore, K. L. and A. F. Dalley (2006), *Clinically Oriented Anatomy, 5th Edition*, Lippincott, Philadelphia

Purves, D., G. J. Augustine, D. Fitzpatrick, W. C. Hall, A. S. LaMantia, J. O. McNamara, and S. M. Williams (2004), *Neuroscience*, Sinauer Associates, Sunderland, MA

Turlough Fitzgerald, M. J., G. Gruener, and E. Mtui (2007), *Clinical Neuroanatomy and Neuroscience, 5th Edition*, Saunders, Philadelphia

Chapter 2
Brain Metabolic Adaptations to Hypoxia

Michelle A. Puchowicz, Smruta S. Koppaka, and Joseph C. LaManna

Introduction

Oxygen, Brain, and Energy Metabolism

The mammalian brain depends totally on a continuous supply of oxygen to maintain its function. It is well known that in the brain, adaptation to hypoxia occurs through both systemic and vascular changes, which may include metabolic changes. However, the local metabolic changes related to energy metabolism that occur within the cell are not well described (Harik et al., 1994; Harik et al., 1995; LaManna and Harik, 1997). Investigating the metabolic adaptations of the central nervous system to mild hypoxia provides an understanding of the key components responsible for regulating cell survival. This chapter concerns itself with the metabolic responses of the brain to mild hypoxia, that is, to physiological hypoxia. This is the range of hypoxia that can be compensated for with physiological mechanisms that directly or indirectly involve energy related metabolic pathways.

The brain's metabolic response is dependent on the severity and/or length of time of exposure (acute and chronic). During prolonged exposure to a low oxygen environment, systemic adaptations, such as increased pulmonary minute volume and packed red cell volume, result in maintained oxygen delivery to the brain, whereas the central cerebrovascular and metabolic adaptations preserve tissue oxygen and energy supply to support neuronal function. Other changes in the brain tissue include a decrease in the volume density of neuronal mitochondria (Stewart et al., 1997) and cytochrome oxidase activity (Chávez et al., 1995; LaManna et al., 1996) that probably reflect an overall decrease in resting cerebral metabolic rate for oxygen ($CMRO_2$) of about 15%; although this has not been measured directly, it is compatible with the slight decrease in body temperature (Mortola, 1993; Wood and Gonzales, 1996; LaManna et al., 2004). The tendency for hypoxia to decrease brain activity may be related to the idea of central respiratory depression (Neubauer et al., 1990) or the activation of survival pathways that regulate energy metabolism, such as hypoxia-inducible factor-1 (HIF-1) (Chávez et al., 2000). More severe hypoxia leads to pathology and cell death due to the failure of these compensatory mechanisms and subsequent energy depletion, which will not be the focus of this chapter.

D.W. McCandless (ed.) *Metabolic Encephalopathy*,
doi: 10.1007/978-0-387-79112-8_2, © Springer Science+Business Media, LLC 2009

Hypoxia

Hypoxia per se is a decrease in the partial pressure of oxygen in the ambient air from the mean sea level value of about 160 Torr (dry gas). In normal physiology, the pO_2 in the pulmonary vein is about 105 Torr due to the contribution of water vapor and carbon dioxide. Most of the oxygen is carried by hemoglobin and at these partial pressures hemoglobin is fully saturated. When the arterial pO_2 falls below 90 Torr, which can occur through decreased fraction inspired oxygen (FiO_2), decreased barometric pressure due to increasing altitude, or through lung pathology (pulmonary hypoxia), then a condition of hypoxemia (lower blood oxygen) occurs. A decrease in the oxygen availability to the tissues occurs in uncompensated hypoxemia, anemia (anemic hypoxia, i.e., low hemoglobin and thus lower oxygen carrying capacity), carbon monoxide (toxic hypoxia), or blood flow restriction (ischemic hypoxia), all resulting in the activation of local tissue metabolic response mechanisms. Adequate brain oxygenation requires both sufficient delivery and uptake by cells. If these two components are not met, such as during hypoxia, then an adaptation process takes place. Adaptations which occur in tissue include increased capillary surface area that results in increased oxygen delivery, decreased energy demand and increased energy production efficiency. During acute and chronic exposures, adaptations include metabolic responses that tend to stabilize energy metabolism (Harik et al., 1995; Lauro and LaManna, 1997).

Acute and Chronic Exposure

Hypoxia is not necessarily a pathological condition. At the cellular level, limiting the amount of oxygen exposure in tissues to only the amount needed to drive activity-induced metabolism is one protective strategy against oxygen toxicity. Thus, the mammalian brain usually exists in a low tissue oxygen milieu. Local tissue mechanisms ensure that the oxygen environment is controlled, not maximized. These physiological considerations underlie the functional magnetic resonance imaging (fMRI) phenomenon known as BOLD (blood oxygen level dependent) response to a focal activation task, proving to be a valuable tool for understanding brain function and energy metabolism, holding promise for pathological diagnosis and treatment monitoring.

For acute hypoxic exposures, if blood oxygen is below 90 Torr, but above 45 Torr, then the hypoxia is mild and can be compensated for by normal physiological processes, and usually does not lead to any tissue damage. Depending on duration, below 45 Torr, permanent damage including neuronal degeneration will most likely occur. If mild hypoxia persists, such as with chronic exposure, long term compensatory responses are activated. Acclimatizing adaptations involve both systemic and metabolic changes which may take days to weeks to become established, but then allow habitation at moderate hypoxia and brief periods of severe hypoxia with far less damage than before acclimatization.

Metabolic regulators of cerebral tissue PtO_2 are complex and involve both vascular and metabolic adaptations. For the brain, the pattern of adaptation includes sequential responses that raise brain PtO_2 (Xu and LaManna, 2006). The initial response is to increase blood flow, followed by an increase in hematocrit and then microvessel density as a result of angiogenesis (vascular adaptations) (Brown et al., 1985; Beck and Krieglstein, 1987; LaManna et al., 1992; LaManna and Harik, 1997; Dunn et al., 2000; LaManna et al., 2004)

In a study using awake or anesthetized rats, the functional MRI responses to graded hypoxia were investigated. CBF, BOLD and cerebral metabolic rate of oxygen ($CMRO_2$) changes were estimated. Hypoxia in the animals that were awake revealed compensatory responses for sustaining blood pressure and increasing both heart and respiration rates. CBF and BOLD were found to decrease in rats that were awake at low pO_2. $CMRO_2$ estimated using a biophysical BOLD model did not change under mild hypoxia but was reduced under severe hypoxia relative to base-line. The authors concluded that with severe hypoxia brain tissue in conditions of being awake appeared better oxygenated than with anesthesia (Duong, 2007).

It is known that neurodegenerative processes such as Alzheimer's result in altered CBF. Altered CBF during these conditions has been reported to improve with hypoxic exposure through the regulation of nitric oxide (NO) (Sun et al., 2006). A model of neurodegenerative brain disorder (via administration of a toxic fragment of beta-amyloid) in rats showed that preadaptation to hypoxia prevented endothelial dysfunction and improved the efficiency of NO storage (Mashina et al., 2006).

Metabolic Adaptations to Hypoxia

Hypoxia is known to induce adaptive changes in the brain which are related to energy metabolism (Semenza et al., 1994; Harik et al., 1995; Harik and LaManna, 1995; LaManna and Harik, 1997; Jones and Bergeron, 2001; Lu et al., 2002; Shimizu et al., 2004). Cerebral metabolic rate for glucose has been reported to be elevated in both acute (Pulsinelli and Duffy, 1979; Beck and Krieglstein, 1987) and chronic hypoxia (Harik et al., 1995). Regional metabolic rate for glucose (CMR_{glu}) is known to increase up to 40% with hypoxia (Harik et al., 1995). Brain glucose and lactate concentrations are also known to increase (about double), whereas glycogen and cytochrome oxidase activities decrease (40% and 25%, respectively) (Harik et al., 1995; LaManna et al., 1996) compared to normoxic controls. In humans, using positron emission tomography, transient hypermetabolism in the basal ganglia has been observed in newborns who had suffered hypoxic-ischemic encephalopathy (Batista et al., 2007)

Glucose Metabolism

Significantly higher levels of brain lactate and pyruvate concentrations and increased lactate to pyruvate ratios accompany hypocapnic hypoxia (Beck and Krieglstein,

1987), suggesting that glycolysis was stimulated. In support of these findings, the activation of the glycolytic enzymes such as the hexokinase (HK), glucose-6-phosphate (G6P), and phosphofructokinase (PFK) without changes in cerebral glucose content or plasma glucose levels were also observed. The association with the increase in glycolysis and hypoxia and no changes in glucose content were further investigated and confirmed. Using 3-O-methylglucose method to measure cerebral glucose content and 2-deoxyglucose method to determine CMR_{glu}, the authors concluded that there is evidence that hypoxia results in a stimulation of glycolysis. The authors further point out that during short term hypoxia, there was dissociation between the increased CBF and fall in pH without changes in the coupling to glucose utilization (CMRglu), suggesting that under hypoxia, local cerebral blood flow matches local metabolic demand, irrespective of altered pH.

The effect of hypoxia on the developing brain was examined in P10 and P30 postnatal rats. With chronic hypoxia in the P10 rats, lactate dehydrogenase (LDH) activity was found to be significantly increased and the hexokinase activity decreased in certain regions of the brain compared to the controls. Neither 2-ketoglutarate dehydrogenase complex (regulator of TCA cycle) nor citrate synthase activities were significantly altered by hypoxia in the P10 rats, but significant regional changes were observed in the P30 rats. The mechanisms leading to the change in the glycolytic enzyme activities have been reported to be related to the up regulation of HIF-1 (Lai et al., 2003).

Activation of Glycolysis: Ph Paradox

The association of decreased energy production with hypoxia is mostly likely a result of a decrease in oxidatively derived ATP, possibly through the inhibition of electron transport chain activities. Lactic acidosis has been reported to occur during hypoxia (Siesjö, 1978), as a result of increased glycolytic rates via the activation of phosphofructokinase. In chronic severe hypoxia, acidosis can even lead to irreversible cell damage.

Hypoxia induces hyperventilation that results in increased oxygen uptake with a simultaneous decrease in arterial PCO_2. An imbalance due to this response leads to alterations in blood and tissue acid—base balance such as alkalosis. Systemic alkalosis is balanced by secretion of bicarbonate in the kidney, but the CNS alkalosis is offset by increased glycolytic ATP production. In oxidative phosphorylation, a proton is consumed when ATP is synthesized. In glycolysis, there is no net production or consumption of protons. When ATP is hydrolyzed in energy requiring reactions, a proton is produced. Thus, utilization of ATP produced from oxidative phosphorylation is pH neutral and consumption of ATP from glycolysis is acid producing (Dennis et al., 1991). Since glycolysis provides the substrate for oxidative phosphorylation, there are always some protons being produced, but an increase in glycolysis without a corresponding increase in oxidative phosphorylation will increase acid production, and this is demonstrated by an increase in brain lactate concentration.

One could predict that this increase in glycolysis cannot substantively support brain function through glycolytically derived ATP. It is more likely instead that the increase in glycolysis functions to balance tissue acid—base disturbances and maintain brain pH, which are related to hyperventilation-induced decreased $PaCO_2$ which would otherwise tend to become alkalotic (Lauro and LaManna, 1997). The substrate hydrogen is oxidized by oxygen in coupled oxidative-phosphorylation through chemiosmotic mechanisms. Finally, creatine kinase catalyzes the phospho-creatine/creatine: ATP/ADP reaction is in equilibrium with the hydrogen ion concentration. During chronic hypoxia, tissue intracellular pH becomes acidified primarily due to the turnover of glycolytically produced ATP, retention of CO_2 from residual oxidative-phosphorylation, and net breakdown of ATP (Hochachka and Mommsen, 1983; Dennis et al., 1991)

Enzyme Related Changes in Oxidative Metabolism

Cytochrome oxidase activity is an indicator of the energy demands of the cell and has been reported to vary with conditions of hypoxia (LaManna et al., 1996). The energy demand of the cell indicates mitochondrial ATP production. The enzyme cytochrome oxidase (complex IV) is a large transmembrane protein complex found in the mitochondria that is primarily involved in the intracellular defense against oxygen toxicity by the safe metabolism of oxygen metabolism. It is the last protein in the electron transport chain and plays an important role in transferring electrons by binding electrons from each of the four cytochrome c molecules to completely reduce molecular oxygen.

With acute hypoxia (15 h-8% oxygen) the cytochrome oxidase activity in the neuron increases, but is unchanged in glia (Hamberger and Hydén, 1963). With chronic exposure, cytochrome-oxidase activity has been found to decrease (Chávez et al., 1995; LaManna et al., 1996; Caceda et al., 2001). A decrease in cytochrome-oxidase activity indicates a decrease in oxidative metabolism and therefore the apparent increase in CMRglu may be as a result of only increased glycolysis and not oxidative metabolism. The activation of glycolysis together with the decrease in cytochrome oxidase activity is consistent with the hypothesis that pH is maintained through the balance between glycolytically derived ATP and oxidatively derived ATP.

Premature transfer of electrons, either at complex I or complex III, results in increased generation of ROS (reactive oxygen species). Studies suggest that there are adaptations that may function to prevent excessive ROS production in hypoxic cells. Pyruvate dehydrogenase kinase 1(PDK1; PK) via phosphorylation, inactivates pyruvate dehydrogenase, which is responsible for the production of acetyl-CoA from pyruvate. Thus, the inactivation of pyruvate dehydrogenase reduces the delivery of acetyl-CoA to the TCA cycle and together with the hypoxia-induced expression of LDHA (lactate dehydrogenase A), reduces the levels of NADH and FADH2 delivered to the electron transport chain (Semenza, 2007), then reducing ROS production. It has been suggested that the upregulation of PK may play a role in induction of hypoxic tolerance (Shimizu et al., 2004).

Intrinsic Brain Tissue Oxygen Sensors and Regulators

Hypoxic Inducible Factor: Energy Metabolism

A primary participant in hypoxic angiogenesis is hypoxia inducible factor 1 (HIF-1) (Chávez et al., 2000). HIF-1 accumulates in hypoxia and is thought to be one of the crucial signaling pathways that might lead to an understanding of the diseases that are associated with oxygen deprivation and metabolic compromise. A family of dioxygenases called HIF prolyl 4 hydroxylases (PHDs) governs the activation of the HIF pathway (Jaakkola et al., 2001; Epstein et al., 2001; Freeman et al., 2003; Appelhoff et al., 2004). Decreased activity in PHDs has been described to occur with oxygen deprivation, triggering cellular homeostatic responses (Siddiq et al., 2007). HIF-1 is known to play a major role as an oxygen sensor pathway in the brain (LaManna, 2007) (Sharp and Bernaudin, 2004). Vascular endothelial growth factor (VEGF) is a molecule that is upregulated by HIF-1 and initiates capillary angiogenesis (Pichiule and LaManna, 2002; Pichiule et al., 2004). Hypoxia is not the only agent resulting in HIF-1 accumulation. For example, non-hypoxic HIF-1 accumulation can occur by growth factors such as IGF-1, iron chelation, cobalt chloride, and alteration of substrate availability. Likewise, overproduction or stimulation of PHD would be expected to prevent HIF-1 accumulation even in moderate hypoxia. Recently, we have found that the presence of ketone bodies in normoxic brain induces HIF-1. In normoxic cell culture, others have found that intermediates of energy metabolism also induce HIF, such as pyruvate (Dalgard et al., 2004) and succinate (Selak et al., 2005).

HIF-1α remains responsive to tissue pO_2 which it does not adapt. The HIF-1 signal is thus maintained until the tissue pO_2 is restored by angiogenesis. HIF-1 is a transcription factor that activates over 40 known genes that have a hypoxic response element (HRE) in their promoter region. Almost all the enzymes of glycolysis have a HRE and are upregulated in prolonged mild hypoxia (LaManna et al., 2007). Thus, the role that HIF-1 plays in the regulation of energy metabolism in response to hypoxia is most likely to aid in the restoration of energy homeostasis.

There are also iron-containing molecules that might act as sensors because they bind oxygen in the physiological range. For example, neuroglobin, a heme protein, is upregulated by sustained hypoxia and may play a protective role (Sun et al., 2003; Li et al., 2006). The effector mechanisms of these iron containing proteins have not been assessed as yet (Brunori and Vallone, 2006).

Brain Metabolic Indicators of Hypoxia: Glucose and Ketone Body Transporters

Not only does the glycolytic rate increase with hypoxia, but the transport of glucose at the blood—brain barrier (BBB) also increases (Harik et al., 1994). The GLUT-1 transporter, responsible for carrier-mediated facilitated diffusion at the BBB is associated with HIF-1 (Chávez et al., 2000). In rats, the increase in glucose transporter and capillary density results in three times increase in glucose flux rate capacity in hypoxic adapted rats. The large disparity in the increased transport capacity compared to the increased cerebral metabolic rate for glucose (CMR_{glu}) can be explained by the net reduction of the arterial glucose delivery by half. In species such as the rat (but not humans) that lack GLUT-1 transporters in circulating red blood cells, the source of glucose available for transport is limited to the glucose in the plasma. After hypoxic adaptation, the rate of whole blood flow through the tissue re-normalizes, but this means that the plasma flow rate is half the pre-exposure rate. The increased transport capacity compensates for the decreased plasma flow rate, and supports the increased resting CMR_{glu}, allowing for transient increases in energy demand due to focal neuronal activation.

The increased density of glucose transporters at the blood—brain barrier together with the increase in the glucose influx is consistent with increased glucose concentrations in the brain. However, the glucose consumption (CMRglu) is only slightly elevated (about 15%) (Harik et al., 1995). The concomitant findings of decreased brain glycogen and increased brain lactate suggest that glycolysis increases with hypoxia (Lauro and LaManna, 1997). This slight increase in CMRglu would contribute a relatively small amount to the energy needs of the cells. Glycolysis produces almost twenty times less ATP per glucose molecule than oxidative phosphorylation, so a 15% increase would be negligible.

Recently, we have found that the moncarboxylate transporter (MCT 1) is upregulated at the BBB with 3 weeks of exposure to hypoxia in rat brain. The MCT family is the primary transporter for short chain acids, such as ketone bodies, lactate and pyruvate. The relative increase in MCT1 and GLUT1 at the BBB following exposure of hypobaric-hypoxia at 10% (0.5 atm. oxygen) is shown in Fig. 2.1. Consistent with previous studies, a 35% increase in capillary density, as measured by GLUT1, is indicative of a hypoxic response (Harik et al., 1996). Similarly, a 20% increase in MCT1 is also observed. These data indicate that the ratio of GLUT1 to MCT1 remains about threefold with hypoxic exposure, suggesting that MCT1 is associated with capillary density at the BBB. This study also shows that GLUT1 remains the more abundant substrate transporter in the brain.

Glutamate Transporters

Glutamate excitotoxicity is associated with ischemia, oxidative stress, seizures, hypoxia and neurodegenerative diseases that can result in neuronal death (Nilsson and Lutz, 1991; Dallas et al., 2007). Disruption in glutamate homeostasis as a result of the release of glutamate and the subsequent increase in cellular levels are known to activate inotropic NMDA (*N*-methyl-D-aspartate) receptors resulting in neuronal damage. Recently, prolonged hypoxia in whole-cell cortical astrocytes using patch-clamp measurements was shown to significantly reduce glutamate uptake via loss

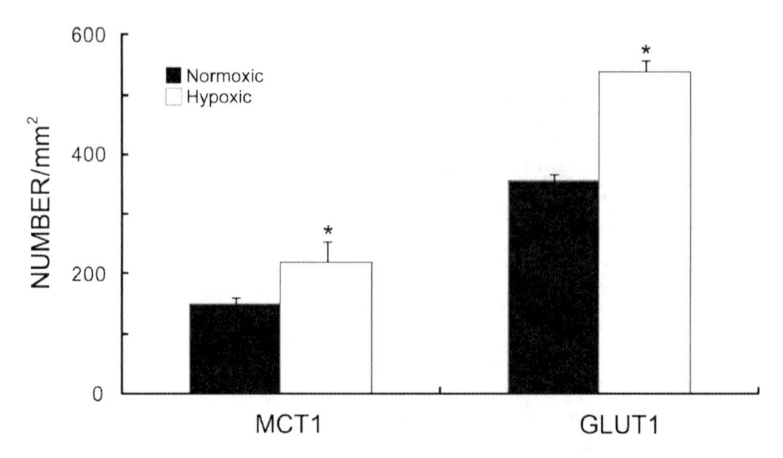

Fig. 2.1 Glucose and ketone body transporters with hypoxia. The relative increase in moncarboxylate (MCT1) and glucose (GLUT1) transporters at the blood—brain barrier following 10% (0.5 atm. oxygen) exposure of hypobaric-hypoxia. Quantification of cerebral capillary density, as measured by GLUT1 immunostaining, and MCT1 immunoreactivities (number of counts per mm²) showed about a 30% upregulation of MCT 1 and GLUT1 transporters with 3 week hypoxic exposure in rat brain. *P < 0.05, hypoxic vs. normoxic

of the activity of the glutamate transporter, EAAT (Dallas et al., 2007). It was further concluded that the down-regulation of the EAAT with chronic hypoxic exposure was directly related to hypoxia and the activation of the nuclear factor, NF-kB, and not the transcriptional regulator, HIF-1.

Fetal guinea pig and newborn piglet model studies have demonstrated that brain tissue hypoxia results in brain cell membrane damage as evidenced by increased membrane lipid peroxidation and decreased Na$^+$, K$^+$-ATPase activity. Brain hypoxia was found to increase the NMDA receptor agonist-dependent Ca^{2+} in synaptosomes of hypoxic as compared to normoxic fetuses (Mishra and Delivoria-Papadopoulos, 1999).

Ketosis and Hypoxia

Ketosis has been considered to improve hypoxic tolerance by improving the metabolic energy state as a result of imbalance in glucose metabolism and energy insufficiency. Ketone bodies are suggested to have beneficial applications in both mitochondrial energy metabolism as well as non-oxidative metabolism.

The classic study by Owen et al. based on arterio-venous differences suggested that the adult brain uses ketone bodies as a principle substrate during starvation, resulting in a remarkable decrease in brain glucose consumption (Owen et al., 1967). Review of the current literature has shown inconsistencies in the magnitude/level at which the mammalian brain uses ketones as an oxidative substrate to glucose (alternate energy substrate to glucose). In a study of fasted rats exposed to altitude, it was reported that improved survival was a result of elevated blood ketone bodies (Myles, 1976). Recent studies have suggested the use of ketones as a therapy for both non-pathological and pathological conditions.

Ketone bodies are known to supplement the brain energy metabolism through the oxidation (utilization) of beta-hydroxybutyrate (BHB) and acetoacetate (AcAc) especially when glucose availability is minimal. Ketosis is induced in most mammals by fasting, starvation or by feeding a high fat-low carbohydrate diet. Under conditions of glucose sparing, the majority of the ketone bodies are supplied by the periphery via liver metabolism (Balasse and Fery, 1989). The liver produces ketone bodies BHB and AcAc via ketogenesis as a result of the partial beta-oxidation of free fatty acids, which are then taken up by peripheral tissues such as those of the brain and utilized.

Ketone bodies are thought to have therapeutic implications that involve their effects on pathological conditions, redox state, diabetes (insulin resistance), non-mitochondrial (glycolysis) and mitochondrial metabolism (glucose metabolism e.g. anaplerosis, oxidative metabolism of glucose, enzyme activities of TCA cycle and oxidative phosphorylation) (Veech, 2004). The potential benefits of ketone bodies on improving overall physical and cognitive performance as well as protection from oxidative stress is thought to be linked to improved metabolic efficiency, but remains to be explored. One could speculate that during conditions of limited oxygen supply, ketone bodies might be beneficial in limiting tissue damage.

Hypoxic-Tolerance with Ketosis

Though not well investigated, the majority of research on hypoxia and the effects of ketosis on cerebral function and metabolism has primarily studied conditions of severe hypoxia. Hypoxia has been described to elevate blood-plasma ketone levels following severe hypoxic exposure. In one study, a sequential severe exposure to hypoxia (4.5% O_2) in mice induced metabolic changes that protected against the lethal effects of hypoxia as measured by hypoxic survival time. The rationale for improved survival time was proposed by the authors to be through the alteration of substrate utilization and mobilization. The combination of three successive pretreatments with hypoxia and intra-peritoneal bolus of BHB dramatically and significantly increased the hypoxia survival time. Hypoxia survival time was not improved with glucose-pretreatment. Hypometabolic hypothermia was also reported, most likely as a consequence of depression of oxidative metabolism (Rising and D'Alecy, 1989). These results suggest that ketones provide a beneficial effect through altering glucose metabolism by possibly improving redox state through the regulation of glycolysis.

Protection from hypoxia in fasted animals (mildly ketotic) resulted in improved survival time and a reduction in lactic acid production without a generalized reduction in cerebral energy metabolism. It was suggested that this protection was due to a shift toward ketone-metabolism with a subsequent reduction of glucose oxidation. The increase in hypoxia survival time could not be accounted for by blood-glucose levels. The causal relation responsible for the increased hypoxia survival time was presumed to be ketosis (Kirsch and D'Alecy, 1979; Eiger et al., 1980).

In a neonatal rodent model of hypoxia-ischemia, ketosis was reported to limit brain damage after 3 h of hypoxia exposure and preserve cerebral energy metabolism. Increased levels of BHB in rat pups (7-day old) might provide a critical and supplemental energy source particularly under times of neuropathological damage (Dardzinski et al., 2000). These data are consistent with the bilateral carotid occlusion rodent model where the authors report that BHB administered exogenously resulted in amelioration of the disruption of cerebral energy metabolism with hypoxia. The mechanism was described to be possibly through the feedback inhibition of pyruvate dehydrogenase complex via increased acetyl-CoA availability. The increased availability of acetyl-CoA could also result in decreased lactate production through feedback inhibition of phosphofructokinase (PFK), the key rate-limiting enzyme in glycolysis (Suzuki et al., 2001). The aspect of the inhibition of the allosteric enzyme PFK was previously proposed in a dog model where A—V differences of glucose and BHB across brain were measured following acute venous infusion of BHB during hypoxia (Chang and D'Alecy, 1993). With acute hypoxia, BHB was found to exhibit neuro-protection by the mechanism of depressing glucose uptake and consumption instead of acting as a cerebral energy substrate presumably through feedback inhibition of PFK. However, this mechanism remains to be understood.

BHB treatment was reported to protect hippocampal cells for 2 h. In cultured hypoxic-exposed hippocampal rat neurons treated with BHB, a concomitant decrease of cytochrome-C release, caspace-3 activation and poly (ADP-ribose) polymerase was observed. Mitochondrial transmembrane potential was maintained during a 2 h exposure to hypoxia (Masuda et al., 2005).

In our rodent model of diet-induced ketosis, we found with chronic mild hypoxic exposure (0.5 atm.) that adaptation to hypoxia did not interfere with ketosis induced by feeding a high fat ketogenic diet. This was indicated by the sustained elevated levels of ketones present in the brain and plasma following the 3-week duration of diet and hypoxia. The reduced lactate in the ketotic groups was thought to be a consequence of ketosis. Hypoxia and ketosis did not result in metabolic acidosis. Plasma lactate was significantly reduced both in normoxic and hypoxic groups relative to non-ketotic standard diet fed groups. The lower plasma lactate levels in the non-ketotic standard diet group suggests the alteration of glucose metabolism by the ketone bodies, possibly through the inhibition of glycolysis or by increased lactate disposal (Puchowicz et al., 2005). Additionally, we have shown that ketosis results in the elevation of HIF-1 in brains of normoxic rats fed the ketogenic diet for 3 weeks (Fig. 2.2) and it remained elevated through week six.

One mechanism explaining the hypoxic response observed with ketosis is through the elevation of HIF-1 (see the section on hypoxia and HIF-1). However the exact biochemical mechanism remains to be explored. Together with the associated changes in glucose metabolism, one could speculate that there is a relationship

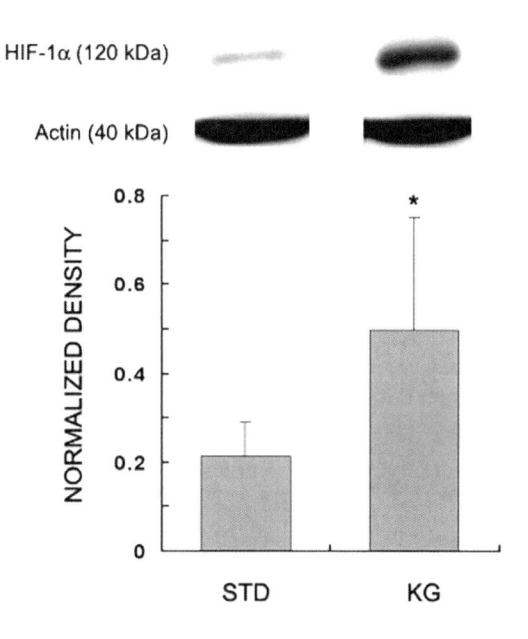

Fig. 2.2 Ketosis induced hypoxic tolerance. Western Blot protein analysis of HIF-1α in cortical brain of diet induced ketotic rats. A threefold upregulation of HIF-1α was evident with feeding a ketogenic diet (KG) for 3 weeks during normoxic conditions compared to standard fed (STD). * < 0.05, KG vs. STD

between the utilization and regulation of energy substrates and hypoxic tolerance (Semenza et al., 1994; Jones and Bergeron, 2001; Lu et al., 2002). The stabilization of HIF-1 has been described to occur by two pathways, either metabolic or hypoxia. In both cases the reaction that results in the stabilization of HIF-1 requires 2-oxoglutarate of which succinate is the product. We hypothesize that the metabolic side of HIF-1 regulation is through the inhibition of the PHD reaction by an intermediate of energy metabolism, such as succinate. The metabolism of ketones has been known to result in an increase in citric acid cycle intermediates such as citrate and succinate as compared to glucose metabolism. Figure 2.3 shows the relationship of the metabolic pathways of glucose and ketone bodies entering the citric acid cycle and HIF-1. With the metabolism of ketones, mitochondrial succinate, at elevated levels, is then transported out of the mitochondria into the cytosol resulting in the inhibition of PHD and thus stabilization of HIF-1. The inhibition of PHD is most likely through product inhibition of the reaction of HIF-1 to the hydroxylated form via PHD.

Conclusions

Mild, prolonged hypoxia evokes systemic and CNS mechanisms that result in successful acclimatization. The CNS response includes increased glucose metabolism, decreased oxidative capacity, and microvascular remodeling by capillary angiogenesis which results in decreased diffusion distances for oxygen from erythrocytes to mitochondria. The changes induced by hypoxia are reversible upon return to a normal

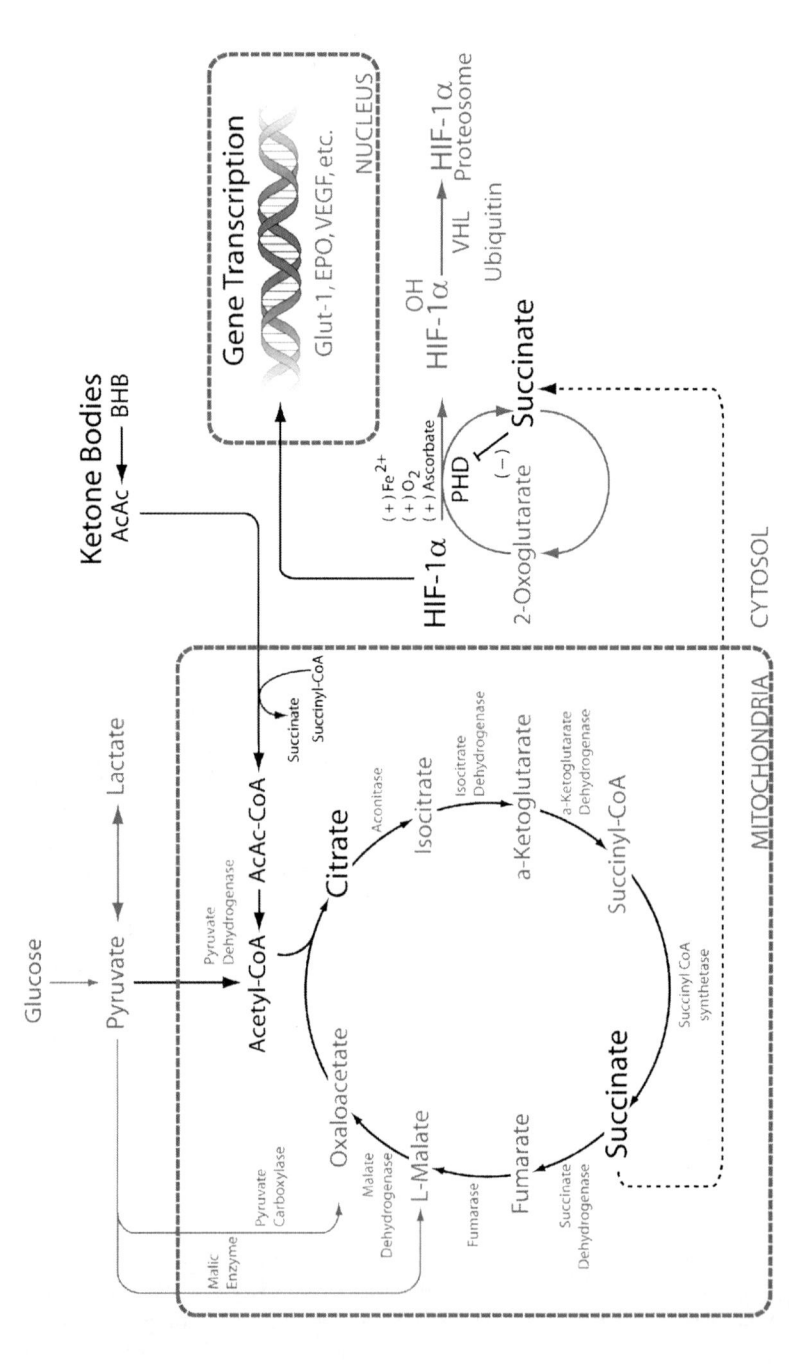

Fig. 2.3 Proposed scheme of the metabolic side of HIF-1 regulation through ketosis. The inhibition of the PHD reaction by an intermediate of energy metabolism, succinate, as a result of ketone body utilization by brain. The relationship of the metabolic pathways of glucose and ketone bodies entering the citric acid cycle and HIF-1 is illustrated. In contrast to glucose metabolism, increased ketone metabolism results in elevated levels of mitochondrial succinate, which is transported out of the mitochondria into the cytosol resulting in the inhibition of PHD and thus stabilization of HIF-1

oxygen environment. The transcription factor, HIF-1, plays a major role in orchestrating the metabolic and vascular responses to hypoxia. Cerebral glycolytic ATP utilization contributes to tissue acid—base balance. Ketones invoke a hypoxia—like response and are protective in hypoxic conditions.

Acknowledgments The authors would like to thank Constantinos Tsipis for his technical support in the art work and in the preparation of this review. This research has been supported by the National Institutes of Health, R01-NS38632 and P50 GM066309.

References

Appelhoff, R.J., Tian, Y.M., Raval, R.R., Turley, H., Harris,A.L., Pugh, C.W., Ratcliffe, P.J., and Gleadle, J.M. 2004. Differential function of the prolyl hydroxylases PHD1, PHD2, and PHD3 in the regulation of hypoxia-inducible factor. Journal of Biological Chemistry 279:38458–38465

Balasse, E.O. and Fery, F. 1989. Ketone body production and disposal: Effects of fasting, diabetes, and exercise. Diabetes Metabolic Reviews 5:247–270

Batista, C.E., Chugani, H.T., Juhasz, C., Behen, M.E., and Shankaran, S. 2007. Transient hypermetabolism of the basal ganglia following perinatal hypoxia. Pediatric Neurology 36:330–333

Beck, T. and Krieglstein, J. 1987.Cerebral circulation, metabolism, and blood—brain barrier in rats in hypocapnic hypoxia. American Journal of Physiology 252:H504–H512

Brown, M.M., Wade, J.P.H., and Marshall, J. 1985. Fundamental importance of arterial oxygen content in the regulation of cerebral blood flow in man. Brain 108:81–93

Brunori, M. and Vallone, B. 2006. A globin for the brain. FASEB Journal 20:2192–2197

Caceda, R., Gamboa, J.L., Boero, J.A., Monge, C., and Arregui, A. 2001. Energetic metabolism in mouse cerebral cortex during chronic hypoxia. Neuroscience Letters 301:171–174

Chang, A.S.Y. and D'Alecy, L.G. 1993. Hypoxia and beta-hydroxybutyrate acutely reduce glucose extraction by the brain in anesthetized dogs. Canadian Journal of Physiology and Pharmacology 71:465–472

Chávez, J.C., Pichiule, P., Boero, J., and Arregui, A. 1995. Reduced mitochondrial respiration in mouse cerebral cortex during chronic hypoxia. Neuroscience Letters 193:169–172

Chávez, J.C., Agani, F., Pichiule, P., and LaManna, J.C. 2000. Expression of hypoxic inducible factor 1α in the brain of rats during chronic hypoxia. Journal of Applied Physiology 89:1937–1942

Dalgard, C.L., Lu, H., Mohyeldin, A., and Verma, A. 2004. Endogenous 2-oxoacids differentially regulate expression of oxygen sensors. The Biochemical Journal 380:419–424

Dallas, M., Boycott, H.E., Atkinson, L., Miller, A., Boyle, J.P., Pearson, H.A., and Peers, C. 2007. Hypoxia suppresses glutamate transport in Astrocytes. Journal of Neuroscience 27:3946–3955

Dardzinski, B.J., Smith, S.L., Towfighi, J., Williams, G.D., Vannucci, R.C., and Smith, M.B. 2000. Increased plasma beta-hydroxybutyrate, preserved cerebral energy metabolism, and amelioration of brain damage during neonatal hypoxia ischemia with dexamethasone pretreatment. Pediatric Research 48:248–255

Dennis, S.C., Gevers, W., and Opie, L.H. 1991. Protons in ischemia: Where do they come from; where do they go to? Journal of Molecular and Cellular Cardiology 23:1077–1086

Dunn, J.F., Grinberg, O., Roche, M., Nwaigwe, C.I., Hou, H.G., and Swartz, H.M. 2000. Noninvasive assessment of cerebral oxygenation during acclimation to hypobaric hypoxia. Journal of Cerebral Blood Flow and Metabolism 20:1632–1635

Duong, T.Q. 2007. Cerebral blood flow and BOLD fMRI responses to hypoxia in awake and anesthetized rats. Brain Research 1135:186–194

Eiger, S.M., Kirsch, J.R., and D'Alecy, L.G. 1980. Hypoxic tolerance enhanced by beta-hydroxybutyrate-glucagon in the mouse. Stroke 11:513–517

Epstein, A.C., Gleadle, J.M., McNeill, L.A., Hewitson, K.S., O'Rourke, J., Mole, D.R., Mukherji, M., Metzen, E., Wilson, M.I., Dhanda, A., Tian, Y.M., Masson, N., Hamilton, D.L.,Jaakkola, P., Barstead, R., Hodgkin, J., Maxwell, P.H., Pugh, C.W., Schofield, C.J., and Ratcliffe, P.J. 2001. C. elegans EGL-9 and mammalian homologs define a family of dioxygenases that regulate HIF by prolyl hydroxylation. Cell 107:43–54

Freeman, R.S., Hasbani, D.M., Lipscomb, E.A., Straub, J.A., and Xie, L. 2003. SM-20, EGL-9, and the EGLN family of hypoxia-inducible factor prolyl hydroxylases. Molecules and Cells 16:1–12

Hamberger, A. and Hydén, H. 1963. Inverse enzymatic changes in neurons and glia during increased function and hypoxia. The Journal of Cell Biology 16:521–525

Harik, S.I. and LaManna, J.C. 1995. Adaptation of the brain to prolonged hypobaric hypoxia: Alterations in the microcirculation and in glucose metabolism. Queen Burlington, VT: City Printers, pp. 18–30

Harik, S.I., Behmand, R.A., and LaManna, J.C. 1994. Hypoxia increases glucose transport at blood—brain barrier in rats. Journal of Applied Physiology 77:896–901

Harilk, S.I., Lust, W.D., Jones, S.C., Lauro, K.L., Pundik, S., and LaManna, J.C. 1995. Brain glucose metabolism in hypobaric hypoxia. Journal of Applied Physiology 79:136–140

Harik, N., Harik, S.I., Kuo, N.T., Sakai, K., Przybylski, R.J., and LaManna, J.C. 1996. Time course and reversibility of the hypoxia-induced alterations in cerebral vascularity and cerebral capillary glucose transporter density. Brain Research 737:335–338

Hochachka, P.W. and Mommsen, T.P. 1983. Protons and anaerobiosis. Science 219:1391–1397

Jaakkola, P., Mole, D.R., Tian, Y.M., Wilson, M.I., Gielbert, J., Gaskell, S.J., Kriegsheim, A.A., Hebestreit, H.F., Mukherji, M., Schofield, C.J., Maxwell, P.H., Pugh, C.W., and Ratcliffe, P.J. 2001. Targeting of HIF-alpha to the von Hippel—Lindau ubiquitylation complex by O_2-regulated prolyl hydroxylation. Science 292:468–472

Jones, N.M. and Bergeron, M. 2001. Hypoxic preconditioning induces changes in hif-1 target genes in neonatal rat brain. Journal of Cerebral Blood Flow and Metabolism 21:1105–1114

Kirsch, J.R. and D'Alecy, L.G. 1979. Effect of altered availability of energy-yielding substrates upon survival from hypoxia in mice. Stroke 10:288–291

Lai, J.C., White, B.K., Buerstatte, C.R., Haddad, G.G., Novotny, E.J., Jr., and Behar, K.L. 2003. Chronic hypoxia in development selectively alters the activities of key enzymes of glucose oxidative metabolism in brain regions. Neurochemical Research 28:933–940

LaManna, J.C. 2007. Hypoxia in the central nervous system. Essays in Biochemistry 43:139–152

LaManna, J.C. and Harik, S.I. 1997. Brain metabolic and vascular adaptations to hypoxia in the rat. Review and update. Advances in Experimental Medicine and Biology 428:163–167

LaManna, J.C., Vendel, L.M., and Farrell, R.M. 1992. Brain adaptation to chronic hypobaric hypoxia in rats. Journal of Applied Physiology 72:2238–2243

LaManna, J.C., Kutina-Nelson, K.L., Hritz, M.A., Huang, Z., and Wong-Riley, M.T.T. 1996. Decreased rat brain cytochrome oxidase activity after prolonged hypoxia. Brain Research 720:1–6

LaManna, J.C., Chavez, J.C., and Pichiule, P. 1997. Genetics and gene expression of glycolysis. In: "Handbook of Neurochemistry and Molecular Neurobiology", 3rd edn., "Brain Energetics, Integration of Molecular and Cellular Processes" (vol. eds., GE Gibson and GA Dienel) Springer, 771–788.

LaManna, J.C., Chavez, J.C., and Pichiule, P. 2004. Structural and functional adaptation to hypoxia in the rat brain. Journal of Experimental Biology 207:3163–3169

Lauro, K.L. and LaManna, J.C. 1997. Adequacy of cerebral vascular remodeling following three weeks of hypobaric hypoxia. Examined by an integrated composite analytical model. Advances in Experimental Medicine and Biology 411:369–376

Li, R.C., Lee, S.K., Pouranfar, F., Brittian, K.R., Clair, H.B., Row, B.W., Wang, Y., and Gozal, D. 2006. Hypoxia differentially regulates the expression of neuroglobin and cytoglobin in rat brain. Brain Research 1096:173–179

Lu, H., Forbes, R.A., and Verma, A. 2002. Hypoxia-inducible factor 1 activation by aerobic glycolysis implicates the Warburg effect in carcinogenesis. The Journal of Biological Chemistry 277:23111–23115

Mashina, S.Y., Aleksandrin, V.V., Goryacheva, A.V., Vlasova, M.A., Vanin, A.F., Malyshev, I.Y., and Manukhina, E.B. 2006. Adaptation to hypoxia prevents disturbances in cerebral blood flow during neurodegenerative process. Bulletin of Experimental Biology and Medicine 142:169–172

Masuda, R., Monahan, J.W., and Kashiwaya, Y. 2005. D-beta-hydroxybutyrate is neuroprotective against hypoxia in serum-free hippocampal primary cultures. Journal of Neuroscience Research 80:501–509

Mishra, O.P. and Delivoria-Papadopoulos, M. 1999. Cellular mechanisms of hypoxic injury in the developing brain. Brain Research Bulletin 48:233–238

Mortola, J.P. 1993. Hypoxic hypometabolism in mammals. News in Physiological Sciences 8:79–82

Myles, W.S. 1976. Survial of fasted rats exposed to altitude. Canadian Journal of Physiology and Pharmacology 54:883–886

Neubauer, J.A., Melton, J.E., and Edelman, N.H. 1990. Modulation of respiration during brain hypoxia. Journal of Applied Physiology 68:441–451

Nilsson, G.E. and Lutz, P.L. 1991. Release of inhibitory neurotransmitters in response to anoxia in turtle brain. The American Journal of Physiology (AJP) — Legacy 261:R32–R37

Owen, O.E., Morgan, A.P., Kemp, H.G., Sullivan, J.M., Herrera, M.G., and Cahill, G.F.jr. 1967. Brain metabolism during fasting. Journal of Clinical Investigation 46:1589–1595

Pichiule, P. and LaManna, J.C. 2002. Angiopoietin-2 and rat brain capillary remodeling during adaptation and de-adaptation to prolonged mild hypoxia. Journal of Applied Physiology 93:1131–1139

Pichiule, P., Chavez, J.C., and LaManna, J.C. 2004. Hypoxic regulation of angiopoietin-2 expression in endothelial cells. Journal of Biological Chemistry 279:12171–12180

Puchowicz, M.A., Emancipator, D.S., Xu, K., Magness, D.L., Ndubuizu, O.I., Lust, W.D., and LaManna, J.C. 2005. Adaptation to chronic hypoxia during diet-induced ketosis. Advances in Experimental Medicine and Biology 566:51–57

Pulsinelli, W.A. and Duffy, T.E. 1979. Local cerebral glucose metabolism during controlled hypoxemia in rats. Science 204:626–629

Rising, C.L. and D'Alecy, L.G. 1989. Hypoxia-induced increases in hypoxic tolerance augmented by beta-hydroxybutyrate in mice. Stroke 20:1219–1225

Selak, M.A., Armour, S.M., Mackenzie, E.D., Boulahbel, H., Watson, D.G., Mansfield, K.D., Pan, Y., Simon, M.C., Thompson, C.B., and Gottlieb, E. 2005. Succinate links TCA cycle dysfunction to oncogenesis by inhibiting HIF-alpha prolyl hydroxylase. Cancer Cell 7:77–85

Semenza, G.L. 2007. Oxygen-dependent regulation of mitochondrial respiration by hypoxia-inducible factor 1. The Biochemical Journal 405:1–9

Semenza, G.L., Roth, P.H., Fang, H.M., and Wang, G.L. 1994. Transcriptional regulation of genes encoding glycolytic enzymes by hypoxia-inducible factor 1. Journal of Biological Chemistry 269:23757–23763

Sharp, F.R. and Bernaudin, M. 2004. HIF1 and oxygen sensing in the brain. Nature Reviews Neuroscience 5:437–448

Shimizu, T., Uehara, T., and Nomura, Y. 2004. Possible involvement of pyruvate kinase in acquisition of tolerance to hypoxic stress in glial cells. Journal of Neurochemistry 91:167–175

Siddiq, A., Aminova, L.R., and Ratan, R.R. 2007. Hypoxia inducible factor prolyl 4-hydroxylase enzymes: Center stage in the battle against hypoxia, metabolic compromise and oxidative stress. Neurochemical Research 32:931–946

Siesjö, B.K. 1978. Brain Energy Metabolism. WileyChichester:

Stewart, P.A., Isaacs, H., LaManna, J.C., and Harik, S.I. 1997. Ultrastructural concomitants of hypoxia-induced angiogenesis. Acta Neuropathologica 93:579–584

Sun, Y., Jin, K., Peel, A., Mao, X.O., Xie, L., and Greenberg, D.A. 2003. Neuroglobin protects the brain from experimental stroke invivo. Proceedings of the National Academy of Sciences of the United States of America 100:3497–3500

Sun, X., He, G., Qing, H., Zhou, W., Dobie, F., Cai, F., Staufenbiel, M., Huang, L.E., and Song, W. 2006. Hypoxia facilitates Alzheimer's disease pathogenesis by up-regulating BACE1 gene expression. Proceedings of the National Academy of Sciences of the United States of America 103:18727–18732

Suzuki, M., Suzuki, M., Sato, K., Dohi, S., Sato, T., Matsuura, A., and Hiraide, A. 2001. Effect of beta-hydroxybutyrate, a cerebral function improving agent, on cerebral hypoxia, anoxia and ischemia in mice and rats. Japanese Journal of Pharmacology 87:143–150

Veech, R.L. 2004. The therapeutic implications of ketone bodies: the effects of ketone bodies in pathological conditions: ketosis, ketogenic diet, redox states, insulin resistance, and mitochondrial metabolism. Prostaglandins, Leukotrienes and Essential Fatty Acids 70:309–319

Wood, S.C. and Gonzales, R. 1996. Hypothermia in hypoxic animals: mechanisms, mediators, and functional significance. Comparative Biochemistry and Physiology 113B:37–43

Xu, K. and LaManna, J.C. 2006. Chronic hypoxia and the cerebral circulation. Journal of Applied Physiology 100:725–730

Chapter 3
Hypoglycemic Brain Damage

Roland N. Auer

Historical Aspects of Hypoglycemia

Although the prevalence of hypoglycemia is thought to be high, blood glucose levels rarely substantiate this (Anderson and Lev-Ran, 1985). This situation changes entirely in the context of diabetes, where hypoglycemic episodes occur with a frequency and severity determined by the intensity of insulin treatment. Indeed, diabetes treatment is a balance (The DCCT Research Group, 1993) between the desire to prevent retinopathy, neuropathy and nephrology and the desire to prevent severe hypoglycemia and permanent brain damage.

Knowledge of hypoglycemic brain damage is thus important in the clinical management of diabetes. But profound hypoglycemia also occurs in the context of insulin overdose of either homicidal or suicidal nature. Other clinical contexts include medication error, where insulin is mistakenly given to the wrong, non-diabetic patient or when insulin dose is miscalculated for a diabetic. Oral hypoglycemic mediations releasing endogenous insulin can cause hypoglycemic brain damage. Lastly, tumors of the β-cells of the islets of Langerhans, so-called insulinomas, can cause hypoglycemic brain damage.

While hypoglycemia associated with starvation, or hypoglycemia accompanying countless diseases including widespread cancer, adversely affects brain function, we note that hypoglycemic brain damage is not produced. An artificial stimulus or source of insulin is needed to produce hypoglycemic brain damage. Chronic low blood sugar, however prolonged, does not suffice. Coma and a flat EEG and a subsequent time period are prerequisites for hypoglycemic brain damage. Insulin coma was first produced systematically in attempts to ameliorate psychiatric conditions; however, if coma was reversed prior to 30 min, brain injury was limited. This is very unlike ischemia where 30 min of global ischemia is intolerable.

When insulin was only in its second decade of medical use, it was developed, in desperation, into a therapy to attempt to alleviate the mental devastation wrought by schizophrenia. The therapy involved 30 min of hypoglycemic coma, or as it was then termed "insulin coma" (Sakel, 1937). Our present understanding of hypoglycemic coma allows us to conceive of spreading depression of Leao passing over the brain surface followed by complete loss of the direct current potential of the cerebral cortex thereafter. Thus begins the time of profound hypoglycemia, with brain-damaging potential. The desired period of coma, after some experience with this procedure, was 30 min, since it was discovered that if the patient remained in coma for longer than 30 min, he would be tragically transformed from a "reversible coma" to an "irreversible coma" (Baker, 1938; Fazekas et al., 1951). This, we now know, was due to an accelerating quantity of neuronal necrosis that occurs between 30 and 60 min of hypoglycemic coma (Auer et al., 1984a,b). When treating patients, inattentiveness to the time would prolong coma beyond potential for recovery, since 60 min of hypoglycemic coma leads to virtual decortication, such that a large number of cortical neurons are lost. After the Second World War, the development of the first effective medications in psychiatry led to the phasing out of insulin-induced hypoglycemia as a therapy for psychiatric disease (Mayer-Gross, 1951).

It should be noted that prior to the introduction of Sakel's therapy, indeed prior to the discovery of insulin, hypoglycemia must have been seen in the context of insulin-secreting pancreatic tumors (Terbrüggen, 1932). However, the stimuli for the nosologic recognition of hypoglycemic brain damage were (Abdul-Rahman and Siesjö, 1980) the discovery of insulin in 1921, (Abdul-Rahman et al., 1980) the ensuing widespread use of insulin in the treatment of diabetes and (Agardh and Siesjö, 1981) the use of hypoglycemic coma in the treatment of schizophrenia. Critical reading of this historical literature reveals that the duration of coma, not the blood sugar level, is critical in determining hypoglycemic brain damage. Since a flat EEG is a clinical counterpart of coma, we begin with a brief review of the electroencephalogram.

The EEG in Hypoglycemia

Disappearance of brain electrical activity is an all-or-none phenomenon, and is a necessary prerequisite for hypoglycemic brain damage. Animals allowed to have delta waves for hours, still show no dead neurons (Auer et al., 1984a,b). The electroencephalogram (EEG) is usually not obtained during hypoglycemia in the usual clinical situations encountered, but controlled hypoglycemia and EEG recording have been done in humans (Meyer and Portnoy, 1958). Such studies have shown us that focal neurological deficits can appear as glucose delivery reduced to a particular brain region. Such focal neurological deficits are reversed on glucose administration. Experimentally, it has been established that EEG determines the presence of brain damage over a range of blood sugars that vary by more than a factor of 10. The clinical state thus trumps the absolute level of blood glucose, in importance.

The normal EEG consists of waves in the alpha range of $8 - 13$ Hz (α waves) and beta range of $13 - 25$ Hz (β waves). Waves in the theta range of $4 - 8$ Hz (θ waves) constitute a very minor component of the normal EEG and delta wave activity in the range of $1–4$ Hz (δ waves) is absent.

As the blood glucose levels progressively drop in hypoglycemia to the range of $1–2$ mM, θ waves increase and coarse δ waves appear. These are accompanied by

Table 3.1 Stages of hypoglycemia

Clinical	EEG	Blood glucose (mM)
Normal	Normal	> 3.5
Anxiety (adrenergic discharge)	↑amplitude ↓frequency (θ, δ waves)	2–3.5
Stupor	δ waves	1–2
Coma Cushing response (↑BP)	Flat	< 1.36

clinical stupor or drowsiness (see Table 3.1). Changes in the brain monoamines dopamine, noradrenaline and serotonin already occur at this stage (Agardh et al., 1979), probably explaining at least partly, the changes in mentation that occur in the early, pre-coma stages of hypoglycemia. Free fatty acids increase over six times (Agardh et al., 1980), due to phospholipid breakdown, and there is inhibition of plasma membrane function and contained ion pumps (Agardh et al., 1982). Metabolically, this still pre-lethal stage corresponds also to progressive carbohydrate depletion in cerebral tissue, until partial energy failure occurs in a threshold manner. This usually occurs when brain glucose has fallen by over 97% (Feise et al., 1976), and blood glucose to the range of 1 mM (18 mg%).

The cerebellum suffers a lesser metabolic insult (Agardh and Siesjö, 1981; Agardh et al., 1981a,b), probably due to the greater efficiency of the cerebellar glucose transporter (LaManna and Harik, 1985), explaining the relative resistance of the cerebellum to neuronal death due to hypoglycemic brain damage. The cerebellum and brainstem are so resistant that protein synthesis actually continues during hypoglycemic coma (Kiessling et al., 1986).

As the duration of hypoglycemia increases, coma finally supervenes and this is accompanied electroencephalographically by isoelectricity, or flat EEG. The blood glucose is by now almost always in the range of < 1 mM. The signs and stages of hypoglycemia are outlined in Table 3.1.

The absolute level of the blood sugar is unimportant once it reaches the asymptotic low levels of hypoglycemic coma. It is the fact of cerebral EEG isoelectricity that is the harbinger of neuronal necrosis. Experimentally, a flat EEG was seen over the range of blood glucose levels, from 1.36 down to 0.12 mM (Auer et al., 1984a,b). This is why the clinical state (or EEG) is so much more valuable data than the absolute level of low blood glucose, in assessment of potential brain damage due to hypoglycemia. Controversies surrounding diabetic children, who are potentially hypoglycemic in the classroom, must bear these principles in mind.

Neurochemistry

Glycolytic flux through the Embden—Myerhof pathway is obviously decreased in hypoglycaemia, contributing to a decreased cerebral metabolic rate for glucose (CMRgl) (Abdul-Rahman and Siesjö, 1980). Transamination reactions occur, and the aspartate—glutamate transaminase reaction is shifted to the left (Fig. 3.1).

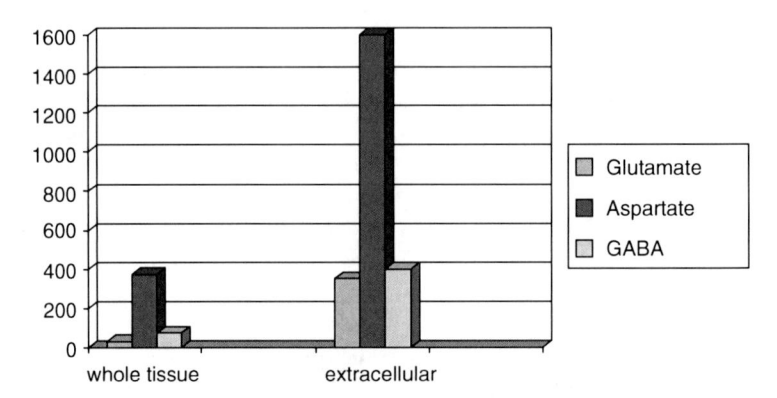

$$aspartate\ +\ \alpha\text{-ketoglutarate}\ \leftrightarrows\ oxaloacetate\ +\ glutamate$$

Fig. 3.1 Altered metabolism of excitatory amino acids during hypoglycemic coma. Oxaloacetate and α-ketoglutarate are the corresponding α-keto-acids to aspartate and glutamate, respectively. During hypoglycemia the aspartate—glutamate transaminase reaction is driven to the left

Fig. 3.2 Hypoglycemia causes an increase in tissue aspartate and decrease in glutamate, while both amino acids flood the extracellular space of the brain. GABA similarly floods the extracellular space, but its inhibitory effects are often insufficient to prevent hypoglycemic convulsions in the face of the excitatory amino acids released. Data from Norberg and Siesjö (1976) for whole tissue and extracellular data from Sandberg et al. (1986) (*See also* Color Insert)

Aspartate in the tissue increases fourfold (Agardh et al., 1978). The increased aspartate spills over from the intracellular to the extracellular space of the brain, where aspartic acid levels increase to 1,600% of control (Sandberg et al., 1985). Excitatory and inhibitory amino acid changes are shown in Fig. 3.2. The balance shifts toward excitation, which is why seizures can be seen in hypoglycemia despite some degree of energy failure (~25–30%).

These are the salient neurochemical features of hypoglycemic brain damage that result in neuronal death. But other biochemical alterations that occur are of interest in that many are the opposite of those that occur in ischemia, which has often been equated with hypoglycemia. One of these perturbations is the consistent development of a profound tissue alkalosis. The cause is twofold. Increased ammonia production as the cell catabolizes protein and delaminates amino acids is one cause. Ammonia is a very strong base and its tissue production powerfully drives up cellular pH. The second reason for alkalosis in hypoglycemia is lactate deficiency. The normally acidifying production of lactic acid is mitigated in profound hypoglycemia. Lactate has a pKa of 3.83, and it tends to pull the tissue pH towards its own pKa. Tonic production of lactate is reduced due to the decreased glycolytic flux in hypoglycemia. One morphologic consequence of this is that infarction is impossible in hypoglycemia due to the impossibility of increasing tissue lactate and lowering pH. Infarction of brain tissue results from profound lactic acidosis or vascular occlusion. Neither occurs in pure hypoglycemic insults to the brain. Thus, selective neuronal necrosis, but not infarction, is seen in hypoglycemia. These events conspire to increase cellular pH to roughly 7.5 from a normal of 7.3 (Pelligrino and Siesjö, 1981). This hypoglycemic alkalosis contrasts with the acidosis engendered by brain tissue in ischemia.

Energy failure occurs in hypoglycemia, with the energy charge falling abruptly to roughly 25–30% of what is normal. Oxidative phosphorylation is decreased and inorganic phosphate is increased (Behar et al., 1985). Adenosine triphosphate (ATP), the chief player in determining the cellular energy state, is reduced. Adenosine monophosphate is increased. Energy equivalents arise through continued turning of the Krebs cycle during hypoglycemia (Sutherland et al., 2008). Conceptually, if the aspartate—glutamate transaminase reaction is written across the Krebs cycle, this aids our understanding of how the Krebs cycle can continue to turn without glucose: the Krebs cycle becomes truncated (Fig. 3.3).

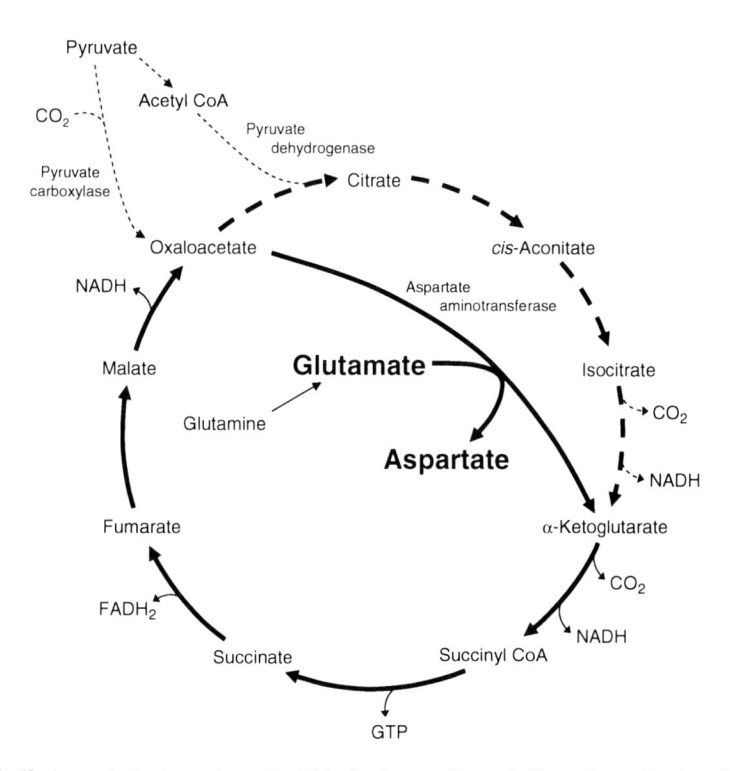

Fig. 3.3 Krebs cycle in hypoglycemia. This fundamental metabolic pathway is altered during hypoglycemic coma. The Krebs cycle continues to turn, however short the glucose supply from Embden—Myerhof glycolysis is, due to a short circuit of metabolites across the circle by amino acid transamination

Brain energy metabolism can be sustained not only by consumption of endogenous substrates such as proteins and fatty acids (Agardh et al., 1980), but also through exogenous molecules which still circulate through the blood in hypoglycemia. These include glycerol (Sloviter et al., 1966), lactate (McIlwain, 1953), and ketone bodies (β-OH butyrate and acetoacetate). Lactate alone can substitute for roughly one-fourth of glucose use (Nemoto and Hoff, 1974).

Oxidation is favored over reduction during hypoglycemia, and all cellular redox pairs tilt their reactions toward oxidation in hypoglycemia. Thus, lactate/pyruvate, NAD/NADH, GSG/GSSG and NADP/NADPH all shift their equilibria toward the oxidized compound of the pair (Agardh et al., 1978). Whether the oxidized cellular state of hypoglycemia leads to oxidative damage to DNA or proteins is still unknown.

Hypoglycemic brain damage is characterized by not only an increase in cerebral blood flow (CBF) (Abdul-Rahman et al., 1980) but interestingly also in a relatively upheld cerebral metabolic rate for oxygen ($CMRO_2$) (Eisenberg and Seltzer, 1962). To account for upheld $CMRO_2$ with decreased CMRgl the use of endogenous substrates by the brain must necessarily be invoked. The use of brain tissue fatty acids and protein catabolic products explain the stoichiometric discrepancy between glucose consumed and CO_2 produced during hypoglycemia. It should be noted that the increased CBF is non-specific, and occurs in many brain insults: the increase in blood pressure represents an attempt by the body to maintain the brain (the Cushing response) in the face of an insult to the brain.

Neuropathology

Once the EEG goes flat, neuronal necrosis appears over 10–30 min (Auer et al., 1984a,b) as aspartate floods the extracellular space (Sandberg et al., 1986). These necrotic neurons can be stained with any acid histological stain, and the increased affinity for acid dyes will cause them to be acidophilic. Since most histologic stains of the brain involve a pink or red acid dye, acidophilic neurons are invariably red in routinely stained tissue sections.

A conspicuous feature of hypoglycemic brain damage in the rat is neuronal necrosis in the dentate gyrus (Fig. 3.4) of the hippocampus (Auer et al., 1985). This seems to be due to the proximity of the NMDA receptors of the molecular layer of the dentate, to the CSF spaces containing the excitatory amino acid aspartate. A similar picture of dentate necrosis is seen sometimes, in human cases of hypoglycemic coma. Although the concept of excitotoxicity was unknown in 1938, toxicity of some kind was postulated by Arthur Weil, when he noticed dentate gyrus neurons near the CSF were necrotic in rabbits (Weil et al., 1938).

One of the mysteries of hypoglycemic brain damage has been its asymmetry. It seems impossible a priori for a metabolic insult to cause an asymmetric pattern of damage in the brain. However, with the understanding that a flat EEG is necessary for brain damage to occur, and the discovery that this is occasionally asynchronous

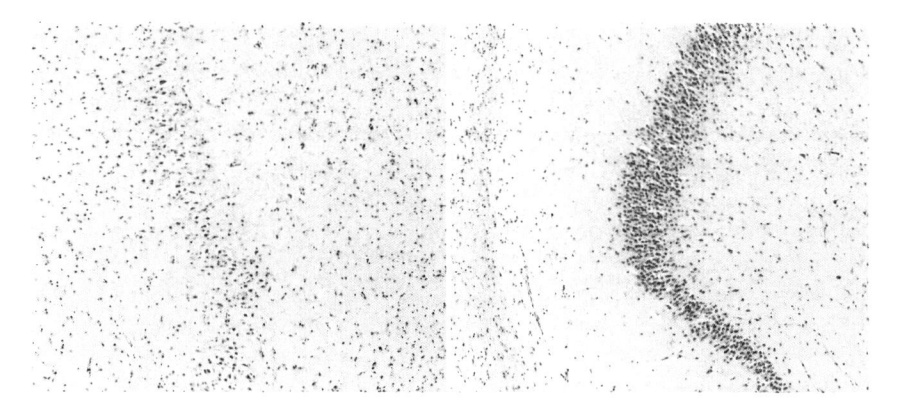

Fig. 3.4 Left and right hippocampus in human hypoglycemia to show asymmetry of brain damage. The dentate granule cells are depleted on the left, but the normal band of well-populated dentate granule cells is seen on the right. Cresyl violet stain (*See also* Color Insert)

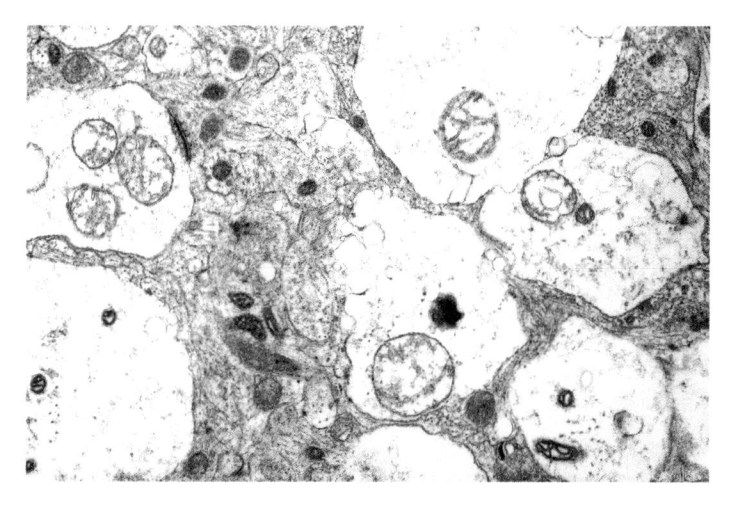

Fig. 3.5 Electron microscopy after only 10 min of EEG silence (see text) reveals already swollen dendrites, and contained mitochondria, sparing intervening axons. This electron microscopic appearance is due to the dendrite location of excitatory amino acid receptors

between the hemispheres (Harris et al., 1984; Wieloch et al., 1984), asynchrony is easily explained. If one hemisphere should develop a flat EEG 10 min before the other, then, a duration of 20 min of flat EEG for the entire brain would really amount in that case to 30 min for one hemisphere but only 20 min for the other, as assessed by an interhemispheric EEG. There is another, more practical implication of this besides theoretically explaining hypoglycemic asymmetry in brain damage. The neuropathologist should never use asymmetry as a criterion favoring either ischemic or hypoglycemic brain damage over the other, since both can be asymmetric.

Electron microscopically (Fig. 3.5), the early lesion of the neuron is marked by dendritic swelling (Auer et al., 1985). This spares the intervening neuropil. The lesion is the electron microscopic hallmark of an excitotoxin. The reason for this is the selective dendritic location of receptors. Thus, amino acids bind to glutamate excitatory receptors on neuronal dendrites, open Ca^{2+} and Na^+ channels which lead also to water fluxes across the membrane, and to the swelling of dendrites, sparing the intervening axons. When cell membrane breaks spread to the soma or perikaryon, the neuron dies.

Hypoglycemia may be summarized as a novel insult that has a number of features unsuspected several decades ago. These include a positive, excitotoxic mechanism of neuronal death, not merely neuronal death by starvation. Asymmetry is sometimes seen, and is explained by the asynchronous onset of electrocerebral silence between the hemispheres. And selective necrosis of the dentate gyrus is not seen in cerebral ischemia, the dentate being the last structure within the hippocampus to be destroyed by ischemia. Based on these principles, it is sometimes possible to tell hypoglycemic from ischemic brain damage in human brains.

References

Abdul-Rahman A, Siesjö BK. 1980. Local cerebral glucose consumption during insulin-induced hypoglycemia, and in the recovery period following glucose administration. Acta Physiol Scand 110:149–159

Abdul-Rahman A, Agardh CD, Siesjö BK. Local cerebral blood flow in the rat during severe hypoglycemia and in the recovery period following glucose injection. Acta Physiol Scand 1980. 109:307–314

Agardh CD, Siesjö BK. Hypoglycemic brain injury: Phospholipids, free fatty acids, and cyclic nucleotides in the cerebellum of the rat after 30 and 60 minutes of severe insulin-induced hypoglycemia. J Cereb Blood Flow Metab 1981. 1:267–275

Agardh CD, Folbergrová J, Siesjö BK. Cerebral metabolic changes in profound insulin-induced hypoglycemia, and in the recovery period following glucose administration. J Neurochem 1978. 31:1135–1142

Agardh CD, Carlsson A, Lindqvist M, Siesjö BK. The effect of pronounced hypoglycemia on monoamine metabolism in rat brain. Diabetes 1979. 28:804–809

Agardh CD, Kalimo H, Olsson Y, Siesjö BK. Hypoglycemic brain injury. I. Metabolic and light microscopic findings in rat cerebral cortex during profound insulin-induced hypoglycemia and in the recovery period following glucose administration. Acta Neuropathol 1980. 50:31–41

Agardh CD, Chapman AG, Nilsson B, Siesjö BK. Endogenous substrates utilized by rat brain in severe insulin-induced hypoglycemia. J Neurochem 1981a 36:490–500

Agardh CD, Kalimo H, Olsson Y, Siesjö BK. Hypoglycemic brain injury: Metabolic and structural findings in rat cerebellar cortex during profound insulin-induced hypoglycemia and in the recovery period following glucose administration. J Cereb Blood Flow Metab 1981b. 1:71–84

Agardh CD, Chapman AG, Pelligrino D, Siesjö BK. Influence of severe hypoglycemia on mitochondrial and plasma membrane function in rat brain. J Neurochem 1982. 38:662–668

Anderson RW, Lev-Ran A. Hypoglycemia: the standard and the fiction. Psychosomatics 1985. 26:38–47

Auer RN, Olsson Y, Siesjö BK. Hypoglycemic brain injury in the rat. Correlation of density of brain damage with the EEG isoelectric time: a quantitative study. Diabetes 1984a 33:1090–1098

Auer RN, Wieloch T, Olsson Y, Siesjö BK. The distribution of hypoglycemic brain damage. Acta Neuropathol (Berl) 1984b. 64:177–191

Auer RN, Kalimo H, Olsson Y, Wieloch T. The dentate gyrus in hypoglycemia: pathology implicating excitotoxin-mediated neuronal necrosis. Acta Neuropathol (Berl) 1985. 67:279–288

Baker AB. Cerebral lesions in hypoglycemia. II. Some possibilities of irrevocable damage from insulin shock. Arch Pathol 1938. 26:765–776

Behar KL, den Hollander JA, Petroff OAC, Hetherington HP, Prichard JW, Shulman RG. Effect of hypoglycemic encephalopathy upon amino acids, high-energy phosphates, and pHi in the rat brain in vivo: Detection by sequential 1H and 31P NMR spectroscopy. J Neurochem 1985. 44:1045–1055

Eisenberg S, Seltzer HS. The cerebral metabolic effects of acutely induced hypoglycemia in human subjects. Metabolism 1962. 11:1162–1168

Fazekas JF, Alman RW, Parrish AE. Irreversible posthypoglycemic coma. Am J Med Sci 1951. 222:640–643

Feise G, Kogure K, Busto R, Scheinberg P, Reinmuth O. Effect of insulin hypoglycemia upon cerebral energy metabolism and EEG activity in the rat. Brain Res 1976. 126:263–280

Harris RJ, Wieloch T, Symon L, Siesjö BK. Cerebral extracellular calcium activity in severe hypoglycemia: Relation to extracellular potassium and energy state. J Cereb Blood Flow Metab 1984. 4:187–193

Kiessling M, Auer RN, Kleihues P, Siesjö BK. Cerebral protein synthesis during long-term recovery from severe hypoglycemia. J Cereb Blood Flow Metab 1986. 6:42–51

LaManna JC, Harik SI. Regional comparisons of brain glucose influx. Brain Res 1985. 326:299–305

Mayer-Gross W. Insulin coma therapy of schizophrenia: Some critical remarks on Dr. Sakel's report. J Ment Sci 1951. 97:132–135

McIlwain H. Glucose level, metabolism, and response to electrical impulses in cerebral tissues from man and laboratory animals. Biochem J 1953. 55:618–624

Meyer JS, Portnoy HD. Localized cerebral hypoglycemia simulating stroke. Neurology 1958. 8:601–614

Nemoto EM, Hoff JT. Lactate uptake and metabolism by brain during hyperlactatemia and hypoglycemia. Stroke 1974. 5:48–53

Norberg K, Siesjö BK. Oxidative metabolism of the cerebral cortex of the rat in severe insulin induced hypoglycemia. J Neurochem 1976. 26:345–352

Pelligrino D, Siesjö BK. Regulation of extra- and intracellular pH in the brain in severe hypoglycemia. J Cereb Blood Flow Metab 1981. 1:85–96

Sakel M. The methodical use of hypoglycemia in the treatment of psychoses. Am J Psychiat 1937. 94:111–129

Sandberg M, Nyström B, Hamberger A. Metabolically derived aspartate – Elevated extracellular levels in vivo in iodoacetate poisoning. J Neurosci Res 1985. 13:489–495

Sandberg M, Butcher SP, Hagberg H. Extracellular overflow of neuroactive amino acids during severe insulin-induced hypoglycemia: in vivo dialysis of the rat hippocampus. J Neurochem 1986. 47:178–184

Sloviter HA, Shimkin P, Suhara K. Glycerol as a substrate for brain metabolism. Nature 1966. 210:1334–1336

Sutherland GR, Tyson RL, Auer RN. 2008. Truncation of the Krebs cycle during hypoglycemic coma. Medicinal Chemistry

Terbrüggen A. Anatomische Befunde bei spontaner Hypoglykämie infolge multipler Pankreasinseladenome. Beitr z path Anat u allg Path 1932. 88:37–59

The DCCT Research Group. The effect of intensive treatment of diabetes on the development and progression of long-term complications in insulin-dependent diabetes mellitus. The Diabetes Control and Complications Trial Research Group. N Engl J Med 1993. 329:977–986

Weil A, Liebert E, Heilbrunn G. Histopathologic changes in the brain in experimental hyperinsulinism. Arch Neurol Psychiat 1938. 39:467–481

Wieloch T, Harris RJ, Symon L, Siesjö BK. Influence of severe hypoglycemia on brain extracellular calcium and potassium activities, energy and phospholipid metabolism. J Neurochem 1984. 43:160–168Auer RN, Hugh J, Cosgrove E, Curry B. Neuropathologic findings in three cases of profound hypoglycemia. Clin Neuropathol 1989. 8:63–68

Chapter 4
Experimental Ischemia: Summary of Metabolic Encephalopathy

W. David Lust, Jennifer Zechel, and Svetlana Pundik

Introduction

Cerebral ischemia refers to a lack of adequate blood flow to the brain, which may be the result of an embolism, blood clot, blood vessel constriction secondary to increased intracranial pressure or a hemorrhage. Why the brain is so susceptible to alterations in Cerebral Blood Flow (CBF) has been extensively studied. The brain is a very demanding organ requiring an uninterrupted supply of nutrients to feed the tens of billions of cells which make up the CNS, necessary for the processing and storing of information and for controlling many vital functions within the organism. Maintaining the structure and function of this complex tissue requires a disproportionately large amount of energy when compared to most other organs of the body. This is clearly demonstrated by the fact that the brain comprises about 2% of total body mass and yet consumes about 20% of the total basal O_2 and receives approximately 15% of the resting cardiac output. An important concept in normal brain metabolism is that energy production is tightly coupled to energy consumption (i.e., work).

Stroke can be categorized as either focal or global ischemia as is the case in stroke and cardiac arrest, respectively. The emphasis of this review will be on the focal strokes, since the metabolic pathophysiology of this type of stroke has made greater advances in recent years and arguably is more important to understanding acute stroke. Stroke in humans, referred to as "Brain Attack, " consists of a focal neurological deficit that develops abruptly, primarily attributable to either cerebral vessel occlusion or to the spontaneous rupture of an intracranial artery with hemorrhage into the brain parenchyma or subarachnoid space (Walker and Marx, 1981). Brain infarction, a localized lesion caused by the occlusion of a brain vessel (usually an artery), accounts for about 75% of the lesions produced by stroke, with brain hemorrhage (11%) and subarachnoid hemorrhage (5%) accounting for most of the rest (Anderson and Whisnant, 1982; Robins and Baum, 1981; Sacco et al., 1982). Thus, human stroke takes many forms depending on the etiology and spatial/temporal characteristics of the lesion and the types of cell damage (see below).

D.W. McCandless (ed.) *Metabolic Encephalopathy*,
doi: 10.1007/978-0-387-79112-8_4, © Springer Science+Business Media, LLC 2009

Energy Metabolism in the Brain

It is the aim of this section to give an overview of nutrient consumption and how the energy produced from oxidative metabolism is absolutely critical for the maintenance of brain functions and the structural integrity of the brain. There are a number of books on the subject and the reader can find more detailed information in both reviews and books (Lipton, 1999; Ginsberg and Bogousslavsky, 1998; Mergenthaler et al., 2004; Welsch et al., 1997).

The brain derives the energy for function and maintenance of structure under normal circumstances from the oxidation of glucose and the formula for the oxidative catabolism of glucose is shown below:

$$C_6H_{12}O_6 + 6O_2 > energy + 6CO_2 + 6H_2O$$

where $C_6H_{12}O_6$ is glucose, O_2 is oxygen, CO_2 is carbon dioxide, H2O is water and ATP is the predominant energy formed. The oxidation of glucose can be considered as a three stage process: glycolysis, tricarboxylic acid cycle terminating in electron transport, and oxidative phosphorylation. Each glucose is metabolized to generate two pyruvates through the glycolytic pathway, yielding two equivalents of both ATP (adenosine triphosphate) and NADH (reduced nicotinamide adenine dinucleotide). The pyruvate then enters the tricarboxylic acid cycle (TCA), generating either additional reduced pyridine nucleotides (NADH) or reduced flavin mononucleotides ($FADH_2$). The NADH and $FADH_2$ serve as substrate for the respiratory chain and the proton gradient generated across the inner mitochondrial membrane drives the phosphorylation of ADP (i.e., oxidative phosphorylation). Each NADH contains sufficient usable energy to phosphorylate 3 ADPs, whereas each $FADH_2$ only has the energy for the phosphorylation of two ADP molecules. Through these three pathways, 36 ATPs theoretically can be produced from each glucose moiety, with the major production of energy occurring during electron transport and oxidative phosphorylation. In the subsequent sections, the specific pathways in the oxidation of glucose will be described in greater detail.

The standard free energy yield (ΔG) for the oxidation metabolism of glucose to carbon dioxide and water at 25°C and pH 7 is approximately −686 kcal mol^{-1} (note that the negative standard free energy favors the forward reaction). Since the brain cannot readily use energy in the form of heat or pressure, the energy produced is chemical in nature and, for the most part, is in the form of ATP. The energy for most of the active processes in the brain comes from the hydrolysis of ATP, which is coupled to reactions where either the adenylate or the phosphate moiety is first covalently bound to an enzyme or a substrate molecule and then one of the moieties is released in the form of either free ADP or Pi. The free-energy change in the brain from the hydrolysis of ATP has been estimated to be −7.86 kcal mole^{-1}, as determined from the following equation:

$$\Delta G = \Delta G° + RT \ln[ADP][Pi]/[ADP]$$

where ΔG is the free-energy change, $\Delta G°'$ is the standard free-energy change, R is the universal gas constant, T is the temperature in degrees Kelvin, [ADP] is the concentration of adenosine diphosphate, [ATP] is the concentration of ATP and [Pi] is the concentration of inorganic phosphate. The actual free-energy change in the cytosol may be as much as -14.1 kcal mole^{-1} based on the observation that free cytosolic ADP is actually only about 5% of the total tissue concentration (i.e., 30 μmol kg^{-1} wet weight) (Veech et al., 1979). The calculated higher free-energy change would increase the efficiency of energy conservation in the brain, which has been estimated to be approximately 20%. ATP can be hydrolyzed to other products such as AMP and inorganic pyrophosphate, but these energy yielding reactions are less common to the cell. Thus, the oxidation of glucose provides cellular energy in the form of ATP, the hydrolysis of which serves most of the active processes in the brain. Generally, the other high-energy phosphates, including P-creatine, phosphoenolpyruvate and 1,3 bisphosphoglycerate, lack enzymes for donating the phosphate from a high-energy source to a lower one without using ATP as an intermediate. For the interested reader, a more extensive description of the various aspects of cerebral bioenergetics are to be found in books and review articles by Erecinska and Silver (1989), Lehninger et al. (1993), and Siesjo (1978).

Biochemical and Physiological Consequences of Experimental Ischemia

In terms of the "Brain Attack" concept, the loss of the TCA and oxidative phosphorylation places the tissue energetically at risk. At best, the anaerobic glycolytic pathway can only minimally reduce the energy debt following ischemia, since the increase in energy demands caused by the disruption of ion gradients overwhelms the glycolytic capacity. Cellular energy reserves including glucose, glycogen, ATP and P-creatine in the brain are limited and the high-rate energy consumption of the brain will exhaust these stores within a minute in the absence of blood flow. An equation for determining the limited endogenous high-energy phosphate (HEP) stores in the brain is presented below (Lowry et al., 1964).

$$HEP = phosphocreatine + 2ATP + ADP + 2glucose + 2.9glycogen$$

Flow Thresholds

Modern electrophysiological techniques and accurate cerebral blood flow determinations have refined our understanding of the relationship between neuronal function, tissue viability, and critical levels of regional cerebral blood flow in both global and focal models of ischemia.

Experimental studies of middle cerebral artery occlusion in various species have demonstrated a blood flow gradient from normal flow in areas outside the affected territory, to modest decreases in the adjacent perifocal region or penumbra, to a profound drop in the ischemic core (Morawetz et al., 1978, 1979; Symon et al., 1974). The slope of this gradient, in part, depends on the extent and functional capacity of collateral blood supply. Since the arteries of the striatum are end vessels with little collateral flow, many events in the ischemic core mimic metabolically those that occur in tissue affected by global ischemia. The tissue in the border zone between moderately reduced flow (i.e., penumbra) and the ischemic core yields a graded range of changes to the transition regions and its homeostatic mechanisms. Different cellular functions, which require specific minimum levels of blood flow, are affected in these regions depending on the level of blood flow reduction (Siesjo, 1992a, 1992b). The various functional perturbations that occur once the relative blood flow decreases below these thresholds are shown in the following schematic representation (Fig. 4.1). The CBF is presented as a percent of control, since interspecies differences are evident for the control flow.

The thresholds described in this section were determined in experimental models using both smaller rodents and primates. A list of various thresholds in a number of species has been summarized (Hossmann, 1998). Similar values have been reported in humans (Siesjo, 1992a, 1992b). While absolute values may vary somewhat between species and the type of anesthesia, the percent reduction from normal flow to these thresholds appears to be uniform and constant. Critical CBF values for loss

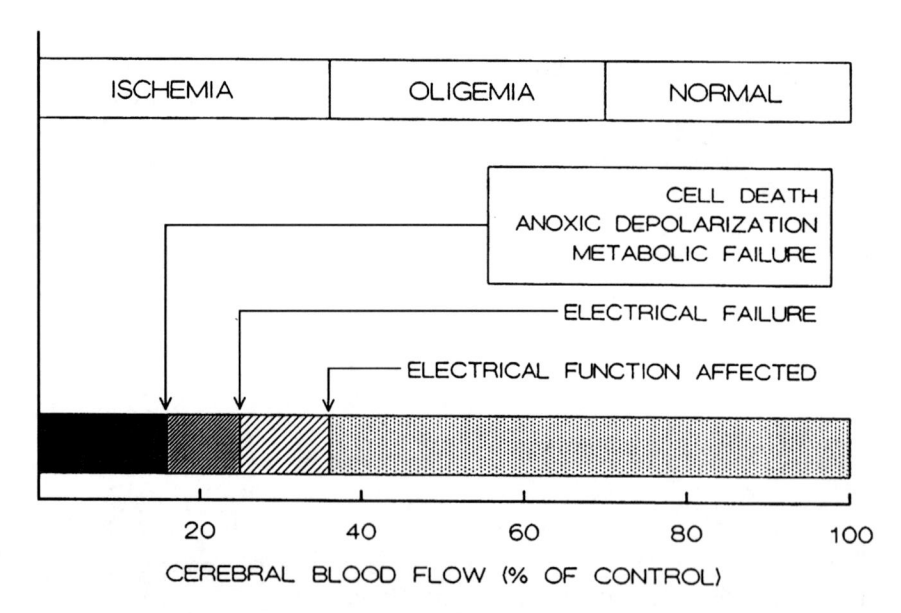

Fig. 4.1 Relative cerebral blood flow thresholds resulting in a range of functional perturbations in the brain. The range of cerebral blood flow on the abscissa range from normal, oligemia, state where total blood volume is reduced, and ischemia

of synaptic transmission in baboons, corresponding to loss of neuronal function, are between 15 and 18 ml/100 g min^{-1} or approximately 23% of control (Branston et al., 1974; Heiss et al., 1976). The threshold for membrane pump failure, and thus for loss of cellular integrity in primates, is approximately 18% of control or 10 ml/100 g min^{-1} (Astrup et al., 1977; Branston et al., 1979). The level of blood flow reduction for ion pump failure appears to be similar to that for energy failure. The presence of these two distinct thresholds implies that some regions in the peri-focal area contain cells that are electrophysiologically quiescent but nonetheless viable. These regions constitute the ischemic penumbra, defined by Astrup, Siesjo, and Symon (1981) as an area with EEG quiescence and low extracellular K$^+$. While flow reduction is one component that determines the severity of an ischemic insult, the duration of flow reduction and age of the animal are part of the equation and are also of paramount importance.

Ischemia Impact on Intermediary Metabolism

The first major comprehensive description of metabolic perturbations following global ischemia was published by Lowry et al. (1964) who described the pronounced changes in energy metabolites upon decapitation of the mouse. We have subsequently examined global ischemia in the gerbil and the high-energy and glucose related metabolite changes are presented in (Fig. 4.2). The findings of a rapid depletion of the high-energy phosphates and glucose within 1 min of bilateral occlusion in the gerbil are essentially the same as those described by Lowry et al. (1964). The consumption of glycogen, a storage form of glucosyl units, was somewhat slower as was the increase in the levels of 5′AMP. Knowing that the brain relies primarily on glucose oxidation for the production for energy, deprivation of oxygen results in anaerobic glycolysis of the endogenous glucosyl moieties, the metabolite changes are rather predictable and are simply explained by the several chemical equations (Siesjo et al., 1998).

$$1.\ ATP + H_2O \rightarrow ADP + Pi + H^+$$
$$2.\ PCr + ADP + H^+ \rightarrow Cr + ATP$$
$$3.\ 2ADP \rightarrow 5'AMP + ATP$$

The first equation is a reflection of the energy needed to maintain the brain function (ATPase) when the production of ATP by oxidative phosphorylation is compromised; the second (creatine phosphokinase) and third (myokinase) reactions are secondary pathways attempting to maintain the ATP levels. In addition, ATP can be produced by anaerobic glycolysis and this explains the relative depletion of the glucose-related metabolites and a tenfold increase in lactate with a resulting acidosis (data not shown.).

In contrast to global ischemia, the time course of energy failure in focal ischemia is slower which, in part, is explained by the localized nature of the insult and the ability

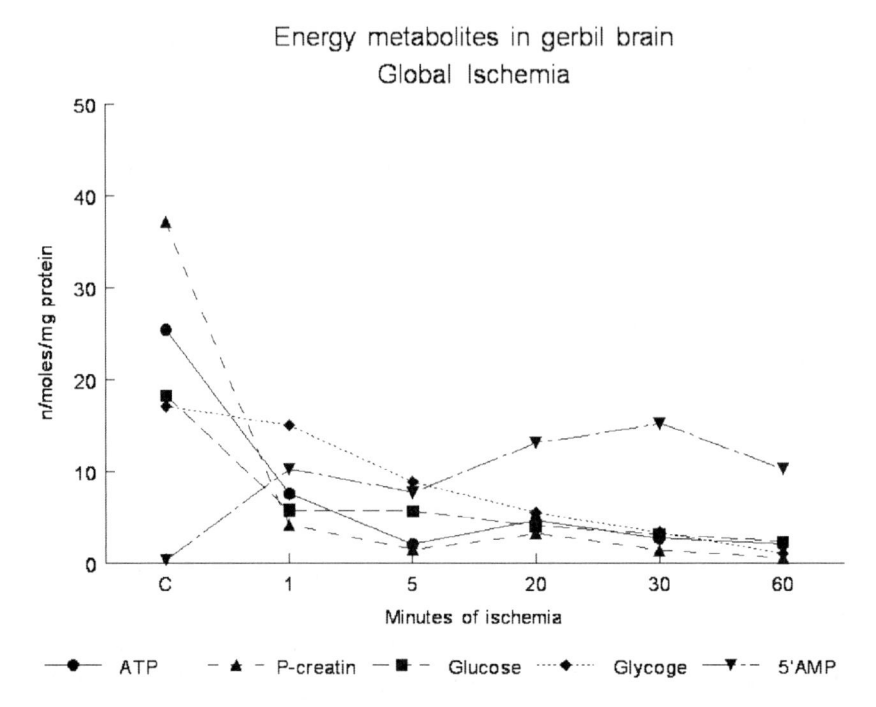

Fig. 4.2 Time course of high-energy phosphates and glucose-related metabolites in a global model of ischemia. The metabolites were determined at various times up to 1 h following bilateral occlusion of the common carotid arteries in the gerbil

of collateral circulation to offset the loss of flow to the ischemic region (Folbergrova et al., 1992). The inset of a coronal section in the ATP graph depicts the various areas sampled and the extent of the ischemia in these four regions is DS > VC > LC > DC over a 6 h of occlusion (abbreviations in legend Fig. 4.3). Clearly, the decline in ATP is attenuated in a focal setting compared to that of global ischemia, but the levels are still detectible after 6 h of occlusion. Somewhat surprising is that all four areas become infarcted after one day of deocclusion of the vessels (i.e., reperfusion) following 6 h of ischemia, even though ATP was reduced, but not depleted (see Fig. 4.9 below).

There are also events that increase the workload on the penumbra following focal ischemia. Cerebral spreading depression (SD) was first described more than 50 years ago and is characterized by a slow transient cellular depolarization moving at 3–4 mm min−1 over the surface of the cortex (Leao, 1944). Evidence of spontaneous SDs has been demonstrated in the penumbra and the trigger to the SD is thought to be the large efflux of potassium and glutamate into the extracellular space from the ischemic core (Strong and Dardis, 2005; Kempski et al., 2000). In a recent study, the extent of the workload in control and non-ischemic rats appears to be similar, but the ability of the bioenergetics to recover was compromised in the penumbra of the ischemic animal. The conclusion is that SD in control animals does not cause cellular

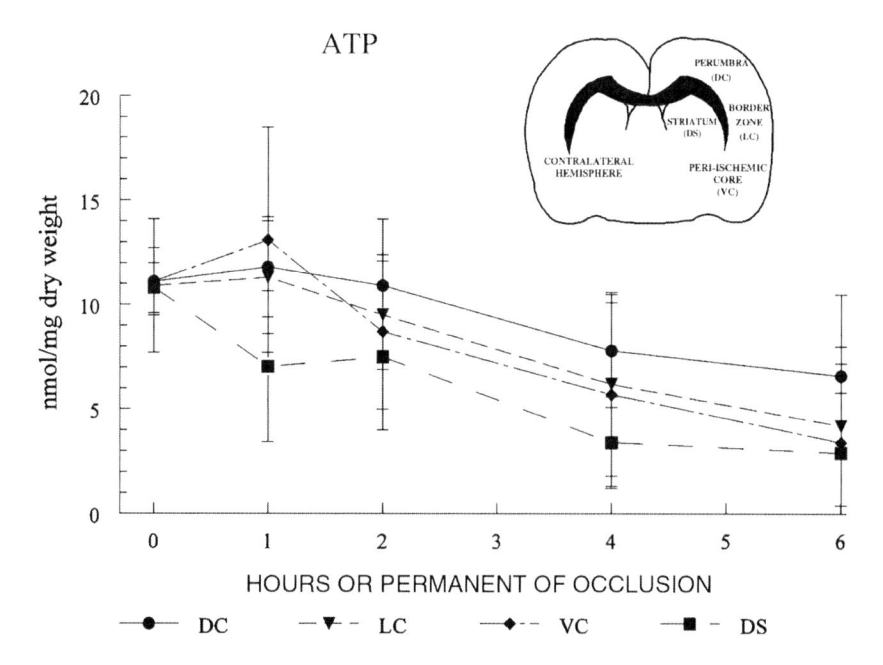

Fig. 4.3 Time course of ATP changes in a focal rat model of permanent ischemia. The inset indicates the four areas within the middle cerebral artery distribution that were dissected from lyophilized coronal sections and measured for ATP. The dorsolateral striatum (DS) is the core and the dorsolateral cortex (DC) is the penumbra with the other regions transitional between these two regions. In this figure, the border zone is the lateral cortex and the per-ischemic cortex (VC). The nomenclature of the transitional area is often variable and can cause problems with the interpretation of the results between studies. Control values were determined from the cerebral cortex of the contralateral hemisphere

damage to the brain, in contrast to the additional workload in compromised tissue which causes an energy imbalance leading to the evolution of cellular damage and eventually terminal anoxic depolarization (Selman et al., 2004).

Triggered Events Secondary to Energy Failure

There are myriad changes in the intracellular milieu following ischemia which result in loss of cellular function and if allowed to persist, will cause irreversible damage or cell death. The schematic representation below shows the dramatic early effects on the brain which tend to deteriorate and a predominance of literature indicate the excessive influx of calcium may be the most devastating consequence (Fig. 4.4).

ENERGY FAILURE

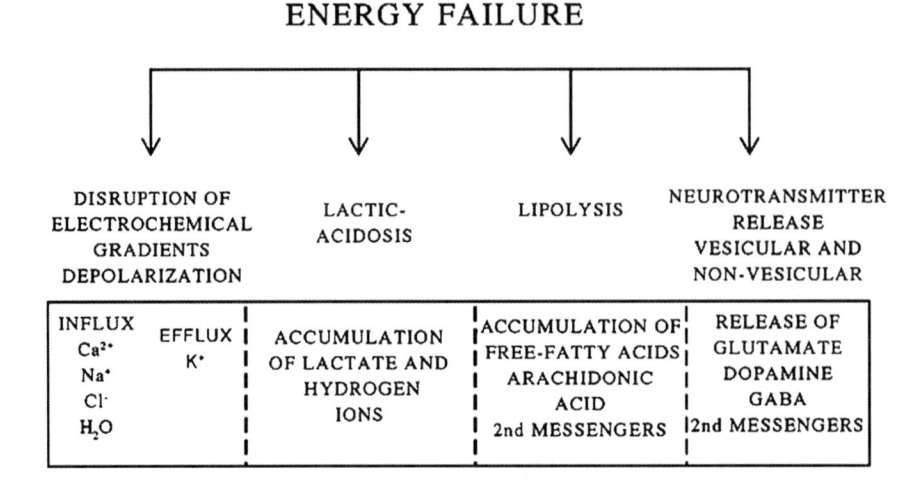

Fig. 4.4 A schematic representation indicating those events that are triggered by energy failure in the brain. In global models, the onset of ionic fluxes, lipolysis, lacticacidosis and neurotransmitter release occur almost immediately and the processes continue with time. The individual resulting events are shown in the lower boxes

Ion Homeostasis

With the onset of ischemia, there is an influx of sodium, calcium and chloride with the concomitant release of potassium. The movement of the sodium and chloride results in the influx of water and brain swelling (Kimelberg, 2005). The influx of calcium is compounded secondarily by the release of endogenous calcium stores from the endoplasmic reticulum (Phillis et al., 2002). The excessive rise in Ca^{2+} that results from an ischemia-induced failure of these homeostatic mechanisms represents a non-physiological stimulus that activates a wide array of intracellular receptors, membrane channels, proteins and enzymes, which lead to the compromise of the cell's functional and structural integrity.

Loss of Ca^{2+} homeostasis leading to an elevated level of intracellular Ca^{2+} has been implicated as a cause of irreversible cell injury in ischemia (Siesjo and Bengtsson, 1989). Both voltage-sensitive and agonist-operated calcium channels control the movement of calcium into the cell, and the latter are predominantly involved in the initiation of the pathophysiological processes resulting from the ischemia.

Since Ca^{2+} plays an important role as an intracellular messenger, the rise in Ca^{2+} may disrupt several intracellular processes and thus compromise the cell's ability to recover from the insult. The importance of Ca^{2+} as an intracellular messenger can be appreciated by the number of different mechanisms employed by the cells to maintain Ca^{2+} homeostasis (Fig. 4.5). Intracellular Ca^{2+} concentration is maintained around $10-^7 M$, while extracellular concentration is in the range of $10-^3 M$ and this electrical chemical gradient exerts a large inward force on Ca^{2+} ions. This gradient takes energy to maintain, requiring the active extrusion of Ca^{2+} from the cell either by a Ca^{2+}activated ATPase or by electrogenic (3:1) Na^+/Ca^{2+} exchange, which uses the membrane Na^+ gradient as the energy source which is lost during ischemia. Regulation of intracellular free Ca^{2+} over the short term can be achieved by the binding or sequestration of calcium. A large portion of the intracellular Ca^{2+} is bound to calcium-binding proteins or other molecules. The remainder of the Ca^{2+} is sequestered in an energy-dependent process, principally in the endoplasmic reticulum and mitochondria. From this description, it is apparent that restoration of calcium homeostasis is a major workload which requires significant amounts of ATP or other related energy sources.

Acidosis

One of the early findings in experimental ischemia was the production of lactate and a concomitant proton which leads to tissue acidification (Lowry et al., 1964). The activation of anaerobic glycolysis produces tissue acidification which led to an

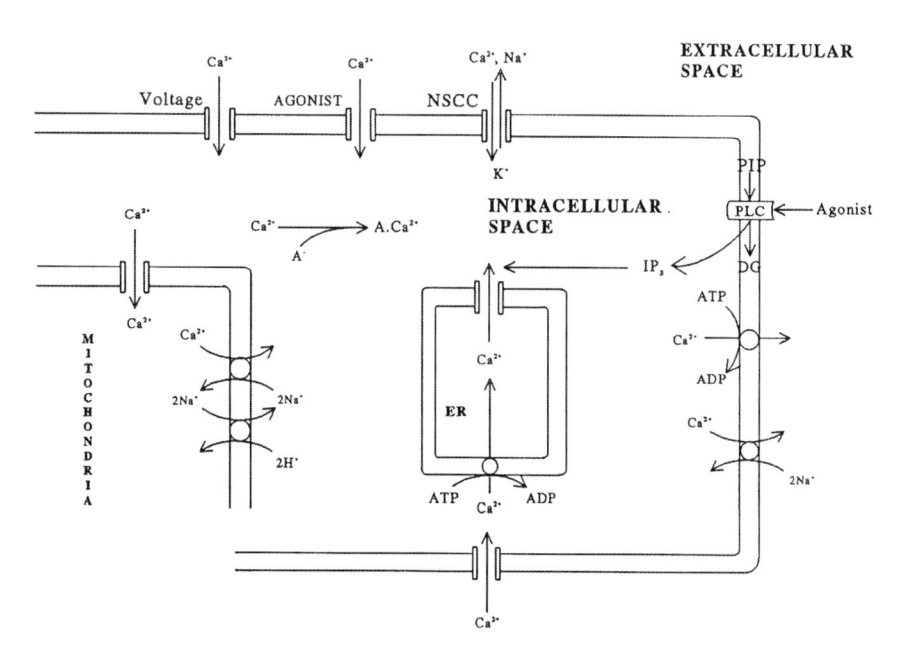

Fig. 4.5 Schematic illustrating the homeostatic mechanisms controlling cellular calcium. Abbreviations: NSCC, non-specific calcium channel; PLC, phospholipase C; PIP, polyphosphoinosides; IP3, inositol triphosphate; DG, diglycerides; A, calcium binding protein; ER, endoplasmic reticulum

early hypothesis that acidification was a major contributor to cell damage. In support of this proposal, preischemic hyperglycemia has been shown to enhance cellular damage (Kagansky et al., 2001). The enhanced acidification in hyperglycemia has been purported to enhance glutamate efflux, brain edema and blood-brain barrier disruption. In global ischemia, the decreases in pH in normoglyemic and hyperglycemic gerbils have been reported to be 0.34 and 0.48 pH units, respectively (Hoffman et al., 1994). While the magnitude of these changes are somewhat modest, it is clear that even small changes of pH can have pronounced effects on cellular homeostasis including a reduction of glycolysis (Trivedi and Danforth, 1966). The exact net effect of lacticacidosis in the pathophysiology of normoglycemic ischemia is somewhat controversial, since it has been shown that a reduction of pH blocks the NMDA glutamate receptor and the anti-excitotoxicity should be beneficial against the ischemic insult (Giffard et al., 1990).

Second Messengers

Another example of brain dysfunction is the changes in secondary messengers owing to the well-documented enhanced release of the primary messengers such as neurotransmitters. Generally, each neurotransmitter or binding ligand has a specific secondary messenger which can be ions, enzymes, or other proteins and they often function through intracellular amplification systems. The cyclic nucleotides have been shown to change significantly during 1 h of global ischemia in the gerbil (Kobayashi et al., 1977). The levels of cyclic AMP synthesized by adenylate cyclase increased almost tenfold in the first min of ischemia and gradually decreased to values fourfold over those of control at 1 h of ischemia (Fig. 4.6). The magnitude of the increased cyclic AMP would undoubtedly further amplify the glycogen phosphorylase cascade (Lust and Passonneau, 1976). In contrast, the concentrations of cyclic GMP formed by guanylate cyclase gradually decreased by about 80% after 1 h of ischemia. To our knowledge, the magnitude of the changes in the cyclic nucleotides in ischemia is far greater than those observed in normal brain function, suggesting that an imbalance between second messenger systems may have marked neuropathological effects on brain function. The pronounced changes in the cyclic nucleotides could be due to the activities of the respective cyclases or to phosphodiesterases. The major perturbations in cyclic AMP and cyclic GMP would have an effect on their targets, protein kinase A and a cyclic GMP-dependent protein kinase, respectively (Churn and DeLorenzo, 1998) which would further disrupt the normal function of the brain.

Another group of second messengers during ischemia is associated with the catabolism of fatty acids (Bazan, 2005). Many of these second messengers are initiated by the action of phospholipase A2 (PLA2) and phospholipase C (PLC). The action of PLC degrades phosphoinositol 4,5 bisphosphate to form inositol 1,4,5-trisphosphate (IP3) and diacylglycerol (DAG). The target for the soluble IP2 is the IP3 receptor in the endoplasmic reticulum which elicits a release of endogenous calcium from this organelle. In contrast, the DAG binds PKC adjacent to the membrane and activates

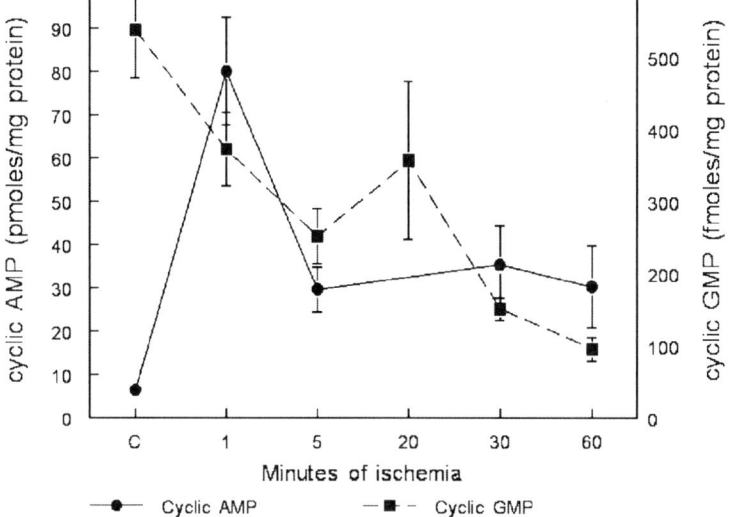

Fig. 4.6 Time course of cyclic nucleotides in global ischemia. The left ordinate indicate the concentrations of cyclic AMP and on the right are those for the cyclic GMP. The concentrations of cyclic AMP and cyclic GMP are expressed in pmol/mg protein and fmol/mg protein, respectively, indicating the order of magnitude lower amounts of cyclic GMP

a process which leads to modulation of neuronal excitability. One of the products of the PLA2 is arachidonic acid (AA) which increases significantly following ischemia. The AA serves as a substrate for the production of a variety of eicosanoids which include prostaglandins, leukotrienes and thromboxanes. These molecules have a number of internal actions and also possess primary messenger characteristics upon being extruded from the cell. The interpretation of the action of each of the messengers is complicated by changes in certain key enzymes during ischemia and the potential of "cross-talk" during the insult (Pelligrino and Wang, 1998). The importance of second messenger changes indicates that ischemia not only profoundly affects primary messengers, but continues to perturb the secondary messenger, causing additional cellular changes which need to be reversed upon reperfusion. Thus, it will become evident in the subsequent sections that reversing energy failure alone does not ensure a restoration of function.

Recirculation Postischemia

In recent years, the focus of ischemia research has been reperfusion, since only the absence of reflow ensures that the cells will become necrotic. Reperfusion alone, however, does not ensure reversibility of cell damage and unless it is performed in

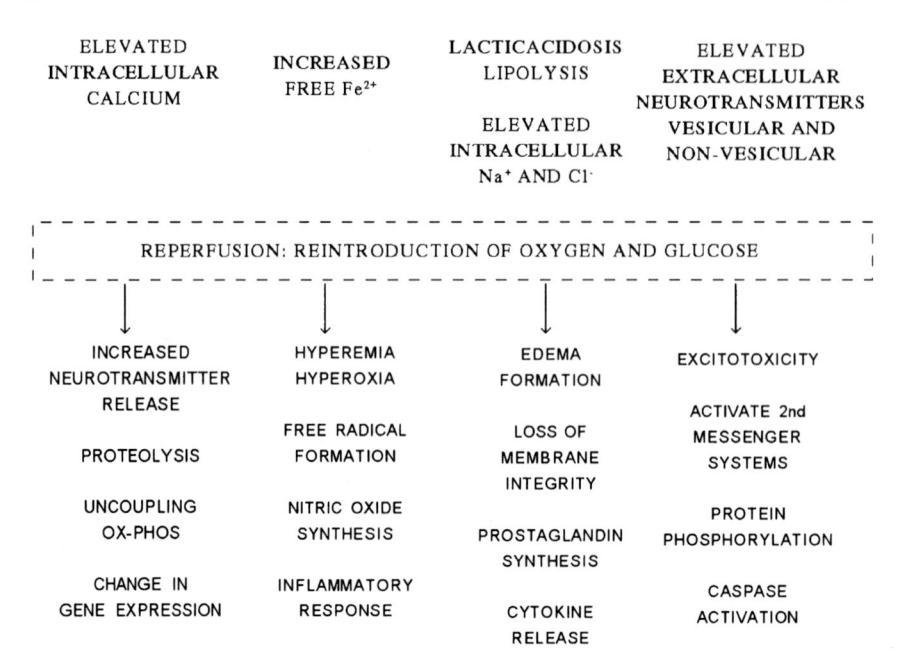

Fig. 4.7 Schematic representation of the myriad reperfusion-induced processes. Many of the events listed may be the consequence of ischemia-induced changes and yet others may be in response to the reintroduction of oxygen and glucose to an ischemic milieu in which many of the cellular mechanisms may have been severely perturbed

a timely manner is not effective. Experimental evidence from both clinical and basic science indicates that the earlier the onset of reflow, the better the outcome. Clinically, NINDS guidelines for the use of tissue plasminogen activator, a thrombolytic agent, indicate that treatment should be instituted within 3 h of the onset of ischemia (Graham, 2003). In experimental animals, earlier reperfusion following focal ischemia resulted in a better outcome (Selman et al., 1990).

Reflow to the perturbed intracellular milieu triggers a multitude of potentially irreversible pathophysiological events (Fig. 4.7). The most glaring example is the burst of free radicals upon reoxygenation and the onset of mitochondrial dysfunction, which has become a major focus in a number of neurodegenerative diseases including stroke. Molecular changes during reperfusion have been well-documented in recent years, as has the emergence of inflammation (see below).

Free Radicals

Since the first demonstration of a reduction in the antioxidant, vitamin C, during ischemia, the importance of free radicals and reactive oxygen species (ROS) has grown in spite of the fact that most of the free radicals are in relatively low concentrations with a short half-life (Flamm et al., 1978). The direct measurement of free radical formation has, for this reason, relied heavily on those oxidized products of free radicals including proteins, DNA and lipids. There are a variety of free radicals, numerous pathways to their formation and far-ranging targets for oxidative damage which has resulted in a somewhat incomplete understanding of the multiplicity of effects (Margaill et al., 2005) (Fig. 4.8). As shown in the schematic representation, it is evident that there are multiple pathways not only for ROS synthesis (i.e., italicized numbers), but also for free radical scavenging (i.e., italicized letters). The entire set of reactions presented in the figure are relevant to the reader, but will not be discussed individually owing to space constraints and the relative impact of each pathway on ischemic pathophysiology. Five major synthetic pathways involve xanthine oxidase, mitochondria, cyclooxygenase 2, leukocytes and nitric oxide synthase. The importance of the role of the conversion of xanthine dehydrogenase to xanthine oxidase and generating O_2^- (superoxide ion) and H_2O_2 (hydrogen peroxide) remains somewhat controversial, owing to the magnitude of the ROS produced and the inability to show the interconversion from the dehydrogenase to the oxidase form (Betz et al., 1991).

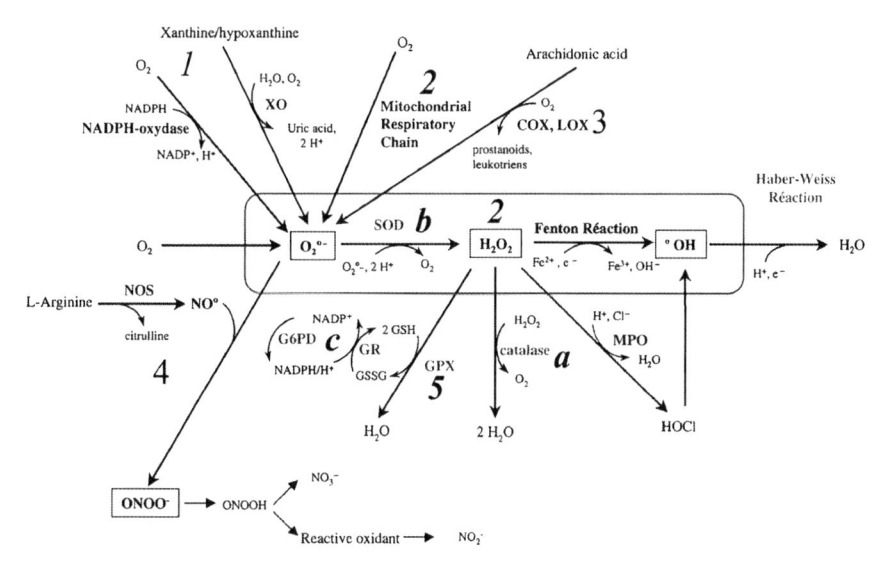

Fig. 4.8 Schematic representation of free radical formation and scavenger system. The comprehensive figure is from Margaill et al., 2005 which is organized to visualize the interaction of the various processes in the production and destruction of free radicals. The numbers represent various free radical synthetic pathway: *1.* xanthine oxidase; *2.* mitochondria production of superoxide ion and hydrogen peroxide; *3.* lipooxygenase and lipoxgenase pathway; *4.* nitric oxide synthesis to peroxynitrite; and *5.* glutathione peroxidase to oxidize glutathione. The italic letters are scavenging systems: a. catalase; b. superoxide dismutase, and c. glutathione reductase. The reactions for these different pathways are clearly indicated in the schematic representation. Abbreviations: GSH, reduced glutathione; GR, glutathione reductase; GPX, glutathione reductases; HOCL, hypochlorous acid; LOX, lipooxygenase; MPO, myeloperoxidase; SOD, superoxide dismutase, other symbols for free radicals are found in the text. Permission to reproduce the schematic representation was granted by Elsevier Limited, Oxford, United Kingdom

The increased production of free radicals in dysfunctional mitochondria occurs during reperfusion when the generation exceeds the capacity of the multiple endogenous free radical scavengers. Nitric oxide is produced by nitric oxide synthase and the NO is thought to interact with O_2^- to form peroxynitrite which readily nitrosylates proteins (Eliasson et al., 1999; Takizawa et al., 1999). The large accumulation of AA is a substrate for the formation of products of lipooxygenase and cyclooxygenases which are also a potential source of free radicals, O_2^- and H_2O_2 (Nogawa et al., 1997). Another source of the free radicals is from the adhesion and infiltration of polynuclear leukocytes. While the evidence for the oxidative stress is convincing, the specific susceptible targets leading to oxidative damage and cell death require additional research.

Metabolic Consequence of Reperfusion: Reversal of Ischemic Process?

The recovery of the bioenergetics following global and focal ischemia is the mandatory first step to restoring the cellular homeostasis to a preischemic state. As previously stated, the time and completeness of reperfusion are probably the most critical components for achieving complete recovery. As the time threshold is exceeded (depending on the ischemia model), reperfusion results in a delayed rate of metabolite restoration, an inability to restore metabolites to normal levels and finally, in long-term ischemia, restoration fails totally, leading to cell death. Many of the reperfusion perturbations cannot be reversed in the absence of energy. Examples of both global and focal ischemia are presented to show how ischemic density elicits a different recovery pattern.

The gerbil model of global ischemia has been extensively studied and its unique characteristic is a delayed neuronal death of the CA 1 pyramidale cells 4 days after reflow, even though the histological profiles of the neurons at day 2 are evident, although somewhat abnormal (Kirino, 1982). The re-introduction of oxygen and glucose to the CA1 neurons, as well as other regions of the hippocampus and cortex, results in a normalized energy profile over a period of 2 days of reflow after 5 min of bilateral occlusion of the common carotid arteries (Kobayashi et al., 1977; Arai et al., 1986). While many initial mechanisms of the CA 1 neuronal death have been proposed (Kirino, 2000), one of the final stages is the decrease of ATP levels by more than 50% which occurs between day 2 and day 4 of reflow during the demise of the neurons. The absence of complete ATP depletion is probably due to the surviving glia within the stratum pyramidale. Selective vulnerability of the CA 1 neurons is typical of most global models including cardiac arrest and resuscitation (Xu et al., 2006).

In focal models of ischemia, reperfusion is rather heterogeneous owing to the intensity of the original ischemic insult. It is important to remember that the various regions within the middle cerebral artery distribution are in a quasi-steady state. As shown in fig. 4.9, the regions of necrosis measured after 1 day of reflow expand to

NORMOTENSIVE MODEL
VOLUME OF INFARCTION, 1 DAY

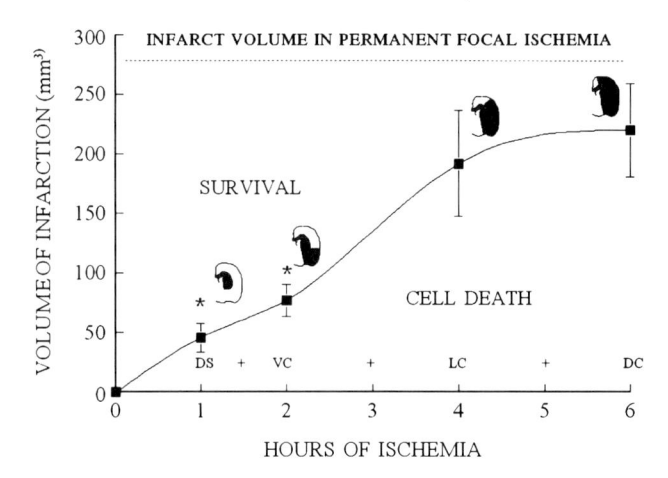

Fig. 4.9 The relationship of infarct volume at one day after various periods of focal ischemia. The black portion of the hemi-section at 1, 2 4 and 6 h of focal ischemia indicates the areas destined for infarction 1 day after reperfusion. This figure clearly shows that the status of a given region is subject to change with increasing periods of ischemia. Clearly, the initial characteristics of the penumbra of being electrical quiescent with normal extracellular K^+ is subject to changes with time and is only a transient condition. The asterisks (*) indicate the infarct volume significantly different from those of permanent focal ischemia

the entire MCA region if the focal ischemia persists for 6 h, which means the ischemic core designated by the dorsal striatum (DS) actually expands sequentially to incorporate the peri-infarct core, the lateral cortex and finally the penumbra (DC). The interpretation of these findings is that all regions are in an unstable steady state which eventually succumbs to the inability of the collateral flow to provide sufficient bioenergetics to maintain the work load of the tissue. Alternatively, the ability of the cells to survive may not only depend on the intensity of the ischemic insult, but also on the tissues' ability to withstand the reperfusion-induced events. It is possible that the time and duration of reduced CBF interact to increase the susceptibility of a given region to reflow-induced damage. The differences in selective vulnerability found between experimental and clinical stroke may also be attributed to different species-specific characteristics of the phenomenon (Hossmann, 1998).

A number of studies have examined the metabolic recovery after 1 or 2 h of focal ischemia and the findings were quite similar (Folbergrova et al., 1995, Hata et al., 2000; Selman et al., 1999; Lust et al., 2002). The high-energy phosphates markedly increase within the first hour of reperfusion and continue to return toward normal values for 2–3 h of reflow. In certain circumstances, however, the pattern of the recovery of nicotinamide adenine nucleotides does display a heterogenous pattern (Welsch, 1998). The short term results would suggest that reperfusion was sufficient to reverse the energy failure following ischemia and restore the function and integrity of the brain. Rather surprisingly, metabolic recovery of the middle cerebral artery following 2 h, subsequently deteriorates by 4 h of reflow and this effect was reversed by the spin trap agent, N-tert-butyl-α-phenynitrone (Folbergrova et al., 1995). In another study, the delayed energy failure after 1 h of focal ischemia was examined in multiple regions of the brain and the effect became evident at 4 h of reperfusion (Hata et al., 2000). In our study, metabolite status during reperfusion in the four regions within the MCA distribution were determined following 2, 4, 6, and 8 h of ischemia (Lust et al., 2002). The results clearly showed that the onset of secondary energy failure occurred earlier after longer periods of ischemia with even the penumbra exhibiting energy failure after 8 h of reperfusion following 8 h of ischemia. Certainly, these collective findings indicate that bioenergetics are critical stages in recovery and further strongly indicate that reperfusion alone may not ensure survival of the tissue.

In a similar paradigm, the metabolic status and response of rats aged 24 months was examined and compared to that in the adult rat in a focal model of cortical ischemia. The concentrations of ATP and P-creatine were measured in the core area adjacent to the core (peri-infarct region), the peri-penumbra region and the penumbra for 8 h of reperfusion following 2 h of cortical ischemia (Fig. 4.10). The results for the core and penumbra have not been presented to emphasize the fate of the transitional zones. Energy failure in the adult brain was only evident in the peri-infarct region, but expanded to include the peri-penumbral region in aging animals. In addition, the recovery of the high-energy phosphates in the aging rat was compromised, indicating that mitochondrial susceptibility to damage increases with age. The results clearly indicate that the aging brain is more susceptible to focal ischemia than that of the adult brain. This is supported by the many changes reported in the aging brain, including marked perturbations of free radical homeostasis and the intense oxidative damage of macromolecules that has been shown within the cell, particularly in the mitochondria where a burst of ROS has also been shown following focal ischemia (Di Lisa and Bernardi, 2005; Lee et al., 2000; Melov, 2002; Sastre et al., 2003). The secondary energy failure during reperfusion suggests that production of energy by mitochondria is perturbed, which may be due to the activation of relatively newly discovered mitochondrial permeability megapore in ischemia (see below). Parenthetically, perinatal brain energy demands are approximately 10% of that of the adult, making it more resistant to an ischemic insult (Pundik et al., 2006). It is concluded that the age of the animal is an important factor in the outcome from an ischemic insult and has to be considered with the intensity and time of ischemia.

Recovery: Mitochondrial Dysfunction

Previous literature has reported on various susceptible sites within the mitochondrion, but much of the data is in isolated reports and has been difficult to interpret in the evolution of the pathophysiology of ischemia. For example, a secondary significant

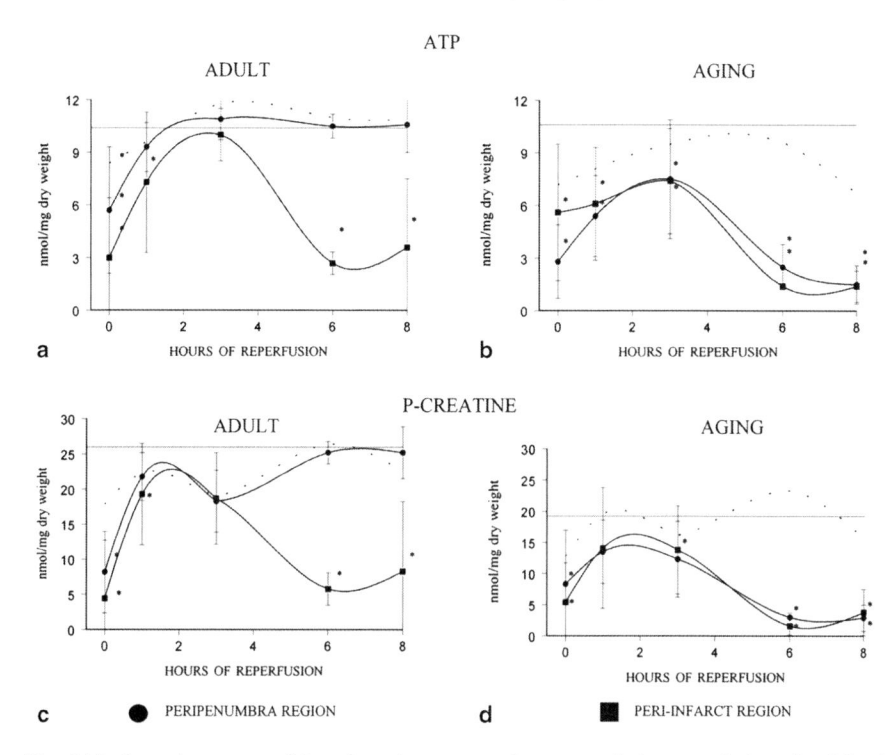

Fig. 4.10 Secondary energy failure is a phenomenon that occurs during reperfusion after 2 h of ischemia when the CBF is essentially normal and the glucose levels are at or above control. The left two panels represent ATP and P-creatine levels in the adult rate, whereas the right panels indicate changes in the aging brain. Note that the core values were not presented, since SEF occured independent of age. The time course for the penumbra is indicated by the dotted lines which do not show SEF over the 8 h reperfusion time-course following 2 h of ischemia even in the core

accumulation of calcium has been reported in the vulnerable dorsolateral striatum at 6 and 24 h of recirculation (Zaidan and Sims, 1994) and this was associated with an earlier depression of mitochondrial respiration and reduction of pyruvate dehydrogenase (Sims and Pulsinelli, 1987). These findings convincingly focused on a potential lesion in the mitochondria (Almeida et al., 1995) and yet mitochondrial dysfunction had not received the attention that other hypotheses have, and this may be related to complexities of studying the production and effect of free radicals in mitochondria upon recirculation. The discovery of the mitochondrial transition pore (MTP) has markedly increased the interest in mitochondria as a major organelle in brain damage (Friberg & Wieloch, 2002).

While alterations in protein synthesis and apoptosis remain as putative elements in the evolution of cell death (Siesjo and Siesjo, 1996), the relatively new discovery of mitochondrial permeability transition pore (MTP) to diminish the bioenergetic capacity of the cell is another focus that is consistent with cell injury and cell death. The MTP is thought to be composed of the interaction of the adenine nucleotide transporter on the outer membrane and a voltage dependent anion channel on the inner membrane, which becomes activated by cyclophilin D located in the matrix (Tanveer et al., 1996). In its simplest terms, the inner membrane of the mitochondria becomes leaky, causing the release of a number of ions and molecules <1500 kD. Most noticeable is the release of protons, the electromotive force driving oxidative phosphorylation and the synthesis of ATP. The pore opening is facilitated by oxidative stress of nicotinamide adenine nucleotides and inorganic phosphate, and inhibited by acidification and elevated ADP. Another interesting aspect of the MTP is that the opening of the megapore is inhibited by immunosuppressants, such as cyclosporine A and FK506 (Folbergrova et al., 1995, 1997). Overall, the existence of MTP has brought a new dimension to the study of energy metabolism both during and after ischemia. More information on the MTP in apoptosis is presented below, revealing the pro- and anti-apoptotic processes of mitochondria.

The MTP is essentially turned off during ischemia in a milieu of depolarized mitochondria, acidification and an inability of mitochondria to sequester calcium (Friberg and Wieloch, 2002). Our observation of secondary energy failure during reperfusion would be consistent with the conditions favoring the opening of MTP. Energy status during reflow had recovered to near-normal values which is compatible with the reversal of the perturbations described above, which prevent the megapore opening. While such a relationship has not been definitively proven, experiments to examine cytochrome c release, among others, could be designed to test this proposal.

Reperfusion: Molecular Events and Cell Death

In the nervous system, cell death in response to ischemia is the result of the complex interplay between the severity of the blood flow reduction and duration of the insult, the responses of neighboring cells and the molecular signaling within the cell. Despite the large number of combinations of these variables present, cellular demise may be in the form of necrotic, autophagic or apoptotic cell death. Whether these cell death pathways share common events is a matter of speculation, but it seems likely that there would be interactive common pathways that are indirectly dependent on the metabolic status of the tissue.

Necrosis

Necrosis generally occurs within the core of the infarct, where the cerebral blood flow drops to <20% of normal levels. This severe decrease in blood flow rapidly depletes energy from cells which have low energy reserves, leaving ATP levels essentially depleted. It is characterized by cell and organelle swelling, followed by the loss of membrane integrity (Festjens et al., 2006). Cells destined to die by necrosis spill their contents, including high levels of glutamate, into the extracellular space. This proves to be a dangerous process, resulting in double jeopardy for neighboring cells, not only from excitotoxic death due to elevated glutamate levels, but also from the effects of an inflammatory response.

Although once believed to be an uncontrolled, disorganized form of cell death, there is increasing evidence that necrosis is, in fact programmed, dependent on signaling pathways within the cell. It is becoming evident that a central player in the necrotic signaling pathway is RIP1, a serine-threonine kinase involved in death receptor-induced necrosis (Harper et al., 2003; Holler et al., 2000; Lewis et al., 2000) and toll-like receptor triggered necrosis (Meylan et al., 2004). As a result of DNA damage, ischemia-reperfusion and glutamate excitotoxicity, poly(ADP-ribose) polymerase-1 (PARP-1) is activated, catabolizing the hydrolysis of adenine dinucleotide (NAD$^+$), resulting in the depletion of ATP (Berger, 1985; Carson et al., 1986; Oleinick and Evans, 1985). RIP1 functions downstream of PARP-1, along with TRAF2, and is necessary for JNK activation. The sustained activation of JNK compromises the integrity of the mitochondrial membrane (Shen et al., 2004; Xu et al., 2006), possibly through Bcl-2 family members (Maundrell et al., 1997) or Bid (Deng et al., 2003), and leads to necrosis. RIP1 is also involved in the activation of NF-κB and MAPKs (Devin et al., 2000; Devin et al., 2003; Kelliher et al., 1998; Ting et al., 1996; Zhang et al., 2000), which in turn release inflammatory cytokines, including IL-6 (Vanden Berghe et al., 2006).

Autophagy

Autophagy appears to be an evolutionarily conserved mechanism, serving to recycle proteins from the cytoplasm or organelles, and has only recently been described as a mode of programmed cell death. It is characterized by the formation of autophagosomes, which use lysosomal hydrolases to degrade contents (Levine and Klionsky, 2004). The presence of autophagy in ischemia has been previously described (Degterev et al., 2005; Nitatori et al., 1995; Zhu et al., 2005), although it is not well characterized. The primary targets of autophagy are depolarized mito-chondria, the endoplasmic reticulum and peroxisomes (Elmore et al., 2001; Hamasaki et al., 2005; Iwata et al., 2006; Yu et al., 2004). In the case of autophagic cell death, catalase has also been reported to be selectively degraded (Yu et al., 2006). In spite of the selective elimination of mitochondria, it has been shown that ATP levels actually increase during autophagy, very likely due to the cessation of protein translation and other energy consuming tasks. Loss of damaged mitochondria may also prevent them from leaking proapoptotic factors into the cytosol, potentially avoiding cell death (Lemasters et al., 1998). There are a number of signaling pathways that have been implicated in autophagic cell death, some of which are also involved in necrotic and apoptotic cell death (Gozuacik and Kimchi, 2007). Bcl2 family members, important in the regulation of the apoptotic pathway based out of mito-chondria (Gross et al., 1999), have been shown to have important roles in the regulation of autophagic cell death. Pro-apoptotic members such as BNIP3 and Bax cause autophagic cell death (Camougrand et al., 2003; Vande et al., 2000), whereas anti-apoptotic members, including Bcl-2 and Bcl-XL, have the opposite effect (Cardenas-Aguayo et al., 2003; Saeki et al., 2000; Vande et al., 2000).

RIP1 also plays a critical role in the induction of autophagic cell death. In response to the application of a pan-caspase inhibitor, zVAD, RIP1 triggered a novel cell death pathway, involving MKK7, JNK and c-Jun (Yu et al., 2004). It is important to note that this mode of cell death is ATP dependent, since protein synthesis is necessary. This may correlate to the biphasic nature of translational arrest in postischemic neurons (DeGracia, 2004), where a reversible translation arrest is seen early in postischemia, followed by a resumption of translation, or the movement to an irreversible arrest state. Protein translation may also be affected at the level of the endoplasmic reticulum, since cerebral ischemia also produces high levels of stress. This activates the translation initiation factor eIF2α kinase, PERK (Koumenis et al., 2002)., which has been shown to regulate autophagy under certain circumstances (Talloczy et al., 2002).

Apoptosis

Apoptosis is generally believed to be an orderly form of cell death, characterized by cell shrinkage, membrane blebbing, chromatin condensation and fragmentation, and the formation of apoptotic bodies. Cells which undergo apoptosis as a result of ischemia have CBF that is greater than 25% of normal values and ATP levels of 50–70% of the normal, generally found within the ischemic penumbra. This amount of blood flow allows for a low level of metabolism, and the generation of ATP via anaerobic respiration for energy. However, not enough ATP can be generated to maintain the transmembrane ion concentration of calcium, potassium and sodium, inhibiting impulse conduction and synaptic function in neurons, and leading to the accumulation of glutamate. The presence of ATP is critical to the complex signaling mechanisms involved in apoptosis.

In response to ischemia and reperfusion, mitochondrial membranes undergo depolarization and elevated levels of Ca^{2+} in the intermembrane space. The control of mitochondrial membrane depolarization depends, in part, on Bcl-2 family member proteins. The proapoptotic group of Bcl-2 members consists of the Bax-subfamily (Bax, Bak, and Bok) and the BH3-only proteins (Bid, Bim, Bik, Bad, Bmf, Hrk, Noxa, Puma, Blk, BNIP3, and Spike) (Cory and Adams, 2002; Mund et al., 2003). It appears that the main function of the Bcl-2 family proteins is to guard mitochondrial integrity and to control the release of mitochondrial proteins into the cytoplasm (Cory and Adams, 2002). It is believed that the proapoptotic Bax and Bak provoke or contribute to the permeabilization of the outer mitochondrial membrane, either by forming channels by themselves (Antonsson et al., 2000) or by interacting with components such as VDAC (Tsujimoto and Shimizu, 2000).

In addition to Bcl-2 itself, there are a number of other prosurvival proteins: Bcl-XL, Bcl-w, A1, and Mcl-1. Prosurvival members of the Bcl-2 family act by binding to and inactivating proapoptotic members, including Bax and Bak, and through the stabilization of the mitochondrial membrane. Upon mitochondrial membrane depolarization, cytochrome c is released and apoptotic signaling is initiated within the cell. Cytochrome c functions in a regulatory manner, preceding the morphological changes that are associated with apoptosis (King, 1997). Once released, cytochrome c binds with Apaf-1 and ATP, which subsequently binds with procaspase-9 to form an apoptosome. The apoptosome cleaves the procaspase to form active caspase 9, which in turn activates the effector caspase-3.

In addition to the release of cytochrome c, SMACs (second mitochondria-derived activator of caspases) are released due to the increase in mitochondrial membrane permeability. These bind to IAPs (inhibitor of apoptosis proteins), preventing them from carrying out their normal function of caspase inhibition, allowing apoptosis to proceed.

Ischemic Neuroprotection

A major omission in this overview is the discussion of interventions to minimize experimental ischemic damage (Hossmann, 2006; Martinez-Vila and Irimia, 2005). There have been so many reports of neuroprotection in experimental models of ischemia and yet, for the most part, the efficacy has not readily been translated to clinical trials. Even though examples of experimental neuroprotection may be informative, a thorough discussion of the subject would be unwieldy and is beyond the scope of this review. Nevertheless, there are new approaches including molecular manipulations and optimizing host defenses which appear to be promising (Dirnagl et al., 2003). Another approach would be to devise a pharmacological cocktail to offset the reflow events (shown in Fig. 4.7), but this is complicated by the multifaceted reperfusion response (Lapchak and Araujo, 2007; Mehta et al., 2007). Until potential therapies have been confirmed in the clinical setting, development of novel ischemic intervention still remains a compelling objective for research.

Summary

Research in the field of experimental ischemia has yielded an enormous amount of information regarding the events that occur within the brain in the absence of cerebral blood flow. Given the myriad pathophysiological events that occur both during and after ischemia, this chapter attempts to provide the reader with a very broad overview of the cascade of events triggered by energy failure. Due to the high energy demands and low energy reserves of the brain, it is obvious that time has to be the primary concern of the physician once human stroke has occurred, and that the restoration of oxygen and glucose is a mandatory first step in salvaging brain function and preventing permanent damage (see subsequent chapter). Early research in experimental stroke focused on energy failure due to the absence of oxygen and glucose, and it is intriguing that current investigations have come a full circle, now focusing on the inability to generate energy due to metabolic dysfunction, even in the presence of these needed nutrients. Metabolic pathways that are operational immediately following the loss of CBF, eventually become dysfunctional and lead to a CBF independent-like ischemic profile due to an inability to oxidize glucose.

References

Almeida, A., Allen, K.L., Bates, T.E., and Clark, J.B. (1995): Effect of reperfusion following cerebral ischaemia on the activity of the mitochondrial respiratory chain in the gerbil brain. J. Neurochem., 65:1698–1703.

Antonsson, B., Montessuit, S., Lauper, S., Eskes, R., and Martinou, J.C. (2000): Bax oligomerization is required for channel-forming activity in liposomes and to trigger cytochrome c release from mitochondria. Biochem. J., 345(Pt 2):271–278.

Anderson GL and Whisnant JP. (1982): A comparison of trends in mortality from stroke in the United States and Rochester, Minnesota. Stroke. (6):804–9.

Arai, H., Passonneau, J.V., and Lust, W.D. (1986): Energy metabolism in delayed neuronal death of CA1 neurons of the hippocampus following transient ischemia in the gerbil. Metab. Brain Dis., 1:263–278.

Astrup, J., Siesjö, B.K., and Symon, L. (1981): Thresholds in cerebral ischemia — The ischemic penumbra. Stroke, 12(6):723–725.

Astrup, J., Symon, L., Branston, N., and Lassen, N. (1977): Cortical evoked potential and extracellular K$^+$ and H$^+$ at critical levels of brain ischemia. Stroke, 8:51–57.

Bazan, N.G. (2005): Synaptic signaling by lipids in the life and death of neurons. Mol. Neurobiol., 31:219–230.

Berger, N.A. (1985): Poly(ADP-ribose) in the cellular response to DNA damage. Radiat. Res., 101:4–15.

Betz, A.L., Randall, J., and Martz, D. (1991): Xanthine oxidase is not a major source of free radicals in focal cerebral ischemia. Am. J. Physiol., 260:H563–H568.

Branston, N.M., Hope, D.T., and Symon, L. (1979): Barbiturates in focal ischemia of primate cortex: Effects on blood flow distribution, evoked potential and extracellular potassium. Stroke, 10:647–653.

Branston, N.M., Symon, L., Crockard, H.A., and Pasztor, E. (1974): Relationship between the cortical evoked potential and local cortical blood flow following acute middle cerebral artery occlusion in the baboon. Exp. Neurol., 45:195–208.

Camougrand, N., Grelaud-Coq, A., Marza, E., Priault, M., Bessoule, J.J., and Manon, S. (2003): The product of the UTH1 gene, required for Bax-induced cell death in yeast, is involved in the response to rapamycin. Mol. Microbiol., 47:495–506.

Cardenas-Aguayo, M.C., Santa-Olalla, J., Baizabal, J.M., Salgado, L.M., and Covarrubias, L. (2003): Growth factor deprivation induces an alternative non-apoptotic death mechanism that is inhibited by Bcl2 in cells derived from neural precursor cells. J. Hematother. Stem Cell Res., 12:735–748.

Carson, D.A., Seto, S., Wasson, D.B., and Carrera, C.J. (1986): DNA strand breaks, NAD metabolism, and programmed cell death. Exp. Cell Res., 164:273–281.

Churn, S.B. and DeLorenzo, R.J. (1998): Intracellular Messengers and Mediators. In: *Cerebrovascular Disease: Pathophysiology, Diagnosis and Management*, Ginsberg MD et al. (eds), pp. 433–439. Blackwell, Malden MA.

Cory, S. and Adams, J.M. (2002): The Bcl2 family: Regulators of the cellular life-or-death switch. Nat. Rev. Cancer, 2:647–656.

DeGracia, D.J. (2004): Acute and persistent protein synthesis inhibition following cerebral reperfusion. J. Neurosci. Res., 77:771–776.

Degterev, A., Huang, Z., Boyce, M., Li, Y., Jagtap, P., Mizushima, N., Cuny, G.D., Mitchison, T.J., Moskowitz, M.A., and Yuan, J. (2005): Chemical inhibitor of nonapoptotic cell death with therapeutic potential for ischemic brain injury. Nat. Chem. Biol., 1:112–119.

Deng, Y., Ren, X., Yang, L., Lin, Y., and Wu, X. (2003): A JNK-dependent pathway is required for TNFalpha-induced apoptosis. Cell, 115:61–70.

Devin, A., Cook, A., Lin, Y., Rodriguez, Y., Kelliher, M., and Liu, Z. (2000): The distinct roles of TRAF2 and RIP in IKK activation by TNF-R1: TRAF2 recruits IKK to TNF-R1 while RIP mediates IKK activation. Immunity, 12:419–429.

Devin, A., Lin, Y., and Liu, Z.G. (2003): The role of the death-domain kinase RIP in tumour-necrosis-factor-induced activation of mitogen-activated protein kinases. EMBO Rep., 4:623–627.

Di Lisa, F. and Bernardi, P. (2005): Mitochondrial function and myocardial aging. A critical analysis of the role of permeability transition. Cardiovasc. Res., 66:222–232.

Dirnagl, U., Simon, R.P., and Hallenbeck, J.M. (2003): Ischemic tolerance and endogenous neuroprotection. Trends Neurosci., 26:248–254.

Eliasson, M.J., Huang, Z., Ferrante, R.J., Sasamata, M., Molliver, M.E., Snyder, S.H., and Moskowitz, M.A. (1999): Neuronal nitric oxide synthase activation and peroxynitrite formation in ischemic stroke linked to neural damage. J. Neurosci., 19:5910–5918.

Elmore, S.P., Qian, T., Grissom, S.F., and Lemasters, J.J. (2001): The mitochondrial permeability transition initiates autophagy in rat hepatocytes. FASEB J., 15:2286–2287.

Erecinska, M. and Silver, I.A. (1989): ATP and brain function. J. Cereb. Blood Flow Metab., 9:2–19.

Festjens, N., Vanden Berghe, T., and Vandenabeele, P. (2006): Necrosis, a well-orchestrated form of cell demise: Signalling cascades, important mediators and concomitant immune response. Biochim. Biophys. Acta, 1757:1371–1387.

Flamm, E.S., Demopoulos, H.B., Seligman, M.L., Poser, R.G., and Ransohoff, J. (1978): Free radicals in cerebral ischemia. Stroke, 9:445–447.

Folbergrova, J., Li, P.A., Uchino, H., Smith, M.L., and Siesjo, B.K. (1997): Changes in the bioenergetic state of rat hippocampus during 2.5 min of ischemia, and prevention of cell damage by cyclosporin A in hyperglycemic subjects. Exp. Brain Res., 114:44–50.

Folbergrova, J., Memzawa, H., Smith, M.L., and Siesjo, B.K. (1992): Focal and perifocal changes in tissue energy state during middle cerebral artery occlusion in normo- and hyerglycemic rats. J. Cereb. Blood Flow Metab., 12:25–33.

Folbergrova, J., Zhao, Q., Katsura, K., and Siesjo, B. (1995): N-tert-Buryl-a-phenylnitrone improves recovery of brain energy state in rats following transient focal ischemia. Proc. Natl. Acad. Sci. U S A., 92:5057–5061.

Friberg, H. and Wieloch, T. (2002): Mitochondrial permeability transition in acute neurodegeneration. Biochimie, 84:241–250.

Giffard, R.G., Monyer, H., Christine, C.W., and Choi, D.W. (1990): Acidosis reduces NMDA receptor activation, glutamate neurotoxicity, and oxygen-glucose deprivation neuronal injury in cortical cultures. Brain Res., 506:339–342.

Ginsberg, M.D. and Bogousslavsky, J. (1998): *Cerebrovascular Disease: Pathophysiology, Diagnosis and Management*. Blackwell, Malden, MA.

Gozuacik, D. and Kimchi, A. (2007): Autophagy and cell death. Curr. Top. Dev. Biol., 78:217–245.

Graham, G.D. (2003): Tissue plasminogen activator for acute ischemic stroke in clinical practice: A meta-analysis of safety data. Stroke, 34:2847–2850.

Gross, A., McDonnell, J.M., and Korsmeyer, S.J. (1999): BCL-2 family members and the mitochondria in apoptosis. Genes Dev., 13:1899–1911.

Hamasaki, M., Noda, T., Baba, M., and Ohsumi, Y. (2005): Starvation triggers the delivery of the endoplasmic reticulum to the vacuole via autophagy in yeast. Traffic, 6:56–65.

Harper, N., Hughes, M., MacFarlane, M., and Cohen, G.M. (2003): Fas-associated death domain protein and caspase-8 are not recruited to the tumor necrosis factor receptor 1 signaling complex during tumor necrosis factor-induced apoptosis. J. Biol. Chem., 278:25534–25541.

Hata, R., Maeda, K., Hermann, D., Mies, G., and Hossmann, K.A. (2000): Evolution of brain infarction after transient focal cerebral ischemia in mice. J. Cereb. Blood Flow Metab., 20:937–946.

Heiss, W.D., Hayakawa, T., and Waltz, A.G. (1976): Cortical neuronal function during ischemia. Effects of occlusion of one middle cerebral artery on single-unit activity in cats. Arch. Neurol., 33:813–820.

Hoffman, T.L., LaManna, J.C., Pundik, S., Selman, W.R., Whittingham, T.S., Ratcheson, R.A., and Lust, W.D. (1994): Early reversal of acidosis and metabolic recovery following ischemia. J. Neurosurg., 81:567–573.

Holler, N., Zaru, R., Micheau, O., Thome, M., Attinger, A., Valitutti, S., Bodmer, J.L., Schneider, P., Seed, B., and Tschopp, J. (2000): Fas triggers an alternative, caspase-8-independent cell death pathway using the kinase RIP as effector molecule. Nat. Immunol., 1:489–495.

Hossmann, K.A. (1998): Thresholds of Ischemic Injury. In: Cerebrovascular Disease: Pathophysiology, Diagnosis, and Management, Ginsberg, M. et al (eds), pp. 193–204. Blackwell, Malden.

Hossmann, K.A. (2006): Pathophysiology and therapy of experimental stroke. Cell. Mol. Neurobiol., 26:1057–1083.

Iwata, J., Ezaki, J., Komatsu, M., Yokota, S., Ueno, T., Tanida, I., Chiba, T., Tanaka, K., and Kominami, E. (2006): Excess peroxisomes are degraded by autophagic machinery in mammals. J. Biol. Chem., 281:4035–4041.

Kagansky, N., Levy, S., and Knobler, H. (2001): The role of hyperglycemia in acute stroke. Arch. Neurol., 58:1209–1212.

Kelliher, M.A., Grimm, S., Ishida, Y., Kuo, F., Stanger, B.Z., and Leder, P. (1998): The death domain kinase RIP mediates the TNF-induced NF-kappaB signal. Immunity, 8:297–303.

Kempski, O., Otsuka, H., Seiwert, T., and Heimann, A. (2000): Spreading depression induces permanent cell swelling under penumbra conditions. Acta Neurochir. Suppl., 76:251–255.

Kimelberg, H.K. (2005): Astrocytic swelling in cerebral ischemia as a possible cause of injury and target for therapy. Glia, 50:389–397.

King, M.A. (1997): Pocket companion to robbins pathologic basis of disease — Robbins, SL, Cotran, RS, Kumar, V. N&Hc-Perspectives on Community, 18:99.

Kirino, T. (1982): Delayed neuronal death in the gerbil hippocampus following ischemia. Brain Res., 239:57–69.

Kirino, T. (2000): Delayed neuronal death. Neuropathology, 20 (Suppl):S95–S97

Kobayashi, M., Lust, W.D., and Passonneau, J.V. (1977): Concentrations of energy metabolites and cyclic nucleotides during and after bilateral ischemia in the gerbil cerebral cortex. J. Neurochem., 29:53–59.

Koumenis, C., Naczki, C., Koritzinsky, M., Rastani, S., Diehl, A., Sonenberg, N., Koromilas, A., and Wouters, B.G. (2002): Regulation of protein synthesis by hypoxia via activation of the endoplasmic reticulum kinase PERK and phosphorylation of the translation initiation factor eIF2alpha. Mol. Cell Biol., 22:7405–7416.

Lapchak, P.A. and Araujo, D.M. (2007): Advances in ischemic stroke treatment: Neuroprotective and combination therapies. Expert Opin. Emerg. Drugs, 12:97–112.

Leao, A.A.P. (1944): Spreading depression of activity in cerebral cortex. J. Neurophysiol., 7:359–390.

Lee, C.K., Weindruch, R., and Prolla, T.A. (2000): Gene-expression profile of the ageing brain in mice. Nat. Genet., 25:294–297.

Lehninger, A.L., Nelson, D.L., and Cos, M.M. (1993): . Worth Publishers, New York.

Lemasters, J.J., Nieminen, A.L., Qian, T., Trost, L.C., Elmore, S.P., Nishimura, Y., Crowe, R.A., Cascio, W.E., Bradham, C.A., Brenner, D.A., and Herman, B. (1998): The mitochondrial permeability transition in cell death: A common mechanism in necrosis, apoptosis and autophagy. Biochim. Biophys. Acta, 1366:177–196.

Levine, B. and Klionsky, D.J. (2004): Development by self-digestion: Molecular mechanisms and biological functions of autophagy. Dev. Cell, 6:463–477.

Lewis, J., Devin, A., Miller, A., Lin, Y., Rodriguez, Y., Neckers, L., and Liu, Z.G. (2000): Disruption of hsp90 function results in degradation of the death domain kinase, receptor-interacting protein (RIP), and blockage of tumor necrosis factor-induced nuclear factor-kappaB activation. J. Biol. Chem., 275:10519–10526.

Li, P.A., Kristian, T., Shamloo, M., and Siesjo, K. (1996): Effects of preischemic hyperglycemia on brain damage incurred by rats subjected to 2.5 or 5 minutes of forebrain ischemia. Stroke, 27:1592–1601.

Lipton, P. (1999): Ischemic cell death in brain neurons. Physiol. Rev., 79:1431–1568.

Lowry, O.H., Passonneau, J.V., Hasselberger, F.X., and Schulz, D.W. (1964): Effect of ischemia on known substrates and cofactors of the glycolytic pathway in brain. J. Biol. Chem., 239:18–30.

Lust, W.D. and Passonneau, J.V. (1976): Cyclic nucleotides in murine brain: Effect of hypothermia on adenosine 3?,5? monophosphate, glycogen phosphorylase, glycogen synthase and metabolites following maximal electroshock or decapitation. J. Neurochem., 26:11–16.

Lust, W.D., Taylor, C., Pundik, S., Selman, W.R., and Ratcheson, R.A. (2002): Ischemic cell death: Dynamics of delayed secondary energy failure during reperfusion following focal ischemia. Metab. Brain Dis., 17:113–121.

Margaill, I., Plotkine, M., and Lerouet, D. (2005): Antioxidant strategies in the treatment of stroke. Free Radic. Biol. Med., 39:429–443.

Martinez-Vila, E. and Irimia, P. (2005): Challenges of neuroprotection and neurorestoration in ischemic stroke treatment. Cerebrovasc. Dis., 20 (Suppl 2):148–158.

Maundrell, K., Antonsson, B., Magnenat, E., Camps, M., Muda, M., Chabert, C., Gillieron, C., Boschert, U., Vial-Knecht, E., Martinou, J.C., and Arkinstall, S. (1997): Bcl-2 undergoes phosphorylation by c-Jun N-terminal kinase/stress-activated protein kinases in the presence of the constitutively active GTP-binding protein Rac1. J. Biol. Chem., 272:25238–25242.

Mehta, S.L., Manhas, N., and Raghubir, R. (2007): Molecular targets in cerebral ischemia for developing novel therapeutics. Brain Res. Rev., 54:34–66.

Mergenthaler, P., Dirnagl, U., and Meisel, A. (2004): Pathophysiology of stroke: Lessons from animal models. Metab. Brain Dis., 19:151–167.

Melov S. (2002) Animal models of oxidative stress, aging, and therapeutic antioxidant interventions. Int J Biochem Cell Biol. 34(11):1395–400.

Meylan, E., Burns, K., Hofmann, K., Blancheteau, V., Martinon, F., Kelliher, M., and Tschopp, J. (2004): RIP1 is an essential mediator of Toll-like receptor 3-induced NF-kappa B activation. Nat. Immunol., 5:503–507.

Morawetz, R.B., Crowell, R.H., DeGirolami, U., Marcoux, F.W., Jones, T.H., and Halsey, J.H. (1979): Regional cerebral blood flow thresholds during cerebral ischemia. Fed. Proc., 38:2493–2494.

Morawetz, R.B., DeGirolami, U., Ojemann, R.G., Marcoux, F.W., and Crowell, R.M. (1978): Cerebral blood flow determined by hydrogen clearance during middle cerebral artery occlusion in unanesthetized monkeys. Stroke, 9:143–149.

Mund, T., Gewies, A., Schoenfeld, N., Bauer, M.K., and Grimm, S. (2003): Spike, a novel BH3-only protein, regulates apoptosis at the endoplasmic reticulum. FASEB J., 17:696–698.

Nitatori, T., Sato, N., Waguri, S., Karasawa, Y., Araki, H., Shibanai, K., Kominami, E., and Uchiyama, Y. (1995): Delayed neuronal death in the CA1 pyramidal cell layer of the gerbil hippocampus following transient ischemia is apoptosis. J. Neurosci., 15:1001–1011.

Nogawa, S., Zhang, F., Ross, M.E., and Iadecola, C. (1997): Cyclo-oxygenase-2 gene expression in neurons contributes to ischemic brain damage. J. Neurosci., 17:2746–2755.

Oleinick, N.L. and Evans, H.H. (1985): Poly(ADP-ribose) and the response of cells to ionizing radiation. Radiat. Res., 101:29–46.

Pelligrino, D.A. and Wang, Q. (1998): Cyclic nucleotide crosstalk and the regulation of cerebral vasodilation. Prog. Neurobiol., 56:1–18.

Phillis, J.W., Diaz, F.G., O'Regan, M.H., and Pilitsis, J.G. (2002): Effects of immunosuppressants, calcineurin inhibition, and blockade of endoplasmic reticulum calcium channels on free fatty acid efflux from the ischemic/reperfused rat cerebral cortex. Brain Res., 957:12–24.

Pundik S, Robinson S, Lust WD, Zechel J, Buczek M, Selman WR. (2006) Regional metabolic status of the E-18 rat fetal brain following transient hypoxia/ischemia. Metab Brain Dis. 21(4):309–17.

Robins M, Baum HM. (1981) The National Survey of Stroke. Incidence. Stroke. (Suppl 1):I45–57.

Sacco RL, Wolf PA, Kannel WB, McNamara PM. (1982). Survival and recurrence following stroke. The Framingham study. Stroke. 13(3):290–5.

Saeki, K., Yuo, A., Okuma, E., Yazaki, Y., Susin, S.A., Kroemer, G., and Takaku, F. (2000): Bcl-2 down-regulation causes autophagy in a caspase-independent manner in human leukemic HL60 cells. Cell Death Differ., 7:1263–1269.

Sastre, J., Pallardo, F.V., and Vina, J. (2003): The role of mitochondrial oxidative stress in aging. Free Radic. Biol. Med., 35:1–8.

Selman, W.R., Crumrine, R.C., Ricci, A.J., LaManna, J.C., Ratcheson, R.A., and Lust, W.D. (1990): Impairment of metabolic recovery with increasing periods of middle cerebral artery occlusion in rats. Stroke, 21:467–471.

Selman, W.R., Lust, W.D., Pundik, S., Zhou, Y., and Ratcheson, R.A. (2004): Compromised metabolic recovery following spontaneous spreading depression in the penumbra. Brain Res., 999:167–174.

Selman, W.R., Zhou, Y., Pundik, S., Ratcheson, R.A., and Lust, W.D. (1999): A role for secondary energy failure in the evolution of ischemic cell death following reversible focal ischemia. J. Cereb. Blood Flow Metab., 19:S618(Abstract)

Shen, H.M., Lin, Y., Choksi, S., Tran, J., Jin, T., Chang, L., Karin, M., Zhang, J., and Liu, Z.G. (2004): Essential roles of receptor-interacting protein and TRAF2 in oxidative stress-induced cell death. Mol. Cell Biol., 24:5914–5922.

Siesjo, B.K. (1978): Brain Energy Metabolism. Wiley, Chichester.

Siesjo, B.K. (1992a): Pathophysiology and treatment of focal cerebral ischemia. Part II: Mechanisms of damage and treatment. J. Neurosurg., 77:337–354.

Siesjo, B.K. (1992b): Pathophysiology and treatment of focal cerebral ischemia. Part I: Pathophysiology. J. Neurosurg., 77:169–184.

Siesjo, B.K. and Bengtsson, F. (1989): Calcium fluxes, calcium antagonists, and calcium-related pathology in brain ischemia, hypoglycemia, and spreading depression: A unifying hypothesis. J. Cereb. Blood Flow Metab., 9:127–140.

Siesjo, B.K., Elmer, E., Janelidze, S., Keep, M., Kristian, T., Ouyang, Y.B., and Uchino, H. (1999): Role and mechanisms of secondary mitochondrial failure. Acta Neurochir. Suppl. (Wien), 73:7–13.

Siesjo, B.K., Kristian, T., and Katsura, K. (1998): Overview of Bioenergetic Failure and Metabolic Cascades in Brain Ischemia. In: Cerebrovascular Disease: Pathophysiology, Diagnosis and Management, Ginsberg M.D. et al. (eds), pp. 3–13. Blackwell, Malden, MA.

Siesjo, B.K. and Siesjo, P. (1996): Mechanisms of secondary brain injury. Anasethesiology, 13:247–268.

Sims, N.R. and Pulsinelli, W.A. (1987): Altered mitochondrial respiration in selectively vulnerable brain subregions following transient forebrain ischemia in the rat. J. Neurochem., 49:1367–1374.

Strong, A.J. and Dardis, R. (2005): Depolarisation phenomena in traumatic and ischaemic brain injury. Adv. Tech. Stand. Neurosurg., 30:3–49.

Symon, L., Pasztor, E., and Branston, N.M. (1974): The distribution and density of reduced cerebral blood flow following acute middle cerebral artery occlusion: An experimental study by the technique of hydrogen clearance in baboons. Stroke, 5:355–364.

Takizawa, S., Fukuyama, N., Hirabayashi, H., Nakazawa, H., and Shinohara, Y. (1999): Dynamics of nitrotyrosine formation and decay in rat brain during focal ischemia-reperfusion. J. Cereb. Blood Flow Metab., 19:667–672.

Talloczy, Z., Jiang, W., Virgin, H.W., Leib, D.A., Scheuner, D., Kaufman, R.J., Eskelinen, E.L., and Levine, B. (2002): Regulation of starvation- and virus-induced autophagy by the eIF2alpha kinase signaling pathway. Proc. Natl. Acad. Sci. U S A, 99:190–195.

Tanveer, A., Virji, S., Andreeva, L., Totty, N.F., Hsuan, J.J., Ward, J.M., and Crompton, M. (1996): Involvement of cyclophilin D in the activation of a mitochondrial pore by Ca2– and oxidant stress. Eur. J. Biochem., 238:166–172.

Ting, A.T., Pimentel-Muinos, F.X., and Seed, B. (1996): RIP mediates tumor necrosis factor receptor 1 activation of NF-kappaB but not Fas/APO-1-initiated apoptosis. EMBO J., 15:6189–6196.

Trivedi, B. and Danforth, W.H. (1966): Effect of pH on the kinetics of frog muscle phosphofructokinase. J. Biol. Chem., 241:4110–4112.

Tsujimoto, Y. and Shimizu, S. (2000): VDAC regulation by the Bcl-2 family of proteins. Cell Death Differ., 7:1174–1181.

Vande, V.C., Cizeau, J., Dubik, D., Alimonti, J., Brown, T., Israels, S., Hakem, R., and Greenberg, A.H. (2000): BNIP3 and genetic control of necrosis-like cell death through the mitochondrial permeability transition pore. Mol. Cell Biol., 20:5454–5468.

Vanden Berghe, T., Kalai, M., Denecker, G., Meeus, A., Saelens, X., and Vandenabeele, P. (2006): Necrosis is associated with IL-6 production but apoptosis is not. Cell Signal., 18:328–335.

Veech, R.L., Lawson, J.W., Cornell, N.W., and Krebs, H.A. (1979): Cytosolic phosphorylation potential. J. Biol. Chem., 254:6538–6547.

Walker, G. and Marx, J.L. (1981): The national survey of stroke: Clinical findings. Stroke, 12(supp 1):1–13.

Wallace, D. (2001): A mitochondrial paradigm for degenerative diseases and ageing. Novartis Found Symp., 235:247–63.

Welsch, K.M., Caplan, L., Reis, D., Siesjo, B.K., and Weir, R. (1997): *Primer on Cerebrovascular Disease*. Academic Press, San Diego, CA.

Xu, Y., Huang, S., Liu, Z.G., and Han, J. (2006): Poly(ADP-ribose) polymerase-1 signaling to mitochondria in necrotic cell death requires RIP1/TRAF2-mediated JNK1 activation. J. Biol. Chem., 281:8788–8795.

Xu, K., Puchowicz, M.A., Lust, W.D., and LaManna, J.C. (2006): Adenosine treatment delays postischemic hippocampal CA1 loss after cardiac arrest and resuscitation in rats. Brain Res., 1071:208–217.

Yang, H. and Smith, D.L. (1997): Kinetics of cytochrome c folding examined by hydrogen exchange and mass spectrometry. Biochemistry, 36:14992–14999.

Yu, L., Alva, A., Su, H., Dutt, P., Freundt, E., Welsh, S., Baehrecke, E.H., and Lenardo, M.J. (2004): Regulation of an ATG7-beclin 1 program of autophagic cell death by caspase-8. Science, 304:1500–1502.

Yu, L., Wan, F., Dutta, S., Welsh, S., Liu, Z., Freundt, E., Baehrecke, E.H., and Lenardo, M. (2006): Autophagic programmed cell death by selective catalase degradation. Proc. Natl. Acad. Sci. U S A, 103:4952–4957.

Zaidan, E. and Sims, N.R. (1994): The calcium content of mitochondria from brain subregions following short-term forebrain ischemia and recirculation in the rat. J. Neurochem., 63:1812–1819.

Zhang, S.Q., Kovalenko, A., Cantarella, G., and Wallach, D. (2000): Recruitment of the IKK signalosome to the p55 TNF receptor: RIP and A20 bind to NEMO (IKKgamma) upon receptor stimulation. Immunity, 12:301–311.

Zhu, C., Wang, X., Xu, F., Bahr, B.A., Shibata, M., Uchiyama, Y., Hagberg, H., and Blomgren, K. (2005): The influence of age on apoptotic and other mechanisms of cell death after cerebral hypoxia-ischemia. Cell Death. Differ., 12:162–176.

Chapter 5
Metabolic Encephalopathy
Stroke – Clinical Features

Svetlana Pundik and Jose I. Suarez

Introduction

Stroke research has been the subject of increased attention in recent years. As a result, significant progress has been made in the understanding of the mechanisms of ischemic injury and development of potential methods of treatment. Unfortunately, only one therapy has been shown to be of benefit for acute ischemic stroke patients in a randomized controlled clinical trial: intravenous administration of recombinant tissue plasminogen activator (rt-PA) (The National Institute of Neurological Disorders and Stroke rt-PA Stroke Study Group (NINDS), 1995). Clinical and basic scientists have been contemplating the reasons for the failure to translate a large number of potential treatments into clinical practice. There are a number of possible explanations that include heterogeneity of human disease, and inherent species-related differences. In contrast to animal models where subjects have similar genetic material and undergo identical procedures, human stroke patients vary in their co-morbidities, age, risk factors, type and location of stroke, duration and severity of ischemia, among other things. Interestingly, the factors that contribute to the heterogeneity of human stroke play a significant role in the reversibility of ischemic injury (Table 5.1). We will discuss in this chapter some of the aspects that are associated with the ability of the brain tissue to improve its function following ischemia as it is seen in clinical practice.

Predictors of Good Outcome

Duration of Ischemia

Restoration of cerebral blood flow (CBF) to normal remains the mainstay of acute stroke treatment. The main challenge of acute stroke treatment is timely evaluation and administration of thrombolytic therapy for qualified patients. Animal studies have demonstrated that there is a very short window of opportunity when brain tissue is salvageable if CBF is reinstated. Clinical studies have demonstrated similar

D.W. McCandless (ed.) *Metabolic Encephalopathy*,   69
doi: 10.1007/978-0-387-79112-8_5, © Springer Science + Business Media, LLC 2009

Table 5.1 Predictors of favorable outcome after ischemic stroke

Predictors	References
Modifiable	
Duration of ischemia	1995; Alexandrov et al., 2001; Christou et al.,2000; Kammersgaard et al., 2004
Glycemic control	Alvarez-Sabin et al.,2003; Beghi et al., 1989; Harik and LaManna, 1988; Kamada et al., 2007; Leigh et al., 2004; Martini and Kent, 2007; Nedergaard, 1987; Nordt et al., 1993, 2000; Parsons et al., 2002; Selman et al., 1991
Optimizing cerebral perfusion, collateral circulation	Alexandrov et al., 1997; Kidwell et al., 2002; Lee et al., 2002; Suarez et al., 2002
Appropriate blood pressure management	Castillo et al., 2004; Vemmos et al., 2004; Yong et al., 2005
Hyperthermia	Azzimondi et al., 1995; Noor et al., 2003; Reith et al., 1996; Weimar et al., 2002
Non-modifiable	
Younger age	Engelter et al., 2006; Harik and LaManna, 1988; Kamada et al., 2007; Liebeskind, 2005; Wegener et al., 2004
Size of ischemic territory	Harik and LaManna, 1988; Johnston et al., 2000; Ribo et al., 2005
Severity of symptoms	Harik and LaManna, 1988; Leigh et al., 2004; Toni et al., 1997; Wegener et al., 2004
Preconditioning	Alexandrov et al., 2001; Ghosh and Galinanes, 2003; Gidday, 2006; Johnston, 2004; Kharbanda et al., 2002; Kirino, 2002; Moncayo et al., 2000; Schaller, 2005; Sitzer et al., 2004w; Wegener et al., 2004; Weih et al., 1999; Wu et al., 2003

observations. In fact, safety and efficacy of thrombolysis directly depends on time from onset of symptoms to recanalization. The National Institute of Neurological Disorders (NINDS) acute stroke trial showed that the best outcome is achieved for patients who are treated within the first 90 min of ischemia (NINDS, 1995). Several studies since then have described that earlier reperfusion is associated with improved outcome (Alexandrov et al., 2001; Christou et al., 2000). In addition, therapy delay leads to further complications, especially intracranial hemorrhage (Kidwell et al., 2002). Therefore, the shorter the time of ischemia, the better the outcome will be.

Infarct Size and Stroke Severity

Functional recovery following stroke depends on stroke location, stroke size and the severity of resulting neurological deficits. Patients with severe neurological deficits, reflected in higher National Institutes of Health Stroke Score (NIHSS)

(Lyden et al., 1994), fare worse than those with lesser deficits (Johnston et al., 2000; Tseng and Chang, 2006; Weimar et al., 2002). While infarct volume is a main outcome measure of stroke in laboratory animals, it plays a role in clinical practice as a predictor only when used in combination with other factors such as age, stroke type, or medical co-morbodities (Johnston et al., 2000). Severity of neurological symptoms only mildly correlates with infarct size (Johnston et al., 2002; Saver et al., 1999). Strokes that involve so called "silent areas" may not result in large NIHSS scores. However, by using detailed neurological and neuropsychological testing and neuroimaging, greater clinical abnormalities can be better documented than when using a gross assessment with NIHSS alone (Menezes et al., 2007). Detailed clinical evaluation, however, remains a main predictor of functional outcome following stroke (Johnston et al., 2002).

Collateral Flow

Focal ischemia produces a gradient of CBF in the territory of the occluded vessel. There is a central area or core with the lowest CBF, and a surrounding area with higher CBF values (albeit critically low compared to normal levels) called the penumbra. The latter may be nourished by existing collateral vasculature during acute stages of cerebral ischemia. Thus, in addition to the duration of ischemia, the amount of collateral circulation determines the outcome in reversible ischemia (Alexandrov et al., 1997; Kucinski et al., 2003; Liebeskind, 2005). As seen in animal stroke studies, the amount of CBF is directly correlated with the degree of damage at different durations of ischemia (Jones et al., 1981). The presence of collateral CBF during the first hours of stroke may allow a longer time for recanalization and lead to recovery of functions. In fact, early improvements in acute stroke may be due to collateral CBF (Toni et al., 1997).

Visualization of collateral circulation can be achieved with several neuroradiological tools. One could use conventional cerebral angiography or perfusion computer tomography or Magnetic Resonance Imaging (MRI) or positron emission tomography (PET) to document collateral CBF. Modern neuroimaging allows for the observation of the effect of collateral CBF in humans. A difference between Diffusion Weighted Images (DWI) and Perfusion Weighted Images (PWI) MRI sequences suggests salvageable tissue (Albers et al., 2006; Johnston et al., 2002) (Fig. 5.1). DWI identifies ischemic tissue with the highest risk of irreversible ischemic damage. PWI are obtained after a rapid injection of a paramagnetic contrast agent and qualitatively describe CBF. Therefore, the difference between the lesions obtained from these two sequences may represent the penumbra. Currently, clinical trials are attempting to determine if this information could be used to extend the window of thrombolytic treatment in a certain group of patients. Although it is unclear whether and to what extent acute DWI lesion translates into irreversibly damaged area, it is obvious that the presence of DWI/PWI mismatch indicates salvageable ischemic brain.

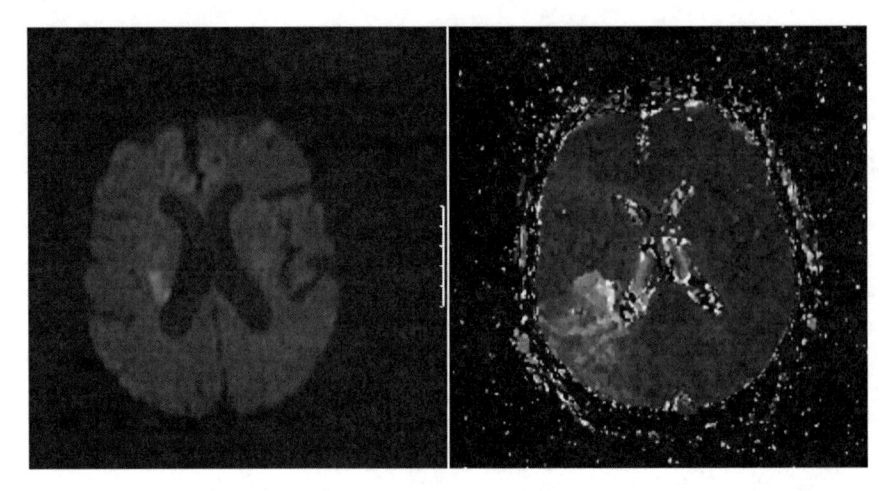

Fig. 5.1 Diffusion Weighted (*left panel*) and Perfusion Weighted (*right panel*) MRI sequences of acute ischemic stroke. Lighter grey area on the perfusion MRI is larger than the diffusion restriction area suggesting the presence of hypoperfused but salvageable tissue

Patient's Age

Recovery following stroke is compromised in the elderly (Johnston et al., 2000; Macciocchi et al., 1998; Weimar et al., 2002). The reasons for the age-dependent discrepancy are still poorly understood. There are several factors that could be contributing to it. One of them is that older patients have a higher incidence of co-morbidities that compromise an efficient rehabilitation process. Older patients have increased rates of hypertension, vascular disease, cardiac dysfunction, and depression among other factors that may also contribute to poor recovery. At the cellular level, ischemic injury affects structures and processes that are already compromised in the aged tissues. The biological process of aging is associated with mitochondrial dysfunction, failure of the free radical scavenging system, and deregulation of protein synthesis (Toescu (2005). However, in spite of higher morbidity and mortality in the elderly after a stroke than in younger patients, older stroke victims benefit from early thrombolytic therapy (Engelter et al., 2006; Kammersgaard et al., 2004). Therefore, while older patients present a challenge in rehabilitation they, nevertheless, should receive aggressive medical care in the acute stages of a stroke.

Hyperglycemia and History of Diabetes

Elevated serum glucose in animal models of reversible ischemic stroke is associated with up to 50% larger volume of infarction (Gisselsson et al., 1999; Nedergaard, 1987; Selman et al., 1991). The mechanisms of hyperglycemic injury are thought to be due to lactic acidosis and changes in CBF (Harik and LaManna, 1988;

Nedergaard, 1987). In observational clinical studies, patients with hyperglycemia at the time of stroke have up to 8 times larger infarct volumes and higher risks of unfavorable outcome including increased mortality rates (Alvarez-Sabin et al., 2003; Beghi et al., 1989; Leigh et al., 2004; Parsons et al., 2002). Hyperglycemia acts as an independent predictor of hemorrhagic transformation after thrombolytic therapy with rt-PA (Beghi et al., 1989; Broderick et al., 1995; Leigh et al., 2004). Of interest is that the association of hyperglycemia and poor outcome is unrelated to a history of diabetes. Acute hyperglycemia interferes with rt-PA mediated thrombolysis (Ribo et al., 2005). In fact, hyperglycemia is known to inhibit endogenous fibrinolysis (Nordt et al., 1993) and inhibit activity of exogenous rt-PA (Pandolfi et al., 2001). Treatment of hyperglycemia partially restores fibrinolytic function of vascular endothelium and exogenous rt-PA (Nordt et al., 2000; Pandolfi et al., 2001). Furthermore, reperfusion injury to the brain vasculature is accelerated with hyperglycemia due to pro-inflammatory, pro-thrombotic and vasoconstrictive vascular processes (Kamada et al., 2007; Martini and Kent, 2007).

Hypertension

Hypertension is one of the modifiable risk factors for stroke. Blood pressure control reduces the risk of stroke by 38% (Whisnant, 1996). Functional outcomes following stroke are also dependent on whether the patients' blood pressure is regulated (Castillo et al., 2004; Yong et al., 2005). Favorable outcomes and blood pressure association is complex. Blood pressure regulation varies in acute versus subacute versus late stages of ischemic stroke. Extremely high blood pressures (SBP >= 185 and DBP >= 110) are unequivocally associated with poor outcomes and hemorrhagic transformation for patients that receive thrombolytic therapy (Vemmos et al., 2004). However, moderately elevated BP (SBP >= 140 and DBP >= 90) is associated with improved functional outcomes (Yong et al., 2005). At the same time, patients who have untreated low systemic pressures during acute stroke do poorly (Castillo et al., 2004). The need for such tight blood pressure regulation in the acute stroke period is due to a loss of microvascular cerebral autoregulation and the existence of pressure-dependent penumbra that is supplied by collateral CBF. In order to maximize CBF in the penumbra, blood pressure is allowed to run high in the first few hours of stroke. Because of the lack of autoregulation, any sudden changes in pressure appear to be deleterious. Increased BP variability during the first 72 h of stroke is associated with mildly increased risk of unfavorable outcome, especially after thrombolysis (Yong et al., 2005).

Role of Preconditioning

The phenomenon of preconditioning has been investigated in various animal stroke models. A vast number of research studies present similar conclusions. A mild reversible metabolic stress that precedes a stroke reduces the amount of damage produced by this stroke, i.e. what does not kill you makes you stronger. The list of metabolic stresses that stimulate endogenous neuroprotection is growing every day. Preconditioning can be achieved with one of the following methods: sublethal hypoxia, brief episodes of ischemia, spreading depression, hyperoxia, hyper or hypothermia, inflammatory stimuli, seizures, metabolic inhibitors, ischemia of other organs such as heart or skeletal muscle (Gidday, 2006; Kirino, 2002). The mechanisms of ischemic tolerance include molecular processes involved in membrane stabilization, inhibition of apoptosis, nitric oxide related events, protein phosphorylation, anti-inflamatory processes, and glial proliferation as well as augmentation of stress responses (Kirino, 2002). A thorough understanding of the mechanisms of stimulation of endogenous neuroprotection will hopefully lead to the development of new stroke therapies.

It is usually a challenge to translate processes observed in experimental stroke models into clinical practice. Endogenous preconditioning is not an exception. Some have investigated transient ischemic attacks (TIA) as an ischemic preconditioning factor. The findings are controversial. The majority of clinical observational studies on this topic reflect a purported association between improved outcome in patients who had a TIA within a few days prior to the stroke (Moncayo et al., 2000; Schaller, 2005; Sitzer et al., 2004; Wegener et al., 2004; Weih et al., 1999). However, a recent report showed a lack of this association (Johnston, 2004). Among the explanations for such discrepancies are small sample size, variability in outcome measure, heterogeneity of stroke type and location and observational character of the studies. It is also encouraging to see that other human system organs such as cardiac and skeletal muscles and liver exhibit effects of ischemic preconditioning (Clavien et al., 2003; Ghosh and Galinanes, 2003; Kharbanda et al., 2002; Wu et al., 2003).

Acute Stroke Treatment

Recanalization remains the most important step in the treatment of ischemic stroke. In fact, without recanalization further treatment is very limited. However, recanalization alone does not restore all the function especially when ischemia has been allowed to last for several hours. Restoration of normal CBF to an ischemic brain region is beneficial only within the first 3–6 h. In very rare situations and mostly in the brain stem ischemias, recanalization procedures are undertaken beyond the 6-h limit. Recanalization into a severely ischemic tissue generates a cascade of untoward processes. Reperfusion injury leads to further deterioration of tissue that is already compromised. In general, shorter and less severe ischemia leads to fewer reperfusion injury consequences. In clinical practice the most significant and most visible complication of reperfusion into a severely injured brain parenchyma is hemorrhage. It is mostly the increased risk of hemorrhagic transformation that limits the administration of thrombolytic therapy beyond 3 h. The key to success in thrombolysis remains a shorter duration of ischemia.

Recanalization

Clot removal can be achieved in several ways. Intravenous (IV) rt-PA is the most widely used and to this date the only medication that is approved by the Food and Drug Administration (FDA) for acute stroke treatment as previously mentioned. Treatment protocol is based on the design of the NINDS Stroke Trial that was published in 1995 (NINDS, 1995). When patients present within 3 h from symptom onset and all the exclusion and inclusion criteria are met, thrombolysis is initiated by giving 0.9 mg kg^{-1} IV over 1 h with 10% given as a bolus. In the NINDS trial, favorable outcome was achieved by at least 30% more patients in the treatment arm group. This group also experienced higher rates of symptomatic intracranial hemorrhage, 6.4% versus 0.6% in the placebo group. Other thrombolytic agents such as urokinase, prourokinase, streptokinase, reteplase, antistreplase, and staphylokinase have also been tested. They are either not used because of serious side effects (streptokinase) or only utilized in experimental protocols. Several clinical trials attempted to test the utility of intravenous thrombolytic therapy with rt-PA or other agents at later ischemia durations (Clark et al., 1999; Hacke et al., 1998). Unfortunately, higher risk of symptomatic intracranial hemorrhage and the lack of significant benefit have demonstrated that there is no safety or efficacy of intravenous therapy past the 3-h window. In addition, a defibrinogenating agent, ancrod, derived from viper venom, is being tested in a larger trial after pilot study showed a favorable risk-benefit ratio (Sherman et al., 2000).

At selected institutions, thrombolytic therapy is also given via the intraarterial (IA) route for patients who present within 6 h from the time of onset. The main advantage of IA therapy is a direct visualization of cerebral vasculature. IA rt-PA therapy is given in the smallest effective doses and directly into the clot and has higher rates of recanalization (Qureshi, 2004; Suarez et al., 2002). The main disadvantage of IA thrombolysis is the requirement of additional resources, such as trained angiographers, who may not be available in a majority of emergency departments. Cerebral angiography is a highly technical and labor intensive procedure. Because of that, IA therapy is associated with an unwanted delay in treatment administration. In order to minimize the delay and maximize the efficacy of treatment thrombolytics can be administered as a combination of IV and IA. In the Interventional Management of Stroke study patients with significant neurological deficits, as indicated by NIHSS >= 10, received rt-PA IV 0.6 mg kg^{-1} (10% as a bolus) of rt-PA within the 3 h of symptoms, which was immediately followed by an angiography and IA therapy, if necessary, based on angiographical findings (IMS Study Investigators, 2004). Subjects who received IV/IA therapy had significantly better outcomes. Further studies of IV/IA treatment are in progress.

Another advantage of angiography during acute stroke treatment is that it allows mechanical disruption of the clot. There are several methods available today. Among them are angioplasty balloons, clot extraction devices (corkscrew retriever MERCI device), ultrasound aided catheters and simple manipulation of the clot with a microcatheter that is used for delivery of the thrombolytic agent. Studies have shown that new devices are efficient in clot removal, but it is still unclear if their use results in improvement of outcome (Smith et al., 2005). Therefore, their use remains limited but is nonetheless growing as more and more angiographers become familiar with the techniques.

Platelet glycoprotein IIB-IIIA inhibitors are used as adjunct to thrombolysis, angioplasty or stenting. Several studies have shown favorable outcomes in a selected group of patients (Eckert et al., 2005; Lee et al., 2002; Qureshi et al., 2006).

Neuroprotection

There have been many attempts to develop a neuroprotective agent that could be administered alone or in addition to thrombolytic therapy. Unfortunately, there are no FDA approved neuroprotective treatments for acute stroke patients. To this date clinical trials have tested glutamate receptor (NMDA) antagonist, MK-801; Ca^{++} channel blockers, free radical scavangers, anti-inflammatory agents (for review see, (Turley et al., 2005; Weinberger, 2006). In some trials the effects were even negative. Many other studies are underway. One promising approach in neuroprotective treatment is the use of recombinant erythropoietin in the acute hours of cerebral ischemia. A small study demonstrated significant improvement in outcome measures in treated patients (Ehrenreich et al., 2002).

Beyond Clot Removal

Maintenance of Perfusion Pressure

As mentioned earlier, survival of ischemic penumbra is highly dependent on the efficiency of collateral vasculature to provide adequate CBF in the periphery of the ischemic territory. The viability of the penumbra depends on the degree and the duration of ischemia. In animal stroke models the reduction of CBF in the periphery of the ischemic territory is down to about 20% (Jones et al., 1981). Similar levels of diminished CBF are observed in humans (Bandera et al., 2006). In a healthy brain, CBF is maintained between 60 and 150 mmHg by vasoconstriction or vasodilatation in response to changes of perfusion pressure. However, during ischemia this autoregulation is lost, which makes the penumbra highly dependent on systemic blood pressures.

Blood pressure in the early hours of stroke is above 160/90 mmHg in 80% of patients and generally normalizes without antihypertensive treatment in a majority of them (Oppenheimer and Hachinski, 1992; Phillips, 1994). Acute blood pressure management depends on whether thrombolytic therapy is being administered.

According to the American Heart Association recommendations, patients who do not meet criteria for thrombolysis are allowed to have pressures as high as 230 mmHg systolic and 130 mmHg diastolic unless evidence of other organ failure is present (Adams et al., 2007). However, in order to reduce the risk of hemorrhagic complications from thrombolysis, blood pressure of over 185 mmHg systolic and 110 mmHg diastolic on more than two readings requires anti-hypertensive therapy. Systolic hypertension is usually treated with labetelol, nicardipine or hydralyzine. Nitroglycerin is usually given for diastolic hypertension. These agents are chosen for their rapid and relatively reliable action with low risk of overshoot and minimal potential to increase intracranial pressure (Rose and Mayer, 2004). It is important to note that current recommendations for blood pressure control during acute stroke are controversial. However, it is essential to avoid sudden drops in blood pressure (Powers, 1993). Overall, maintenance of adequate and yet not dangerously high perfusion pressure is one of the most challenging tasks during the first hours of stroke.

Management of Hyperthermia, Hypothermic Treatments

Therapeutic hypothermia has shown about a 30% reduction in brain infarction volume and may be due to decreased excitotoxicity, improved anti-inflammatory effects, free radical scavenging, and anti-apoptotic processes (Maher and Hachinski, 1993; Onesti et al., 1991; Pabello et al., 2005; Toyoda et al., 1996; Van Hemelrijck et al., 2005; Yanamoto et al., 1996). Hypothermic therapy is neuroprotective for cardiac arrest survivors (Hypothermia after Cardiac Arrest Study Group, 2002; Nolan et al., 2003). Deep and mild hypothermia is used during vascular and cardiac surgeries and traumatic brain injury. While mild hypothermia is recommended and is being used in many medical centers throughout the world for cardiac arrest, its use for patients with acute ischemic stroke has not been shown efficacious. In addition, benefits of hypothermia are not without problems. First, even a mild hypothermia is associated with such side effects as cardiac arrhythmias, infections and hypotension (Olsen et al., 2003). Second, discomfort from the temperatures and reflex shivering require induction of a coma for these patients. Therefore, to deal with the side effects of hypothermia, patients need to be placed on mechanical ventilation, heavily sedated and paralyzed. These procedures carry a significant level of their own risks, and therefore halt its implementation into clinical practice.

Just as hypothermia is shown to reduce ischemic brain damage, elevated body temperature produces greater injury (Azzimondi et al., 1995; Weimar et al., 2002). Clinical (Reith et al., 1996) and animal studies (Noor et al., 2003) described a larger volume of infarction and worse functional outcomes for those with fevers during early times of cerebral ischemia. While induction of hypothermia presents a challenging task, maintaining normothermia in acute stroke patients is usually manageable. It is routine in a stroke care unit to investigate and treat causes of temperatures above 37.5°C by administering antipyretic measures.

Management of Hyperglycemia

Treatment of hyperglycemia has not been directly studied in acute stroke patients. However, there is a vast literature on treatment of hyperglycemia in the setting of acute illness (van den et al., 2001, 2006). Elevation of blood glucose might be secondary to stress in these situations and may resolve spontaneously. It is proposed that the aim of intensive insulin therapy should be 140–180 mg dl^{-1}. There are clinical trails underway, that are testing strategies on achieving nomoglycemia in acute stroke.

Future in Stroke Therapies

Stroke is the third leading cause of morbidity and mortality in the United States. Therefore, efforts that are being made in stroke research to develop novel therapies are very important. Early recanalization should remain a key element in acute stroke treatment. The fact that only a very limited number of findings in the laboratories have been translated into clinical practice is on the minds of many basic and clinical investigators. There have been many proposals for addressing this problem. Suggestions of multimodal approaches have been made. Perhaps attending to several pathological events induced by ischemia is necessary to demonstrate the clinical effect. It is also important that putative neuroprotectants are subjected to vigorous investigation prior to taking a given agent to clinical trials. Perhaps more careful population selection that might involve additional imaging might be a solution.

References

Adams, H.P., Jr., del Zoppo, G., Alberts, M.J., Bhatt, D.L., Brass, L., Furlan, A., Grubb, R.L., Higashida, R.T., Jauch, E.C., Kidwell, C., Lyden, P.D., Morgenstern, L.B., Qureshi, A.I., Rosenwasser, R.H., Scott, P.A., and Wijdicks, E.F. (2007): Guidelines for the early management of adults with ischemic stroke: a guideline from the American Heart Association/American Stroke Association Stroke Council, Clinical Cardiology Council, Cardiovascular Radiology and Intervention Council, and the Atherosclerotic Peripheral Vascular Disease and Quality of Care Outcomes in Research Interdisciplinary Working Groups: the American Academy of Neurology affirms the value of this guideline as an educational tool for neurologists. Stroke, 38:1655–1711.
Albers, G.W., Thijs, V.N., Wechsler, L., Kemp, S., Schlaug, G., Skalabrin, E., Bammer, R., Kakuda, W., Lansberg, M.G., Shuaib, A., Coplin, W., Hamilton, S., Moseley, M., and Marks, M.P. (2006): Magnetic resonance imaging profiles predict clinical response to early reperfusion: the diffusion and perfusion imaging evaluation for understanding stroke evolution (DEFUSE) study. Ann. Neurol., 60:508–517.
Alexandrov, A.V., Black, S.E., Ehrlich, L.E., Caldwell, C.B., and Norris, J.W. (1997): Predictors of hemorrhagic transformation occurring spontaneously and on anticoagulants in patients with acute ischemic stroke. Stroke, 28:1198–1202.

Alexandrov, A.V., Burgin, W.S., Demchuk, A.M., El Mitwalli, A., and Grotta, J.C. (2001): Speed of intracranial clot lysis with intravenous tissue plasminogen activator therapy: sonographic classification and short-term improvement. Circulation, 103:2897–2902.

Alvarez-Sabin, J., Molina, C.A., Montaner, J., Arenillas, J.F., Huertas, R., Ribo, M., Codina, A., and Quintana, M. (2003): Effects of admission hyperglycemia on stroke outcome in reperfused tissue plasminogen activator—treated patients. Stroke, 34:1235–1241.

Azzimondi, G., Bassein, L., Nonino, F., Fiorani, L., Vignatelli, L., Re, G., and D'Alessandro, R. (1995): Fever in acute stroke worsens prognosis. A prospective study. Stroke, 26:2040–2043.

Bandera, E., Botteri, M., Minelli, C., Sutton, A., Abrams, K.R., and Latronico, N. (2006): Cerebral blood flow threshold of ischemic penumbra and infarct core in acute ischemic stroke: a systematic review. Stroke, 37:1334–1339.

Beghi, E., Bogliun, G., Cavaletti, G., Sanguineti, I., Tagliabue, M., Agostoni, F., and Macchi, I. (1989): Hemorrhagic infarction: risk factors, clinical and tomographic features, and outcome. A case-control study. Acta Neurol. Scand., 80:226–231.

Brekenfeld, C., Remonda, L., Nedeltchev, K., Bredow, F., Ozdoba, C., Wiest, R., Arnold, M., Mattle, H.P., and Schroth, G. (2005): Endovascular neuroradiological treatment of acute ischemic stroke: techniques and results in 350 patients. Neurol.Res., 27 Suppl 1:S29–S35

Broderick, J.P., Hagen, T., Brott, T., and Tomsick, T. (1995): Hyperglycemia and hemorrhagic transformation of cerebral infarcts. Stroke, 26:484–487.

Castillo, J., Leira, R., Garcia, M.M., Serena, J., Blanco, M., and Davalos, A. (2004): Blood pressure decrease during the acute phase of ischemic stroke is associated with brain injury and poor stroke outcome. Stroke, 35:520–526.

Christou, I., Alexandrov, A.V., Burgin, W.S., Wojner, A.W., Felberg, R.A., Malkoff, M., and Grotta, J.C. (2000): Timing of recanalization after tissue plasminogen activator therapy determined by transcranial doppler correlates with clinical recovery from ischemic stroke. Stroke, 31:1812–1816.

Clark, W.M., Wissman, S., Albers, G.W., Jhamandas, J.H., Madden, K.P., and Hamilton, S. (1999): Recombinant tissue-type plasminogen activator (Alteplase) for ischemic stroke 3 to 5 hours after symptom onset. The ATLANTIS Study: a randomized controlled trial. Alteplase Thrombolysis for Acute Noninterventional Therapy in Ischemic Stroke. JAMA, 282:2019–2026.

Clavien, P.A., Selzner, M., Rudiger, H.A., Graf, R., Kadry, Z., Rousson, V., and Jochum, W. (2003): A prospective randomized study in 100 consecutive patients undergoing major liver resection with versus without ischemic preconditioning. Ann. Surg., 238:843–850.

Eckert, B., Koch, C., Thomalla, G., Kucinski, T., Grzyska, U., Roether, J., Alfke, K., Jansen, O., and Zeumer, H. (2005): Aggressive therapy with intravenous abciximab and intra-arterial rtPA and additional PTA/stenting improves clinical outcome in acute vertebrobasilar occlusion: combined local fibrinolysis and intravenous abciximab in acute vertebrobasilar stroke treatment (FAST): results of a multicenter study. Stroke, 36:1160–1165.

Ehrenreich, H., Hasselblatt, M., Dembowski, C., Cepek, L., Lewczuk, P., Stiefel, M., Rustenbeck, H.H., Breiter, N., Jacob, S., Knerlich, F., Bohn, M., Poser, W., Ruther, E., Kochen, M., Gefeller, O., Gleiter, C., Wessel, T.C., De Ryck, M., Itri, L., Prange, H., Cerami, A., Brines, M., and Siren, A.L. (2002): Erythropoietin therapy for acute stroke is both safe and beneficial. Mol. Med., 8:495–505.

Engelter, S.T., Bonati, L.H., and Lyrer, P.A. (2006): Intravenous thrombolysis in stroke patients of > or = 80 versus <80 years of age — A systematic review across cohort studies. Age Ageing, 35:572–580.

Ghosh, S. and Galinanes, M. (2003): Protection of the human heart with ischemic preconditioning during cardiac surgery: role of cardiopulmonary bypass. J. Thorac. Cardiovasc. Surg., 126:133–142.

Gidday, J.M. (2006): Cerebral preconditioning and ischaemic tolerance. Nat. Rev. Neurosci., 7:437–448.

Gisselsson, L., Smith, M.L., and Siesjo, B.K. (1999): Hyperglycemia and focal brain ischemia. J. Cereb. Blood Flow Metab., 19:288–297.

Hacke, W., Kaste, M., Fieschi, C., von Kummer, R., Davalos, A., Meier, D., Larrue, V., Bluhmki, E., Davis, S., Donnan, G., Schneider, D., Diez-Tejedor, E., and Trouillas, P. (1998): Randomised double-blind placebo-controlled trial of thrombolytic therapy with intravenous alteplase in acute ischaemic stroke (ECASS II). Second European-Australasian Acute Stroke Study Investigators. Lancet, 352:1245–1251.

Harik, S.I. and LaManna, J.C. (1988): Vascular perfusion and blood-brain glucose transport in acute and chronic hyperglycemia. J. Neurochem., 51:1924–1929.

Hypothermia after Cardiac Arrest Study Group (2002): Mild therapeutic hypothermia to improve the neurologic outcome after cardiac arrest. N. Engl. J. Med., 346:549–556.

IMS Study Investigators (2004): Combined intravenous and intra-arterial recanalization for acute ischemic stroke: The Interventional Management of Stroke Study. Stroke, 35:904–911.

Johnston, K.C., Connors, A.F., Jr., Wagner, D.P., Knaus, W.A., Wang, X., and Haley, E.C., Jr. (2000): A predictive risk model for outcomes of ischemic stroke. Stroke, 31:448–455.

Johnston, K.C., Wagner, D.P., Haley, E.C., Jr., andConnors, A.F., Jr. (2002): Combined clinical and imaging information as an early stroke outcome measure. Stroke, 33:466–472.

Johnston, S.C. (2004): Ischemic preconditioning from transient ischemic attacks? Data from the Northern California TIA Study. Stroke, 35:2680–2682.

Jones, T.H., Morawetz, R.B., Crowell, R.M., Marcoux, F.W., FitzGibbon, S.J., DeGirolami, U., and Ojemann, R.G. (1981): Thresholds of focal cerebral ischemia in awake monkeys. J. Neurosurg., 54:773–782.

Kamada, H., Yu, F., Nito, C., and Chan, P.H. (2007): Influence of hyperglycemia on oxidative stress and matrix metalloproteinase-9 activation after focal cerebral ischemia/reperfusion in rats: relation to blood-brain barrier dysfunction. Stroke, 38:1044–1049.

Kammersgaard, L.P., Jorgensen, H.S., Reith, J., Nakayama, H., Pedersen, P.M., and Olsen, T.S. (2004): Short- and long-term prognosis for very old stroke patients. The Copenhagen Stroke Study. Age Ageing, 33:149–154.

Kharbanda, R.K., Mortensen, U.M., White, P.A., Kristiansen, S.B., Schmidt, M.R., Hoschtitzky, J.A., Vogel, M., Sorensen, K., Redington, A.N., and MacAllister, R. (2002): Transient limb ischemia induces remote ischemic preconditioning in vivo. Circulation, 106:2881–2883.

Kidwell, C.S., Saver, J.L., Carneado, J., Sayre, J., Starkman, S., Duckwiler, G., Gobin, Y.P., Jahan, R., Vespa, P., Villablanca, J.P., Liebeskind, D.S., and Vinuela, F. (2002): Predictors of hemorrhagic transformation in patients receiving intra-arterial thrombolysis. Stroke, 33:717–724.

Kirino, T. (2002): Ischemic tolerance. J. Cereb. Blood Flow Metab., 22:1283–1296.

Kucinski, T., Koch, C., Eckert, B., Becker, V., Kromer, H., Heesen, C., Grzyska, U., Freitag, H.J., Rother, J., and Zeumer, H. (2003): Collateral circulation is an independent radiological predictor of outcome after thrombolysis in acute ischaemic stroke. Neuroradiology, 45:11–18.

Lee, D.H., Jo, K.D., Kim, H.G., Choi, S.J., Jung, S.M., Ryu, D.S., and Park, M.S. (2002): Local intraarterial urokinase thrombolysis of acute ischemic stroke with or without intravenous abciximab: a pilot study. J. Vasc. Interv. Radiol., 13:769–774.

Leigh, R., Zaidat, O.O., Suri, M.F., Lynch, G., Sundararajan, S., Sunshine, J.L., Tarr, R., Selman, W., Landis, D.M., and Suarez, J.I. (2004): Predictors of hyperacute clinical worsening in ischemic stroke patients receiving thrombolytic therapy. Stroke, 35:1903–1907.

Liebeskind, D.S. (2005): Collaterals in acute stroke: beyond the clot. Neuroimaging Clin. N. Am., 15:553–573, x.

Lyden, P., Brott, T., Tilley, B., Welch, K.M., Mascha, E.J., Levine, S., Haley, E.C., Grotta, J., and Marler, J. (1994): Improved reliability of the NIH stroke scale using video training. NINDS TPA Stroke Study Group. Stroke, 25:2220–2226.

Macciocchi, S.N., Diamond, P.T., Alves, W.M., and Mertz, T. (1998): Ischemic stroke: relation of age, lesion location, and initial neurologic deficit to functional outcome. Arch.Phys.Med.Rehabil., 79:1255–1257.

Maher, J. and Hachinski, V. (1993): Hypothermia as a potential treatment for cerebral ischemia. Cerebrovasc. Brain Metab. Rev., 5:277–300.

Martini, S.R. and Kent, T.A. (2007): Hyperglycemia in acute ischemic stroke: a vascular perspective. J. Cereb. Blood Flow Metab., 27:435–451.

Menezes, N.M., Ay, H., Wang, Z.M., Lopez, C.J., Singhal, A.B., Karonen, J.O., Aronen, H.J., Liu, Y., Nuutinen, J., Koroshetz, W.J., and Sorensen, A.G. (2007): The real estate factor: quantifying the impact of infarct location on stroke severity. Stroke, 38:194–197.

Moncayo, J., de Freitas, G.R., Bogousslavsky, J., Altieri, M., and van Melle, G. (2000): Do transient ischemic attacks have a neuroprotective effect? Neurology, 54:2089–2094.

Nedergaard, M. (1987): Transient focal ischemia in hyperglycemic rats is associated with increased cerebral infarction. Brain Res., 408:79–85.

Nolan, J.P., Morley, P.T., Vanden Hoek, T.L., Hickey, R.W., Kloeck, W.G., Billi, J., Bottiger, B.W., Morley, P.T., Nolan, J.P., Okada, K., Reyes, C., Shuster, M., Steen, P.A., Weil, M.H., Wenzel, V., Hickey, R.W., Carli, P., Vanden Hoek, T.L., and Atkins, D. (2003): Therapeutic hypothermia after cardiac arrest: an advisory statement by the advanced life support task force of the International Liaison Committee on Resuscitation. Circulation, 108:118–121.

Noor, R., Wang, C.X., and Shuaib, A. (2003): Effects of hyperthermia on infarct volume in focal embolic model of cerebral ischemia in rats. Neurosci. Lett., 349:130–132.

Nordt, T.K., Klassen, K.J., Schneider, D.J., and Sobel, B.E. (1993): Augmentation of synthesis of plasminogen activator inhibitor type-1 in arterial endothelial cells by glucose and its implications for local fibrinolysis. Arterioscler. Thromb., 13:1822–1828.

Nordt, T.K., Peter, K., Bode, C., and Sobel, B.E. (2000): Differential regulation by troglitazone of plasminogen activator inhibitor type 1 in human hepatic and vascular cells. J. Clin. Endocrinol. Metab., 85:1563–1568.

Olsen, T.S., Weber, U.J., and Kammersgaard, L.P. (2003): Therapeutic hypothermia for acute stroke. Lancet Neurol., 2:410–416.

Onesti, S.T., Baker, C.J., Sun, P.P., and Solomon, R.A. (1991): Transient hypothermia reduces focal ischemic brain injury in the rat. Neurosurgery, 29:369–373.

Oppenheimer, S. and Hachinski, V. (1992): Complications of acute stroke. Lancet, 339:721–724.

Pabello, N.G., Tracy, S.J., Snyder-Keller, A., and Keller, R.W., Jr. (2005): Regional expression of constitutive and inducible transcription factors following transient focal ischemia in the neonatal rat: influence of hypothermia. Brain Res., 1038:11–21.

Pandolfi, A., Giaccari, A., Cilli, C., Alberta, M.M., Morviducci, L., De Filippis, E.A., Buongiorno, A., Pellegrini, G., Capani, F., and Consoli, A. (2001): Acute hyperglycemia and acute hyperinsulinemia decrease plasma fibrinolytic activity and increase plasminogen activator inhibitor type 1 in the rat. Acta Diabetol., 38:71–76.

Parsons, M.W., Barber, P.A., Desmond, P.M., Baird, T.A., Darby, D.G., Byrnes, G., Tress, B.M., and Davis, S.M. (2002): Acute hyperglycemia adversely affects stroke outcome: a magnetic resonance imaging and spectroscopy study. Ann. Neurol., 52:20–28.

Phillips, S.J. (1994): Pathophysiology and management of hypertension in acute ischemic stroke. Hypertension, 23:131–136.

Powers, W.J. (1993): Acute hypertension after stroke: the scientific basis for treatment decisions. Neurology, 43:461–467.

Qureshi, A.I. (2004): Endovascular treatment of cerebrovascular diseases and intracranial neoplasms. Lancet, 363:804–813.

Qureshi, A.I., Harris-Lane, P., Kirmani, J.F., Janjua, N., Divani, A.A., Mohammad, Y.M., Suarez, J.I., and Montgomery, M.O. (2006): Intra-arterial reteplase and intravenous abciximab in patients with acute ischemic stroke: an open-label, dose-ranging, phase I study. Neurosurgery, 59:789–796.

Reith, J., Jorgensen, H.S., Pedersen, P.M., Nakayama, H., Raaschou, H.O., Jeppesen, L.L., and Olsen, T.S. (1996): Body temperature in acute stroke: relation to stroke severity, infarct size, mortality, and outcome. Lancet, 347:422–425.

Ribo, M., Molina, C., Montaner, J., Rubiera, M., Delgado-Mederos, R., Arenillas, J.F., Quintana, M., and Alvarez-Sabin, J. (2005): Acute hyperglycemia state is associated with lower tPA-induced recanalization rates in stroke patients. Stroke, 36:1705–1709.

Rose, J.C. and Mayer, S.A. (2004): Optimizing blood pressure in neurological emergencies. Neurocrit. Care, 1:287–299.

Saver, J.L., Johnston, K.C., Homer, D., Wityk, R., Koroshetz, W., Truskowski, L.L., and Haley, E.C. (1999): Infarct volume as a surrogate or auxiliary outcome measure in ischemic stroke clinical trials. The RANTTAS Investigators. Stroke, 30:293–298.

Schaller, B. (2005): Ischemic preconditioning as induction of ischemic tolerance after transient ischemic attacks in human brain: its clinical relevance. Neurosci. Lett., 377:206–211.

Selman, W.R., Crumrine, R.C., Rosenstein, C.C., Jenkins, C., LaManna, J.C., Ratcheson, R.A., and Lust, W.D. (1991): Rapid metabolic failure in spontaneously hypertensive rats after middle cerebral artery ligation. Metab. Brain Dis., 6:57–64.

Sherman, D.G., Atkinson, R.P., Chippendale, T., Levin, K.A., Ng, K., Futrell, N., Hsu, C.Y., and Levy, D.E. (2000): Intravenous ancrod for treatment of acute ischemic stroke: the STAT study: a randomized controlled trial. Stroke Treatment with Ancrod Trial. JAMA, 283:2395–2403.

Sitzer, M., Foerch, C., Neumann-Haefelin, T., Steinmetz, H., Misselwitz, B., Kugler, C., and Back, T. (2004): Transient ischaemic attack preceding anterior circulation infarction is independently associated with favourable outcome. J. Neurol. Neurosurg. Psychiatry, 75:659–660.

Smith, W.S., Sung, G., Starkman, S., Saver, J.L., Kidwell, C.S., Gobin, Y.P., Lutsep, H.L., Nesbit, G.M., Grobelny, T., Rymer, M.M., Silverman, I.E., Higashida, R.T., Budzik, R.F., and Marks, M.P. (2005): Safety and efficacy of mechanical embolectomy in acute ischemic stroke: results of the MERCI trial. Stroke, 36:1432–1438.

Suarez, J.I., Zaidat, O.O., Sunshine, J.L., Tarr, R., Selman, W.R., and Landis, D.M. (2002): Endovascular administration after intravenous infusion of thrombolytic agents for the treatment of patients with acute ischemic strokes. Neurosurgery, 50:251–259.

The National Institute of Neurological Disorders and Stroke rt-PA Stroke Study Group (1995): Tissue plasminogen activator for acute ischemic stroke. N. Engl. J. Med., 333:1581–1587.

Toescu, E.C. (2005): Normal brain ageing: models and mechanisms. Philos. Trans. R. Soc. Lond. B Biol. Sci., 360:2347–2354.

Toni, D., Fiorelli, M., Bastianello, S., Falcou, A., Sette, G., Ceschin, V., Sacchetti, M.L., and Argentino, C. (1997): Acute ischemic strokes improving during the first 48 hours of onset: predictability, outcome, and possible mechanisms. A comparison with early deteriorating strokes. Stroke, 28:10–14.

Toyoda, T., Suzuki, S., Kassell, N.F., and Lee, K.S. (1996): Intraischemic hypothermia attenuates neutrophil infiltration in the rat neocortex after focal ischemia-reperfusion injury. Neurosurgery, 39:1200–1205.

Tseng, M.C. and Chang, K.C. (2006): Stroke severity and early recovery after first-ever ischemic stroke: results of a hospital-based study in Taiwan. Health Policy, 79:73–78.

Turley, K.R., Toledo-Pereyra, L.H., and Kothari, R.U. (2005): Molecular mechanisms in the pathogenesis and treatment of acute ischemic stroke. J. Invest. Surg., 18:207–218.

van den, B.G., Wilmer, A., Hermans, G., Meersseman, W., Wouters, P.J., Milants, I., Van Wijngaerden, E., Bobbaers, H., and Bouillon, R. (2006): Intensive insulin therapy in the medical ICU. N. Engl. J. Med., 354:449–461.

van den, B.G., Wouters, P., Weekers, F., Verwaest, C., Bruyninckx, F., Schetz, M., Vlasselaers, D., Ferdinande, P., Lauwers, P., and Bouillon, R. (2001): Intensive insulin therapy in the critically ill patients. N. Engl. J. Med., 345:1359–1367.

Van Hemelrijck, A., Hachimi-Idrissi, S., Sarre, S., Ebinger, G., and Michotte, Y. (2005): Post-ischaemic mild hypothermia inhibits apoptosis in the penumbral region by reducing neuronal nitric oxide synthase activity and thereby preventing endothelin-1-induced hydroxyl radical formation. Eur. J. Neurosci., 22:1327–1337.

Vemmos, K.N., Spengos, K., Tsivgoulis, G., Zakopoulos, N., Manios, E., Kotsis, V., Daffertshofer, M., and Vassilopoulos, D. (2004): Factors influencing acute blood pressure values in stroke subtypes. J. Hum. Hypertens., 18:253–259.

Wegener, S., Gottschalk, B., Jovanovic, V., Knab, R., Fiebach, J.B., Schellinger, P.D., Kucinski, T., Jungehulsing, G.J., Brunecker, P., Muller, B., Banasik, A., Amberger, N., Wernecke, K.D., Siebler, M., Rother, J., Villringer, A., and Weih, M. (2004): Transient ischemic attacks before ischemic stroke: preconditioning the human brain? A multicenter magnetic resonance imaging study. Stroke, 35:616–621.

Weih, M., Kallenberg, K., Bergk, A., Dirnagl, U., Harms, L., Wernecke, K.D., and Einhaupl, K.M. (1999): Attenuated stroke severity after prodromal TIA: a role for ischemic tolerance in the brain? Stroke, 30:1851–1854.

Weimar, C., Ziegler, A., Konig, I.R., and Diener, H.C. (2002): Predicting functional outcome and survival after acute ischemic stroke. J. Neurol., 249:888–895.

Weinberger, J.M. (2006): Evolving therapeutic approaches to treating acute ischemic stroke. J. Neurol. Sci., 249:101–109.

Whisnant, J.P. (1996): Effectiveness versus efficacy of treatment of hypertension for stroke prevention. Neurology, 46:301–307.

Wu, Z.K., Laurikka, J., Saraste, A., Kyto, V., Pehkonen, E.J., Savunen, T., and Tarkka, M.R. (2003): Cardiomyocyte apoptosis and ischemic preconditioning in open heart operations. Ann. Thorac. Surg., 76:528–534.

Yanamoto, H., Hong, S.C., Soleau, S., Kassell, N.F., and Lee, K.S. (1996): Mild postischemic hypothermia limits cerebral injury following transient focal ischemia in rat neocortex. Brain Res., 718:207–211.

Yong, M., Diener, H.C., Kaste, M., and Mau, J. (2005): Characteristics of blood pressure profiles as predictors of long-term outcome after acute ischemic stroke. Stroke, 36:2619–2625.

Chapter 6
The Role of Animal Models in the Study of Epileptogenesis

Kate Chandler, Pi-Shan Chang, and Matthew Walker

Introduction

Epileptogenesis is the process that leads to the development of epilepsy: the propensity to have recurrent, spontaneous seizures. During epileptogenesis, brain excitability increases due to molecular, cellular and network alterations. These changes are thought to be initiated by one or more brain insults which may be naturally occurring events such as traumatic brain injury, but can also be modeled in animals, using insults such as chemically induced status epilepticus (SE: a prolonged seizure).

The study of epileptogenesis is critical for (a) identifying patients who are at risk of developing epilepsy and (b) targeting drugs that can modify the epileptogenic process and could therefore prevent the development of the disease.

The interpretation of many of the pathologic and electrophysiologic changes in human brain tissue is confounded by (1) the influence of treatment; (2) the difficulty in differentiating cause from effect (i.e. it is possible that the changes are the result not the cause of the seizures) and (3) the lack of adequate control tissue for comparison. In order to overcome these handicaps, animal models of mesial temporal lobe epilepsy (TLE) are used to study epileptogenesis – the two most studied being the kindling model and the post-SE model.

Epileptogenesis in Human Epilepsies and Experimental Models

Epilepsy is commonly categorised as idiopathic, symptomatic, probable symptomatic (previously known as cryptogenic) and reactive. Idiopathic epilepsies are presumed to have an underlying genetic cause; indeed ion channelopathies have been identified in several human idiopathic epilepsy syndromes (Gardiner, 2005). Symptomatic epilepsies, which are thought to account for up to 50% of all epilepsy cases (Delorenzo et al., 2005), arise secondary to an underlying identifiable brain insult or lesion such as stroke, intracranial neoplasia or encephalitis. Probable

D.W. McCandless (ed.) *Metabolic Encephalopathy*,
doi: 10.1007/978-0-387-79112-8_6, © Springer Science + Business Media, LLC 2009

symptomatic epilepsies are likely to be symptomatic but have no cause that can be identified using current methods (Engel, 2006). Finally, reactive seizures arise secondary to an underlying metabolic disturbance that has an indirect effect on brain excitability, such as hepatic encephalopathy.

Naturally occurring epileptogenic insults in patients with symptomatic epilepsy include encephalitis, SE, traumatic brain injury and stroke. However, only a small proportion of people who suffer such insults develop epilepsy. The process can be considered in three stages: (1) the insult; (2) the latent period (epileptogenesis occurs during this period, before epilepsy develops); (3) epilepsy (Fig. 6.1). In some people the latent period can be extremely long. Approximately ninety per cent of SE patients, who develop epilepsy, do so within 7 years (Hesdorffer et al., 1998). This period tends to be shorter in patients who have had a traumatic brain injury or stroke (within 2 years; Hesdorffer et al., 1998), suggesting that the length of the epileptogenic period varies depending on the epileptogenic insult. Once a patient starts to have recurrent seizures, the epileptic disease state probably continues to progress, as seizures may induce additional neuronal alteration that leads to lowering of the seizure threshold.

The prolonged latent period in people suggests that either epileptogenesis is a long process, or that a second insult is necessary for epilepsy to occur. This "second hit hypothesis" proposes that an initial insult results in lowered seizure threshold, and then a later insult, the 'second hit', results in the expression of epilepsy (Walker et al., 2002).

Rodent models of epilepsy are commonly produced by stimulating SE chemically or electrically. Following this insult, the latent period ensues (which usually lasts

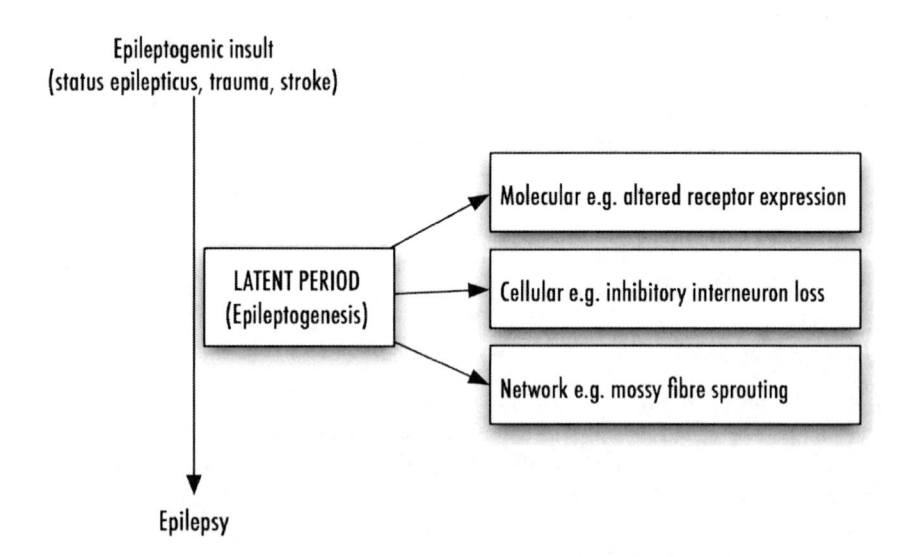

Fig. 6.1 Illustration of the development of epilepsy following a specific brain insult

days to weeks in animal models) and the state of recurrent seizures then occurs. In these animal models, epileptogenesis is actively occurring in the latent period, and therefore modification of this process can potentially change the outcome i.e. whether that animal goes on to develop epilepsy. The onset of epilepsy, and therefore presumably the process of epileptogenesis, is more rapid in SE models compared to traumatic brain injury or stroke models.

Hippocampal Sclerosis

The hippocampus plays a prominent part in epilepsy research, as it is an area of the brain that not only is vulnerable to damage by seizures but also commonly acts as the substrate for seizures. The hippocampus (literally "sea horse" due to its curled appearance) is part of the limbic lobe — a structure originally defined by Broca as a ring of grey matter on the medial aspect of each hemisphere. Bouchet and Cazauvieilh were the first to make an association between pathological changes in the hippocampus and epilepsy in 1825 during their work on potential links between mental disorders and epilepsy. During these studies they observed pathological changes in the hippocampus in five epileptic patients. The hippocampi had become firm in texture. They termed this pathological change "sclerosis." In 1880, Sommer reviewed the literature on the role of the hippocampus in epilepsy. He was probably the first person to associate the hippocampus with temporal lobe seizures. He postulated that hippocampal cells were vulnerable to insults and that the resulting Ammon's horn sclerosis was the cause of epilepsy (Fisher et al., 1998). In 1899 Bratz described the microscopic appearance of hippocampal sclerosis (HS):

1. Destruction of pyramidal neurons in Ammon's Horn (hippocampus proper), particularly in CA1
2. Loss of hilar neurons and CA4 pyramidal neurons (endfolium sclerosis)
3. Granule cell loss

He also showed that epilepsy is not always associated with HS. He thought that HS caused certain types of seizures rather than being a result of seizures. Certainly there is evidence that HS could indeed cause seizures. Most patients with refractory mesial TLE have HS (Babb and Brown, 1986). There is usually cell loss in both hippocampi although it tends to be asymmetrical. Unilateral hippocampal atrophy is often seen in these patients, which is characterised on MRI by reduced volume and increased signal on T2-weighted images. The location of the seizure focus frequently corresponds to the side of the atrophic hippocampus (Fish and Spencer, 1995). Removal of the sclerotic tissue during temporal lobe resection successfully "cures" 80% of the patients (Arruda et al., 1996). However, among patients with hippocampal atrophy, around 10% have bilateral atrophy. In these patients surgical success is achieved not by removing the more or less atrophied hippocampus, but by removing the seizure focus as identified by EEG recordings (Fish and Spencer, 1995). Later in the twentieth century an alternative explanation developed.

Scheibel et al. (1974) demonstrated a positive correlation between the chronicity of the seizure disorder and the severity of the hippocampal damage which led many to believe that HS was an eventual consequence of chronic seizures rather than a cause. From the 1950s onwards, workers started to associate HS with prior cerebral injury early in life. Retrospective studies frequently show that HS is associated with prolonged febrile seizures in childhood (Davies et al., 1996; Lewis, 1999) but see Tarkka et al. (2003). However HS is sometimes present even in the absence of any apparent initial precipitating injury. Despite the link between febrile seizures and TLE, many children who have had febrile seizures do not develop HS or TLE (Tarkka et al., 2003). Clearly, there is a complex relationship between early cerebral injury, TLE and HS. The most commonly used animal models of epilepsy have similar pathology to HS, which is induced by an insult such as SE.

Animal Models of Epilepsy

Epilepsy can be modeled in diverse species from *Drosophila* to more complex organisms such as non-human primates. These can be separated into models of seizures and models of epilepsy, although some models can be considered to be both. For example, if flurothyl is administered to rodents, they have a generalised seizure but do not develop spontaneous seizures. In contrast, if SE is induced by administration of pilocarpine, the animal develops spontaneous seizures after a latent period of days to weeks. The former case is a model of seizure activity but not of epilepsy, and the latter is a model of an acutely precipitated seizure leading to a chronic epileptic state. Some studies refer to models as "acute", e.g. maximal electroshock, pentylenetetrazol – models that produce immediate seizure activity – and "chronic" models that involve an insult such as SE, which is followed by spontaneous seizures following a latent period. Study of the latent period may reveal changes associated with epileptogenesis. There are also many species in which epilepsy occurs spontaneously, often due to a known or suspected genetic predisposition e.g. dogs, rats, mice, hamsters.

Transgenic mice with seizure disorders have also been developed (Upton and Stratton, 2003). These become important when considering that at least 40–50% of all forms of human epilepsy are idiopathic generalised epilepsies, which are characterised by the lack of antecedent disease and a presumed genetic origin. Transgenic models include those engineered with functionally identical mutations to those in human inherited epilepsy, models which have single altered genes to determine whether those genes are involved in epileptogenesis and lastly, spontaneous mutations in mice, which have provided a source of potential candidate genes.

Epileptiform activity can also be induced in a brain slice *in vitro*. This can be achieved by applying drugs, electrical stimuli, manipulating ion concentrations, or combinations of these. Such drugs include 4-aminopyridine, pilocarpine and the $GABA_A$ receptor antagonist bicuculline. Raising the concentration of potassium or lowering the concentration of magnesium can also induce epileptiform activity.

The activity in these *in vitro* models can then be measured by making extracellular field potential recordings in the desired region or using multiple electrode arrangements. Propagation of epileptiform activity through a slice can also be monitored by measuring intrinsic optical signal (IOS). Intrinsic optical signals are generated *in vitro* by changes in light scattering properties through refraction and reflection and/or by changes in light absorption. These phenomena are mainly secondary to cell swelling leading to alterations in the extracellular space volume (D'Arcangelo et al., 2001). The relevance of synchronised bursts in the limited network of the slice preparation to the *in vivo* situation is uncertain, especially since certain synchronised bursts *in vivo* have been proposed to be very different from seizures and even to have an antiepileptic effect (de Curtis and Avanzini, 2001).

Since the methods of modeling epilepsy are so diverse, it is imperative to select the most appropriate models for studying epileptogenesis. The most common methods of studying this process are in rodents (Table 6.1), in particular the kindling or post-SE models.

Post-status Epilepticus Models

The probability of SE patients subsequently developing epilepsy within 2 years is 41% compared with 13% of those with acute symptomatic seizures but no SE (Hesdorffer et al., 1998). This suggests a relationship between the prolonged seizures of SE and subsequent epileptogenesis, although a relationship between the length of seizure and the nature and severity of the precipitant cannot be discounted.

Table 6.1 Examples of Commonly used Rodent Models of Epilepsy, Seizures, and Epileptiform Activity. The Models Marked in Bold are used as Models of Epilepsy – Spontaneous Seizures can Occur Later. SE (Status Epilepticus)

A) In vivo	i) Chemical	**Pilocarpine (+/– lithium) SE**
		Kainic acid SE
		Tetanus toxin
		Pentylenetetrazol
		Fluorothyl
		Penicillin
	ii) Electrical	**Kindling**
		Perforant path stimulation SE
		Maximal Electroshock
	iii) Genetic	**Spontaneous**
		Transgenic
B) In vitro (slice preparations)		4-aminopyridine
		Low magnesium
		Pilocarpine

In humans, SE has been shown to result in hippocampal damage and subsequent HS. The hippocampus thus has a dichotomous role as the substrate for epilepsy, and as the structure susceptible to damage by prolonged seizures.

Animal models of generalised convulsive as well as limbic SE have supported these findings. Limbic SE has been induced by the systemic or local administration of kainic acid (Ben-Ari, 1985), systemic administration of pilocarpine (a muscarinic receptor agonist) (Turski et al., 1989) or protocols using electrical stimulation of limbic areas (Lothman et al., 1990). SE in these models results in hippocampal damage similar to that observed in humans. Following these acute episodes of limbic SE, many of the animals go on to develop spontaneous limbic seizures after the latent period of days to weeks (Ben-Ari, 1985; Turski et al., 1989; Lothman et al., 1990).

Pilocarpine Model

Although there are many convulsants that can be given to induce SE and later epilepsy, the two most studied are SE induced by kainite acid or pilocarpine. The latter is perhaps the most commonly used, and results in SE and then the later development of spontaneous seizures in the majority of animals. Turski et al. (1989) were the first to demonstrate that systemic administration of pilocarpine, an agonist of muscarinic acetylcholine receptors, induced limbic seizures in rats. The seizures consisted of staring spells, olfactory and gustatory automatisms and motor limbic seizures that developed over 1–2 h and built up progressively into limbic SE. The earliest electrographic alterations were in the hippocampus; these then propagated to the amygdala and cortex. A pattern of cell loss occurred in the hippocampus that is similar to human patients with TLE. The pilocarpine model can be used to ask different questions depending on the time period studied: a) the immediate consequences of SE, b) the changes that occur during epileptogenesis (i.e. during the latent period) and c) the period of spontaneous recurrent seizures.

Perforant Path Stimulation

Sloviter and Damiano (1981) developed a model of seizure-induced neuronal damage which involved continuous electrical stimulation of the perforant path of urethane anaesthetised rats for 24 h. The stimulation evoked granule cell population spikes (due to the summation of action potentials from many neurons), epileptiform discharges and reduced dentate inhibition, which could be quantified by measuring loss of paired pulse inhibition. Hilar interneurons and CA3 pyramidal neurons were damaged. McIntyre et al. (1982) then modified the model by stimulating the perforant path in unanaesthetised previously kindled rats for 60 min at a frequency of 60 Hz. Since the process of kindling itself induces epilepsy (see below), models were then developed in which the perforant path was stimulated in na rats and self-sustaining SE was eventually induced (Mazarati et al., 1998). Several different stimulating protocols have been used, including either intermittent or continuous stimulation. Walker et al. (1999) modified previous protocols described by Sloviter and Damiano (1981). This modification consisted of continuous stimulation of the perforant path at 20 Hz for 2 h via electrodes implanted into the angular bundle. Activity is recorded in the dentate gyrus via another electrode throughout the procedure. In this model, electrographic and behavioural seizure activity becomes self-sustaining and persists after the stimulation has stopped.

Tetanus Toxin Model

Intrahippocampal injection of tetanus toxin also results in spontaneous seizures, even after the clearance of the toxin, and this model has also contributed to our understanding of the pathophysiology of mesial TLE (Mellanby et al., 1977). However, this model does not result in HS (Jefferys et al., 1992), and the seizures usually abate in contrast to the human condition.

The Kindling Model

Kindling is the repetition of subthreshold stimuli that initially evoke afterdischarges but not seizures (Goddard, 1967). Repetition of these stimuli results in a gradual lengthening of the afterdischarges, eventually leading to progressively more severe seizures. Once an animal has been kindled, the heightened response to the stimulus seems to be permanent, and spontaneous seizures can occur (McNamara et al., 1993). The hippocampus and amygdala are easily kindled, resulting in a well-described progression of limbic seizures. Kindling shares several characteristics with NMDA-dependent long-term potentiation (LTP) of excitatory synaptic transmission. This has led to the suggestion that kindling and LTP have similar underlying mechanisms. In support of this, the rate at which kindling occurs is retarded in rodents treated with NMDA receptor antagonists. There are, however, several differences between kindling and LTP. Although NMDA receptor antagonists can completely block the induction of LTP, they are unable to block kindling completely (Cain et al., 1992). Perhaps a more fundamental difference is that the kindling process requires afterdischarges; the repeated induction of LTP without afterdischarges does not induce kindling. LTP of glutamatergic synaptic transmission may contribute to kindling by increasing the excitatory synaptic drive and the likelihood of evoking afterdischarges, but is alone insufficient to explain the cellular mechanisms of kindling (Cain et al., 1992).

The kindling model is particularly well-suited to the study of epileptogenesis. Brain regions can be precisely activated by focal stimulation, the interictal, ictal and post-ictal periods are easily manipulated (Sato et al., 1990) and the seizure types are similar to those seen in naturally occurring TLE (Loscher, 2002).

However, the kindling model rarely results in the development of spontaneous seizures unless a large number of stimulations are applied ("over-kindling") (Pinel and Rovner, 1978).

Kindling alone is unlikely to explain the occurrence of HS in association with other pathology, because kindling itself usually results in no or minimal hippocampal damage and sclerosis (Tuunanen and Pitkanen, 2000). Kindling could, however, explain the progression of mesial temporal epilepsy. Spontaneous seizures in the kindling model can result in progressive neuronal loss within the hippocampus (Cavazos et al., 1994). Indeed, even following single seizures, there is evidence of both apoptotic cell death and also neurogenesis in the dentate granule cell layer (Bengzon et al., 1997). This would suggest that recurrent seizures may cause further structural and functional changes in the hippocampus. Human evidence for this has mainly been indirect. Epilepsy duration correlates with hippocampal volume loss and progressive neuronal loss and dysfunction (Theodore and Gaillard, 1999). There has also been a case report demonstrating that hippocampal volumes decrease with time in HS (Van Paesschen et al., 1998) and the appearance of HS de novo has been observed following secondary generalised brief tonic-clonic seizures (Briellmann et al., 2001), although whether this is directly the cause of the seizures rather than resultant hypoxia during seizures or some other underlying pathological process remains uncertain (Sutula and Pitkanen, 2001).

Cellular Alterations During Epileptogenesis

Epileptogenic insults often result in an immediate selective neuronal loss, which is most prominent in areas CA1 and CA3 of the hippocampus. The latent period of variable length then follows and it is presumed that structural and functional changes occur during this time, which eventually lead to epilepsy. These changes involve not only neuronal loss, but also rearrangement of neuronal circuits, release of growth factors, failure to recruit inhibitory interneurons, gliogenesis, neurogenesis, altered inhibition, altered non-synaptic transmission, and changes in receptors and ion channels. All these processes may play a part in epileptogenesis.

Thus multifarious processes are involved in epileptogenesis, and this has complicated the identification of drugs that can be given in the latent period, which will prevent the later development of epilepsy.

Neuronal Loss

Both humans with TLE and rodents following SE demonstrate substantial hippocampal neuronal loss, which is specific to particular subfields including CA1, CA3 and the hilus (Norman, 1964; DeGiorgio et al., 1992; Turski et al., 1983, 1989; Clifford et al., 1987). There are a number of factors during prolonged seizures that may lead to neuronal death, including disruption of ion homeostatis resulting in cell swelling (necrosis) and a large influx of calcium into neurons via predominantly NMDA receptors, which results in a cascade of biochemical events, eventually leading to cell death (apoptosis).

After brain injury, such as ischemic injury, stroke, or traumatic brain injury, apoptosis plays a critical role in neuronal death (Liou et al., 2003). Apoptosis is a physiological process for killing cells and is important for the normal development and function of multicellular organisms (Strasser et al., 2000). Apoptosis results from a highly ordered molecular cascade, including inflammatory/cytokine-processing Caspases - 1, 5, and 11 and apoptosis-regulatory caspases - 2, 3, 6, 7, 8, 9, and 10. It may also involve changes in gene transcription (Henshall and Simon, 2005). Apoptosis after seizures may involve extrinsic pathway activation which involves caspases 2 and 8, intrinsic pathway activation which relates to mitochondrial calcium loading, cytochrome c release and calpain-mediated release of apoptosis-inducing factor (AIF), and executioner caspase activation which includes caspase 3, 6 and 7 (Henshall and Simon, 2005).

Other mechanisms that may play a crucial role are mitochondrial dysfunction and the production of free radicals. The brain is sensitive to oxidative damage because of its high aerobic metabolic demand, high polyunsaturated fatty acid content, poor repair capacity and high iron load. Free radicals disrupt membrane lipids and cause membrane failure. They may also interact with, damage and fragment DNA and inhibit mitochondrial respiration. Free radicals may induce modification of nucleotide bases by glycation or oxidation. Free radicals may also be involved in the development of seizures. During epileptogenesis, free radicals rise markedly (Gupta et al., 2003). The mechanisms underlying free radical formation during seizures are unclear, but mitochondrial dysfunction may play a critical part in the production of the free radical superoxide (O_2^-). The autooxidation of catecholamines, xanthine oxidase, phospholipase A2 and NAD(P)H oxidases, may also contribute to O_2^- production. Superoxide can combine with nitric oxide to produce peroxynitrite, which is neurotoxic.

Mitochondrial dysfunction during seizures can also alter neuronal excitability. The inhibition of the mitochondrial respiratory chain enzymes, such as cytochrome c oxidase and succinate dehydrogenase, evokes seizures. This may be due to an intracellular decrease in ATP levels and alterations in neuronal calcium homeostasis (Kunz, 2002). Alternatively, free radicals may attack mitochondria, inhibit the activity of the respiratory chain and induce a transient permeability. This results in a decline of ATP production and excessive release of free radicals, consequently causing cell death (Arzimanoglou et al., 2002).

The Creation of New Circuits: Mossy Fibre Sprouting

Hippocampal mossy fibres, which are the axons of dentate granule cells, converge in the dentate hilus and innervate hilar and CA3 neurons (Frotscher et al., 1994). In the rodent, the main mossy fibre axons leave the hilus and travel through CA3 in stratum lucidum, to the level of the apical dendrites of CA3 pyramidal cells (Henze et al., 2000).

Mossy fibre axons form synapses with excitatory and inhibitory cells of the hilus and area CA3.

There are inhibitory presynaptic G-protein coupled receptors in mossy fibre synapses for a number of transmitters, such as glutamate, γ-aminobutyric acid (GABA), adenosine and dynorphin. The release of glutamate and GABA can therefore have a depressant effect on mossy fibre synaptic transmission. The mossy fibre pathway predominantly synapses onto inhibitory interneurons, rather than excitatory hilar mossy cells and CA3 pyramidal cells.

The presynaptic terminals of mossy fibres are large and complex and completely surround the branched dendritic spines with up to 20 independent release sites. The synapses of mossy fibre demonstrate strong activity-dependent plasticity, and a presynaptic form of LTP, which does not require NMDA receptor activation (Nicoll and Malenka, 1995). A CA3 pyramidal neuron receives only approximately fifty synapses from mossy fibres, whereas it receives about 12,000 synapses from other CA3 neurons located in the ipsilateral and contralateral hippocampus. Mossy fibre synapses are however very efficient and are able to bring CA3 pyramidal cells to the firing threshold. Therefore, stimulating mossy fibres can lead to activation of recurrent excitatory synapses by firing of CA3 pyramidal neurons, resulting in epileptiform activity (Miles and Wong, 1983).

Re-organisation of the mossy fibre axons ("sprouting") occurs in both animal models and human patients with TLE into the inner molecular layer of the dentate gyrus. The role of these aberrant fibres has been a subject of controversy; sprouting is considered to be a response to the loss of neuronal targets. The two lines of thought are that, either sprouted fibres synapse onto dentate granule cells and form recurrent excitatory synapses, or they synapse mainly onto interneurons, thus increasing inhibition in the epileptic hippocampus. There are several lines of evidence showing that sprouted fibres make aberrant connections with granule cells both in animal models of epilepsy (Okazaki et al., 1995; Buckmaster et al., 2002) and human patients (Zhang and Houser, 1999), resulting in functional recurrent excitatory synapses (Tauck and Nadler, 1985). More recently, Buckmaster et al. (2002) provided evidence that most new synapses made by sprouted mossy fibres were with GABA-negative dendritic spines. In contrast to these findings Sloviter (1992) suggested that mossy fibre sprouting restores inhibition by innervation of interneurons by the sprouts and electron microscopic studies have revealed that sprouted mossy fibres probably synapse onto interneurons as well (Kotti et al., 1997).

What are the functional consequences of sprouting and is it necessary for epileptogenesis to occur? Longo and Mello (1997) showed that if sprouting is inhibited by cycloheximide (a protein synthesis inhibitor) in laboratory models of epilepsy, epileptogenesis still occurs. In contrast with this work however, Williams et al. (2002) failed to prevent sprouting with cycloheximide although they did show that spontaneous seizures could develop in the absence of Timm-stain positive mossy fibre sprouting. Similarly, spontaneous recurrent seizures develop even in the absence of sprouting in a modified kainic acid model of epilepsy (Zhang and Houser, 1999). But, in this last study, the presence of sprouting and neuronal cell loss was associated with more severe seizures. These results do not necessarily mean that mossy fibre sprouting is not important in increasing hippocampal excitability; they demonstrate that sprouting is not required for recurrent seizures to develop.

Growth Factors

Growth factors regulate a variety of cellular processes. They bind to receptors on the cell surface, activating cellular proliferation and differentiation. Growth factors have a role in the development of epilepsy. There is evidence that nerve growth factor (NGF) accelerates epileptogenesis in the kindling model and increases mossy fibre sprouting in the CA3 region and inner molecular layer (Adams et al., 1997). Brain derived neurotrophic factor (BDNF) mRNA was markedly increased in the hippocampus and the neocortex in fully kindled rats (Simonato et al., 1998). In the kindling model, the expression of BDNF mRNA was increased in the dentate granule cell layer, and was influenced by neurotrophin-3 (Elmer et al., 1996). Further evidence demonstrates that synthetic peptides designed to prevent neuro-trophin binding to their receptors, significantly retard kindling-induced epilepsy and inhibit mossy fibre sprouting (Rashid et al., 1995).

Tyrosine kinase B (TrkB), the primary target of BDNF, plays a vital role in epi-leptogeneis. Characteristic neuropathologic changes result from the interaction of BDNF with TrkB, such as neuronal loss and axonal sprouting, plastic changes in neuronal networks and synapses, neurogenesis, and dendritic outgrowth. Increased BDNF causes neuronal hyperexcitability and promotes LTP of excitatory synaptic transmission. Further studies suggest that BDNF redistributes (e.g. to distal dendrites) during the development of epilepsy.

The Dormant Basket Cell Hypothesis

Sloviter proposed the "dormant basket cell hypothesis" to account for data obtained in rats that had been exposed to prolonged perforant path stimulation under urethane anesthesia (Sloviter, 1983, 1987, 1991b). He observed that the principal cell disinhibition and hyperexcitability that immediately followed prolonged seizure discharges, correlated closely with selective neuronal injury to hilar neurons and CA3 pyramidal cells, and were similar to those changes produced acutely by bicuculline (Sloviter, 1991b). He suggested that principal cell disinhibition might be caused by the loss of mossy cell activation of dentate basket cells and the loss of CA3 pyramidal cell excitation of the area CA1 basket cells. Findings by Bekenstein and Lothman (1993) supported the hypothesis, as have several subsequent studies (Mangan et al., 1995; Jefferys and Traub, 1998; Doherty and Dingledine, 2001;

Denslow et al., 2001). In contrast, Esclapez et al. (1997) argued that in CA1, inhibition remained intact. Sloviter suggested that if inhibitory interneurons do indeed survive, perhaps a weak glutamate receptor agonist might restore inhibition by selectively activating surviving inhibitory interneurons (Sloviter, 1991a). This has subsequently been supported by Khalilov et al. (2002) who describe that a GluR5 subunit-containing glutamate receptor agonist selectively excites hippocampal interneurons and evokes hippocampal principal cell inhibition. The dormant basket cell hypothesis was tested experimentally by Ratzliff et al. (2004) who demonstrated that specific ablation of mossy cells decreased the excitability of granule cells in response to perforant path stimulation. In contrast, when the interneurons themselves were ablated, it led to an increase in excitability as would be expected. The "irritable mossy cell hypothesis" was then proposed, stating that it was the survival of the mossy cells that caused an increase in dentate gyrus hyperexcitability (Ratzliff et al., 2004; Santhakumar et al., 2000).

Gliogenesis

The supportive network of glia, particularly astrocytes, has been suspected for many years to be involved in epileptogenesis (Scharfman and Gray, 2007). It has been implicated for several reasons: (1) gliosis is required to induce post traumatic epilepsy (Penfield, 1927); (2) astrocytes assist ion homeostasis and hence neuronal excitability (Orkand et al., 1966); (3) Interplay between glial cell volume and extracellular volume affect neuronal hypersynchrony (Lux et al., 1986; Hochman et al., 1995). Historically, astrocytes have been referred to as passive bystanders that merely provide support to neuronal networks (Halassa et al., 2007). This view is changing, because astrocytes can generate oscillatory intracellular calcium waves, which can propagate through the astrocytic network and release neurotransmitters such as glutamate (Bezzi et al., 2004). Further, astrocytic calcium waves have properties that could facilitate seizures.

Astroglial gliosis is frequently observed in the epileptic foci of patients and animal models of epilepsy. In particular, kindling results in structural changes in astrocytes in the brain regions known to propagate kindled seizures. However, this has been suspected to be related to neuronal loss. Khurgel et al. (1995) demonstrated astrocyte activation following kindling, in the absence of significant neuronal degeneration. It was therefore hypothesized that neuronal loss is not a prerequisite for epileptogenesis.

Recent work by Tian et al. (2005) demonstrated that activation of single astrocytes by photolytic uncaging of intracellular calcium, could induce paroxysmal depolarizing shifts (PDSs) in rat hippocampal slices, while neuronal action potentials and synaptic transmission were blocked. They also observed an increase in glutamate release in the rat hippocampus during seizures, which was thought to be released from astrocytes (Tian et al., 2005). Finally they also provided evidence that some antiepileptic drugs appeared to act directly on astrocytes.

Neurogenesis

The view that new neurons are not produced in the adult brain has been slowly changing over the past century. There has been experimental evidence, since the 1960s, that neurogenesis occurs in the adult brain (Altman, 1962; Altman and Das, 1965) and this evidence has been broadened with the use of the thymidine analogue BrDU to identify newly born cells in the brain (Gage, 2002). Neurogenesis in the adult brain mainly occurs in three areas: the subventricular zone, olfactory bulb and dentate gyrus, although other sites have been reported. Clearly new neurons in the dentate gyrus could have a key role in epileptogenesis, depending on their effects on the hippocampal circuitry. Newly born granule cells send axonal projections to their usual anatomical pathway, the mossy fibres. They appear to develop electrophysiological properties similar to other granule cells (van Praag et al., 2002).

Adult neurogenesis is profoundly modified by environment and disease. Seizure activity increases neurogenesis in the dentate gyrus (Gray and Sundstrom, 1998). Although this activity-induced neurogenesis is interesting, what effect does it have on epileptogenesis? Here, the evidence is conflicting. It has been argued that newly born neurons do not survive for long (Bengzon et al., 1997) although this is not always the case (Scharfman et al., 2000). Other studies suggest that new neurons have a marked effect on epileptogenesis. Jung et al., (2004) showed that reduction of new neurons after pilocarpine induced SE was associated with reduced seizure frequency. Further, ectopic granule cells and hippocampal slice recordings discharge spontaneous action with potential bursts which were synchronised with area CA3 activity, suggesting abnormal activity (Scharfman et al., 2000). Although these lines of evidence suggest that new neurons increase excitability, it is unclear whether new neurons could also either innervate GABAergic cells or develop into granule cells with a GABAergic phenotype (Gutierrez, 2005). In summary, current evidence suggests that new dentate granule cells increase excitability and one could speculate that reduction of new granule cell birth could reduce the likelihood of the development of epilepsy. However, this hypothesis should be viewed with caution, as Hattiangady et al., (2004) have shown a decline in neurogenesis in chronic epilepsy. This could theoretically lead to cognitive dysfunction and depression, which might be worsened by drugs that inhibit neurogenesis.

Altered Inhibition

Blocking GABAergic inhibition produces seizures. Loss of inhibition is frequently observed in epilepsy models and tissue from patients. Paradoxically, other studies have reported increased inhibition in epilepsy. Despite a lot of evidence that there is decreased inhibition in the epileptic hippocampus, there have been several conflicting findings. In the dentate gyrus, both decreased and increased inhibition have been observed. An increase in paired-pulse inhibition of granule cells has been observed in several epilepsy models. This was observed in the amygdala kindling model (Tuff et al., 1983) and in the kainic acid model (Haas et al., 1996; Buckmaster and Dudek, 1997). Granule cells from epileptic rats have increased $GABA_A$ receptor current density when compared with controls (Gibbs et al., 1997) and there are more $GABA_A$ receptors per synapse in kindled animals when compared with the controls (Otis et al., 1994; Nusser et al., 1998). There are a number of explanations for these contrasting findings. First, studies are often not directly comparable because they use different animal models. Second, inhibition is measured in different ways, including paired pulse inhibition, miniature inhibitory post-synaptic current (mIPSC), spontaneous inhibitory post-synaptic current (IPSC) frequency, and evoked inhibitory post-synaptic potential (IPSP). Third, it has been proposed that results may differ depending on the septotemporal level of the hippocampus examined (Bernard et al., 2000). In vivo studies in rats tend to concentrate on the more accessible septal level of the hippocampus; however, tissue resected from humans includes the anterior (the human equivalent to the rat temporal) region (NB in the human the hippocampal poles are termed anterior and posterior, in contrast to the rat that is described as having temporal and septal poles). It is here that the hilar neuron loss is most severe (Babb et al., 1984) and the number of GAD-positive (Buckmaster and Jongen-Relo, 1999) and the somatostatin- and parvalbumin-immunoreactive interneurons are most reduced (Buckmaster and Dudek, 1997). Therefore, reduced granule cell inhibition and interneuron loss in the temporal dentate gyrus may be missed in the studies of the septal dentate gyrus.

Non-synaptic Mechanisms in Epileptogenesis

Non-synaptic mechanisms refer to those mechanisms that are independent of active chemical synaptic transmission in the synchronization of neuronal activity during seizures, and may contribute to chronic epileptogenesis. These "non-synaptic" pileptogenesis. These "non-synaptic" mechanisms include electronic coupling via gap junctions, electrical field effects, and ionic interactions.

First, electronic coupling via gap junctions involves a specialized membrane structure. Gap junctions are intercellular channels composed of connexin proteins, which can be modulated by a number of intracellular and extracellular factors. The function of gap junctions is to allow certain molecules and ions to move freely between neurons. This permits direct electrical transmission between cells, chemical transmission between cells through small second messengers, and the passage of small molecules (1,000 D). Gap junction channels contribute to neuronal synchronization in the brain. In addition, they may also play a role in hypersynchrony in epilepsy models (Kohling et al., 2001; Perez et al., 2000). Recent studies indicate that gap junctions can significantly modify the expression, the duration, and the propagation of seizures in vitro (Gajda et al., 2003) and in vivo in epilepsy models.

Second, electrical-field effects depend on neuronal orientation, polarity and the size of the extracellular space (Ghai et al., 2000). Tightly packed neurons that are arranged in parallel are susceptible to the effect of activity-induced electrical fields.

Electric fields can modulate neuronal activity in the central nervous system. They influence a variety of cellular events, such as membrane differentiation, neurite growth during both development and structural regeneration, organization of local neuronal circuits, and even receptor localization (Faber and Korn, 1989). When synaptic activity is blocked in hippocampal slices by removing extracellular Ca^{2+}, recurrent spontaneous paroxysms, termed seizure-like events, occur in the CA1 region. CA1 neurons discharge in synchrony and result in massive neuronal excitation (Konnerth et al., 1986).

The third mechanism is ionic interaction, which involves activity-dependent shifts in the intracellular and extracellular concentration of ions. Intense electrical activity during seizures is associated with transmembrane ionic currents, causing K^+ and Cl^- redistribution. Increasing K^+ concentration in the extracellular milieu increases membrane excitability and has a slow synchronizing effect on neuronal networks (Jensen and Yaari, 1997). Extracellular chloride may also play a critical role in neuronal synchronization (Hochman et al., 1999).

Receptor/Acquired Channel Changes

Several studies have observed changes in the expression of genes encoding various ion channels following SE. These include $GABA_A$ receptors, sodium, potassium and calcium channels and hyperpolarisation-activated cyclic nucleotide-gated (HCN) channels (Brooks-Kayal et al., 1998; Chen et al., 2001; Ellerkmann et al., 2003; Bernard et al., 2004).

$GABA_A$ Receptors

Altered expression and function of $GABA_A$ receptors in epilepsy play a key role in epileptogenesis. There is an increase in membrane $GABA_A$ receptors in the dentate gyrus which is reflected in an increase in the efficacy of GABA in activating whole cell currents and increase in the mIPSC amplitude (Otis et al., 1994; Gibbs et al., 1997; Nusser et al., 1998). As well as an increase in receptor density, $GABA_A$ receptor function is also altered in TLE. $GABA_A$ receptors on dentate granule cells become sensitive to blockade by zinc in contrast to $GABA_A$ receptors in controls which are zinc insensitive (Buhl et al., 1996; Gibbs et al., 1997). This is due to altered subunit composition from receptors containing predominantly alpha1 and alpha2 subunits to receptors containing much higher levels of the alpha4 subunit (Brooks-Kayal et al., 1998). This potentially leads to a zinc-induced collapse of inhibition in the dentate gyrus and a breakdown of the barrier function of the dentate gyrus (Heinemann et al., 1992; Lothman et al., 1992; Buhl et al., 1996; Gibbs et al., 1997).

Historically the activity of GABA at $GABA_A$ receptors has been considered to be inhibitory due to the hyperpolarization of the post-synaptic neuron. However, GABA acts as an excitatory neurotransmitter under certain circumstances and depolarizing $GABA_A$ receptor responses are implicated in spontaneous interictal discharges in human TLE (Cohen et al., 2002). There is a negative shift in the equilibrium potential of chloride during neuronal development. Internal chloride is regulated by NKCC (Na^+, K^+, $2Cl^-$ co- transporters), which tend to pump chloride into neurons and the KCl transporter KCC2, which tends to extrude chloride; the expression of these two transporters is under developmental regulation with NKCC being expressed early and later in the expression of KCC2. During epileptogenesis, there may be the recapitulation of early chloride transporter expression leading to depolarizing GABA responses (Munoz et al., 2007). In addition, the neuronal expression of the carbonic anhydrase isoform VII leads to transient excitatory GABAergic transmission in adult neurons.

$GABA_B$ Receptors

$GABA_B$ receptor changes are also implicated in epileptogenesis. $GABA_B$ receptors undergo changes in TLE in both human patients and experimental models. Haas et al. (1996) measured paired pulse suppression of recurrent IPSPs (disinhibition) in rats two weeks after kainic acid-induced SE and reported a downregulation of $GABA_B$ receptors in the polysynaptic recurrent inhibitory circuit in the dentate gyrus. They proposed that this is one mechanism that could induce an enhancement of dentate inhibition after seizures. Wasterlain et al. (1996) showed loss of $GABA_B$ mediated slow inhibitory postsynaptic potentials (IPSPs) recorded from dentate granule cells following unilateral stimulation of the perforant path. Wu and Leung (1997) reported reduced paired pulse depression of IPSCs recorded in CA1 after kindling that persisted for at least 21 days due to downregulation of presynaptic $GABA_B$ receptors. A decrease in the efficacy of presynaptic $GABA_B$ receptors in glutamatergic terminals was shown in the basolateral amygdala after amygdala kindling (Asprodini et al., 1992). Kokaia and Kokaia (2001) reported changes in $GABA_B$ immunoreactivity in the hippocampus after kindling. A loss of $GABA_B$ receptor mediated heterosynaptic depression was observed following SE in two different rodent models that were accompanied by reduced $GABA_B$ receptor binding in stratum lucidum (Chandler et al., 2003). Alterations in $GABA_B$ receptor expression have also been reported in hippocampal tissue from human patients with TLE (Munoz et al., 2002; Billinton et al., 2001; Princivalle et al., 2002), with both increased and decreased expression depending upon the hippocampal subfield.

Metabotropic Glutamate Receptors (mGluRs)

Metabotropic glutamate receptors play an important role in the central nervous system, regulating both neuronal excitability (Colwell and Levine, 1999) and the release of neurotransmitters (Cartmell and Schoepp, 2000). mGluRs may also play an important role in synaptic plasticity (e.g. long term potentiation and long-term depression; (Riedel and Reymann, 1996)). Recent studies show that group one metabotropic glutamate receptors may be involved in epileptogeneis. Expression of mGluR5 increases in hippocampus in TLE patients (Notenboom et al., 2006).

Activation of group one mGluRs is proconvulsant, and activation of group two and three mGluRs is anticonvulsant (Alexander and Godwin, 2006). During epileptogenesis, mGluR expression and function are rapidly altered in several animal models. Epileptogenesis can be induced by the direct activation of mGluR5, and the maintenance of seizures involves mGluR1 and mGluR5 (mGluR1 appears to play a more dominant role than mGluR5 (Wong et al., 2005)).

NMDA Receptors

The N-methyl D-aspartate (NMDA) receptor is ionotropic and allows flow of Na^+, K^+ and $Ca2^+$ ions. Calcium entering via NMDA receptors during SE may initiate a cascade of cellular events, including activation of catabolic enzymes, impairment of energy metabolism, generation of reactive oxygen species and ultimately cell death. NMDA receptor activation also leads to permanent alterations in neuronal circuits and excitability. There is some evidence that NMDA receptor expression and function is altered in epilepsy. Mathern et al. (1998) found that NMDAR2b mRNA was upregulated in the dentate gyrus of rats after self-sustaining limbic SE and in human TLE patients. Behr et al. (2001) showed a transient NMDA receptor mediated facilitation of high frequency input in the rat dentate gyrus after kindling. Altered expression of NMDA subunits was also recently reported in pentylenetetrazol-kindled animals (Zhu et al., 2004). Recent evidence suggests that NMDA receptor currents are increased following SE (Scimemi et al., 2006).

Kainate Receptors

The kainate receptor consists of four subunits composed from the subunits GluR5, GluR6, GluR7, KA1 and KA2 (Dingledine et al., 1999). Kainate receptors are permeable to sodium and potassium ions. The kainate receptor plays a role in the regulation of excitatory synaptic transmission (Lauri et al., 2001), regulation of inhibitory synaptic transmission (Clarke et al., 1997), excitatory synaptic transmission (Vignes and Collingridge, 1997) and LTP (Lauri et al., 2001; Vissel et al., 2001).

Changes in editing of mRNA encoding for AMPA receptor subunit GluR2 has been observed in hippocampi from epileptic humans (Vollmar et al., 2004). GluR2 is increased and GluR1 is decreased following Fe^{3+} induced epileptogenesis (Doi et al., 2001). However, GluR2 was reduced in limbic forebrain and amygdala 24 h after amygdala kindling (Prince et al., 1995). Enhanced expression of GluR1 flip AMPA receptor subunits has been observed in hippocampal astrocytes from epilepsy patients (Seifert et al., 2004). After lithium pilocarpine induced SE, the expression of different AMPA receptor subunits changes (e.g., GluR2 increases and GluR3 decreases) in dentate granule neurons (Porter et al., 2006). Interestingly, the AMPA receptor antagonist YM90K markedly retarded the evolution of kindling (Kodama et al., 1999).

Channelopathies

Recently, mutations in genes encoding for receptors and ion channels have been associated with naturally occurring human epilepsies. These include subunits of the $GABA_A$ receptor (Baulac et al., 2001) sodium channel (Wallace et al., 2001), chloride channel (Haug et al., 2003), potassium channels (Eunson et al., 2000), calcium channels (Jouvenceau et al., 2001), and nicotinic acetylcholine receptors (De Fusco et al., 2000).

There is also burgeoning evidence of acquired channelopathies occurring in experimental models. There is increased excitability of CA1 pyramidal neuron dendrites in pilocarpine-treated rats due to decreased availability of A-type potassium ion channels (Bernard et al., 2004). A reduction in the cationic current, I_h, has been described in the entorhinal cortex neurons leading to increased dendritic excitability 24 h following a single seizure episode in rats (Shah et al., 2004). I_h is also altered following experimental complex febrile seizures which lead to hyperexcitability (Chen et al., 2001). An increase in the density of a 'T-type' calcium current has been observed in CA1 neurons, which leads to intrinsic bursting behaviour (Su et al., 2002).

Interventions to Prevent Epileptogenesis

The multiple processes of epileptogenesis provide a number of sites for potential interventions to prevent epilepsy. Elucidation of the epileptogenic process should provide significantly more effective methods of treatment. The opportunities for intervention can be divided into four strategic approaches: (1) initial insult modification (which can be achieved by early pharmacological intervention and neuro-surgery), (2) neuroprotection (3) antagonism of epileptogenesis and (4) disease modification . Interventions of the epileptogenic cascade may be able to prevent neuronal injury or death, preserving or restoring neuronal function, and neuronal recovery or regeneration (Walker et al., 2002). Epileptogenesis depends on processes different from those involved in seizure generation. Drugs can be designed to inhibit the initial damage which is produced by brain insults, such as excitotoxic cell death, and to prevent or reverse alterations in neuronal circuits that contribute to lowered seizure thresholds.

Initial Insult Modification

Initial insult modification may have an antiepileptogenic and disease-modifying effect. The effect of phenobarbital (80 mg kg^{-1}), MK-801 (4 mg kg^{-1}), or phenytoin (100 mg kg^{-1}) at 1, 2, and 4 h after initiation of "continuous" hippocampal stimulation on the development of epilepsy was investigated. Both phenobarbital and MK-801 reduced the percentage of animals developing epilepsy. The mechanism may have involved inhibition of the cascade of events that leads to the death of critical neuronal populations and cell death-induced reactive gliosis, sprouting, and other types of damage-induced network reorganization underlying the development of spontaneous seizures (Pitkanen, 2002).

Neuroprotection to Prevent Epileptogenesis

Neuronal injury and death may play an important role in epileptogenesis. Pharmacological neuroprotection in epileptogenesis can be considered as primary and secondary. Antiepileptic drugs and compounds designed to act on voltage-sensitive Na^+ and Ca^{2+} channels or on glutamate receptors are considered primary neuroprotection. Secondary neuroprotection refers to interventions that act on the cascade leading to necrosis or apoptosis (Artemowicz and Sobaniec, 2005).

Treatment during or after brain injury in chronic epilepsy models, such as SE and kindling, reduce cell loss and neuronal death. Such neuroprotection could contribute to modification of or delay in the development of epilepsy.

Antiepileptic drugs (AEDs) are aimed at preventing and suppressing seizure activity. Some may ameliorate necrotic and apoptotic neuronal death. Injection of dizocilpine (MK-801) and retigabine after kainite-induced SE prevented neurodegeneration and expression of markers of apoptosis in limbic brain regions and significantly reduced the damage in the limbic regions (Ebert et al., 2002) (Brandt et al., 2003). Topiramate also can act as a potent neuroprotectant following SE (Fisher et al., 2004). Valproate has a powerful neuroprotective effect in the hippocampal formation and the dentate hilus in SE models (Brandt et al., 2006). Vigabatrin protected the hippocampus from brain damage efficiently in Ammon's horn, and to a lesser extent in the hilus following pilocarpine-induced SE (Andre et al., 2001). Carbamazepine reduced damage to the hippocampal formation in a pilocarpine-induced SE model (Capella and Lemos, 2002). The group treated with atipamezole had milder hilar cell damage in amygdala-kindled rats (Pitkanen et al., 2004). The severity of hippocampal cell loss was milder after diazepam treatment in amygdala-kindled rats (Pitkanen et al., 2005).

However, it is still questionable whether neuroprotection during or after brain injury will alter or prevent the process of epileptogenesis. Indeed, preventing neurodegeneration or neuronal cell loss often does not prevent epileptogenesis (Ebert et al., 2002; Rigoulot et al., 2004).

Antiepileptogenesis

The definition of an antiepileptogenic compound is one that prevents or slows the process of developing epilepsy (Cole and Dichter, 2002). Several experimental studies have aimed at preventing epileptogenesis in animal models, including SE and kindling models, by administering AEDs. Some have mild or questionable effects, and others are without effect even if they neuroprotect (Walker et al., 2002).

This is probably because AEDs were designed and tested to prevent seizures rather than for modifying the changes described above.

An alternative approach has been to target NMDA receptors. NMDA receptors play a critical role in neuronal death and may also be important for the initiation of many of the changes described above. MK-801, a potent NMDA receptor antagonist, prevented the development of seizures following pilocarpine-induced SE (Raza et al., 2004), and has been shown to have a neuroprotective and antiepileptogenic effect in other epilepsy models (Prasad et al., 2002), but not in the kainite SE model (Brandt et al., 2003). This may be because by the time the drug is administered many of the pathways determining cellular and network alterations are already active. Undoubtedly the future lies in identifying downstream targets that may prevent the epileptogenic process.

Conclusion

There is a vast array of changes in organisation, connectivity, receptors, neuronal properties and astrocyte function that can also contribute to epileptogenicity, and much of this information has been acquired by the study of animal models. The great challenges are to differentiate pro-epileptogenic changes from anti-epileptogenic, compensatory changes, and to determine which the critical processes are. It is likely that epileptogenicity is not one single process, but that many diverse processes can result in the expression of epilepsy. Further understanding of epileptogenesis will enable us to target patients with the appropriate antiepileptogenic drugs and to reduce the likelihood of developing epilepsy.

References

Adams, B, Sazgar, M, Osehobo, P, Van der Zee, CE, Diamond, J, Fahnestock, M, Racine, RJ (1997) Nerve growth factor accelerates seizure development, enhances mossy fiber sprouting, and attenuates seizure-induced decreases in neuronal density in the kindling model of epilepsy. J Neurosci, 17:5288–5296.

Alexander, GM, Godwin, DW (2006) Metabotropic glutamate receptors as a strategic target for the treatment of epilepsy. Epilepsy Res, 71:1–22.

Altman, J (1962) Are new neurons formed in the brains of adult mammals? Science, 135:1127–1128.

Altman, J, Das, GD (1965) Post-natal origin of microneurones in the rat brain. Nature, 207:953–956.

Andre, V, Ferrandon, A, Marescaux, C, Nehlig, A (2001) Vigabatrin protects against hippocampal damage but is not antiepileptogenic in the lithium-pilocarpine model of temporal lobe epilepsy. Epilepsy Res, 47:99–117.

Arruda, F, Cendes, F, Andermann, F, Dubeau, F, Villemure, JG, Jones-Gotman, M, Poulin, N, Arnold, DL, Olivier, A (1996) Mesial atrophy and outcome after amygdalohippocampectomy or temporal lobe removal. Ann Neurol, 40:446–450.

Artemowicz, B, Sobaniec, W (2005) Neuroprotection possibilities in epileptic children. Rocz Akad Med Bialymst, 50 (Suppl 1):91–95.

Arzimanoglou, A, Hirsch, E, Nehlig, A, Castelnau, P, Gressens, P, Pereira de Vasconcelos, A (2002) Epilepsy and neuroprotection: an illustrated review. Epileptic Disord, 4:173–182.

Asprodini, EK, Rainnie, DG, Shinnick-Gallagher, P (1992) Epileptogenesis reduces the sensitivity of presynaptic γ-aminobutyric acidB receptors on glutamatergic afferents in the amygdala. J Pharmacol Exp Ther, 262:1011–1021.

Babb, TL, Brown, WJ (1986) Neuronal, dendritic, and vascular profiles of human temporal lobe epilepsy correlated with cellular physiology in vivo. Adv Neurol, 44:949–966.

Babb, TL, Brown, WJ, Pretorius, J, Davenport, C, Lieb, JP, Crandall, PH (1984) Temporal lobe volumetric cell densities in temporal lobe epilepsy. Epilepsia, 25:729–740.

Baulac, S, Huberfeld, G, Gourfinkel-An, I, Mitropoulou, G, Beranger, A, Prud'homme, JF, Baulac, M, Brice, A, Bruzzone, R, LeGuern, E (2001) First genetic evidence of GABA(A) receptor dysfunction in epilepsy: a mutation in the γ2-subunit gene. Nat Genet, 28:46–48.

Behr, J, Heinemann, U, Mody, I (2001) Kindling induces transient NMDA receptor-mediated facilitation of high-frequency input in the rat dentate gyrus. J Neurophysiol, 85:2195–2202.

Bekenstein, JW, Lothman, EW (1993) Dormancy of inhibitory interneurons in a model of temporal lobe epilepsy. Science, 259:97–100.

Ben-Ari, Y (1985) Limbic seizure and brain damage produced by kainic acid: mechanisms and relevance to human temporal lobe epilepsy. Neuroscience, 14:375–403.

Bengzon, J, Kokaia, Z, Elmer, E, Nanobashvili, A, Kokaia, M, Lindvall, O (1997) Apoptosis and proliferation of dentate gyrus neurons after single and intermittent limbic seizures. Proc Natl Acad Sci U S A, 94:10432–10437.

Bernard, C, Anderson, A, Becker, A, Poolos, NP, Beck, H, Johnston, D (2004) Acquired dendritic channelopathy in temporal lobe epilepsy. Science, 305:532–535.

Bernard, C, Cossart, R, Hirsch, JC, Esclapez, M, Ben-Ari, Y (2000) What is GABAergic inhibition? How is it modified in epilepsy? Epilepsia, 41 (Suppl 6):S90–S95.

Bezzi, P, Gundersen, V, Galbete, JL, Seifert, G, Steinhauser, C, Pilati, E, Volterra, A (2004) Astrocytes contain a vesicular compartment that is competent for regulated exocytosis of glutamate. Nat Neurosci, 7:613–620.

Billinton, A, Baird, VH, Thom, M, Duncan, JS, Upton, N, Bowery, NG (2001) GABA(B) receptor autoradiography in hippocampal sclerosis associated with human temporal lobe epilepsy. Br J Pharmacol, 132:475–480.

Brandt, C, Gastens, AM, Sun, M, Hausknecht, M, Loscher, W (2006) Treatment with valproate after status epilepticus: effect on neuronal damage, epileptogenesis, and behavioral alterations in rats. Neuropharmacology, 51:789–804.

Brandt, C, Glien, M, Potschka, H, Volk, H, Loscher, W (2003) Epileptogenesis and neuropathology after different types of status epilepticus induced by prolonged electrical stimulation of the basolateral amygdala in rats. Epilepsy Res, 55:83–103.

Brandt, C, Potschka, H, Loscher, W, Ebert, U (2003) N-methyl-D-aspartate receptor blockade after status epilepticus protects against limbic brain damage but not against epilepsy in the kainate model of temporal lobe epilepsy. Neuroscience, 118:727–740.

Briellmann, RS, Newton, MR, Wellard, RM, Jackson, GD (2001) Hippocampal sclerosis following brief generalized seizures in adulthood. Neurology, 57:315–317.

Brooks-Kayal, AR, Shumate, MD, Jin, H, Rikhter, TY, Coulter, DA (1998) Selective changes in single cell GABA(A) receptor subunit expression and function in temporal lobe epilepsy. Nat Med, 4:1166–1172.

Buckmaster, PS, Dudek, FE (1997a) Network properties of the dentate gyrus in epileptic rats with hilar neuron loss and granule cell axon reorganization. J Neurophysiol, 77:2685–2696.

Buckmaster, PS, Dudek, FE (1997b) Neuron loss, granule cell axon reorganization, and functional changes in the dentate gyrus of epileptic kainate-treated rats. J Comp Neurol, 385:385–404.

Buckmaster, PS, Jongen-Relo, AL (1999) Highly specific neuron loss preserves lateral inhibitory circuits in the dentate gyrus of kainate-induced epileptic rats. J Neurosci, 19:9519–9529.

Buckmaster, PS, Zhang, GF, Yamawaki, R (2002) Axon sprouting in a model of temporal lobe epilepsy creates a predominantly excitatory feedback circuit. J Neurosci, 22:6650–6658.

Buhl, EH, Otis, TS, Mody, I (1996) Zinc-induced collapse of augmented inhibition by GABA in a temporal lobe epilepsy model. Science, 271:369–373.

Cain, DP, Boon, F, Hargreaves, EL (1992) Evidence for different neurochemical contributions to long-term potentiation and to kindling and kindling-induced potentiation: role of NMDA and urethane-sensitive mechanisms. Exp Neurol, 116:330–338.

Capella, HM, Lemos, T (2002) Effect on epileptogenesis of carbamazepine treatment during the silent period of the pilocarpine model of epilepsy. Epilepsia, 43 (Suppl 5):110–111.

Cartmell, J, Schoepp, DD (2000) Regulation of neurotransmitter release by metabotropic glutamate receptors. J Neurochem, 75:889–907.

Cavazos, JE, Das, I, Sutula, TP (1994) Neuronal loss induced in limbic pathways by kindling: evidence for induction of hippocampal sclerosis by repeated brief seizures. J Neurosci, 14:3106–3121.

Chandler, KE, Princivalle, AP, Fabian-Fine, R, Bowery, NG, Kullmann, DM, Walker, MC (2003) Plasticity of GABA(B) receptor-mediated heterosynaptic interactions at mossy fibers after status epilepticus. J Neurosci, 23:11382–11391.

Chen, K, Aradi, I, Thon, N, Eghbal-Ahmadi, M, Baram, TZ, Soltesz, I (2001) Persistently modified h-channels after complex febrile seizures convert the seizure-induced enhancement of inhibition to hyperexcitability. Nat Med, 7:331–337.

Clarke, VR, Ballyk, BA, Hoo, KH, Mandelzys, A, Pellizzari, A, Bath, CP, Thomas, J, Sharpe, EF, Davies, CH, Ornstein, PL, Schoepp, DD, Kamboj, RK, Collingridge, GL, Lodge, D, Bleakman, D (1997) A hippocampal GluR5 kainate receptor regulating inhibitory synaptic transmission. Nature, 389:599–603.

Clifford, DB, Olney, JW, Maniotis, A, Collins, RC, Zorumski, CF (1987) The functional anatomy and pathology of lithium-pilocarpine and high-dose pilocarpine seizures. Neuroscience, 23:953–968.

Cohen, I, Navarro, V, Clemenceau, S, Baulac, M, Miles, R (2002) On the origin of interictal activity in human temporal lobe epilepsy in vitro. Science, 298:1418–1421.

Cole, AJ, Dichter, M (2002) Neuroprotection and antiepileptogenesis: overview, definitions, and context. Neurology, 59:S1–S2.

Colwell, CS, Levine, MS (1999) Metabotropic glutamate receptor modulation of excitotoxicity in the neostriatum: role of calcium channels. Brain Res, 833:234–241.

D'Arcangelo, G, Tancredi, V, Avoli, M (2001) Intrinsic optical signals and electrographic seizures in the rat limbic system. Neurobiol Dis, 8:993–1005.

Davies, KG, Hermann, BP, Dohan, FCJ, Foley, KT, Bush, AJ, Wyler, AR (1996) Relationship of hippocampal sclerosis to duration and age of onset of epilepsy, and childhood febrile seizures in temporal lobectomy patients. Epilepsy Res, 24:119–126.

de Curtis, M, Avanzini, G (2001) Interictal spikes in focal epileptogenesis. Prog Neurobiol, 63:541–567.

De Fusco, M, Becchetti, A, Patrignani, A, Annesi, G, Gambardella, A, Quattrone, A, Ballabio, A, Wanke, E, Casari, G (2000) The nicotinic receptor β 2 subunit is mutant in nocturnal frontal lobe epilepsy. Nat Genet, 26:275–276.

DeGiorgio, CM, Tomiyasu, U, Gott, PS, Treiman, DM (1992) Hippocampal pyramidal cell loss in human status epilepticus. Epilepsia, 33:23–27.

Delorenzo, RJ, Sun, DA, Deshpande, LS (2005) Cellular mechanisms underlying acquired epilepsy: the calcium hypothesis of the induction and maintainance of epilepsy. Pharmacol Ther, 105:229–266.

Denslow, MJ, Eid, T, Du, F, Schwarcz, R, Lothman, EW, Steward, O (2001) Disruption of inhibition in area CA1 of the hippocampus in a rat model of temporal lobe epilepsy. J Neurophysiol, 86:2231–2245.

Dingledine, R, Borges, K, Bowie, D, Traynelis, SF (1999) The glutamate receptor ion channels. Pharmacol Rev, 51:7–61.

Doherty, J, Dingledine, R (2001) Reduced excitatory drive onto interneurons in the dentate gyrus after status epilepticus. J Neurosci, 21:2048–2057.

Doi, T, Ueda, Y, Tokumaru, J, Mitsuyama, Y, Willmore, LJ (2001) Sequential changes in AMPA and NMDA protein levels during Fe(3 +)-induced epileptogenesis. Brain Res Mol Brain Res, 92:107–114.

Draguhn, A, Traub, RD, Schmitz, D, Jefferys, JG (1998) Electrical coupling underlies high-frequency oscillations in the hippocampus in vitro. Nature, 394:189–192.

Ebert, U, Brandt, C, Loscher, W (2002) Delayed sclerosis, neuroprotection, and limbic epileptogenesis after status epilepticus in the rat. Epilepsia, 43 (Suppl 5):86–95.

Ellerkmann, RK, Remy, S, Chen, J, Sochivko, D, Elger, CE, Urban, BW, Becker, A, Beck, H (2003) Molecular and functional changes in voltage-dependent Na(+) channels following pilocarpine-induced status epilepticus in rat dentate granule cells. Neuroscience, 119:323–333.

Elmer, E, Kokaia, M, Kokaia, Z, Ferencz, I, Lindvall, O (1996) Delayed kindling development after rapidly recurring seizures: relation to mossy fiber sprouting and neurotrophin, GAP-43 and dynorphin gene expression. Brain Res, 712:19–34.

Engel, JJ (2006) Report of the ILAE classification core group. Epilepsia, 47:1558–1568.

Esclapez, M, Hirsch, JC, Khazipov, R, Ben-Ari, Y, Bernard, C (1997) Operative GABAergic inhibition in hippocampal CA1 pyramidal neurons in experimental epilepsy. Proc Natl Acad Sci USA, 94:12151–12156.

Eunson, LH, Rea, R, Zuberi, SM, Youroukos, S, Panayiotopoulos, CP, Liguori, R, Avoni, P, McWilliam, RC, Stephenson, JB, Hanna, MG, Kullmann, DM, Spauschus, A (2000) Clinical, genetic, and expression studies of mutations in the potassium channel gene KCNA1 reveal new phenotypic variability. Ann Neurol, 48:647–656.

Faber, DS, Korn, H (1989) Electrical field effects: their relevance in central neural networks. Physiol Rev, 69:821–863.

Fish, DR, Spencer, SS (1995) Clinical correlations: MRI and EEG. Magn Reson Imaging, 13:1113–1117.

Fisher, A, Wang, X, Cock, HR, Thom, M, Patsalos, PN, Walker, MC (2004) Synergism between topiramate and budipine in refractory status epilepticus in the rat. Epilepsia, 45:1300–1307.

Fisher, PD, Sperber, EF, Moshe, SL (1998) Hippocampal sclerosis revisited. Brain Dev, 20:563–573.

Frotscher, M, Soriano, E, Misgeld, U (1994) Divergence of hippocampal mossy fibers. Synapse, 16:148–160.

Gage, FH (2002) Neurogenesis in the adult brain. J Neurosci, 22:612–613.

Gajda, Z, Gyengesi, E, Hermesz, E, Ali, KS, Szente, M (2003) Involvement of gap junctions in the manifestation and control of the duration of seizures in rats in vivo. Epilepsia, 44:1596–1600.

Gajda, Z, Szupera, Z, Blazso, G, Szente, M (2005) Quinine, a blocker of neuronal cx36 channels, suppresses seizure activity in rat neocortex in vivo. Epilepsia, 46:1581–1591.

Gardiner, M (2005) Genetics of idiopathic generalized epilepsies. Epilepsia, 46 (Suppl 9):15–20.

Ghai, RS, Bikson, M, Durand, DM (2000) Effects of applied electric fields on low-calcium epileptiform activity in the CA1 region of rat hippocampal slices. J Neurophysiol, 84:274–280.

Gibbs, JWr, Shumate, MD, Coulter, DA (1997) Differential epilepsy-associated alterations in postsynaptic GABA(A) receptor function in dentate granule and CA1 neurons. J Neurophysiol, 77:1924–1938.

Goddard, GV (1967) Development of epileptic seizures through brain stimulation at low intensity. Nature, 214:1020–1021.

Gray, WP, Sundstrom, LE (1998) Kainic acid increases the proliferation of granule cell progenitors in the dentate gyrus of the adult rat. Brain Res, 790:52–59.

Gupta, YK, Veerendra Kumar, MH, Srivastava, AK (2003) Effect of Centella asiatica on pentylenetetrazole-induced kindling, cognition and oxidative stress in rats. Pharmacol Biochem Behav, 74:579–585.

Gutierrez, R (2005) The dual glutamatergic-GABAergic phenotype of hippocampal granule cells. Trends Neurosci, 28:297–303.

Haas, KZ, Sperber, EF, Moshe, SL, Stanton, PK (1996) Kainic acid-induced seizures enhance dentate gyrus inhibition by downregulation of GABA(B) receptors. J Neurosci, 16:4250–4260.

Halassa, MM, Fellin, T, Haydon, PG (2007) The tripartite synapse: roles for gliotransmission in health and disease. Trends Mol Med, 13:54–63.

Hattiangady, B, Rao, MS, Shetty, AK (2004) Chronic temporal lobe epilepsy is associated with severely declined dentate neurogenesis in the adult hippocampus. Neurobiol Dis, 17:473–490.

Haug, K, Warnstedt, M, Alekov, AK, Sander, T, Ramirez, A, Poser, B, Maljevic, S, Hebeisen, S, Kubisch, C, Rebstock, J, Horvath, S, Hallmann, K, Dullinger, JS, Rau, B, Haverkamp, F, Beyenburg, S, Schulz, H, Janz, D, Giese, B, Muller-Newen, G, Propping, P, Elger, CE, Fahlke, C, Lerche, H, Heils, A (2003) Mutations in CLCN2 encoding a voltage-gated chloride channel are associated with idiopathic generalized epilepsies. Nat Genet, 33:527–532.

Heinemann, U, Beck, H, Dreier, JP, Ficker, E, Stabel, J, Zhang, CL (1992) The dentate gyrus as a regulated gate for the propagation of epileptiform activity. Epilepsy Res Suppl, 7:273–280.

Henshall, DC, Simon, RP (2005) Epilepsy and apoptosis pathways. J Cereb Blood Flow Metab, 25:1557–1572.

Henze, DA, Urban, NN, Barrionuevo, G (2000) The multifarious hippocampal mossy fiber pathway: a review. Neuroscience, 98:407–427.

Hesdorffer, DC, Logroscino, G, Cascino, G, Annegers, JF, Hauser, WA (1998) Risk of unprovoked seizure after acute symptomatic seizure: effect of status epilepticus. Ann Neurol, 44:908–912.

Hochman, DW, Baraban, SC, Owens, JW, Schwartzkroin, PA (1995) Dissociation of synchronization and excitability in furosemide blockade of epileptiform activity. Science, 270:99–102.

Hochman, DW, D'Ambrosio, R, Janigro, D, Schwartzkroin, PA (1999) Extracellular chloride and the maintenance of spontaneous epileptiform activity in rat hippocampal slices. J Neurophysiol, 81:49–59.

Houser, CR, Esclapez, M (1996) Vulnerability and plasticity of the GABA system in the pilocarpine model of spontaneous recurrent seizures. Epilepsy Res, 26:207–218.

Jefferys, JG, Evans, BJ, Hughes, SA, Williams, SF (1992) Neuropathology of the chronic epileptic syndrome induced by intrahippocampal tetanus toxin in rat: preservation of pyramidal cells and incidence of dark cells. Neuropathol Appl Neurobiol, 18:53–70.

Jefferys, JG, Traub, RD (1998) 'Dormant' inhibitory neurons: do they exist and what is their functional impact? Epilepsy Res, 32:104–113.

Jensen, MS, Yaari, Y (1997) Role of intrinsic burst firing, potassium accumulation, and electrical coupling in the elevated potassium model of hippocampal epilepsy. J Neurophysiol, 77:1224–1233.

Jouvenceau, A, Eunson, LH, Spauschus, A, Ramesh, V, Zuberi, SM, Kullmann, DM, Hanna, MG (2001) Human epilepsy associated with dysfunction of the brain P/Q-type calcium channel. Lancet, 358:801–807.

Jung, KH, Chu, K, Kim, M, Jeong, SW, Song, YM, Lee, ST, Kim, JY, Lee, SK, Roh, JK (2004) Continuous cytosine-b-D-arabinofuranoside infusion reduces ectopic granule cells in adult rat hippocampus with attenuation of spontaneous recurrent seizures following pilocarpine-induced status epilepticus. Eur J Neurosci, 19:3219–3226.

Khalilov, I, Hirsch, J, Cossart, R, Ben-Ari, Y (2002) Paradoxical anti-epileptic effects of a GluR5 agonist of kainate receptors. J Neurophysiol, 88:523–527.

Khurgel, M, Switzer, RCr, Teskey, GC, Spiller, AE, Racine, RJ, Ivy, GO (1995) Activation of astrocytes during epileptogenesis in the absence of neuronal degeneration. Neurobiol Dis, 2:23–35.

Kodama, M, Yamada, N, Sato, K, Kitamura, Y, Koyama, F, Sato, T, Morimoto, K, Kuroda, S (1999) Effects of YM90K, a selective AMPA receptor antagonist, on amygdala-kindling and long-term hippocampal potentiation in the rat. Eur J Pharmacol, 374:11–19.

Kohling, R, Gladwell, SJ, Bracci, E, Vreugdenhil, M, Jefferys, JG (2001) Prolonged epileptiform bursting induced by 0-Mg(2 +) in rat hippocampal slices depends on gap junctional coupling. Neuroscience, 105:579–587.

Kokaia, Z, Kokaia, M (2001) Changes in GABA(B) receptor immunoreactivity after recurrent seizures in rats. Neurosci Lett, 315:85–88.

Konnerth, A, Heinemann, U, Yaari, Y (1986) Nonsynaptic epileptogenesis in the mammalian hippocampus in vitro. I. Development of seizurelike activity in low extracellular calcium. J Neurophysiol, 56:409–423.

Kotti, T, Riekkinen, PJS, Miettinen, R (1997) Characterization of target cells for aberrant mossy fiber collaterals in the dentate gyrus of epileptic rat. Exp Neurol, 146:323–330.

Kunz, WS (2002) The role of mitochondria in epileptogenesis. Curr Opin Neurol, 15:179–184.

Lauri, SE, Bortolotto, ZA, Bleakman, D, Ornstein, PL, Lodge, D, Isaac, JT, Collingridge, GL (2001) A critical role of a facilitatory presynaptic kainate receptor in mossy fiber LTP. Neuron, 32:697–709.

Lewis, DV (1999) Febrile convulsions and mesial temporal sclerosis. Curr Opin Neurol, 12:197–201.

Lieberman, DN, Mody, I (1999) Properties of single NMDA receptor channels in human dentate gyrus granule cells. J Physiol, 518:55–70.

Liou, AK, Clark, RS, Henshall, DC, Yin, XM, Chen, J (2003) To die or not to die for neurons in ischemia, traumatic brain injury and epilepsy: a review on the stress-activated signaling pathways and apoptotic pathways. Prog Neurobiol, 69:103–142.

Longo, BM, Mello, LE (1997) Blockade of pilocarpine- or kainate-induced mossy fiber sprouting by cycloheximide does not prevent subsequent epileptogenesis in rats. Neurosci Lett, 226:163–166.

Loscher, W (2002) Animal models of epilepsy for the development of antiepileptogenic and disease-modifying drugs. A comparison of the pharmacology of kindling and post-status epilepticus models of temporal lobe epilepsy. Epilepsy Res, 50:105–123.

Lothman, EW, Bertram, EH, Kapur, J, Stringer, JL (1990) Recurrent spontaneous hippocampal seizures in the rat as a chronic sequela to limbic status epilepticus. Epilepsy Res, 6:110–118.

Lothman, EW, Stringer, JL, Bertram, EH (1992) The dentate gyrus as a control point for seizures in the hippocampus and beyond. Epilepsy Res Suppl, 7:301–313.

Lux, HD, Heinemann, U, Dietzel, I (1986) Ionic changes and alterations in the size of the extracellular space during epileptic activity. Adv Neurol, 44:619–639.

Mangan, PS, Rempe, DA, Lothman, EW (1995) Changes in inhibitory neurotransmission in the CA1 region and dentate gyrus in a chronic model of temporal lobe epilepsy. J Neurophysiol, 74:829–840.

Mathern, GW, Pretorius, JK, Mendoza, D, Lozada, A, Leite, JP, Chimelli, L, Fried, I, Sakamoto, AC, Assirati, JA, Adelson, PD (1998) Increased hippocampal AMPA and NMDA receptor subunit immunoreactivity in temporal lobe epilepsy patients. J Neuropathol Exp Neurol, 57:615–634.

Mazarati, AM, Wasterlain, CG, Sankar, R, Shin, D (1998) Self-sustaining status epilepticus after brief electrical stimulation of the perforant path. Brain Res, 801:251–253.

McIntyre, DC, Nathanson, D, Edson, N (1982) A new model of partial status epilepticus based on kindling. Brain Res, 250:53–63.

McNamara, JO, Bonhaus, W, Shin, C. The kindling model of epilepsy. In: Schwartzkroin PA, editor. Epilepsy: models, mechanisms, and concepts. Cambridge: Cambridge University Press, 1993:21–47

Mellanby, J, George, G, Robinson, A, Thompson, P (1977) Epileptiform syndrome in rats produced by injecting tetanus toxin into the hippocampus. J Neurol Neurosurg Psychiatry, 40:404–414.

Miles, R, Wong, RK (1983) Single neurones can initiate synchronized population discharge in the hippocampus. Nature, 306:371–373.

Munoz, A, Arellano, JI, DeFelipe, J (2002) GABABR1 receptor protein expression in human mesial temporal cortex: changes in temporal lobe epilepsy. J Comp Neurol, 449:166–179.

Munoz, A, Mendez, P, DeFelipe, J, Alvarez-Leefmans, FJ (2007) Cation-chloride cotransporters and GABA-ergic innervation in the human epileptic hippocampus. Epilepsia, 48:663–673.

Nicoll, RA, Malenka, RC (1995) Contrasting properties of two forms of long-term potentiation in the hippocampus. Nature, 377:115–118.

Norman, RM (1964) The neuropathology of status epilepticus. Med Sci Law, 14:46–51.

Notenboom, RG, Hampson, DR, Jansen, GH, van Rijen, PC, van Veelen, CW, van Nieuwenhuizen, O, de Graan, PN (2006) Up-regulation of hippocampal metabotropic glutamate receptor 5 in temporal lobe epilepsy patients. Brain, 129:96–107.

Nusser, Z, Hajos, N, Somogyi, P, Mody, I (1998) Increased number of synaptic GABA(A) receptors underlies potentiation at hippocampal inhibitory synapses. Nature, 395:172–177.

Okazaki, MM, Evenson, DA, Nadler, JV (1995) Hippocampal mossy fiber sprouting and synapse formation after status epilepticus in rats: visualization after retrograde transport of biocytin. J Comp Neurol, 352:515–534.

Orkand, RK, Nicholls, JG, Kuffler, SW (1966) Effect of nerve impulses on the membrane potential of glial cells in the central nervous system of amphibia. J Neurophysiol, 29:788–806.

Otis, TS, De Koninck, Y, Mody, I (1994) Lasting potentiation of inhibition is associated with an increased number of γ-aminobutyric acid type A receptors activated during miniature inhibitory postsynaptic currents. Proc Natl Acad Sci USA, 91:7698–7702.

Penfield, W (1927) The mechanism of cicatricial contraction in the brain. Brain, 50:499–517.

Perez Velazquez, JL, Carlen, PL (2000) Gap junctions, synchrony and seizures. Trends Neurosci, 23:68–74.

Pinel, JP, Rovner, LI (1978) Experimental epileptogenesis: kindling-induced epilepsy in rats. Exp Neurol, 58:190–202.

Pitkänen, A (2002) Drug-mediated neuroprotection and antiepileptogenesis: animal data. Neurology. 59(9 Suppl 5):S27–33.

Pitkanen, A, Kharatishvili, I, Narkilahti, S, Lukasiuk, K, Nissinen, J (2005) Administration of diazepam during status epilepticus reduces development and severity of epilepsy in rat. Epilepsy Res, 63:27–42.

Pitkanen, A, Narkilahti, S, Bezvenyuk, Z, Haapalinna, A, Nissinen, J (2004) Atipamezole, an α(2)-adrenoceptor antagonist, has disease modifying effects on epileptogenesis in rats. Epilepsy Res, 61:119–140.

Porter, BE, Cui, XN, Brooks-Kayal, AR (2006) Status epilepticus differentially alters AMPA and kainate receptor subunit expression in mature and immature dentate granule neurons. Eur J Neurosci, 23:2857–2863.

Prasad, A, Williamson, JM, Bertram, EH (2002) Phenobarbital and MK-801, but not phenytoin, improve the long-term outcome of status epilepticus. Ann Neurol, 51:175–181.

Prince, HK, Conn, PJ, Blackstone, CD, Huganir, RL, Levey, AI (1995) Down-regulation of AMPA receptor subunit GluR2 in amygdaloid kindling. J Neurochem, 64:462–465.

Princivalle, AP, Duncan, JS, Thom, M, Bowery, NG (2002) Studies of GABA(B) receptors labelled with [(3)H]-CGP62349 in hippocampus resected from patients with temporal lobe epilepsy. Br J Pharmacol, 136:1099–1106.

Rashid, K, Van der Zee, CE, Ross, GM, Chapman, CA, Stanisz, J, Riopelle, RJ, Racine, RJ, Fahnestock, M (1995) A nerve growth factor peptide retards seizure development and inhibits neuronal sprouting in a rat model of epilepsy. Proc Natl Acad Sci USA, 92:9495–9499.

Ratzliff, AH, Howard, AL, Santhakumar, V, Osapay, I, Soltesz, I (2004) Rapid deletion of mossy cells does not result in a hyperexcitable dentate gyrus: implications for epileptogenesis. J Neurosci, 24:2259–2269.

Raza, M, Blair, RE, Sombati, S, Carter, DS, Deshpande, LS, DeLorenzo, RJ (2004) Evidence that injury-induced changes in hippocampal neuronal calcium dynamics during epileptogenesis cause acquired epilepsy. Proc Natl Acad Sci USA, 101:17522–17527.

Riedel, G, Reymann, KG (1996) Metabotropic glutamate receptors in hippocampal long-term potentiation and learning and memory. Acta Physiol Scand, 157:1–19.

Rigoulot, MA, Koning, E, Ferrandon, A, Nehlig, A (2004) Neuroprotective properties of topiramate in the lithium-pilocarpine model of epilepsy. J Pharmacol Exp Ther, 308:787–795.

Santhakumar, V, Bender, R, Frotscher, M, Ross, ST, Hollrigel, GS, Toth, Z, Soltesz, I (2000) Granule cell hyperexcitability in the early post-traumatic rat dentate gyrus: the 'irritable mossy cell' hypothesis. J Physiol, 524 (Pt 1):117–134.

Sato, M, Racine, RJ, McIntyre, DC (1990) Kindling: basic mechanisms and clinical validity. Electroencephalogr Clin Neurophysiol, 76:459–472.

Scharfman, HE, Goodman, JH, Sollas, AL (2000) Granule-like neurons at the hilar/CA3 border after status epilepticus and their synchrony with area CA3 pyramidal cells: functional implications of seizure-induced neurogenesis. J Neurosci, 20:6144–6158.

Scharfman, HE, Gray, WP (2007) Relevance of seizure-induced neurogenesis in animal models of epilepsy to the etiology of temporal lobe epilepsy. Epilepsia, 48 (Suppl 2):33–41.

Scheibel, ME, Crandall, PH, Scheibel, AB (1974) The hippocampal-dentate complex in temporal lobe epilepsy. A Golgi study. Epilepsia, 15:55–80.

Scimemi, A, Schorge, S, Kullmann, DM, Walker, MC (2006) Epileptogenesis is associated with enhanced glutamatergic transmission in the perforant path. J Neurophysiol, 95:1213–1220.

Seifert, G, Huttmann, K, Schramm, J, Steinhauser, C (2004) Enhanced relative expression of glutamate receptor 1 flip AMPA receptor subunits in hippocampal astrocytes of epilepsy patients with Ammon's horn sclerosis. J Neurosci, 24:1996–2003.

Shah, MM, Anderson, AE, Leung, V, Lin, X, Johnston, D (2004) Seizure-induced plasticity of h channels in entorhinal cortical layer III pyramidal neurons. Neuron, 44:495–508.

Simonato, M, Molteni, R, Bregola, G, Muzzolini, A, Piffanelli, M, Beani, L, Racagni, G, Riva, M (1998) Different patterns of induction of FGF-2, FGF-1 and BDNF mRNAs during kindling epileptogenesis in the rat. Eur J Neurosci, 10:955–963.

Sloviter, RS (1983) "Epileptic" brain damage in rats induced by sustained electrical stimulation of the perforant path. I. Acute electrophysiological and light microscopic studies. Brain Res Bull, 10:675–697.

Sloviter, RS (1987) Decreased hippocampal inhibition and a selective loss of interneurons in experimental epilepsy. Science, 235:73–76.

Sloviter, RS (1991a) Feedforward and feedback inhibition of hippocampal principal cell activity evoked by perforant path stimulation: GABA-mediated mechanisms that regulate excitability in vivo. Hippocampus, 1:31–40.

Sloviter, RS (1991b) Permanently altered hippocampal structure, excitability, and inhibition after experimental status epilepticus in the rat: the "dormant basket cell" hypothesis and its possible relevance to temporal lobe epilepsy. Hippocampus, 1:41–66.

Sloviter, RS (1992) Possible functional consequences of synaptic reorganization in the dentate gyrus of kainate-treated rats. Neurosci Lett, 137:91–96.

Sloviter, RS, Damiano, BP (1981) Sustained electrical stimulation of the perforant path duplicates kainate-induced electrophysiological effects and hippocampal damage in rats. Neurosci Lett, 24:279–284.

Stefan, H, Lopes da Silva, FH, Loscher, W, Schmidt, D, Perucca, E, Brodie, MJ, Boon, PA, Theodore, WH, Moshe, SL (2006) Epileptogenesis and rational therapeutic strategies. Acta Neurol Scand, 113:139–155.

Strasser, A, O'Connor, L, Dixit, VM (2000) Apoptosis signaling. Annu Rev Biochem, 69:217–245.

Su, H, Sochivko, D, Becker, A, Chen, J, Jiang, Y, Yaari, Y, Beck, H (2002) Upregulation of a T-type Ca2 + channel causes a long-lasting modification of neuronal firing mode after status epilepticus. J Neurosci, 22:3645–3655.

Sutula, TP, Pitkanen, A (2001) More evidence for seizure-induced neuron loss: is hippocampal sclerosis both cause and effect of epilepsy? Neurology, 57:169–170.

Tarkka, R, Paakko, E, Pyhtinen, J, Uhari, M, Rantala, H (2003) Febrile seizures and mesial temporal sclerosis: no association in a long-term follow-up study. Neurology, 60:215–218.

Tauck, DL, Nadler, JV (1985) Evidence of functional mossy fiber sprouting in hippocampal formation of kainic acid-treated rats. J Neurosci, 5:1016–1022.

Theodore, WH, Gaillard, WD (1999) Association between hippocampal volume and epilepsy duration. Ann Neurol, 46:800.

Tian, GF, Azmi, H, Takano, T, Xu, Q, Peng, W, Lin, J, Oberheim, N, Lou, N, Wang, X, Zielke, HR, Kang, J, Nedergaard, M (2005) An astrocytic basis of epilepsy. Nat Med, 11:973–981.

Traub, RD, Pais, I, Bibbig, A, LeBeau, FE, Buhl, EH, Hormuzdi, SG, Monyer, H, Whittington, MA (2003) Contrasting roles of axonal (pyramidal cell) and dendritic (interneuron) electrical coupling in the generation of neuronal network oscillations. Proc Natl Acad Sci USA, 100:1370–1374.

Tuff, LP, Racine, RJ, Adamec, R (1983) The effects of kindling on GABA-mediated inhibition in the dentate gyrus of the rat. I. Paired-pulse depression. Brain Res, 277:79–90.

Turski, L, Ikonomidou, C, Turski, WA, Bortolotto, ZA, Cavalheiro, EA (1989) Review: cholinergic mechanisms and epileptogenesis. The seizures induced by pilocarpine: a novel experimental model of intractable epilepsy. Synapse, 3:154–171.

Turski, WA, Cavalheiro, EA, Schwarz, M, Czuczwar, SJ, Kleinrok, Z, Turski, L (1983) Limbic seizures produced by pilocarpine in rats: behavioural, electroencephalographic and neuropathological study. Behav Brain Res, 9:315–335.

Tuunanen, J, Pitkanen, A (2000) Do seizures cause neuronal damage in rat amygdala kindling? Epilepsy Res, 39:171–176.

Upton, N, Stratton, S (2003) Recent developments from genetic mouse models of seizures. Curr Opin Pharmacol, 3:19–26.

Van Paesschen, W, Duncan, JS, Stevens, JM, Connelly, A (1998) Longitudinal quantitative hippocampal magnetic resonance imaging study of adults with newly diagnosed partial seizures: one-year follow-up results. Epilepsia, 39:633–639.

van Praag, H, Schinder, AF, Christie, BR, Toni, N, Palmer, TD, Gage, FH (2002) Functional neurogenesis in the adult hippocampus. Nature, 415:1030–1034.

Vignes, M, Collingridge, GL (1997) The synaptic activation of kainate receptors. Nature, 388:179–182.

Vissel, B, Royle, GA, Christie, BR, Schiffer, HH, Ghetti, A, Tritto, T, Perez-Otano, I, Radcliffe, RA, Seamans, J, Sejnowski, T, Wehner, JM, Collins, AC, O'Gorman, S, Heinemann, SF (2001) The role of RNA editing of kainate receptors in synaptic plasticity and seizures. Neuron, 29:217–227.

Vollmar, W, Gloger, J, Berger, E, Kortenbruck, G, Kohling, R, Speckmann, EJ, Musshoff, U (2004) RNA editing (R/G site) and flip-flop splicing of the AMPA receptor subunit GluR2 in nervous tissue of epilepsy patients. Neurobiol Dis, 15:371–379.

Walker, MC, Perry, H, Scaravilli, F, Patsalos, PN, Shorvon, SD, Jefferys, JG (1999) Halothane as a neuroprotectant during constant stimulation of the perforant path. Epilepsia, 40:359–364.

Walker, MC, White, HS, Sander, JW (2002) Disease modification in partial epilepsy. Brain, 125:1937–1950.

Wallace, RH, Scheffer, IE, Barnett, S, Richards, M, Dibbens, L, Desai, RR, Lerman-Sagie, T, Lev, D, Mazarib, A, Brand, N, Ben-Zeev, B, Goikhman, I, Singh, R, Kremmidiotis, G, Gardner, A, Sutherland, GR, George, ALJ, Mulley, JC, Berkovic, SF (2001) Neuronal sodium-channel α1-subunit mutations in generalized epilepsy with febrile seizures plus. Am J Hum Genet, 68:859–865.

Wasterlain, CG, Shirasaka, Y, Mazarati, AM, Spigelman, I (1996) Chronic epilepsy with damage restricted to the hippocampus: possible mechanisms. Epilepsy Res, 26:255–265.

Williams, PA, Wuarin, JP, Dou, P, Ferraro, DJ, Dudek, FE (2002) Reassessment of the effects of cycloheximide on mossy fiber sprouting and epileptogenesis in the pilocarpine model of temporal lobe epilepsy. J Neurophysiol, 88:2075–2087.

Wong, RK, Bianchi, R, Chuang, SC, Merlin, LR (2005) Group I mGluR-induced epileptogenesis: distinct and overlapping roles of mGluR1 and mGluR5 and implications for antiepileptic drug design. Epilepsy Curr, 5:63–68.

Wu, C, Leung, LS (1997) Partial hippocampal kindling decreases efficacy of presynaptic GABAB autoreceptors in CA1. J Neurosci, 17:9261–9269.

Zhang, N, Houser, CR (1999) Ultrastructural localization of dynorphin in the dentate gyrus in human temporal lobe epilepsy: a study of reorganized mossy fiber synapses. J Comp Neurol, 405:472–490.

Zhu, LJ, Chen, Z, Zhang, LS, Xu, SJ, Xu, AJ, Luo, JH (2004) Spatiotemporal changes of the N-methyl-D-aspartate receptor subunit levels in rats with pentylenetetrazole-induced seizures. Neurosci Lett, 356:53–56

Chapter 7
Seizure-Induced Neuronal Plasticity and Metabolic Effects

Monisha Goyal

Introduction

Epilepsy, the most common acquired chronic neurological disease, occurs in 1% of the human population. Despite treatment with the newest antiepileptic medications, almost one-third of the individuals continue having seizures (Kwan and Brodie, 2000). Many of those with seizure persistence and even some with seizure remittance suffer often from under-appreciated co-morbidities including cognitive deficits and psychopathology such as anxiety, depression, and poor attention.

Our understanding of epileptogenesis and its concurrent effects is based mainly on animal models. Using humans with epilepsy to study effects of human epilepsy is fraught with multiple problems including ethics, medication effects, and reproducibility. More recently, however, human brain tissue from surgical resections has been studied (this represents only a small subgroup of patients with epilepsy). Modern imaging techniques have also helped unveil widespread metabolic abnormalities associated with epilepsy.

This chapter highlights our current understanding of the age-dependent spectrum of seizure effects in both animal models and humans, and wherever possible, incorporates the contribution of other disciplines such as pathology and modern neuroimaging.

Historical Perspective

It is widely accepted today that prolonged seizures can selectively kill vulnerable neurons. This selective neuronal loss (diffuse cortical atrophy, unilateral hippocampal sclerosis, cerebellar atrophy) was described in the 1800s in institutionalized patients with refractory epilepsy and was assumed to be secondary to hypoxia/ischemia. In the 1970s, after experiments in primates, Meldrum and colleagues proposed that selective hippocampal and neuronal loss resulted from ischemic changes due to local seizure activity lasting 82–120 min (Meldrum, 2002).

This idea that selective neuronal damage was caused by abnormal electrical discharges remained controversial until it was shown that the perforant path

D.W. McCandless (ed.) *Metabolic Encephalopathy*,
doi: 10.1007/978-0-387-79112-8_7, © Springer Science+Business Media, LLC 2009

stimulation caused damage in the hilus and CA3 in the hippocampus (Sloviter, 1983). By this time, there was an increasing body of evidence showing that though oxygen and glucose consumption increased during seizures, there was reactive compensation by increase in blood flow (Meldrum, 1983).

So how then did cell death occur if there was adequate compensation? In the 1980s, it was hypothesized that the link between electrical discharges and ischemic cell change was mitochondrial injury with resulting energy depletion mediated by calcium overload. Thus the role of excitotoxicity, initially described by Olney, was expanded to include neurodegeneration in epilepsy (Olney and Sharpe, 1969; Olney and de Gubareff, 1978; Meldrum, 1983).

These concepts have been further developed in the last 25 years with advances in neuroimaging and the elucidation of cellular mechanisms and neural circuitry in experimental paradigms.

Insights from Experimental Models and Human Studies

Animal models have proved invaluable in providing the foundation for basic research on epileptogenesis. The relevance of ideas generated from these models is then tested on human brain slices from surgical specimens when possible (Sarkisian, 2001; Cortez et al., 2006). Commonly used animal models are summarized in Table 7.1.

Table 7.1 Common animal models[a]

Model	Mechanism	Seizure type
Chemoconvulsants		
Kainate	Glutamate agonist	Temporal lobe
Pilocarpine (+/− lithium)	Acetylcholine agonist	
Picrotoxin	GABA antagonist	
Bicuculline		
4-amino pyridine		Partial
Tetanus toxin	Neurotoxin	
Pertussis toxin		
Electrical/chemical Stimulation		
Kindling		
Perforant-path stimulation (PPS)		Temporal lobe
Chemoconvulsants		
Kainate	Glutamate agonist (maximal dosages)	
Pentylenetetrazol (PTZ)	GABA antagonist (maximal dosages)	
Bicuculline		Generalized
Picrotoxin		
Flurothyl	Other agents	
Ouabain		
Penicillin	GABA antagonist	Absence-like
Electrical stimulation		
Maximal electroshock (MES)		Generalized

[a]Adapted from Sarkisian (2001), with permission from Elsevier

Experimental models for seizures can be broadly divided into two groups: models for acute seizures and for chronic focal epilepsy. The acute models are responsible for much of our understanding of the mechanisms of epilepsy. They involve the administration of drugs such as picrotoxin, bicuculline, pentylenetetrazol (PTZ), and 4-aminopyridine to previously healthy animals. Since they are models of convulsions rather than epilepsy, they relate to seizures induced by drugs or by metabolic derangements in humans.

Models for partial epilepsy employ intracerebral tetanus toxin and kindling (repetitive, initially subconvulsive electrical or chemical stimulation to a focal target, resulting in repetitive seizures).

Other methods such as chemoconvulsants (kainic acid, systemic pilocarpine) and prolonged electrical stimulation induce an initial episode of status epilepticus (SE). After 2–3 weeks, there is relapse into recurrent partial seizures. The main criticism of models triggered by SE is that they do not mimic most human epilepsy. Rather than an acute event triggering epilepsy, the initial seizure in most humans is induced by a longstanding pathology such as cortical dysplasia, which is associated with a cascade of events leading to epileptogenesis. Otherwise, seizures in these animal models and human epilepsies show similar latency periods from initial injury to the emergence of recurrent partial seizures and share both EEG and histopathological features.

It is well established that SE is associated with adverse outcomes in humans and both cognitive and motor dysfunction in animals. There may also be an increased risk for subsequent seizures as well as a typical neuronal loss pattern predominantly in the hippocampus. However, it is becoming abundantly clear that these changes occur not only after SE but also after brief seizures.

The change involves a sequence of events after an initial insult encompassing biochemical, anatomic, and functional changes. During this latency period of epileptogenesis and subsequently with repeated seizures, there is activity-dependent reorganization of neural circuitry affecting gene expression, synaptic physiology, and activation of late cell death pathways, which culminates in neuronal loss, neurogenesis, axonal sprouting and dendritic changes, astrocytic dysfunction, altered receptors and ion channels (See Fig. 7.1). Of note, the most commonly studied human and experimental models (kindling, for example) involve temporal lobe onset seizures and extrapolation from these paradigms to other epilepsy subtypes should therefore be done with caution.

Neuronal Loss

Many models of recurrent seizures do not demonstrate neuronal loss, and as in many, spontaneous seizures are preceded by SE, the initial neuronal loss may be attributed to SE rather than subsequent seizures (Gorter et al., 2003). Longstanding debate has centered on whether hippocampal sclerosis (neuronal loss and gliosis in CA1, CA3, and hilus of the dentate gyrus) results from seizures or is the cause of seizures.

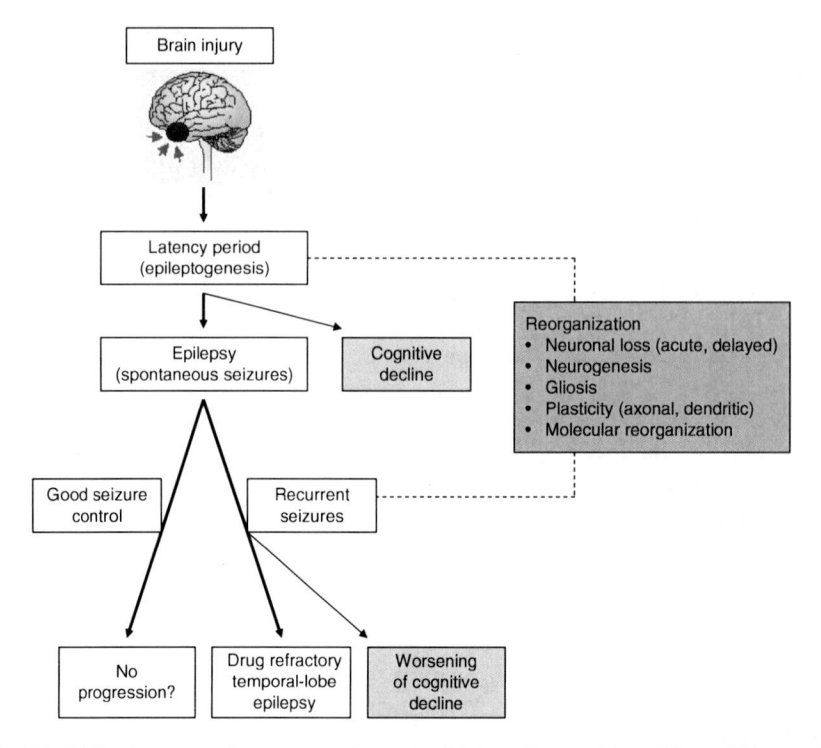

Fig. 7.1 Epileptic process in symptomatic temporal-lobe epilepsy. Adapted from Pitkanen, A., and Sutula, T.P. 2002. Is epilepsy a progressive disorder? Prospects for new therapeutic approaches in temporal-lobe epilepsy. The Lancet Neurology 1:173, with permission from Elsevier (*See also* Color Insert)

Multiple experimental studies, however, support the viewpoint that both SE and recurrent seizures induce neuronal loss. Neuronal damage is seen in primates with prolonged bicuculline-induced seizures (Meldrum et al., 1973) and in chemically or electrographically induced SE (Sankar et al., 1998). The most cited experimental paradigm demonstrating neuronal loss with recurrent seizures is the kindling model. Studies in this rodent model have demonstrated neuronal loss of 20–50% in a pattern similar to hippocampal sclerosis. This loss of neurons correlates with the number of seizures (Cavazos and Sutula, 1990; Cavazos et al., 1994; Frantseva et al., 2000a; Kotloski et al., 2002; Sankar et al., 2002). There is also a corresponding increase in neuron-specific enolase, a marker for neuronal injury (Hansen et al., 1990) and injury to inhibitory interneurons has also been shown (Gorter et al., 2001). Other models, including genetically mutated animals, also demonstrate neuronal loss with recurrent seizures (Qiao and Noebels, 1993).

Cell loss in the kindling model is not only limited to the hippocampus but also includes the amygdala and the entorhinal cortex and is dependent on the stimulation site (Kotloski et al., 2002).

In human epilepsy, Neuronal loss is well known, especially in the hippocampus. Studies have also found increased concentrations of enolase in human serum and CSF with spontaneous and provoked seizure in subgroups of both children and adults with epilepsy (DeGiorgio et al., 1995).

The mechanism of cell loss involves excitotoxicity provoked by intense neuronal firing (Grisar, 1986). There is excessive excitatory neurotransmitter release by activation of N-methyl-D-aspartate (NMDA) receptors and voltage-activated calcium channels which enables intracellular calcium influx in the neurons and glia. This calcium influx leads to a cascade of biochemical processes that ultimately lead to cell death: mitochondrial dysfunction with uncoupling of oxidative phosphorylation, generation of reactive oxygen species, and activation of many proteolytic and catabolic enzymes that adversely affect cell function.

Emerging evidence indicates that calcium functions as a major second messenger system and is involved in epileptogenesis and maintenance of persistent seizures. Prolonged seizures and spontaneous recurrent seizures cause prolonged elevations in calcium levels that then mediate second messenger effects on neuronal plasticity (Pal et al., 2001; Raza et al., 2004). As a second messenger, calcium affects gene transcription directly by modulating transcription factors or indirectly through calcium dependent kinases and phosphatases (Mellstrom and Naranjo, 2001).

Mitochondrial Dysfunction

The concept of mitochondrial dysfunction in epilepsy has emerged with the recognition that vast metabolic and bioenergetic changes are seen not only with isolated seizures but also with chronic epilepsy. The increase in metabolic demand results in increased blood flow and the increased rate of glycolysis exceeds pyruvate utilization by pyruvate hydrogenase, which then results in the build up of lactate.

During glutamate toxicity, mitochondria help sequester excessive calcium. However, as the calcium concentration increases, the electron transport chain is inhibited with resultant production of superoxide. This, in turn, leads to the opening of the mitochondrial permeability transition pore which results in the collapse of the mitochondrial proton gradient and ATP production (Bernardi, 1999), as well as in the release of cytochrome C, which triggers apoptosis (Kluck et al., 1997).

Dysfunction of Complex I respiratory chain enzyme and ultrastructural damage in the hippocampal mitochondria is seen in kainite-induced SE in rats (Chuang et al., 2004). A key player in mitochondrial oxidative phosphorylation, Complex I is a major source of superoxide and its dysfunction may increase mitochondrial reactive oxygen species (ROS) production and redox signaling (Taylor et al., 2003). The perforant path stimulation model shows mitochondrial dysfunction and decreased brain glutathione (Cock et al., 2002). Pilocarpine-treated rats show selective decline in Complexes I and IV activity in hippocampal CA1and CA3 subfields (Kudin et al., 2002). This pattern of complex I deficiency in CA3 region is also seen in humans (Kunz et al., 2000).

Seizure-induced neuronal death involves both apoptotic and necrotic cell death; both processes are intricately involved with mitochondrial function (Sloviter et al., 1996). Oxidative stress not only causes cell death but alters cellular biophysical characteristics, such as the band-pass frequency responses of neurons (Frantseva et al., 1998), possibly indicating a rapid reorganization of cellular networks into different functional assemblies (Morimoto et al., 2004).

Role of Free Radicals

Prolonged seizure activity in animals results in increased production of reactive oxygen species via activation of phospholipase A2 (Dennis, 1994) and nitric oxide synthase (Lipton et al., 1996) leading to excess arachidonic acid and nitric oxide synthesis respectively. The resulting free radicals destroy cytoskeletons, nucleic acids, and membrane lipids (Hall et al., 1999).

Free radical overproduction is seen in vivo and in vitro using the kindling model (Frantseva et al., 2000a, b). Studies show that in the kainate animal model, vital components of the cellular antioxidant defense mechanism (glutathione, mitochondrial aconitase) are depleted following seizures. These changes are seen 8 and 16 h post-SE. Neuronal cell loss in hippocampal CA3 region is generally seen 2–7 days after kainate administration.

Activation of Proteolytic and Catabolic Enzymes

Numerous enzyme systems are impacted by NMDA receptor-mediated intracellular calcium increase. Caspases, a family of endonucleases, trigger excitotoxicity-induced apoptosis Zipfel et al., 2000). Another family of proteases, the calpains, degrades the cytoskeleton, receptors, G proteins, and calcium binding proteins when activated (Emerich, 1999).

Prolonged seizures of around 30 min produce necrotic and apoptotic cell death in the same cell population with different time courses or in different cell populations. The current concept of the fate of a neuron is determined by many factors following an initial seizure including early gene expression with alteration of mRNA and proteins (enzymes, receptors, and ion channel subunits) modulated by calcium mediated effects. These changes alter the vulnerability of that neuron to subsequent excitotoxic stresses. Therefore, genetics influences seizure susceptibility (manifested in animal models and febrile seizures in humans) and consequences of seizures (hippocampal sclerosis).

Neurogenesis

Neurogenesis has been demonstrated in rodents, triggered by stimuli such as brief seizures. Secondarily generalized kindled seizures result in two- to threefold increase in granule cell neurogenesis in the hippocampus (Parent et al., 1998). This phenomenon is also seen with other experimental paradigms including partial seizures with kindling and electroconvulsive shocks in adult animals (Madsen et al., 2000; Nakagawa et al., 2000).

Axonal Sprouting and Dendritic Changes

Persistent sprouting of granule cell axons or mossy fibers in the dentate gyrus with recurrent seizures occurs with kindling and in stargazer and tottering mice (Stanfield, 1989; Cavazos et al., 1991). Sprouting is also seen in other regions of the hippocampus and with other animal models. This synaptic reorganization produces enhanced frequency of glutaminergic spontaneous synaptic currents (Esclapez et al., 1999). The phenomenon of axonal plasticity occurs with both partial and generalized seizures and is seen throughout the life span.

Multiple studies in animal models document loss of dendritic spines and occurrence of varicose swelling including pilocarpine- and tetanus-induced seizures (Isokawa, 1998; Jiang et al., 1998). Dendritic spines are felt to be the main site of excitatory glutamatergic synaptic inputs into neurons. Since each spine makes a synaptic connection with a single presynaptic terminal, a decrease in dendritic spines leads to alterations in both pre- and postsynaptic components of neuronal circuitry. Several studies suggest an upregulation of dendritic spines with particular learning tasks (Moser et al., 1994). The findings of hippocampal and neocortical dendritic spine loss in epilepsy implies a causal relationship with cognitive and memory deficits.

In humans with epilepsy, both mossy fiber sprouting, as well as decreased dendritic branching and spine density are seen (Babb and Brown, 1986; Swann et al., 2000; Wong, 2005).

Astrocyte Dysfunction

Interest is growing in the role of glial cells in epilepsy (Binder and Steinhauser, 2006). Direct stimulation of astrocytes causes neuronal synchronization in acute epilepsy models (Tian et al., 2005). Like neurons, astrocytes exhibit calcium-induced release of glutamate which provides excitation to neurons in the vicinity (Volterra and Meldolesi, 2005). Glial cells show plasticity with structural alterations in receptors, membrane channels, and transporters in both experimental epilepsy and human tissue.

Glial cell activation and proliferation with upregulation of glial-fibrillary acidic protein is seen with kindled seizures in the hippocampus. This phenomenon is time-dependent and even reversible (Torre et al., 1993; Adams et al., 1998).

Primarily responsible for glutamate uptake, glial cells may contribute to the increased extracellular levels of glutamate seen in epileptogenic tissues (Glass and

Dragunow, 1995). Dysfunctional glial glutamate transporters are implicated in seizures with MTS (During and Spencer, 1993). In addition, the MTS tissue has decreased concentrations of glutamine synthetase, an enzyme that converts glutamate to glutamine after reuptake by astrocytes. This may contribute to increased glutamate concentrations in astrocytes and extracellular space seen in sclerotic tissue (Eid et al., 2004) and in patients with MTS using MRS (Petroff et al., 2002).

Alterations in both astrocytic potassium and water channels have been associated with epilepsy. Increased extracellular potassium concentration enhances epileptiform activity (Feng and Durand, 2006) and this finding has been reproduced in human hippocampal slices (Gabriel et al., 2004). In human sclerotic tissue, down regulation of potassium channels suggests impairment in potassium buffering and contribution to hyperexcitability (Hinterkeuser et al., 2000; Steinhauser and Seifert, 2002).

Brain tissue excitability is closely related to osmolarity and the size of the extracellular space (Schwartzkroin et al., 1998), with hypo-osmolar states lowering seizure threshold (Andrew et al., 1989). The aquaporins function as water channels and are expressed by glial cells especially at astroglial endfeet adjacent to blood vessels and astrocytic membranes ensheathing glutamatergic synapses (Nagelhus et al., 2004). Altered expression of these channels is seen with an overall increase in sclerotic hippocampi but with a decrease in perivascular expression. This decrease in perivascular expression may alter water flux and increase seizure propensity (Lee et al., 2004; Eid et al., 2005).

Astrocytic alterations are implicated in seizures associated with tumors. Microdialysis studies of gliomas show increased glutamate concentrations at the periphery in peri-tumoral tissue (Bianchi et al., 2004). Ye et al. have shown decreased glutamate uptake and decreased glutamate transporter expression in glioma cells (Ye et al., 1999). Mislocalization of potassium channels (Olsen and Sontheimer, 2004) and reduced potassium currents (Bordey and Sontheimer, 1998) are seen in malignant astrocytes. Glial cell changes have also been found in post traumatic animal models of epilepsy with reduction in astrocytic glutamate transporter expression and impairment in potassium homeostasis (D'Ambrosio et al., 1999; Samuelsson et al., 2000).

Altered Gene Expression

Genetic background influences seizure-induced plasticity. The number of genes affected in hippocampal and cerebellar tissue after a chemically induced seizure varies depending on the strain of mice (Schauwecker, 2002). Seizures themselves increase expression of certain genes such as *fos* and *jun*, transcription factors involved in activity-dependent hippocampal signaling and neuronal injury. Kindled seizures also induce changes in gene and protein expression (Hughes et al., 1999). More recently, DNA microarrays have been used to study alteration in gene expression in animal models and epileptogenic tissue. Up and down regulation of genes occurs in a time-dependent fashion during both epileptogenesis and recurrent seizures. These gene products regulate transcription, protein synthesis and degradation, metabolism, structural proteins, and receptors (Lukasiuk and Pitkanen, 2004).

Altered Neurotransmitter Receptors Within Epileptic Foci

Recent work has shown that rather than inhibitory effects, GABAergic synapses may have excitatory effects in basal conditions in immature and adult tissue (Coulter, 1999; Cossart et al., 2005). In human temporal lobe epilepsy, altered GABA-A receptor function is associated with changes in subunit composition (Loup et al., 2000).

Animal models show dynamic changes in receptor numbers and subunit composition during the development of epileptogenesis after the initial drug- or electrical-induced SE (Gibbs et al., 1997; Nusser et al., 1998). How this impacts seizure generation needs further study but change in receptor subunit composition may enhance that potential (McIntyre et al., 2002).

Changes in excitatory amino acid receptors have been documented by multiple laboratories (Mody and Heinemann, 1987; Kohr and Mody, 1994; Kraus et al., 1994). Increased mRNA levels of NMDA receptor subunits have been measured in dentate-granule cells of patients with hippocampal sclerosis and mesial temporal lobe epilepsy (Mathern et al., 1999).

Altered Ion Channel Function

Synchronization of hyperexcitable neurons is dependent on ionic currents that flow through sodium, potassium, and calcium channels. Alterations in ion channels are seen in both animal models and surgically resected human tissue (Lombardo et al., 1996; Beck et al., 1997; Vreugdenhil et al., 1998; Straub et al., 2000).

Altered Neurochemistry

Animal studies demonstrate increased extracellular glutamate during SE (Ueda et al., 2002). Reduced GABA and increased lactate have been measured in vivo in a brain dialysis study in human temporal lobe epilepsy (During and Spencer, 1993; During et al., 1994). Using microdialysis prior to neurological surgery, neurotoxic levels of glutamate have been associated with seizure progression (Wilson et al., 1996). This increase in glutamate is implicated in neuronal death because the antagonism of certain glutamate receptors during SE is neuroprotective (Ebert et al., 2002).

Do Seizures Cause More Seizures?

Some evidence supports seizures begetting seizures. This phenomenon is seen in electrically/chemically induced SE which results in spontaneous seizures after a few weeks, but is dependent on the developmental stage of the animal. Seizure-induced cell death may increase the chance of seizure recurrence because of the loss of inhibition from injured interneurons which may lead to a lower inhibitory drive. Seizure-induced sprouting of nerve terminals results in synaptic rearrangements that promotes hyperexcitability in the dentate (Glass and Dragunow, 1995; Ben-Ari, 2001) and is associated with increased seizure activity (Ratzliff et al., 2002). After generalized convulsive seizures in rodents, increases in the power of specific frequency bands (Mackenzie et al., 2002), changes in excitation/inhibition (Lopes da Silva et al., 1994), and calcium homeostatic mechanisms (Pal et al., 2000) also support this idea.

Kindling by electrical and chemical induction is seen in many species including primates, suggesting it is an effect of neural plasticity and not merely an experimental model. Other manifestations of progression include increase in seizure frequency and duration, change in seizure type, and progressive cognitive and memory impairment (Sutula et al., 1995; Jiang et al., 1999; Nissinen et al., 2000).

However, other seizure-induced changes do not support seizure recurrence. The presence of interictal activity decreases the probability of seizure occurrence in several in vitro and in vivo studies (Barbarosie and Avoli, 1997). If seizures induce seizures, seizure control would imply improvement in the natural history of epilepsy but this is not the case.

Seizure-Associated Changes in the Developing Brain

During the early postnatal period in animal models, the developing brain is particularly susceptible to seizures. In humans, this correlates with increased risk of seizures in childhood, especially in the neonate, and is associated with hyperexcitability in the immature brain from relatively slow maturation of inhibitory neurotransmission in comparison with the more robust development of the excitatory system (Brooks-Kayal, 2005). Gamma-aminobutyric acid (GABA), the main inhibitory neurotransmitter in the adult brain, is excitatory in the immature brain because of higher intracellular chloride ion content. With GABA receptor activation, chloride ions exit the cell resulting in depolarization (Cossart et al., 2005). GABA-mediated excitation has been associated with the development of a secondary epileptogenic foci in the contralateral immature rat hippocampi (Khalilov et al., 1999; Khalilov et al., 2003).

The excitatory effect is also mediated by the potentiation of glutamate, the main excitatory neurotransmitter, in part because of low expression of the primary glutamate transporter. In addition, post-synaptic transmission is enhanced by NMDA receptors which depolarize more easily and AMPA receptors which stay open longer (Monyer et al., 1994; Pickard et al., 2000).

Prolonged or repetitive early life seizures increase the subsequent risk of developing epilepsy by affecting hippocampal excitability. Animal studies confirm this in the lithium-pilocarpine model (Zhang et al., 2004).

Unlike seizures in mature animals, seizures in early life do not cause cell loss but there is reduced neurogenesis in the dentate gyrus (McCabe et al., 2001) with extensive synaptic reorganization and mossy fiber sprouting (Huang et al., 1999). Injury following prolonged seizures does occur including long-term adverse effects on learning and memory (Holmes, 2004). Impairment of visual-spatial memory has also been seen (Lynch et al., 2000). Although it is well established that cognitive impairment occurs in young animals with recurring seizures, the exact mechanism(s) resulting in this effect remain unclear.

Contributions of Neuroimaging

Seizure-induced changes detected on MRI are time sensitive, evolving from a transient peri-ictal phenomenon to permanent structural change. Animal studies suggest that a seizure initiates a local cellular disturbance leading to cell swelling and fluctuations in extracellular water. This results in transient increase in the volume of a brain structure, with an increase in T2-weighted signal intensity and changes in the acute diffusion coefficient (Briellmann et al., 2005).

These acute changes on MRI typically show complete resolution even after SE (Salmenpera et al., 2000). However, in both children and adults, there may be evolution to atrophy in the region that showed initial swelling with an increased T2-weighted signal intensity, attributed to gliosis and neuronal cell loss (Tien and Felsberg, 1995; Meierkord et al., 1997; Perez et al., 2000). This phenomenon is typically seen with the development of hippocampal sclerosis.

Contributions of volumetric MRI and functional neuroimaging techniques are discussed below. These modalities have helped elucidate pathophysiological mechanisms of epilepsy, showing more widespread seizure-induced metabolic changes than previously appreciated.

Volumetric MRI

With increasing resolution due to higher Tesla images, volumetric studies show hippocampal and cerebellar volume loss of 3% and neocortical volume loss of 1.6% with an interscan interval of 3.5 years (Lemieux et al., 2000; Liu et al., 2001). Hippocampal atrophy, identified with hippocampal volume measurements, correlates well with hippocampal neuron loss, especially in the CA1 sub-region. Decreased ipsilateral thalamic volume has also been documented in temporal lobe epilepsy (Dreifuss et al., 2001; Natsume et al., 2003).

Positron Emission Topography (PET)

A variety of available PET ligands allow the measurement of glucose metabolism, central benzodiazepine receptors, and opioid and dopamine receptors. 18f-fluoro-deoxyglucose PET shows regional cerebral glucose utilization that is sensitive but has nonspecific etiology. Regional hypometabolism may be seen in 90% of patients with medial temporal lobe epilepsy (Gaillard et al., 1995) and in 70% of patients with neocortical epilepsy (Engel et al., 1995). Unlike the clear-cut correlation between MRI volume loss and neuronal loss, PET hypometabolism is caused not only by neuronal loss but also by a metabolic disturbance.

In intractable nonlesional frontal and temporal epilepsy, PET has shown hypometabolism in the ipsilateral hippocampus and thalamus. Secondarily, generalized seizures and long duration of epilepsy correlate with lower glucose metabolism (Benedek et al., 2004). These studies highlight the progressive involvement of cortical-subcortical networks in seizure propagation. This phenomenon of progressive metabolic activation of cortical-subcortical limbic seizure networks is described with increasing stages of electric kindling in rats using autoradiography of glucose metabolism (Handforth and Ackermann, 1995).

Magnetic Resonance Spectroscopy (MRS)

MRS allows evaluation of neuronal integrity and function by measuring N-acetylaspartate (NAA), a normal byproduct of neuronal cellular metabolism. NAA is a marker of neuronal cell dysfunction, not just volume loss, but has nonspecific etiology. Other metabolites including choline (marker for cellular proliferation), creatinine, lactate, GABA, glutamate, and glutamine can also be measured. MRS may detect metabolic abnormalities not seen on MRI, such as an increase in lactate during SE in humans, even after a single seizure (Castillo et al., 2001; Mueller et al., 2001). Focal decreases in NAA are seen with nonlesional temporal lobe epilepsy and extratemporal partial epilepsies (Matthews et al., 1990; Stanley et al., 1998). NAA abnormalities are described in ipsilateral and contralateral normal-volume hippocampi in temporal lobe epilepsy with subsequent normalization after surgery if the seizures resolve (Cendes et al., 1997). A recent study detected neuronal dysfunction in thalami of patients with absence epilepsy (Fojtikova et al., 2006).

Diffusion Tensor Imaging (DTI)

DTI identifies the motion of water that is quantified by voxel and region-based methods measuring diffusivity and fractional anisotropy. Diffusivity, a measurement of amplitude of diffusional motion, is increased with neuronal loss and gliosis.

Fractional anisotropy reflects directional motion of water. Fluid motion in the brain is restricted to the same axis as the axon or myelin sheath. With neuronal damage, fractional anisotropy decreases secondary to less restricted motion in various axes.

Both peri-ictal and post-ictal changes in DTI occur in animal models and in human epilepsy after SE and brief seizures. In the peri-ictal phase, reduced diffusivity is seen in animal models and in up to 50% of patients after seizures in one series (Diehl et al., 2005). This may be secondary to cellular swelling and reduced extracellular space due to the failure of ATPase which results in the accumulation of intracellular sodium and water (Wang et al., 1996). Transient decreases in diffusivity are seen post-ictally in kainic acid-induced SE in rodents. Increase in diffusivity is seen with chronic pilocarpine-induced seizures thought to be due to the loss of cellular and structural integrity.

In humans with temporal lobe seizures, DTI abnormalities correspond with histopathological changes such as gliosis (Rugg-Gunn et al., 2002). DTI in patients with chronic epilepsy show increased diffusivity and decreased anisotropy in sclerosed hippocampi. This increase in diffusivity extends beyond the seizure origin into normal appearing brain tissue (Yoo et al., 2002). This increase has also been observed in normal tissue on the contralateral side and does not correspond to PET abnormalities (Kimiwada et al., 2006).

DTI thus far has shown previously unappreciated metabolic abnormalities involving water diffusion and serial images with this modality hold promise in detecting continuing cerebral damage in epilepsy.

Functional MRI

The most common method of fMRI is blood oxygenation level-dependent (BOLD) imaging. Using hemoglobin as an endogenous contrast agent, this technique measures magnetization difference between oxy- and deoxy hemoglobin to create the fMRI signal. fMRI, then, is an indirect measure of neuronal activity, which is accompanied by its hemodynamic correlate, also known as neurovascular coupling.

Much work has been done on neurovascular coupling during seizures (Suh et al., 2006b). It is widely accepted that an increase in neuronal activity increases the cerebral metabolic rate of oxygen consumption, resulting in an increase in cerebral blood flow and blood volume. This increase in cerebral blood flow occurs 1–2 s after the onset of neuronal activity as shown by PET scan (with a temporal resolution of seconds). This oversupply of oxygenated hemoglobin compared with deoxygenated hemoglobin produces the blood-oxygen-level-dependent (BOLD) signal imaged with fMRI.

More recently, techniques with higher spatial and temporal resolution, such as ORIS (optical recording of intrinsic signals), imaging spectroscopy, oxygen-sensitive electrodes, and fMRI at 1.5 and 4-T, have shown neuronal and blood oxygenation changes in the first few hundred milliseconds after neuronal firing. These changes manifest as an "initial dip" or a rapid decrease in tissue oxygenation (increase in deoxygenated hemoglobin) before the well-known increase in cerebral blood flow. This "initial dip", though not shown by some studies, indicates that after neuronal firing, there is a brief ischemic environment until arterioles dilate due to cerebral autoregulation.

In the rat model of interictal spikes from focal iontophoresis of bicuculline, a GABA-A antagonist, each interictal event produces a focal increase in deoxygenated hemoglobin lasting as long as 3–4 s using ORIS. The onset latency appears within 100 ms of the event. Increase in cerebral blood volume, however, also occurs within 100 ms after the interictal spike. However, the slope of the rise in deoxygenated hemoglobin is steeper than the slope of the rise in blood volume. This suggests that despite the rapid increase in blood flow, metabolic demand for oxygen is not met. Eventually the slope for blood volume becomes steeper than that for deoxygenated blood, indicating hyperoxygenation resulting in the BOLD signal. When interictal spikes occur frequently, there is a more persistent increase in deoxygenated hemoglobin, indicating persistent unmet demand for oxygen (Suh et al., 2005).

This hypothesis has also been confirmed using ORIS in rat models with aminopyridine (4-AP). With ictal events of 60–90 s, increase in deoxygenated hemoglobin lasts the duration of the seizure despite a rapid increase in blood volume (Bahar et al., 2006). The epileptic dip has also been demonstrated by measuring partial pressure of oxygen using oxygen sensitive electrodes in rats with 4-AP focus (Zhao et al., 2005). These studies suggest that focal seizures are associated with an increased metabolic demand for oxygen which remains unmet for at least some time despite autoregulatory mechanisms.

Recently ORIS was used in humans with refractory epilepsy, who underwent craniotomy for cortical resection. Cortical stimulation elicited the initial dip in all subjects and showed a linear relationship between stimulus amplitude and the area of the optical signal with the initial dip and early blood volume increase. More than 3 s later, the decrease in deoxygenated hemoglobin (relevant to the BOLD signal) was no longer localized to the electrophysiologic activation but involved several centimeters of surrounding cortex (Suh et al., 2006a).

Besides ictal events, BOLD signal is also measured in human interictal discharges (Benar et al., 2002). It is unknown how this period of relative ictal/interictal ischemia impacts neuronal function or causes permanent damage. However, in humans, interictal spikes may be associated with transient negative impact on cognition (Shewmon and Erwin, 1988; Binnie and Marston, 1992).

Clinical Perspective and Future Considerations

In general, childhood epilepsy is associated with behavior problems and cognitive decline. Though many factors like anti-seizure drugs, genetics, and seizure duration may be contributing to these effects, lower intelligence quotients (IQ) scores, learning disabilities, and psychopathology are more common than in the normal population. Increased risk of learning disabilities is present despite well-controlled seizures (Bailet and Turk, 2000).

There is marked heterogeneity in epilepsy syndromes with lack of inevitable progression of seizures in some and predictable remission in others. Examples of predictable remission include benign familial neonatal convulsions and benign focal epilepsies. However, what has been traditionally thought of as "benign" may not be so. Subtle cognitive abnormalities are coming to light in the so-called "benign" epilepsies (Yung et al., 2000; Sanchez-Carpintero and Neville, 2003).

Do chronic recurrent seizures lead to cognitive abnormalities or does the underlying brain dysfunction cause both? The latter is supported by the fact that despite seizure control and normalization of EEG, individuals continue having problems with learning, behavioral, and psychosocial issues, even in comparison with other chronically ill children. Though adults also continue to have a lower quality of life, the adverse effect on cognition is seen especially in children and may, therefore, be due, in part, to the interference of intellectual abilities during a period of rapid childhood development (Bjornaes et al., 2001).

It is becoming increasingly clear that an epileptogenic tissue undergoes some fundamental metabolic changes. There are altered neurotransmitter receptors in the remaining surviving neurons that may have altered circuitry. Screening of potential anti-seizure drugs usually occurs in acute seizure models such as pentylenetetrazol, maximal electroshock, fluorothyl (rather than epileptic). Screening drugs in these epileptic animals targeting these altered receptors may be more beneficial and have a wider therapeutic index.

Conclusion

Clinically, it is well accepted that seizures are associated with brain damage and neuronal loss. Experimentally and from surgically excised human tissue studies, the emerging perspective is that epileptogenesis and subsequent seizures trigger activity-dependent structural and functional remodeling of neuronal circuitry. This neuronal plasticity shows marked age dependence and exerts long term and at times progressive cognitive and behavioral sequelae.

Advancement in functional imaging and volumetric MRI studies has helped elucidate progressive metabolic changes in subcortical structures in epilepsy, perhaps indicative of a growing maladaptive network manifesting as neurological and behavioral pathology.

Better understanding of these effects and their underlying mechanisms expands the spectrum of potential therapeutic intervention targets that aim not only for seizure suppression but also to influence and modify seizure-induced neuronal plasticity.

References

Adams, B., Von Ling, E., Vaccarella, L., Ivy, G.O., Fahnestock, M., and Racine, R.J. 1998. Time course for kindling-induced changes in the hilar area of the dentate gyrus: reactive gliosis as a potential mechanism. Brain Res 804(2):331–336

Andrew, R.D., Fagan, M., Ballyk, B.A., and Rosen, A.S. 1989. Seizure susceptibility and the osmotic state. Brain Res 498(1):175–180

Babb, T.L., and Brown, W.J. 1986. Neuronal, dendritic, and vascular profiles of human temporal lobe epilepsy correlated with cellular physiology in vivo. Adv Neurol 44:949–966

Bahar, S., Suh, M., Zhao, M., and Schwartz, T.H. 2006. Intrinsic optical signal imaging of neocortical seizures: the 'epileptic dip '. Neuroreport 17(5):499–503

Bailet, L.L., and Turk, W.R. 2000. The impact of childhood epilepsy on neurocognitive and behavioral performance: a prospective longitudinal study. Epilepsia 41(4):426–431

Barbarosie, M., and Avoli, M. 1997. CA3-driven hippocampal-entorhinal loop controls rather than sustains in vitro limbic seizures. J Neurosci 17(23):9308–9314

Beck, H., Steffens, R., Heinemann, U., and Elger, C.E. 1997. Properties of voltage-activated Ca2+ currents in acutely isolated human hippocampal granule cells. J Neurophysiol 77(3):1526–1537

Benar, C.G., Gross, D.W., Wang, Y., Petre, V., Pike, B., Dubeau, F., and Gotman, J. 2002. The BOLD response to interictal epileptiform discharges. Neuroimage 17(3):1182–1192

Ben-Ari, Y. 2001. Cell death and synaptic reorganizations produced by seizures. Epilepsia 42(Suppl 3):5–7

Benedek, K., Juhasz, C., Muzik, O., Chugani, D.C., and Chugani, H.T. 2004. Metabolic changes of subcortical structures in intractable focal epilepsy. Epilepsia 45(9):1100–1105

Bernardi, P. 1999. Mitochondrial transport of cations: channels, exchangers, and permeability transition. Physiol Rev 79(4):1127–1155

Bianchi, L., De Micheli, E., Bricolo, A., Ballini, C., Fattori, M., Venturi, C., Pedata, F., Tipton, K.F., and Della Corte, L. 2004. Extracellular levels of amino acids and choline in human high grade gliomas: an intraoperative microdialysis study. Neurochem Res 29(1):325–334

Binder, D.K., and Steinhauser, C. 2006. Functional changes in astroglial cells in epilepsy. Glia 54(5):358–368

Binnie, C.D., and Marston, D. 1992. Cognitive correlates of interictal discharges. Epilepsia 33(Suppl 6):S11–S17

Bjornaes, H., Stabell, K., Henriksen, O., and Loyning, Y. 2001. The effects of refractory epilepsy on intellectual functioning in children and adults. A longitudinal study. Seizure 10(4):250–259

Bordey, A., and Sontheimer, H. 1998. Electrophysiological properties of human astrocytic tumor cells in situ: enigma of spiking glial cells. J Neurophysiol 79(5):2782–2793

Briellmann, R.S., Wellard, R.M., and Jackson, G.D. 2005. Seizure-associated abnormalities in epilepsy: evidence from MR imaging. Epilepsia 46(5):760–766

Brooks-Kayal, A.R. 2005. Rearranging receptors. Epilepsia 46(Suppl 7):29–38

Castillo, M., Smith, J.K., and Kwock, L. 2001. Proton MR spectroscopy in patients with acute temporal lobe seizures. AJNR Am J Neuroradiol 22(1):152–157

Cavazos, J.E., and Sutula, T.P. 1990. Progressive neuronal loss induced by kindling: a possible mechanism for mossy fiber synaptic reorganization and hippocampal sclerosis. Brain Res 527(1):1–6

Cavazos, J.E., Golarai, G., and Sutula, T.P. 1991. Mossy fiber synaptic reorganization induced by kindling: time course of development, progression, and permanence. J Neurosci 11(9):2795–2803

Cavazos, J.E., Das, I., and Sutula, T.P. 1994. Neuronal loss induced in limbic pathways by kindling: evidence for induction of hippocampal sclerosis by repeated brief seizures. J Neurosci 14(5 Pt 2):3106–3121

Cendes, F., Andermann, F., Dubeau, F., Matthews, P.M., and Arnold, D.L. 1997. Normalization of neuronal metabolic dysfunction after surgery for temporal lobe epilepsy. Evidence from proton MR spectroscopic imaging. Neurology 49(6):1525–1533

Chuang, Y.C., Chang, A.Y., Lin, J.W., Hsu, S.P., and Chan, S.H. 2004. Mitochondrial dysfunction and ultrastructural damage in the hippocampus during kainic acid-induced status epilepticus in the rat. Epilepsia 45(10):1202–1209

Cock, H.R., Tong, X., Hargreaves, I.P., Heales, S.J., Clark, J.B., Patsalos, P.N., Thom, M., Groves, M., Schapira, A.H., Shorvon, S.D., and Walker, M.C. 2002. Mitochondrial dysfunction associated with neuronal death following status epilepticus in rat. Epilepsy Res 48(3):157–168

Cortez, M.A., Perez Velazquez, J.L., and Snead, O.C., III 2006. Animal models of epilepsy and progressive effects of seizures. Adv Neurol 97:293–304

Cossart, R., Bernard, C., and Ben-Ari, Y. 2005. Multiple facets of GABAergic neurons and synapses: multiple fates of GABA signalling in epilepsies. Trends Neurosci 28(2):108–115

Coulter, D.A. 1999. Chronic epileptogenic cellular alterations in the limbic system after status epilepticus. Epilepsia 40(Suppl 1):S23–S33; discussion S40–S21

D'Ambrosio, R., Maris, D.O., Grady, M.S., Winn, H.R., and Janigro, D. 1999. Impaired $K(^+)$ homeostasis and altered electrophysiological properties of post-traumatic hippocampal glia. J Neurosci 19(18):8152–8162

DeGiorgio, C.M., Correale, J.D., Gott, P.S., Ginsburg, D.L., Bracht, K.A., Smith, T., Boutros, R., Loskota, W.J., and Rabinowicz, A.L. 1995. Serum neuron-specific enolase in human status epilepticus. Neurology 45(6):1134–1137

Dennis, E.A. 1994. Diversity of group types, regulation, and function of phospholipase A2. J Biol Chem 269(18):13057–13060

Diehl, B., Symms, M.R., Boulby, P.A., Salmenpera, T., Wheeler-Kingshott, C.A., Barker, G.J., and Duncan, J.S. 2005. Postictal diffusion tensor imaging. Epilepsy Res 65(3):137–146

Dreifuss, S., Vingerhoets, F.J., Lazeyras, F., Andino, S.G, Spinelli, L., Delavelle, J., and Seeck, M. 2001. Volumetric measurements of subcortical nuclei in patients with temporal lobe epilepsy. Neurology 57(9):1636–1641

During, M.J., and Spencer, D.D. 1993. Extracellular hippocampal glutamate and spontaneous seizure in the conscious human brain. Lancet 341(8861):1607–1610

During, M.J., Fried, I., Leone, P., Katz, A., and Spencer, D.D. 1994. Direct measurement of extracellular lactate in the human hippocampus during spontaneous seizures. J Neurochem 62(6):2356–2361

Ebert, U., Brandt, C., and Loscher, W. 2002. Delayed sclerosis, neuroprotection, and limbic epileptogenesis after status epilepticus in the rat. Epilepsia 43(Suppl 5):86–95

Eid, T., Thomas, M.J., Spencer, D.D., Runden-Pran, E., Lai, J.C., Malthankar, G.V., Kim, J.H., Danbolt, N.C., Ottersen, O.P., and de Lanerolle, N.C. 2004. Loss of glutamine synthetase in the human epileptogenic hippocampus: possible mechanism for raised extracellular glutamate in mesial temporal lobe epilepsy. Lancet 363(9402):28–37

Eid, T., Lee, T.S., Thomas, M.J., Amiry-Moghaddam, M., Bjornsen, L.P., Spencer, D.D., Agre, P., Ottersen, O.P., and de Lanerolle, N.C. 2005. Loss of perivascular aquaporin 4 may underlie deficient water and K^+ homeostasis in the human epileptogenic hippocampus. Proc Natl Acad Sci USA 102(4):1193–1198

Emerich, D.F. 1999. Intracellular events associated with cerebral ischemia. In: L.P. Miller (Ed.), Stroke therapy: basic, preclinical, and clinical directions. Wiley-Liss, New York, pp. 195–218

Engel, J., Jr., Henry, T.R., and Swartz, B.E. 1995. Positron emission tomography in frontal lobe epilepsy. Adv Neurol 66:223–238; discussion 238–241

Esclapez, M., Hirsch, J.C., Ben-Ari, Y., and Bernard, C. 1999. Newly formed excitatory pathways provide a substrate for hyperexcitability in experimental temporal lobe epilepsy. J Comp Neurol 408(4):449–460

Feng, Z., and Durand, D.M. 2006. Effects of potassium concentration on firing patterns of low-calcium epileptiform activity in anesthetized rat hippocampus: inducing of persistent spike activity. Epilepsia 47(4):727–736

Fojtikova, D., Brazdil, M., Horky, J., Mikl, M., Kuba, R., Krupa, P., and Rektor, I. 2006. Magnetic resonance spectroscopy of the thalamus in patients with typical absence epilepsy. Seizure 15(7):533–540

Frantseva, M.V., Perez Velazquez, J.L., and Carlen, P.L. 1998. Changes in membrane and synaptic properties of thalamocortical circuitry caused by hydrogen peroxide. J Neurophysiol 80(3):1317–1326

Frantseva, M.V., Perez Velazquez, J.L., Tsoraklidis, G., Mendonca, A.J., Adamchik, Y., Mills, L.R., Carlen, P.L., and Burnham, M.W. 2000a. Oxidative stress is involved in seizure-induced neurodegeneration in the kindling model of epilepsy. Neuroscience 97(3):431–435

Frantseva, M.V., Velazquez, J.L., Hwang, P.A., and Carlen, P.L. 2000b. Free radical production correlates with cell death in an in vitro model of epilepsy. Eur J Neurosci 12(4):1431–1439

Gabriel, S., Njunting, M., Pomper, J.K., Merschhemke, M., Sanabria, E.R., Eilers, A., Kivi, A., Zeller, M., Meencke, H.J., Cavalheiro, E.A., Heinemann, U., and Lehmann, T.N. 2004. Stimulus and potassium-induced epileptiform activity in the human dentate gyrus from patients with and without hippocampal sclerosis. J Neurosci 24(46):10416–10430

Gaillard, W.D., Bhatia, S., Bookheimer, S.Y., Fazilat, S., Sato, S., and Theodore, W.H. 1995. FDG-PET and volumetric MRI in the evaluation of patients with partial epilepsy. Neurology 45(1):123–126

Gibbs, J.W., III, Shumate, M.D., and Coulter, D.A. 1997. Differential epilepsy-associated alterations in postsynaptic GABA(A) receptor function in dentate granule and CA1 neurons. J Neurophysiol 77(4):1924–1938

Glass, M., and Dragunow, M. 1995. Neurochemical and morphological changes associated with human epilepsy. Brain Res Brain Res Rev 21(1):29–41

Gorter, J.A., van Vliet, E.A., Aronica, E., and Lopes da Silva, F.H. 2001. Progression of spontaneous seizures after status epilepticus is associated with mossy fibre sprouting and extensive bilateral loss of hilar parvalbumin and somatostatin-immunoreactive neurons. Eur J Neurosci 13(4):657–669

Gorter, J.A., Goncalves Pereira, P.M., van Vliet, E.A., Aronica, E., Lopes da Silva, F.H., and Lucassen, P.J. 2003. Neuronal cell death in a rat model for mesial temporal lobe epilepsy is induced by the initial status epilepticus and not by later repeated spontaneous seizures. Epilepsia 44(5):647–658

Grisar, T.M. 1986. Neuron-glia relationships in human and experimental epilepsy: a biochemical point of view. Adv Neurol 44:1045–1073

Hall, E.D., Kupina, N.C., and Althaus, J.S. 1999. Peroxynitrite scavengers for the acute treatment of traumatic brain injury. Ann N Y Acad Sci 890:462–468

Handforth, A., and Ackermann, R.F. 1995. Mapping of limbic seizure progressions utilizing the electrogenic status epilepticus model and the 14C-2-deoxyglucose method. Brain Res Brain Res Rev 20(1):1–23

Hansen, A., Jorgensen, O.S., Bolwig, T.G., and Barry, D.I. 1990. Hippocampal kindling alters the concentration of glial fibrillary acidic protein and other marker proteins in rat brain. Brain Res 531(1–2):307–311

Hinterkeuser, S., Schroder, W., Hager, G., Seifert, G., Blumcke, I., Elger, C.E., Schramm, J., and Steinhauser, C. 2000. Astrocytes in the hippocampus of patients with temporal lobe epilepsy display changes in potassium conductances. Eur J Neurosci 12(6):2087–2096

Holmes, G.L. 2004. Effects of early seizures on later behavior and epileptogenicity. Ment Retard Dev Disabil Res Rev 10(2):101–105

Huang, L., Cilio, M.R., Silveira, D.C., McCabe, B.K., Sogawa, Y., Stafstrom, C.E., and Holmes, G.L. 1999. Long-term effects of neonatal seizures: a behavioral, electrophysiological, and histological study. Brain Res Dev Brain Res 118(1–2):99–107

Hughes, P.E., Alexi, T., Walton, M., Williams, C.E., Dragunow, M., Clark, R.G., and Gluckman, P.D. 1999. Activity and injury-dependent expression of inducible transcription factors, growth factors and apoptosis-related genes within the central nervous system. Prog Neurobiol 57(4):421–450

Isokawa, M. 1998. Remodeling dendritic spines in the rat pilocarpine model of temporal lobe epilepsy. Neurosci Lett 258(2):73–76

Jiang, M., Lee, C.L., Smith, K.L., and Swann, J.W. 1998. Spine loss and other persistent alterations of hippocampal pyramidal cell dendrites in a model of early-onset epilepsy. J Neurosci 18(20):8356–8368

Jiang, W., Duong, T.M., and de Lanerolle, N.C. 1999. The neuropathology of hyperthermic seizures in the rat. Epilepsia 40(1):5–19

Khalilov, I., Dzhala, V., Medina, I., Leinekugel, X., Melyan, Z., Lamsa, K., Khazipov, R., and Ben-Ari, Y. 1999. Maturation of kainate-induced epileptiform activities in interconnected intact neonatal limbic structures in vitro. Eur J Neurosci 11(10):3468–3480

Khalilov, I., Holmes, G.L., and Ben-Ari, Y. 2003. In vitro formation of a secondary epileptogenic mirror focus by interhippocampal propagation of seizures. Nat Neurosci 6(10):1079–1085

Kimiwada, T., Juhasz, C., Makki, M., Muzik, O., Chugani, D.C., Asano, E., and Chugani, H.T. 2006. Hippocampal and thalamic diffusion abnormalities in children with temporal lobe epilepsy. Epilepsia 47(1):167–175

Kluck, R.M., Bossy-Wetzel, E., Green, D.R., and Newmeyer, D.D. 1997. The release of cytochrome c from mitochondria: a primary site for Bcl-2 regulation of apoptosis. Science 275(5303):1132–1136

Kohr, G., and Mody, I. 1994. Kindling increases N-methyl-D-aspartate potency at single N-methyl-D-aspartate channels in dentate gyrus granule cells. Neuroscience 62(4):975–981

Kotloski, R., Lynch, M., Lauersdorf, S., and Sutula, T. 2002. Repeated brief seizures induce progressive hippocampal neuron loss and memory deficits. Prog Brain Res 135:95–110

Kraus, J.E., Yeh, G.C., Bonhaus, D.W., Nadler, J.V., and McNamara, J.O. 1994. Kindling induces the long-lasting expression of a novel population of NMDA receptors in hippocampal region CA3. J Neurosci 14(7):4196–4205

Kudin, A.P., Kudina, T.A., Seyfried, J., Vielhaber, S., Beck, H., Elger, C.E., and Kunz, W.S. 2002. Seizure-dependent modulation of mitochondrial oxidative phosphorylation in rat hippocampus. Eur J Neurosci 15(7):1105–1114

Kunz, W.S., Kudin, A.P., Vielhaber, S., Blumcke, I., Zuschratter, W., Schramm, J., Beck, H., and Elger, C.E. 2000. Mitochondrial complex I deficiency in the epileptic focus of patients with temporal lobe epilepsy. Ann Neurol 48(5):766–773

Kwan, P., and Brodie, M.J. 2000. Early identification of refractory epilepsy. N Engl J Med 342(5):314–319

Lee, T.S., Eid, T., Mane, S., Kim, J.H., Spencer, D.D., Ottersen, O.P., and de Lanerolle, N.C. 2004. Aquaporin-4 is increased in the sclerotic hippocampus in human temporal lobeepilepsy. Acta Neuropathol (Berl) 108(6):493–502

Lemieux, L., Liu, R.S., and Duncan, J.S. 2000. Hippocampal and cerebellar volumetry in serially acquired MRI volume scans. Magn Reson Imaging 18(8):1027–1033

Lipton, S.A., Choi, Y.B., Sucher, N.J., Pan, Z.H., and Stamler, J.S. 1996. Redox state, NMDA receptors and NO-related species. Trends Pharmacol Sci 17(5):186–187; discussion 187–189

Liu, R.S., Lemieux, L., Bell, G.S., Bartlett, P.A., Sander, J.W., Sisodiya, S.M., Shorvon, S.D., and Duncan, J.S. 2001. A longitudinal quantitative MRI study of community-based patients with chronic epilepsy and newly diagnosed seizures: methodology and preliminary findings. Neuroimage 14(1 Pt 1):231–243

Lombardo, A.J., Kuzniecky, R., Powers, R.E., and Brown, G.B. 1996. Altered brain sodium channel transcript levels in human epilepsy. Brain Res Mol Brain Res 35(1–2):84–90

Lopes da Silva, F.H., Pijn, J.P., and Wadman, W.J. 1994. Dynamics of local neuronal networks: control parameters and state bifurcations in epileptogenesis. Prog Brain Res 102:359–370

Loup, F., Wieser, H.G., Yonekawa, Y., Aguzzi, A., and Fritschy, J.M. 2000. Selective alterations in GABAA receptor subtypes in human temporal lobe epilepsy. J Neurosci 20(14): 5401–5419

Lukasiuk, K., and Pitkanen, A. 2004. Large-scale analysis of gene expression in epilepsy research: is synthesis already possible? Neurochem Res 29(6):1169–1178

Lynch, M., Sayin, U., Bownds, J., Janumpalli, S., and Sutula, T. 2000. Long-term consequences of early postnatal seizures on hippocampal learning and plasticity. Eur J Neurosci 12(7):2252–2264

Mackenzie, L., Medvedev, A., Hiscock, J.J., Pope, K.J., and Willoughby, J.O. 2002. Picrotoxin-induced generalised convulsive seizure in rat: changes in regional distribution and frequency of the power of electroencephalogram rhythms. Clin Neurophysiol 113(4):586–596

Madsen, T.M., Treschow, A., Bengzon, J., Bolwig, T.G., Lindvall, O., and Tingstrom, A. 2000. Increased neurogenesis in a model of electroconvulsive therapy. Biol Psychiatry 47(12):1043–1049

Mathern, G.W., Pretorius, J.K., Mendoza, D., Leite, J.P., Chimelli, L., Born, D.E., Fried, I., Assirati, J.A., Ojemann, G.A., Adelson, P.D., Cahan, L.D., and Kornblum, H.I. 1999. Hippocampal N-methyl-D-aspartate receptor subunit mRNA levels in temporal lobe epilepsy patients. Ann Neurol 46(3):343–358

Matthews, P.M., Andermann, F., and Arnold, D.L. 1990. A proton magnetic resonance spectroscopy study of focal epilepsy in humans. Neurology 40(6):985–989

McCabe, B.K., Silveira, D.C., Cilio, M.R., Cha, B.H., Liu, X., Sogawa, Y., and Holmes, G.L. 2001. Reduced neurogenesis after neonatal seizures. J Neurosci 21(6):2094–2103

McIntyre, D.C., Hutcheon, B., Schwabe, K., and Poulter, M.O. 2002. Divergent GABA(A) receptor-mediated synaptic transmission in genetically seizure-prone and seizure-resistant rats. J Neurosci 22(22):9922–9931

Meierkord, H., Wieshmann, U., Niehaus, L., and Lehmann, R. 1997. Structural consequences of status epilepticus demonstrated with serial magnetic resonance imaging. Acta Neurol Scand 96(3):127–132

Meldrum, B.S. 1983. Metabolic factors during prolonged seizures and their relation to nerve cell death. Adv Neurol 34:261–275

Meldrum, B.S. 2002. Implications for neuroprotective treatments. Prog Brain Res 135:487–495

Mellstrom, B., and Naranjo, J.R. 2001. Mechanisms of Ca^{2+}-dependent transcription. Curr Opin Neurobiol 11(3):312–319

Meldrum, B.S., Vigouroux, R.A., and Brierley, J.B. 1973. Systemic factors and epileptic brain damage. Prolonged seizures in paralyzed, artificially ventilated baboons. Arch Neurol 29(2):82–87

Mody, I., and Heinemann, U. 1987. NMDA receptors of dentate gyrus granule cells participate in synaptic transmission following kindling. Nature 326(6114):701–704

Monyer, H., Burnashev, N., Laurie, D.J., Sakmann, B., and Seeburg, P.H. 1994. Developmental and regional expression in the rat brain and functional properties of four NMDA receptors. Neuron 12(3):529–540

Morimoto, K., Fahnestock, M., and Racine, R.J. 2004. Kindling and status epilepticus models of epilepsy: rewiring the brain. Prog Neurobiol 73(1):1–60

Moser, M.B., Trommald, M., and Andersen, P. 1994. An increase in dendritic spine density on hippocampal CA1 pyramidal cells following spatial learning in adult rats suggests the formation of new synapses. Proc Natl Acad Sci USA 91(26):12673–12675

Mueller, S.G., Kollias, S.S., Trabesinger, A.H., Buck, A., Boesiger, P., and Wieser, H.G. 2001. Proton magnetic resonance spectroscopy characteristics of a focal cortical dysgenesis during status epilepticus and in the interictal state. Seizure 10(7):518–524

Nagelhus, E.A., Mathiisen, T.M., and Ottersen, O.P. 2004. Aquaporin-4 in the central nervous system: cellular and subcellular distribution and coexpression with KIR4.1. Neuroscience 129(4):905–913

Nakagawa, E., Aimi, Y., Yasuhara, O., Tooyama, I., Shimada, M., McGeer, P.L., and Kimura, H. 2000. Enhancement of progenitor cell division in the dentate gyrus triggered by initial limbic seizures in rat models of epilepsy. Epilepsia 41(1):10–18

Natsume, J., Bernasconi, N., Andermann, F., and Bernasconi, A. 2003. MRI volumetry of the thalamus in temporal, extratemporal, and idiopathic generalized epilepsy. Neurology 60(8):1296–1300

Nissinen, J., Halonen, T., Koivisto, E., and Pitkanen, A. 2000. A new model of chronic temporal lobe epilepsy induced by electrical stimulation of the amygdala in rat. Epilepsy Res 38(2–3):177–205

Nusser, Z., Hajos, N., Somogyi, P., and Mody, I. 1998. Increased number of synaptic GABA(A) receptors underlies potentiation at hippocampal inhibitory synapses. Nature 395(6698): 172–177

Olney, J.W., and de Gubareff, T. 1978. Glutamate neurotoxicity and Huntington's chorea. Nature 271(5645):557–559

Olney, J.W., and Sharpe, L.G. 1969. Brain lesions in an infant rhesus monkey treated with monsodium glutamate. Science 166(903):386–388

Olsen, M.L., and Sontheimer, H. 2004. Mislocalization of Kir channels in malignant glia. Glia 46(1):63–73

Pal, S., Limbrick, D.D., Jr., Rafiq, A., and DeLorenzo, R.J. 2000. Induction of spontaneous recurrent epileptiform discharges causes long-term changes in intracellular calcium homeostatic mechanisms. Cell Calcium 28(3):181–193

Pal, S., Sun, D., Limbrick, D., Rafiq, A., and DeLorenzo, R.J. 2001. Epileptogenesis induces long-term alterations in intracellular calcium release and sequestration mechanisms in the hippocampal neuronal culture model of epilepsy. Cell Calcium 30(4):285–296

Parent, J.M., Janumpalli, S., McNamara, J.O., and Lowenstein, D.H. 1998. Increased dentate granule cell neurogenesis following amygdala kindling in the adult rat. Neurosci Lett 247(1):9–12

Perez, E.R., Maeder, P., Villemure, K.M., Vischer, V.C., Villemure, J.G., and Deonna, T. 2000. Acquired hippocampal damage after temporal lobe seizures in 2 infants. Ann Neurol 48(3):384–387

Petroff, O.A., Errante, L.D., Rothman, D.L., Kim, J.H., and Spencer, D.D. 2002. Glutamate-glutamine cycling in the epileptic human hippocampus. Epilepsia 43(7):703–710

Pickard, L., Noel, J., Henley, J.M., Collingridge, G.L., and Molnar, E. 2000. Developmental changes in synaptic AMPA and NMDA receptor distribution and AMPA receptor subunit composition in living hippocampal neurons. J Neurosci 20(21):7922–7931

Qiao, X., and Noebels, J.L. 1993. Developmental analysis of hippocampal mossy fiber outgrowth in a mutant mouse with inherited spike-wave seizures. J Neurosci 13(11):4622–4635

Ratzliff, A.H., Santhakumar, V., Howard, A., and Soltesz, I. 2002. Mossy cells in epilepsy: rigor mortis or vigor mortis? Trends Neurosci 25(3):140–144

Raza, M., Blair, R.E., Sombati, S., Carter, D.S., Deshpande, L.S., and DeLorenzo, R.J. 2004. Evidence that injury-induced changes in hippocampal neuronal calcium dynamics during epileptogenesis cause acquired epilepsy. Proc Natl Acad Sci USA 101(50): 17522–17527

Rugg-Gunn, F.J., Eriksson, S.H., Symms, M.R., Barker, G.J., Thom, M., Harkness, W., and Duncan, J.S. 2002. Diffusion tensor imaging in refractory epilepsy. Lancet 359(9319):1748–1751

Salmenpera, T., Kalviainen, R., Partanen, K., Mervaala, E., and Pitkanen, A. 2000. MRI volumetry of the hippocampus, amygdala, entorhinal cortex, and perirhinal cortex after status epilepticus. Epilepsy Res 40(2–3):155–170

Samuelsson, C., Kumlien, E., Flink, R., Lindholm, D., and Ronne-Engstrom, E. 2000. Decreased cortical levels of astrocytic glutamate transport protein GLT-1 in a rat model of posttraumatic epilepsy. Neurosci Lett 289(3):185–188

Sanchez-Carpintero, R., and Neville, B.G. 2003. Attentional ability in children with epilepsy. Epilepsia 44(10):1340–1349

Sankar, R., Shin, D.H., Liu, H., Mazarati, A., Pereira de Vasconcelos, A., and Wasterlain, C.G. 1998. Patterns of status epilepticus-induced neuronal injury during development and long-term consequences. J Neurosci 18(20):8382–8393

Sankar, R., Shin, D., Liu, H., Wasterlain, C., and Mazarati, A. 2002. Epileptogenesis during development: injury, circuit recruitment, and plasticity. Epilepsia 43(Suppl 5):47–53

Sarkisian, M.R. 2001. Overview of the current animal models for human seizure and epileptic disorders. Epilepsy Behav 2(3):201–216

Schauwecker, P.E. 2002. Complications associated with genetic background effects in models of experimental epilepsy. Prog Brain Res 135:139–148

Schwartzkroin, P.A., Baraban, S.C., and Hochman, D.W. 1998. Osmolarity, ionic flux, and changes in brain excitability. Epilepsy Res 32(1–2):275–285

Shewmon, D.A., and Erwin, R.J. 1988. The effect of focal interictal spikes on perception and reaction time. I. General considerations. Electroencephalogr Clin Neurophysiol 69(4): 319–337

Sloviter, R.S. 1983. "Epileptic" brain damage in rats induced by sustained electrical stimulation of the perforant path. I. Acute electrophysiological and light microscopic studies. Brain Res Bull 10(5):675–697

Sloviter, R.S., Dean, E., Sollas, A.L., and Goodman, J.H. 1996. Apoptosis and necrosis induced in different hippocampal neuron populations by repetitive perforant path stimulation in the rat. J Comp Neurol 366(3):516–533

Stanfield, B.B. 1989. Excessive intra- and supragranular mossy fibers in the dentate gyrus of tottering (tg/tg) mice. Brain Res 480(1–2):294–299

Stanley, J.A., Cendes, F., Dubeau, F., Andermann, F., and Arnold, D.L. 1998. Proton magnetic resonance spectroscopic imaging in patients with extratemporal epilepsy. Epilepsia 39(3):267–273

Steinhauser, C., and Seifert, G. 2002. Glial membrane channels and receptors in epilepsy: impact for generation and spread of seizure activity. Eur J Pharmacol 447(2–3):227–237

Straub, H., Kohling, R., Frieler, A., Grigat, M., and Speckmann, E.J. 2000. Contribution of L-type calcium channels to epileptiform activity in hippocampal and neocortical slices of guinea-pigs. Neuroscience 95(1):63–72

Suh, M., Bahar, S., Mehta, A.D., and Schwartz, T.H. 2005. Temporal dependence in uncoupling of blood volume and oxygenation during interictal epileptiform events in rat neocortex. J Neurosci 25(1):68–77

Suh, M., Bahar, S., Mehta, A.D., and Schwartz, T.H. 2006a. Blood volume and hemoglobin oxygenation response following electrical stimulation of human cortex. Neuroimage 31(1):66–75

Suh, M., Ma, H., Zhao, M., Sharif, S., and Schwartz, T.H. 2006b. Neurovascular coupling and oximetry during epileptic events. Mol Neurobiol 33(3):181–197

Sutula, T., Lauersdorf, S., Lynch, M., Jurgella, C., and Woodard, A. 1995. Deficits in radial arm maze performance in kindled rats: evidence for long-lasting memory dysfunction induced by repeated brief seizures. J Neurosci 15(12):8295–8301

Swann, J.W., Al-Noori, S., Jiang, M., and Lee, C.L. 2000. Spine loss and other dendritic abnormalities in epilepsy. Hippocampus 10(5):617–625

Taylor, E.R., Hurrell, F., Shannon, R.J., Lin, T.K., Hirst, J., and Murphy, M.P. 2003. Reversible glutathionylation of complex I increases mitochondrial superoxide formation. J Biol Chem 278(22):19603–19610

Tian, G.F., Azmi, H., Takano, T., Xu, Q., Peng, W., Lin, J., Oberheim, N., Lou, N., Wang, X., Zielke, H.R., Kang, J., and Nedergaard, M. 2005. An astrocytic basis of epilepsy. Nat Med 11(9):973–981

Tien, R.D., and Felsberg, G.J. 1995. The hippocampus in status epilepticus: demonstration of signal intensity and morphologic changes with sequential fast spin-echo MR imaging. Radiology 194(1):249–256

Torre, E.R., Lothman, E., and Steward, O. 1993. Glial response to neuronal activity: GFAP-mRNA and protein levels are transiently increased in the hippocampus after seizures. Brain Res 631(2):256–264

Ueda, Y., Yokoyama, H., Nakajima, A., Tokumaru, J., Doi, T., and Mitsuyama, Y. 2002. Glutamate excess and free radical formation during and following kainic acid-induced status epilepticus. Exp Brain Res 147(2):219–226

Volterra, A., and Meldolesi, J. 2005. Astrocytes, from brain glue to communication elements: the revolution continues. Nat Rev Neurosci 6(8):626–640

Vreugdenhil, M., Faas, G.C., and Wadman, W.J. 1998. Sodium currents in isolated rat CA1 neurons after kindling epileptogenesis. Neuroscience 86(1):99–107

Wang, Y., Majors, A., Najm, I., Xue, M., Comair, Y., Modic, M., and Ng, T.C. 1996. Postictal alteration of sodium content and apparent diffusion coefficient in epileptic rat brain induced by kainic acid. Epilepsia 37(10):1000–1006

Wilson, C.L., Maidment, N.T., Shomer, M.H., Behnke, E.J., Ackerson, L., Fried, I., and Engel, J., Jr. 1996. Comparison of seizure related amino acid release in human epileptic hippocampus versus a chronic, kainate rat model of hippocampal epilepsy. Epilepsy Res 26(1):245–254

Wong, M. 2005. Modulation of dendritic spines in epilepsy: cellular mechanisms and functional implications. Epilepsy Behav 7(4):569–577

Ye, Z.C., Rothstein, J.D., and Sontheimer, H. 1999. Compromised glutamate transport in human glioma cells: reduction-mislocalization of sodium-dependent glutamate transporters and enhanced activity of cystine-glutamate exchange. J Neurosci 19(24):10767–10777

Yoo, S.Y., Chang, K.H., Song, I.C., Han, M.H., Kwon, B.J., Lee, S.H., Yu, I.K., and Chun, C.K. 2002. Apparent diffusion coefficient value of the hippocampus in patients with hippocampal sclerosis and in healthy volunteers. AJNR Am J Neuroradiol 23(5):809–812

Yung, A.W., Park, Y.D., Cohen, M.J., and Garrison, T.N. 2000. Cognitive and behavioral problems in children with centrotemporal spikes. Pediatr Neurol 23(5):391–395

Zhang, G., Raol, Y.H., Hsu, F.C., Coulter, D.A., and Brooks-Kayal, A.R. 2004. Effects of status epilepticus on hippocampal GABAA receptors are age-dependent. Neuroscience 125(2):299–303

Zhao, M., Ma, H., Suh, M., and Schwartz, T.H. 2005. Decrease in brain tissue oxygenation in spite of an increase in cerebral blood flow during acute focal 4-aminopyridine seizures in rat neocortex, Abstract for Society of Neuroscience Annual Conference, November 12–16

Zipfel, G.J., Babcock, D.J., Lee, J.M., and Choi, D.W. 2000. Neuronal apoptosis after CNS injury: the roles of glutamate and calcium. J Neurotrauma 17(10):857–869

Chapter 8
Metabolic Encephalopathies in Children

Joseph DiCarlo

Metabolic Encephalopathies in Children

Encephalopathies in children arise from an array of sources. Intoxication from accidental ingestion can produce a profound encephalopathy that clears almost as quickly as it appears. The encephalopathy of septic shock is under-appreciated, yet the brain should be counted among organs shut down, almost always temporarily, in children with multiple organ failure. The child's brain is fairly resilient in the face of many of these entities, except rarely in the face of meningitis, but even here severe long-term damage seems to result only from the rare misdiagnosed or under-treated case.

However, encephalopathies with a metabolic basis tend to be the most problematic for infants or children, with functional outcomes dependent upon timely and prudent interventions. Three varieties of metabolic encephalopathy in children are discussed here. The first two are closely related. Inborn (genetic) errors of metabolism can present in the newborn as severe encephalopathy from hyperammonemia alone. When a metabolic error presents months to years later, a degree of hepatic insufficiency may complicate the metabolic derangement. In acute or fulminant hepatic failure of any etiology (i.e., infections, drug-induced, toxin-related), the rise in serum ammonia may be only moderate but other factors contribute to the ensuing encephalopathy, which may be devastating within days.

The third variety, the severe encephalopathy produced by diabetic ketoacidosis (DKA), is entirely another matter. The cerebral edema associated with DKA was commonly thought of as being the result of osmotic shifts during therapy to restore hydration; there is now evidence that edema is the endpoint of a vasogenic phenomenon – i.e., it might be due to capillary leak and/or membrane disruption, the result of profoundly deficient circulation, and it is present before rehydration takes place. The severe case of DKA – associated cerebral edema is likely to be accompanied by disruptions in other organ systems as well, including gastrointestinal perforations and necrosis. The therapeutic challenge lies in balancing the restoration of perfusion and substrate delivery, while limiting the accumulation of fluid.

D.W. McCandless (ed.) *Metabolic Encephalopathy*,
doi: 10.1007/978-0-387-79112-8_8, © Springer Science + Business Media, LLC 2009

Inborn Errors of Metabolism

In neonates, serious inborn errors of metabolism may present as an overwhelming illness characterized most dramatically by encephalopathy, while direct involvement of the liver is modest. An infant with branched-chain organic aciduria presents with ketosis or ketoacidosis, and hyperammonemia. The most common abnormal organic acidurias are maple syrup urine disease, isovaleric acidemia, propionic aciduria and methylmalonic aciduria. Specific diagnoses are made by detecting acylcarnitine and other organic acid compounds in plasma and urine by gas chromatography mass spectrometry (MS) or tandem MS-MS (Ogier de Baulny and Saudubray, 2002).

Urea cycle defects usually present with hyperammonemia alone. The most common hereditary urea cycle disorder, ornithine transcarbamoylase (OTC) deficiency, is an X-linked recessive disorder. The gene responsible for the enzyme is located on Xp21.1, and is expressed in the liver and gut. Males presenting in the neonatal period carry the most severe version of the disease (Gordon, 2003). Presentations outside the newborn period consist of intermittent vomiting and encephalopathy developing over days to weeks. By this time laboratory examination might reveal hyperammonemia and all the parameters that would indicate hepatic failure, including elevated transaminases, coagulopathy and decreased blood urea nitrogen (Mustafa and Clarke, 2006). Certain defects might not present until adulthood. In heterozygous adults, OTC deficiency can cause seizures and hyperammonemic coma especially after a large protein load (Legras et al., 2002).

Valproate: Induced Hyperammonemic Encephalopathy

Valproic acid, a common anticonvulsant, can accumulate and induce a hyperammonemic encephalopathy, presenting as acutely impaired consciousness, focal neurologic symptoms, and increased seizure frequency. Moreover, valproate-induced hyperammonemic encephalopathy can occur more readily in the child (or adult) with carnitine deficiency or with congenital urea cycle enzymatic defects. In fact, the occult presence of a urea cycle enzymatic defect is occasionally uncovered by the development of valproate-induced encephalopathy; an occult enzyme deficiency can even make its first presentation, in the absence of valproate, in the generically stressed critical care patient, young or old (Verrotti et al., 2002; Thakur et al., 2006; Summar et al., 2005).

Valproate may contribute to hyperammonemia by inhibiting carbamoylphosphate synthetase-I, the enzyme that begins the urea cycle. Phenobarbital may potentiate the toxic effect of valproate. The electroencephalogram in severe hyperammonemic encephalopathy exhibits continuous generalized slowing, a predominance of theta and delta waves, bursts of frontal intermittent rhythmic delta activity, and triphasic waves (Segura-Bruna et al., 2006).

Hepatic Encephalopathy

Previous theories, centered on induced increases in gamma-aminobutyric acid (GABA) 'tone' and the generation of benzodiazepine-like substances, have less traction presently. As ammonia is a central component in the development of hepatic encephalopathy, the mechanism is similar to that found in encephalopathy due to inborn errors, with a possible osmotic derangement in astrocytes or changes in cellular metabolism and alterations in cerebral blood flow. Cerebral edema in hepatic encephalopathy may be due to an increase in cerebral blood volume and cerebral blood flow, in part due to inflammation, to glutamine and to toxic products of a diseased liver (Mattarozzi et al., 2005). In a prospective series of 121 adults with cirrhosis, ammonia levels correlated with the severity of hepatic encephalopathy. Venous sampling was as reliable as arterial for ammonia measurement (Ong et al., 2003). Mild hypothermia has been shown to counteract some of these changes in the experimental setting; clinical trials of mild cooling in hepatic encephalopathy may be on the horizon (Mattarozzi et al., 2005).

Astrocytes and Ammonia

Swollen astrocytes constitute a major component of the brain edema resulting from hyperammonemia or fulminant hepatic failure. Intracerebral ammonia is detoxified in the cytoplasm of astrocytes, where glutamine synthetase catalyzes the conversion of glutamate and ammonia to glutamine. The elevated glutamine may increase intracellular osmolarity, promoting an influx of water with resultant astrocytic swelling (Bachmann, 2002). An alternative explanation posits that as it accumulates in mitochondria, the glutamine is metabolized by phosphate-activated glutaminase back into glutamate and ammonia. In the mitochondria the glutamine-derived ammonia provokes the excessive production of free radicals and the induction of the mitochondrial 'membrane permeability transition', rendering the astrocytes swollen and dysfunctional (Albrecht and Norenberg, 2006). The mitochondrial membrane permeability transition is a calcium ion-dependent increase in mitochondrial membrane permeability that occurs after the opening of a voltage-dependent anion channel, the permeability transition pore (Jones, 2002). High concentrations of calcium ions, along with oxidative stress, cause the transition pore to open, flooding and uncoupling the mitochondrion. If the pore remains open, ATP is depleted and cell necrosis ensues (Halestrap, 2006).

Therapy for Hyperammonemic and Hepatic Encephalopathy

First line therapy for hyperammonemia is directed at reducing the production and absorption of gut-derived ammonia. Non-absorbable antibiotics are better than non-absorbable disaccharides (e.g., lactulose) in reducing the risk of no improvement

(1.24, 1.02–1.50, 10 trials) and modestly better at lowering the blood ammonia concentration (weighted mean difference 2.35 μmol L^{-1}, 0.06–13.45μmol L^{-1}, 10 trials) (Als-Nielsen et al., 2004). The ideal oral antibiotic would have a wide spectrum of antibacterial activity against aerobic and anaerobic Gram-negative and Gram-positive bacteria, and a very low rate of systemic absorption (Festi et al., 2006).

In an infant with hyperammonemia and a suspected inborn metabolism error it is now common to start treatment empirically with a metabolic cocktail of intravenous sodium phenylacetate (NaPh) and sodium benzoate (NaBz). Sodium benzoate provides an alternative pathway for the disposal of waste nitrogen, interacting with glycine to form hippurate. The subsequent renal excretion of hippurate results in the elimination of ammonia ions. If hyperammonemia is severe, hemodialysis is initiated. In some centers hemodialysis will be repeated once or twice daily in the acute phase; in others the initial hemodialysis is followed by continuous veno-venous hemofiltration. In one case study involving empiric treatment with NaPh and NaBz, the dialytic and convective (hemofiltration) clearance of ammonia, NaPh, and NaBz was measured, first by hemodialysis and then by hemofiltration. The clearance of ammonia was 57 mL min^{-1} by hemodialysis and 37 mL min^{-1} by hemofiltration. Clearance of the therapeutic agents, NaBz and NaPh, was about 37 mL min^{-1} by hemodialysis and 12 mL min^{-1} by hemofiltration. Despite fairly efficient clearance of both NaPh and NaBz by hemofiltration or dialysis, the hyperammonemia was corrected (Legras et al., 2002). Sodium benzoate may be used on a short or long-term basis to stabilize or improve encephalopathy in liver failure of any etiology (Mattarozzi et al., 2005).

Plasma Filtration or Removal

Hemodialysis and/or hemofiltration should be employed proactively in the child with evolving hyperammonemic encephalopathy. In a series of eighteen children undergoing dialysis/filtration for hyperammonemia due to urea cycle defects, organic acidemias, and Reye syndrome, initial therapy with hemodialysis was associated with improved survival. About half of the cases were transitioned to continuous hemofiltration for an average of six days to maintain metabolic control. Delays of more than 24 h from diagnosis to initiation of therapy were associated with an increased risk of mortality (Ogier de Baulny, 2002).

As the etiology in most cases of hepatic failure is irreversible short of transplantation and the target for removal or modulation is more complex than simply ammonia alone, continuous hemofiltration is more commonly chosen than hemodialysis for the child who needs extracorporeal support. Hemofiltration is effective at direct removal of water and unbound small and middle molecular weight solutes. Because it involves the exchange of large amounts of extracellular water, hemofiltration may also result in the movement, metabolism or removal of larger molecules as well (Di Carlo and Alexander, 2005). However, this latter effect would take time to accomplish. By itself, plasmapheresis (plasma exchange) is effective in normalizing coagulation parameters and controlling volume status in small children. However it may have no effect on the neurologic complications of liver failure nor does it impact the ability of the liver to regenerate. But the addition of sequential or simultaneous plasmapheresis therapy to a regimen of hemofiltration is helpful in some cases of fulminant hepatic failure, thrombotic thrombocytopenic purpura, and the hemolytic uremic syndrome (Yorgin et al., 2000). Plasmapheresis, or more accurately in the case of liver 'replacement' therapy, plasma exchange, removes larger intravascular components, including activated factors involved in the coagulation cascade. The multi-organ instability induced by fulminant hepatic failure might be more easily managed with this dual approach; such therapeutic complexity is usually unnecessary in the encephalopathy induced by hyperammonemia alone (Biancofiore et al., 2003).

Charcoal Hemoperfusion

Charcoal hemoperfusion can also remove the toxin in hepatic failure. A model of galactosamine-induced fulminant hepatic failure in a rat with grade 3 hepatic encephalopathy, demonstrated the potential utility of multisorbent plasma perfusion over uncoated spherical charcoal, and an endotoxin removing adsorbent (polymyxin B-sepharose). Timing, duration and frequency of treatment impacted liver cell proliferative response as compared to untreated fulminant hepatic failure paired controls (Ryan et al., 2001).

MARS

By dialyzing across a membrane with albumin added to the dialysis solution, one might be able to attract toxins otherwise tightly bound to proteins in the serum and thus unavailable to conventional dialysis or hemofiltration. However, once the binding sites are occupied on the albumin in the dialysis solution, no further capture can occur. A continuous supply of fresh albumin would be prohibitively expensive. However a method was devised to refresh the albumin in solution by passing the dialysate through a sorbent column, removing the toxin from albumin binding sites. The on-line real-time version of this device has been termed the Molecular Absorbing Recirculating System, or MARS (Stange et al., 2002). The MARS technology can remove both water-soluble and albumin-bound toxins, and can provide renal support in case of renal failure.

Hepatocyte Columns

The 'non-biological' systems discussed above depend on nonspecific mechanisms for detoxification, i.e., detoxification by removal rather than metabolism. Intriguing work is in progress with 'biological' systems that can provide detoxification, biotransformation and biosynthetic functions. Biological systems generally employ an extracorporeal circuit that allows blood to come in contact with hepatocytes. Thus far there is no effective highly differentiated human hepatocyte line available for clinical use. Clinically tested systems presently use either porcine hepatocytes or human hepatoma cell lines (Demetriou, 2005).

Hepatectomy

In fulminant hepatic failure the necrosing liver not only defaults on its synthetic and filtrative functions, but also provokes an intense inflammatory reaction, triggering the cytokine cascade and its consequences. In the most dramatic cases it may be prudent to remove the liver in its entirety even if a donor organ is not yet available. In this situation transplant hepatectomy with portocaval shunting begins the bridge to transplantation; while awaiting procurement, the critical care physician must then attempt to replace the hepatic synthetic function (with plasma infusion or exchange) and filtrative function (with hemofiltration, dialysis or hemadsorption). A surprising degree of stability can be achieved in the first day or two following total hepatectomy. By the third day an evolving metabolic acidosis heralds impending instability (Hammer et al., 1996).

Common Strategies in Hepatic Encephalopathy

MARS has been slow to capture the imagination of critical care physicians, at least in pediatrics. The majority of critical care units employ either hemofiltration or plasma exchange when supporting the child with a failed liver. There is a widely varying degree of comfort with hemofiltration or plasma exchange devices as well, particularly with regard to the safety of active anticoagulation in the face of severe coagulopathy. For hemofiltration, the past decade has witnessed the gradual conversion from heparin- based systemic anticoagulation (or occasionally, no anticoagulant) to citrate-based 'regional' or 'circuit' anticoagulation. Citrate infused into the circuit binds ionized calcium, rendering the coagulation cascade ineffective. Calcium is simultaneously infused remotely in the patient, normalizing peripheral blood ionized calcium, with no net effect on system coagulation parameters.

Probably more important than the choice of technology is the timing of its application and to some extent the 'prescription' applied (i.e., the clearance rate in hemofiltration or MARS and the frequency and volume in the case of plasma exchange). Timing cannot be over-emphasized, but it is rarely described. We will institute hemofiltration as the child with acute hepatic failure is actively evolving from West Haven Criteria grade 2 encephalopathy (drowsiness, lethargy, obvious personality changes, inappropriate behavior, and intermittent disorientation, usually regarding time) to grade 3 (somnolent but can be aroused, unable to perform mental tasks, disorientation to time and place, marked confusion, amnesia, occasional fits of rage, incomprehensible speech). Employing this strategy, one might reasonably expect to be able to support the child for up to two to three weeks if necessary while waiting for organ procurement. On occasion, the native liver might even recover within this window, particularly if the etiology of hepatic failure was hepatitis A or a drug or toxin.

Diabetic Ketoacidosis

Clinically apparent cerebral edema occurs in less than 1% of all episodes of diabetic ketoacidosis, (Dunger et al., 2004) but its subclinical presence is far more common. In a series of children with DKA, 22/41 had narrowing of the cerebral ventricles as measured by magnetic resonance imaging(MRI) (Glaser et al., 2006). Initial pCO2 correlates well with the presence of cerebral edema (relative risk of cerebral edema for each decrease of 7.8 mm Hg, 3.4; 95% confidence interval, 1.9–6.3); initial serum urea nitrogen concentrations correlate as well (relative risk of cerebral edema for each increase of 9 mg dL^{-1} [3.2 mmol L^{-1}], 1.7; 95% confidence interval, 1.2–2.5) (Glaser et al., 2001).

Subtle findings support the association of hypoperfusion with DKA encephalopathy. N-acetylaspartate (NAA) is a neuronal-axonal marker that can be used as a dynamic marker of neuronal dysfunction and integrity. The ratio of NAA to creatine in the basal ganglia decreases during acute DKA, suggesting that neuronal integrity is compromised. Children at highest risk for cerebral edema during DKA are those with greater dehydration and greater hypocapnia at presentation (Jones, 1979; Muir et al., 2001; Glaser, 2001). Volume depletion may lead to hypoperfusion and brain ischemia, especially as the hyperventilation that accompanies DKA can induce cerebral vasoconstriction. This can occur particularly within more vulnerable areas like the basal ganglia. Proton magnetic resonance spectroscopy has detected lactate peaks within the basal ganglia, suggestive of anaerobic cerebral metabolism (Wooton-Gorges et al., 2005).

Fourteen children were studied using both diffusion and perfusion weighted MRI, during DKA treatment and after recovery. The apparent diffusion coefficients (ADCs) were elevated during DKA treatment in all regions except the occipital gray matter, suggesting an increase in water diffusion. Perfusion MRI during DKA treatment demonstrated shorter mean transit times and higher peak tracer concentrations, indicative of increased cerebral blood flow (CBF). Elevated ADC values during DKA treatment suggest that a vasogenic process, rather than osmotic cellular swelling, is responsible for edema formation (Glaser et al., 2004). Transcranial Doppler evaluations of middle cerebral artery flow velocities and cerebral autoregulation demonstrated normal to increased cerebral blood flow, elevated regional cerebral oxygenation, impaired autoregulation, and changes in brain volume in diabetic ketoacidosis. This suggests that a transient loss of cerebral autoregulation might allow a paradoxical increase in CBF and the development of vasogenic cerebral edema (Roberts et al., 2006).

There is something unsettling about the vasogenic theories, however. If hypoper fusion and then reperfusion were the sole culprits, one would expect a similar incidence of cerebral edema in profound septic shock. Yet, but for the subpopulation with concomitant meningitis, the brain is free of edema in sepsis, except perhaps in toxic shock syndrome (Churchwell et al., 1995; Smith and Gulinson, 1988). Even in the most dramatic and persistent of cytokine-driven illnesses, the hemophagocytic syndrome, the cerebral cortex is spared until the disease runs its course; at that point there is evidence of a perivascular, not cytotoxic, event. Early on, only the meninges are involved, with infiltration of lymphocytes and macrophages. Edema appears later as a perivascular infiltrate. Only in advanced disease is there diffuse infiltration into the tissue (Henter and Nennesmo, 1997).

Conversely, cerebral edema in DKA is always distributed globally. There is understandable reluctance to abandon the osmotic theories for the genesis DKA – related cerebral edema. The hypertonicity in untreated DKA and the rapidly changing osmotic gradient in resolving DKA may indeed contribute something to fluid movement between neuron and interstitium; and in the treatment phase the rate of change of this gradient is nearly impossible to control.

To complicate matters further, in childhood diabetes, in recent years, type 2 diabetes, characterized by insulin-resistance instead of insulin absence, has become more prevalent among children. Just one decade ago, it would not have been included in the differential diagnosis of pediatric encephalopathy. Recognition of type 2 diabetes in children, and its potential lethality, is very recent. A report from Florida in 2004 described a cluster of seven obese African American youth who were initially thought to have died from DKA due to type 1 diabetes, despite meeting the criteria for the hyperglycemic hyperosmolar state characteristic of type 2 diabetes and not for DKA. All had previously unrecognized type 2 diabetes (Morales and Rosenbloom, 2004).

As obesity and type 2 diabetes in childhood grow in prevalence, such related complications may also increase. Diagnostic criteria for the hyperglycemic hyperosmolar non-ketotic syndrome include blood glucose above 600 mg dL^{-1} and serum osmolality above 330 mOsm L^{-1} with only mild acidosis (serum bicarbonate 15–20 mmol L^{-1} and mild ketonuria 15 mg dL^{-1} or less). In a subsequent series it seems that the diagnosis was entertained early: the Glasgow Coma Scale at presentation was 13 (range 9–15); mean body mass index (BMI) at presentation exceeded the 97th percentile; mean serum osmolality was 393 mOsm L^{-1}; and mean blood glucose was an unavoidable 1,604 mg dL^{-1}. One of these children died, from multisystem organ failure (Fourtner et al., 2005).

In another series with hyperglycemic hyperosmolar nonketotic (HHNK) syndrome, eight children (seven male and one female, all African-American) from 11 to 17 years of age each had a body mass index exceeding the 97th percentile. Presentation ranged from confusion to coma. Serum bicarbonate was less than 14 mmol L^{-1}. Corrected sodium in all patients was in the hypernatremic range in conjunction with high effective serum osmolality. Metabolic control was achieved in all patients within 36 h of admission. All but one recovered with fluid replacement and intravenous insulin therapy. One child succumbed to massive pulmonary embolism Bhowmick et al., 2005).

Cerebral edema is not a prominent issue in complicated insulin-resistant diabetes in adults, and perhaps not in children as well. In fact, while the experience with pediatric type 2 diabetes is still modest, it appears that the pitfalls of therapy are similar to those encountered in the treatment of the adult, where the consequences of undertreatment include vascular occlusion, rhabdomyolysis and multiple organ failure (Magee and Bhatt, 2001). Hypercoagulability has been described in type 1 DKA, although its consequences have been relatively minor (e.g., deep venous thromboses); (Davis et al., 2007) undertreatment (in terms of fluid repletion) is otherwise thought of as advantageous in DKA. It is tempting to suggest the routine use of gentle anticoagulation if central venous catheters are employed in the stabilization of DKA, and in all cases of pediatric HHNK with or without invasive catheters.

How then should one support the child with hyperglycemic hyperosmolar non-ketotic syndrome? In adults with HHNK, computed tomography reveals an increase in brain tissue density, suggesting that the brain is dehydrated, not edematous (Azzopardi et al., 2002). Serum glucose concentrations in HHNK can be as high as in the most extreme case of DKA, but of course ketonemia is minimal or absent. Ketones have been shown to exert osmotic pressure; (Puliyel and Bhambhani, 2003) could it be the presence (in DKA) or absence (in HHNK) of ketone bodies that distinguishes the two milieu in terms of risk for cerebral edema? Or, perhaps another easily measured variable, the serum (and by inference the tissue) pH, is most important? A large surveillance study of DKA in the United Kingdom determined that in the absence of cerebral edema, the level of consciousness is directly related to the serum pH, with confusion/agitation associated with a pH of 6.96 ± 0.11 and coma ensuing at pH 6.88 ± 0.09 (Edge et al., 2006a, b). In children with DKA and cerebral edema (n = 43), acidosis was more severe, and serum potassium and blood urea nitrogen were higher, than in those with DKA without cerebral edema (Edge et al., 2006a, b). Neither serum glucose concentration nor serum osmolality were associated with degree of consciousness or presence of edema.

In fact, the evidence may be mounting that it is the degree of dehydration, and therefore the severity of malperfusion, that predisposes to eventual cerebral edema. It may be that if dehydration in type 2 diabetes is pushed far enough, one might encounter cerebral edema in that population as well. The simple answer might be sought in a regression analysis that includes body mass index, degree of dehydration (in terms of kg lost as a percentage of body weight), ketosis and pH.

There are only serendipitous reports of hemofiltration used in the patient with DKA, and in all cases the indication was renal failure, not encephalopathy (Kawata et al., 2006). In the child with intact renal function during DKA, urine output often remains excessive during the first day of therapy until serum glucose concentrations come under control. With care, one can control the rate of change in total body water, though its distribution will remain elusive. Would there be a role, therapeutically or experimentally, for hemofiltration in the setting of DKA and encephalopathy? It is hard to imagine one, as removal of the putative toxins (glucose or ketone bodies) proceeds easily upon administration of fluid and insulin, and rapid removal of either one might induce an unfavorable shift of fluid into cells. In recovering DKA, bathing the interstitium with many liters of crystalloid may create more problems than it would solve. For the time being it is prudent to maintain a strategy of controlled rehydration in the child with DKA encephalopathy.

Conclusion

Metabolic encephalopathies present a serious and often rewarding challenge that tests the limits of critical care therapeutics. The neuropharmacologist may someday devise a discrete control mechanism for the mitochondrial membrane permeability transition, or there may eventually exist an elegant way to stabilize membranes interrupted by profound malperfusion. In the meantime, in the support of the child with severe encephalopathy, the intensive care physician will continue to modulate the interstitial milieu with broad strokes: simply, with controlled or restricted input of fluids and perhaps diuretic encouragement of its egress; or elaborately, with any of a number of therapeutically distinct technologies, including intermittent hemodialysis to remove target small solutes, continuous hemofiltration to remove water and solute and fundamentally alter the interstitial milieu, or plasmapheresis or exchange to eliminate larger vasoactive contributors to multiple organ (and therefore brain) failure.

References

Albrecht, J. and Norenberg, M.D. 2006. *Glutamine: A Trojan horse in ammonia neurotoxicity* Hepatology **44**(4): pp. 788–794

Als-Nielsen, B., L.L. Gluud, and C. Gluud, 2004. *Non-absorbable disaccharides for hepatic encephalopathy: Systematic review of randomised trials.* Bmj, **328**(7447): p. 1046

Azzopardi, J., et al.2002. *Lack of evidence of cerebral oedema in adults treated for diabetic ketoacidosis with fluids of different tonicity.* Diabetes Res Clin Pract, **57**(2): pp. 87–92

Bachmann, C., 2002. *Mechanisms of hyperammonemia.* Clin Chem Lab Med, **40**(7): pp. 653–662

Bhowmick, S.K., Levens, K.L. andRettig, K.R. 2005. *Hyperosmolar hyperglycemic crisis: An acute life-threatening event in children and adolescents with type 2 diabetes mellitus.* Endocr Pract, **11**(1): pp. 23–29

Biancofiore, G., et al.2003. *Combined twice-daily plasma exchange and continuous veno-venous hemodiafiltration for bridging severe acute liver failure.* Transplant Proc, **35**(8): pp. 3011–3014

Churchwell, K.B., et al., 1995. *Intensive blood and plasma exchange for treatment of coagulopathy in meningococcemia.* J Clin Apher, **10**(4): pp. 171–177

Davis, J., et al.2007. *DKA, CVL and DVT. Increased risk of deep venous thrombosis in children with diabetic ketoacidosis and femoral central venous lines.* Ir Med J, **100**(1): p. 344

Demetriou, A.A., 2005. *Hepatic assist devices.* Panminerva Med, **47**(1): pp. 31–37

Di Carlo, J.V. and S.R. Alexander, 2005. *Hemofiltration for cytokine-driven illnesses: The mediator delivery hypothesis.* Int J Artif Organs, **28**(8): pp. 777–786

Dunger, D.B., et al.2004. *ESPE/LWPES consensus statement on diabetic ketoacidosis in children and adolescents.* Arch Dis Child, **89**(2): pp. 188–194

Edge, J.A., et al.2006a. *Conscious level in children with diabetic ketoacidosis is related to severity of acidosis and not to blood glucose concentration.* Pediatr Diabetes, **7**(1): pp. 11–15

Edge, J.A., et al., et al.2006b. *The UK case-control study of cerebral oedema complicating diabetic ketoacidosis in children.* Diabetologia, **49**(9): pp. 2002–2009

Festi, D., et al., 2006. *Management of hepatic encephalopathy: Focus on antibiotic therapy.* Digestion, **73**(Suppl 1): pp. 94–101

Fourtner, S.H., S.A. Weinzimer, and L.E. Levitt Katz, 2005. *Hyperglycemic hyperosmolar non-ketotic syndrome in children with type 2 diabetes.* Pediatr Diabetes, **6**(3): pp. 129–135

Glaser, N., et al., et al.2001. *Risk factors for cerebral edema in children with diabetic ketoacidosis. The Pediatric Emergency Medicine Collaborative Research Committee of the American Academy of Pediatrics.* N Engl J Med, **344**(4): pp. 264–269

Glaser, N.S., et al.2004. *Mechanism of cerebral edema in children with diabetic ketoacidosis. J Pediatr*, **145**(2): pp. 164–171

Glaser, N.S., et al.2006. *Frequency of sub-clinical cerebral edema in children with diabetic ketoacidosis.* Pediatr Diabetes, **7**(2): pp. 75–80

Gordon, N., 2003. *Ornithine transcarbamylase deficiency: A urea cycle defect.* Eur J Paediatr Neurol, **7**(3): pp. 115–121

Halestrap, A.P., 2006. *Calcium, mitochondria and reperfusion injury: A pore way to die.* Biochem Soc Trans, **34**(Pt 2): pp. 232–237

Hammer, G.B., et al.1996. *Continuous venovenous hemofiltration with dialysis in combination with total hepatectomy and portocaval shunting. Bridge to liver transplantation.* Transplantation, **62**(1): pp. 130–132

Henter, J.I. and I. Nennesmo, 1997. *Neuropathologic findings and neurologic symptoms in twenty-three children with hemophagocytic lymphohistiocytosis.* J Pediatr, **130**(3): pp. 358–365

Jones, M., 1979. *Energy metabolism in the developing brain.* Seminars in Perinatology, **3:** pp. 121–129

Jones, E.A., 2002. *Ammonia, the GABA neurotransmitter system, and hepatic encephalopathy.* Metab Brain Dis, **17**(4): pp. 275–281

Kawata, H., et al.2006. *The use of continuous hemodiafiltration in a patient with diabetic ketoacidosis.* J Anesth, **20**(2): pp. 129–131

Legras, A., et al.2002. *Late diagnosis of ornithine transcarbamylase defect in three related female patients: Polymorphic presentations.* Crit Care Med, **30**(1): pp. 241–244

Magee, M.F. and B.A. Bhatt, 2001. *Management of decompensated diabetes. Diabetic ketoacidosis and hyperglycemic hyperosmolar syndrome.* Crit Care Clin, **17**(1): pp. 75–106

Mattarozzi, K., et al.2005. *Distinguishing between clinical and minimal hepatic encephalopathy on the basis of specific cognitive impairment.* Metab Brain Dis, **20**(3): pp. 243–249

Morales, A.E. and A.L. Rosenbloom, 2004. *Death caused by hyperglycemic hyperosmolar state at the onset of type 2 diabetes.* J Pediatr, **144**(2): pp. 270–273

Muir, A., et al., *Cerebral edema in childhood diabetic ketoacidosis: Natural history, radiographic findings and early identification.* 2001

Mustafa, A. and J.T. Clarke, 2006. *Ornithine transcarbamoylase deficiency presenting with acute liver failure.* J Inherit Metab Dis, **29**(4): p. 586

Ogier de Baulny, H., 2002. *Management and emergency treatments of neonates with a suspicion of inborn errors of metabolism.* Semin Neonatol, **7**(1): pp. 17–26

Ogier de Baulny, H. and J.M. Saudubray, 2002. *Branched-chain organic acidurias.* Semin Neonatol, **7**(1): pp. 65–74

Ong, J.P., et al., et al.2003. *Correlation between ammonia levels and the severity of hepatic encephalopathy.* Am J Med, **114**(3): pp. 188–193

Puliyel, J.M. and V. Bhambhani, 2003. *Ketoacid levels may alter osmotonicity in diabetic ketoacidosis and precipitate cerebral edema.* Arch Dis Child, **88**(4): p. 366

Roberts, J.S., et al.2006. *Cerebral hyperemia and impaired cerebral autoregulation associated with diabetic ketoacidosis in critically ill children.* Crit Care Med, **34**(8): pp. 2217–2223

Ryan, C.J., et al.2001. *Multisorbent plasma perfusion in fulminant hepatic failure: effects of duration and frequency of treatment in rats with grade III hepatic coma.* Artif Organs, **25**(2): pp. 109–118

Segura-Bruna, N., et al.2006. *Valproate-induced hyperammonemic encephalopathy.* Acta Neurol Scand, **114**(1): pp. 1–7

Smith, D.B. and J. Gulinson, 1988. *Fatal cerebral edema complicating toxic shock syndrome.* Neurosurgery, **22**(3): pp. 598–599

Stange, J., et al.2002. *The molecular adsorbents recycling system as a liver support system based on albumin dialysis: A summary of preclinical investigations, prospective, randomized, controlled clinical trial, and clinical experience from 19 centers.* Artif Organs, **26**(2): pp. 103–110

Summar, M.L., 2005. *Unmasked adult-onset urea cycle disorders in the critical care setting.* Crit Care Clin, **21**(Suppl 4): pp. S1–S8

Thakur, V., et al.2006. *Fatal cerebral edema from late-onset ornithine transcarbamylase deficiency in a juvenile male patient receiving valproic acid.* Pediatr Crit Care Med, **7**(3): pp. 273–276

Verrotti, A., et al.2002. *Valproate-induced hyperammonemic encephalopathy.* Metab Brain Dis, **17**(4): pp. 367–373

Wooton-Gorges, S., et al., 2005. *Detection of cerebral beta-hydroxybutyrate, acetoacetate and lactate on proton MR spectroscopy in children with diabetic ketoacidosis.* Am J Neuroradiol, **26**: pp. 1286–1291

Yorgin, P.D., et al.2000. *Concurrent centrifugation plasmapheresis and continuous venovenous hemodiafiltration.* Pediatr Nephrol, **14**(1): pp. 18–21

Chapter 9
Pathophysiology of Hepatic Encephalopathy: Studies in Animal Models

Roger F. Butterworth

Introduction

Hepatic Encephalopathy (HE) is a serious neuropsychiatric complication of both acute and chronic liver failure. A study group concluded in 2002 that "HE is a spectrum of neuropsychiatric abnormalities seen in patients with liver dysfunction after exclusion of other known brain diseases" (Ferenci et al., 2002). A multiaxial definition of HE was proposed that defines both the type of hepatic abnormality and the characteristics of the neurological manifestations. Three types of hepatic abnormalities were defined, namely:

Type A: HE associated with Acute Liver Failure (ALF)
Type B: HE associated with portal-systemic bypass with no intrinsic hepatocellular disease (in practice this type is rare)
Type C: HE associated with cirrhosis and portal hypertension or portal-systemic shunts.

In the case of chronic liver disease, episodic HE and persistent HE were defined and the term "minimal hepatic encephalopathy" (MHE) was coined to replace the hitherto inappropriate term "subclinical encephalopathy". It is expected that this new system of classification will help to dispel the confusion frequently present in current textbook definitions and to facilitate multi-centre clinical trials in HE.

Neurological symptoms characteristic of HE include shortened attention span, sleep abnormalities and motor incoordination progressing through lethargy to stupor and coma. In Type C HE, symptoms take an undulating course progressing relatively slowly whereas in Type A HE, the neurological disorder may progress from altered mental status to coma within days. Seizures are not uncommon and mortality rates are high in the acute (Type A) form of HE where death invariably results from brain herniation as a consequence of intracranial hypertension brought on by brain swelling (edema).

D.W. McCandless (ed.) *Metabolic Encephalopathy*,
doi: 10.1007/978-0-387-79112-8_9, © Springer Science+Business Media, LLC 2009

Neuropathology of HE

The Astrocyte

The principal neuropathologic finding in HE is altered astrocyte morphology. The cellular characteristics of the astrocyte changes in HE are a direct function of the nature of the liver failure (acute vs. chronic, ie. Type A vs. Type C) as well as the severity of HE.

Von Hösslin and Alzheimer in 1912 described morphological abnormalities of astrocytes in a disorder known as Westphal–Strümpell pseudosclerosis, a disorder shown subsequently to be identical to acquired hepatocerebral degeneration (Wilson's Disease). The term "Alzheimer Type II astrocyte" is now used to describe these characteristic morphological features manifested by astrocytes that consist of large, pale (watery-looking) nuclei, margination of the chromatin and prominent nucleoli as shown in Fig. 9.1. Intranuclear glycogen inclusions are also evident in these cells.

Hepatic Encephalopathy (HE) in chronic liver failure, regardless of the etiology of liver disease, is characterized by the presence of Alzheimer Type II astrocytes. The number of the cells showing the Alzheimer Type II phenotype is significantly correlated with the severity of encephalopathy (Adams and Foley, 1953; Butterworth et al., 1987). Alzheimer Type II astrocytes are found in grey and white matter of HE brains where they show regional selectivity (Butterworth et al., 1987). The nuclei take on a variety of shapes from round (in cerebral cortex) to irregular or lobulated forms (in basal ganglia) and in both cases may occur in pairs or triplets suggestive of hyperplasia (Norenberg, 1987).

Magnetic Resonance Imaging (MRI) studies reveal alterations of pathophysiological significance in HE patients. Bilateral signal hyperintensities are observed in globus pallidus on T_1-weighted MRI (see Sect. "Manganese") in over 80% of cirrhotic patients and T_2-weighted Fast-FLAIR MRI reveals white matter lutencies along the corticospinal tract of cirrhotic patients with overt HE (Rovira et al., 2002). Both of these MRI alterations resolve following liver transplantation.

Studies in experimental animal models of HE continue to help to characterize the early morphologic changes in astrocytes. For example, feeding of ammonia cation exchange resins to rats following end-to-side portacaval anastomosis results in severe encephalopathy (Norenberg, 1987). In early stages of HE in these animals, astrocytes exhibit evidence of hypertrophy (increased mitochondria and endoplasmic reticulum) (Fig. 9.2). Later stages of encephalopathy (coma) are accompanied by contractions of mitochondria and degenerative changes.

Mixed glial-neuronal cultures exposed to sera from HE patients and from animals with experimental HE develop morphological changes characteristic of Alzheimer Type II astrocytes (Mossakowski et al., 1970). Exposure of cultured rat cortical astrocytes to ammonia, the principal putative neurotoxin generated in liver failure (Sect.Ammonia), results in changes that mimic the in vivo findings consisting, at the light microscopic level, of increased cytoplasmic basophilia, vacuolization and cellular disintegration (Norenberg, 1987). Ultrastructural studies show that the initial response consists of proliferation of mitochondria and smooth endoplasmic reticulum

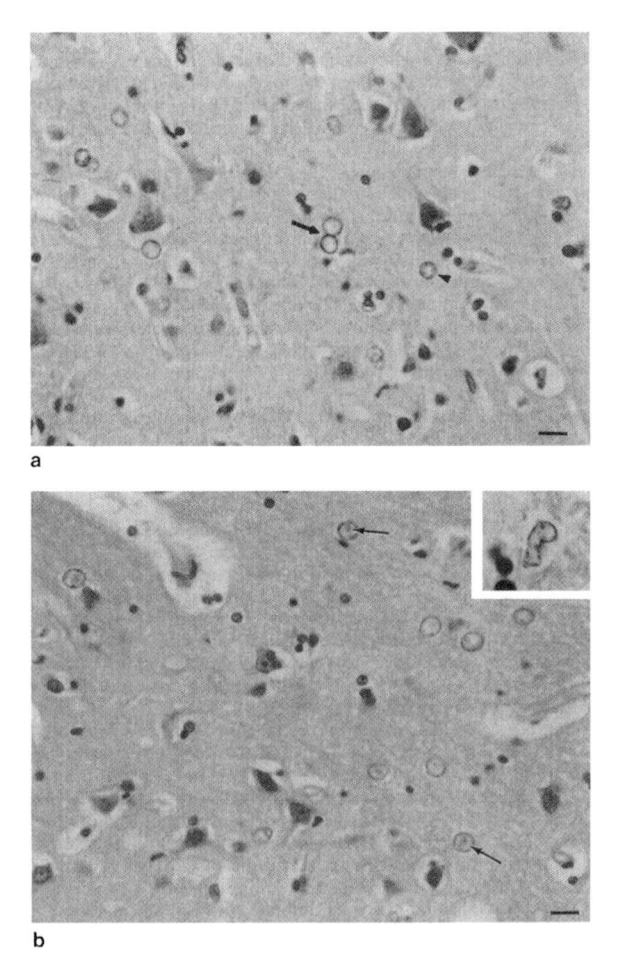

Fig. 9.1 Alzheimer Type II astrocytosis in HE (chronic liver failure) (**a**) Light micrograph of cerebral cortex from a cirrhotic patient who died in hepatic coma. Note prominence of pale, enlarged astroglial nuclei frequently occurring in pairs (*arrow*) suggestive of hyperplasia. A normal astrocyte nucleus is shown for comparison purposes (*arrowhead*). Bar = 20 μM. (**b**) Similar section showing intranuclear glycogen inclusions (*arrow*). *Inset*: irregular lobular astrocyte in pallidum. Reproduced from Norenberg (1987), with permission from Humana Press

with appearance of dense bodies resembling lipofuscin granules. Loss of intermediate filaments have been described in both human HE (Sobel et al., 1981) and in ammonia-exposed astrocytes in culture (Norenberg, 1987).

Neuronal Cell Death in HE

It is generally assumed that neuronal cell death is minimal in liver failure and is insufficient to account for the neurological complications characteristic of HE. A careful review of the literature, however, reveals that neuronal cell death and

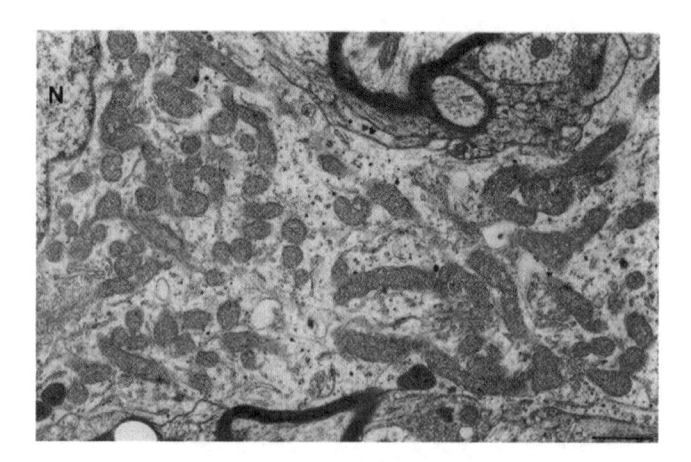

Fig. 9.2 Early changes in astrocytic morphology in experimental HE (chronic liver failure) Electron micrograph of an astrocyte process showing mitochondrial proliferation from a portacaval-shunted rat with mild HE resulting from feeding of ammonia resins. N: nucleus, Bar = 1 μM Reproduced from Norenberg (1987), with permission from Humana Press

severe neuronal dysfunction are well documented in liver failure. Several distinct clinical entities have been described including acquired (non-Wilsonian) hepatocerebral degeneration, post-shunt myelopathy, cerebellar degeneration and extrapyramidal disorders (Parkinsonism) (see Butterworth, 2007 for review). Furthermore, studies in experimental animals reveal that multiple cell death mechanisms occur in the brain with liver failure. Such mechanisms include lactic acidosis, oxidative/nitrosative stress, NMDA receptor-mediated excitotoxicity together with evidence of an inflammatory response and the presence of proinflammatory cytokines. These mechanisms are discussed in more details in Sects. "Astrocyte Metabolism and Function in HE" and "Neurotransmitter Function in HE" of this review. It has been proposed that the extent of neuronal cell death in liver failure is attenuated by the presence of compensatory mechanisms including down-regulation of NMDA receptors, the occurrence of hypothermia and the increased synthesis of neuroprotective compounds such as the neurosteroid allopregnanolone (Butterworth, 2007). The occurrence of neuronal cell death in liver failure suggests that some of the so-called "sequellae" of liver transplantation (gait ataxia, memory loss and confusion) could reflect pre-existing neuropathology (Kril and Butterworth, 1996).

Pathogenesis of HE: Role of Blood-Borne Toxins

Ammonia

Evidence of an association between HE and ammonia dates back over a century to the studies of Eck, who described the effects of portacaval anastomosis in dogs. Feeding of meat to shunted dogs led to severe neurological impairment progressing to coma. Subsequently, in the 1950s, attempts were made to treat ascites in cirrhotic patients using ammonium ion-exchange resins. This treatment led to reduction of ascitic volume but precipitated neurological symptoms that were indistinguishable from HE (Gabuzda et al., 1952). Arterial blood ammonia concentrations are frequently increased in patients with HE and brain ammonia may reach millimolar concentrations at coma stages of encephalopathy. Studies using Positron Emission Tomography (PET) and $^{13}NH_3$ have been used to investigate brain ammonia metabolism in cirrhotic patients. In one study, a significant increase in the cerebral metabolic rate for ammonia (CMR_A) was observed in patients with mild HE (Lockwood et al., 1991) accompanied by an increase in blood—brain barrier permeability to ammonia (Table 9.1). It was suggested that the increased ease with which ammonia appears to move into the brain in HE patients could account for (a) the hypersensitivity of cirrhotic patients to ammoniagenic conditions such as protein loading, gastrointestinal bleeding and constipation, (b) the lack of a tight correlation between blood and brain ammonia concentrations in liver failure and (c) the common occurrence of HE in some patients with near normal arterial ammonia levels. Results of a second $^{13}NH_3$ PET study suggested that increased cerebral trapping of ammonia in cirrhotic patients with HE was also the result of increased circulating ammonia in addition to altered brain ammonia kinetics (Keiding et al., 2006).

There is evidence of a pathogenetic link between hyperammonemia and the phenomenon of Alzheimer type II astrocytosis; this astrocytic phenotype has been described in a wide range of hyperammonemic syndromes associated with congenital urea cycle enzyme defects, as well as in experimental animals with urease-induced hyperammonemia and in primary cultures of astrocytes exposed to ammonia (Butterworth et al., 1987).

Patients with ALF who developed intracranial hypertension and cerebral herniation manifest arterial ammonia levels in the 300–600 μM range, significantly higher than patients who did not herniate (Clemmesen et al., 1999) and there is evidence to suggest that increased brain ammonia concentrations are causally related to the pathogenesis of brain edema in ALF. Brain ammonia concentrations are reported to be in the low millimolar range at coma and edematous stages of encephalopathy in experimental animal models of ALF (Swain et al., 1992). Furthermore, treatment of cerebral cortical slices with ammonia in concentrations equivalent to those reported in brain in experimental ALF results in significant cell swelling (Norenberg, 1987).

Table 9.1 Blood–brain ammonia kinetics in HE patients

	Control subjects (5)	Patients (5)
Arterial NH_3 (mM)	0.030 ± 0.007	$0.062 \pm 0.02*$
Cerebral metabolic rate (NH_3)	0.35 ± 0.15	$0.91 \pm 0.36*$
BBB permeability to NH_3 (ml g^{-1} min^{-1})	0.13 ± 0.03	$0.22 \pm 0.07*$

Data from studies using $^{13}NH_3$ PET in cirrhotic patients with mild HE. Values indicated represent mean ± S.E. from 5 patients per group. Significant differences indicated by * $p < 0.01$ by Student's t test (Adapted from Lockwood et al., 1991 with permission)

There is a convincing body of evidence to suggest that hyperammonemia in liver failure results from altered inter-organ trafficking of ammonia (Chatauret and Butterworth, 2004) as shown in Fig. 9.3.

The intestines have a high glutaminase activity that converts glutamine to glutamate and ammonia. Studies in healthy animals reveal that intestinal glutamine metabolism accounted for 50% of the ammonia produced by the portal-drained viscera, the remaining 50% being produced in the colon, most of which was derived from urea uptake from arterial blood (Weber and Veach, 1979). In liver failure, the contribution of the intestine to hyperammonemia results primarily from reduced hepatic urea synthesis rather than increased intestinal ammonia production (Olde Damink et al., 2002). Liver ammonia removal is highly compartmentalized, involving two distinct but functionally related cell types; urea is synthesized from ammonia via the urea cycle in periportal hepatocytes whereas perivenous hepatocytes transform ammonia into glutamine via glutamine synthetase. Cirrhotic patients commonly have intra and extra-hepatic portal-systemic shunts that may account for a large portion of portal blood flow. This portal-systemic shunting, in addition to the loss of hepatocytes and impaired residual hepatocyte function, results in decreased ammonia removal by the liver. Increased renal ammonia synthesis reported in liver failure is offset by increased urinary ammonia excretion (Olde Damink et al., 2002).

Unlike the liver, skeletal muscle and brain are devoid of an effective urea cycle and consequently must rely on glutamine synthesis for ammonia removal. In liver failure, muscle becomes the major route for ammonia detoxification. Evidence consistent with this notion includes reports of increased glutamine production by skeletal muscle in chronic liver insufficiency (Ganda and Ruderman, 1976). More recently it has been shown that chronic hyperammonemia resulting from portacaval

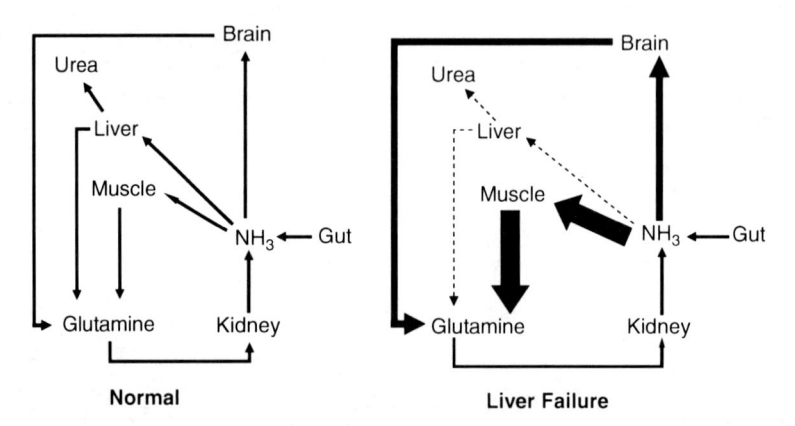

Normal   **Liver Failure**

Fig. 9.3 Effect of liver failure on inter-organ trafficking of ammonia. Under normal physiological conditions, ammonia produced by the gut is removed by the liver as urea (periportal hepatocytes) or glutamine (perivenous hepatocytes). Increased ammonia synthesis by the kidney is offset by increased urinary ammonia excretion. In liver failure, skeletal muscle becomes the major route for ammonia detoxification as a result of a post-translational increase of glutamine synthetase. Unlike muscle, the brain does not adapt to liver failure by induction or glutamine synthetase

anastomosis in rats results in increased glutamine synthetase (GS) activity due to a post-translational increase of GS (Desjardins et al., 1999). ALF also results in increased muscle GS mRNA and protein leading to increased enzyme activity (Rama Rao et al., 1999).

Unlike skeletal muscle, brain does not adapt its ammonia removal capacity by induction of GS. On the contrary, GS activities have consistently been shown to be decreased both in the brains of animals with experimental chronic liver failure (Desjardins et al., 1999) as well as in autopsied brain tissue from cirrhotic patients who died in hepatic coma (Lavoie et al., 1987a). Decreased GS activity in brain in liver failure may be explained by oxidative/nitrosative stress mechanisms leading to increased protein tyrosine nitration of GS as demonstrated in both in vitro and in vivo models of hyperammonemia (Schliess et al., 2002). In spite of the reduction of GS, brain glutamine concentrations are invariably increased in liver failure and increased CSF glutamine concentrations correlate well with the severity of neurological impairment in HE patients. Glutamine concentrations are elevated two to fivefold in CSF and brain tissue from experimental animals with HE and increased brain glutamine has also been described in NMR studies in both human (Laubenberger et al., 1997) and experimental (Sonnewald et al., 1996) HE. Glutamine concentrations in autopsied brain tissue from cirrhotic patients who died in hepatic coma are increased two to fivefold (Lavoie et al., 1987b). However, NMR spectroscopic studies have been unable to demonstrate increased glutamine synthesis in the brain (Zwingmann et al., 2003) suggesting that the increased brain glutamine in liver failure could result primarily from decreased glutamine release from the astrocyte or decreased glutaminase activity. This issue requires further study.

Concentrations of ammonia equivalent to those reported in the brain in HE are known to cause deleterious effects on the CNS functions by both direct and indirect mechanisms (Szerb and Butterworth, 1992, review). Direct effects of the NH_4^+ ion on both inhibitory and excitatory neurotransmission have been reported. Millimolar concentrations of ammonia impair postsynaptic inhibition in the brain by inactivation of the extrusion of Cl^- from neurons (Raabe, 1987). This inactivation of Cl^- extrusion abolishes the concentration gradient for Cl^- across the neuronal membrane. Consequently, the opening of Cl^- channels by the inhibitory neurotransmitter no longer takes place and the inhibitory postsynaptic potential (IPSP) is abolished. It has been demonstrated that brain ammonia concentrations as low as 0.5 mM may exert this adverse effect on inhibitory neurotransmission.

Ammonia also has deleterious effects on excitatory neurotransmission (Szerb and Butterworth, 1992, review) where effects on both presynaptic and postsynaptic neuronal membranes have been reported. NH_4^+ ions, in pathophysiologically relevant concentrations, interfere with excitatory neurotransmission by preventing the action of glutamate at the postsynaptic receptor. In addition, NH_4^+ depolarizes neurons to a variable degree without consistently changing membrane resistance, probably by reducing K^+ concentrations.

Portacaval-shunted rats administered ammonium salts to precipitate coma manifest impaired brain energy metabolism (Hindfelt et al., 1977). Possible mechanisms responsible for this include inhibitory effects of ammonia on the tricarboxylic acid cycle enzyme, α-ketoglutarate dehydrogenase (Lai and Cooper, 1986) and on the malate–aspartate shuttle (Hindfelt et al., 1977). In addition, ammonia stimulates glycolysis at the level of phosphofructokinase. These metabolic effects of ammonia result in increased lactate production in the brain. CSF lactate concentrations correlate well with the deterioration and subsequent recovery of neurological status in portacaval-shunted rats in which HE was precipitated by ammonia administration (Therrien et al., 1991) as well as in cirrhotic patients with mild to moderate HE (Yao et al., 1987).

Manganese

A highly consistent finding on MRI of cirrhotic patients is bilateral symmetrical hyperintensities in globus pallidus on T_1-weighted imaging (Spahr et al., 1996). A representative example of these pallidal hyperintensities is shown in Fig. 9.4.

A convincing body of evidence suggests that these images are the result of pallidal manganese deposition. Manganese is normally eliminated via the hepatobiliary route and blood manganese concentrations are increased in cirrhotic patients who manifest pallidal signal hyperintensities on MRI. Similar pallidal signals have been reported in other conditions including Alagille's Syndrome (Devenyi et al., 1994) (a disorder characterized by cholestasis and intrahepatic bile duct paucity) as well

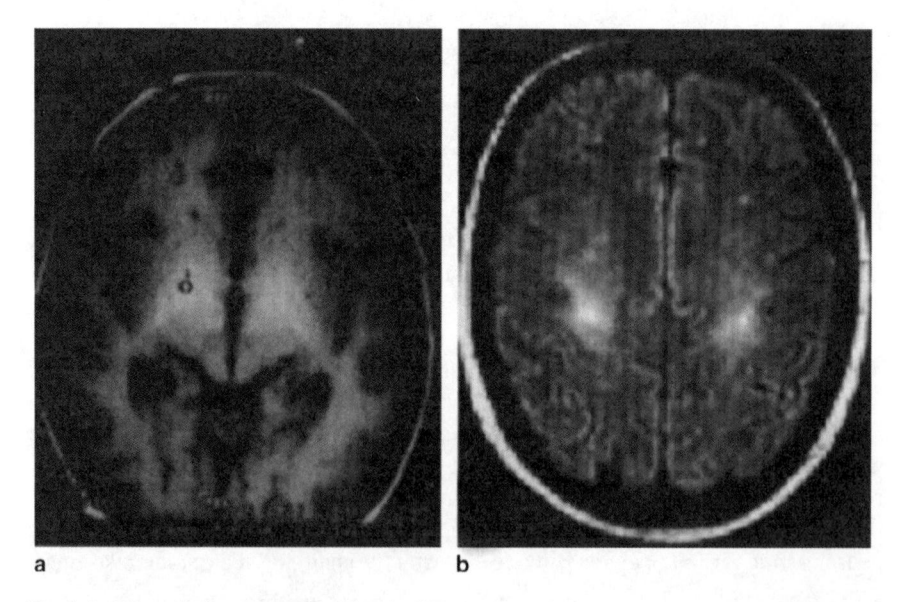

a   b

Fig. 9.4 Magnetic Resonance Imaging in HE. (**a**) T_1-weighted Magnetic Resonance Imaging shows bilateral signal hyperintensities in globus pallidus (*arrow*) of a cirrhotic patient with mild HE. (**b**) T_2-weighted Fast-FLAIR images in white matter along the corticospinal tract of a cirrhotic patient with overt HE (adapted from Rovira et al., 2002)

Table 9.2 Brain manganese concentrations in dissected pallidal tissue from cirrhotic patients who died in hepatic coma and from rats with experimental acute or chronic liver failure

	Pallidal manganese concentration (μg g^{-1})
Patients	
Controls	1.41 ± 0.91 (8)
Cirrhotics	4.04 ± 1.54** (8)
Rats	
Controls (Sham-operated)	0.54 ± 0.02 (6)
Portacaval-shunted	1.05 ± 0.07** (6)
Bile-duct ligated (cirrhosis)	0.85 ± 0.06* (6)
Hepatic devascularization (ALF)	0.72 ± 0.07 (6)

Values represent mean ± S.E. of determinations in dissected pallidal tissue. Number of patient samples in parentheses. Values significantly different from the appropriate control group indicated by *$p < 0.05$, **$p < 0.01$ by Analysis of Variance

as in patients during total parenteral nutrition (Mirowitz et al., 1991). Both conditions are associated with increased blood manganese. Direct measurements using neutron activation analysis reveal up to sevenfold increases in manganese content of dissected pallidum obtained post mortem from cirrhotic patients who died in hepatic coma (Pomier Layrargues et al., 1995) and from experimental animals with surgical portacaval shunts where a selective accumulation of manganese was reported in pallidum and, to a lesser extent, in other basal ganglia structures (Rose et al., 1999) (Table 9.2).

Brain Glucose Metabolism in HE

There is no convincing evidence that HE is primarily the consequence of a reduction in brain levels of high energy phosphates whether assessed biochemically in animal models (Hindfelt et al., 1977; Mans et al., 1994) or spectroscopically (Bates et al., 1989) in animal models or in humans. On the other hand, alterations of brain glucose utilization, of cerebral blood flow and of brain glucose metabolic pathways have been consistently reported. These alterations of brain metabolism are dependent upon the severity of HE, the type of liver failure (acute vs. chronic) and on the brain region under investigation. PET studies using ^{18}F-deoxyglucose reveal significant decreases in glucose utilization localized to the anterior cingulate cortex in cirrhotic patients with mild HE (Lockwood et al., 2002). This brain structure is known to be implicated in the control of the anterior attention system responsible for the monitoring of, for example, the selection of responses to visual stimuli. A significant correlation was observed between the magnitude of the decreased glucose utilization in anterior cingulate and the reduced performance in attention-demanding tasks such as trail making and digit symbol psychometric tests.

Measurement of cerebral blood flow using ^{15}O–H_2O–PET reveals a redistribution of flow from regions such as the anterior cingulate cortex to subcortical structures including the thalamus (Lockwood et al., 1997(review)).

Precipitation of severe encephalopathy and coma in experimental animals with chronic liver failure following the administration of ammonium salts results in severe alterations of brain glucose metabolism leading to accumulation of lactate and alanine (Therrien et al., 1991) and reduced brain concentrations of glutamate and aspartate sufficiently severe to result in impairment of the malate–aspartate shuttle (Hindfelt et al., 1977) but without alterations of high energy phosphates. Increased brain lactate and alanine concentrations result from impaired oxidation of pyruvate in the brain with liver failure and interestingly, it has been demonstrated that ammonium ion, in concentrations equivalent to those encountered in brain at coma stages of HE, is a potent inhibitor of the tricarboxylic acid cycle enzyme α-ketoglutarate dehydrogenase (Lai and Cooper, 1986).

It has been postulated that the clinical features of brain edema in ALF are the consequences of partial brain ischemia rather than of brainstem compression. Consistent with this notion, the cerebral metabolic rate for oxygen ($CMRO_2$) was found to be low in all of 30 patients with ALF in grade IV HE and it was calculated that this level of $CMRO_2$ was inappropriately low even to meet the reduced metabolic requirements of deep coma (Wendon et al., 1994). Furthermore, 21 of the 30 ALF patients showed increased brain lactate production indicative of an underlying brain oxygen deficit. It is well established that brain ischemia is associated with an enhancement of anaerobic metabolism leading to intra and extracellular lactic acidosis and exposure of cultured cortical astrocytes to 20 μM lactate results in significant cell swelling by mechanisms that are both pH-dependent and pH-independent (Staub et al., 1990).

Increased brain and cerebrospinal fluid lactate concentrations have consistently been reported in animal models of ALF (Zwingmann et al., 2003; Mans et al., 1994). ^{13}C-Nuclear Magnetic Resonance studies confirm that experimental ALF results in increased de novo synthesis of lactate from glucose (Zwingmann et al., 2003) in the brain. Experimental ALF also results in increased expression of the endothelial cell/astrocyte glucose transporter GLUT-1 (Bélanger et al., 2006). These changes likely occur in an attempt to maintain brain levels of high energy phosphates in the face of diminished capacity for pyruvate oxidation.

Astrocyte Metabolism and Function in HE

Astrocyte Structural Proteins

Glial Fibrillary Acidic Protein (GFAP) is the major protein of intermediate filaments in differentiated astrocytes. Results of a recent study reveal that GFAP mRNA and protein expression are significantly reduced in frontal cortex of rats with ALF resulting from hepatic devascularization (Bélanger et al., 2002). These findings were selective for GFAP; expression of a second glial neurofilament protein S-100β was unchanged in the brain of these animals. It was suggested that the loss of GFAP and the resulting impairment of visco-elastic properties of the astrocyte could facilitate cell swelling leading to brain edema and its complications, which are characteristic of ALF. Exposure of cultured cortical astrocytes to millimolar concentrations of ammonia results in a loss of GFAP expression (Neary et al., 1994; Bélanger et al., 2002) and it was suggested that ammonia exposure under these conditions led to a destabilization of GFAP mRNA.

GFAP expression in brain has also been studied in both experimental and human *chronic* liver failure where it was reported to be decreased or unchanged, depending upon the brain region under investigation. GFAP-immunolabelling of cerebral cortical astrocytes was reportedly decreased following end-to-side portacaval anastomosis in rats (Norenberg, 1987) and in the cerebrum of patients with chronic liver failure (Sobel et al., 1981). On the other hand, GFAP immunolabelling of cerebellar Bergmann glia in human chronic liver failure was unaltered (Kril and Butterworth, 1996).

Glutamine Synthesis, the Glutamate-Glutamine Cycle

Glutamine Synthetase (GS) is the enzyme primarily responsible for ammonia removal by the brain and is almost exclusively localized in the astrocytes. Not surprisingly, in both acute and chronic liver failure, brain glutamine concentrations are increased (Lavoie et al., 1987a; Laubenberger et al., 1997). However, despite these findings of increased brain glutamine, there is little convincing evidence to suggest that GS gene or protein expression is induced in the brain with liver failure. For example, portacaval anastomosis in rats results in unaltered or even decreased expression and activities of GS in the brain (Girard et al., 1989). These findings again suggest that the increases in concentrations of glutamine consistently observed in the brain with liver failure may result primarily from decreased glutamine release and/or degradation.

Glutamate and Glycine Transporters

The effective and rapid removal of neuronally-released glutamate from the synaptic cleft is achieved by high affinity, energy-dependent glutamate transporters. Evidence from in vitro and in vivo studies suggests that both liver failure and ammonia exposure result in reduced expression and activity of these transporters in the brain (Butterworth, 2001).

Rat hippocampal slices perfused with low millimolar concentrations of ammonia show decreased capacity for uptake of the non-metabolized glutamate analogue D-aspartate (Schmidt et al., 1990) and exposure of these preparations to serum extracts from patients with chronic liver failure and HE leads to reductions in high affinity D-aspartate uptake. Furthermore, a significant inverse correlation was observed between the D-aspartate uptake inhibition and ammonia content of the serum extract from these patients. A significant inhibition by ammonia of high affinity uptake of glutamate by synaptosomal preparations from normal rats has been reported (Mena and Cotman, 1985) and studies of glutamate uptake by synaptosomes from experimental animals with ALF manifest a significant decrease in high affinity glutamate uptake capacity (Oppong et al., 1995).

Astrocytes express high affinity glutamate transporters EAAT-1 and EAAT-2 in the forebrain. The neuronally localized glutamate transporter EAAT-3 does not appear to be localized on nerve terminals and is therefore not considered to play a major role in the removal of synaptic glutamate, at least in cortical regions. EAAT-4 is a neuronal transporter expressed by cerebellar Purkinje cells whereas EAAT-5 is confined to the retina. Thus, most of the brain, particularly the cerebral cortex, hippocampus and midbrain structures rely primarily on astrocytic transporters for the effective removal of glutamate from the synapse. Exposure of cultured rat cortical astrocytes to ammonia results in a significant loss in expression of EAAT-1 mRNA accompanied by a parallel reduction in capacity to transport the non-metabolizeable glutamate analogue d-aspartate (Chan et al., 2000). Expression of a second astrocytic glutamate transporter, EAAT-2 mRNA and protein are decreased in frontal cortical extracts from rats with ALF resulting from hepatic devascularization (Knecht et al., 1997) (Fig. 9.5) and in the brains of mice with ALF due to thioacetamide hepatotoxicity (Norenberg et al., 1997). Loss of EAAT-2 expression in the brains of ALF rats is accompanied by a significant loss of high affinity d-aspartate uptake capacity by brain preparations from these animals

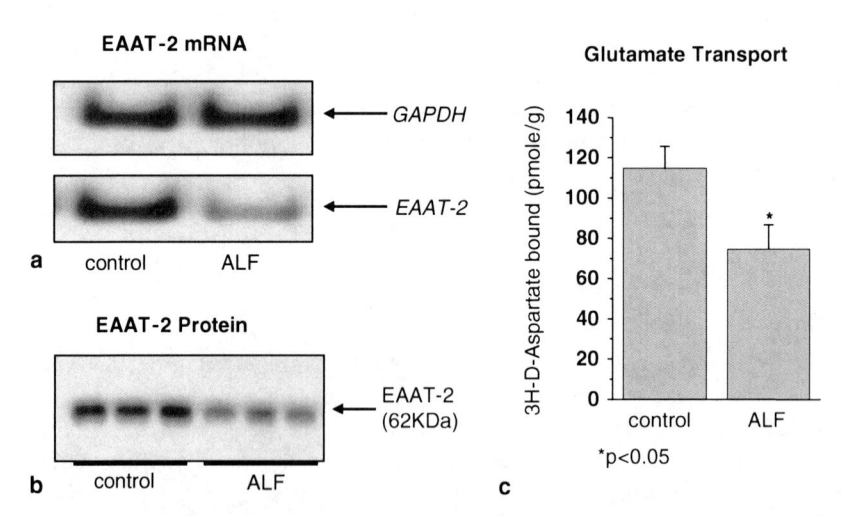

Fig. 9.5 Decreased expression of EAAT-2 in brain in experimental ALF. Decreased EAAT-2 mRNA (panel A) and EAAT-2 protein (panel B) in rats with ALF resulting from hepatic devascularization (portacaval anastomosis followed by hepatic artery ligation). Rats had severe HE and brain edema. Panel C shows decreased uptake of the glutamate analogue ^{3}H-D-Aspartate by cortical slices from ALF rats (Data from Knecht et al., 1997)

(Butterworth, 2001) and the loss of transporter capacity results in increased extracellular brain concentrations of glutamate in the frontal cortex of these animals (Michalak et al., 1996). Increased extracellular brain glutamate has been confirmed in a wide range of models of ALF (Felipo and Butterworth, 2002). In contrast to ALF, chronic liver failure does not consistently result in a loss of glutamate transport capacity (Raghavendra Rao et al., 1995).

Ischemic ALF leads to a significant loss of expression of the astrocytic glycine transporter GLYT-1 in cerebral cortex (Zwingmann et al., 2002) and a concomitant increase in concentration of glycine in brain extracellular fluid of comatose animals (Michalak et al., 1996). Exposure of cultured astrocytes to ammonia likewise leads to a significant reduction in expression of GLYT-1. A major function of glycine in the frontal cortex is the positive allosteric modulation of the glutamate (NMDA) receptor complex on which there is a glycine modulatory site. Stimulation of this site by increased extracellular glycine offers an alternative (or additional) explanation for the increased glutamatergic transmission in HE.

"Peripheral-Type" Benzodiazepine Receptors, Neurosteroids

The mitochrondrial "peripheral-type" benzodiazepine receptor (PTBR) is a multimeric complex comprising three subunits, namely an 18 KDa isoquinoline carboxamine-binding protein (IBP), a 34 KDa voltage-dependent anion channel and a 30 KDa adenine nucleotide carrier (McEnery et al., 1992).

Chronic liver failure resulting from end-to-side portacaval-anastomosis in rats leads to a significant increase in IBP mRNA in frontal cortical extracts of these animals (Desjardins et al., 1997). Concomitant with these changes in gene expression is a significant increase in binding sites for the PTBR ligand ^{3}H-PK11195 (Giguère et al., 1992; Desjardins and Butterworth, 2002). Increased densities of binding sites for ^{3}H-PK11195 have also been reported in autopsied frontal cortex and caudate nuclei of cirrhotic patients who died in hepatic coma (Lavoie et al., 1990). More recently, increased densities of binding of the PET ligand ^{11}C-PK11195 were reported in basal ganglia and frontal cortical regions of the brain of cirrhotic patients (Cagnin et al., 2001) where the magnitude of these increases was positively correlated with the severity of cognitive impairment in these patients.

As the PTBR is localized predominantly on astrocytic mitochondria in mammalian brain, it is not surprising that alterations of PTBR expression in HE are associated with altered mitochondrial function. For example, portacaval anastomosis in rats leads to astrocytic mitochondrial proliferation (Fig. 9.2). Mitochondrial proliferation also results from exposure of both cultured astrocytes (Norenberg and Lapham, 1974) and C6 glioma cells (Shiraishi et al., 1990) to PTBR ligands.

There is evidence to suggest that changes in expression of the PTBR are an integral part of the Alzheimer type II changes that are characteristic of astrocytes in chronic liver failure (Fig. 9.1). Expression of the PTBR IBP is significantly correlated with the presence of Alzheimer type II changes in autopsied brain tissue from cirrhotic patients who died in hepatic coma (Bélanger et al., 2004).

Astrocyte–Astrocyte "Crosstalk"

Astrocytes from the cerebral cortex and hippocampus express functional NMDA receptors which are implicated in both astrocyte-neuron and astrocyte-astrocyte metabolic coupling (Verkhratsky and Kirchhoff, 2007). Astrocyte NMDA receptors respond to glutamate ligands with increases in intracellular Ca^{2+} (Porter and McCarthy, 1995), NO production (Mollace et al., 1995) and protein tyrosine nitration (Schliess et al., 2002). Furthermore, glutamate induces astrocytic swelling due to stimulation of Ca^{2+}-dependent K^+ uptake that is sensitive to NMDA receptor antagonists (Bender et al., 1998).

Exocytotic release of glutamate from astrocytes is well established (Verkhratsky and Kirchhoff, 2007) and astrocytic release of glutamate can trigger activation of NMDA receptors on neighbouring astrocytes giving rise to astrocyte-astrocyte "crosstalk".

In hyperammonemic conditions and ALF, loss of capacity of the astrocytic glutamate transporters (see Sect. Glutamate Transporters) coupled with ammonia-induced exocytotic release of glutamate from astrocytes (Rose et al., 2005) suggest that glutamatergic synaptic regulation is impaired at multiple loci involving both neuronal and astrocytic NMDA receptors as shown in a simplified schematic manner in Fig. 9.6. Activation of astrocytic NMDA receptors by increased synaptic glutamate resulting in increased Ca^{2+}-dependent uptake of K^+ could explain the astrocytic swelling and consequent cytotoxic brain edema in ALF.

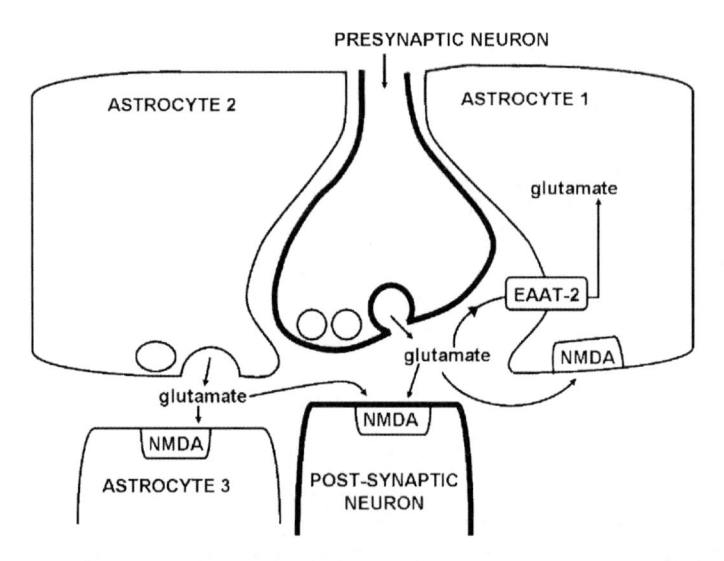

Fig. 9.6 Astrocyte-astrocyte, neuron-astrocyte and astrocyte-neuron "crosstalk" involving exocytotic glutamate release from astrocytes and activation of neuronal and astrocytic NMDA receptors. Exposure to ammonia results in exocytotic (Ca^{2+}-dependent) release of glutamate from both astrocytes and nerve terminals and a down regulation of astrocytic glutamate transporters (EAAT-2 shown here). Activation of NMDA receptors on astrocytes by ammonia results in oxidative/nitrosative stress and nitration of key brain proteins including the ammonia-metabolizing enzyme glutamine synthetase

Neurotransmitter Function in HE

Several lines of evidence suggest that neurological dysfunction particularly in HE, results from neurotransmission failure. Neurotransmission defects in HE include direct neurotoxic effects of the ammonium ion (NH_4^+) on inhibitory and excitatory neurotransmission, the accumulation of neuroactive and neurotoxic metabolites of tryptophan, altered glutamatergic synaptic regulation and activation of "peripheral-type" benzodiazepine receptors resulting in the synthesis of neuroactive steroids with high affinity for the GABA-A receptor.

Glutamate

Brain glutamate concentrations are significantly reduced in experimental ischemic liver failure as well as in toxic liver failure in rats, in parallel with the deterioration of neurological status (Swain et al., 1992). CSF glutamate concentrations in these animals are concomitantly increased and using the technique of in vivo brain dialysis, increased extracellular brain concentrations of glutamate have consistently been reported in experimental ALF (Bosman et al., 1992; Michalak et al., 1996). Evidence for glutamatergic synaptic dysfunction includes reductions in the expression of the high affinity astrocytic glutamate transporter EAAT-2 in the brain of rats with ALF (Knecht et al., 1997) (see Sect. Glutamate Transporters) and alterations of both NMDA and non-NMDA subclasses of glutamate receptors have been reported in experimental animal models of chronic liver failure (Peterson et al., 1990; Maddison et al., 1991). Reduction in the non-NMDA subclass of glutamate receptors was reported in brain in experimental ALF (Michalak and Butterworth, 1997) where it was suggested that this reduction in receptor sites and the consequent relative increase in the other subclass (NMDA) of glutamate receptors could be implicated in the pathogenesis of HE in ALF (Butterworth, 1997).

GABA

Throughout the 1980s a great deal of attention was focussed on the theory that HE was the result of increased activity of the GABA neurotransmitter system in brain. Abnormalities of the brain GABA system were initially reported in experimental animal models of ALF (Schafer and Jones, 1982). However, in animal models of chronic liver failure, no alterations of the GABA system have been reported, whether reflected by GABA content, its related enzymes or receptor sites (Butterworth and Giguère, 1986; Roy et al., 1988; Mans et al., 1992, 1994). Furthermore, autopsied brain tissue from cirrhotic patients who died in hepatic coma contains normal activities of GABA-related enzymes and unchanged densities and affinities of GABA binding sites (Lavoie et al., 1987a,b; Butterworth et al., 1988).

Benzodiazepine binding sites in the central nervous system form part of the GABA-A receptor complex and stimulation of the benzodiazepine binding site "facilitates" the action of GABA on the functionally linked GABA-A site of the complex. In this way, chloride channel opening is increased with resulting hyperpolarization and inhibition. Amelioration of neurological status in cirrhotic patients with HE has been reported following administration of the benzodiazepine antagonist flumazenil (Pomier Layrargues et al., 1994; Gyr et al., 1996). It was suggested that the ameliorative action of flumazenil in HE was due either to inhibition of increased densities of benzodiazepine binding sites or by inhibition of the action of an "endogenous" ligand at these sites. Subsequent investigations revealed no alterations of densities or affinities of these sites in either experimental or human HE (Butterworth et al., 1988) leaving open the possibility that the beneficial effects of flumazenil were the result of the blocking of the action of "endogenous" benzodiazepines (Mullen et al., 1990). Initial reports demonstrated that CSF from patients with advanced HE contained significant amounts of "benzodiazepine-like" substances (Olasmaa et al., 1990) and a known pharmaceutical benzodiazepine, diazepam and its NN-desmethyl metabolite, both of which are positive allosteric modulators of GABA neurotransmission, were isolated from serum and CSF of cirrhotic patients with HE. The interest generated by these reports was tempered by the fact that the concentrations of benzodiazepines reported in blood, CSF and brain extracts of HE patients are very low (well below levels associated with their sedative actions).

An alternative theory to explain increased GABAergic neurotransmission in HE has recently emerged. Two distinct types of benzodiazepine receptors are expressed in the brain, namely the type referred to in the previous section of this chapter, forming part of the GABA-benzodiazepine receptor complex, situated on the postsynaptic neuronal membrane and a second type, the "peripheral-type" benzodiazepine receptor (PTBR), localized on the outer mitochondrial membrane of astrocytes and other glial cells (see Sect. "Peripheral-Type" Benzodiazepine Receptors, Neurosteroids. Endogenous ligands for the PTBR include the neuropeptide diazepam binding inhibitor (DBI) and its processing peptide octadecaneuropeptide (ODN) and increases of these peptides have been reported in HE (Rothstein et al., 1989). Using an immunocytochemical technique and an antibody of high specific activity to synthetic ODN, it was demonstrated that portacaval anastomosis results in increased ODN-immunolabelling in astrocytes and other non-neuronal elements (Butterworth et al., 1991) in several brain regions.

Activation of PTBRs in the brain in liver failure results in increased synthesis of a novel class of compounds known as neurosteroids that are synthesized in the brain mainly by astrocytes independent of peripheral steroidal sources (adrenals and gonads). Neurosteroids bind and modulate different types of neural receptors; effects on the GABA-A receptor complex are the most extensively studied. For example, the neurosteroid tetrahydroprogesterone (allopregnanolone), and tetrahydrodeoxy-corticosterone (THDOC) are potent positive allosteric modulators of the GABA-A receptor complex (Fig. 9.7). As a consequence, neurosteroids stimulate inhibitory neurotransmission in the CNS, and it has been suggested that neuroinhibitory

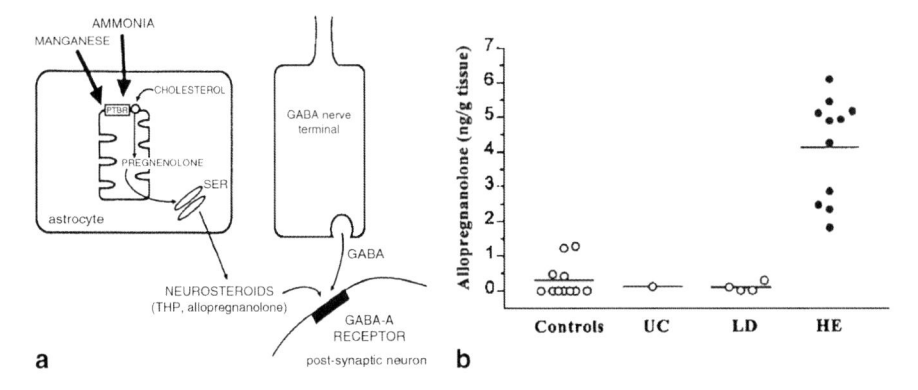

Fig. 9.7 (a) Activation of the astrocytic mitochondrial PTBR results in the synthesis of neuros-teroids such as allopregnanolone with potent agonist actions at the GABA-A receptor complex. **(b)** Allopregnanolone levels are increased in autopsied brain tissue from cirrhotic patients who died in hepatic coma (HE) but not in non-encephalopathic cirrhotic patients (LD), or in a patient who died in uremic coma (UC). Modified from Ahboucha et al., 2005)

changes including "increased GABA-ergic tone" are the consequences of increased brain concentrations of allopregnanolone (Ahboucha and Butterworth, 2007). In fact, increased brain concentrations of allopregnanolone have been reported in autopsied brain tissue from cirrhotic patients who died in hepatic coma (Ahboucha et al., 2005) (Fig. 9.7).

Serotonin

CSF tryptophan concentrations are increased in patients in hepatic coma (Young et al., 1975) and portacaval anastomosis results in increases in its blood–brain barrier transport (Huet et al., 1981). Since tryptophan hydroxylation (the rate-limiting step in brain 5HT synthesis) is not saturated at normal blood and brain tryptophan concentrations, increased availability of tryptophan has the potential to result in increased 5HT synthesis in brain. Evidence consistent with increased 5HT synthesis and turnover in the brain in human and experimental HE includes the report of increased concentrations of the 5HT metabolite 5-hydroxyindoleacetic acid (5HIAA) in CSF of cirrhotic patients in hepatic coma (Young et al., 1975), in autopsied brain tissue from HE patients (Bergeron et al., 1989) and in the brains of portacaval-shunted rats (Bergeron et al., 1990). Moreover, these studies showed that the 5HIAA/5HT concentration ratio (considered to reflect 5HT turnover) is elevated in brain tissue from cirrhotic patients who died in hepatic coma and in the brains of portacaval-shunted rats administered ammonium salts to precipitate severe encephalopathy. Direct measurement of 5HT turnover in the brain, using an in vivo decarboxylase inhibition assay, also reveals increased 5HT turnover (Bengtsson et al., 1991) in the brains of portacaval-shunted animals. Increased brain

5HT turnover correlates with the degree of portal-systemic shunting and level of hyperammonemia in rats following variable portal vein stenosis (Lozeva et al., 2004), an animal model of minimal HE. Other evidence for altered 5HT metabolism and function in human HE includes the report of increased activities of monoamine oxidase MAO_A and of MAO_A mRNA in autopsied brain tissue from cirrhotic patients who died in hepatic coma (Mousseau et al., 1997) (in human brain, MAO_A is responsible for 5HT breakdown) and concomitant increases in concentrations of the 5HT metabolite, 5HIAA (Bergeron et al., 1989). Increased densities of postsynaptic serotonin ($5HT_2$) receptors, measured using 3H-ketanserin, were also described in this material (Raghavendra Rao and Butterworth, 1994). Together, these findings suggest a 5HT synaptic *deficit* in HE. Serotonin plays a key role in the regulation of sleep mechanisms and altered 5HT neurotransmission has been implicated in a range of psychiatric disorders in humans. Thus, the changes in 5HT turnover in brain could relate to the early signs and symptoms characteristic of HE such as altered sleep patterns and depression.

Other neuroactive and neurotoxic metabolites of tryptophan are also reportedly increased in the brain with chronic liver failure. One such example is quinolinic acid (QUIN) synthesized from tryptophan via the kynurenine pathway. QUIN synthesis is particularly sensitive to increased availability of tryptophan. QUIN has been identified in both rodent and human brain extracts and cerebral cortical QUIN concentrations are elevated in rats following portacaval anastomosis (Moroni et al., 1986a). Furthermore, QUIN concentrations are increased up to sevenfold in CSF and autopsied frontal cortex of cirrhotic patients who died in hepatic coma (Moroni et al., 1986b).

Another neuroactive tryptophan metabolite is the trace amine tryptamine. Tryptamine content of CSF is increased in HE patients (Young and Lal, 1980) and a significant loss of ^{3}H-tryptamine binding sites (consistent with down regulation of these sites) has been described in autopsied brain tissue from these patients (Mousseau et al., 1994).

Dopamine

The presence of extrapyramidal signs and symptoms in patients with end-stage chronic liver failure has prompted, by analogy with the well-established dopamine (DA) deficit in Parkinson's Disease, evaluation of the DA system in HE. Studies in autopsied brain tissue from cirrhotic patients who died in hepatic coma reveal severalfold increases of the DA metabolite homovanillic acid (Bergeron et al., 1989). Similar findings, consistent with increased DA metabolism or turnover, were reported in the brains of rats following portacaval anastomosis (Bergeron et al., 1995). A possible explanation for these findings could relate to the report of increased activities of the monoamine metabolizing enzyme MAO-A (Mousseau et al., 1997). Densities of the post-synaptic DA receptor (D_2 receptor) are reduced in autopsied pallidum of cirrhotic patients (Mousseau et al., 1993), a finding that could result from manganese deposition in this material (Pomier Layrargues et al., 1995).

Histamine

Histamine is involved in a wide range of physiological functions including sleep, arousal and circadian rhythmicity. Alterations of the brain histamine system in HE include increases in concentrations of both histamine and its principle metabolite tele-methylhistamine in the hypothalamus of portacaval-hunted rats together with increased histamine release from the hypothalamus of these animals (Lozeva-Thomas et al., 2004), suggesting increased histaminergic activity. Increased brain histamine could result from increased brain availability of the precursor amino acid L-histidine. Activation of postsynaptic histamine H_1 receptors suppresses deep slow-wave sleep (Tasaka, 1994) and these receptors are implicated in the entrainment of circadian rhythms to the light-dark cycle in mammals (Jacobs et al., 2000). HE in chronic liver failure is characterized by decreased total duration of slow-wave sleep and disorganization of the normal sleep cycle (Cordoba et al., 1998). Similar changes in sleep patterns and diurnal rhythms have been reported in portacaval-shunted rats. Reports of upregulation of histamine H_1 receptors in the cerebral cortex of HE patients and of portacaval-shunted rats together with changes in serotonin receptors in HE patients' brains could account for the accompanying sleep disturbances. Histamine H_1 receptor blockers improve sleep quality in humans and significant improvements in circadian rhythmicity have been reported in portacaval-shunted rats administered the H_1 receptor blocker mepyramine (Lozeva et al., 2000).

Opioid System

Cirrhotic patients are sensitive to morphine and portacaval shunting is known to lead to increased pain sensitivity, phenomena that involve the endogenous opioid neurotransmitter system. Increased circulating levels of the endogenous opioid met-enkephalin have been reported in patients with primary biliary cirrhosis (Thornton and Losowsky, 1988) and brain levels of β-endorphin are increased in experimental chronic liver failure (Panerai et al., 1982). Portacaval anastomosis in rats results in region-selective increases of μ and δ opioid receptor sites in the brain (DeWaele et al., 1996) a finding that has been linked to increased ethanol consumption in a free-choice drinking paradigm in these animals (DeWaele et al., 1997). Extrapolation of these findings to humans suggests the intriguing possibility that significant liver disease resulting in activation of the brain opioid system could result in increased alcohol consumption, resulting in a vicious cycle and acceleration of alcoholic cirrhosis and its complications.

Oxidative/Nitrosative Stress and Inflammation

Nitric Oxide

Both acute hyperammonemia (Kosenko et al., 1995) and chronic hyperammonemia resulting from portacaval anastomosis (Raghavendra Rao et al., 1997) result in increased expression of nitric oxide synthase (NOS-I isoform) in the brain. Nitric oxide synthase activities are also increased by an ammonia-induced stimulation of L-arginine uptake into neuronal preparations both in vitro and in vivo (Raghavendra Rao et al., 1997). Selective NOS-I inhibitors, such as nitroarginine, inhibit many of the metabolic and toxic effects of acute hyperammonemia (Kosenko et al., 1995). However, NOS-I inhibitors were ineffective in the prevention of intracranial hypertension in portacaval shunted rats administered ammonium salts (Larsen et al., 2001).

Exposure of cultured astrocytes to millimolar concentrations of ammonia results in oxidative stress and in protein tyrosine nitration (Schliess et al., 2002), a process that was dependent upon stimulation of the inducible isoform of NOS (NOS-2). Astrocytic protein tyrosine nitration was also described in the brain of rats following portacaval anastomosis where tyrosine nitration of GS was described (Schliess et al., 2002). Reduced GS enzyme activity resulting from tyrosine nitration offers a cogent explanation for the consistent reports of decreased capacity for ammonia removal in the brain of animals following portacaval anastomosis.

Inflammation

It is becoming increasingly evident that ammonia neurotoxicity is not the only pathophysiologic process with the potential to adversely affect cerebral function in ALF. In particular, there is evidence to suggest that infection and inflammation also play a significant role. The development of infection in patients with ALF is associated with a more rapid progression of HE and cerebral edema (Vaquero et al., 2003) and large clinical studies have convincingly shown a higher prevalence of infection together with the "systemic inflammatory response syndrome" (SIRS) in ALF patients. SIRS is a response to the presence of proinflammatory cytokines such as the interleukins IL-1 and IL-6 and Tumor Necrosis Factor (TNFα) (Blei, 2004) and increased circulating levels of cytokines have consistently been reported in ALF patients (Nagaki et al., 2000; Jalan et al., 2002). Cytokines are also formed and released by the necrotic liver. Jalan et al. (2002) reported a beneficial effect of hepatectomy in an ALF patient with uncontrolled intracranial hypertension and showed that hepatectomy reduced ICP and concomitantly reduced circulating cytokines.

It is well established that clinical conditions characterized by high cytokine levels and cytokine activation of the vascular endothelium manifest altered *blood–brain barrier permeability* (Sharief and Thompson, 1992). The presence of endotoxin, known to stimulate cytokine production, leads to increased blood–brain barrier permeability in experimental ALF (Tominaga et al., 1991). Furthermore, arteriovenous-difference studies by Jalan et al. (2002) reveal a net production of pro-inflammatory cytokines in the brain of patients with ALF.

Therapeutic Advances

Therapy for HE continues to be a challenge. The reasons for this include a lack of standardization of assessment of mental status, the resistance of some ethics boards to sanction placebo-controlled trials, confounding issues such as the presence of precipitating factors like infection, hyponatremia and unidentified neuroactive drugs and difficulties related to the effective blinding of patients and clinical investigators to the treatment regimen. Attention to these issues is urgently needed, the resolution of which would undoubtedly provide a stimulus for translational research in HE.

Clinical studies continue to focus upon the reduction of circulating ammonia with regrettably little attention paid to the therapeutic targeting of the brain.

Ammonia-Lowering Strategies

Restriction of dietary protein is no longer recommended as a means of prevention of HE in cirrhotic patients. Long-term nitrogen restriction is potentially harmful and a positive nitrogen balance is essential in order to promote liver regeneration and to increase the capacity for ammonia removal by skeletal muscle. Protein intake in the 1–2 g kg^{-1} per day range is now generally recommended in order to maintain an adequate nitrogen balance.

Non-absorbable disaccharides such as lactulose and lactitol are still routinely used to decrease ammonia production in the gut despite a lack of adequate controlled clinical trials. In the case of lactulose, the ammonia-lowering effect appears to involve increased fecal nitrogen excretion by facilitation of the incorporation of ammonia into bacteria as well as a cathartic effect. Antibiotics such as neomycin have traditionally been used to lower blood ammonia by inhibition of ammonia production by intestinal bacteria. However, neomycin therapy is associated with significant toxic side effects and is increasingly being replaced by alternative antibiotics such as rifaximin.

A recent study demonstrated significant lowering of blood ammonia and concomitant improvement in neuropsychiatric status in cirrhotic patients with grade I–II HE following administration of the novel hypoglycaemic agent acarbose (Gentile et al., 2005). One suggested mechanism of action of acarbose involves inhibition of proteolytic flora responsible for gut ammonia production. Renewed interest (after a 40 year hiatus) has recently begun in the area of lowering of gut ammonia production involving oral supplementation with probiotic gut flora. The objective of these probiotics is to increase the intestinal production of urease-negative bacteria, thereby decreasing gut ammonia production. Lactic acid producing probiotics have the advantage of reducing gut ammonia absorption. Symbiotic preparations consist of both lactic acid-producing probiotics and fermentable fibre. Initial controlled clinical trials with these agents led to significant improvement of neurological status and lowering of venous ammonia (Liu et al., 2004) in cirrhotic patients.

An alternative means of lowering blood ammonia in acute or chronic liver failure involves the stimulation of ammonia fixation. Ammonia is normally removed by urea formation (periportal hepatocytes) as well as by glutamine formation (perivenous hepatocytes, muscle and brain). Strategies aimed at stimulating residual urea or glutamine synthesis have emerged in recent years and one of the most successful, L-ornithine L-aspartate, has been shown in controlled clinical trials to be effective in lowering blood ammonia and concomitantly improving neuropsychiatric status in cirrhotic patients (Kircheis et al., 1997). Studies in experimental animals reveal that, in chronic liver failure, the beneficial effect of L-ornithine L-aspartate is primarily due to the stimulation of urea formation in the liver and in the stimulation of glutamine formation by skeletal muscle. In experimental ALF, the ammonia-lowering effect of L-ornithine L-aspartate appears to be due predominantly to the stimulation of muscle glutamine synthetase (Rose et al., 1998). However, no studies to date have evaluated the effects of L-ornithine L-aspartate in ALF patients. Successful lowering of blood ammonia concentrations in cirrhotic patients has also been accomplished using sodium benzoate.

A particularly novel approach to lowering blood and brain ammonia concentrations in liver failure involves the use of l-carnitine and results of a recent study, following up on studies in experimental animals (Therrien et al., 1997) demonstrated a protective effect of L-carnitine agonist ammonia-precipitated encephalopathy in cirrhotic patients (Malaguarnera et al., 2005).

Neuropharmacologic Advances

As the precise pathophysiologic mechanisms and alterations in neurotransmitter systems responsible for the pathogenesis of HE become more clearly defined, novel pharmacological approaches are starting to emerge.

A number of controlled clinical trials have been performed to evaluate the efficacy of the benzodiazepine receptor antagonist flumazenil in HE patients. In a subset of these patients, improvements following flumazenil have been spectacular. However, enthusiasm for this approach has been tempered by the possible confounding effects of prior exposure of patients to pharmaceutical benzodiazepines (used as sedatives or as part of an endoscopic work-up in cirrhotic patients) and by the poor correlation between the clinical response and blood levels of benzodiazepines in these patients. Adding to these difficulties is the short half-life and lack of an oral formulation for flumazenil.

Benzodiazepine partial inverse agonists display weak negative intrinsic activity and act as mild GABA-A receptor antagonists (Haefely, 1994). One such agent, Ro15–4513, ameliorates the symptoms of HE in rats with thioacetamide-induced

(Yurdaydin et al., 1993) or ischemic (Bosman et al., 1991) liver failure. Similar results were obtained with a second benzodiazepine partial inverse agonist sarmazenil. It was recently determined that a probable mechanism responsible for the beneficial effect of these agents most likely involves attenuation of the effects of GABA agonist neuro-steroids such as allopregnanolone (Ahboucha et al., 2006). Surprisingly, no studies have so far been initiated to test the efficacy of benzodiazepine partial inverse agonists in patients with either acute or chronic liver failure.

Both L-DOPA and the dopamine receptor agonist bromocriptine have been used in clinical trials in patients with PSE. While results were not encouraging in terms of overall cognitive improvement, it is possible that they have a beneficial effect on motor performance.

Studies in experimental animals with toxic or ischemic liver failure demonstrate that antagonists of certain serotonin receptor subtypes and administration of the glutamate (NMDA) receptor antagonist memantime improve neurological function and/or prevent brain edema in these animals (Vogels et al., 1997).

Following up on the demonstration of a beneficial effect of histamine H_1 antagonist on neurological function in an experimental animal model of HE (Lozeva et al., 2000) results of a controlled clinical trial revealed improvements in sleep quality in cirrhotic patients with minimal HE following administration of the H_1 blocker hydroxyzine (Spahr et al., 2007).

Hypothermia

Hypothermia extends the survival time and prevents the development of brain edema in rats with ALF, caused by hepatic devascularization and mild hypothermia (33—35°C), reduces ammonia-induced brain swelling and increased intracranial pressure in portacaval-shunted rats administered ammonium salts. These findings have led to the successful use of mild hypothermia for the treatment of uncontrolled intracranial hypertension in patients with ALF (Jalan et al., 1999). Mechanisms so far identified that underlie the beneficial effect of hypothermia in ALF include reduced blood–brain transfer of ammonia, improved cerebrovascular hemodynamics and normalization of extracellular brain amino acid patterns (for review of these mechanisms, see Vaquero et al., 2005). Mild hypothermia also improves hepatic function in experimental toxic liver injury (Vaquero et al., 2007) Mild-to-moderate hypothermia has the potential to serve as an important strategy in the management of patients with ALF awaiting liver transplantation.

Liver Support Systems

Artificial liver support systems aimed at the normalization of plasma composition of essential compounds and removal of toxins continue to be evaluated for the management and treatment of HE. Biological systems are generally based upon bioreactors loaded with hepatocytes. However, controlled clinical trials using such systems are limited. Non-biological systems such as hemodialysis and albumin dialysis remove water soluble or protein bound substances including ammonia, manganese and benzodiazepines as well as certain cytokines (Chamuleau et al., 2006).

Summary

There can be little doubt, given the overwhelming body of evidence from both clinical and experimental studies, that ammonia plays a major role in the pathogenesis of HE in both acute and chronic liver failure. Ammonia is increased in the brain in both acute and chronic liver failure to attain millimolar concentrations at coma stages of encephalopathy. Levels of the ammonia-detoxification product, glutamine, are increased in brain and correlate with the degree of neurological impairment in chronic liver disease. The neuropathological feature characteristic of HE in chronic liver failure, Alzheimer type II astrocytosis, is also encountered in chronic hyper-ammonemic syndromes of non-hepatic origin. Millimolar concentrations of ammonia have direct toxic effects on both CNS inhibition and excitation. Prolonged exposure of brain to increased ammonia results in disruption of inter-cellular metabolic coupling and in a defect in glutamatergic synaptic regulation. Accumulation of brain glutamine leads to increased brain uptake of tryptophan, which in turn results in increased synthesis of neuroactive (tryptamine) and neurotoxic (quinolinic acid) metabolites of serotonin. Increased activities of monoamine oxidase and of serotonin receptor densities are consistent with a serotonin synaptic deficit in human HE. Precipitous increases of brain ammonia are implicated in the pathogenesis of brain edema and its complications in ALF.

It is unlikely, however, that ammonia's direct and indirect effects on brain functions are the sole cause of neurologic dysfunction in liver failure. Other neurotoxic substances such as manganese and proinflammatory cytokines could gain entry to brain in liver failure and act synergistically with ammonia-related mechanisms.

Studies of cerebral energy metabolism using either biochemical or spectroscopic techniques have provided little convincing evidence that HE is the consequence of primary cellular energy failure. However, brain glucose utilization is altered in liver failure being selectively decreased in anterior cingulate cortex, a brain structure associated with visual processing and attention. Brain glucose oxidation is decreased in acute and chronic liver failure resulting in lactate accumulation.

In addition to alterations of astrocyte morphology, liver failure leads to alterations in expression of genes coding for important astrocyte proteins including the structural protein GFAP, glutamate and glycine transporters as well as the mitochondrial PTBR.

Liver failure leads to significant changes in synthesis, degradation and/or function of a wide range of neurotransmitters including glutamate, GABA, the monoamines (serotonin, dopamine, histamine) and the opioid system. In the case of glutamate, modifications include decreased glutamate transport capacity resulting in increased extracellular glutamate and NMDA receptor activation. The discovery of neurosteroids with potent agonist modulatory action at the GABA-A receptor may explain the notion of "increased GABAergic tone" in HE. Alterations of monoaminergic (serotonin, histamine) systems have been implicated in the pathogenesis of the sleep disturbances and depression that is prevalent in early HE resulting from chronic liver failure.

There is evidence to suggest that oxidative and nitrosative stress play a role in the pathogenesis of HE. Induction of NOS isoforms have been described in the brain in liver failure and the consequent nitration of key brain proteins such as glutamine synthetase has been reported.

Management and treatment of HE continues to rely heavily on reduction of circulating ammonia using a range of agents including those aimed at reducing gut ammonia production (lactulose, acarbose, antibiotics, probiotics) and those aimed at increased ammonia fixation (l-ornithine l-aspartate, benzoate). As cerebral mechanisms become better elucidated, therapies aimed at the brain are starting to appear. There is evidence from studies in animal models of liver failure for a beneficial action of a wide range of neuropharmacologic agents that include the NMDA receptor antagonist memantine, histamine H_1 blockers, opioid receptor antagonists, and benzodiazepine receptor inverse agonists as well as agents such as Lconiinee. There is convincing evidence for a beneficial effect of mild hypothermia in the management of patients with ALF, the beneficial effects being multiple, including reduction of blood–brain ammonia transfer, normalization of brain glucose oxidation, attenuation of oxidative stress and of altered gene expression.

References

Adams RD, Foley JM. The neurological disorder associated with liver disease. In: *Metabolic and Toxic Diseases of the Nervous System.* (H.H. Merritt, and C.C. Hare, eds.)Vol. 32. Williams and Wilkins, Baltimore, USA, pp. 198–237, 1953

Ahboucha S, Butterworth RF. The neurosteroid system: Implication in the pathophysiology of hepatic encephalopathy. *Neurochem. Int.*, 52, 575–587, 2008

Ahboucha S, Pomier-Layrargues G, Mamer O, Butterworth RF. Increased brain concentrations of a neuroinhibitory steroid in human hepatic encephalopathy. *Ann. Neurol.*, 58, 169–170, 2005

Ahboucha S, Coyne L, Hirakawa R, Butterworth RF, Halliwell RF. An interaction between benzodiazepines and neuroactive steroids at $GABA_A$ receptors in cultured hippocampal neurons. *Neurochem. Int.*, 48, 703–707, 2006

Bates TE, Williams SR, Kauppinen RA, Godian DG. Observation of cerebral metabolites in an animal model of ALF in vivo: ^1H and ^{31}P nuclear magnetic resonance study. *J. Neurochem.*, 53, 102–110, 1989

Bélanger M, Desjardins P, Chatauret N, Butterworth RF. Loss of expression of glial fibrillary acidic protein in acute hyperammonemia. *Neurochem. Int.*, 41(2–3), 155–160, 2002

Bélanger M, Ahboucha S, Desjardins P, Butterworth RF. Upregulation of peripheral-type (mitochondrial) benzodiazepine receptors in hyperammonemic syndromes: Consequences for neuronal excitability. *Adv. Mol. Cell Biol.*, 31(3), 983–997, 2004

Bélanger M, Desjardins P, Chatauret N, Butterworth RF. Selectively increased expression of the astrocytic/endothelial glucose transporter protein GLUT1 in acute liver failure (p NA). *Glia*, 53(5), 557–562, 2006

Bender AS, Schousboe A, Reichelt W, Norenberg MD. Ionic mechanisms in glutamate-induced astrocyte swelling: Role of K^+ influx. *J. Neurosci. Res.*, 52, 307–321, 1998

Bengtsson F, Bugge M, Johansen KH, Butterworth RF. Brain tryptophan hydroxylation in the portacaval shunted rat: A hypothesis for the regulation of serotonin turnover in vivo. *J. Neurochem.*, *56*, 1069–1074, 1991

Bergeron M, Reader TA, Pomier Layrargues G, Butterworth RF. Monoamines and metabolites in autopsied brain tissue from cirrhotic patients with hepatic encephalopathy. *Neurochem. Res.*, *14*, 853–859, 1989

Bergeron M, Swain MS, Reader TA, Grondin L, Butterworth RF. Effect of ammonia on brain serotonin metabolism in relation to function in the portacaval shunted rat. *J. Neurochem.*, *55*, 222–229, 1990

Bergeron M, Reader TA, Pomier Layrargues G, Butterworth RF. Monoamines and metabolites in autopsied brain tissue from cirrhotic patients with hepatic encephalopathy. *Neurochem. Res.*, *20*, 79–86, 1995

Blei AT. Infection, inflammation and hepatic encephalopathy, synergism redefined. *J. Hepatol.*, *40*, 327–330, 2004

Bosman DK, van den Buijs CA, de Haan JG, Maas MA, Chamuleau RA. The effects of benzodiazepine-receptor antagonists and partial inverse agonists on acute hepatic encephalopathy in the rat. *Gastroenterology*, *101*, 772–781, 1991

Bosman DK, Deutz NEP, Maas MAW, van Eijk MHH, Smit JJH, de Haan JG, Chamuleau RAFM. Amino acid release from cerebral cortex in experimental ALF, studied by in vivo cerebral cortex microdialysis. *J. Neurochem.*, *59*, 591–599, 1992

Butterworth RF. Hepatic encephalopathy and brain edema in acute hepatic failure: Does glutamate play a role? *Hepatology*, *25*, 1032–1034, 1997

Butterworth RF. Neurotransmitter dysfunction in hepatic encephalopathy: New approaches and new findings. *Metab. Brain Dis.*, *16*, 55–65, 2001

Butterworth RF. Neuronal cell death in hepatic encephalopathy. *Metab. Brain Dis.*, *22*, 309–320, 2007

Butterworth RF, Giguère JF. Cerebral amino acids in portal-systemic encephalopathy: Lack of evidence for altered γ-aminobutyric acid (GABA) function. *Metab. Brain Dis.*, *1*, 221–228, 1986

Butterworth RF, Giguère JF, Michaud J, Lavoie J, Pomier Layrargues G. Ammonia: Key factor in the pathogenesis of hepatic encephalopathy. *Neurochem. Pathol.*, *6*, 1–12, 1987

Butterworth RF, Lavoie J, Giguère JF, Pomier Layrargues G. Affinities and densities of high affinity ^{3}H-muscimol (GABA-A) binding sites and central benzodiazepine receptors are unchanged in autopsied brain tissue from cirrhotic patients with hepatic encephalopathy. *Hepatology*, *8*, 1084–1088, 1988

Butterworth RF, Tonon MC, Désy L, Giguère JF, Vaudry H, Pelletier G. Increased brain content of the endogenous benzodiazepine receptor ligand octadecaneuropeptide (ODN) following portacaval anastomosis in the rat. *Peptides*, *12*, 119–125, 1991

Cagnin A, Taylor-Robinson SD, Forton DM, Banati RB. In vivo quantification of cerebral "peripheral benzodiazepine binding site" in minimal hepatic encephalopathy: A [^{11}C]R-PK11195 positron emission tomography study. *J. Hepatol.*, *34*, 58, 2001

Chamuleau RA, Poyck PP, van de Kerkhove MP. Bioartificial liver: Its pros and cons. *Ther. Apher. Dial.*, *10*(2), 168–174, 2006

Chan H, Hazell AS, Desjardins P, Butterworth RF. Effects of ammonia on glutamate transporter (GLAST) protein and mRNA in cultured rat cortical astrocytes. *Neurochem. Int.*, *37*, 243–248, 2000

Chatauret N, Butterworth RF. Effects of liver failure on inter-organ trafficking of ammonia: Implications for the treatment of encephalopathy. *J. Gastroenterol. Hepatol.*, *19*, S219–S223, 2004

Clemmesen JO, Larsen FS, Kondrup J, Hansen BA, Ott P. Cerebral herniation in patients with ALF is correlated with arterial ammonia concentration. *Hepatology*, *29*, 648–653, 1999

Cordoba J, Cabrera J, Lataif L, Penev P, Zee P, Blei AT. High prevalence of sleep disturbance in cirrhosis. *Hepatology*, *27*(2), 339–345, 1998

Desjardins P, Butterworth RF. The "peripheral-type" benzodiazepine (3) receptor in hyperammonemic disorders. *Neurochem. Int.*, *41*, 109–114, 2002

Desjardins P, Bandeira P, Raghavendra Rao VL, Ledoux S, Butterworth RF. Increased expression of the peripheral-type benzodiazepine receptor-isoquinoline carboxamide binding protein in mRNA brain following portacaval anastomosis. *Brain Res.*, *758*, 255–258, 1997

Desjardins P, Rama Rao VK, Michalak A, Rose C, Butterworth RF. Effect of portacaval anastomosis on glutamine synthetase protein and gene expression in brain, liver and skeletal muscle. *Metab. Brain Dis.*, *14*, 273–282, 1999

Devenyi AG, Barron TF, Mamourian AC. Dystonia, hyperintense basal ganglia, and high whole blood manganese levels in Alagilles's syndrome. *Gastroenterology*, *106*, 1068–1071, 1994

DeWaele JP, Audet RM, Leong DK, Butterworth RF. Portacaval anastomosis results in region-selective changes of β-endorphin content and of μ and δopiod receptor densities in rat brain. *Hepatology*, *24*, 895–901, 1996

DeWaele JP, Audet RM, Rose C, Butterworth RF. The portacaval-shunted rat: A new model for the study of the mechanisms controlling voluntary ethanol consumption and ethanol dependence. *Alcohol Clin. Exp. Res.*, *21*, 305–310, 1997

Felipo V, Butterworth RF. Neurobiology of ammonia. *Prog. Neurobiol.*, *67*, 259–279, 2002

Ferenci P, Lockwood A, Mullen K, Tarter R, Weissenborn K, Blei AT, and the Members of the Working Party. Hepatic encephalopathy – definition, nomenclature, diagnosis and quantification: Final report of the working party at the 11th World Congress of Gastroenterology, Vienna 1998. *Hepatology*, *35*, 716–721, 2002

Gabuzda G Jr., Philips GB, Davidson CS. Reversible toxic manifestations in patients with cirrhosis of the liver given cation-exchange resins. *New Engl. J. Med.*, *246*, 124–130, 1952

Ganda OP, Ruderman NB. Muscle nitrogen metabolism in chronic hepatic insufficiency. *Metabolism*, *25*(4), 427–435, 1976

Gentile S, Guarino G, Romano M, Alagia IA, Fierro M, Annunziata S, Magliano PL, Gravina AG, Torella R. A randomized controlled trial of acarbose in hepatic encephalopathy. *Clin. Gastroenterol. Hepatol.*, *3*, 184–191, 2005

Giguère JF, Hamel E, Butterworth RF. Increased densities of binding sites for the "peripheral-type" benzodiazepine receptor ligand ^{3}H-PK11195 in rat brain following portacaval anastomosis. *Brain Res.*, *585*, 295–298, 1992

Girard G, Giguère JF, Butterworth RF. Effect of portacaval anastomosis on ammonia metabolism in brain and liver. In: *Hepatic Encephalopathy: Pathophysiology and Treatment*. (R.F. Butterworth, and G. Pomier Layrargues, eds.), Humana Press, New Jersey., pp. 79–89, 1989

Gyr K, Meier R Haussler J, Bouletreau P, Fleig WE, Gatta A, Holstege A, Pomier Layrargues G, Schalm SW, Groeneweg M, Scollo-Lavizzari G, Ventura E, Zeneroli ML, Williams R, Yoo Y, Amrein R. Evaluation of the efficacy and safety of flumazenil in the treatment of portal systemic encephalopathy: A double blind, randomised, placebo controlled multicentre study. *Gut*, *39*, 319–324, 1996

Haefely WE. Allosteric modulation of the GABA$_A$ receptor channel: A mechanism for interaction with a multitude of central nervous system function. In: *The Challenge of Neuropharmacology*. (H. Möhler, and M. Da Prada, eds.), F. Hoffman-La Roche , Basel, Switzerland, pp. 15–40, 1994

Hindfelt B, Plum F, Duffy TE. Effects of acute ammonia intoxication on cerebral metabolism in rats with portacaval shunts. *J. Clin. Invest.*, *59*, 386–396, 1977

Huet PM, Pomier Layrargues G, Duguay L, Du Souich P. Blood—brain barrier transport of tryptophan and phenylalanine: Effect of portacaval shunt in dogs. *Am. J. Physiol.*, *4*, 163–169, 1981

Jacobs EH, Yamatodani A, Timmerman H. Is histamine the final neurotransmitter in the entrainment of circadian rhythms in mammals? *Trends Pharmacol. Sci.*, *21*(8), 293–298, 2000

Jalan R, Damink SW, Deutz NE, Lee A, Hayes PC. Moderate hypothermia for uncontrolled intracranial hypertension in acute liver failure. *Lancet*, *354*(9185), 1164–1168, 1999

Jalan R, Pollok A, Shah SH, Madhavan K, Simpson KJ. Liver derived pro-inflammatory cytokines may be important in producing intracranial hypertension in ALF. *J. Hepatol.*, *37*, 536–538, 2002

Keiding S, Sørensen M, Bender D, Munk OL, Ott P, Vilstrup H. Brain metabolism of ^{13}N-ammonia during acute hepatic encephalopathy in cirrhosis measured by positron emission tomography. *Hepatology*, *43*, 42–50, 2006

Kircheis G, Nilius R, Held C, Berndt H, Buchner M, Gortelmeyer R, Hendricks R, Kruger B, Kuklinski B, Meister H, Otto HJ, Rink C, Rosch W, Stauch S. Therapeutic efficacy of l-ornithine-l-aspartate infusions in patients with cirrhosis and hepatic encephalopathy: Results of a placebo-controlled, double-blind study. *Hepatology*, *25*(6), 1351–1360, 1997

Knecht K, Michalak A, Rose C, Rothstein JD, Butterworth RF. Decreased glutamate transporter (GLT-1) expression in frontal cortex of rats with ALF. *Neurosci. Lett.*, *229*, 201–203, 1997

Kosenko E, Kaminsky Y, Grau E, Minana MD, Grisolia S, Felipo V. Nitroarginine, an inhibitor of nitric oxide synthetase, attenuates ammonia toxicity and ammonia-induced alterations in brain metabolism. *Neurochem. Res.*, *20*(4), 451–456, 1995

Kril JJ, Butterworth RF. Diencephalic and cerebellar pathology in alcoholic and non-alcoholic patients with end-stage liver disease. *Hepatology*, *24*(4), 1303, 1996

Lai JCK, Cooper AJL. Brain α-ketoglutarate dehydrogenase: Kinetic properties, regional distribution and effects of inhibitors. *J. Neurochem.*, *47*, 1376–1386, 1986

Larsen FS, Gottstein J, Blei AT. Cerebral hyperemia and nitric oxide synthase in rats with ammonia-induced brain edema. *J. Hepatol.*, *34*(4), 548–554, 2001

Laubenberger J, Haussinger D, Boyer S, Guffer H, Henning J, Lange M. Proton magnetic resonance spectroscopy of brain in symptomatic and asymptomatic patients with liver cirrhosis. *Gastroenterology*, *112*, 1610–1616, 1997

Lavoie J, Giguère JF, Pomier Layrargues G, Butterworth RF. Activities of neuronal and astrocytic marker enzymes in autopsied brain tissue from patients with hepatic encephalopathy. *Metab. Brain Dis.*, *2*, 283–290, 1987a

Lavoie J, Giguère JF, Pomier Layrargues G, Butterworth RF. Amino acid changes in autopsied brain tissue from cirrhotic patients with hepatic encephalopathy. *J. Neurochem.*, *49*, 692–697, 1987b

Lavoie J, Pomier Layrargues G, Butterworth RF. Increased densities of "peripheral-type" benzodiazepine receptors in autopsied brain tissue from cirrhotic patients with hepatic encephalopathy. *Hepatology*, *11*, 874–878, 1990

Liu Q, Duan ZP, Ha DK, Bengmark S, Kurtovic J, Riordan SM. Synbiotic modulation of gut flora: Effect on minimal hepatic encephalopathy in patients with cirrhosis. *Hepatology*, *39*(5), 1441–1449, 2004

Lockwood AH, Yap EWH, Wong W-H. Cerebral ammonia metabolism in patients with severe liver disease and minimal hepatic encephalopathy. *J. Cereb. Blood Flow Metab.*, *11*, 337–341, 1991

Lockwood AH, Weissenborn K, Butterworth RF. An image of the brain in patients with liver disease. *Curr. Opin. Neurol.*, *10*, 525–533, 1997

Lockwood AH, Weissenborn K, Bokemeyer M, Tietge U, Burchert W. Correlations between cerebral glucose metabolism and neuropsychological test performance in non-alcoholic cirrhotics. *Metab. Brain Dis.*, *17*, 29–40, 2002

Lozeva V, Valjakka A, Lecklin A, Olkkonen H, Hippelainen M, Itkonen M, Plumed C, Tuomisto L. Effects of the histamine H(1) receptor blocker, pyrilamine, on spontaneous locomotor activity of rats with long-term portacaval anastomosis. *Hepatology*, *31*(2), 336–344, 2000

Lozeva V, Montgomery JA, Tuomisto L, Rochelean B, Pannunzio M, Huet PM, Butterworth RF. Increased brain serotonin turnover correlates with the degree of shunting and hyperammonemia in rats following variable portal vein stenosis. *J. Hepatol.*, *40*, 742–748, 2004

Lozeva-Thomas V, Ahonen P, Chatauret N, Tuomisto L, Butterworth RF. Brain histamine in experimental acute liver failure: Effects of l-histidine loading. *Inflamm. Res.*, *53*(Suppl. 1), S55–S56, 2004

Maddison JE, Watson WEJ, Dodd PR, Johnston GAR. Alterations in cortical ^{3}H-kainate and α-^{3}H-amino-3-hydroxy-5-methyl-4-isoxazolepropionic acid binding in a spontaneous canine model of chronic hepatic encephalopathy. *J. Neurochem.*, *56*, 1881–1888, 1991

Malaguarnera M, Pistone G, Elvira R, Leotta C, Scarpello L, Liborio R. Effects of l-carnitine in patients with hepatic encephalopathy. *World J. Gastroenterol.*, *11*(45), 7197–7202, 2005

Mans AM, Kukulka KM, McAvoy KJ, Rokosz NC. Regional distribution and kinetics of three sites on the $GABA_A$ receptor: Lack of effect of portacaval shunting. *J. Cereb. Blood Flow Metab 12*, 334–346, 1992

Mans AM, De Joseph MR, Hawkins RA. Metabolic abnormalities and grade of encephalopathy in acute hepatic failure. *J. Neurochem.*, *63*, 1829–1838, 1994

McEnery MW, Snowman AM, Trifiletti RR, Snyder SH. Isolation of the mitochondrial benzodiazepine receptor: Association with the voltage-dependent anion channel and the adenine nucleotide carrier. *Proc. Natl. Acad. Sci. USA*, *89*, 3170–3174, 1992

Mena EE, Cotman CW. Pathologic concentrations of ammonium ions block l-glutamate uptake. *Exp. Neurol.*, *59*, 259–263, 1985

Michalak A, Butterworth RF. Selective loss of binding sites for the glutamate receptor ligands [^{3}H]kainate and (S)-[^{3}H]5-fluorowillardiine in the brains of rats with acute liver failure. *Hepatology*, *24*, 631–635, 1997

Michalak A, Rose C, Butterworth J, Butterworth RF. Neuroactive amino acids and glutamate (NMDA) receptors in frontal cortex of rats with experimental acute liver failure. *Hepatology*, *24*, 908–913, 1996

Mirowitz SA, Westrich TJ, Hirsch JD. Hyperintense basal ganglia on T_1-weighted MR images in patients receiving parenteral nutrition. *Radiology*, *191*, 117–120, 1991

Mollace V, Colasanti M, Rodino P, Lauro GM, Rotiroti D, Nistico G. NMDA-dependent prostaglandin E2 release by human cultured astroglial cells is driven by nitric oxide. *Biochem. Biophys. Res. Commun.*, *215*, 793–799, 1995

Moroni F, Lombardi G, Carla V, Pellegrini D, Carassale GL, Cortesini C. Content of quinolinic acid and other tryptophan metabolites increases in brain regions of rats used as experimental models of hepatic encephalopathy. *J. Neurochem.*, *46*, 869–874, 1986a

Moroni F, Lombardi G, Carla V, Lal S, Etienne P, Nair NPV. Increase in the content of quinolinic acid in cerebrospinal fluid and frontal cortex of patients with hepatic failure. *J. Neurochem.*, *47*, 1667–1671, 1986b

Mossakowski MJ, Renkawek K, Krasnicka Z, Smialek M, Pronaszko A. Morphology and histochemistry of Wilsonian and hepatogenic gliopathy in tissue culture. *Acta Neuropathol.*, *16*, 1–16, 1970

Mousseau DD, Perney P, Pomier Layrargues G, Butterworth RF. Selective loss of pallidal dopamine D2 receptor density in hepatic encephalopathy. *Neurosci. Lett.*, *162*, 192–196, 1993

Mousseau DD, Pomier Layrargues G, Butterworth RF. Region-selective decreases in densities of [^{3}H]-tryptamine binding sites in autopsied brain tissue from cirrhotic patients with hepatic encephalopathy. *J. Neurochem.*, *61*, 621–625, 1994

Mousseau DD, Baker GB, Butterworth RF. Increased density of catalytic sites and expression of brain monoamine oxidase A in humans with hepatic encephalopathy. *J. Neurochem.*, *68*, 1200–1208, 1997

Mullen KD, Szauter KM, Kaminsky-Russ K. "Endogenous" benzodiazepine activity in body fluids of patients with hepatic encephalopathy. *Lancet*, *336*, 81–83, 1990

Nagaki M, Iwai H, Naiki T, Ohnishi H, Muto Y, Moriwaki H. High levels of serum interleukin-10 and tumor necrosis factor-α are associated with fatality in fulminant hepatitis. *J. Infect. Dis.*, *182*, 1103–1108, 2000

Neary JT, Whittemore SR, Zhu Q, Norenberg MD. Destabilization of glial fibrillary acidic protein mRNA in astrocytes by ammonia and protection by extracellular ATP. *J. Neurochem.*, *63*(6), 2021–2027, 1994

Norenberg MD. The role of astrocytes in hepatic encephalopathy. *Neurochem. Pathol.*, *6*, 13–33, 1987

Norenberg MD, Lapham LW. The astrocyte response in experimental portal-systemic encephalopathy: An electron microscope study. *J. Neuropathol. Exp. Neurol.*, *33*, 422–435, 1974

Norenberg MD, Huo Z, Neary JT, Roig-Cantesano A. The glial glutamate transporter in hyperammonemia and hepatic encephalopathy: Relation to energy metabolism and glutamatergic neurotransmission. *Glia*, *21*(1), 124–133, 1997

Olasmaa M, Rothstein JD, Guidotti A, Weber RJ, Paul SM, Spector S, Zeneroli ML, Baraldi M, Costa E. Endogenous benzodiazepine receptor ligands in human and animal hepatic encephalopathy. *J. Neurochem.*, *55*, 2015–2023, 1990

Olde Damink SW, Deutz NE, Dejong CH, Soeters PB, Jalan R. Interorgan ammonia metabolism in liver failure. *Neurochem. Int.*, *41*(2–3), 177–188, 2002

Oppong KNW, Bartlett K, Record CO, Al Mardini H. Synaptosomal glutamate transport in thioacetamide-induced hepatic encephalopathy in the rat. *Hepatology*, *22*, 553–558, 1995

Panerai AE, Salerno F, Baldissera F, Martini A, Di Giulio AM, Mantegazza P. Brain β-endorphin concentrations in experimental chronic liver failure. *Brain Res.*, *247*, 188–190, 1982

Peterson C, Giguère JF, Cotman CW, Butterworth RF. Selective loss of *N*-methyl-D-aspartate-sensitive l-[3H]-glutamate binding sites in rat brain following portacaval anastomosis. *J. Neurochem.*, *55*, 386, 1990

Pomier Layrargues G, Giguère JF, Lavoie J, Perney P, Gagnon S, D'Amour M, Wells J, Butterworth RF. Clinical efficacy of benzodiazepine antagonist RO 15—1788 (flumazenil) in cirrhotic patients with hepatic coma: Results of a randomized double-blind placebo-controlled cross-over trial. *Hepatology*, *19*, 32–37, 1994

Pomier Layrargues G, Spahr L, Butterworth RF. Increased manganese concentrations in pallidum of cirrhotic patients: Cause of magnetic resonance hyperintensity? *Lancet*, *345*, 735, 1995

Porter JT, McCarthy KD. GFAP-positive hippocampal astrocytes *in situ* respond to glutamatergic neuroligands with increases in [Ca^{2+}]i. *Glia*, *13*, 101–112, 1995

Raabe W. Synaptic transmission in ammonia intoxication. *Neurochem. Pathol.*, *6*, 145–166, 1987

Raghavendra Rao VL, Butterworth RF. Alterations of [3H]8-OH-DPAT and [3H]-ketanserin binding sites in autopsied brain tissue from cirrhotic patients with hepatic encephalopathy. *Neurosci. Lett.*, *182*, 69–72, 1994

Raghavendra Rao VL, Audet RM, Butterworth RF. Selective alterations of extracellular brain amino acids in relation to function in experimental portal-systemic encephalopathy: Results of an in vivo microdialysis study. *J. Neurochem.*, *65*, 1221–1228, 1995

Raghavendra Rao VL, Audet RM, Butterworth RF. Increased neuronal nitric oxide synthase expression in brain following portacaval anastomosis. *Brain Res.*, *765*, 169–172, 1997

Rama Rao VK, Desjardins P, Rose C, Therrien G, Butterworth RF. Increased glutamine synthetase expression in skeletal muscle: An important alternative pathway for ammonia removal in liver failure. *Hepatology*, *30*(4), 162A, #7, 1999

Rose C, Michalak A, Pannunzio P, Therrien G, Quack G, Kircheis G, Butterworth RF. L-ornithine-L-aspartate in experimental portal-systemic encephalopathy: Therapeutic efficacy and mechanism of action. *Metab. Brain Dis.*, *13*(2), 24–32, 1998

Rose C, Butterworth RF, Zayed J, Normandin L, Todd K, Spahr L, Huet P-M, Pomier Layrargues G. Manganese deposition in basal ganglia structures results from both portal-systemic shunting and liver dysfunction. *Gastroenterology*, *117*, 640–644, 1999

Rose C, Kresse W, Kettenmann H. Acute ammonia insult results in calcium-dependent glutamate release from cultured astrocytes: An effect of pH. *J. Biol. Chem.*, *280*(22), 20937–20944, 2005

Rothstein JD, McKhann G, Guarneri P, Barbaccia ML, Guidotti A, Costa E. Hepatic encephalopathy and cerebrospinal fluid content of diazepam binding inhibitor (DBI). *Ann. Neurol.*, *26*, 57–62, 1989

Rovira A, Cordoba J, Sanpedro F, Grive E, Rovira-Gols A, Alonso J. Normalization of T2 signal abnormalities in hemispheric white matter with liver transplant. *Neurology*, *59*, 335–341, 2002

Roy S, Pomier Layrargues G, Butterworth RF, Huet PM. Hepatic encephalopathy in cirrhotic and portacaval shunted dogs: Lack of changes in brain GABA uptake, brain GABA levels, brain glutamic acid decarboxylase and brain postsynaptic GABA receptors. *Hepatology*, *8*, 845–849, 1988

Schafer DF, Jones EA. Hepatic encephalopathy and the γ-aminobutyric acid system. *Lancet, 1,* 18–20, 1982

Schliess F, Görg B, Fischer R, Desjardins P, Bidmon HJ, Herrmann A, Butterworth RF, Zilles K, Häussinger D. Ammonia induces MK-801 sensitive nitration and phosphorylation of protein tyrosine residues in rat astrocytes. *FASEB J., 16*(7), 739–741, 2002

Schmidt W, Wolf G, Grungreiff K, Meier M, Reum T. Hepatic encephalopathy influences high affinity uptake of transmitter glutamate and aspartate into the hippocampal formation. *Metab. Brain Dis., 5,* 19–32, 1990

Sharief MK, Thompson EJ. In vivo relationship of tumor necrosis factor-α to blood—brain barrier damage in patients with active multiple sclerosis. *J. Neuroimmunol., 38*(1–2), 27–33, 1992

Shiraishi T, Black KL, Ikesaki K. Peripheral benzodiazepine receptor ligands induce morphological changes in mitochondria of cultured glioma cells. *Soc. Neurosci. Abs., 16,* 214–216, 1990

Sobel RA, De Armond SJ, Forno LS, Eng LF. Glial fibrillary acidic protein in hepatic encephalopathy: An immunohistochemical study. *J. Neuropathol. Exp. Neurol., 40,* 625–632, 1981

Sonnewald U, Therrien G, Butterworth RF. Portal-systemic encephalopathy, disorder of neuron-astrocytic metabolic trafficking: Evidence from ^{13}C-NMR studies. *J. Neurochem., 67,* 1711–1717, 1996

Spahr L, Butterworth RF, Fontaine S, Bui L, Therrien G, Millette P, Lebrun L-H, Zayed J, Leblanc A, Pomier Layrargues G. Increased blood manganese in cirrhotic patients: Relationship to pallidal magnetic resonance signal hyperintensity and neurological symptoms. *Hepatology, 24,* 1116–1120, 1996

Spahr L, Coeytaux A, Giostra E, Hadengue A, Annoni JM. Histamine H₁ blocker hydroxyzine improves sleep in patients with cirrhosis and minimal hepatic encephalopathy: A randomized controlled pilot trial. *Am. J. Gastroenterol., 202,* 744–753, 2007

Staub F, Baethmann A, Peters J, Weight H, Keupski O. Effects of lactacidosis on glial cell volume and viability. *J. Cereb. Blood Flow Metab., 10,* 866–876, 1990

Swain MS, Bergeron M, Audet R, Blei AT, Butterworth RF. Monitoring of neurotransmitter amino acids by means of an indwelling cisterna magna catheter. A comparison of two rodent models of fulminant hepatic failure. *Hepatology, 16,* 1028–1035, 1992

Szerb JC, Butterworth RF. Effect of ammonium ions on synaptic transmission in the mammalian central nervous system. *Prog. Neurobiol., 39,* 135–153, 1992

Tasaka K. New advances in histamine research. Spinger, Tokyo, 1994

Therrien G, Giguère JF, Butterworth RF. Increased cerebrospinal fluid lactate reflects deterioration of neurological status in experimental portal-systemic encephalopathy. *Metab. Brain Dis., 6,* 225–231, 1991

Therrien G, Butterworth J, Rose C, Butterworth RF. Protective effect of l-carnitine in ammonia-precipitated encephalopathy in portacaval shunted rats: Evidence for a central mechanism of action. *Hepatology, 25,* 551–556, 1997

Thornton JR, Losowsky MS. Plasma methionine enkephalin concentration and prognosis in primary biliary cirrhosis. *Br. Med. J., 297,* 1241–1242, 1988

Tominaga S, Watanabe A, Tsuji T. Synergistic effect of bile acid, endotoxin, and ammonia on brain edema. *Metab. Brain Dis., 6*(2), 93–105, 1991

Vaquero J, Polson J, Chung C, Helenowski I, Schiodt FV, Reisch J, et al. Infection and the progression of hepatic encephalopathy in acute liver failure. *Gastroenterology, 125,* 755–764, 2003

Vaquero J, Rose C, Butterworth RF. Keeping cool in acute liver failure: Rationale for the use of mild hypothermia. *J. Hepatol., 43,* 1067–1077, 2005

Vaquero J, Bélanger M, James L, Herrero R, Desjardins P, Cote J, Blei AT, Butterworth RF. Mild hypothermia attenuates liver injury and improves survival in mice with acetaminophen toxicity. *Gastroenterology, 132*(1), 372–383, 2007

Verkhratsky A, Kirchhoff F. NMDA receptors in glia. *The Neuroscientist, 13*(1), 28–37, 2007

Vogels BA, Maas MA, Daalhuisen J, Quack G, Chamuleau RA. Memantine, a noncompetitive NMDA receptor antagonist improves hyperammonemia-induced encephalopathy and acute hepatic encephalopathy in rats. *Hepatology, 25,* 820–827, 1997

Weber FL Jr, Veach GL. The importance of the small intestine in gut ammonium production in the fasting dog. *Gastroenterology, 77,* 235–240, 1979

Wendon JA, Harrison PM, Keays R, Williams R. Cerebral blood flow and metabolism in fulminant liver failure. *Hepatology, 19,* 1407–1413, 1994

Yao H, Sadoshima S, Fujii K, Kusada K, Ishitsuka T, Tamaki K, Fujishima M. Cerebrospinal fluid lactate in patients with hepatic encephalopathy. *Eur. Neurol., 27,* 182–187, 1987

Young SN, Lal S. CNS tryptamine metabolism in hepatic coma. *J. Neural. Transm., 47,* 153–161, 1980

Young SN, Lal S, Feldmuller F, Aranoff A, Martin JB. Relationships between tryptophan in serum and CSF and 5-hydroxyindoleacetic acid in CSF of man: Effects of cirrhosis of the liver and probenecid administration. *J. Neurol. Neurosurg. Psychiat., 38,* 322–330, 1975

Yurdaydin C, Gu ZQ, Nowak G, Fromm C, Holt AG, Basile AS. Benzodiazepine receptor ligands are elevated in an animal model of hepatic encephalopathy: Relationship between brain concentration and severity of encephalopathy. *J. Pharmacol. Exp. Ther., 265,* 565–571, 1993

Zwingmann C, Desjardins P, Chatauret N, Michalak A, Hazell AS, Butterworth RF. Reduced astrocytic glycine transporter Glyt-1 in acute liver failure. *Metab. Brain Dis., 17*(4), 263–274, 2002

Zwingmann C, Chatauret N, Leibfritz D, Butterworth RF. Selective increase of brain lactate synthesis in experimental acute liver failure: Results of a [^{1}H-^{13}C] nuclear magnetic resonance study. *Hepatology, 37,* 420–428, 2003

Chapter 10
Hepatic Encephalopathy

Karin Weissenborn

Hepatic encephalopathy (HE) is a neuropsychiatric syndrome associated with severe, acute and chronic liver failure. According to the recommendations of a working party (Ferenci et al., 2002) held at the 4th World Congress of Gastroenterology in Vienna in 1998, 3 types of HE should be differentiated, on principle:

1. Encephalopathy associated with acute liver failure (type A)
2. Encephalopathy associated with portal-systemic bypass and no intrinsic hepatocellular disease (type B)
3. Encephalopathy associated with cirrhosis and portal hypertension (type C).

Especially with regard to type B and C encephalopathy, further differentiation is recommended, based on the duration and characteristics of the clinical manifestations, as (a) episodic HE (precipitated or spontaneous), (b) persistent HE and (c) minimal HE. The term persistent HE includes treatment-dependent persistent encephalopathy, e.g. a subgroup of patients who develop overt symptoms of HE if appropriate medication is discontinued.

Clinical Features of Hepatic Encephalopathy

Type A encephalopathy and type B and C encephalopathy share many clinical symptoms. However, there are also a considerable number of clinical features which are significantly different. Generally the brain's reaction to a distinct metabolic disturbance differs depending on the rate at which the metabolic alteration occurs, since there is little scope for adaptation to acute severe metabolic alterations.

Acute Liver Failure

Accordingly the neurological clinical status of patients with acute liver failure may rapidly deteriorate. Encephalopathy is a hallmark symptom in these patients, and while the grade of HE is stable over long time periods in cirrhotic patients, patients with

D.W. McCandless (ed.) *Metabolic Encephalopathy*,
doi: 10.1007/978-0-387-79112-8_10, © Springer Science+Business Media, LLC 2009

acute liver failure may present with irritability, insomnia and slight concentration deficits and progress within hours to deep coma.

The interval between the first symptom of liver disease and the first symptoms of hepatic encephalopathy in patients with acute liver failure ranges from hours to weeks, and this interval often predicts the clinical course and prognosis. According to O'Grady et al. (1993) patients with acute liver failure should be sub-classified depending on the jaundice-to-encephalopathy time interval into: (a) hyperacute (onset within 1 week), (b) acute (between 8 and 28 days), and (c) subacute (between 29 days and 12 weeks). Interestingly prognosis is best in the hyperacute sub-group. However, this may be due to the fact that most of the cases in this group are aceta-minophen-induced (Ostapowicz et al., 2002).

Further differences between HE in acute liver failure and in cirrhosis relate to the kind of mood alterations and psycho-motor functions. In acute liver failure irri-tability, insomnia and concentration deficits are followed by confusion and disori-entation, and often by agitation and manic behaviour, before drowsiness occurs and the patient passes to stupor or even coma. In contrast, irritability, agitation and manic behaviour are rarely observed in HE in cirrhotics, where lethargy, psychomo-tor slowing and depression predominate.

Seizures and Intracranial Hypertension are Frequent in Acute Liver Failure

Also in contrast to cirrhotics and patients with porto-systemic bypass without hepato-cellular disease, altered mental status in patients with acute liver failure may be due to the presence of seizure activity or hypoglycemia. A status of complex partial seizures may be misinterpreted as bizarre behaviour in the course of acute hepatic encephalopathy. Seizures may even continue under neuroleptanalgesia, and then cannot be detected by clinical observation.

Ellis et al. (2000) reported a 45% frequency of seizure activity in electively paralyzed and ventilated acute liver failure patients without antiepileptic therapy. Forty-two patients were enrolled in this trial when signs of grade III HE became evident and they had been electively paralyzed and ventilated. Twenty patients were given phenytoin, and 22 acted as controls. Subclinical seizure activity was recorded in 3 of the treated patients and 10 of the control group. Pupillary abnor-malities were observed in 5 and 11 patients, respectively. ICP elevation was found in 3 out of 10 treated patients and 7 out of 9 patients in the control group. Autopsy was performed in 9 patients of the phenytoin group and 10 of the control group. Brain edema was found in 2 of the prophylaxis group and 7 of the control group. Survival was 83% in the treated and 75% in the control group with regard to the patients who had undergone liver transplantation (OLT), and in 100% com-pared to 33% in the control group in those patients who did not fulfill the OLT criteria.

Seizures can be induced by hyperammonemia, relative ischemia and hypogly-caemia in patients with acute liver failure. In addition, they can be induced by the development of brain edema, while on the other hand they increase brain edema.

Since a more recent study published by Bhatia et al. (2004) was unable to reproduce the findings of Ellis and co-workers, a prophylactic treatment with phenytoin is not recommended at present. But, recurrent electroencephalography (EEG) if not continuous EEG monitoring should be performed in these patients.

In a further study Bhatia et al. (2006) reported an incidence of seizures of 22.5% in a group of 80 patients with acute liver failure. All patients had clinical features of brain edema when seizures occurred. Seizures were more frequent in patients with ammonia levels of $\geq$ 124 μmol/l (35%) compared to those with ammonia levels below 124 μmol/l (8%). In this study seizures were uniformly fatal and did not respond to standard antiepileptic medication.

Brain Edema: the Most Severe Complication of Acute Liver Failure

Brain edema is the most severe neurological complication of acute liver failure. It complicates approximately 50–80% of cases with acute liver failure and grade III to IV HE (Ostapowicz et al., 2002). The pathogenesis of brain edema in acute liver failure is complex. For details see the paper of R.F. Butterworth (this volume). In summary, high arterial ammonia levels contribute to the development of brain edema through the accumulation of glutamine and alanine in astrocytes. In addition, vasodilatation and an increased cardiac output are observed, which lead to increased cerebral blood volume and flow and thus increased intra-cranial pressure. To maintain sufficient cerebral perfusion pressure (CPP) of 50 to 65 mm Hg systemic hypotension must be counteracted by the use of vasopressive agents. Sustained CPP of < 40 mm Hg is associated with a high likelihood of ischemic brain damage and poor neurological outcome after transplantation (Rinella and Sanyal, 2006).

Unfortunately, the development of brain edema can hardly be monitored in patients with acute liver failure. Brain edema mostly occurs when hepatic encephalopathy has progressed to grades III to IV. In grades I-II encephalopathy, it has rarely been observed. With progression to grade III HE cerebral edema occurs in up to 35% and with grade IV HE in 65 to 75% of the patients (Munoz, 1993).

With grade III HE, mechanical ventilation is recommended to prevent aspiration and to control agitation which can lead to increases in intracranial pressure. Choice of sedation is done on an individual basis at the moment. Often propofol is used as it is considered to reduce intracranial pressure in ALF patients (Raghavan and Marik, 2006). Barbiturates are not consistently recommended (Polson and Lee, 2005; Rinella and Sanyal, 2006).

Due to sedation, clinical judgment is severely limited in those patients who are most at risk to develop brain edema. Clinical signs such as pupillary dilatation or signs of decerebration are evident only later in the course, and hypertension, bradycardia or irregular respirations may even be masked by concomitant therapy.

Direct intracranial pressure monitoring is the most reliable modality for the measurement of ICP. In practice, however, in several institutions the placement of ICP monitoring devices is avoided because of severe coagulation deficits in the patients. A recent multi-center study in the United States showed that ICP monitoring was performed in only 28% of the patients with acute liver failure and hepatic encephalopathy (Vaquero et al., 2005). The frequency of monitoring differed between the centers involved in the study. It also differed between those patients who were listed for urgent liver transplantation and those who were not. In most patients subdural transducers were used (63.8%). Ten percent of the patients (n = 6) had intracranial bleeding as a result of the ICP monitor, half of them showed clinical deterioration, and 2 patients died. Overall the frequency of complications was significantly less than assumed several years ago by Blei et al. (1993) who performed a survey of transplant centers across the United States. They had estimated that about 20% of the ICP monitoring resulted in intracranial bleeding. In this former study the lowest bleeding rate had been observed with 3.8% in patients with epidural catheters, in contrast to about 20% in patients with subdural or parenchymal catheters.

According to Shami et al. (2003) the bleeding complications can be controlled if recombinant activated factor VII is given directly before the procedure.

If ICP monitoring is performed, rapid and abrupt increases of intracranial pressure can be directly addressed by the infusion of mannitol, for example, or the application of ranitidine. In most patients, however, the management of raised intracranial pressure has to be performed without actual information on its severity. Thus, general rules are followed, such as elevation of the head of the bed to 30 degrees, sedation, minimal stimulation or the administration of mannitol three times a day. Mannitol has been shown to improve survival in a group of patients with acute liver failure (Canalese et al., 1982). If renal failure accomplishes acute liver failure, however, a paradoxical effect can occur with mannitol treatment as serum osmolality increases and volume overload may occur. Therefore plasma osmolality must be checked at least twice a day to assure that it remains <320 mosmol/l (Polson and Lee, 2005; Rinella and Sanyal, 2006).

Interestingly in the recent study of Vaquero et al. (2005) patients who underwent ICP monitoring before liver transplantation received more treatments for elevated ICP values, including mannitol and barbiturates, than those in whom monitoring was not undertaken. However, the 30-day outcomes were similar in both groups. This could be due to an inclusion bias, since ICP monitoring was more often done in those patients who were listed for urgent transplantation. A true evaluation of the treatment effects could have been done only if in some of the patients the management was unchanged, irrespective of the ICP data.

Besides mannitol, hypertonic saline (30%) has also been used to prevent brain edema in ALF patients. A recent trial demonstrated that induction and maintenance of hypernatremia (145–155 mmol/l) in patients with grade 3 and 4 HE resulted in a decreased incidence and severity of intracranial hypertension (Murphy et al., 2004).

Considering the results of recent studies in experimental animals, some groups studied the effect of moderate hypothermia (32–33 °C) on ICP in patients with ALF. Jalan et al. (2004) were able to show that cooling to 32 °C decreased ICP to <20 mm Hg even in patients with refractory intracranial hypertension. Although their results were quite impressive other groups call for a randomized, controlled study before the use of hypothermia can be recommended as a routine measure to treat brain edema in ALF.

Recently Tofteng and Larsen (2004) reported on their experience with indomethacin in ALF patients with severe intracranial hypertension. Indomethacin induces cerebral vasoconstriction. It was applied in twelve cases with ALF and cerebral edema in this study. Indomethacin reduced ICP from 30 mm Hg to 12 mm Hg (mean; $p < 0.05$) and increased the cerebral perfusion pressure from 48 to 65 mm Hg (mean; $p < 0.05$).

As indomethacin has several undesirable side effects such as gastrointestinal bleeding, platelet dysfunction and nephrotoxicity, it is not routinely recommended for use in ALF patients in spite of the positive effects observed.

Since direct ICP monitoring in patients with acute liver failure is prone to bleeding complications, alternative methods are evaluated. Brain imaging is not sensible. Computed tomography brain scanning may fail to demonstrate cerebral edema in ALF patients with elevated intracranial pressure. In addition, imaging methods cannot be used for ICP monitoring. Continuous transcranial Doppler monitoring of cerebral blood flow may be used as an alternative. Transcranial Doppler (TCD) monitoring reveals information about changes in cerebral perfusion and, thereby, indirectly about changes in intracranial pressure. With ICP increases flow resistance also increases, followed by a decrease in diastolic flow, predominantly. TCD monitoring studies in head-trauma patients have shown that there is a positive exponential correlation between pulsatility index (PI) or resistance index (RI) and ICP> 15–25 mmHg if the arterial CO_2 partial pressure is kept constant (Homburg et al., 1993). Data on patients with fulminant hepatic failure are sparse. Larsen et al. (1995a), however, have shown that cerebral blood flow in patients with acute liver failure can be measured reasonably well by continuous TCD monitoring. In 1993, Aggarwal et al. (1993) reported on their experiences with continuous TCD and ICP monitoring during liver transplantation in 5 patients with acute liver failure compared to 5 cirrhotic patients. While they did not observe any changes in TCD parameters in the group of cirrhotics, they found a decrease in the mean flow velocity (MFV) and the diastolic flow velocity (DFV) coincident with a decrease in cerebral perfusion pressure and also an increase in ICP and PI during reperfusion of the grafted liver.

We were able to show a significant relationship between increases in ICP and alteration of the Doppler spectrum and also a significant correlation between CPP and Doppler parameters in a patient who was continuously monitored over a period of 15 hours before transplantation and had severe ICP increase and CPP decrease during the last 6 hours before transplantation. In contrast, the TCD data were normal in 6 patients with acute liver failure and grades I–III HE, who did not present with increased intracranial pressure (Schnittger et al., 1997).

Further, data of Larsen et al. (1995b) show that early in the course of the disease increases in ICP may also be combined with increases in the flow velocity. This finding is in accordance with the current hypotheses that an increase in CBF is of critical importance for the development of brain edema and intracranial hypertension in acute liver failure. In the study of Larsen et al. an increase of the median flow velocity (V_{mean}) to more than 100 cm/sec could be shown to be a predictor of fatal cerebral damage. In patients who developed clinical signs of brain death V_{mean} rose from about 85 cm/sec to 135 cm/sec initially but then abruptly declined after about 10 hours to a mean of 5 cm/sec due to further increase in ICP and resulting decrease of the cerebral perfusion pressure.

Thus, TCD monitoring in patients with acute liver failure may be helpful in the detection of ICP alterations and may also provide cues to estimate the prognosis of an individual patient. Further studies, however, are needed to evaluate the prognostic significance of TCD monitoring in acute liver failure.

Prognosis of Acute Liver Failure

Prognosis of acute liver failure depends on several factors like etiology of liver disease, age, grade of liver disease, grade of encephalopathy, presence or absence of systemic inflammatory response syndrome, etc. Currently available prognostic scoring systems do not adequately predict outcomes or determine candidacy for liver transplantation (Polson and Lee, 2005). In the study of Bhatia et al. (2006), however, three variables were found to be significantly predictive of death: arterial ph $\geq$ 7.40, presence of clinical cerebral edema, and arterial ammonia levels $\geq$ 124 μmol/l.

Survival rates have improved from 15% in the pre-transplant era to $\geq$ 60%. Spontaneous survival rates are now around 40% (Bhatia et al., 2006). Post-transplant survival rates have been reported to be about 80–90% (Sass and Shakil, 2005).

Encephalopathy Associated with Cirrhosis and Portal Hypertension (Type C)

In contrast to acute liver failure, hepatic encephalopathy associated with cirrhosis and portal hypertension progresses over long time periods (from months to years) from psychomotor slowing to severe clouding of consciousness, in principle. Bouts of severe HE however can be precipitated in patients with so far undetected HE by several factors such as gastrointestinal bleeding, infection, hyponatremia, etc. In contrast to acute liver failure, HE in cirrhosis does not cause the patient's demise although it carries a poor prognosis as an indicator of the severity of liver disease.

As in acute liver failure, the main symptoms of HE are alterations in consciousness, cognitive dysfunction and motor disturbances. While patients with ALF often appear irritable and restless in the very beginning, psychomotor slowing is characteristic for type C HE. The alteration of consciousness is the basis for the 4-stage grading system of hepatic encephalopathy used world wide (West Haven classification) (Atterbury et al., 1978).

Grade I HE is characterized by mild confusion, grade II by drowsiness, lethargy and disorientation, grade III by somnolence to sopor, and grade IV by coma with or without response to painful stimuli (a and b) (Table 10.1). This grading system

Table 10.1 Stages of Hepatic Encephalopathy

Stage	Clinical features
Minimal	No clinical abnormalities, pathological psychometric and/or neurophysiological findings
I	Reversal of sleep rhythm, psychomotor slowing, attention deficits, irritability, untidiness, slight disturbances of mood and personality
II	Drowsiness, lethargy, major mental disabilities, intermittent disorientation, inappropriate behaviour
III	Somnolent but rousable, persistent disorientation, pronounced confusion, incoherent speech with perseveration, unable to perform mental tasks
IV	Coma with (IVa) or without (IVb) response to painful stimuli

attributes characteristic disturbances of motor function and speech to distinct grades of HE. Asterixis (flapping tremor), for example, is often associated with grade II, as are dysarthria and ataxia or extra-pyramidal features like tremor, muscular rigidity or posture abnormalities. The neuropsychiatric changes which occur in patients with cirrhosis, however, are much more likely to form a continuum with no natural steps, and the various neuropsychiatric symptoms may arise in various combinations. In practice, for example, disturbed motor function is detectable by subtle neurological assessment even in patients who appear clinically unaffected at first sight (Krieger et al., 1996; Spahr et al., 1996). Asterixis can be observed in patients without cognitive dysfunction or alteration of consciousness, and speech disturbances are one of the first symptoms of HE. In the early stages, speech is monotonous and slow, similar to patients with Parkinson's disease. With increasing ataxia speech becomes slurred and dysarthric. In stuporous patients dysphasia occurs combined with perseveration. Writing disabilities also occur early in the course of the disease. They are traditionally used as diagnostic markers. In the early phase omission of single letters, reversal of order and misspellings are common. With deterioration writing becomes more tremulous, letters are superimposed, and lines of writing converge. The patients become unable to sign their names or to move the pencil from left to right. The patient's diary thus can be effectively used for monitoring of the grade of HE.

It must be emphasized that the combination of extra-pyramidal and cerebellar dysfunction is quite characteristic for hepatic encephalopathy, and cannot be found in any other kind of metabolic encephalopathy, except Wilson's disease. Asterixis, however, which is considered a classic sign of HE, is also seen in other metabolic or toxic encephalopathies, such as uremia, CO_2 retention, hypomagnesemia or intoxication with antiepileptic drugs, for example.

Hyperreflexia and ankle clonus are characteristic in grade III HE, but like the other motor signs, it can be observed also in patients with minimal HE.

In the early grades of HE, personality changes and mood disorders are frequent. Most of the patients show varying degrees of remittent personality change. In some patients, behavior is uninhibited during acute exacerbations of HE, in others previous personality trends are intensified. Strikingly, many patients begin to keep a diary where every daily event is noted meticulously. Mood fluctuates. In some patients depression and euphoria abruptly follow each other. In others only depression or euphoria can be observed interchanging with a stable mood.

Paranoid symptoms are common with grade III HE due to perceptual difficulties and misinterpretations. In addition, hallucinations, mostly visual, may be observed (Sherlock et al., 1954).

Focal neurological signs are uncommon in hepatic encephalopathy and point to complications such as intracranial bleeding. Cadranel et al. (2001), however, showed that HE may even present with focal neurological signs. They reported a series of 34 cirrhotics with 48 episodes of suspected HE. Ten episodes were characterized by focal neurological signs such as hemiparesis. In 2 cases, cerebral bleeding was diagnosed. In the remaining patients cerebral computer tomography or magnetic resonance imaging and cerebrospinal fluid were normal, and the symptoms resolved with therapy for HE.

Minimal Hepatic Encephalopathy

Minimal hepatic encephalopathy is the mildest form of HE. Subtle behavioral changes that are only apparent to the patient's family or old friends and sleep disturbances are the first signs of minimal hepatic encephalopathy (Gitlin et al., 1986; Cordoba et al., 1998). The behavioral changes are predominantly due to a slight impairment of cognitive function, especially attention, resulting from bilateral forebrain and parieto-occipital dysfunction (Lockwood et al., 1993, 2002). Since verbal abilities are preserved in this stage of HE, cerebral dysfunction does not become apparent during the routine clinical examination. The diagnosis can be made only by the application of neuropsychological or neurophysiological measures (Schomerus and Hamster, 1998; Weissenborn et al., 1990, 2001). Patients with minimal HE do worse than healthy controls, especially in tests of psychomotor speed, visual perception and attention. Some of them also show a pathological slowing of the EEG and prolonged latencies of exogenous and endogenous evoked potentials (Amodio and Gatta, 2005; Kullmann et al., 1995; Weissenborn et al., 1990). The prevalence of this 'minimal' form of HE is estimated to be about 30–60% in cirrhotics without overt clinical signs of HE (Bajaj et al., 2007a; Schomerus and Schreiegg, 1993). Although undetectable on clinical grounds, minimal HE is undoubtedly of clinical significance for the patients. It has been shown to interfere with the patients' ability to drive a car (Schomerus et al., 1981; Srivastava et al., 1994; Watanabe et al., 1995; Wein et al., 2004), with their earning capability (Schomerus and Hamster, 2001) and their quality of life (Groeneweg et al., 1998). In addition, mHE predicts the development of overt HE, and is associated with a poor prognosis. The probability of clinically manifest HE after a follow-up of 3 years was estimated to be about 50% in patients with minimal HE compared to 8% in cirrhotics without minimal HE (Hartmann et al., 2000). Thus a few years ago, experts in the field encouraged the use of the term "minimal HE" instead of the terms "subclinical HE" or "latent HE", which have been widely used in the past for this setting, and such a connotation runs the risk of considering this mildest degree of HE void of clinical significance (Ferenci et al., 2002; Lockwood, 2000). Recently it was suggested that minimal HE be combined with grade I HE to a sub-class called "mild HE". This recommendation is sensible because the detection of grade I HE depends on the experience and accuracy of the examiner. In other words grade I HE is often overlooked and misclassified.

Due to growing evidence that minimal HE has a significant impact on the patient's well-being and daily living activities, consensus has emerged that patients with minimal HE should be treated (Lockwood, 2000; Quadri et al., 2007). Recently Prasad et al. (2007) were able to show that treatment of minimal HE with lactulose resulted in an improvement of cognitive function. Whether treatment of minimal HE also affects the prognosis with regard to the development of overt HE needs to be studied.

At present most of the patients with minimal HE remain undiagnosed. Bajaj et al. (2007a) reported upon a survey regarding testing for minimal HE in the United States. An anonymous questionnaire querying among other things the personal opinions regarding mHE, diagnostic strategies, frequency of diagnosis, reasons for testing/non testing was sent to 246 AASLD members. Of the 137 replies received, 84% believed that mHE was a problem and 74% agreed that mHE should be tested for. But 38% did not test for mHE. And only 28% tested more than 50%, only 14% more than 80% of their patients. The main reason for not testing was the time and resources needed for testing, missing test standards in the United States, and questionable therapeutic effects in mHE. Considering the growing evidence on the clinical significance of mHE and successful therapeutic strategies, this survey elucidates the need for a consensus on the diagnostic procedure for mHE.

Chronic Persistent Hepatic Encephalopathy

Rare complications of liver cirrhosis are chronic persistent hepatic encephalopathy and hepatic myelopathy. Both occur in less than 1% of the patients, and both are accompanied by extensive porto-systemic shunts. Chronic persistent hepatic encephalopathy is also known as acquired hepato-lenticular degeneration (Victor, Adams and Cole, 1965). In contrast to the usual cirrhotic patient with HE, these patients show obvious neuronal alterations: a patchy, spongy degeneration most consistently observed in the deep layers of the cerebral cortex and subcortical white matter, particularly in the parieto-occipital cortex, basal ganglia and cerebellum.

Clinically the patients present with chronic progressive extra-pyramidal and cerebellar symptoms, often combined with hyperreflexia and extensor plantar reflexes. Besides motor functions, the patients' cognition and consciousness may also be impaired, but are less impressive than the motor functions. The effect of the classical therapeutic interventions aimed at the lowering of plasma ammonia levels is limited in chronic persistent HE. Liver transplantation has been reported to result in an improvement of the neurological symptoms (Powell et al., 1990). Today, however, controlled studies on this subject are missing.

Hepatic Myelopathy

Hepatic myelopathy is characterized by a rapidly progressing spastic paraparesis without sensible deficits or bladder dysfunction. Brisk tendon reflexes and increased tone of the lower extremities are the predominant findings. Plantar reflexes may be flexor, even in patients with ankle clonus. Usually the patients are bound to a wheelchair or bedridden within months.

In most cases, hepatic myelopathy will develop after several episodes of HE. For unknown reasons, most patients described are men. Imaging of the spinal cord by MRI or myelography reveals normal results, as does the examination of CSF. The main pathological finding in HM is demyelination of the cortico-spinal tracts predominantly in the lower part of the cervical and the thoracic spinal cord. Occasionally demyelination has also been found in the ventral pyramidal tracts and in the posterior columns and spinocerebellar tracts (Campellone et al., 1996, Weissenborn et al., 2003).

Since hepatic myelopathy does not respond to the classical ammonia lowering therapy of hepatic encephalopathy, transplantation should be considered as soon as the diagnosis is made. With successful liver transplant, the patient has a chance of improvement (Weissenborn et al., 2003).

Diagnosis and Differential Diagnosis of Type C HE

The diagnosis of type C hepatic encephalopathy has to be made on clinical grounds. Although the clinical symptoms are characteristic for HE, they are not specific. Thus, other possible causes of CNS dysfunction in a patient with liver cirrhosis have to be excluded. Neither biochemical examinations, nor neuro-imaging or neurophysiological methods supply specific findings that enable one to make the diagnosis without doubt. Nevertheless these diagnostic means have to be used for differential diagnostic purposes to exclude other causes of brain dysfunction or — in the case of hepatic myelopathy — other causes of paraparesis.

Imaging

The diagnostic work-up of a patient with suspected hepatic encephalopathy should include in any case a cerebral computed tomography. Subdural haematoma and also parenchymal bleeding are rare but significant complications of liver cirrhosis that must be ruled out before the diagnosis of HE is made.

HE per se does not induce any structural or functional changes that can be detected by neuro-imaging — neither by CCT nor by magnetic resonance imaging. The characteristic bilateral symmetric pallidal hyperintensities which are present in up to 90% of the patients with liver cirrhosis indicate porto-caval shunting and deposition of manganese within the basal ganglia, but do not correlate with the presence and extent of HE (Lockwood, Butterworth and Weissenborn, 1997).

Laboratory Findings

Alterations of the electrolyte levels are also clinically relevant with regard to differential diagnosis, especially hyponatremia, hyper- and hypoglycemia and renal dysfunction. Hyponatremia may induce an episode of HE, but it may also cause brain dysfunction per se.

Plasma ammonia levels are frequently measured if HE is suspected. An elevated ammonia level at diagnosis indeed points to a potential hepatic origin to the change in brain function. But, although serum levels correlate broadly with stages of encephalopathy, there is substantial overlap between values at different stages. An ammonia level within the normal range is no proof against the diagnosis of HE.

EEG and Evoked Potentials

The majority of patients with clinically overt HE show alterations of the EEG with an increasing amount of theta and delta rhythms with increase in the grade of HE (Davies et al., 1991; Montagnese, Amodio and Morgan, 2004; Parsons-Smith et al., 1957). Like the clinical symptoms, these EEG alterations, however, are unspecific. Thus, the EEG cannot be used to prove the diagnosis of HE. This also holds true for other neurophysiological methods such as evoked potentials.

Wernicke's Disease is the Most Important Differential Diagnosis of HE

In clinical practice, the most important differential diagnosis of HE, especially in patients with alcoholic cirrhosis, is Wernicke's disease. Wernicke's disease shares major symptoms with hepatic encephalopathy: disturbance of consciousness, ataxia, and dysarthria. A differentiation based on clinical symptoms is impossible. According to Kril and Butterworth (1997) the diagnosis of Wernicke's encephalopathy is frequently missed in patients with alcoholic cirrhosis. In a neuropathological examination of the brain of 36 patients who died with the clinical signs of HE, histopathological characteristics of Wernicke's disease were found in 9 patients, while this diagnosis had been made on clinical grounds only in 2 of the patients.

Magnetic resonance imaging may show the characteristic alterations present in Wernicke's encephalopathy: hemorrhage in the region of the mammillary bodies, the thalamus and the inferior olivary nucleus (White et al., 2005; Zhong et al., 2005). In practice, however, it is often impossible to achieve a good quality MRI in these patients due to their pronounced organic brain syndrome. Therefore, thiamine (100 mg/day intravenously for up to ten days, and thereafter changed to oral application) should be administered even in cases with suspected or possible Wernicke's encephalopathy.

Diagnosis of Minimal Hepatic Encephalopathy

Minimal hepatic encephalopathy is defined as cerebral dysfunction in cirrhotics that remains undetected on clinical grounds but can be proven via neuropsychological and/or neurophysiological methods. At present no measure has been defined as "gold standard" for diagnosing minimal HE. From neuropsychological studies it is well known that minimal HE is characterized by deficits in attention, visuo-spatial perception, motor speed and accuracy. Thus, psychometric tests that evaluate exactly these cognitive domains are recommended to be used for making the diagnosis of mHE. In practice the Number Connection Tests A and B are frequently used, although they have been shown to be less sensitive than other tests. Also, the block design test and the digit symbol test from the WAIS —R test battery are frequently used. Recently the Inhibitory Control Test has been recommended (Bajaj et al., 2007). A battery that has been especially developed for the diagnosis of minimal hepatic encephalopathy is the PSE-Syndrom-Test (Schomerus et al., 1999). The test combines the results of the Number Connection tests A and B, the serial dotting test, the digit symbol test and the line tracing test. The battery has been shown to be sensitive, reliable and easy to apply. Four different versions are provided for follow-up studies. The test has been standardized in a German population. Meanwhile, however, it has also been validated in other countries (Romero-Gomez et al., 2007). The PSE-Syndrome-Test has been recommended by the HE-Working group at the 11th World Congress of Gastroenterology held in Vienna in 1998 (Ferenci et al., 2002).

Whether neuropsychological or neurophysiological methods should be preferred for diagnosing mHE is not known. EEG was one of the first methods applied for this purpose. The EEG shows a characteristic generalized slowing which increases with increasing grade of HE in the individual patient. Unfortunately, however, normal EEGs have been found in patients with pronounced clinical symptoms of HE, and at least some of the investigators were unable to find a clear correlation between the amount of EEG alterations and the clinical grade of HE. For example, we were able to show that the EEG is pathological in only about 35% of the patients with grade I HE (Weissenborn et al., 1990). Several studies describe a prevalence of EEG alterations of 40–80%, in clinically not encephalopathic cirrhotic patients (Montagnese et al., 2007; Penin, 1967; Van der Rijt et al., 1990). The sensitivity is increased by the use of automatic-computerized analysis methods (Amodio et al., 1999; Montagnese et al., 2007). Whether the sensitivity of modern EEG analysis exceeds that of other methods has not yet been systematically analyzed.

Besides EEG, special visual evoked potentials (VEP) have been intensively studied in cirrhotic patients with and without HE. They have been proven to be useful for follow-up examinations, as they show an increase or decrease of latency in close correlation to the clinical course of the disease. As a diagnostic tool, however, they cannot be reliably used as they fail minimum requirements on sensitivity and specificity (Weissenborn, 1991).

Recently the critical flicker frequency test (Kircheis et al., 2002; Romero-Gomez et al., 2007) has been recommended for the diagnosis of minimal HE. With this tool, the threshold frequencies at which light pulses are perceived as fused or flickering can be recorded. In two different labs, values between 38 and 39 Hz for the flicker frequency distinguished low-grade HE from normal subjects. Further studies are needed to assess the specificity and sensitivity of this measure more reliably.

Therapy for Hepatic Encephalopathy

Most HE episodes in patients with chronic liver disease are precipitated by events such as oral protein load, gastrointestinal bleeding, obstipation, infection, especially peritonitis, hypokalemia and alkalosis complicating the use of diuretic drugs, administration of sedative drugs, for example for diagnostic procedures or induction of portal-systemic shunt via shunt operation or the TIPSS implantation.

The elimination of these precipitating factors is the first therapeutic step and leads to an improvement without need for any further therapy in many cases. If the symptoms do not resolve or no precipitating factor can be identified more specific therapeutic efforts must be applied:

As hyperammonemia is considered to be the main cause of hepatic encephalopathy, "specific" therapy for HE is aimed at the reduction of ammonia production and resorption. Protein restriction has been recommended for a long time to reduce ammonia production. But patients with cirrhosis are hypercatabolic. They may require up to 1.5 g/kg protein per day. Therefore protein restriction has been limited to patients with severe hepatic encephalopathy, and a reduction to less than 1 g protein per kg body weight has been discouraged in the past. A recent clinical study even showed no benefits of protein restriction (Cordoba et al., 2004). 30 cirrhotics who were admitted to the hospital with hepatic encephalopathy were randomized to a low-protein diet or normal protein diet. After two weeks of treatment, the groups did not significantly differ with regard to the course of hepatic encephalopathy.

Vegetable protein is preferred over animal protein as the dietary source as it seems advantageous (Bianchi et al., 1993). Part of the protein needed to maintain the nitrogen balance can be replaced by a mixture of branched chain amino acids (BCAA) which can produce a positive nitrogen balance to approximately the same order of magnitude as a corresponding amount of food protein, without precipitating HE (Marchesini et al., 2003; Muto et al., 2005).

The reduction of intestinal ammonia synthesis can also be achieved by the administration of non-absorbable disaccharides-like lactulose and lactitol or antibiotics like neomycin, paramomycin, metronidazole or rifaximin. Lactulose exerts several effects: (1). acidification of the intestinal content resulting in a reduction of ammonia absorption and net movement of ammonia from the blood into the bowel and (2). reduced bacterial production of ammonia in the colonic lumen due to environmental changes with promotion of the growth of non-urease producing bacteria and (3). the cathartic effect. The daily dose of lactulose is between 30 and 60 g per day. The goal is to obtain 2–3 soft bowel movements per day. Recently lactulose has been proven to be effective even in patients with minimal HE (Prasad et al., 2007).

Lactitol has been shown to be as effective as lactulose for treating hepatic encephalopathy (Morgan et al., 1989; Morgan and Hawley 1987). Common side effects of the disaccharides are flatulence and abdominal cramping.

Antibiotics like neomycin or metronidazole are administered either when an episode of HE arises or chronically to avoid the development of clinically overt HE. In the past, neomycin was the most used antibiotic. There is increasing evidence, however, that rifaximin has a more favorable benefit : risk ratio in the treatment of HE (Leevy and Phillips, 2007; Mas et al., 2003). While neuro- and nephrotoxicity have to be considered in neomycin rifaximin is well-tolerated. Recently Leevy and Phillips (2007) showed in 145 patients that the frequency and grade of HE was lower with rifaximin therapy compared to lactulose therapy, and that with rifaximin therapy the patients were less often hospitalized than with lactulose.

The supplementation with zinc has been debated for years (Marchesini et al., 1996; Riggio et al., 1991). Zinc is a cofactor in all reactions of the urea cycle. The precipitation of overt hepatic encephalopathy in the presence of zinc deficiency and the improvement of HE after supplementation of zinc have been imposingly demonstrated (Van der Rijt et al., 1991). Patients with cirrhosis and zinc deficiency will improve their ability to synthesize urea after provision of zinc. Long term zinc supplementation may be useful for patients with mild chronic forms of encephalopathy.

Ammonia metabolism within the muscle may be improved by the administration of L-ornithine-L- aspartate (LOLA). Controlled trials suggest that enteral and parenteral formulations of ornithine aspartate significantly reduce blood ammonia levels and have useful therapeutic effects in patients with cirrhosis and encephalopathy (Kircheis et al., 1997; Stauch et al., 1998). Poo et al. (2006) recently showed in a randomized, lactulose-controlled study of oral ornithine-aspartate in 20 patients with clinically overt HE that LOLA in contrast to lactulose significantly improved parameters of mental status, number connection test scores, asterixis scores and EEG. Both lactulose and LOLA reduced serum ammonia levels and improved quality of life scores. Since only 20 patients were included in this study the results cannot be considered proof for the superiority of LOLA compared to lactulose, while the data underscore the effect of both agents. Ornithine aspartate is well tolerated in general. From a theoretical point of view the combination of disaccharides and ornithine aspartate may be useful in patients with insufficient efficacy of only one of the two drugs.

Medical management of hepatic encephalopathy is mainly used for patients who do not yet meet the criteria for liver transplantation or for patients awaiting liver transplantation. It must be underscored that severe neuropsychiatric symptoms in patients with severe liver disease must not be considered as contraindications but instead as clear indications for liver transplantation.

Overt hepatic encephalopathy is accompanied with a bad prognosis (Bustamante et al., 1999; Hui et al., 2002). According to the results of a recent study the survival probability after the first episode of overt encephalopathy in patients with chronic liver disease is 42% at 1-year of follow-up and 23% at 3 years. Considering the survival probability reported for patients undergoing liver transplantation, which is approximately 80% at 1 year and up to 70% at 5 years, patients developing their first episode of HE should be considered as potential candidates for liver transplantation.

References

Aggarwal, S., Kang, Y., DeWolf, A., Scott, V., Martin, M., Policare, R. 1993. Transcranial doppler monitoring of cerebral blood flow velocity during liver transplantation. Transplant Proc 25 (2): 1799–1800

Amodio, P., Gatta, A. 2005. Neurophysiological investigation of hepatic encephalopathy. Metabol Brain Dis 20 (4): 369–379

Amodio, P., Marchetti, P., Del Piccolo, F., de Tourtchaninoff, M., Varghese, P., Zuliani, C., Campo, G., Gatta, A., Guerit, J.M. 1999. Spectral versus visual EEG analysis in mild hepatic encephalopathy. Clin Neurophysiol 110: 1334–1344

Atterbury, C.E., Maddrey, W., Conn, H.O. 1978. Neomycin-sorbitol and lactulose in the treatment of acute portal-systemic encephalopathy. A controlled, double-blind clinical trial. Am J Dig Dis. 23: 398–406

Bajaj, J.S., Etemadian, A., Hafeezullah, M., Saeian, K. 2007a. Testing for minimal hepatic encephalopathy in the United States: An AASLD survey. Hepatology 45 (3): 833–834

Bajaj, J.S., Saeian, K., Verber, M.D., Hischke, D., Hoffmann, R.G., Franco, J., Varma, R.R., Rao, S.M. 2007b. Inhibitory control test is a simple method to diagnose minimal hepatic encephalopathy and predict development of overt hepatic encephalopathy. Am J Gastroenterol 102: 1–7

Bhatia, V., Batra, Y., Acharya, S.K. 2004. Prophylactic phenytoin does not improve cerebral edema or survival in acute liver failure — a controlled clinical trial. J Hepatology 41: 89–96

Bhatia, V., Singh, R., Acharya, S.K. 2006. Predictive value of arterial ammonia for complications and outcome in acute liver failure. GUT 55 (1): 98–104

Bianchi, G.P., Marchesini, G., Fabbri, A., Rondelli, A., Bugianesi, E., Zoli, M., Pisi, E. 1993. Vegetable versus animal protein diet in cirrhotic patients with chronic encephalopathy: A randomized cross-over comparison. J Intern Med 233: 385–392

Blei, A.T., Olafsson, S., Webster, S., Levy, A. 1993. Complications of intracranial pressure monitoring in fulminant hepatic failure. Lancet 341: 157–158

Bustamante, J., Rimola, A., Ventura, P.J., Navasa, M., Cirera, I., Reggiardo, V., Rodes, J. 1999. Prognostic significance of hepatic encephalopathy in patients with cirrhosis. J Hepatol 30: 890–895

Cadranel, J.F., Lebiez, E., Di Martino, V., Bernard, B., El Koury, S., Tourbah, A., Pidoux, B., Valla, D., Opolon, P. 2001. Focal neurological signs in hepatic encephalopathy in cirrhotic patients: An underestimated entity ? Am J Gastroenterol 96: 515–518

Campellone, J.V., Lacomis, D., Giuliani, M.J., Kroboth, F.J. 1996. Hepatic myelopathy. Case report with review of the literature. Clin Neurol Neurosurg 98: 242–246

Canalese, J., Gimson, A.E., Davis, C., Mellon, P.J., Davis, M., Williams, R. 1982. Controlled trial of dexamethasone and mannitol for the cerebral edema of fulminant hepatic failure. Gut 23: 625–629

Cordoba, J., Cabrera, J., Lataif, L., Penev, P., Zee, P., Blei, A.T. 1998. High prevalence of sleep disturbances in cirrhosis. Hepatology 27: 339–345

Cordoba, J., Lopez-Hellin, J., Planas, M., Sabin, P., Sanpedro, F., Castro, F., Esteban, R., Guardia, J. 2004. Normal protein diet for episodic hepatic encephalopathy: Results of a randomized study. J Hepatol 41: 38–43.

Davies, M.G., Rowan, M.J., Feely, J. 1991. EEG and event related potentials in hepatic encephalopathy. Metab Brain Dis 1991; 6: 175–186.

Ellis, A.J., Wendon, J.A., Williams, R. 2000. Subclinical seizure activity and prophylactic phenytoin infusion in acute liver failure: A controlled clinical trial. Hepatology 32: 536–541

Ferenci, P., Lockwood, A.H., Mullen, K., Tarter, R., Weissenborn, K., Blei, A.T. 2002. Hepatic encephalopathy — definition, nomenclature, diagnosis, and quantification: final report of the working party at the 11th World Congress of Gastroenterology, Vienna 1998; Hepatology 35(3): 716–721

Gitlin, N., Lewis, D.C., Hinkley, L. 1986. The diagnosis and prevalence of subclinical hepatic encephalopathy in apparently healthy ambulant non-shunted patients with cirrhosis. J Hepatol 3: 75–82

Groeneweg, M., Quero, J.C., De Bruijn, I., Hartmann, I.J., Essink-bot, M.L., Hop, W.C., Schalm, S.W. 1998. Subclinical hepatic encephalopathy impairs daily functioning. Hepatology 28: 45–49

Hartmann, I.J.C., Groeneweg, M., Quero, J.C., Beijeman, S.L., de Man, R.A., Hop, W.C., Schalm, S.W. 2000. The prognostic significance of subclinical hepatic encephalopathy. Am J Gastroenterol 2000; 95: 2029–2034

Homburg, A.M., Jacobsen, M., Enevoldsen, E. 1993. Transcranial doppler recordings in raised intracranial pressure. Acta Neurol Scand 87: 488–493

Hui, A.Y., Chan, H.L., Leung, N.W., Hung, L.C., Chan, F.K., Sung, J.J. 2002. Survival and prognostic indicators in patients with hepatitis B virus-related cirrhosis after onset of hepatic decompensation. J Clin Gastroenterol 34(5): 569–572

Jalan, R., Olde Damink, S.W., Deutz, N.E., Hayes, P.C., Lee, A. 2004. Moderate hypothermia in patients with acute liver failure and uncontrolled intracranial hypertension. Gastroenterology 127: 1338–1346

Kircheis, G., Nilius, R., Held, C., Berndt, H., Buchner, M., Görtelmeyer, R., Hendricks, R., Kruger, B., Kuklinski, B., Meister, H., Otto, H.J., Rink, C., Rosch, W., Stauch, S. 1997. Therapeutic efficacy of L-Ornithine-L-Aspartate infusions in patients With cirrhosis and hepatic encephalopathy: Results of a placebo-controlled, double-blind study. Hepatology 25: 1351–1360

Kircheis, G., Wettstein, M., Timmermann, L., Schnitzler, A., Haussinger, D. 2002. Critical flicker frequency for quantification of low-grade hepatic encephalopathy. Hepatology. 35: 357–366.

Krieger, S., Jauss, M., Jansen, O., Theilmann, L., Geissler, M., Krieger, D. 1996. Neuropsychiatric profile and hyperintense globus pallidus on T1-weighted magnetic resonance images in liver cirrhosis. Gastroenterology 111: 147–155

Kril, J.J., Butterworth, R.F. 1997. Diencephalic and cerebellar pathology in alcoholic and non-alcoholic patients with end-stage liver disease. Hepatology 26 (4): 837–841

Kullmann, F., Hollerbach, S., Holstege, A., Schölmerich, J. 1995. Subclinical hepatic encephalopathy: The diagnostic value of evoked potentials. J Hepatol 22: 101–110

Larsen, F.S., Ejlersen, E., Hansen, B.A., Knudsen, G.M., Tygstrup, N., Secher, N.H. 1995 a. Functional loss of cerebral blood flow autoregulation in patients with fulminant hepatic failure. J Hepatol 23: 212–217

Larsen, F.S., Pott, F., Hansen, B.A., Ejlersen, E., Knudsen, G.M., Clemmesen, J.D., Secher, N.H. 1995b. Transcranial doppler sonography may predict brain death in patients with fulminant hepatic failure. Transpl Proc 27 (6): 3510–3511

Leevy, C.B., Phillips, J.A. 2007. Hospitalizations during the use of rifaximin versus lactulose for the treatment of hepatic encephalopathy. Dig Dis Sci; 52: 737–741

Lockwood, A.H. 2000. "What's in a name?" Improving the care of cirrhotics. J Hepatol 32: 859–861

Lockwood, A.H., Butterworth, R.F., Weissenborn, K. 1997. An image of the brain in liver disease. Curr Opin Neurol 10(6): 525–533

Lockwood, A.H., Murphy, B.W., Donnelly, K.Z., Mahl, T.C., Perini, S. 1993. Positron-emission tomographic localization of abnormalities of brain metabolism in patients with minimal hepatic encephalopathy. Hepatology 18 (5): 1061–1068

Lockwood, A.H., Weissenborn, K., Bokemeyer, M., Tietge, U., Burchert, W. 2002. Correlations between cerebral glucose metabolism and neuropsychological test performance in non-alcoholic cirrhotics. Metab Brain Dis 17(1): 29–40

Marchesini, G., Bianchi, G., Merli, M., Amodio, P., Panella, C., Loguercio, C., Rossi Fanelli, F., Abbiati, R., and Italian BCAA Study Group. 2003. Nutritional supplementation with branched-chain amino acids in advanced cirrhosis: A double-blind, randomized trial. Gastroenterology 124: 1792–1801

Marchesini, G., Fabbri, A., Bianchi, G., Brizi, M., Zoli, M. 1996. Zinc supplementation and amino-acid-nitrogen metabolism in patients with advanced cirrhosis. Hepatology 23: 1084–1092

Mas, A., Rodes, J., Sunyer, L., Rodrigo, L., Planas, R., Vargas, V., Castells, L., Rodriguez-Martinez, D., Fernandez-Rodriguez, C., Coll, I., Pardo, A.;, Spanish Association for the study of the Liver Hepatic Encephalopathy Cooperative Group. 2003. Comparison of rifaximin and lactitol in the treatment of acute hepatic encephalopathy: Results of a randomized, double-blind, double-dummy, controlled clinical trial. J Hepatol. 38: 51–58.

Montagnese, S., Amodio, P., Morgan, M.Y. 2004. Methods for diagnosing hepatic encephalopathy in patients with cirrhosis: A multidimensional approach. Metab Brain Dis 19: 281–312

Montagnese, S., Jackson, C., Morgan, M.Y. 2007. Spatio-temporal decomposition of the electroencephalogram in patients with cirrhosis. J Hepatol 46: 447–458

Morgan, M.Y., Alonso, M., Stanger, L.C. 1989. Lactitol and lactulose for the treatment of sub-clinical hepatic encephalopathy in cirrhotic patients. J Hepatol 8: 208–217

Morgan, M.Y., Hawley, K.E. 1987. Lactitol versus lactulose in the treatment of acute hepatic encephalopathy in cirrhotic patients: a double-blind, randomized trial. Hepatology 7(6): 1278–1284

Munoz, S.J. 1993. Difficult management problems in fulminant hepatic failure. Semin Liver Disease 13: 395–413

Murphy, N., Auzinger, G., Bernal, W., Wendon, J. 2004. The effect of hypertonic sodium chloride on intracranial pressure in patients with acute liver failure. Hepatology 39: 464–470

Muto, Y., Sato, S., Watanabe, A., Moriwaki, H., Suzuki, K., Kato, A., Kato, M., Nakamura, T.,; Higuchi, K., Nishiguchi, S.;, Kumada, H., Ohashi, Y.;, for the Long-Term-Survival Study (LOTUS) Group. 2005. Effects of oral branched-chain amino acid granules on event-free survival in patients with liver cirrhosis. Clin Gastroenterol Hepatol. 3: 705–713.

O'Grady, J.G., Schalm, S.W., Williams, R. 1993. Acute liver failure: Redefining the syndromes. Lancet 342: 273–275

Ostapowicz, G.A., Fontana, R.J., Schiodt, F.V., Larson, A., Davern, T.J., Han, S.H., McCashland, T.M., Shakil, A.O., Hay, J.E., Hynan, L., Crippin, J.S., Blei, A.T., Samuel, G., Reisch, J., Lee, W.M.;, U.S. Acute Liver Failure Study Group. 2002. Results of a prospective study of acute liver failure at 17 tertiary care centers in the United States. Ann Intern Med 137: 947–954

Parsons-Smith, B.G., Summerskill, W.H.J., Dawson, A.M., Sherlock, S. 1957. The electroencephalograph in liver disease. Lancet ii: 867–871

Penin, H. 1967. Über den diagnostischen Wert des Hirnstrombildes bei der hepato-portalen Encephalopathie. Fortschr Neurol Psychiatr 35 (4): 174–234

Polson, J., Lee, W.M. 2005. AASLD position paper: The management of acute liver failure. Hepatology 41(5): 1179–1197

Poo, J.L., Gongora, J., Sanchez-Avila, F., Aguilar-Castillo, S., Garcia-Ramos, G., Fernandez-Zertuche, M., Rodriguez-Fragoso, L., Uribe, M. 2006. Efficacy of oral L-ornithine-L-aspartate in cirrhotic patients with hyperammonemic hepatic encephalopathy. Results of a randomized, lactulose-controlled study. Ann Hepatol 5(4): 281–288

Powell, E.E., Pender, M.P., Chalk, J.B., Parkin, P.J., Strong, R., Lynch, S., Kerlin, P., Cooksley, W.G., Cheng, W., Powell, L.W. 1990. Improvement in chronic hepatocerebral degeneration following liver transplantation. Gastroenterology 98: 1079–1082

Prasad, S., Dhiman, R.K., Duseja, A., Chawla, Y.K., Sharma, A., Agarwal, R. 2007. Lactulose improves cognitive functions and health-related quality of life in patients with cirrhosis who have minimal hepatic encephalopathy. Hepatology 45: 549–559

Quadri, A.M., Ogunwale, B.O., Mullen, K.D. 2007. Can we ignore minimal hepatic encephalopathy any longer? Hepatology 45: 547–548

Raghavan, M., Marik, P.E. 2006. Therapy of intracranial hypertension in patients with fulminant hepatic failure. Neurocrit Care 4: 179–189

Riggio, O., Ariosto, F.,, Merli, M., Caschera, M., Zullo, A., Balducci, G., Ziparo, V., Pedretti, G., Fiaccodori, F., Bottari, E. et al. 1991. Short-term oral zinc supplementation does not improve chronic hepatic encephalopathy: results of a double-blind crossover trial. Dig Dis Sci 36: 1204–1208

Rinella, M.E., Sanyal, A. 2006. Intensive management of hepatic failure. Sem Respiratory Crit Care Med 27(3): 241–261

Romero-Gomez, M., Cordoba, J., Jover, R., Del Olmo, J.A., Ramirez, M., Rey, R., de Madaria, E., Montoliu, C., Nunez, D., Flavia, M., Company, L., Rodrigo, J.M., Felipo, V. 2007. Value of the critical flicker frequency in patients with minimal hepatic encephalopathy. Hepatology 45(4): 879–885

Sass, D.A., Shakil, A.O. 2005. Fulminant hepatic failure. Liver Transplantation 11(6): 594–605

Schnittger, C., Weissenborn, K., Böker, K., Kolbe, H., Dengler, R., Manns, M.P. 1997. Continuous noninvasive cerebral perfusion monitoring in fulminant hepatic failure and brain oedema. In Advances in Hepatic Encephalopathy & Metabolism in Liver Disease. eds C. Record and H. Al Mardini (eds.)., pp. 515–519. New Castle upon Tyne: Medical Faculty of the University of Newcastle upon Tyne

Schomerus, H., Hamster, W. 1998. Neuropsychological aspects of portal-systemic encephalopathy. Metab Brain Dis 13(4): 361–377

Schomerus, H., Hamster, W. 2001. Quality of life in cirrhotics with minimal hepatic encephalopathy. Metab Brain 16 (1/2): 37–42

Schomerus, H., Hamster, W., Blunck, H., Reinhard, U., Mayer, K., Dolle, W. 1981. Latent portasystemic encephalopathy: I. Nature of cerebral functional defects and their effect on fitness to drive. Dig Dis Sci 26: 622–630

Schomerus, H., Schreiegg, J. 1993. Prevalence of latent portasystemic encephalopathy in an unselected population of patients with liver cirrhosis in general practice. Z Gastroenterol 31: 231–234

Schomerus, H., Weissenborn, K., Hamster, W., Rueckert, N., Hecker, H. 1999. PSE-syndrom-test. Psychodiagnostisches Verfahren zur quantitativen Erfassung der (minimalen) portosystemischen Enzephalopathie. Frankfurt: Swets Test Services.

Shami, V.M., Caldwell, S.H., Hespenheide, E.E., Arseneau, K.O., Blickston, S.J., Macik, B.G. 2003. Recombinant activated factor VII for coagulopathy in fulminant hepatic failure compared with conventional therapy. Liver Transpl 9: 138–143

Sherlock, S., Summerskill, W.H.J., White, L.P., Phear, E.A. 1954. Portal-systemic encephalopathy. Neurological complications of liver disease. Lancet I: 453–457

Spahr, L., Butterworth, R.F., Fontaine, S., Bui, L., Therrien, G., Milette, P.C., Lebrun, L.H., Zayed, J., Leblanc, A., Pomier-Layrargues, G. 1996. Increased blood manganese in cirrhotic patients: Relationship to pallidal magnetic resonance signal hyperintensity and neurological symptoms. Hepatology 1996; 24: 1116–1120

Srivastava, A., Mehta, R., Rothke, S.P., Rademaker, A.W., Blei, A.T. 1994. Fitness to drive in patients with cirrhosis and portal systemic shunting: A pilot study evaluating driving performance. J Hepatol 21: 1023–1028

Stauch, S., Kircheis, G., Adler, G., Beckh, K., Ditschuneit, H., Görtelmeyer, R., Hendricks, R., Heuser, A., Karoff, C., Malfertheiner, P., Mayer, D., Rosch, W., Steffens, J. 1998. Oral L-ornithine-L-aspartate therapy of chronic hepatic encephalopathy: Results of a placebo-controlled double-blind study. J Hepatol 28: 856–864

Tofteng, F., Larsen, F.S. 2004. The effect of indomethacin on intracranial pressure, cerebral perfusion and extracellular lactate and glutamate concentrations in patients with fulminant hepatic failure. J Cereb Blood Flow Metab 24: 798–804

Van der Rijt, C.D.C., Schalm, S., De Groot, G.H., De Vlieger, M. 1990. Objective measurement of hepatic encephalopathy by means of automated EEG analysis. Electroencephalogr Clin Neurophysiol 75: 289–295

Van der Rijt, K., Schalm, S.W., Schat, H., Foeken, K., De Jong, G. 1991. Overt hepatic encephalopathy precipitated by zinc deficiency. Gastroenterology 100: 114–118

Vaquero, J., Fontana, R.J., Larson, A.M., Bass, N.M.T., Davern, T.J., Shakil, A.O., Han, S., Harrison, M.E., Stravitz, T.R., Munoz, S., Brown, R., Lee, W.M., Blei, A.T. 2005. Complications and use of intracranial pressure monitoring in patients with acute liver failure and severe encephalopathy. Liver Transpl 12: 1581–1589

Victor, M., Adams, R.D., Cole, M. 1965. The acquired (non-Wilsonina) type of chronic hepatocerebral degeneration. Medicine (Baltimore) 44: 345–394

Watanabe, A., Tuchida, T., Yata, Y., Kuwabara, J. 1995. Evaluation of neuropsychological function in patients with liver cirrhosis with special reference to their driving ability. Metab Brain Dis. 10: 239–248

Wein, C., Koch, H., Popp, B., Oehler, G., Schauder, P. 2004. Minimal hepatic encephalopathy impairs fitness to drive. Hepatology 39: 739–745

Weissenborn, K. 1991. Neurophysiological methods in the diagnosis of Eearly hepatic encephalopathy. In Progress in Hepatic Encephalopathy and Metabolic Nitrogen Exchange. Bengtsson F, Jeppsson B, Almdal T, Vilstrup H (eds.)., pp 27–39. Boca Raton: CRC Press

Weissenborn, K., Ennen, J.C., Schomerus, H., Ruckert, N., Hecker, H. 2001. Neuropsychological characterization of hepatic encephalopathy. J Hepatol. 34: 768–773.

Weissenborn, K., Scholz, M., Hinrichs, H., Wiltfang, J., Schmidt, F.W., Künkel, H. 1990. Neurophysiological assessment of early hepatic encephalopathy. Electroencephalogr Clin Neurophysiol 1990; 75: 289–295

Weissenborn, K., Tietge, U.J., Bokemeyer, M., Mohammadi, B., Bode, U., Manns, M.P., Caselitz, M. 2003. Liver transplantation improves hepatic myelopathy: Evidence by three cases. Gastroenterology. 2003; 124: 346–351

White, M.L., Zhang, Y., Andrew, L.G., Hadley, W.L. 2005. MR imaging with diffusion-weighted imaging in acute and chronic wernicke encephalopathy. AJNR 26: 2306–2310

Zhong, C., Jin, L., Fei, G. 2005. MR imaging of nonalcoholic wernicke encephalopathy: A follow-up study. AJNR 26: 2301–2305

Chapter 11
Uremic and Dialysis Encephalopathies

Allen I. Arieff

Introduction

Because of advancements in dialysis therapy and renal transplantation, and their medical management, patients with end-stage renal disease (ESRD) rarely develop severe clinical manifestations before the institution of therapy. In the USA, there are currently about 450,000 patients receiving dialysis therapy (Meyer & Hostetter, 2007). Although uncommon in the USA and Canada, patients who have chronic renal failure (GFR below 30 ml/min) and have not yet received dialysis therapy may develop a symptom complex characterized by fatigue, mild sensorial clouding, decreased mental acuity, tremor, anorexia, nausea and sleep disturbances (Gusbeth-Tatomir et al., 2007; Perl et al., 2006;). However, the institution of apparently adequate maintenance dialysis therapy does not eliminate central nervous system (CNS) manifestations of uremia. CSN disorders of both untreated renal failure and those persisting despite dialysis are referred to as uremic encephalopathy. The treatment of end stage renal disease with dialysis has itself been associated with the emergence of several distinct disorders of the central nervous system. These disorders are: dialysis disequilibrium syndrome, dialysis dementia, stroke, sexual dysfunction and a new syndrome of chronic dialysis-dependent encephalopathy (Arieff, 2004).

The dialysis disequilibrium syndrome has been rarely observed since about 1970. It is usually a consequence of the initiation of dialysis therapy in a minority of patients. Dialysis dementia is a progressive, generally fatal encephalopathy which can affect patients on chronic hemodialysis as well as children with chronic renal failure who have not been treated with dialysis. Cardiovascular disorders are the major cause of death in hemodialysis patients, accounting for 40% of deaths (Go et al., 2004; Herzog et al., 2007). These include myocardial infarction, cardiomyopathy, ischemic heart disease and stroke (Shik & Parfrey, 2005; Toyoda et al., 2005). The factors associated with uremia which lead to an increased incidence of mortality from stroke are not well known, but are beginning to be elucidated (Koren-Morag et al., 2006; Shik & Parfrey, 2005). Additionally, parathyroid hormone may be a major factor in the pathogenesis of stroke in patients with ESRD (De Boer et al., 2002; Sato et al., 2003). In addition to the above manifestations of neurologic dysfunction which are specifically related to uremia, dialysis, or both, a number of other neurologic disorders occur with increased frequency in patients who have end stage renal disease and are being

D.W. McCandless (ed.) *Metabolic Encephalopathy*,
doi: 10.1007/978-0-387-79112-8_9, © Springer Science+Business Media, LLC 2009

treated with chronic hemodialysis. Subdural hematoma, acute stroke, certain electro-lyte disorders (hyponatremia, hypernatremia, phosphate depletion and hypercal-cemia), vitamin deficiencies, Wernickes encephalopathy, drug intoxication, hypertensive encephalopathy and acute trace element intoxication must be considered in patients with chronic renal failure who manifest an altered mental state. Patients with renal failure are also at risk to develop organic brain disease and other metabolic encephalopathies, which might afflict the general population (Schulz & Arieff, 2003). Therefore, when a patient with end-stage kidney disease presents with altered mental status, a thorough and complete evaluation is necessary.

Uremic Encephalopathy

Uremic encephalopathy is an acute or subacute organic brain syndrome that regularly occurs in patients with acute or chronic renal failure when the glomer-ular filtration rate declines below about 15% of the normal. As with other organic brain syndromes, these patients display variable disorders of conscious-ness, psychomotor behavior, thinking, memory, speech, sleep, perception and emotion (Teschan & Arieff, 1985). The term uremic encephalopathy refers to the nonspecific neurologic symptoms of uremia. The symptoms may include sluggishness and easy fatigue, daytime drowsiness and insomnia with a tendency toward sleep-inversion (Gusbeth-Tatomir et al., 2007; Perl et al., 2006), itching, inability to perform mental (cognitive) tasks, slurring of speech, anorexia, nausea, and vomiting. Other symptoms may include restlessness, diminished sexual interest and performance, myoclonus, asterixis, disorientation and confusion, ataxia, and rarely, with the current ready availability of dialysis, coma and convulsions (Arieff, 2004).

There are certain characteristics of uremic encephalopathy which are espe-cially noteworthy. The symptoms and their severity and overall rate of progres-sion vary directly with the rate at which renal insufficiency develops. Uremic symptoms are generally more severe and progress more rapidly in patients with acute renal failure than in those with chronic renal failure. The symptoms are readily ameliorated by dialysis procedures or renal transplantation and suppressed by maintenance dialysis regimens. Thus the encephalopathy of renal failure is important to recognize precisely because it is promptly and decisively treatable by generally available clinical methods. The causes of uremic encephalopathy are doubtless multiple and complex. However, since the widespread introduction of recombinant human erythropoietin (EPO) as a therapeutic agent in patients with varying degrees of renal insufficiency, including those being treated with hemo-dialysis, it is now clear that brain functions and quality of life are improved by correction of the anemia associated with ESRD with EPO (Ayus et al., 2005; Moreno et al., 2000).

Diagnosis of Uremic Encephalopathy

The presenting symptoms of uremia are similar to many other encephalopathic states. The differential diagnosis is even more complex, since patients with renal failure are subjected to other intercurrent illnesses that may also induce other encephalopathic effects. In patients with renal failure, treatment with dialysis will restore more normal body fluid composition. Despite the possibility that multiple causes of encephalopathy might occur simultaneously, uremic encephalopathy may be successfully differentiated in most instances by means of the usual clinical methods.

In patients who have other medical problems such as advanced liver disease with hepatic insufficiency, it is often difficult to differentiate whether the encephalopathy is due to hepatic or renal causes. In patients with renal failure, the major route for elimination of urea is not available; thus, there is an increase in blood urea. The amount of urea that enters the colon is increased because of the elevated plasma urea. Urea is then acted on by colonic bacteria and mucosal enzymes in a manner similar to that of protein and amino acids. This leads to increased ammonia production in uremic subjects that may either increase plasma ammonia levels or lead to misinterpretation of this test.

If the patient with kidney failure also has cirrhosis or some other form of liver failure, this additional ammonia load may present a stress that cannot be adequately handled by the diseased liver. The result may be increased blood and central nervous system ammonia levels with development of encephalopathy (Fraser & Arieff, 1985). Thus, patients with cirrhosis and end-stage kidney disease are at particular risk for developing encephalopathy since both conditions act synergistically to increase both blood and central nervous system ammonia. It should also be noted that plasma urea and serum creatinine do not always adequately reflect renal function in patients with severe liver disease. Recent studies suggest that many patients who have cirrhosis, ascites, and normal plasma urea and creatinine may in fact have severe renal functional impairment (Gines et al., 1988; Papadakis & Arieff, 1987; Takabatake et al., 1988). In such individuals, differentiation of hepatic from uremic encephalopathy on clinical grounds may be difficult.

Acute Renal Failure

The clinical manifestations of acute renal failure have been studied in several large patient series (Cooper & Arieff, 1979; Teschan et al., 1979). Abnormalities of mental status have been noted as early and sensitive indices of a neurologic disorder, which progressed rapidly into disorientation and confusion (Fraser & Arieff, 1988a, 1993). When uremia is untreated and allowed to progress, coma often supervenes. Cranial nerve signs such as nystagmus and mild facial asymmetries are common, though usually transient. There can be visual field defects and papilledema of the optic fundi. About half the patients have dysarthria, and many have diffuse weakness and fasciculations. Marked variation of deep tendon reflexes is noted in most patients, often in an asymmetrical pattern. Progression of hyperreflexia, with sustained clonus at the patella or ankle, is common (Moe & Sprague, 1994).

Chronic Renal Failure

The incidence of ESRD in the United States is 260 cases per million of the population (Schulman & Himmelfarb, 2004). Among patients with ESRD treated with chronic hemodialysis, comorbid diseases, mainly infectious and cardiovascular, continue to increase. Among patients with ESRD treated with chronic hemodialysis in 2000, there were 13.9 hospital days per patient year. Among such patients, the mortality in the USA is 182 deaths per 1000 patient-years at risk, or 247 deaths per 1000 patient-years at risk for hemodialysis patients (Schulman & Himmelfarb, 2004). In an overview, the 5-year mortality among chronic hemodialysis patients in the USA was about 50%, although most deaths were due to co-morbid factors (Meyer & Hostetter, 2007). The most frequent causes of ESRD (in the USA) are diabetes, hypertension, glomerulonephritis, arteriosclerotic cardiovascular disease and polycystic kidney disease (Schulman & Himmelfarb, 2004).

There are multiple neurological complications reported in patients with chronic renal failure (CRF) (Brouns & De Deyn, 2004; Fraser & Arieff, 1999; Meyer & Hostetter, 2007; Ogura et al., 2001; Palmer, 2003). Many are not improved by chronic dialysis.

After about 2002, the electroencephalogram (EEG) has rarely been used as a clinical tool for evaluation of patients with renal nsufficiency. The findings which follow are not currently widely applicable. When the diagnosis of renal failure was first established, EEGs in patients with acute renal failure (Cooper et al., 1978) were found to be generally grossly abnormal. Generally, the percentage of EEG power less than 5 Hz and less than 7 Hz, which are standard measurements of the percentage of EEG power devoted to abnormal (delta) slow wave activity, are over 20 times the normal value. The percentage of EEG frequencies above 9 Hz and below 5 Hz are not affected by dialysis for 6 to 8 weeksand return to normal with recovery of renal function. Similar findings have been shown in experimental animals with renal failure (Guisado et al., 1975). In patients with acute renal failure, the EEG is abnormal within 48 hours of the onset of renal failure (Cooper et al., 1978) and is generally not affected by dialysis within the first 3 weeks.

The EEG findings in patients with chronic renal failure are well described and generally less severe than those observed in patients with acute renal failure (Teschan & Arieff, 1985). After the initiation of dialysis, there may be an initial period of clinical stabilization, during which time the EEG deteriorates, (up to six months) but after that it approaches normal values (Kiley et al., 1976). However, improvement is seen after renal transplantation (Bolton, 1976; Teschan et al., 1983). Cognitive functions have also been shown to be impaired in uremia. These include global cognitive functions, memory, sustained attention, selective attention, speed of decision making, short-term memory, language and mental manipulation of symbols (Murray et al., 2006, 2007; Pereira et al., 2005). The causes of the EEG abnormalities observed in uremic patients are probably multifactorial but there is evidence that a very important element may be the effect of the parathyroid hormone (PTH) on the brain. In experimental animals with either acute or chronic renal failure, many of the EEG abnormalities can be shown to be related to a direct effect of PTH on the brain, which leads to an elevated brain content of calcium Ca^{2+} (Guisado et al., 1975). Studies in patients with either acute renal failure or chronic renal failure suggest a similar pathogenesis (Fraser & Arieff, 1995; Mahoney & Arieff, 1982).

Psychological Testing

Several different types of psychological tests have been applied to subjects with chronic renal failure. These have been designed to measure the effects of either dialysis, renal transplantation or parathyroidectomy (Cogan et al., 1978; Murray et al., 2006; Pereira et al., 2007; Pliskin et al., 1996; Sehgal et al., 1997). The Trailmaking Test has been administered to a number of uremic subjects. In general, their performance was less effective than that of the normal ones; improvement with practice limits repeated use of this test. The Continuous Memory Test correlates quite well with the degree of renal failure, as does the Choice Reaction Time. Scores in both tests improved with treatment by dialysis or renal transplantation. Similar but less impressive results were obtained with the Continuous Performance Test. Of all these tests, it appeared that the CRT was best correlated with renal function and with improvement in the patient's condition as a result of dialysis or transplantation (Fraser & Arieff, 1994; Teschan & Arieff, 1985).

Patients with chronic renal failure, who were maintained on dialysis, have been evaluated as to the possible effects of PTH on psychological function (Cogan et al., 1978). After establishment of baseline values, patients with chronic renal failure underwent parathyroidectomy for other medical reasons (e.g. bone disease, soft tissue calcification, and persistent hypercalcemia, all of which were unresponsive to medical management). In these patients, parathyroidectomy resulted in a significant improvement in several areas of psychological testing. They showed significant improvement in Raven's Progressive Matrices percentile scores and visual motor index (VMI) raw and percentage scores. These are tests of general cognitive function, nonverbal problem solving and visual-motor or visual spatial skills (Fraser & Arieff, 1994; Teschan & Arieff, 1985).

In addition, they manifested significantly fewer errors on the Trailmaking Test as well as significantly lower raw and T-score values on the Profile of Mood States Fatigue Scale postoperatively, in which they reported feeling significantly less fatigue, weariness, and inertia after undergoing surgery. Control subjects who underwent neck surgery for other reasons generally did not demonstrate significant postoperative improvement in tests of cognition (Cogan et al., 1978). Other studies have shown that in general there is intellectual impairment in most patients with chronic renal failure being treated with dialysis, virtually all of whom have secondary hyperparathyroidism (Brouns & De Deyn, 2004; Gilli & Bastiani, 1983; Murray et al., 2006, 2007; Pereira et al., 2005). Cognitive data were compared with other information, such as age, sex, length of dialysis, and biochemical variables by a multiple regression technique. The analysis suggested that of the cognitive data, those obtained with the BDLT bore the strongest relation to duration of dialysis. Other studies have suggested that the WAIS full-scale IQ in dialysis patients is below that of the general population (Murray et al., 2006; Pereira et al., 2007).

Biochemical Changes in the Brain

To determine the possible causes of the EEG abnormalities and clinical manifestations observed in patients with either acute renal failure or chronic renal failure, in vivo biochemical studies have been carried out in the brains of both patients and laboratory animals. Measurements have included brain intracellular pH and concentrations of Na^+, K^+, Cl^-, Al^{3+}, Ca^{2+}, Mg^{2+}, urea, adenine nucleotides (creatine phosphate, ATP, ADP, AMP), lactate, and ($Na^+ + K^+$)-activated adenosine triphosphatase (ATPase) enzyme activity (Arieff et al., 1975, 1977; Cooper et al., 1978; Fraser & Arieff, 1988a; Mahoney & Arieff, 1982; Mahoney et al., 1984; Minkoff et al., 1972; Perry et al., 1985; Van den Noort et al., 1968). In patients with acute renal failure, the brain content of water, K^+ and Mg^{2+} is normal, while Na^+ is modestly decreased and Al^{3+} is slightly elevated (Cooper et al., 1978). However, cerebral cortex Ca^{2+} content is almost twice the normal value (Cooper et al., 1978; Mahoney & Arieff, 1982; Mahoney et al., 1984). Similar findings have been observed in dogs with acute renal failure (Arieff et al., 1976; Guisado et al., 1975;). Alterations of cerebral metabolism that might be related to the changes in permeability mentioned above have also been studied in animals (Deferrari, 1981; Mahoney et al., 1983, 1984; Verkman & Fraser, 1986). In the brain of rats with acute renal failure, creatine phosphate, ATP, and glucose were increased, but there were corresponding decreases in AMP, ADP, and lactate. Total brain adenine nucleotide content and ($Na+ - K+$)- activated ATPase were normal to low. The uremic brain utilized less ATP and thus failed to produce ADP, AMP, and lactate at normal rates. The brain energy charge was normal, as was the redox state, and these findings were not altered by hypoxia (Fraser & Arieff, 1988a; Mahoney et al., 1984). There was a corresponding decrease in brain metabolic rate, along with elevated glucose and low lactate levels (Mahoney et al., 1984). Patients with chronic renal insufficiency (glomerular filtration rate below 20 ml/min) have decreased brain uptake of glutamine and increased ammonia uptake. The relevance of these findings, in terms of neurotransmitters or other brain function, is unknown (Deferrari, 1981).

In animals with either acute or chronic renal failure, both urea concentration and osmolality are similar in brain, cerebrospinal fluid, and plasma. The solute content of brain in animals with acute renal failure is such that essentially all of the increase in brain osmolality is due to an increase of brain urea concentration. However, in animals with chronic renal failure, about half of the increase in brain osmolality is due to the presence of undetermined solute (idiogenic osmoles) with the other half due to an increase of urea concentration (Arieff et al., 1975, 1977; Mahoney et al., 1983).

In dogs with chronic renal failure, brain content of Na+, K+, Cl– and water are not different from control values. Similarly, the extracellular space was not different from control (Mahoney et al., 1983). Calcium content was measured in eight parts of the brain in dogs that had chronic renal failure for four months. Calcium content was found to be normal in the subcortical white matter, pons, medulla, cerebellum, thalamus and caudate nucleus. However, calcium was about 60% above control values in both cortical gray matter and hypothalamus (Mahoney et al., 1983). Magnesium content was normal in all eight parts of the brain, as was the water content (Mahoney et al., 1983). Other investigators have also found the cerebral cortex calcium content elevated in dogs with chronic renal failure (Akmal et al., 1984).

In patients with acute renal failure, as well as those with chronic renal failure who had not been treated with dialysis, brain calcium (cerebral cortex) was significantly increased versus control values. The brain samples were obtained from autopsy material.

In animals that have acute renal failure and metabolic acidosis, the intracellular pH (pHi) of brain and skeletal muscle is normal (Arieff et al., 1977). In dogs with chronic renal failure, intracellular pH is normal in brain, liver and skeletal muscle (Mahoney et al., 1983). In patients with renal failure, intracellular pH has been reported to be normal in both skeletal muscle and leukocytes, as well as in the "whole body" (Levin & Baron, 1977; Maschio et al., 1970; Tizianello et al., 1977). The pH of CSF has also been shown to be normal in both patients and laboratory animals with renal failure (Arieff et al., 1977; Mahoney et al., 1983). Thus, despite the presence of extracellular metabolic acidemia in patients or laboratory animals with either acute renal failure or chronic renal failure, the intracellular pH is normal in the brain, white cells, liver and skeletal muscle.

In general then, studies of brain tissue from both intact animal models of uremia and humans with renal failure have revealed many different biochemical abnormalities associated with the uremic state. However, such investigations have not as yet revealed much about the fundamental mechanisms that might induce such abnormalities. Such studies probably will have to be done in isolated cell systems or subcellular systems from the brain. These systems have the advantage of permitting one to study isolated manifestations of the uremic state while removing the numerous potential confounding influences present in an in vivo model.

Central Nervous System Pathology with Uremia

Pathologic studies of brain have been reported in over 400 patients dying with chronic renal failure (Arieff, 1986; Rotter & Roettger, 1974). It has been shown that there is some necrosis of the granular layer of the cerebral cortex. Small intracerebral hemorrhages and necrotic foci are seen in about 10% of uremic patients and focal glial proliferation is seen in about 2%. In general, changes in the brain of patients who died with chronic renal failure are probably non-specific and related more to any of a number of concomitant underlying disease states (Burks et al.,

1976). Cerebral edema has not been observed in the brains of humans and laboratory animals with either acute renal failure or chronic renal failure (Cogan et al., 1978; Cooper et al., 1978). However, recent studies using MR imaging technique have shown that ESRD may be associated with interstitial brain edema (Chen et al., 2007). The initial hemodialysis often leads to increased interstitial brain edema (Chen et al., 2007). About 33% of adult patients with chronic renal failure have subcortical and periventricular lesions in brain white matter (Martinez-Vea et al., 2006). These appear to represent ischemic brain damage in middle aged uremic patients (Martinez-Vea et al., 2006). Silent cerebral infarcts are present in about 50% of hemodialysis patients (Naganuma et al., 2005). Such lesions represent a major risk factor for a subsequent stroke. Among adults maintained on chronic hemodialysis, stroke was common, usually with infarction in the vertibrobasilar territory (Toyoda et al., 2005).

The recent use of advanced neuroimaging techniques has led to substantial in vivo study of uremic brain in humans (Arieff et al., 1994). Acute and subacute movement disorders have been observed in patients with ESRD. These have been associated with bilateral basal ganglia and internal capsule lesions (Wang et al., 2004). Cerebral atrophy has been observed in chronic hemodialysis patients, and it tends to worsen as dialysis therapy continues (Savazzi, 1988; Savazzi et al., 1995). Cerebral atrophy had previously been thought to be associated with dialysis dementia, but this is apparently not the case (Mahurkar et al., 1978). ESRD has also been reported to lead to deterioration of vision. Some cases are associated with uremic pseudotumor cerebri, and in these cases, surgical optic nerve fenestration may improve visual loss (Guy et al., 1990; Korzets et al., 1998).

Pathophysiology of Uremic Encephalopathy

Although there are many factors that contribute to uremic encephalopathy, most investigations have shown no correlation between encephalopathy and any of the commonly measured indicators of renal failure. In recent years, there has been considerable discussion of the possible role of PTH as a uremic toxin. There is a substantial amount of evidence to suggest that PTH may exert an adverse effect on the central nervous system (Cogan et al., 1978; Cooper et al., 1978; Guisado et al., 1975; Mahoney & Arieff, 1982).

Role of PTH: In patients dying with acute or chronic renal failure, the calcium content in brain cerebral cortex is significantly elevated (Cogan et al., 1978; Cooper et al., 1978; Guisado et al., 1975). Dogs with acute or chronic renal failure show increases of brain gray matter calcium and EEG changes similar to those seen in humans with acute renal failure (Arieff & Massry, 1974; Arieff et al., 1975; Mahoney et al., 1983, 1984). In dogs, both the EEG and brain calcium abnormalities found with uremia can be prevented by parathyroidectomy. Conversely, these abnormalities can be reproduced by the administration of PTH to normal animals while maintaining serum calcium and phosphate in the normal range. Thus, PTH is essential to produce some of the central nervous system manifestations in the canine model of uremia (Guisado et al., 1975; Akmal et al., 1984).

PTH is known to have central nervous system effects in humans even in the absence of impaired renal function. Neuropsychiatric symptoms have been reported to be among the most common manifestations of primary hyperparathyroidism (Heath et al., 1980; Luxenberg et al., 1984). Patients with primary hyperparathyroidism also have EEG changes similar to those observed in patients with acute renal failure (Cooper et al., 1978; Goldstein & Massry, 1980). The common denominator appears to be elevated plasma levels of PTH (Cogan et al., 1978; Cooper et al., 1978; Guisado et al., 1975; Mahoney & Arieff, 1982).

In patients with acute renal failure the EEG is abnormal within 18 hours of the onset of renal failure and is generally not affected by dialysis for periods of up to 8 weeks (Cooper et al., 1978). In patients with either primary or secondary hyperparathyroidism, parathyroidectomy results in an improvement of both EEG and psychological testing, suggesting a direct effect of PTH on the central nervous system. Similarly, dialysis results in a decrement of brain (cerebral cortex) calcium toward normal in both patients and laboratory animals with renal failure concomitant with improvement of the EEG (Cogan et al., 1978; Cooper et al., 1978; Guisado et al., 1975). In uremic patients, both EEG changes and psychological abnormalities are improved by parathyroidectomy or medical suppression of PTH. PTH, high brain calcium content or both are probably responsible, at least in part, for some of the encephalopathic manifestations of renal failure.

The mechanisms by which PTH might impair central nervous system function are only partially understood. The increased calcium content in such diverse tissues as skin, cornea, blood vessels, brain, and heart in patients with hyperparathyroidism suggests that PTH may somehow facilitate the entry of Ca^{+2} into such tissues. The finding of increased calcium in the brains of both dogs and humans with either acute or chronic renal disease and secondary hyperparathyroidism is consistent with the conception that part of the central nervous system dysfunction and EEG abnormalities found in acute renal failure or chronic renal failure may be due in part to a PTH-mediated increase in brain calcium. Calcium is essential for the function of neurotransmission in the central nervous system and for a large number of intracellular enzyme systems. Thus, increased brain calcium content could disrupt cerebral function by interfering with any of these processes (Rasmussen, 1986). It is also possible that PTH itself may have a detrimental effect on the central nervous system.

Uremic Neurotoxins

Central Nervous System

The number of compounds retained by the body in patients with renal failure, either singly or in combination, is substantial [May, 1996 #5799]. Numerous studies have been carried out in order to attempt to indentify which of the many compounds which are elevated in uremic subjects is truly a uremic toxin. Criteria established by Bergstom and Furst for uremic toxins are as follows [Bergstrom, 1983 #5801]: (a) The compound should be chemically indentified and quantifiable in biologic fluids; (b) The concentration of the substance in plasma from uremic subjects should be higher than that found in subjects who do not have renal insufficiency; (c) The concentration of the substance in plasma should somehow correlate with specific uremic symptoms, and these should be alleviated with reduction of the substance to normal; (d) The toxic effects of the substance should be demonstrable at concentrations found in plasma from uremic patients.

Uremic neurotoxins would imply retention of solutes which have specific detremental effects upon the nervous system function, whether it is the peripheral nervous sytem or central nervous system [Vanholder, 1998 #5605; Vanholder, 1998 #5599].

There are at least three different types of uremic solutes which are potentially toxic and which can be characterized [Miyata, 2000 #5576]. These include; (a) small water soluble compounds, such as urea and creatinine [Vanholder, 1998 #5605]; (b) middle molecules; (c) protein-bound compounds. Most of the small water soluble compounds, such as urea and creatinine, are not particularly toxic and are easily removed with dialysis.

Guanidine Compounds

Guanidine compounds have been postulated to be "uremic toxins" for many years [Giovannetti, 1973 #4049], based upon possible detremental effects upon the central nervous system. Recent studies have demonstrated that several guanidino compounds are present in uremic brain [De Deyn, 1995 #5591] and may be important in the etiology of uremic encephalopathy [De Deyn, 2002 #5725]. There are at least 4 guanidino compounds which are experimental convulsants. These guanidino compounds appear to work by activation of NMDA receptors by guanidinosuccinic acid (GSA). Activation of the NMDA receptor is a major pathologic mechanism in the etiology of several types of brain damage, including head trauma [Faden, 1989 #535] and stroke [Lipton, 1994 #4838] [Beal, 1992 #4547].

In addition, guanidine compounds have a depressant effect on the mitochondrial function [Davidoff, 1973 #3936]. In the brain of uremic patients, guanidino compounds were measured in 28 different regions [De Deyn, 1995 #5591]. Guanidinosuccinic acid levels were elevated by up to 100 fold in uremic brains vs control brains, and levels increased with increasing extent of uremia. The brain levels of guanidinosuccinic acid in ureic brain were similar to those observed in normal animal brain following injection to blood levels which cause convulsions [De Deyn, 1995 #5591]. Guanidines inhibit neutrophil superoxide production, can induce seizures, and supress natural killer cell response to interleukin-2 [Dhondt, 2000 #5592]. Other guanidines, which are arginine analogues, are competitive inhibitors of NO synthetase, which impairs removal of AGEs [Asahi, 2000 #5803] and can lead to vasoconstriction, hypertension, ischemic globmerular injury, immmune dysfunction and neurological changes [Dhondt, 2000 #5592].

Middle molecules are large molecular weight compounds (300 to 12000 Daltons) which have in the past been felt to be responsible for some of the manifestations of uremia[Scribner, 1975 #4047]. Despite the fact that at one time, dialysis membranes were designed with the specific intent of removing more middle molecules, evidence of their toxicity is generally lacking [Man, 1973 #4708; Kjellstrand, 1979 #1202; Scribner, 1975 #4047]. Although there has recently been renewed interest in these molecules [Kjellstrand, 1979 #1202; Vanholder, 1994 #5593], evidence of their toxicity is still conjectural [Dhondt, 2000 #5592] (Winchester & Audia, 2006).

With established renal insufficiency, guanidines, which are competitive inhibitors of NO synthetase, will rapidly accumulate in blood, and their presence will impair removal of AGEs [Asahi, 2000 #5803] and can lead to worsening hypertension, immmune dysfunction [Cohen, 2001 #5741] and neurological changes [Dhondt, 2000 #5592], such as stroke [Wanner, 2002 #5785].

Advanced Glycation End Products

AGEs can modify tissues, enzymes and proteins and may play a role in the pathogenesis of dialysis-associated amyloidosis [Miyata, 2000 #5576]. AGEs may also play a role in the pathogenesis of diabetic nephropathy [Makita, 1991 #2182]. AGEs are markedly elevated in plasma of patients with ESRD (Haag-Weber, 1998 #5607). The AGEs react with vascular cells to inactivate endothelial NO and may increase the propensity of ESRD patients to develop hypertension. Current dialysis therapy is relatively ineffective in removal of AGEs, so that there is accumulation of AGEs in patients with ESRD, particularly those with diabetes mellitus (Haag-Weber, 1998 #5607). The AGEs are "middle molecules" and have the potential to cause tissue damage and lead to hypertension. Thus, at least some "middle molecules" may actually be deleterious in patients with ESRD, and they are poorly removed with conventional dialysis (Winchester & Audia, 2006). There is evidence that angiotensin converting enzyme antagonists decrease the formation of AGEs [Miyata, 2002 #5823]. Protein- bound compounds (toxins) are not substantially removed by dialysis, and almost all are lipophilic. Such compounds include polyamines such as spermine (Koenig et al., 1988). Spermine is postulated to be a uremic toxin and appears to react with the NMDA receptor, which affects calcium and sodium permeability in brain cells (Koenig et al., 1988; Yu et al., 1999). Stimulation of the NMDA receptor in brain is the final common pathway for brain cell death in a number of pathological pathways (Beal, 1992 #4547; Lipton, 1994 #4838). The uremic state is associated with increased oxidative stress, resulting in protein oxidation products in plasma and cell membranes. There is eventual alteration of proteins with the formation of oxidized amino acids, including glutamine and glutamate (Miyata, 2000 #5576). Such reactions may eventually lead to the stimulation of the NMDA receptor in the brain, with brain cell damage or death (Wratten et al., 2000; O'Hooge, 2003 #6049).

Neurologic Complications of End-Stage Renal Disease and Its Therapy

Dialysis Disequilibrium Syndrome

In patients with ESRD, there are several central nervous system disorders that may occur as a consequence of dialytic therapy. Dialysis disequilibrium syndrome (DDS) is a clinical syndrome that occurs in patients being treated with hemodialysis. The syndrome was first described in 1962 and may include symptoms such as headache, nausea, emesis, blurring of vision, muscular twitching, disorientation, hypertension, tremors and seizures (Arieff, 1983, 1989). The syndrome of DDS has been expanded to include milder symptoms, such as muscle cramps, anorexia, restlessness, and dizziness (Arieff, 1994). Although DDS has been reported among all age groups, it is more common among younger patients, particularly the pediatric age group (Grushkin et al., 1972). The syndrome is most often associated with rapid hemodialysis of patients with acute renal failure, but it also has occasionally been reported following maintenance hemodialysis of patients with chronic renal failure (Porte et al., 1973). The pathogenesis of DDS has been extensively investigated and the findings are summarized elsewhere (Arieff, 1982, 1994). The symptoms are usually self-limiting but recovery may take several days. It appears that present methods of dialysis have altered the clinical picture of DDS. Most reports of seizures, coma, and death were reported prior to 1970. The symptoms of DDS as reported in the last 25 years (1982–2007) have generally been mild, consisting of nausea, weakness, headache, fatigue, and muscle cramps. Almost all cases have occurred in patients undergoing their initial few hemodialyses. It is also unclear whether any patient ever actually died from DDS or in fact from other neurological complications associated with dialysis, such as acute stroke, subdural hematoma, subarachnoid hemorrhage or hyponatremia (Arieff, 1994). Recently, the diagnosis of DDS has become a "wastebasket" for a number of disorders that can occur in patients with renal failure and may affect the central nervous system (Arieff, 1994). It is important to recognize that DDS is rare and the diagnosis should be one of exclusion.

DDS has been treated either by addition of osmotically active solute (glucose, glycerol, albumin, urea, fructose, NaCl, mannitol) to the dialysate, or by intravenous infusion of mannitol or glycerol. With the technique of pure ultrafiltration the patient is subjected to ultrafiltration without dialysis. The net result is loss of fluid without the patient undergoing dialysis. Ultrafiltration followed by dialysis does not appear to be associated with DDS (Ronco et al., 1998). Additionally, DDS can be prevented by decreasing the time on dialysis and increasing the frequency of dialysis at the initiation of hemodialysis in patients. Mannitol infusion accompanying the initial three hemodialyses has been successful in prevention of DDS (Rodrigo et al., 1977). Administration of 50 ml of 50% mannitol both at the initiation of dialysis and after two hours of dialysis has generally been successful in preventing symptoms of DDS. Chronic peritoneal dialysis is currently in use worldwide. Different types of peritoneal dialysis are carried out 'in-center', at home, or in combinations of ambulatory plus home (Sharad et al., 1998). Patients undergo continuous low-volume peritoneal dialysis for as long as 24 hours per day. Symptoms of DDS have not been presently reported in patients utilizing this mode of dialysis.

Chronic Dialysis Dependent Encephalopathy

Radiologic and Pathologic Examination of Uremic Brain

There does not currently exist a large prospective study of the radiology of uremic brain in humans. Many pathologic studies of the brains of patients who died with chronic renal failure are old, and there does not exist an extensive study of uremic brain which has utilized more modern pathologic methodology (Olsen, 1961; Rotter & Roettger, 1974). Prior to 1974, subdural hemorrhages were felt to be very common in dialyzed subjects, and were reported in about 1–3% of such autopsies. In addition, intracerebral hemorrhages were said to be present in about 6% of dialysis patients who expired. Cerebral edema is not found in the brain of patients or laboratory animals with chronic renal failure, either by biochemical or histologic criteria. However, recent MR imaging studies suggest that interstitial edema is present in the brain of some patients with ESRD (Chen et al., 2007). Generalized but variable neuronal degeneration is often present but its anatomical location is quite variable. Small intracerebral hemorrhages and necrotic foci are seen in about 10% of uremic patients, and focal glial proliferation is found in about 2%. White matter lesions, including infarction, are present in about 33% of patients with chronic renal insufficiency (Martinez-Vea et al., 2006). Among patients with maintenance hemodialysis for at least five years, the majority appear to have suffered from stroke, usually small and ischemic (Mattana et al., 1998; Toyoda et al., 2005). Cognitive function is frequently impaired in chronic hemodialysis patients and is often so subtle that it is not diagnosed (Murray et al., 2006). There are pathological lesions which are frequent in chronic hemodialysis patients. Silent cerebral infarction and ischemic white matter lesions are present in over 50% of such individuals (Naganuma et al., 2005; Martinez-Vea et al., 2006). Additionally, recent studies demonstrate a very high frequency of ischemic stroke in chronic hemodialysis patients (Koren-Morag et al., 2006; Monk & Bennett, 2006; Toyoda et al., 2005). It is not known how many of this population have diabetes as the cause of their renal failure. The increase of stroke would be expected in such individuals (Koren-Morag et al., 2006; Toyoda et al., 2005; Zoccali et al., 2003).

Neuroimaging studies demonstrate bilateral basal ganglia lesions in diabetic patients with ESRD (Wang et al., 2004). Pet scaning demonstrates decreased glucose utilization in these lesions. A few cases of optic neuropathy have been reported in uremic individuals (Winkelmayer et al., 2001). White matter lesions may be present in one-third of the patients with chronic renal failure (Martinez-Vea et al., 2006).

Uremic patients are prone to several risk factors which have a tendency to increase the incidence of stroke. These risk factors include diabetes mellitus, a smoking history, hypertension, chronic infection (often catheter related), chronic inflammation, elevated cholesterol and triglycerides and advanced atherosclerosis (Zoccali et al., 2003). When uremic patients maintained with chronic dialysis suffer a possible stroke, brain neuroimaging, using CT or MRI is likely to be carried out for diagnostic purposes. However, as yet, a series of such neuroimaging cases has not been assembled for research purposes. The few available studies suggest that the brain of patients with uremia have a very high incidence of cerebral atrophy, which is disproportionately high for the age of the individuals studied (Savazzi et al., 1995). Earlier studies had also found a high incidence of cerebral atrophy among chronic hemodialysis patients (Papageorgiou et al., 1982). There appears to be subtle brain damage, not detectable by standard neuroimaging techniques or the EEG, but often manifested by deterioration of intellectual capability (Murray et al., 2006, 2007).

It is currently not unusual for patients to survive on hemodialysis for 25 years (Dean & Allegretti, 2003). However, among patients who have been on hemodialysis for over a decade, there is often mental deterioration, with markedly decreased intellectual capability, even without medical evidence of an actual stroke (Murray et al., 2006, 2007).

The collective syndrome of Chronic Dialysis Dependant Encephalopathy is a combination of probable organic mental disorders plus psychiatric disorders commonly associated with hemodialysis (Arieff, 2004; Murray et al., 2006, 2007; Pereira et al., 2005, 2007) (Table 1). The clinical manifestations include impaired intellectual capability and cognition, decreased exercise capability, sexual dysfunction and psychiatric disorders ranging from depression to psychois, sleep disorders (Perl et al., 2006), and suicidal behavior (Murray et al., 2006, 2007).The recent use of advanced neuroimaging techniques has led to an increase in our understanding of changes in the uremic brain in humans. Acute and subacute movement disorders have been

Table 1 Clinical Manifestations of Chronic Dialysis-Dependant Encephalopathy

Decreased intellectual capability
Impaired cognition
Chronic depression
Decreased capability for physical activity
Myopathy
Deterioration of vision
Suicidal behavior
Sexual dysfunction
Sleep disturbances
Pruritis
Psychosis

observed in patients with ESRD (Okada et al., 1991). These have been associated with bilateral basal ganglia and internal capsule lesions (Okada et al., 1991; Wang et al., 1998). Cerebral atrophy has been observed in chronic hemodialysis patients and it tends to worsen as dialysis therapy continues (Savazzi et al., 1995). Cerebral atrophy had previously been thought to be associated with dialysis dementia, but this is apparently not the case (Mahurkar et al., 1978). ESRD has also been reported to lead to deterioration of vision (Korzets et al., 1998). Some cases are associated with uremic pseudotumor cerebri (Guy et al., 1990).

Dialysis Dementia

Dialysis dementia (also called dialysis encephalopathy) is a progressive, frequently fatal neurologic disease which was initially described in several reports from 1970–1973 (Alfrey et al., 1972; Mahurkar et al., 1973; Siddiqui et al., 1970). Existence of the syndrome was then independently confirmed worldwide by several different groups in the early to mid-1970s (Arieff & Mahoney, 1983; Dunea et al., 1978;). In adults, the disease is seen almost exclusively in patients being treated with chronic hemodialysis. The early literature focused on the distinctive neurologic findings (Alfrey et al., 1976; Arieff & Mahoney, 1983; Dunea et al., 1978). However, more recent reports from both Europe and the USA suggest that some forms of dialysis dementia may be a part of a multi-system disease which may include encephalopathy, osteomalacic bone disease, proximal myopathy, and anemia (Fraser & Arieff, 1988b; Pierides et al., 1980).

The etiology of this syndrome remains controversial (Arieff, 1990; Arieff & Mahoney, 1983). Although an increase in brain aluminum content has been strongly implicated in some cases of dialysis dementia, the evidence is far less convincing in others. Based upon current progress, it now appears useful to subdivide dialysis dementia into three categories: (a) an epidemic form which is related to contamination of the dialysate, often with aluminum; (b) sporadic cases in which aluminum intoxication is less likely to be a contributory factor; and (c) dementia associated with congenital or early childhood renal disease. This entity has been reported in several children who were never dialyzed or exposed to aluminum compounds. These early childhood cases may represent developmental neurologic defects resulting from exposure of the growing brain to a uremic environment (Greenberg, 1978).

The initial reports of dialysis dementia in the 1970's were soon followed by reports throughout the world (Fraser & Arieff, 1993). These patients all had the endemic form and usually had been on chronic hemodialysis for over 2 years before the onset of symptoms. Early manifestations consisted of a mixed dysarthria-apraxia of speech with slurring, stuttering and hesitancy. Personality changes, including psychoses, led to dementia, myoclonus, and seizures. Symptoms initially were intermittent and were often worse during dialysis, but generally became constant. In most cases, the disease progressed to death within six months. Speech disturbances were found in 90% of patients, affective disorders culminating in dementia in 80%, motor disturbances in 75%, and convulsions in 60–90%. In contrast to this fairly distinct clinical picture, brain histology has generally been normal or nonspecific.

Early in the disease, the EEG shows multifocal bursts of high amplitude delta activity with spikes and sharp waves, intermixed with runs of background activity appearing more normal. These EEG abnormalities may precede overt clinical symptoms by 6 months. As the disease progresses, the normal background activity also deteriorates to slow frequencies (Grushkin et al., 1972). The EEG has been said to be pathognomic, but a similar pattern may also be seen in other metabolic encephalopathies. The diagnosis depends on the presence of the typical clinical picture and is confirmed by the characteristic EEG pattern (Cooper et al., 1978). Magnetoencephalography (MEG) has only recently been used in the evaluation of uremic patients (Thodis et al., 1992). MEG has not yet been used in the evaluation of patients with dialysis dementia.

Aluminum intoxication was first implicated in this disorder by Alfrey and associates (Alfrey et al., 1976). The aluminum content of brain gray matter was elevated to eleven times the normal value in patients with dialysis dementia, versus an increase of three times the normal in patients on chronic hemodialysis without dialysis dementia. Aluminum content was also increased in bone and other soft tissue. Oral phosphate binders containing aluminum [$Al(OH)_3$ and $Al_2(CO_3)_2$] were originally suspected to be the source of the aluminum.

Most of the aluminum in blood is bound to transferrin, so that there is very little free aluminum in the blood (Farrar et al., 1990). The brain contains few transferrin receptors, so that normally, aluminum uptake into the brain is negligible. Any free aluminum in the blood, usually in the form of aluminum-citrate, can readily enter the central nervous system. Normally, there is an excess of gallium-binding sites in plasma, so that even in situations where blood aluminum is increased, there is still almost no free aluminum in blood. Aluminum binding can be studied with the aluminum analogue gallium (Farrar et al., 1988). In studies of gallium-transferrin binding in blood of patients having either Alzheimer's disease, Down syndrome or renal failure treated with chronic hemodialysis, gallium binding to transferrin was significantly reduced in patients with either Down syndrome or Alzheimer's disease (Farrar et al., 1990). However, gallium binding to transferrin was normal in patients with chronic renal failure treated with hemodialysis. In such patients, there was accumulation of aluminum in those brain regions with high densities of transferrin receptors (Farrar et al., 1990).

The aforementioned findings involve studies in less than 20 patients with dialysis dementia (Farrar et al., 1990) and five with chronic renal failure treated with hemodialysis. More such studies are needed before it can be conclusively stated that the distribution of aluminum in brains of patients with dialysis dementia is not similar to that in patients with Alzheimer's disease. In patients with chronic renal failure without dialysis dementia, neurofibrillary changes have not been found.

Recent studies have further added to our knowledge of the possible effects of aluminum on the central nervous system in patients with chronic renal failure. A possible pathophysiologic basis for detrimental effects of aluminum on the central nervous system has been described by Altmann and associates (Altmann et al., 1987).

Dihydropteridine reductase is an important enzyme in the synthesis of several important neurotransmitters, such as tyrosine and acetyl choline. They found that erythrocyte levels of dihydropteridine reductase activity were less than predicted values, and correlated with plasma aluminum levels (Altmann et al., 1987). After treatment with desferrioxamine, red cell dihydropteridine reductase activity levels doubled. Although brain levels of dihydropteridine reductase activity were not evaluated, it was suggested that high brain aluminum levels might lead to decreased availability of dihydropteridine reductase in the brain. It has been suggested that the mere presence of an increased body aluminum burden has an adverse effect on overall mortality (Chazan et al., 1988). More specifically, an increased body aluminum burden (estimated by the desferrioxamine infusion test) has been associated with memory impairment and increased severity of myoclonus with decreased motor strength (Sprague et al., 1988).

Altmann and associates evaluated patients with chronic renal failure and apparently normal cerebral function (Altmann et al., 1989). They found that when compared to a control group with similar IQ, the patients with chronic renal failure had abnormalities in six tests of psychomotor function. Plasma aluminum levels were only mildly elevated (59 ± 9 µgm/l). When 15 of these patients were treated for three months with desferrioxamine, anemia improved and the erythrocyte activity of dihydropteridine reductase rose significantly. Changes in erythrocyte dihydropteridine reductase activity correlated significantly with changes in psychomotor performance (Altmann et al., 1987, 1989). Even at high blood Al levels, most Al is bound to transferrin (Farrar et al., 1990) and thus cannot bind to the cerebral transferrin receptors. It may be that patients who develop dialysis dementia have less transferrin binding capacity, less transferrin, or a greater density of transferrin receptors in the brain.

Thus, aluminum in the bloodstream may be potentially toxic in patients with chronic renal failure, possibly leading to both dialysis dementia and osteomalacia (Dunea et al., 1978). Most Nephrologists would agree that the potential hazards of poor control of plasma phosphate are worse than the potential toxicity of aluminum accumulation from oral aluminum containing antiacids. However, progress in the control of hyperphosphatemia in hemodialysis patients by other means has almost eliminated the use of aluminum containing compounds (in 2005). Calcium carbonate (or acetate) was found to be more effective for the control of hyperphosphatemia than is aluminum hydroxide (Mai et al., 1989). More recently, sevelamer (Renagel®, Genzyme Corp, Cambridge, MA), a polymeric phosphate binder, is widely used for control of phosphate in chronic dialysis patients (Slatopolsky, 1999). Renagel® has been found to be more effective than either calcium carbonate, calcium acetate or aluminum hydroxide for the treatment of hyperphosphatemia in dialysis patients (Slatopolsky, 1999), and it does not introduce aluminum into the body. Lanthanum carbonate (Fosrenol ™) has only recently been introduced in the United States for the control of hyperphosphatemia in patients with ESRD (Albaaj & Hutchison, 2005; Joy & Finn, 2003). Its place in the management of dialysis patients with hyperphosphatemia remains to be determined, but it is more efficacious than either calcium acetate or any of the aluminum-containing antiacids. However, aluminum is still the second most prevalent element in the earth's crust, and a substantial quantity will enter the body, even without administration of aluminum containing antiacids (Perl & Good, 1990; Rifat et al., 1990).

Deionization of the water used to prepare dialysate is now a standard preventive measure (Good & Perl, 1988). However, deionization may be beneficial by removing any number of other agents. Other trace elements may be present in water which can result in central nervous system toxicity. Such elements include silicon, cadmium, mercury, lead, manganese, copper, nickel, thallium, boron and tin (Hershey et al., 1983). Among these potentially neurotoxic elements, no one has measured brain content of cadmium, mercury, nickel, thallium, vanadium, or boron. Manganese was found to have increased in cortical white matter in the eight encephalopathic patients in whom it was measured (Cartier et al., 1978). These patients also had elevated aluminum levels in gray matter.

Most of the controversy over the etiology of dialysis dementia has involved those cases which occur sporadically. As noted previously, dialysate aluminum levels are not always elevated. The use of aluminum containing antacids is no different in patients with dialysis dementia than in unaffected patients, and brain aluminum levels in patients with dialysis dementia may overlap with those of unaffected patients (Arieff, 1990). The largest group of "sporadic cases" has been reported from Nashville, Tennessee (Ward et al., 1978). The reported incidence of dialysis dementia in the area is 5%, despite the use of deionized water with aluminum levels below 5 µg/L for dialysate. Osteomalacic bone disease was not clinically apparent in this group. Serum aluminum levels in the encephalopathic group were 3–4 times higher than other dialyzed patients, despite equivalent prescribed doses of aluminum containing phosphate binders. These results suggest greater absorption and/or retention of aluminum or other trace metal contamination in this group of encephalopathic patients. No other metals were measured in the Nashville study.

The evidence available thus far indicates that aluminum is elevated in the brain (cortical gray matter) of patients with dialysis dementia. However, the actual contribution of aluminum to the encephalopathy remains unclear. Aluminum content has been reported to be elevated in the brains of patients with other disorders, including senile dementia and Alzheimer's syndrome, and might actually be a nonspecific finding associated with dementia. Aluminum is also elevated in the brains of patients who have other disorders associated with altered blood-brain barrier. Such disorders include renal failure, hepatic encephalopathy and metastatic cancer. Other evidence suggests that brain aluminum content may also increase as a function of the aging process.Blood-brain barrier abnormalities can result in increased brain aluminum content (Banks & Kastin, 1983).

Despite these unresolved questions, most outbreaks of the epidemic form of dialysis dementia have been associated with high levels of aluminum in the dialysate (Berkseth & Shapiro, 1980). Lowering the dialysate aluminum to below 20 µg/L, usually by deionization, appears to prevent the onset of the disease in patients who are beginning dialysis. New cases may continue to appear in those patients who were previously exposed to the high aluminum dialysate, although the course is milder and mortality is somewhat decreased. In patients with overt disease, eliminating the source of aluminum has resulted in improvement in some but not all patients. Renal transplantation has generally not been helpful in patients with established dialysis dementia. Diazepam or clonazepam are useful in controlling seizure activity associated with the disease, but become ineffective later on and do not alter the final outcome. Treatment of sporadic cases, in which the etiology is not clear, is more difficult. Every effort should be made to identify a treatable cause. Dialysis dementia must be differentiated from other metabolic encephalopathies, such as hypercalcemia and hypophosphatemia, hyperparathyroidism, acute heavy metal intoxications and structural neurologic lesions, such as subdural hematoma (Arieff, 1990). Because of the low incidence, the uncertain etiology, and the poor correlation of plasma with tissue aluminum levels, screening tests have not generally been employed.

The source of excess Al^{+3} in the brain is not entirely clear. Some Al^{+3} apparently is absorbed after oral administration of aluminum-containing antacids (Arieff, 1990). Significant absorption of oral aluminum can occur in patients with chronic renal failure but the weight of evidence is against oral aluminum as the major source. The retention of Al^{+3} after oral administration of Al^{+3} salts is greater in patients with renal failure than in normal subjects (Graf et al., 1981). However, with increased use of Fosrenol and Renagel to control plasma phosphate in patients with renal insufficiency, as well as discovery of other toxicities associate with oral aluminum use (Goodman et al., 1984), the clinical use of aluminum-containing antacids has become reduced considerably in the USA and Western Europe.

The typical daily dietary Al^{+3} intake is 10 to 100 mg (Campbell et al., 1957), although absorption is normally minimal. This quantity of dietary Al^{+3} is more than enough to account for the entire increase of brain Al^{+3} observed in patients with dialysis dementia. Among 22 such patients, the mean brain Al^{+3} content was 22 mg/kg dry weight. The normal human brain weighs about 1500 gm and is about 80% water, or 300 gm dry weight. Thus, the total increase in Al^{+3} content for the whole brain is less than 7 mg in patients with dialysis dementia. Therefore, the entire increase of brain Al^{+3} in such patients can theoretically be accounted for by dietary aluminum. The increase in body aluminum stores may also be, in part, the result of Al^{+3} contamination from other sources, such as Al^{+3} in dialysate water, dialysis system aluminum pipes, or aluminum leaked from anodes.

There are a large number of children who have renal insufficiency and also require hospitalization with intravenous therapy. Such children may receive large quantities of intravenous aluminum (Al^{3+}) from contamination of intravenous solutions with aluminum salts (Andreoli et al., 1984). Thus, even in the absence of hemodialysis therapy, children with chronic renal failure may receive large quantities of intravenous aluminum, which may explain the development of dialysis dementia even in the absence of dialysis (Andreoli et al., 1984). The location of the aluminum in the brain of patients with dialysis dementia has not been well established. In Alzheimer's disease, it initially appeared that the aluminum was localized only in the nuclear regions of neurofilrillary tangles (Perl & Brody, 1980). More recent investigations reveal that in Alzheimer's disease, aluminum accumulates in at least four different sites: DNA containing structures of the nucleus, protein moieties of neurofilrillary tangles, amyloid cores of senile plaques, and cerebral ferritin (Garruto et al., 1983; Candy et al., 1992).

Senile plaques and neurofilrillary tangles are of course diagnostic features of Alzheimer's disease. It is not generally appreciated that in dialysis dementia, the brain also contains senile plaques and neurofilrillary tangles in the majority of cases (Brun & Dictor, 1981). However, in dialysis dementia, aluminum was not located in the neurons but rather, in glial cells and the walls of blood vessels (Good & Perl, 1988). The aforementioned findings involve studies in less than 20 patients (Brun & Dictor, 1981; Farrar et al., 1990). More such studies are needed before it can be conclusively stated that the distribution of aluminum in the brain of patients with dialysis dementia is not similar to that in patients with Alzheimer's disease.

Although the source of the increased Al^{+3} in the brain of patients with dialysis encephalopathy can theoretically be accounted for on the basis of increased Al^{+3} intake,it is unclear as to how the Al^{+3} enters the brain. The increased body aluminum burdens present in uremic subjects may contribute to increased Al^{+3} content in the brain of such individuals. To clarify the role of oral ingestion of aluminum salts in the causation of increased brain Al^{+3} content, it would be instructive to examine brain tissue from patients without renal failure who had ingested large quantities of $Al(OH)_3$, i.e., patients with chronic peptic ulcer disease. However since such material is not likely to be available, studies in laboratory animals given large quantities of aluminum salts should provide similar information. In both rats and dogs receiving oral aluminum salts, there is a significant increment in brain Al^{+3} content (Arieff et al., 1979). Administration of PTH to rats receiving aluminum salts results in an additional increment of brain Al^{+3} content (Graf et al., 1981). Thus, in laboratory animals, both a chronic increase in oral Al^{+3} ingestion, or PTH excess, can lead to an increase of cerebral cortex Al^{+3}, even in the absence of renal failure.

Alternative Etiologies

Many other possible causes of dialysis dementia have been proposed. These include other trace element contaminants, normal pressure hydrocephalus, slow virus infection of the central nervous system, and regional alterations in cerebral blood flow (Arieff, 1990). Slow virus infection of the nervous system is a possible etiology for dialysis dementia. The clinical manifestations resemble those of other slow virus infections, such as Kuru or Creutzfeldt-Jakob disease (Selkoe, 1978 #4714; Gajdusek, 1985 #1662).

In summary, dialysis dementia probably represents an end point in a disease of multiple etiology. There are at least three subgroups and in two of them the etiology of dialysis encephalopathy must be regarded as unknown. The possible role of aluminum, or other trace element abnormalities, is unclear. At this time, there is no known satisfactory treatment for patients with dialysis encephalopathy. Most patients reported in the literature thus far have not survived, usually dying within 18 months of the time of diagnosis. The syndrome is not alleviated by increased frequency of dialysis, and usually not by renal transplantation (Burks, 1976 #461; Arieff, 1983 #497). Definitive therapy must await a better understanding of the pathogenesis of this disorder. The use of deferoxamine to chelate aluminum or other trace elements is experimental and has not been productive. There has been some improvement in patients with dialysis dementia treated with deferoxamine (Malluche et al., 1984). Deferoxamine can be used to remove aluminum from the body, but this approach has not been shown to improve the clinical status of patients with dialysis dementia.

Other Central Nervous System Complications of Dialysis

In addition to dialysis dementia and DDS, there are several other neurological disorders which have been reported in patients being treated with dialysis. In most instances, patients have initially presented with headache, nausea, emesis or hypotension, while some have had seizures. Most such patients have initially been diagnosed as having DDS, while others, particularly those with chronic subdural hematoma, have been suspected of having dialysis dementia. The disorders include: copper intoxications, subdural hematoma, muscle cramps, nonketotic hyperosmolar coma with hyperglycemia, cerebral embolus secondary equipment malfunction, acute cerebrovascular accident, depletion syndrome, malfunction of fluid proportioning system, excessive ultrafiltration with hypotension and seizures, hypoglycemia, and Wernicke's Encephalopathy (Fraser & Arieff, 1993; Mahoney & Arieff, 1982). Subdural hematoma is not an infrequent cause of death in patients maintained with chronic hemodialysis, although much less after 1990 (Leonard & Shapiro, 1975). This condition may initially present with headache, drowsiness, nausea, and vomiting. If the patient loses consciousness or develops signs of increased intracranial pressure, a diagnosis of subarachnoid bleeding should be considered. Such episodes in uremic patients are usually fatal unless the patients are operated upon. If the above symptoms persist between hemodialysis, or progressively worsen, subdural hematoma is likely, particularly if the patient is taking anticoagulants. On physical examination, there is often evidence of localized neurological disease; there may be signs of meningeal irritation and somnolence and focal seizures may be observed. The diagnosis can usually be made by modern neuroimaging techniques {computed axial tomography (CAT scan) or magnetic resonance imaging (MRI)} (Arieff et al., 1994; Kucharczyk et al., 1985).

Improper proportioning of dialysate, due to either human or mechanical error, is still an important cause of neurological abnormality in dialysis patients (Bleumle, 1968). The usual effect of such dialysate abnormalities is the production of either hypo- or hypernatremia. Both of these abnormalities of body fluid osmolality can lead to seizures and coma, although different mechanisms are involved. In acute hypernatremia, there will be excessive thirst, lethargy, irritability, seizures, and coma, with spasticity and muscle rigidity. In acute hyponatremia there is weakness, fatigue, and dulled sensorium, which may also progress to seizures and coma, respiratory arrest, and death. Such symptoms developing soon after initiation of hemodialysis should alert the physician to the possibility of such an error. A check of the dialysate osmolality or sodium concentration is the most rapid means of detecting this problem. Death has been reported as a consequence of either hyper- or hyponatremia (Arieff, 1985).

About one liter of fluid per hour can be removed by ultrafiltration hemodialysis and about 300 ml/hr by peritoneal dialysis using hypertonic dialysate. Such a rate of fluid removal from the intravascular space may be faster than the rate at which fluid can be replaced from the interstitial compartment, and hypotension may develop. Symptoms of hypotension may include seizures which, although actually due to cerebrovascular insufficiency, may be mistaken for DDS, particularly in diabetic subjects.

Most of the neurologic complications of renal transplantation relate to secondary afflictions, such as infection and neoplasia (Chan et al., 2001). As already discussed, most of the neurologic complications of the uremic state tend to improve following renal transplantation. These include the neuropathy, encephalopathy and EEG changes (Bolton, 1976; Chan et al., 2001).

Stroke in Patients Treated with Chronic Hemodialysis

Cerebrovascular disease is a common cause of serious disability and death in chronic hemodialysis patients, and stroke represents the second most frequent cause of death (Go et al., 2004; Monk & Bennett, 2006) (the three most frequent are heart attack, stroke and infection) (Mazzuchi et al., 2000). In the USA and Western Europe (including Israel), cardiovascular disease is far more common in dialysis patients than in the rest of the population (Go et al., 2004; Shik & Parfrey, 2005). A part of the reason may be that the major cause of ESRD in the USA is diabetes mellitus (27% in 2005), far greater than in Europe (19%) and Japan (10%) (Shik & Parfrey, 2005; Mauer et al., 2001). Among the factors which doubtless contribute to the high incidence of stroke in patients with ESRD treated with hemo-dialysis are the high incidence of hypertension, the large number of such patients who have diabetes mellitus, and the accelerated arteriosclerosis in such patients (Herzog et al., 2007). In addition, uremic patients tend to have high cholesterol levels and a high incidence of obesity, and they tend to smoke cigarettes excessively (Bronner et al., 1995). There is a high incidence of chronic infection in dialysis patients, which leads to elevated blood levels of atherogenic risk factors, such as cytokines, which appear to contribute to the increased incidence of stroke in such patients (Ayus & Sheikh-Hamad, 1998). The elevated cytokines are largely due to the high incidence of chronic inflamatory conditions in chronic hemodialysis patients (Ayus & Sheikh-Hamad, 1998).

There is substantial recent evidence that chronic inflamation plays a role in the pathogenesis of cardiovascular disease (Panichi et al., 2000; Takaki et al., 2003). Cytokines released from involved tissues stimulate the liver to synthesize acute phase proteins, including C-reactive protein (CRP). Elevated levels of CRP and other cytokines constitute an independent risk factor for cardiovascular disease. AGEs can modify tissues, enzymes and (Miyata et al., 2000, 2001). AGEs are markedly elevated in plasma of patients with ESRD, particularly if they also have diabetes mellitus (Breyer et al., 1996; Haag-Weber, 1998; Makita et al., 1991). These AGEs react with vascular cells to inactivate endothelial NO and may increase the propensity of ESRD patients to develop arteriosclerosis and hypertension (Haag-Weber, 1998). Patients with uremia have an accumulation of proinflamatory compounds, including AGEs, and these probably impair defense mechanisms against oxidative injury (Kaysen, 2001). There is evidence for increased cytokine production secondary to blood interaction with bioincompatable dialysis components. In particular, blood-dialyzer interaction can activate mononuclear cells, leading to production of inflamatory cytokines (Vanholder et al., 2001). Synthetic high-flux dialyzer membranes are permeable to the pro-inflamatory cytokines, and are capable of removing interleukin-1B (IL-1B), tumor necrosis factor alpha and interleukin-6, thus offering a potential therapeutic approach (Lonnemann, 2000). It is unclear whether cytokine removal by continuous renal replacement therapy will decrease the incidence of stroke. The use of sorbents with continuous plasma filtration offers another possibility for a novel therapeutic approach (Ronco et al., 1998; Wratten et al., 2000). Some of the cytokines, such as IL-1B, tumor necrosis factor alpha, and interleukin-6, may induce an inflamatory state, and are believed to play an important role in dialysis-related mortality (Zimmerman & Herringer, 1999). Recent prospective studies have demonstrated that patients with ESRD and higher blood levels of certain cytokines have a greater mortality and have a larger number of cardiovascular events (Kaysen, 2001). Contaminated dialysate water can result in pyogenic substances of bacterial origin being absorbed into the dialysis membrane (Lonnemann, 2000). The consequence could be induction of an inflammatory response in some dialysis patients. Substances of bacterial origin activate circulating mononuclear cells to produce proinflamatory cytokines. The cytokines include IL-1B, tumor necrosis factor alpha, and interleukin-6, and they mediate the acute phase response resulting in elevated levels of acute phase proteins, including CRP (Lonnemann, 2000). The effects of dialysis reuse on cytokine production has not been evaluated, but may be important, as reuse could theoretically lead to more contamination of the dialysate (Ayus & Sheikh-Hamad, 1998; Lonnemann, 2000). Reactive carbonyl compounds and AGEs, which tend to modify proteins in a deleterious manner, can be decreased by the use of a peritoneal dialysate containing icodextrin and amino acids instead of glucose (Miyata et al., 2000, 2001).

The cardiovascular disease and cerebrovascular disease can lead to cerebral ischemia. Cerebral ischemia initiates a number of processes which can lead to progressive brain damage (Brott & Bogousslavsky, 2000). Cerebral ischemia can lead to activation of free radicals, NMDA and apoptosis, in the brain, all potential mechanisms of brain damage in patients with hypoperfusion or stroke (Vexler et al., 1997). Anoxic injury to brain endothelial cells can increase production of NO, which can lead to free radical formation (Kumar et al., 1996). Apoptosis, or programed cell death, is another mode of destruction of brain cells in stroke (Vexler et al., 1997). Glutamate activates the NMDA receptor complex and can also lead to later activation of apoptosis (Honig & Rosenberg, 2000). In general, the ischemic event of a stroke only serves to initiate the biochemical events that may lead to brain damage. Interventions which counter these biochemical events may decrease the brain damage associated with acute stroke. Recent knowledge of the pathogenesis of stroke has led to a major expansion in the opportunities for prevention of stroke. High grade carotid stenosis can lead to stroke, although the exact numbers of patients who will suffer stroke when they have carotid stenosis is not known. Screening patients who have renal failure for the presence of carotid stenosis will diagnose a substantial number of such patients, although at considerable cost. However, because of noninvasive diagnostic techniques such as duplex Doppler ultrasonography, screening for carotid stenosis involves essentially no morbidity. Studies of the aortic arch for the presence of large atherosclerotic plaques (more than 4 mm thick) is an important predictor for the possibility of stroke in the future (Amarenco et al., 1992), as is the presence of atrial fibrillation (Snow et al., 2003). Other common preventive measures include treatment of hypertension, cessation of smoking, lowering of plasma cholesterol, control of plasma glucose (in diabetic pateints), weight loss, increased exercise, and decreased alcohol consumption. Other possible preventive measures include dietary antioxidants, low dose aspirin, and a decrease of intake of saturated fatty acids (Bronner et al., 1995). Although treatment of hypertension is known to decrease the incidence of stroke, not all antihypertensive agents are of equal efficacy. In general, only beta-blockers, thiazide diuretics and angiotensin-converting enzyme (ACE) inhibitors have been shown to reduce the incidence of stroke, whereas, alpha-adrenergic blocking agents and short acting calcium channel blockers may not (ALLHAT, 2002). Given that the aforementioned are the likely mechanisms of brain damage in a stroke, a whole new field is opened in terms of potential therapeutic agents for decreasing brain damage associated with stroke. Such agents include calcium channel blockers, inhibitors of NMDA receptors, and agents which scavenge free radicals (Albers et al., 1995). It is now clear that in many cases acute stroke can often be successfully treated, but only if physicians realize that stroke should now be considered a medical emergency where timely therapy can make the difference in the functional survival of the brain. Therapies for acute stroke which are now being administered in teaching hospitals in the USA start with acute neuroimaging in the emergency room. An initial CT scan will usually reveal acute stroke and if present, serves to differentiate occlusive from hemorrhagic stroke. If a non-hemorrhagic stroke is present, treatment prospects can be examined with magnetic resonance angiography (MRA), which is non-invasive. Contrast should not be administered to patients with impaired renal function, but can be given to dialysis patients. When acute stroke is diagnosed within the appropriate time window (within three hours of the onset of symptoms) current therapies may include intravenous thrombolytic therapy (Study Group, 1996), intraarterial thrombolytic therapy, antithrombotic and antiplatelet drugs, defibrinogenating agents and neuroprotective drugs (Brott & Bogousslavsky, 2000; del Zoppo, 1995; rt-PA Stroke Study Group, 1995; Sherman et al., 2000). Administraton of the defibrinogenating agent ancrod to patients with acute ischemic stroke resulted in a better functional status after a 3-month follow-up (Sherman et al., 2000). Nizofenone can scavenge free radicals and inhibit glutamate release, and may prove useful as a cerebroprotective agent (Yasuda & Nakajiima, 1993).

Some cases of acute stroke will be due to dissection of the carotid or vertebral artery systems. These patients have lesions which are not amenable to dissolution of clot, as the obstructing lesion is in fact a hemorrhage in the arterial wall (Schievink, 2001). Dissection of the carotid or vertebral artery system can be initiated by chiropractic manipulation of the cervical spine; such maneuvers should probably be avoided in dialysis patients (Schievink, 2001). In addition, some cases of apparent acute stroke in dialysis patients will be due to subdural hematoma, which must always be considered in the differential diagnosis of stroke in dialysis patients.

Sexual Dysfunction in Uremia

Disturbances in sexual function are a common complication of chronic renal failure (Palmer, 2003). These complications occur in both genders and may include erectile dysfunction, decreased libido and decreased frequency of intercourse (Palmer, 2003). Studies in uremic rats showed that erectile impairment was associated with a disturbance in NO synthetase gene expression (Abdel-Gawad et al., 1999). Sexual dysfunction in men with ESRD treated with maintainance hemodialysis is common, and previously impotence was observed in at least 50% of such patients. A number of abnormalities associated with renal failure appear to be important in the genesis of impotence. There are abnormalities in autonomic nervous system function (Campese et al., 1981; Campese & Liu, 1990), impairment in arterial and venous systems of the penis (along with vascular pathology in other vascular beds), hypertension (many drugs used to treat hypertension cause secondary impotence), and other associated endocrine abnormalities. There are also the associated effects of aging, with impotence observed in over 50% of men over 60 years old who do not have renal failure (Lamberts et al., 1997). There are a variety of approaches to the evaluation of impotency in uremic men (Palmer, 2003). Patients with ESRD have a high incidence of cardiovascular disease, which impairs penile vessels along with those of the rest of the body (Meeus et al., 2000). The incidence of hypertension is also higher in ESRD patients than in the rest of the population, and hypertension is a major contributor to vascular disease (Meeus et al., 2000). Many drugs used to treat hypertension can lead to impotence (calcium channel blockers, beta blockers, thiazides, guanethidine). The incidence of depression is high in patients with ESRD, and many drugs used to treat depression can lead to impotence (phenothiazines, tricyclics, fluxetine). Although appreciaton of the aforementioned abnormalities may increase our understanding, until very recently, there was little that could be done other than to discontinue certain drugs used to treat hypertension or depression (Palmer, 1999). There are now a number of drugs which can successfully treat impotence (Leland, 1997). Alprostadil was successful, but had to be delivered transurethrally (Nathan et al., 1997). In particular, sildenafil can be administered orally and is highly effective, even in men who have cardiovascular disease (Herrmann et al., 2000) or uremia (Palmer, 2003). Other treatments for impotence among men with ESRD include penile prostheses, direct injection of alpha blocking agents or other vasodilators (papaverine, phentolamine, alprostadil) into the penis, and vacuum constrictive devices (Leland, 1997).

References

Abdel-Gawad M, Huynh H & Brock GB. (1999). Experimental chronic renal failure-associated erectile dysfunction: Molecular alterations in nitric oxide synthetase pathway and IGF-I system. Molecular Urol 3, 117–125.

Akmal M, Goldstein DA, Multani S & Massry SG. (1984). Role of uremia, brain calcium and parathyroid hormone on changes in electroencephalogram in chronic renal failure. Am J Physiol 246 (Renal, Fluid, Electrolyte Physiol. 15), F575–F579.

Albaaj F & Hutchison AJ. (2005). Lanthanum carbonate for the treatment of hyperphosphatemia in renal failure and dialysis patients. Expert Opin Pharmacother 6, 319–328.

Albers GW, Atkinson RP, Kelley RE & Rosenbaum DM. (1995). Safety, tolerability and pharmacokinetics of the NMDA antagonist dextrorphan in patients with acute stroke. Stroke 26, 254–258.

Alfrey AC, LeGendre GR & Kaehny WD. (1976). The dialysis encephalopathy syndrome: Possible aluminum intoxication. N Engl J Med 294, 184–188.

Alfrey AC, Mishell J, Burks SR, Contiguglia SR, Rudolph H, Lewin E & Holmes JH. (1972). Syndrome of dysphaxia and multifocal seizures associated with chronic hemodialysis. Trans Amer Soc Artif Intern Organs 18, 257–261.

ALLHAT. (2002). Major outcomes in high-risk hypertensive patients randomized to angiotensin-converting enzyme inhibitor or calcium channel blocker vs diuretic. JAMA 288, 2981–2997.

Altmann P, Al-Salihi F, Butter K, Cutler P, Blair J, Leeming R, Cunningham J & Marsh F. (1987). Serum aluminum levels and erythrocyte dihydropteridine reductase activity in patients on hemodialysis. N Engl J Med 317, 80–84.

Altmann P, Hamon C, Blair J, Dhanesha U, Cunningham J & Marsh F. (1989). Disturbance of cerebral function by aluminum in haemodialysis patients without overt aluminum toxicity. Lancet ii, 7–12.

Amarenco P, Duyckaerts C, Tzourio C, Henin D, Bousser MG & Hauw JJ. (1992). The prevalence of ulcerated plaques in the aortic arch in patients with stroke. New Engl J Med 326, 221–225.

Andreoli SP, Bergstein JM & Sherrard DJ. (1984). Aluminum intoxication from aluminum-containing phosphate binders in children with azotemia not undergoing dialysis. N Engl J Med 310, 1079.

Arieff AI. (1982). Dialysis disequilibrium syndrome: Current concepts on pathogenesis. In Controversies in Nephrology, Schreiner GE & Winchester JF, pp. 367–376. George Washington University Press, Washington, D.C.

Arieff AI. (1983). Dialysis disequilibrium syndrome. In Textbook of Nephrology, 1st edn, Massry SG & Glassock RJ, pp. 7.24–27.26. Williams & Wilkins, Baltimore.

Arieff AI. (1985). Effects of water, acid base, and electrolyte disorders on the central nervous system. In Fluid, Electrolyte and Acid-Base Disorders, Arieff AI & DeFronzo RA, pp. 969–1040. Churchill Livingstone, New York.

Arieff AI. (1986). Neurological manifestations of uremia. In The Kidney, 3rd edn, Brenner BM & Rector FC, Jr., pp. 1731–1756. W. B. Saunders, Philadelphia.

Arieff AI. (1989). Dialysis disequilibrium syndrome. In Textbook of Nephrology, 2nd edn, Massry SG & Glassock RJ, pp. 1168–1170. Williams & Wilkins, Baltimore.

Arieff AI. (1990). Aluminum and the pathogenesis of dialysis dementia. Environ Geochem Health 12, 89–93.

Arieff AI. (1994). Dialysis disequilibrium syndrome: Current concepts on pathogenesis and prevention. Kidney Int 45, 629–635.

Arieff AI. (2004). Neurological complications of renal iInsufficiency. In Brenner & Rector's The Kidney, 7th edn, Brenner BM, pp. 2227–2254. W.B. Saunders, Philadelphia.

Arieff AI, Cooper JD, Armstrong D & Lazarowitz VC. (1979). Dementia, renal failure and brain aluminium. Ann Intern Med 90, 741–747.

Arieff AI, Fraser CL, Rowley H, Truwit C & Kucharczyk J. (1994). Metabolic encephalopathy. In Magnetic Resonance Neuroimaging, 1st edn, Kucharczyk J, Moseley M & Barkovich AJ, pp. 319–349. CRC Press, Boca Raton, FL.

Arieff AI, Guisado R & Massry SG. (1975). Uremic encephalopathy: Studies on biochemical alterations in the brain. Kidney Int 7, S194–S200.

Arieff AI, Guisado R & Massry SG. (1977). Central nervous system pH in uremia and the effects of hemodialysis. J Clin Invest 58, 306.

Arieff AI, Kerian A, Massry SG & DeLima J. (1976). Intracellular pH of brain: alterations in acute respiratory acidosis and alkalosis. Am J Physiol 230, 804–812.

Arieff AI & Mahoney CA. (1983). Pathogenesis of dialysis encephalopathy. Neurobehav Toxicol Teratol 5, 641–644.

Arieff AI & Massry SG. (1974). Calcium metabolism of brain in acute renal failure. Effects of uremia, hemodialysis, and parathyroid hormone. J Clin Invest 53, 387–392.

Asahi K, Ichimori K, Nakaawa H & Izuhara Y. (2000). Nitric oxide inhibits the formation of advanced glycation end products. Kidney Int 58, 1780–1787.

Ayus JC, Go AS, Valderrabano F, Verde E, De Vinuesa SG, Achinger SG, Lorenzo V, Arieff AI & Luno J. (2005). Effects of erythropoietin on left ventricular hypertrophy in adults with severe chronic renal failure and hemoglobin <10 g/dl. Kidney Int 68, 788–795.

Ayus JC & Sheikh-Hamad D. (1998). Silent infection in clotted hemodialysis access grafts. J Am Soc Nephrol 9, 1314–1317.

Banks WA & Kastin AJ. (1983). Aluminium increases permeability of the blood-brain barrier to labelled DSIP and b-endorphin: Possible implications for senile and dialysis dementia. Lancet, 1227–1229.

Beal MF. (1992). Mechanisms of excitotoxicity in neurologic disease. FASEB J 6, 3338–3344.

Bergstrom J & Furst P. (1983). Uremic toxins. In Replacement of Renal Function by Dialysis, Drukker W, Parsons FM & Maher JF, pp. 354–377. Martinus Nijhoff, Boston.

Berkseth RO & Shapiro FL. (1980). An epidemic of dialysis encephalopathy and exposure to high aluminum dialysate. In Controversies in Nephrology, Schriener GE & Winchester JF, p. 42. Georgetown University Press, Georgetown, MD.

Bleumle LW. (1968). Current status of chronic hemodialysis. Am J Med 44, 749.

Bohlender JM, Franke S, Stein G & Wolf G. (2005). Advanced glycation end products and the kidney. Am J Physiol: Renal Physiol 289, F645–F659.

Bolton CF. (1976). Electrophysiologic changes in uremic neuropathy after successful renal transplantation. N Engl J Med 284, 1170.

Breyer JA, Bain RP, Evans JK, Nahman NS & Lewis EJ. (1996). Predictors of the progression of renal insufficiency in patients with insulin-dependent diabetes and overt diabetic nephropathy. Kidney Int 50, 1651–1658.

Bronner IL, Kanter DS & Manson JE. (1995). Primary prevention of stroke. New Engl J Med 333, 1392–1400.

Brott T & Bogousslavsky J. (2000). Treatment of acute ischemic stroke. New Engl J Med 343, 710–722.

Brouns R & De Deyn PP. (2004). Neurological complications in renal failure: A review. Clin Neurol Neurosurg 107, 1–16.

Brun A & Dictor M. (1981). Senile plaques and tangles in dialysis dementia. Acta Path Microbiol Scand Sect. A 89, 193–198.

Burks JS, Alfrey AC, Huddlestone J, Norenberg MD & Lewin E. (1976). A fatal encephalopathy in chronic haemodialysis patients. Lancet 1, 764–768.

Campbell IR, Cass JF & Cholak J. (1957). Aluminum in the enviornment of man. AMA Arch Indust Health 15, 359–361.

Campese VM & Liu CL. (1990). Sexual dysfunction in uremia. Contrib Nephrol 77, 1–14.

Campese VM, Romoff MS & Levitan D. (1981). Mechanisms of autonomic nervous system dysfunction in uremia. Kidney Int 20, 246.

Candy JM, McArthur FK, Oakley AE, Taylor GA & Chen CP. (1992). Aluminum accumulation in relation to senile plaque and neurobibrillary tangle formation in the brains of patients with renal failure. J Neurol Sci 107, 210–218.

Cartier F, Allain P, Gary J, Chatel M, Menault F & Pecker S. (1978). Encephalopathie myoclonique progressive des dialyses: Role de l'eau utilisee pour l'hemodialyse. Nouv Presse Med 7, 97–102.

Chan L, Wang W & Kam I. (2001). Outcomes and complications of renal transplantation. In Diseases of the Kidney, 7th edn, Schrier RW & Gottschalk CW, pp. 2871–2938. J.B. Lippincott, Philadelphia.

Chazan JA, Blonsky SL, Abuelo JG & Pezzullo JC. (1988). Increased body aluminum: An independent risk factor in patients undergoing long-term hemodialysis? Arch Intern Med 148, 1817–1820.

Chen CL, Lai PH, Chou KJ, Chung HM & Fang HC. (2007). A preliminary report of brain edema in pateints with uremia at first hemodialysis: Evaluation by diffusion-weighted MR imaging. Am J Neuroradiol 28, 68–71.

Clark WR & Gao D. (2002). Low-molecular weight proteins in end-stage renal disease: Potential toxicity and dialytic removal mechanisms. J Am Soc Nephrol 13, S41–S47.

Cogan MG, Covey C, Arieff AI, Wisniewski A, Clark OH, Lazorowitz VC & Leach W. (1978). Central nervous system manifestations of hyperparathyroidism. Amer J Med 65, 963–970.

Cohen H, Rdnicki M & Horl WH. (2001). Uremic toxins modulate the spontaneous apoptotic cell death and essential functions of neutrophils. Kidney Int 59 (Suppl. 78), S48–S52.

Cooper JD & Arieff AI. (1979). Lindau disease treated by bilateral nephrectomy and hemodialysis. West J Med 130, 456–458.

Cooper JD, Lazorowitz VC & Arieff AI. (1978). Neurodiagnostic abnormalities in patients with acute renal failure. Evidence for neurotoxicity of parathyroid hormone. J Clin Invest 61, 1448–1455.

D'Hooge R, Van de Vijver G, Van Bogaert PP, Marescau B & Vanholder R. (2003). Involvement of voltage- and ligand-gated Ca2+ channels in the neuroexcitatory and synergistic efects of putative uremic neurotoxins. Kidney Int 63, 1764–1775.

Davidoff F. (1968). Effects of guanidine derivatives on mitochrondial function I. J Clin Invest 47, 2331–2343.

Davidoff F. (1973). Guanidine derivatives in medicine. N Engl J Med 289, 141–146.

De Boer IH, Gorodetskaya I, Young B, Hsu CY & Chertow GM. (2002). The severity of secondary hyperparathyroidism in chronic renal insufficiency is GRF-dependent, race dependent, and associated with cardiovascular disease. J Am Soc Nephrol 13, 2762–2769.

Dean S & Allegretti C. (2003). 25 years of good health. In Patient Line, ed. McKenna CA, pp. 1–3. FMC Medical Services, Lexington, MA.

De Deyn PP, D'Hooge R, Van Bogaert P & Marescau B. (2002). Endogenous guanidino compounds as uremic neurotoxins. Kidney Int 59 (suppl. 78), S77–S83.

De Deyn PP, Marescau B, D'Hodge R, Possemiers I, Nagler J & Mahler C. (1995). Guanidino compound levels in brain regions of non-dialyzed uremic patients. Neurochem Int 27, 227–237.

Deferrari G. (1981). Brain metabolism of amino acids and ammonia in patients with chronic renal insufficiency. Kidney Int 20, 505.

del Zoppo GJ. (1995). Acute stroke – on the theshold of a therapy. New Engl J Med 333, 1632–1633.

Dhondt A, Vanholder R, van Beisen W & Lameire N. (2000). The removal of uremic toxins. Kidney Int 58 (Suppl. 76), S47–S59.

Dunea G, Mahurkar SD & Mamdami B. (1978). Role of aluminum in dialysis dementia. Ann Intern Med 88, 502–504.

Evans RW, Rader B & Manninen DL. (1990). The quality of life of hemodialysis reciepients treated with recombinant human erythropoietin. JAMA 263.

Faden AI, Demediuk P, Panter SS & Vink R. (1989). The role of excitatory amino acids and NMDA receptors in traumatic brain injury. Science 244, 798–800.

Farrar G, Altmann P, Welch S, Wychrij O, Ghose B, Lejeune J, Corbett J, Prasher J & Blair JA. (1990). Defective gallium-transferrin binding in Alzheimer disease and Down syndrome: Possible mechanism for accumulation of aluminum in brain. Lancet 335, 747–750.

Farrar G, Morton AP & Blair JA. (1988). The intestinal spectiation of gallium: Possible models to describe the bioavailability of aluminum. In Trace Element Analytical Chemistry in Medicine and Biology, Bratter P & Schramel P, pp. 343–347. Walter de Gruyter, Berlin.

Fraser CL & Arieff AI. (1985). Hepatic encephalopathy. N Engl J Med 313, 865–873.

Fraser CL & Arieff AI. (1988a). Nervous system complications in uremia. Ann Intern Med 109, 143–153.

Fraser CL & Arieff AI. (1988b). Nervous system manifestations of renal failure. In Diseases of the Kidney, 4th edn, Schrier RW & Gottschalk CW, pp. 3063–3092. Little, Brown, Boston, MA.

Fraser CL & Arieff AI. (1993). Nervous system manifestations of renal failure. In Diseases of the Kidney, 5th edn, Schrier RW & Gottschalk CW, pp. 2789–2816. Little, Brown, Boston, MA.

Fraser CL & Arieff AI. (1994). Metabolic encephalopathy as a complication of renal failure: Mechanisms and mediators. In New Horizons: The Science and Practice of Acute Medicine, 5th edn, Matuschak GM, pp. 518–526. Williams & Wilkins, Baltimore.

Fraser CL & Arieff AI. (1995). Metabolic encephalopathy as a complication of acid base, and electrolyte disorders. In Fluid, Electrolyte and Acid-Base Disorders, 2nd edn, Arieff AI & DeFronzo RA, pp. 685–740. Churchill Livingstone, New York.

Fraser CL & Arieff AI. (1999). Neuropsychiatric complications of uremia. In Therapy in Nephrology and Hypertension, Brady HR & Wilcox CS, pp. 488–490. W.B. Saunders, Philadelphia.

Garruto RM, Fukatsu R, Yanagihara R, Gajdusek DC, Hook G & Fiori CE. (1983). Imaging of calcium and aluminum in neurofibrillary tangle-bearing neurons in parkinsonism-dementia of Guam. Proc Natl Acad Sci USA 81, 1875–1879.

Gilli P & Bastiani P. (1983). Cognitive function and regular dialysis treatment. Clin Nephrol 19, 188–192.

Gines P, Tito L, Arroyo V & Planas J. (1988). Randomized comparative study of therapeutic paracentesis with and without intravenous albumin in cirrhosis. Gastroenterology 94, 1493–1502.

Giovannetti S, Balestri PL & Barsotti G. (1973). Methylguanidine in uremia. Arch Intern Med 131, 709.

Go AS, Chertow GM, Fan D, McCulloch CE & Hsu C. (2004). Chronic kidney disease and the risks of death, cardiovascular events, and hospitalization. New Engl J Med 351, 1296–1305.

Goldstein DA & Massry SG. (1980). The relationship between the abnormalities in EEG and blood levels of parathyroid hormone in dialysis patients. J Clin Endocrinol Metab 51, 130.

Good PF & Perl DP. (1988). A lasar microprobe mass analysis study of aluminum distribution in the cerebral cortex of dialysis encephalopathy. J Neuropathol Exp Neurol 47, 321.

Goodman WG, Gilligan J & Horst R. (1984). Short term aluminum administration in the rat: Effects on bone formation and relationship to renal osteomalacia. J Clin Invest 73, 171.

Graf H, Stummvoll HK & Messinger V. (1981). Aluminum removal by hemodialysis. Kidney Int 19, 587.

Greenberg MD. (1978). Brain damage in hemodialysis patients. Dialy Transplant 7, 238.

Grimm G, Stockenhuber F, Schneeweiss B, Madl C, Zeitlhofer J & Schneider B. (1990). Improvement of brain fuunction in hemodialysis patients treated with erythropoietin. Kidney Int 38, 480–486.

Grushkin CM, Korsch B & Fine RN. (1972). Hemodialysis in small children. JAMA 221, 869.

Guisado R, Arieff AI, Massry SG, Lazarowitz V & Kerian A. (1975). Changes in the electroencephalogram in acute uremia. Effects of parathyroid hormone and brain electrolytes. J Clin Invest 55, 738–745.

Gusbeth-Tatomir P, Boisteanu D, Seica A, Buga C & Covic A. (2007). Sleep disorders: A systematic review of an emerging major clinical issue in renal patients. Int Urol Nephrol 39, 1217–1226.

Guy J, Johnston PK, Corbett JJ, Day AL & Glaser JS. (1990). Treatment of visual loss in pseudotumor cerebri associated with uremia. Neurology 40, 28–32.

Haag-Weber M. (1998). AGE-modified proteins in renal failure. In Critical Care Nephrology, Ronco C & Bellomo R, pp. 878–883. Kluwer, Hingham, MA.

Heath H, Hodgson SF & Kennedy MA. (1980). Primary hyperparathyroidism: Incidence, morbidity, and potential impact in a community hospital. New Engl J Med 302, 189–193.

Herrmann HC, Chang G, Klugherz BD & Mahoney PD. (2000). Hemodynamic effects of sildenafil in men with severe coronary artery disease. New Engl J Med 342, 1622–1626.

Hershey CO, Ricanati ES, Hershey LA, Varnes AW, Lavin PJM & Strain WH. (1983). Silicon as a potential uremic neurotoxin: Trace element analysis in patients with renal failure. Neurology 33, 786–789.

Herzog CA, Littrell K, Arko C, Frederick PD & Blaney M. (2007). Clinical characteristics of dialysis patients with acute myocardial infarction in the United States. Circul 116, 1465–1472.

Honig LS & Rosenberg RN. (2000). Apoptosis and neurologic disease. Am J Med 108, 317–330.

Joy MS & Finn WF. (2003). Randomized, double-blind, placebo-controlled, dose-titratio, phase III study assessing the efficacy and tolerability of lanthanum carbonate: A new phosphate binder for the treatment of hyperphosphatemia. Am J Kidney Dis 42, 96–107.

Kaysen GA. (2001). The microinflamatory state in uremia: Causes and potential consequences. J Am Soc Nephrol 12, 1549–1557.

Kiley JE, Woodruff MW & Pratt KL. (1976). Evaluation of encephalopathy by EEG frequency analysis in chronic dialysis patients. Clin Nephrol 5, 245.

Kjellstrand CM, Arieff AI, Friedman EA, Furst P, Henderson LW & Massry SG. (1979). Inadequacy of dialysis: Why patients are not well. Trans Am Soc Artif Intern Organs XXV, 518–520.

Koenig H, Goldstone AD, Lu CY & Trout JJ. (1988). Polyamines: Transducers of osmotic signals at the blood-brain barrier. Proc Amer Soc Neurochem 19, 79.

Koren-Morag N, Goldbourt U & Tanne D. (2006). Renal dysfunction and risk of ischemic stroke or TIA in patients with cardiovascular disease. Neurol 67, 224–228.

Korzets Z, Zeltzer E, Rathaus M, Manor R & Bernheim J. (1998). Uremic optic neuropathy. A uremic manifestation mandating dialysis. Am J Nephrol 18, 240–242.

Kucharczyk W, Brant-Zawadzki M & Norman D. (1985). Magnetic resonance imaging of the central nervous system–An update. West J Med 142, 54–62.

Kumar M, Liu GJ, Floyd RA & Grammas P. (1996). Anoxic inujury of endothelial cells increses production of nitric oxide and hydroxyl radicals. Biochem Biophys Res Comm 219, 497–501.

Lamberts SWJ, van den Beld AW & van der Lely A. (1997). The endocrinology of aging. Science 278, 419–424.

Leland J. (1997). A pill for impotence? Newsweek Nov. 17, 62–68.

Leonard A & Shapiro FL. (1975). Subdural hematoma in regularly hemodialyzed patients. Ann Intern Med 82, 650–658.

Levin GE & Baron DN. (1977). Leucocyte intracellular pH in patients with metabolic acidosis or renal failure. Clin Sci (Oxford) 52, 325.

Lipton SA & Rosenberg PA. (1994). Excitatory amino acids as a final common pathway for neurologic disorders. New Engl J Med 330, 613–622.

Lonnemann G. (2000). Chronic inflamation in hemodialysis: The role of contaminated dialysate. Blood Purif 18, 214–223.

Luxenberg J, Feigenbaum LZ & Aron JM. (1984). Reversible long-standing dementia with normocalcemic hyperparathyroidism. J Am Geriat Soc 32, 546–547.

Mahoney CA & Arieff AI. (1982). Uremic encephalopathies: Clinical, biochemical and experimental features. Amer J Kidney Dis 2, 324–336.

Mahoney CA, Arieff AI, Leach WJ & Lazarowitz VC. (1983). Central and peripheral nervous system effects of chronic renal failure. Kidney Int 24, 170–177.

Mahoney CA, Sarnacki P & Arieff AI. (1984). Uremic encephalopathy: Role of brain energy metabolism. Am J Physiol 247 (Renal Fluid Electrolyte Physiol. 16), F527–F532.

Mahurkar SD, Dkar SK, Salta R, Myers L, Smith LC & Dunea G. (1973). Dialysis dementia. Lancet 1, 1412–1415.

Mahurkar SD, Myers L, Cohen J, Kamath RV & Dunea G. (1978). Electroencephalographic and radionucleotide studies in dialysis dementia. Kidney Int 13, 306–315.

Mai ML, Emmett M, Sheikh MS, Santa Ana CA, Schiller L & Fordtran JS. (1989). Calcium acetate, an effective phosphorus binder in patients with renal failure. Kidney Int 36, 690–695.

Makita Z, Radoff S, Rayfield EJ, Yang Z, Skolnik E, Delaney V, Friedman EA, Cerami A & Vlassara H. (1991). Advanced glycosylation end products in patients with diabetic nephropathy. N Engl J Med 325, 836–842.

Malluche HH, Smith AJ & Abreo K. (1984). The use of deferoxamine in the management of aluminum accumulation in bone in patients with renal failure. N Engl J Med 311, 140.

Martinez-Vea A, Salvado E, Bardaji A, Gutierrez C, Ramos A & Garcia C. (2006). Silent cerebral white matter lesions and their relationship with vascular risk factors in middle-aged predialysis patients with CKD. Am J Kidney Dis 47, 241–250.

Maschio G, Bazzato G, Bertaglia E, Sardini D & Mioni G. (1970). Intracellular pH and electrolyte content of skeletal muscle in patients with chronic renal acidosis. Nephron 7, 481–487.

Mattana J, Effiong C, Gooneratne R & Singhal PC. (1998). Outcome of stroke in patients undergoing hemodialysis. Arch Intern Med 158, 537–541.

Mauer M, Fioretta P, Woredekal Y & Friedman EA. (2001). Diabetic nephropathy. In Diseases of the Kidney, 7th edn, Schrier RW & Gottschalk CW, pp. 2083–2127. J.B. Lippincott, Philadelphia.

Mazzuchi N, Carbonell E & Fernandez-Caen J. (2000). Importance of blood pressure control in hemodialysis patient survival. Kidney Int 58, 2147–2154.

Meeus F, Kourilsky O, Guerin AP, Gaudry C, Marchais SJ & London GM. (2000). Pathophysiology of cardiovascular disease in hemodialysis patients. Kidney Int 58 (Suppl. 76), S140–S147.

Meyer TW & Hostetter TH. (2007). Uremia. New Engl J Med 357, 1316–1325.

Minkoff L, Gaertner M, Darah C, Mercier C & Levin ML. (1972). Inhibition of brain sodium-potassium ATPase in uremic rats. J Lab Clin Med 80, 71–78.

Miyata T, Kirokawa K & van Ypersele de Strihou C. (2000). Relevance of oxidative and carbonyl stress to long-term uremic complications. Kidney Int 58 (Suppl. 76), S120–S125.

Miyata T, Sugiyama S, Saito A & Kirokawa K. (2001). Reactive carbonyl compounds related uremic toxicity (carbonyl stress). Kidney Int 59 (Suppl. 78), S25–S31.

Miyata T, van Ypersele de Strihou C, Ueda Y, Ichimori K & Kirokawa K. (2002). Angiotensin II receptor antagonists and angiotensin-converting enzyme inhibitors lower in vitro the formation of advanced glycation end products: Biochemical mechanisms. J Am Soc Nephrol 13, 2478–2487.

Moe SM & Sprague SM. (1994). Uremic encephalopathy. Clin Nephrol 42, 251–256.

Monk RD & Bennett DA. (2006). Reno-cerebrovascular disease? Neurol 67, 196–198.

Moreno F, Sanz-Guajardo D, Lopez-Gomez J, Jofre R & Valderrabano F. (2000). Increasing the hematocrit has a beneficial effect on quality of life and is safe in selected hemodialysis patients. J Am Soc Nephrol 11, 335–342.

Murray AM, Pederson SI, Tupper DE, Hochhalter AK, Miller WA, Zaun D, Collins AJ, Kane R & Foley RN. (2007). Acute variation in cognitive function in hemodialysis patients. Am J Kidney Dis 50, 270–278.

Murray AM, Tupper DE, Knopman DS & Gilbertson DT. (2006). Cognitive impairment in hemodialysis patients is common. Neurology 67, 216–223.

Naganuma T, Uchida J & Tsuchida K. (2005). Silent cerebral infarction predicts vascular events in hemodialysis patients. Kidney Int 67, 2434–2439.

Nathan HP, Hellstrom WJG & Kaiser FE. (1997). Treatment of men with erectile dysfunction with transurethral alprostadil. New Engl J Med 336, 1–7.

Ogura T, Makinodan A, Kubo T, Hayashida T & Hirasawa Y. (2001). Electrophysiological course of uraemic neuropathy in haemodialysis patients. Postgrad Med J 77, 451–454.

Okada J, Yoshikawa K, Matsuo H & Oouchi M. (1991). Reversible MRI and CT findings in uremic encphalopathy. Neuoradiology 33, 524–526.

Olsen S. (1961). The brain in uremia. Acta Psychiat Scand 36 (Suppl. 156), 1–128.

Palmer BF. (1999). Sexual dysfunction in uremia. J Am Soc Nephrol 10, 1381–1388.

Palmer BF. (2003). Sexual dysfunction in men and women with chronic kidney disease and end-stage kidney disease. Adv Ren Replace Ther 10, 48–60.

Panichi V, Migliori M, De Pietro S, Taccola D & Andreini B. (2000). The link of biocompatibility to cytokine production. Kidney Int 58 (Suppl. 76), S96–S103.

Papadakis MA & Arieff AI. (1987). Unpredictability of clinical evaluation of renal function in cirrhosis. Amer J Med 82, 945–952.

Papageorgiou C, Ziroyannis P & Vathylakis J. (1982). A comparative study of brain atrophy by computerized tomography in chronic renal failure and chronic hemodialysis. Acta Neurol Scand 66, 378–384.

Pereira AA, Weiner DE, Scott T, Chandra P, Bluestein R, Griffith J & Sarnak MJ. (2007). Subcortical cognitive impairment in dialysis patients. Hemodial Int 11, 309–314.

Pereira AA, Weiner DE, Scott T & Sarnak MJ. (2005). Cognitive function in dialysis patients. Am J Kidney Dis 45, 448–462.

Perl DP & Brody AR. (1980). Alzheimer's disease: X-ray spectrometric evidence of aluminum accumulation in neurofibrillary tangle-bearing neurons. Science 208, 297–299.

Perl DP & Good PF. (1990). Microprobe studies of aluminum accumulation in association with human central nervous system disease. Environ Geochem Health 12, 97–102.

Perl J, Unruh ML & Chan CT. (2006). Sleep disorders in end-stage renal disease; 'Markers of inadequate dialysis?' Kidney Int 70, 1687–1693.

Perry TL, Yong VW, Kish SJ, Ito M, Foulks JG, Godolphin WJ & Sweeney VP. (1985). Neurochemical abnormalities in brain of renal failure patients treated by repeated hemodialysis. J Neurochem 45, 1043–1048.

Pierides AM, Edwards WG Jr, Cullum UX Jr, Mc call JT & Ellis HA. (1980). Hemodialysis encephalopathy with osteomalacic fractures and muscle weakness. Kidney Int 18, 115.

Pliskin NH, Yurk HM, Ho LT & Umans JG. (1996). Neurocognitive function in chronic hemodialysis patients. Kidney Int 49, 1435–1440.

Porte FK, Johnson WJ & Klass DW. (1973). Prevention of dialysis disequilibrium syndrome by use of high sodium concentration in the dialysate. Kidney Int 3, 327.

Rasmussen H. (1986). The calcium messenger system, parts I and II. New Engl J Med 314, 1094–1101 and 1164–1170.

Rifat SL, Eastwood MR, McLachlan DRC & Corey PN. (1990). Effect of exposure of miners to aluminum powder. Lancet 336, 1162–1165.

Rodrigo F, Shideman J, McHugh R, Buselmeier T & Kjellstrand C. (1977). Osmolality changes during hemodialysis: Natural history, clinical correlations, and influence of dialysate glucose and intravenous mannitol. Ann Intern Med 86, 554–561.

Ronco C, Brendolan A & Bellomo R. (1998). Current technology for continuous renal replacement therapies. In Critical Care Nephrology, Ronco C & Bellomo R, pp. 1269–1308. Kluwer, Hingham, MA.

Rotter W & Roettger P. (1974). Comparative pathologic-anatomic study of cases of chronic global renal insufficiency with and without preceding hemodialysis. Clin Nephrol 1, 257.

rt-PA Stroke Study Group N. (1995). Tissue plasminogen activator for acute ischemic stroke. New Engl J Med 333, 1581–1587.

Sato Y, Kaji M, Metoki N, Satoh K & Iwamoto J. (2003). Does compensatory hyperparathyroidism predispose to ischemic stroke? Neurology 60, 626–629.

Savazzi GM. (1988). Pathogenesis of cerebral atrophy in uraemia. Nephron 49, 94–103.

Savazzi GM, Cusamo F, Vinci S & Allegri L. (1995). Progression of cerebral atrophy in patients on regular hemodialysis treatment: Long-term follow-up with cerebral computed tomography. Nephron 69, 29–33.

Schievink WI. (2001). Spontaneous dissection of the carotid and vertebral arteries. New Engl J Med 344, 898–906.

Schrier RW & Wang W. (2004). Acute renal failure and sepsis. New Engl J Med 351, 159–169.

Schulman G & Himmelfarb J. (2004). Hemodialysis. In Brenner & Rector's The Kidney, 7th edn, Brenner BM, pp. 2563–2624. W.B. Saunders, Philadelphia, PA.

Schulz JB & Arieff AI. (2003). Metabolic and toxic encephalopathies. In Neurological Disorders: Course and Treatment, 2nd edn, eds. Brandt T, Caplan LR, Dichgans J, Diener HC & Kennard C, pp. 991–1010. Academic Press, London.

Scribner BH & Babb AL. (1975). Evidence for toxins of "middle" molecular weight. Kidney Int 7 (Suppl 3), S349.

Sehgal AR, Grey SF, DeOreo PB & Whitehouse PJ. (1997). Prevalence, recognition, and implications of mental impairment among hemodialysis patients. Am J Kidney Dis 30, 41–49.

Sharad G, Saran R & Nolph KW. (1998). Indications, contraindications and complications of peritoneal dialysis in the critically ill. In Critical Care Nephrology, Ronco C & Bellomo R, pp. 1373–1382. Kluwer, Hingham, MA.

Sherman DG, Atkinson RP & Chippendale T. (2000). Intravenous ancrod for treatment of acute ischemic stroke. JAMA 283, 2395–2403.

Shik J & Parfrey PS. (2005). The clinical edidemiology of cardiovascular disease in chronic kidney disease. Curr Opin Nephrol Hypertens 14, 550–557.

Siddiqui JY, Fitz AE & Lawton RL. (1970). Causes of death in patients receiving long-term hemodialysis. JAMA 212, 1350.

Slatopolsky E, Martin K & Hruska K. (1980). Parathyroid hormone metabolism and its potential as a uremic toxin. Am J Physiol 239, F1–F12.

Slatopolsky EA. (1999). Renagel®, a nonabsorbed calcium- and aluminum-free phosphate binder, lowers serum phosphorus and parathyroid hormone. Kidney Int 55, 299–307.

Snow V, Weiss KB & LeFevre M. (2003). Management of newly detected atrial fibrillation. Ann Intern Med 139, 1009–1017.

Sprague SM, Corwin HL, Tanner CM, Wilson RS, Green BJ & Goetz CG. (1988). Relationship of aluminum to neurocognitive dysfunction in chronic dialysis patients. Arch Intern Med 148, 2169–2172.

Study Group E. (1996). Thrombolytic therapy with streptokinase in acute ischemic stroke. New Engl J Med 335, 145–150.

Takabatake T, Ohta H, Ishida Y, Hara H, Ushiogi Y & Hattori N. (1988). Low serum creatinine levels in severe hepatic disease. Arch Intern Med 148, 1313–1315.

Takaki J, Nishi T, Nangaku M, Shimoyama H, Inada T, Matsuyama N, Kumano H & Kuboki T. (2003). Clinical and psychological aspects of restless legs syndrome in uremic patients on hemodialysis. Am J Kidney Dis 41, 833–839.

Taki K, Takayama F, Tsuruta Y & Niwa T. (2006). Oxidative stress, advanced glycation end product and coronary artery calcification in hemodialysis patients. Kidney Int 70, 218–224.

Teschan PE & Arieff AI. (1985). Uremic and dialysis encephalopathies. In Cerebral Energy Metabolism and Metabolic Encephalopathy, ed. McCandless DW, pp. 263–285. Plenum, New York.

Teschan PE, Bourne JR & Reed RB. (1983). Electrophysiological and neurobehavioral responses to therapy: The national cooperative dialysis study. Kidney Int 23, 558.

Teschan PE, Ginn HE, Bourne JR, Ward JW, Hamel B, Nunnally JC, Musso M & Vaughn WK. (1979). Quantitative indices of clinical uremia. Kidney Int 15, 676–697.

Thodis E, Anninos PA, Pasadakis P, Adamopoulos AV, Panagoutsos S & Vargemezis V. (1992). Evaluation of CNS function in CAPD patients using magnetoencephalography (MEG). Adv Perit Dial 8, 181–184.

Tizianello A, Deferrari G, Gurreri G & Acquarone N. (1977). Effects of metabolic alkalosis, metabolic acidosis and uraemia on whole-body intracellular pH in man. Clin Sci Mol Med (Oxford) 52, 125–135.

Toyoda K, Fujii K, Fujimi S, Kumai Y, Tsuchimochi H, Ibayashi S & Lida M. (2005). Stroke in patients on maintainance hemodialysis: A 22-year single center study. Am J Kidney Dis 45, 1058–1066.

Van den Noort S, Eckel RE, Brine K & Hrdlicka JT. (1968). Brain metabolism in uremic and adenosine-infused rats. J Clin Invest 47, 2133–2142.

Vanholder R. (1998a). Low molecular weight uremic toxins. In Critical Care Nephrology, Ronco C & Bellomo R, pp. 855–868. Kluwer, Hingham, MA.

Vanholder R. (1998b). Pathogenesis of uremic toxicity. In Critical Care Nephrology, Ronco C & Bellomo R, pp. 845–853. Kluwer, Hingham, MA.

Vanholder R, De Smet R & Glorieux G. (2003). Review on uremic toxins: Classification, concentration, and interindividual variability. Kidney Int 63, 1934–1943.

Vanholder R, De Smet R & Lameire N. (2001). Protein-bound uremic solutes: The forgotten toxins. Kidney int 59 (Suppl. 78), S266–S270.

Verkman AS & Fraser CL. (1986). Water and non-electrolyte permeability in brain synaptosomes isolated from normal and uremic rats. Am J Physiol 250, R306–R312.

Vexler ZS, Roberts TPL, Bollen AW, Derugin N & Arieff AI. (1997). Transient cerebral ischemia: Association of apoptosis induction with hypoperfusion. J Clin Invest 99, 1453–1459.

Wang HC, Brown P & Lees AJ. (1998). Acute movement disorders with bilateral basal ganglia lesions in uremia. Movement Dis 13, 952–957.

Wang HC, Hsu JL & Shen YY. (2004). Acute bilateral basal ganglia lesions in patients with diabetic uremia: An FDG-PET study. Clin Nucl Med 29, 475–478.

Ward MK, Feest TG, Ellis HA, Parkinson IS & Keer DN. (1978). Osteomalacia dialysis osteodystrophy: Evidence for a water-borne aetiological agent, probably aluminum. Lancet 1, 841.

Winchester JF & Audia P. (2006). Extracorporeal strategies for the removal of middle molecules. Semin Dial 19, 110–114.

Winkelmayer WC, Eigner M, Berger O, Grisold W & Leithner C. (2001). Optic neuropathy in uremia. Am J Kidney Dis 37, E23.

Wratten ML, Tetta C, Ursini F & Sevanian A. (2000). Oxidant stress in hemodialysis: Prevention and treatment strategies. Kidney Int 58 (Suppl. 76), S126–S132.

Yasuda H & Nakajiima A. (1993). Brain protection against ischemic injury by nizofenone. Cerebrovasc Brain Metabol Rev 5, 264–268.

Yu SP, Yeh CH, Strasser M, Tian M & Choi DW. (1999). NMDA receptor-mediated K+ efflux and neuronal apoptosis. Science 284, 336–338.

Zimmerman JS & Herringer JS. (1999). Inflamation enhances cardiovascular risk and mortality in hemodialysis patients. Kidney Int 55, 648–658.

Zoccali C, Mallamaci F & Tripipepi G. (2003). Inflamation and atherosclerosis in end-stage renal disease. Blood Purif 21, 29–36.

Chapter 12
Thiamine Deficiency: A Model of Metabolic Encephalopathy and of Selective Neuronal Vulnerability

Saravanan Karuppagounder and Gary E. Gibson*

Summary Thiamine (vitamin B1) deficiency (TD) is a unique example of a nutritional deficit that produces a generalized impairment in oxidative metabolism and leads to metabolic encephalopathy or delirium, memory deficits and selective neuronal death in particular brain regions. Experimental TD is a classical model of a nutritional deficit associated with a generalized impairment of oxidative metabolism and selective cell loss in the brain. The response to TD is altered by the genetic background (i.e., strain) and the age of the animal. Changes in thiamine-dependent processes have also been implicated in ischemia (stroke), diabetes and multiple neurodegenerative disorders. An understanding of the mechanism by which TD leads to brain dysfunction and eventually to selective neuronal death is likely to facilitate our understanding of the role of thiamine in all these disorders. In addition, the results are likely to help our understanding of the fundamental mechanisms leading to altered neuronal functions and neuronal death in these other disorders.

Introduction

Thiamine-dependent processes play key roles in the pentose shunt, the linkage of glycolysis to the tricarboxylic acid (TCA) cycle and within the TCA cycle. Thiamine is rapidly converted to the biologically active form, thiamine pyrophosphate (TPP), which is an essential coenzyme for critical metabolic pathways. The TPP-dependent enzymes in the brain include transketolase (EC 2.2.1.1), a key enzyme of the pentose phosphate shunt. The pentose shunt provides NADPH, which is important for maintenance of the redox state of glutathione and for NO$^\bullet$ production. Although the pentose shunt is generally regarded as cytosolic, evidence suggests that a complete pentose shunt also exists in the endoplasmic reticulum. This cellular localization suggests that the thiamine-dependent pentose shunt may play a role in protein folding. There are also three thiamine-dependent mitochondrial enzyme complexes [pyruvate dehydrogenase (PDHC) (EC 1.2.4.1, EC

---

* To whom Correspondence should be addressed.

2.3.1.12, EC 1.6.4.3), α-ketoglutarate dehydrogenase (KGDHC) (EC 1.2.4.2, EC 1.2.4.4, EC 2.3.1.61, EC 1.6.4.3) and the branched chain α-keto acid dehydrogenase]. PDHC is the enzyme that links glycolysis and the TCA cycle. PDHC can exist in both active and inactive forms. KGDHC is a key and arguably a rate-limiting enzyme of the TCA cycle. Classically, KGDHC and PDHC enzymes provide reducing equivalents that are critical for ATP formation. These enzymes are sensitive to oxidants which regulate their activity (Jeitner et al., 2005). Recent data suggests that PDHC and KGDHC can also produce H_2O_2 (Starkov et al., 2004). Thiamine may play a role in Na^+/Ca^{2+} exchange and in $NO^•$ oxidant signaling (Gibson and Blass, 2007).

For decades, TD has been known to cause neurological disease in humans and animals. The classical disease related to thiamine deficiency in humans is Wernicke Korsakoff syndrome. The symptoms include delirium, severe memory deficits and selective neurodegeneration. The neurological consequences of TD in animals have been known for nearly a century and reduced activities of thiamine-dependent enzymes in brains from TD animals have been documented since Sir Rudolph Peters' classical studies on "biochemical lesions" in 1929. Recent interest in thiamine-dependent processes has surged because of reductions in thiamine-dependent enzymes and the expected consequences, including a decline in brain metabolic rates and increased oxidative stress accompanying several neurodegenerative disorders (see Table 12.1). A regionally selective loss of neurons occurs in all these disorders, but the molecular and cellular basis is unknown. TD provides a highly reproducible model in which mild interruption of oxidative metabolism

Table 12.1 Thiamine Deficiency Has Been Implicated in the Symptoms and Pathophysiology of Numerous Disorders (Only one to three references of many references are listed. For example Pubmed lists 305 references for thiamine and diabetes.)

Aquired immune deficiency (AIDS) (Alcaide et al., 2003)
Dialysis (Ueda et al., 2006)
Diabetes (Karachalias et al., 2005) (Hammes et al., 2003)
Chemo-therapy (Coy et al., 2005; Foldi et al., 2007; Lee et al., 2005; McLure et al., 2004)
Genetic disorders of thiamine transporter (Neufeld et al., 2001)
Nutrition (Wernicke Korsakoff Syndrome) (Victor et al., 1971)
Age related neurodegenerative disorders
Alzheimer's Disease (Gibson et al., 1988)
Parkinson's disease (Gibson et al., 2003; Jimenez-Jimenez et al., 1999)
Huntington disease (Klivenyi et al., 2004; Yates et al., 1990)
Down's (Yates et al., 1990)
Picks (Yates et al., 1990)
ALS (Sheline et al., 2002)
Progressive supranuclear palsy (Park et al., 2001a)
Spinal cerebellar ataxia (Mastrogiacomo and Kish, 1994)
Peripheral neuropathy after gastrectomy (Alves et al., 2006)
Lead poisoining (Wang et al., 2007a)
Stroke (Martin et al., 2005; Sheline and Wei, 2006; Shin et al., 2004; Shneider, 1991)

leads to interruption of neuronal function and eventually to neuronal loss in selective brain regions.

Thiamine-Dependent Processes are Altered in Multiple Human Diseases

Analysis of post-mortem samples from humans who died from various diseases confirms that thiamine-dependent processes are altered in human patients with disease. The majority of the observations related to age-related neurodegenerative diseases have been made on autopsy tissues especially of the brain but measures on blood components and cultured fibroblasts confirm that thiamine dependent enzymes are altered in stroke and neurodegenerative diseases. The diseases include Wernicke-Korsakoff Syndrome, stroke/ischemia, Alzheimer's disease (AD), Parkinson's disease, Huntington's disease, Progressive Supranuclear Palsy and at least five adult-onset neurodegenerative diseases that are caused by genes containing a variably increased CAG repeat within their coding region. The resulting polyglutamine domains bind to enzymes of energy metabolism including the thiamine-dependent enzyme KGDHC (Table 12.1). There is also a large literature on in-born-errors of metabolism in which thiamine dependent processes are altered, and the findings have been reviewed recently (Gibson and Blass, 2007).

Wernicke Korsakoff syndrome (WKS) is the neurological disorder most clearly linked to thiamine deficiency in humans. WK develops in a subset of chronic alcoholics, who are vitamin deficient because so many calories are consumed as alcohol instead of normal diet, and a diet rich in carbohydrates increases the metabolic demand for thiamine. Thiamine dependent enzymes were diminished in the brains of patients who died with WKS, but not in alcoholic controls (Butterworth et al., 1993). Transketolase in fibroblasts from those patients who develop WKS syndrome binds TPP more avidly than the control lines. The Km was nearly ten times higher in patients with WKS. Thus, these patients have an abnormality of transketolase that would be clinically unimportant if the diet was adequate (Blass and Gibson, 1977, 1979). The latter demonstrate a predisposing biochemical mutation to a neurological diseases that is only revealed by inadequate diet.

The results that implicate alterations in thiamine-dependent processes in an age-related disorder are strongest for AD. In post-mortem brains from AD patients, the activities of KGDHC are reduced more than 50–75% and those of transketolase more than 45%. PDHC decreases about 50%. Decreases occur in histologically damaged, and in relatively undamaged areas. These decreases occur even though the activities of other mitochondrial enzymes increase in brains of AD patients (Bubber et al., 2005; Gibson et al., 1988; Hazell and Wang, 2005; Heroux et al., 1996). Alterations in thiamine-dependent processes also occur in peripheral tissues from patients with AD. Small (-12%), but statistically significant abnormalities of TPP stimulation of transketolase (a sensitive measure of thiamine status) were identified in red blood cells and of transketolase activities in cultured fibroblasts (−15%) from patients with AD (Gibson et al., 1988). Furthermore, KGDHC is more sensitive to stress in fibroblasts from controls than from patients with AD (Gibson et al., 1999).

The mechanisms underlying the reductions in thiamine-dependent processes in AD brain are still unknown. In general, thiamine-dependent enzymes are diminished without an alteration in protein levels. This suggests that the proteins have been modified. This could alter the interaction of thiamine with the protein as appears to occur in transketolase from patients with WKS. Since KGDHC is particularly sensitive to oxidants and considerable oxidative stress accompanies AD, oxidant dependent modification of the thiamine dependent enzymes has been hypothesized. However, convincing data is still lacking. A better understanding of mechanisms underlying the deficit in thiamine-dependent processes and of the consequences of reducing thiamine-dependent processes will likely lead to the development of better treatment paradigms for patients.

Thiamine-Dependent Processes are also Diminished in Animal Models of Neurological Disease

Behavioral deficits related to reduction in thiamine-dependent processes occur in animal models of various neurodegenerative diseases. The percentage of PDHC in the active form is significantly reduced in R6/2 mice at 12 weeks of age. Huntington's disease is a neurodegenerative illness caused by expansion of CAG repeats at the N-terminal end of the protein huntingtin. In striatum and cortex in mice with 150 CAG repeats (R6/2 strain) thiamine dependent processes are altered. Dichloroacetate (DCA) stimulates PDHC activity and lowers cerebral lactate concentrations. DCA significantly increases survival, improves motor function, delays loss of body weight, attenuates the development of striatal neuron atrophy, and prevents diabetes, and DCA ameliorates the deficit. These results provide further evidence for a role of energy dysfunction including deficits in thiamine-dependent processes in HD pathogenesis and suggest that DCA may exert therapeutic benefits in HD (Andreassen et al., 2001).

Mice that are deficient in dihydrolipoamide dehydrogenase (Dld+/-), the E3 component of KGDHC and PDHC show increased vulnerability to 1-methyl-4-phenyl-1,2,3,6-tetrahydropyridine, malonate and 3-nitropropionic acid, which have been proposed for use in models of Parkinson's Disease and Huntington's Disease. MPTP cause significantly greater depletion of tyrosine hydroxylase-positive neurons in the substantia nigra of Dld+/- mice than that seen in the wild-type littermate controls. Striatal lesion volumes produced by malonate and 3-NP were significantly increased in Dld+/- mice. KGDHC activity was also found to be reduced in putamen from patients with HD. These findings provide further evidence that thiamine-dependent defects may contribute to the pathogenesis of neurodegenerative diseases (Klivenyi et al., 2004).

TD Models Both the Acute and Chronic Effects of Mild Impairment of Oxidative Metabolism

TD models both acute and long term mechanisms for how mild impairment of oxidative metabolism alters brain function. Thiamine-dependent processes may induce acute changes in function that are not reflected in neuron loss. Such changes include altered neurological function through changes in myelination (Kark et al., 1975), axonal conduction (Pawlik et al., 1977) and synaptic transmission (Barclay et al., 1982). TD can also model the chronic changes that result from mild impairment of oxidative metabolism. TD produces selective neuronal loss that induces permanent functional deficits. The TD model is powerful to study both. In the early stages, there is a clear loss of brain function (both motor performance mediated by CNS and memory skills) and no neuronal loss. At later stages, there is neuronal loss in specific brain regions, especially the thalamus (Falk et al., 1976; Ke et al., 2003). Both the acute and chronic abnormalities in response to mild impairment of oxidative metabolism may model similar changes in diseases such as AD or other neurodegenerative disorders that are accompanied by mild impairment of oxidative metabolism. The acute and chronic changes may occur sequentially or simultaneously. A better understanding of the mechanism underlying both the acute and chronic deficits in thiamine-dependent processes will probably lead to the development of a better treatment paradigm for patients with diseases that are accompanied by mild impairment of oxidative metabolism.

A variety of models have been developed to study TD. TD in rodents can be produced in multiple ways. Simple deprivation of thiamine will deplete thiamine and thiamine-dependent processes. However, this expands the time until the symptoms occur, and increases the variability for time of the onset of the symptoms. Injection of inhibitors of thiamine utilization in conjunction with the thiamine deficient diet shortens the time until onset of symptoms and provides a remarkably reproducible model. Pyrithiamine, which is structurally similar to thiamine, blocks the thiamine pyrophospho kinase, which catalysis the phosphorylation of thiamine to thiamine pyrophosphate (TPP) so that the production of the metabolically active form of thiamine, TPP, is impaired. Pyrithiamine readily crosses the blood brain barrier so that TD is produced in the brain and in the periphery. On the other hand, oxythiamine does not cross the BBB and only produces TD in the periphery. The precise timing of the acute and chronic changes in TD varies with the model. All the models lead to diminished food intake, so, often paired fed controls are used. These have never shown that pathology is related to TD.

The TD model can be used to ask mechanistic questions by using either the temporal response of changes or by preventing the neuronal loss either with various pharmacological treatments or different transgenic mice. One can provide excess thiamine to determine when irreversible changes have occurred, which provides further insight into mechanism. In addition to manipulation with various pharmacological agents (e.g., cholinergic drugs), compounds that bypass the metabolic step or block can be administered. For example, ketone bodies can bypass deficits in thiamine dependent processes and improve patients' well being (Blass et al.,

1975; Falk et al., 1976). Thus, the highly reproducible acute and chronic changes make TD an attractive model to ask mechanistic questions about how mild impairment of oxidative metabolism impairs brain function.

Behavioral Deficits due to TD Precede Neuronal Death (Acute Treatment)

The ease of assessing behavioral deficits in TD increases the power of the model to ask mechanistic questions. Thiamine was originally discovered and isolated by a bioassay of neurological function. The loss of righting reflex, opistotonus or even seizures were used as end points in early studies. These markers have proved useful and are reasonably definitive but are of limited value for mechanistic studies, because by the time these changes occur there is neuronal death, and many of the observed changes may be secondary to other severe neurological changes (e.g., seizures) or other independent pathways.

To test the causes of neurological dysfunction and to detect changes in behavior at early stages of TD that precede neuronal death in TD mice, several behavioral measures have been standardized. The behavioral tasks have also been used to determine treatments for the reversal of TD-induced changes or to determine when they become irreversible.

The string test or tight rope test measures the ability of the mouse or rat to move along an elevated taut string. The task is very sensitive to TD and other metabolic encephalopathies such as hypoxia. However, successful use requires considerable skill and time. After only one day of TD, 38% of rats perform poorly. By day 5 of TD, 50% of rats have persistently decreased scores. This occurs before the onset of weight loss or appearance of neurological deficits such as seizures. The abnormalities are related to deficits in the central nervous system (CNS) because no alterations occur in rats treated with oxythiamine that does not act in the brain (Barclay et al., 1981a,b). Nor are the deficits related to a TD-induced weight loss because pair fed controls whose diet is restricted do not have these deficits.

TD alters spontaneous open field behavior (Barclay and Gibson, 1982; Barclay et al., 1982). TD mice increase "staring" after just three days of treatment, and this is not observed with TD that is restricted to non-brain tissues (i.e., oxythiamine treatment) (Barclay and Gibson, 1982; Barclay et al., 1982). Open field behavior can also be followed by the mice breaking a light beam which is tracked with a computer. In general, TD increases locomotor activity and this is followed by a gradual decrease (Freeman et al., 1987).

Performance on the rotarod task declines with thiamine deficiency in a highly reproducible manner. Impaired motor performance on rotarod is apparent by 8 days of TD (−32%) and is severe by 10 days of TD (−97%) (Karuppagounder et al., 2007; Shi et al., 2007). The decline in performance on this task approximates changes in neuronal death. Thus, compounds or conditions that improve rotarod performance may delay neuron death.

TD also induces memory loss. TD reduces performance on a passive avoidance task and the deficit is proportional to the severity of TD (Nakagawasai et al., 2000a,b). The percentage of mice making an inappropriate choice increases from about 15% in controls to greater than 60% with TD. If thiamine is administered at early times of TD, the deficits are reversible (Nakagawasai et al., 2000a,b). TD reduces the learning ability (Y maze performance based on coupling the correct choice to light to avoid shock) at a stage of TD when mice do not exhibit regular pathological lesions, the loss of cholinergic neurons, decreases of NeuN-positive hippocampal neurons, or abnormal long-term potentiation of hippocampal CA1 and CA3. Re-administering thiamine reverses the weakened learning ability (Zhao et al., 2007).

TD-Induced Behavioral Deficits are Altered by Age and Genetics

The genetic background of both humans and mice alters their response to both genetic and "environmentally" induced neurodegenerative diseases. Changes in open field behavior reveal that genetic background alters the response to TD. Open field behavior was determined during thiamin deficiency in two strains of young mice. In CD-1 mice, TD reduces total distance traveled and vertical movements after 7 days and the decline is more than 50% by day 9. The open field behavior of untreated Balb/c mice is about 40% less than in CD-1 mice which respond to TD in a qualitatively different manner. The activity of the Balb/c mice increases, and then decreases with TD (Freeman et al., 1987). Thus, TD provides a convenient and reproducible model in which to study the interaction of genetics with mild impairment of oxidative metabolism.

TD-induced changes in open field behavior are altered by the age of the mice. The locomotor activity of 3 month old mice peaks on day 6 (126% of initial score), whereas 10 and 30 month old mice show a much greater increase (about 175% of initial scores), and the peak is on day 7. Although the activity of the thiamine-dependent enzyme transketolase (TK) is affected similarly at all ages, the activity of KGDHC in the brain of aged mice is more sensitive to thiamine deficiency than in the brain of young mice. KGDHC activity declines 41%, 57% and 74% at 3, 10, and 30 months, respectively. Thus, the current mouse model is an attractive one to study the interaction of a mild reduction in metabolism with aging (Freeman et al., 1987).

Reversal of TD-Induced Behavioral Deficits Provides Insight into Underlying Mechanisms

Reversal of both the acute and long-term effects of TD has been used to test the mechanism in TD. The behavioral tasks described in previous paragraphs, especially the tight rope task, are ideal for studying the effects of drugs on TD-induced changes. A series of studies using the tight rope task demonstrate that TD induces cholinergic deficiency. Systematic manipulation with cholinergic drugs (agonists and cholinesterase inhibitors) reveals that behavioral deficits are related to muscarinic, cholinergic abnormalities in the CNS. Indeed, TD-induced abnormalities can be reversed as well by the cholinesterase inhibitor physostigmine or the muscarinic agonist arecoline as with thiamine itself. TD-induced increases in "staring" are also ameliorated with central cholinergic muscarinic drugs as effectively as with thiamine (Barclay and Gibson, 1982; Barclay et al., 1982). TD-induced decline in the turnover of acetylcholine supports the behavioral evidence of a decline in cholinergic function (Barclay et al., 1981a).

TD-induced behavioral deficits and neuronal loss can be reversed by the administration of thiamine if the treatment is provided early enough. For example, in mice, thiamine administration by day 7 of TD prevents TD-induced neuron loss and deficits in rotarod performance. If thiamine is administered later, the neuron loss is permanent (Ke et al., 2003). These straightforward studies are important from a mechanistic point of view because they indicate that certain irreversible steps in cell death occur by a defined time in the treatment.

TD-Induced Neurological Deficits that are Not Reversed by Thiamine Administration

Thiamine administration after varying periods of TD can be used to determine the long term effects of the neuronal loss (or permanent synaptic deficits) due to mild impairment of oxidative metabolism. Deficits that persist after administration of thiamine are due to neuronal death or induction of permanent synaptic deficits. Although TD is used as a tool to create a lesion, the animals are not TD during the task. In-vivo acetylcholine efflux, a marker of memory-related activation, was measured in the hippocampus and the amygdala of TD and control rats while they were tested on a spontaneous alternation task. During behavioral testing, all animals display increases in acetylcholine efflux in both the hippocampus and amygdala. However, during spontaneous alternation testing acetylcholine efflux in the hippocampus and the alternation scores are higher in control rats than in TD-treated rats. In contrast, acetylcholine efflux in the amygdala is not suppressed in TD treated rats, relative to control rats, prior to or during behavioral testing (Roland and Savage, 2007; Savage et al., 2007). This approach allows assessment of specific pathways that are permanently altered by TD and their role in memory independent of the acute effects of mild impairment of oxidative metabolism (Langlais and Savage, 1995; Roland and Savage, 2007; Savage et al., 2007).

Impairing Thiamine-Dependent Processes Alters Cholinergic Function and Behavioral Performance, and Induces Cell Death

Behavioral deficits related to reduction in thiamine-dependent process occur in animal models of delirium. Hypoxia is a classical model of delirium. Hypoxia, like TD, diminishes performance on the tight rope task. The behavioral abnormalities can be ameliorated by cholinomimetics that act in the CNS (Gibson et al., 1983). Whether this is directly related to thiamine-dependent processes is not clear but inhibitors of PDHC or KGDHC diminish acetylcholine synthesis. Any impairment of pyruvate dehydrogenase leads to a corresponding reduction in acetylcholine synthesis (Gibson et al., 1975). Impairing flux through KGDHC also impairs cholinergic function (Gibson and Blass, 1976). The direct link of these thiamine dependent enzymes to cholinergic function may be a component of all metabolic encephalopathies including hypoxia and hypoglycemia (Gibson and Blass, 1976) as well as TD.

Thus, TD is a typical metabolic encephalopathy, like hypoxia and hypoglycemia. All appear to intervene with the same thiamine-dependent steps in metabolism and lead to a reduction in cholinergic function. These changes can be mediated independent of neuronal death and can occur in the absence of any neuronal death.

The Temporal Sequence of Selective Neuronal Death in TD Can Be Used to Study Mechanism

TD, like all neurological diseases, exhibits selective vulnerability. TD causes selective neuronal death in specific brain regions, while microglia, endothelial cells and astrocytes are activated (Ke and Gibson, 2004). The brain regions in which the neurons die in TD have been described extensively. Neuropathologic evaluation of the brains of patients with WKS reveals a highly selective and reproducible pattern of neuronal cell loss involving primarily diencephalic and brainstem structures. Bilateral symmetrical lesions are consistently observed in thalamic nuclei, mammillary bodies, inferior colliculi, inferior olivary nuclei, and lateral vestibular nuclei (Victor et al., 1971). The changes during TD in rodents mimic those in WKS but differ from those in AD and other neurodegenerative diseases. Nevertheless, the processes that make these neurons may also underlie the selective vulnerability of neuronal populations in other diseases that are accompanied by mild impairment of oxidative metabolism. Even if different details are involved in each disease, an understanding of selective vulnerability in TD will probably shed light on the molecular events underlying all neurodegenerative diseases.

The time frame of the neuron death in TD makes it particularly valuable for studies of selective cell death in neurodegenerative disorders. In mice, the time course of TD-induced changes in neurons was determined in the most sensitive brain region, the submedial thalamic nucleus (SmTN). Significant neuronal loss

(29%) occurs after 8 or 9 days of TD (TD8-9) and increases to 90% neuron loss by days 10–11. At this time the damage spreads to other regions of the thalamus. To test the duration of TD critical for irrevocable changes, mice received thiamine after various durations of TD. Thiamine administration on day 8 blocks further neuronal loss, partially reverses effects on day 9, and is ineffective on days 10-11. These studies indicate that irreversible steps leading to neuronal death are active by day 9. This model provides a unique paradigm for elucidating the molecular mechanisms involved in neuronal commitment to neuronal death cascades following mild impairment of oxidative metabolism (Ke et al., 2003). Selective vulnerability occurs in all neurodegenerative disorders and the mechanisms underlying the neuronal death are still unknown. Determining the mechanism for selective vulnerability in TD will probably help in an understanding of these other disorders.

Immunocytochemical, Histochemical and Message Studies to Evaluate Selective Vulnerability

One possibility for selective vulnerability is the distribution of thiamine-dependent enzymes in the brain. Thus, the thiamine-dependent enzymes in the brain were mapped immunochemically, with histochemistry and by message levels.

KGDHC occurs at low levels in neurons, glia and neutrophil throughout the rat brain. Some regions including those that are enriched with the cholinergic neuronal marker, choline acetyltransferase (ChAT), show relatively high perikaryal enrichment of KGDHC. In several regions, virtually all cholinergic neurons are enriched with KGDHC (Calingasan et al., 1994a,b). In the cerebral cortex, high immunoreactivity occurs mostly in layers III, V, and VI. The hippocampal pyramidal layer in CA1 and CA2 exhibits more intense staining than CA3. In the mammillary body, intensely labeled cells occur in the supramammillary and lateral nuclei. The basal forebrain, basal ganglia, reticular and midline thalamic nuclei, red nucleus, pons, cranial nerve nuclei, inferior and superior colliculi, and cerebellar nuclei also contain highly immunoreactive neurons. The distribution of KGDHC overlaps with that of PDHC (Calingasan et al., 1994b). Thus, there is a regional distribution and high overlap of KGDHC with cholinergic neurons that may account for the acute effects of TD and the link with cholinergic function. However, there is no apparent relation between distribution of the thiamine-dependent enzymes KGDHC or PDHC and TD-induced neuronal death. Furthermore, there is no relation between the regional selective neuron death and the response of these enzymes to thiamine deficiency. The enrichment of thiamine-dependent enzymes in neurons may account for neuronal vulnerability, but it does not account for the loss of neurons in different brain regions.

Analysis of the individual KGDHC subunits does not reveal a regional selective change either. Western blots and immunocytochemistry reveal different aspects of the changes in protein levels. By Western blot analysis of samples from the vulnerable region (SmTN), the immunoreactivity of E1k and E3 increases by 34% and 40%, respectively, but only at mid stage TD. In the non-vulnerable cortex, the immunoreactivity of the three subunits is not altered. Immunocytochemical staining of brain sections from mice in late stages indicates a reduction in the immunoreactivity of all subunits in SmTN, but not in the cortex. Reductions in the E2k and E3 mRNA in SmTN occur before neuronal death (−28% and −18%, respectively) and are more severe at the time of neuronal death (−61% and −66%, respectively). On the other hand, the level of E1k mRNA did not decline in SmTN until TD10 (−48%). In contrast, TD did not alter mRNA levels of the subunits in the cortex at late stages (Shi et al., 2007).

Immunocytochemical measures only represent protein levels, so a unique method was devised to assess the activities of thiamine dependent enzymes. Since enzyme activities measured by traditional methods in a test tube do not necessarily reflect activities in the brain, methods were developed to assess activities in more complicated and life-like environments. These histochemical methods provide a way to estimate the activity of KGDHC in living cells and in frozen sections. By this method, a loss in activity occurs in the selectively vulnerable region but only at later stages when there is already a decline in the number of neurons. At late stages, the overall KGDHC activity declined 52% in vulnerable regions but only 20% in relatively spared regions (Shi et al., 2007). Thus, decreased KGDHC activities as measured by histochemical methods cannot account for regional selectivity (Shi et al., 2007).

The heterogeneous distribution of TK may reflect a variety of metabolic activities among different brain regions but does not provide a simple molecular explanation for selective cell death in either thiamine deficiency or other conditions where TK is reduced (Calingasan et al., 1995c). Parallel distribution of all TPP-dependent enzymes occurs in many regions. But the distribution does not correlate with the predilection of particular brain regions to pathological or biochemical lesions in some neurodegenerative disorders. Thus, there is no correlation between selective vulnerability and regional distribution of TK (Calingasan et al., 1995c). Furthermore, the response of TK to TD does not parallel selective vulnerability. TK activity declines in both vulnerable and spared regions in TD. Immunoblots show a parallel reduction of TK protein. With a few exceptions, immunocytochemistry indicates an overall decline of TK immunoreactivity and the decrease is not specific to vulnerable areas. In contrast to the pronounced, general decline of TK protein, in situ hybridization reveals a regional decrease of 0–25% of TK mRNA in TD. Northern blots indicate a similar level of TK mRNA in whole brain in TD. These results show that the decline of TK activity results from a proportional decrease of TK protein, and the deficiency may be due to an instability of TK protein or an inhibition of TK mRNA translation. The lack of correlation of the distribution and the absence of specific alteration of TK in affected regions suggest that the reduced TK may not be linked directly to selective vulnerability in thiamine deficiency (Sheu et al., 1996).

Ex Vivo Studies to Evaluate Selective Vulnerability

An alternative strategy to determine the molecular basis of selective vulnerability is to make the animal thiamine deficient and then remove the tissue from vulnerable and non-vulnerable regions (e.g., brain slices) and use the tissue for in vitro studies. This allows assessment of much more dynamic measures such as metabolic fluxes to be monitored. To clarify the enzymatic mechanisms of brain damage in TD, glucose oxidation, acetylcholine synthesis, and the activities of the three major TPP-dependent brain enzymes were compared in untreated controls, in symptomatic pyrithiamine-induced thiamin-deficient rats, and in animals in which the symptoms had been reversed by treatment with thiamine. Although brain slices from symptomatic animals produce $^{14}CO_2$ and ^{14}C-acetylcholine from [U-^{14}C] glucose at rates similar to controls under resting conditions, the K^+-induced-increase in these variables is reduced by 50 and 75%, respectively. The activities of transketolase and KGDHC decrease 60–65% and 36%, respectively. The activity of PDHC did not change nor did the activity of its activator pyruvate dehydrogenase phosphate phosphatase (EC 3.1.3.43). Although treatment with thiamine for 7 days reverses the neurological symptoms and restores glucose oxidation, acetylcholine synthesis and 2-oxoglutarate dehydrogenase activity to normal, transketolase activity remains 30–32% lower than in the controls (Gibson et al., 1984).

To further elucidate the molecular basis of the selective damage to various brain regions by thiamin deficiency, changes in enzymatic activities were compared to carbohydrate flux through various pathways from vulnerable and nonvulnerable regions at late or end stages of TD. The changes in enzyme activities do not parallel the pathological vulnerability of these regions to TD. $^{14}CO_2$ production from ^{14}C-glucose labeled in various positions was utilized to assess metabolic flux. At late stages of TD, $^{14}CO_2$ production in the vulnerable regions declined severely (−46 to 70%) and approximately twice as much as those in non-vulnerable regions. Also, the ratio of enzymatic activity to metabolic flux increased as much as 56% in the vulnerable regions, but decreased 18 to 30% in the non-vulnerable region. These differences reflect a greater decrease in flux than enzyme activities in the vulnerable regions. Thus, selective cellular responses to TD can be demonstrated ex vivo, and these changes can be directly related to alterations in metabolic flux. Since they cannot be related to enzymatic alterations in the three regions, factors other than decreases in the activity of these TPP-dependent enzymes must underlie selective vulnerability in this model of thiamin deficiency (Gibson et al., 1989).

TD-Induced Changes in Inflammation and Oxidative Stress Support their Role in the Selective Neuronal Death During TD

In contrast to the more general changes in thiamine-dependent enzyme markers, TD-induced increases in markers of oxidative stress and inflammation in vivo generally parallel selective vulnerability. Whether these changes cause or follow neuronal death has been tested by following the temporal response in multiple cell types and by testing the consequences of various genetic knockouts. The temporal responses of the changes in TD brains provide insight into the mechanism for cell death in thiamine deficiency. The time course of cell specific TD-induced changes in neurons and microglia were determined in the brain region most sensitive to TD (i.e., the submedial thalamic nucleus (SmTN); Fig. 12.1). Although only neurons will die, changes occur in microglia and endothelial cells before neuronal loss. Significant **neuron** loss (29%) occurs after 8 or 9 days of TD and increases to 90% neuron loss by days 10–11 of TD. The changes in neurons include introduction of oxidative stress as indicated by nitrotyrosine formation and alterations in APP processing including the formation of neuritic clusters, protein nitration, and hydroxynonenal. The number of **microglia** increases 16% by day 8 of TD and by nearly 400% on day 11 of TD. The changes in microglia include induction of CD40, iNOS, hemeoxygenase-1, nitrotyrosine, interleukin-1β, TNFα, increased redox active iron and ferritin as well as altered amyloid precursor protein processing (Calingasan et al., 1999; Karuppagounder et al., 2007). Hemeoxygenase-1 (HO-1) positive microglia are not detectable at day 8 of TD, yet

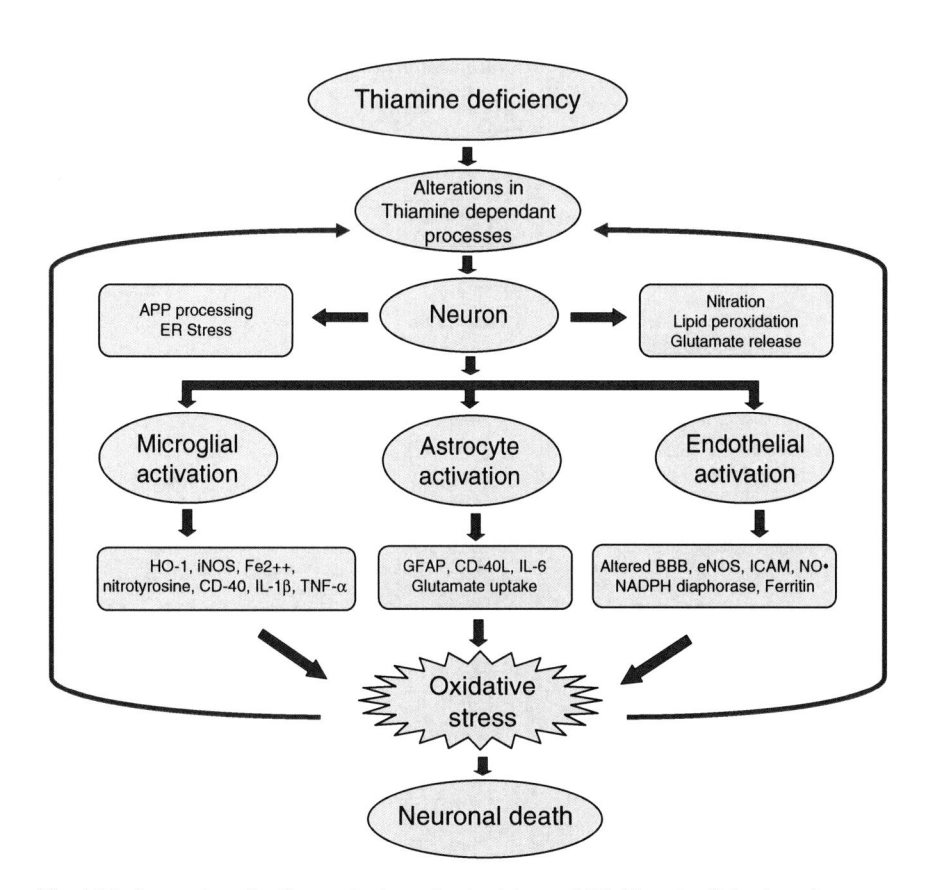

Fig. 12.1 Interaction of cell types in the pathophysiology of TD (*See also* Color Insert)

increase dramatically co-incident with neuron loss. The induction in **endothelial cells** includes NADPH diaphorase, iron and the antioxidant protein ferritin, eNOS and ICAM-1 (Calingasan et al., 1998). Changes in endothelial cells, especially ICAM-1 may allow migration of T-cells, neutrophils and mast cells. These peripheral cells may react with endogenous receptors on microglia and astrocytes to induce an inflammatory response. Changes in **astrocytes** include induction of CD40L, interleukin-6 and at later times increased GFAP (Karuppagounder et al., 2007; Ke et al., 2005a; Todd and Butterworth, 1999) and decreased expression of glutamate transporters (Desjardins and Butterworth, 2005; Hazell et al., 2003). The induction of CD40L on astrocytes appears to be important in the induction of neuron loss. The decrease in glutamate transporters appears important since at least in the late stages of TD, glutamate increases and glutamate blockers protect against neuron loss (Hazell et al., 1993; Langlais and Mair, 1990). Other peripheral cells are also part of the TD-induced response. For example, activated mast cells that release histamines are present in TD brains (McRee et al., 2000; Meng and Okeda, 2003). Peripheral macrophages that release myeloperoxidase are also present in these selectively vulnerable regions (Calingasan et al., 2000).

To test whether the responses of individual cell types reflect inherent properties of the cells, the reaction of different brain cell types to TD and/or inflammation in vitro has been determined. The cells that have been implicated in TD-induced neurotoxicity, including neurons, microglia, astrocytes, and brain endothelial cells, as well as neuroblastoma and BV-2 microglial cell lines, were cultured in either thiamine-depleted media or in normal culture media with amprolium, a thiamine transport inhibitor. The activity levels of a key mitochondrial enzyme, KGDHC, were uniquely distributed among different cell types: The highest activity was in the endothelial cells, and the lowest in primary microglia and neurons. The unique distribution of the activity did not account for the selective response to TD. TD slightly inhibited general cellular dehydrogenases in all cell types, whereas it significantly reduced the activity of KGDHC exclusively in primary neurons and neuroblastoma cells. Among the cell types tested, only in neurons did TD induce apoptosis and cause the accumulation of 4-hydroxy-2-nonenal, a lipid peroxidation product. The results demonstrated that the selective cell changes during TD in vivo reflect inherent properties of the different brain cell types (Park et al., 2000).

To test the interaction of the various cell types during TD, TD primary hippocampal neurons were either cultured alone, or co-cultured with primary astrocytes or microglia. After 7 days of TD, 50% of the neurons died, and the processes of many of the surviving neurons were severely truncated. When TD neurons were co-cultured with astrocytes or microglia, neurons did not die or show decreased neurite outgrowth. This shows that neuronal-glial interactions are critical for maintaining neuronal homeostasis during chronic metabolic impairment (Park et al., 2001b).

The induction of inflammatory markers at the gene level (i.e., mRNA) is region- and cell-specific. A micropunch technique was used to evaluate quantitatively the selective regional changes in mRNA and protein levels. To test whether this method can distinguish between changes in vulnerable and non-vulnerable regions, markers for neuronal loss (NeuN) and endothelial cells (eNOS) and inflammation (IL-1β, IL-6 and TNF-α) in SmTN and cortex of control and TD mice were assessed. TD significantly reduced NeuN and increased CD11b, GFAP and ICAM-1 immunore-activity in the vulnerable SmTN as revealed by immunocytochemistry. When assessed on samples obtained from the SmTN by the micropunch method, NeuN protein declined (−49%), while increases in mRNA levels occurred for eNOS (3.7-fold), IL-1β (43-fold), IL-6 (44-fold) and TNF-α (64-fold) in SmTN with TD. The only TD-induced change that occurred in the cortex with TD was an increase in TNF-α (22-fold) mRNA levels. Immunocytochemical analyses revealed that IL-1β, IL-6 and TNF-α protein levels increased in TD brains and colocalized with glial markers. The results demonstrate that TD induces quantitative, distinct inflamma-tory responses and oxidative stress in vulnerable and non-vulnerable regions that may underlie selective vulnerability (Karuppagounder et al., 2007).

To further test the role of inflammatory/immune mechanisms in TD-induced neurodegeneration, the temporal profile of neurodegeneration was compared to the activation of CD68-positive microglia and ICAM-1-positive endothelial cells dur-ing TD in wild type mice and in CD40L-/- mice. The results demonstrate that CD40 and CD40 ligand (CD40L), two co-stimulatory immune molecules, are involved in TD-induced selective neuronal death. TD induces CD40 immunoreactivity in microglia and CD40L immunoreactivity in astrocytes. Both CD40-positive micro-glia and CD40L-positive astrocytes increase during the progressive TD-induced neuronal death. The number of CD40L-positive astrocytes increase whenever the number of NeuN-positive neurons decrease (Ke et al., 2005a,b).

In general, the results indicate that TD induces alterations in endothelial cells and microglia occur contemporaneously and precede those in astrocytes and neu-rons. This suggests that the initial oxidative stress is amplified in feedback loops and eventually leads to neuronal death (Fig. 12.1).TD provides a unique model for elucidating the molecular mechanisms involved in neuronal commitment to neuro-nal death cascades and contributory microglial activity. Inflammation and oxidative stress may play a major role in TD-induced selective vulnerability (Karuppagounder et al., 2007; Ke et al., 2003; Ke and Gibson, 2004).

Selective Changes in the Blood-Brain Barrier Suggest that they are Involved in Selective Neurodegeneration

Alterations in the blood-brain barrier (BBB) could allow entry of peripheral macro-phages into the brain. This would promote neurodegenerative processes. Late stages of TD are characterized by hemorrhages in the brain indicating a severe disruption of the BBB. Although these changes are restricted to areas of damage, the relation of the breakdown of the BBB to selective neuron loss in TD is not understood. The BBB was examined at different stages of TD using immunoreactivity of immunoglobulin G (IgG) as an indicator of BBB integrity. IgG increases in vulnerable regions prior to the onset of neuronal death and hemorrhage. Non-vulnerable regions display little or no IgG immunoreactivity. Electron microscopy of capillary endothelia in areas of IgG accumulation reveals perivascular edema and intact interendothelial tight junctions. Thus, TD leads to a breakdown of the BBB in vulnerable regions prior to or coinciding with hemorrhage and cell loss (Calingasan et al., 1995a). The presence of mast cells that produce histamine and neutrophils that produce myeloperoxidase suggest migration of cells from the periphery to sites of TD-induced injury. Whether the histamine release precedes of follows the BBB changes is controversial (Calingasan et al., 2000b; McRee et al., 2000; Meng and Okeda, 2003).

Changes in Amyloid Precursor Protein in Response to TD

Mild impairment of oxidative metabolism alters processing of proteins and can possibly lead to the pathology that accompanies neurodegenerative diseases. This has been studied most exhaustively for the pathology related to AD. Amyloid precursor protein (APP) and amyloid-β-peptide (Aβ) are critical in the formation of plaques in AD and these proteins have an important function in calcium regulation and cell death pathways. In rats, TD-induced cell degeneration is accompanied by an accumulation of APP/amyloid precursor-like protein 2 (APLP2) immunoreactivity in abnormal neurites and perikarya along the periphery of, or scattered within, the lesion. While immunocytochemistry and thioflavine S histochemistry fail to show fibrillar beta-amyloid, APP-like immunoreactivity accumulates in aggregates of swollen, abnormal neurites and perikarya along the periphery of the infarct-like lesion in the thalamus. Immunoblotting of the thalamic region around the lesion reveals increased APP-like holoprotein immunoreactivity. Regions without apparent pathological lesions show no alteration in APP-like immunoreactivity. Thus, the oxidative insult associated with cell loss, hemorrhage and infarct-like lesions during TD leads to altered APP metabolism (Calingasan et al., 1995b).

Prompted by these data and our previous findings of a genetic variation in the development of TD symptoms, the studies were extended to mice. In C57BL/6, ApoE knockout, and APP YAC transgenic mice, TD induces abnormal clusters of intensely immunoreactive neurites only in areas of damage. The clusters appear as either irregular clumps or round or oval rosettes that strikingly resemble the neuritic component of Alzheimer amyloid plaques. However, immunostaining using various antisera to synthetic Aβ and thioflavine S histochemistry fail to show evidence of a component of Aβ. This is the first report that chronic oxidative deficits can lead to this novel pathology (Calingasan et al., 1995b, 1996).

To study the effects of TD on plaque pathology, TD was induced in 45, 60 and 90 day-old transgenic mice over expressing a doubly mutated human amyloid precursor protein (APP). TD significantly increases the compact (Thioflavine-S or Congo red) and diffuse (4G8 or 6E10-immunoreactive) plaques. Quantification of plaques in the cortex, hippocampus, olfactory bulb, cerebellum, thalamus, hypothalamus and neostriatum reveals a marked acceleration of amyloid deposition in TD compared to saline treated animals. Thus, the plaques are not associated only with regions with selective neuronal death. These plaques are associated with activated microglia (CD11b-immunoreactive) and are surrounded by reactive astrocytes (GFAP-immunoreactive). These findings demonstrate that the induction of mild impairment of oxidative metabolism, oxidative stress and inflammation induced by TD also promotes accumulation of plaques (Karuppagounder et al., 2006).

Prevention of Cell Death as a Test of Mechanism

Prevention of neuron death at various stages of TD can be used to test theoretical mechanisms. A variety of means have been used to protect against cell loss in TD. To test the duration of TD critical for irrevocable changes, mice received thiamine after various durations of TD. Thiamine administration on day 8 of TD blocks further neuronal loss and induction of HO-1 positive microglia, whereas other microglial changes persist (Ke et al., 2003). Thiamine treatment only partially reverses effects on day 9 of TD, and is ineffective on days 10–11. These studies indicate that irreversible steps leading to neuronal death and induction of HO-1 positive microglia occur on day 9. This suggests that reversal of changes that occur prior to this would prevent neuronal death and the reversal of the cell death by thiamine allows a test of the mechanism.

Another productive way to prevent neuron death is the use of mice that have genetic knockout of various inflammatory/oxidative stress genes. The immunochemical results described in previous sections demonstrate several steps that precede neuronal death. If these indicators are important, then knockout of these molecules should protect against neuronal death. This approach has the obvious limitation that other pathways may compensate. Several paradigms partially protect including knockouts of ICAM-1 (Calingasan et al., 2000b), eNOS (Calingasan et al., 2000a). Each provides protection of about 20%. On the other hand, knockouts of iNOS and nNOS do not provide any protection (Calingasan and Gibson, 2000b). One interpretation of these findings is that the signal is initiated from the neurons and is then exaggerated by the microglial and endothelial cell responses.

In early stages of TD, targeted deletion of CD40 diminishes the number of CD40L-positive astrocytes and reduces neuronal death by 35%. The number of CD40L-positive astrocytes increases whenever the number of NeuN-positive neurons decrease. In early stages of TD, deletion of CD40L diminishes CD40-positive microglia and reduces the neuronal death by 64%. In advanced phases of TD, neither CD40 nor CD40L deletion protects against neuronal death. The results indicate that CD40-CD40L interactions promote neuronal death in early stages of TD, but that at later phases the protective effects of the diminished CD40 or CD40L are over-ridden by other mechanisms (Ke et al., 2005a,b).

The roles of other peripheral inflammatory cascades in TD-induced neuronal death have also been tested (Table 12.3). Deletion of genes for CD4, or CD8 (the

Table 12.2 CD40 and CD40L Alter the Response to Thiamine Deficiency

	Protective	Summary
CD40 -/-	yes	Delays neuronal death;↓ CD40L astrocytes
CD40L -/-	yes	Delays neuronal death; ↓CD40 microglia; ↓activation of microglia and endothelial cells

Table 12.3 Peripheral Inflammatory Cells/Signals Alter the Microglial Response to TD

	Hypertrophy	Proliferation	Activation
Wild type	↑	↔	↑
CD4 -/-	↑	↑	↑
CD8 -/-	↑	↑	↑
IFNγ -/-	↑	↑	↑
NADPH oxidase -/-	↓		↓

co-receptors for T-cells), IFN-gamma (the cytokine produced by T-cell), or NADPH oxidase (the inflammation related oxidase) were tested. None protect against neuronal death in late stages of TD. On the other hand, deletion of the genes for CD4, CD8 and IFN-gamma increase microglial activation, and deletion of the gene for NADPH oxidase decrease microglial activation when compared to control mice. In wild type mice, TD causes hypertrophy of CD68 positive microglia without increasing the number of microglia. However, TD induces hypertrophy and proliferation of CD68-positive microglia in the CD4 (97%), CD8 (57%) or IFN-gamma (96%) genetic knockout mice. In the genetic knockout mice for NADPH oxidase, the microglial activation was 65% less than the wild type mice. The results demonstrate that mice deficient in specific T cells (CD4-/-, CD8-/-) or activated T cell product (IFN-gamma-/-) have increased microglia activation, but mice deficient in NADPH oxidase have decreased microglial activation. However, at the time point tested, the deletions are not neuroprotective. The results suggest that inflammatory responses play a role in TD-induced pathological changes in the brain, and the inflammation appears to be a late event that reflects a response to neuronal damage, which may spread the damage to other brain regions (Ke et al., 2006)(Table 12.2).

Diminishing oxidative stress diminishes TD-induced neuron loss. Dietary restriction prolongs life span in rodents. It is thought to do this by enhancing the ability of tissues to handle oxidative stress. Dietary restriction limits TD-induced oxidative stress as assessed by heme-oxygenase-1 and limits the neuronal loss (Calingasan and Gibson, 2000a). At the symptomatic stage, complexin I and complexin II levels (synapse markers) in the medial thalamus decrease by 63% and 45%, respectively, compared to control animals. Cotreatment with the antioxidant N- acetylcysteine prevents both neuronal loss and down regulation of complexins (Hazell and Wang, 2005).

The Cellular Basis of Cell Death in Response to Mild Impairment of Oxidative Metabolism (Fig. 12.2)

Cell Culture

Cells in culture allow much more detailed studies of mechanism than in vivo. Cultured cerebellar granule neurons were treated with amprolium, a potent inhibitor of thiamine transport. Exposure to amprolium causes apoptosis and the generation of reactive

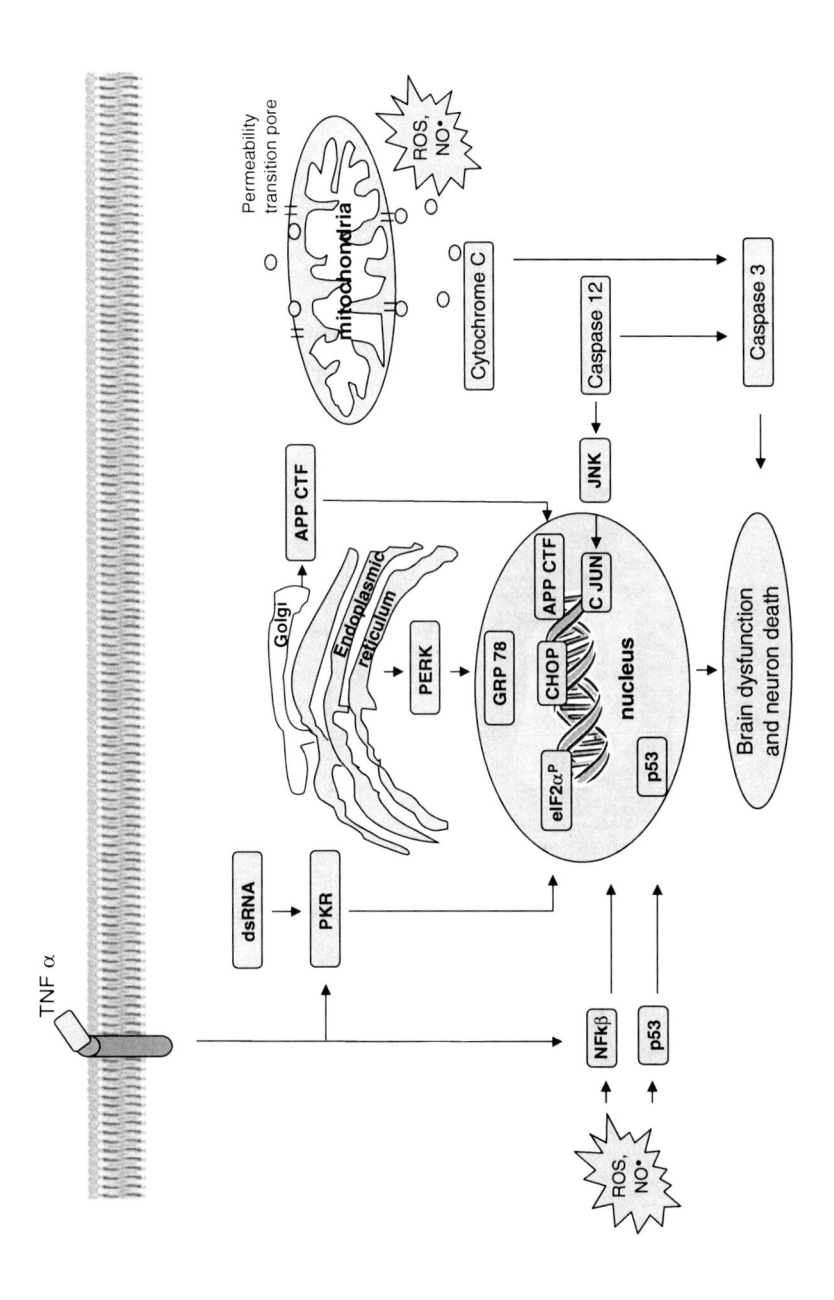

Fig. 12.2 Interactions of protein processing and nuclear translocation in the pathophysiology of TD (*See also* Color Insert)

oxygen species in cerebellar granule neurons. Similar to observations in vivo, TD up-regulates markers for ER stress. Treatment with a selective inhibitor of caspase-12 significantly alleviates amprolium-induced death of the neurons. Thus, ER stress may play a role in TD-induced brain damage (Wang et al., 2007c). In a complementary series of studies, cerebellar granule cells were exposed to thiamine-deficient medium for up to 7 days in the absence or presence of the central thiamine antagonist pyrithiamine. Exposure of cells for 7 days to thiamine-deficient medium alone causes no detectable cell death. On the other hand, pyrithiamine treatment reduces thiamine phosphate esters, decreases activities of the thiamine-dependent enzymes KGDHC and transketolase, causes a two-fold increase in lactate release, a lowering of pH, and significant cell death. DNA fragmentation studies did not reveal evidence of apoptotic cell death. Addition of alpha-tocopherol (vitamin E) or butylated hydroxyanisole to TD cells protects. Furthermore, MK-801, an NMDA receptor antagonist, is not neuro-protective. These results suggest that reactive oxygen species (ROS) play a major role in thiamine deficiency-induced neuronal cell death (Pannunzio et al., 2000).

Specific inhibitors of thiamine-dependent enzymes are useful in determining how reductions in thiamine-dependent enzymes lead to cell death and how these changes may relate to the acute changes induced by the inhibition of KGDHC. As discussed earlier, the acute reduction in KGDHC activity impairs acetylcholine synthesis. Eventually these same processes and/or additional steps lead to neuron death. α-keto-β-methyl-n-valeric acid (KMV), a structural analogue of α-ketoglutarate, was used to inhibit KGDHC activity in order to test effects of reduced KGDHC on mito-chondrial function and cell death cascades in PC12 cells. KMV decreases *in situ* KGDHC activity by 52% (1 hr) or 65% (2 hr). Under the same conditions, KMV does not alter the mitochondrial membrane potential (MMP) nor production of reac-tive oxygen species (ROS). However, KMV increases lactate dehydrogenase (LDH) release from cells by 100%, promotes translocation of mitochondrial cytochrome c to the cytosol, and activates caspase-3. Inhibition of the mitochondrial permeability transition pore (MPTP) by cyclosporin A partially blocks this KMV-induced change in cytochrome c (–40%) and LDH (–15%) release, and prevents cell death. Thus, impairment of this key mitochondrial enzyme in PC12 cells may lead to cytochrome c release and caspase-3 activation by partial opening of the MPTP before the loss of mitochondrial membrane potentials (Huang et al., 2003).

Thiamine alters the genetic response to cell injury to protect cells. TPP inhibits p53 DNA binding, and the thiamine inhibits intracellular p53 activity. Thiamine affects two p53-regulated cellular responses to ionizing radiation: rereplication and apoptosis (McLure et al., 2004). Thiamine also interacts with NFkappaB. Diabetes induces NFKappaB and this induction is diminished by thiamine (Hammes et al., 2003).

Protein Processing

TD alters protein processing and induces endoplasmic reticulum (ER) stress which is a response to unfolded proteins. The microsomal fractions contain 20–24% of the thiamine in the cells (Balaghi and Pearson, 1966). TD up-regulates several markers of ER stress, such as glucose-regulated protein (GRP) 78, growth arrest and DNA-damage inducible protein or C/EBP-homologus protein (GADD153/Chop), and phosphorylation of eIF2alpha and cleavage of caspase-12 (Wang et al., 2007c). Large increases (as much as 5.5 fold) and a subsequent decline precede neuronal death. ER morphology is altered in the thalamus and this occurs before the loss of neurons. The changes include swelling accompanied by enlargement of the lumen. The rough ER displays an irregular packing pattern in the thalamus of TD mice. All these changes precede neuronal death (Wang et al., 2007c).

Translocation of Proteins to the Nucleus

TD induces translocation of proteins to the nucleus that alter transcription. Many of the changes that are described above occur in microglia or endothelial cells before any changes occur in neurons. Recent studies suggest that migration of proteins (transcription factors) to the nucleus may be early markers of TD-induced neuronal death. Migration of a carboxy terminal fragment of amyloid precursor protein from the cytosol to the nucleus occurs at early stages of TD. These fragments could initiate changes in the neurons that then lead to the other inflammatory processes (Karuppagounder, et al., 2008). These possibilities have not been tested. The changes in APP processing probably reflect the altered protein processing described in the previous paragraph.

Nuclear translocation of other factors that alter transcription is also induced by TD. Upon activation, double-stranded RNA-activated protein kinase (PKR) phosphorylates its substrate, the alpha-subunit of eukaryotic initiation factor-2 (eIF2alpha), which inhibits translation. TD in mice induces phosphorylation of PKR and phosphorylation of eIF2alpha in the thalamus. TD also induces nuclear translocation of PKR in primary cultures of cerebellar granule neurons. Both PKR inhibition and dominant-negative PKR mutant protect cerebellar granule neurons against TD-induced cell death. TD promotes the association between RAX and PKR. The antioxidant vitamin E dramatically decreases the RAX/PKR association and ameliorates TD-induced cell death (Wang et al., 2007b).

Other Roles of Thiamine May also Be Critical in TD-Induced Cell Death

Thiamine may also act as antioxidant. It is particularly effective at absorbing NO· species. Chronic treatment with high concentrations of thiamine diminished c-DCF-detectable ROS production by t-BHP. DAF-detectable NO· induced by SIN-1 is neutralized by thiamine at lower concentrations (Gibson and Blass, 2007; Huang et al., 2005).

Several lines of evidence suggest that tranketolase is the rate limiting step in the pentose shunt. A major role of the pentose shunt is to provide NADPH for maintaining the redox state of glutathione. Thus, a depletion of thiamine would result in less NADPH and more oxidized glutathione. This may be particularly important in the endoplasmic reticulum which has a full pentose shunt of its own (Bublitz and Steavenson, 1988) and high glutathione content (Balaghi and Pearson, 1966). Glutathione in the endoplasmic reticulum is involved in protein folding and processing. Thus, abnormalities could explain the production of altered proteins such as those of amyloid precursor protein that occurs in the brain of thiamine deficient mice.

Conclusion

Thiamine deficiency provides a unique model to study the effects of mild impairment of oxidative metabolism on brain function and selective neurodegeneration. The acute changes provide a model in which to study the mechanisms underlying metabolic encephalopathies. In its chronic form it provides a model of selective neurodegeneration in response to mild impairment of oxidative metabolism.

References

Alcaide, M. L., et al., 2003. Wernicke's encephalopathy in AIDS: a preventable cause of fatal neurological deficit. Int J STD AIDS. 14, 712–713.

Alves, L. F., et al., 2006. [Beriberi after bariatric surgery: not an unusual complication. Report of two cases and literature review]. Arq Bras Endocrinol Metabol. 50, 564–568.

Andreassen, O. A., et al., 2001. Dichloroacetate exerts therapeutic effects in transgenic mouse models of Huntington's disease. Ann Neurol. 50, 112–117.

Balaghi, M., Pearson, W. N., 1966. Tissue and intracellular distribution of radioactive thiamine in normal and thiamine-deficient rats. J Nutr. 89, 127–132.

Barclay, L. L., Gibson, G. E., 1982. Spontaneous open-field behavior in thiamin-deficient rats. J Nutr. 112, 1899–1905.

Barclay, L. L., et al., 1981a. Impairment of behavior and acetylcholine metabolism in thiamine deficiency. J Pharmacol Exp Ther. 217, 537–543.

Barclay, L. L., et al., 1981b. The string test: an early behavioral change in thiamine deficiency. Pharmacol Biochem Behav. 14, 153–157.

Barclay, L. L., et al., 1982. Cholinergic therapy of abnormal open-field behavior in thiamin-deficient rats. J Nutr. 112, 1906–1913.

Blass, J. P., Gibson, G. E., 1977. Abnormality of a thiamine-requiring enzyme in patients with Wernicke-Korsakoff syndrome. N Engl J Med. 297, 1367–1370.

Blass, J. P., Gibson, G. E., 1979. Genetic factors in Wernicke-Korsakoff syndrome. Alcohol Clin Exp Res. 3, 126–134.

Blass, J. P., et al., 1975. Clinical and metabolic abnormalities accompanying deficiencies in pyruvate oxidation. F.A. Hommes (ed.) Academic Press, New York, pp. 193

Bubber, P., et al., 2005. Mitochondrial abnormalities in Alzheimer brain: mechanistic implications. Ann Neurol. 57, 695–703.

Bublitz, C., Steavenson, S., 1988. The pentose phosphate pathway in the endoplasmic reticulum. J Biol Chem. 263, 12849–12853.

Butterworth, R. F., et al., 1993. Thiamine-dependent enzyme changes in the brains of alcoholics: relationship to the Wernicke-Korsakoff syndrome. Alcohol Clin Exp Res. 17, 1084–1088.

Calingasan, N. Y., Gibson, G. E., 2000a. Dietary restriction attenuates the neuronal loss, induction of heme oxygenase-1 and blood-brain barrier breakdown induced by impaired oxidative metabolism. Brain Res. 885, 62–69.

Calingasan, N. Y., Gibson, G. E., 2000b. Vascular endothelium is a site of free radical production and inflammation in areas of neuronal loss in thiamine-deficient brain. Ann N Y Acad Sci. 903, 353–356.

Calingasan, N. Y., et al., 1994a. Distribution of the alpha-ketoglutarate dehydrogenase complex in rat brain. J Comp Neurol. 346, 461–479.

Calingasan, N. Y., et al., 1994b. Selective enrichment of cholinergic neurons with the alpha-ketoglutarate dehydrogenase complex in rat brain. Neurosci Lett. 168, 209–212.

Calingasan, N. Y., et al., 1995a. Blood-brain barrier abnormalities in vulnerable brain regions during thiamine deficiency. Exp Neurol. 134, 64–72.

Calingasan, N. Y., et al., 1995b. Accumulation of amyloid precursor protein-like immunoreactivity in rat brain in response to thiamine deficiency. Brain Res. 677, 50–60.

Calingasan, N. Y., et al., 1995c. Heterogeneous expression of transketolase in rat brain. J Neurochem. 64, 1034–1044.

Calingasan, N. Y., et al., 1996. Novel neuritic clusters with accumulations of amyloid precursor protein and amyloid precursor-like protein 2 immunoreactivity in brain regions damaged by thiamine deficiency. Am J Pathol. 149, 1063–1071.

Calingasan, N. Y., et al., 1998. Induction of nitric oxide synthase and microglial responses precede selective cell death induced by chronic impairment of oxidative metabolism. Am J Pathol. 153, 599–610.

Calingasan, N. Y., et al., 1999. Oxidative stress is associated with region-specific neuronal death during thiamine deficiency. J Neuropathol Exp Neurol. 58, 946–958.

Calingasan, N. Y., et al., 2000. Vascular factors are critical in selective neuronal loss in an animal model of impaired oxidative metabolism. J Neuropathol Exp Neurol. 59, 207–217.

Coy, J. F., et al., 2005. Mutations in the transketolase-like gene TKTL1: clinical implications for neurodegenerative diseases, diabetes and cancer. Clin Lab. 51, 257–273.

Desjardins, P., Butterworth, R. F., 2005. Role of mitochondrial dysfunction and oxidative stress in the pathogenesis of selective neuronal loss in Wernicke's encephalopathy. Mol Neurobiol. 31, 17–25.

Falk, R. E., et al., 1976. Ketonic diet in the management of pyruvate dehydrogenase deficiency. Pediatrics. 58, 713–721.

Foldi, M., et al., 2007. Transketolase protein TKTL1 overexpression: A potential biomarker and therapeutic target in breast cancer. Oncol Rep. 17, 841–845.

Freeman, G. B., et al., 1987. Effect of age on behavioral and enzymatic changes during thiamin deficiency. Neurobiol Aging. 8, 429–434.

Gibson, G. E., Blass, J. P., 1976. Impaired synthesis of acetylcholine in brain accompanying mild hypoxia and hypoglycemia. J Neurochem. 27, 37–42.

Gibson, G. E., Blass, J. P., 2007. Thiamine-dependent processes and treatment strategies in neurodegeneration. Antioxid Redox Signal. 9, 1605–1619.

Gibson, G. E., et al., 1975. Decreased synthesis of acetylcholine accompanying impaired oxidation of pyruvic acid in rat brain minces. Biochem J. 148, 17–23.

Gibson, G. E., et al., 1983. Cholinergic drugs and 4-aminopyridine alter hypoxic-induced behavioral deficits. Pharmacol Biochem Behav. 18, 909–916.

Gibson, G. E., et al., 1984. Correlation of enzymatic, metabolic, and behavioral deficits in thiamin deficiency and its reversal. Neurochem Res. 9, 803–814.

Gibson, G. E., et al., 1988. Reduced activities of thiamine-dependent enzymes in the brains and peripheral tissues of patients with Alzheimer's disease. Arch Neurol. 45, 836–840.

Gibson, G., et al., 1989. Regionally selective alterations in enzymatic activities and metabolic fluxes during thiamin deficiency. Neurochem Res. 14, 17–24.

Gibson, G. E., et al., 1999. Oxidative stress and a key metabolic enzyme in Alzheimer brains, cultured cells, and an animal model of chronic oxidative deficits. Ann N Y Acad Sci. 893, 79–94.

Gibson, G. E., et al., 2003. Deficits in a tricarboxylic acid cycle enzyme in brains from patients with Parkinson's disease. Neurochem Int. 43, 129–135.

Hammes, H. P., et al., 2003. Benfotiamine blocks three major pathways of hyperglycemic damage and prevents experimental diabetic retinopathy. Nat Med. 9, 294–299.

Hazell, A. S., Wang, C., 2005. Downregulation of complexin I and complexin II in the medial thalamus is blocked by N-acetylcysteine in experimental Wernicke's encephalopathy. J Neurosci Res. 79, 200–207.

Hazell, A. S., et al., 1993. Cerebral vulnerability is associated with selective increase in extracellular glutamate concentration in experimental thiamine deficiency. J Neurochem. 61, 1155–1158.

Hazell, A. S., et al., 2003. Thiamine deficiency results in downregulation of the GLAST glutamate transporter in cultured astrocytes. Glia. 43, 175–184.

Heroux, M., et al., 1996. Alterations of thiamine phosphorylation and of thiamine-dependent enzymes in Alzheimer's disease. Metab Brain Dis. 11, 81–88.

Huang, H. M., et al., 2003. Inhibition of alpha-ketoglutarate dehydrogenase complex promotes cytochrome c release from mitochondria, caspase-3 activation, and necrotic cell death. J Neurosci Res. 74, 309–317.

Huang, H. M., et al., 2005. Selective antioxidants differentially modify endoplasmic reticulum Ca2+ stores and capacitative calcium entry. Soc Neurosci. 35, 93–100.

Jeitner, T. M., et al., 2005. Inhibition of the alpha-ketoglutarate dehydrogenase complex by the myeloperoxidase products, hypochlorous acid and mono-N-chloramine. J Neurochem. 92, 302–310.

Jimenez-Jimenez, F. J., et al., 1999. Cerebrospinal fluid levels of thiamine in patients with Parkinson's disease. Neurosci Lett. 271, 33–36.

Karachalias, N., et al., 2005. High-dose thiamine therapy counters dyslipidemia and advanced glycation of plasma protein in streptozotocin-induced diabetic rats. Ann N Y Acad Sci. 1043, 777–783.

Kark, R. A., et al., 1975. Experimental thiamine deficiency. Neuropathic and mitochondrial changes induced in rat muscle. Arch Neurol. 32, 818–825.

Karuppagounder, S. S., et al., 2006. Mild impairment of oxidative metabolism exacerbates pathology in a transgenic model of plaque formation. Soc Neurosci. 36, 170–715.

Karuppagounder, S. S., et al., 2007. Changes in inflammatory processes associated with selective vulnerability following mild impairment of oxidative metabolism. Neurobiol Dis. 26, 353–362.

Karuppagounder, S. S., et al., 2008. Translocation of Amyloid Precursor Protein C-terminal Fragment(s) to the Nucleus Precedes Neuronal Death due to Thiamine Deficiency-induced Mild Impairment of Oxidative Metabolism. Neurochemical research, 33, 1365–1372.

Ke, Z. J., Gibson, G. E., 2004. Selective response of various brain cell types during neurodegeneration induced by mild impairment of oxidative metabolism. Neurochem Int. 45, 361–369.

Ke, Z. J., et al., 2003. Reversal of thiamine deficiency-induced neurodegeneration. J Neuropathol Exp Neurol. 62, 195–207.

Ke, Z. J., et al., 2005a. CD40-CD40L interactions promote neuronal death in a model of neurodegeneration due to mild impairment of oxidative metabolism. Neurochem Int. 47, 204–215.

Ke, Z. J., et al., 2005b. CD40L deletion delays neuronal death in a model of neurodegeneration due to mild impairment of oxidative metabolism. J Neuroimmunol. 164, 85–92.

Ke, Z. J., et al., 2006. Peripheral inflammatory mechanisms modulate microglial activation in response to mild impairment of oxidative metabolism. Neurochem Int. 49, 548–556.

Klivenyi, P., et al., 2004. Mice deficient in dihydrolipoamide dehydrogenase show increased vulnerability to MPTP, malonate and 3-nitropropionic acid neurotoxicity. J Neurochem. 88, 1352–1360.

Langlais, P. J., Mair, R. G., 1990. Protective effects of the glutamate antagonist MK-801 on pyrithiamine-induced lesions and amino acid changes in rat brain. J Neurosci. 10, 1664–1674.

Langlais, P. J., Savage, L. M., 1995. Thiamine deficiency in rats produces cognitive and memory deficits on spatial tasks that correlate with tissue loss in diencephalon, cortex and white matter. Behav Brain Res. 68, 75–89.

Lee, B. Y., et al., 2005. Thiamin deficiency: a possible major cause of some tumors? (review). Oncol Rep. 14, 1589–1592.

Martin, E., et al., 2005. Pyruvate dehydrogenase complex: metabolic link to ischemic brain injury and target of oxidative stress. J Neurosci Res. 79, 240–247.

Mastrogiacomo, F., Kish, S. J., 1994. Cerebellar alpha-ketoglutarate dehydrogenase activity is reduced in spinocerebellar ataxia type 1. Ann Neurol. 35, 624–626.

McLure, K. G., et al., 2004. NAD+ modulates p53 DNA binding specificity and function. Mol Cell Biol. 24, 9958–9967.

McRee, R. C., et al., 2000. Increased histamine release and granulocytes within the thalamus of a rat model of Wernicke's encephalopathy. Brain Res. 858, 227–236.

Meng, J. S., Okeda, R., 2003. Neuropathological study of the role of mast cells and histamine-positive neurons in selective vulnerability of the thalamus and inferior colliculus in thiamine-deficient encephalopathy. Neuropathology. 23, 25–35.

Nakagawasai, O., et al., 2000a. Immunohistochemical estimation of brain choline acetyltransferase and somatostatin related to the impairment of avoidance learning induced by thiamine deficiency. Brain Res Bull. 52, 189–196.

Nakagawasai, O., et al., 2000b. Immunohistochemical estimation of rat brain somatostatin on avoidance learning impairment induced by thiamine deficiency. Brain Res Bull. 51, 47–55.

Neufeld, E. J., et al., 2001. Thiamine-responsive megaloblastic anemia syndrome: a disorder of high-affinity thiamine transport. Blood Cells Mol Dis. 27, 135–138.

Pannunzio, P., et al., 2000. Thiamine deficiency results in metabolic acidosis and energy failure in cerebellar granule cells: an in vitro model for the study of cell death mechanisms in Wernicke's encephalopathy. J Neurosci Res. 62, 286–292.

Park, L. C., et al., 2000. Metabolic impairment elicits brain cell type-selective changes in oxidative stress and cell death in culture. J Neurochem. 74, 114–124.

Park, L. C., et al., 2001a. Mitochondrial impairment in the cerebellum of the patients with progressive supranuclear palsy. J Neurosci Res. 66, 1028–1034.

Park, L. C., et al., 2001b. Co-culture with astrocytes or microglia protects metabolically impaired neurons. Mech Ageing Dev. 123, 21–27.

Pawlik, F., et al., 1977. Peripheral nerve changes in thiamine deficiency and starvation. Acta Neuropathologica. 39, 211–218.

Roland, J. J., Savage, L. M., 2007. Blunted hippocampal, but not striatal, acetylcholine efflux parallels learning impairment in diencephalic-lesioned rats. Neurobiol Learn Mem. 87, 123–132.

Savage, L. M., et al., 2007. Selective septohippocampal — but not forebrain amygdalar — cholinergic dysfunction in diencephalic amnesia. Brain Res. 1139, 210–219.

Sheline, C. T., Wei, L., 2006. Free radical-mediated neurotoxicity may be caused by inhibition of mitochondrial dehydrogenases in vitro and in vivo. Neuroscience. 140, 235–246.

Sheline, C. T., et al., 2002. Cofactors of mitochondrial enzymes attenuate copper-induced death in vitro and in vivo. Ann Neurol. 52, 195–204.

Sheu, K. F., et al., 1996. Regional reductions of transketolase in thiamine-deficient rat brain. J Neurochem. 67, 684–691.

Shi, Q., et al., 2007. Responses of the mitochondrial alpha-ketoglutarate dehydrogenase complex to thiamine deficiency may contribute to regional selective vulnerability. Neurochem Int. 50, 921–931.

Shin, B. H., et al., 2004. Thiamine attenuates hypoxia-induced cell death in cultured neonatal rat cardiomyocytes. Mol Cells. 18, 133–140.

Shneider, A. B., 1991. [Anti-ischemic heart protection using thiamine and nicotinamide]. Patol Fiziol Eksp Ter. 9–10.

Starkov, A. A., et al., 2004. Mitochondrial alpha-ketoglutarate dehydrogenase complex generates reactive oxygen species. J Neurosci. 24, 7779–2788.

Todd, K. G., Butterworth, R. F., 1999. Early microglial response in experimental thiamine deficiency: an immunohistochemical analysis. Glia. 25, 190–198.

Ueda, K., et al., 2006. Severe thiamine deficiency resulted in Wernicke's encephalopathy in a chronic dialysis patient. Clin Exp Nephrol. 10, 290–293.

Victor, M., et al., 1971. The Wernicke-Korsakoff syndrome. A clinical and pathological study of 245 patients, 82 with post-mortem examinations. Contemp Neurol Ser. 7, 1–206.

Wang, C., et al., 2007a. Effect of ascorbic acid and thiamine supplementation at different concentrations on lead toxicity in liver. Ann Occup Hyg. 51, 563–569.

Wang, X., et al., 2007b. Activation of double-stranded RNA-activated protein kinase by mild impairment of oxidative metabolism in neurons. J Neurochem. 103, 2380–2390.

Wang, X., et al., 2007c. Thiamine deficiency induces endoplasmic reticulum stress in neurons. Neuroscience. 144, 1045–1056.

Yates, C. M., et al., 1990. Enzyme activities in relation to pH and lactate in postmortem brain in Alzheimer-type and other dementias. J Neurochem. 55, 1624–1630.

Zhao, N., et al., 2007. Impaired hippocampal neurogenesis is involved in cognitive dysfunction induced by thiamine deficiency at early pre-pathological lesion stage. Neurobiol Dis. 29, 176–185.

Chapter 13
Alcohol, Neuron Apoptosis, and Oxidative Stress

George I. Henderson, Jennifer Stewart, and Steven Schenker

Introduction

The aim of this chapter is to present a contemporary overview of ethanol-induced apoptosis in the brain, with a focus on the potential role of oxidative stress and some new concepts related to glia-mediated neuroprotection and selective vulnerability of neurons to ethanol. While ethanol-related oxidative stress and neuron apoptotic death have been documented in the adult brain, the vast majority of reports have centered on the developing organ (Schenker *et al.*, 1990). We address both settings and offer several potential explanations for the high sensitivity of the fetal brain to these toxic responses to ethanol. Of note is that neurotoxic responses to ethanol have been recognized for several decades yet the mechanisms underlying these often devastating effects remain controversial. The following material abundantly illustrates that the setting is multifactorial with multiple ethanol-related perturbations at play, likely with each impacting to different degrees on various brain areas as well as on different neuron and glia populations. Finally, neuron survival and functions are intimately connected to the glia with which neurons are commingled. Such interactions may often be essential to neuron survival and we include a brief overview of recent studies addressing ethanol effects on neuroprotective glia/neuron interactions.

Ethanol Damage and Neuron Loss in the "Developing" Brain

Ethanol Effects on the Developing Brain

The fetotoxic effects of ethanol in humans first appeared in modern literature almost 40 years ago (Lemoine *et al.*, 1968; Jones *et al.*, 1973). Subsequently, a myriad of reports have linked maternal consumption of ethanol to alterations in fetal brain development, both in the human setting and in a variety of animal models (Schenker *et al.*, 1990; Streissguth *et al.*, 1990). The former is often reflected in a variety of behavioral and performance deficits (Streissguth *et al.*, 1990; Wass *et al.*, 2002; Wozniak *et al.*, 2004) which range widely in severity. The most severe presentation

D.W. McCandless (ed.) *Metabolic Encephalopathy*,
doi: 10.1007/978-0-387-79112-8_13, © Springer Science+Business Media, LLC 2009

of the fetotoxicity of ethanol was labeled the Fetal Alcohol Syndrome, which is typically diagnosed (in humans) on the basis of growth retardation, central nervous system disorders, and characteristic craniofacial anomalies (Lemoine *et al.*, 1968; Jones *et al.*, 1973). However, most ethanol exposed children do not express these three criteria and, inexplicably, some present no adverse effects. One factor however, is clear: utero-ethanol exposure can adversely affect the developing brain, probably by multiple mechanisms. Among the untoward effects of ethanol is a loss of neurons from specific brain areas, a phenomenon that can be connected to impaired neuronal migration as well as to neuron death (Miller 1992, 1995a).

Ethanol Elicits Neuron Death

There is unambiguous evidence that approximately 50% of all neurons formed in the developing brain will die via apoptosis and, in the absence of this, brain development can be grossly abnormal (Lossi and Merighi, 2003; Martin, 2001; Sastry *et al.*, 2000; Kuan *et al.*, 2000). Thus, with ethanol exposure, we have a setting where a xenobiotic is enhancing the death of neurons which may be predisposed to apoptotic death. Baseline measures of the degree of normal neuron death vary with the stages of development, specific brain structures and their components, and are likely to vary by neuron type also (Miller, 1995b). Similarly, while ethanol exposure elicits neuron death throughout the developing brain, distinct structures are clearly affected differently and an estimate of enhanced neuron apoptotic death illustrates that most neurons at a given time are not dying. Thus, we are dealing with a complex setting of "selective vulnerability" of neuron populations to ethanol within various brain areas that likewise express differing sensitivities to ethanol. Clearly, an understanding of the basic mechanisms underlying ethanol effects on neurons in the developing brain and related components e.g. glia, is central to clarifying these two linked phenomena. We present below, information supporting a role for oxidative stress in ethanol-mediated apoptotic death of neurons and suggest a mechanism underlying the varying sensitivities of neurons and brain structures to this effect.

Oxidative Stress

Oxidative Stress in Biological Systems

A "sudden" appearance of molecular oxygen in the earth's atmosphere occurred approximately 2.3 billion years ago (Falkowsky, 2006) and generated vast challenges to ensuing oxygen reliant organisms. Among these is the ability to continually defend biological systems from the single electron reduction products of molecular oxygen. These oxygen radicals, essentially oxygen molecules with unpaired electrons, are highly reactive pro-oxidants that are generated in immense quantities in all eukaryotic systems. When the balance between detoxification and production is skewed towards a state where there is an excess expression of reactive oxygen species, a surfeit of oxidative damage to cellular components occurs and this is termed Oxidative Stress.

Central among the toxic responses to oxidative stress is the induction of apoptotic death (Curtin *et al.*, 2002; Fleury *et al.*, 2002; Polster and Fkskum, 2004; Ryter *et al.*, 2007). While it is clear that it can be an initiator as well as a signaling event within the apoptotic process, the specific mechanisms underlying these remain uncertain. Likely, these responses could be related to the damage of cellular components e.g. DNA, lipids, and polysaccharides. One potential pathway by which ethanol-mediated oxidative stress may elicit apoptosis of neurons is associated with the oxidation of polyunsaturated fatty acids within mitochondria (Ramachandran *et al.*, 2001, 2003). Among the variety of oxidation products of these fatty acids are toxic/pro-apoptotic aldehydes, the most potent being 4-hydroxynonenal (Esterbauer *et al.*, 1990; Uchida et al., 1993). This compound readily induces apoptotic death of neurons (Lovell and Markesbery, 2006; Dwivedi et al., 2007) and is produced in neurons secondary to ethanol-related oxidative stress (Ramachandran *et al.*, 2001, 2003)

Ethanol-Related Origins of Oxidative Stress

The subcellular origins of ethanol-related oxidative stress have not been firmly established, but the adult brain contains cytochrome P450s, including the ethanol-inducible Cyp2E1. Inhibition of this enzyme can prevent increased lipid peroxidation in ethanol exposed brain, suggesting that one source of reactive oxygen species (ROS) resides in the smooth endoplasmic reticulum (Montoliu *et al.*, 1995). However, the same studies also illustrated enhanced reactive oxygen species in synaptosomes, far removed from the "microsomal" compartment. Moderate chronic ethanol consumption may enhance superoxide anion radical production in submitochondrial particles isolated from the adult brain (Ribiere *et al.*, 1994), thus the latter finding could reflect a mitochondrial origin as well. There is compelling evidence from many groups that ethanol can elicit oxidative stress at the mitochondrial level (Devi *et al.*, 1993, Dawson *et al.*, 1993, Kukielka *et al.*, 1994, Garcia-Ruiz, 1995) and therefore, it is likely that, in an organ such as the adult brain which possesses cytochrome P450s, there are at least two contributing subcellular sites (rough endoplasmic reticulum and mitochondria) of ethanol—related oxidative stress. In tissues such as the developing brain, which express little to no Cyp2E1, one could expect mitochondrial dysfunction to be a major player, if not the principle source, in ethanol-generated oxidative stress. On a molecular level, the mechanisms by which ethanol elicits oxidative stress remain controversial. The cytochrome P450 cycle is one which entails a sequential two electron reduction of molecular oxygen with measurable leakage of the superoxide anion radical and hydrogen peroxide. Clearly, induction of Cyp2E1 could increase such leakage, but the relevance of this must be evaluated within the context of a mature tissue that is relatively resistant to ethanol-mediated oxidative stress (see below). The majority of molecular oxygen entering tissues, cycles through the mitochondrial respiratory chain and as much as 1 to 2% leaks as reactive oxygen species in a baseline state. Inhibition of respiratory chain components stimulates production of ROS (Dawson *et al.*, 1993) and ethanol does this in fetal cells (complexes I and IV) (Devi *et al.*, 1994). This is associated with lipid peroxidation and decreased ATP synthesis. One mechanism underlying this inhibition of complex IV is mediated by the lipid peroxidation products, HNE and MDA (Chen *et al.*, 1999, 2000). However, some initial event must occur to enhance mitochondrial ROS and the identity of this event remains uncertain. In summary, there is a variety of evidence that, in neuronal cells, ethanol generates oxidative stress and that, at least in the fetus, this may be of mitochondrial origin.

Ethanol and the Adult Brain

The concept of ethanol as an oxidative stress-generating xenobiotic is not a new one, although until relatively recently, this function has not been connected to brain pathology. Over four decades ago, Di Luzio reported signs of ethanol-related oxidative stress in liver (Di Luzio, 1963) and this has been followed by numerous reports illustrating pro-oxidant responses to ethanol in this organ (Shaw *et al.*, 1983; Bailey and Cunningham, 2002; Tsukamoto, 1993; Cederbaum, 1989). In the adult brain, the earliest reports connecting oxidative damage in various brain areas of the rat to acute and chronic ethanol consumption appeared about twenty years ago (Rouach, 1987; Nordmann *et al.*, 1990; Renis *et al.*, 1996; Boveris *et al.*, 1997). This ethanol-related oxidative stress was associated with decreased brain glutathione, lipid peroxidation, DNA strand breaks, and with alterations in mitochondrial respiratory complexes I and IV. The DNA damage was observed with chronic but not acute ethanol exposure in the hippocampus and cerebellum (Renis *et al.*, 1996). Subsequent studies have shown differential effects of chronic ethanol on reduced glutathione status by brain area with the highest changes in hippocampus and cerebellum, followed by cerebral cortex and striatum (Calabrese *et al.*, 2002). The altered glutathione status was correlated with increases in 4-hydroxynonenal (HNE, a highly reactive product of lipid oxidation) and decreased glutathione reductase. These findings are among the first, illustrating regional vulnerabilities to ethanol that could be connected to an underlying mechanism of action e.g. variable alterations in glutathione status (see below).

Ethanol and the Developing Brain

The first reports documenting ethanol-induced oxidative stress in a fetal model appeared about 20 years ago. They linked maternal ethanol intake to evidence of lipid peroxidation products in fetal rat liver (Dreosti, 1987; Dreosti and Partick,

1987). This ethanol response likewise occurs in cultured fetal rat hepatocytes along with replicative blocks, impaired mitochondrial respiration, and enhanced mitochondrial content of 4-hydroxynonenal (Devi *et al.*, 1993, 1994, 1996). A subsequent report demonstrated that chronic maternal intake of ethanol can reduce fetal brain glutathione content (Reyes *et al.*, 1993), which could be either a cause of or a response to toxin-induced oxidative stress. Acute maternal consumption of ethanol in a binge drinking pattern during days 17 and 18 of gestation in a rat model, increased malondialdehyde (MDA) and conjugated dienes in fetal brain while also decreasing glutathione content (Henderson *et al.*, 1995). Importantly, this binge pattern elicited oxidative damage to fetal brain mitochondria, HNE elevation (but not in the maternal brain) and concomitant mitochondrial damage connected to apoptosis, permeability transition and cytochrome c and apoptosis inducing factor (AIF) release (Ramachandran *et al.*, 2001). Treatment of fetal brain mitochondria with HNE mimicked these effects of ethanol. While this correlation is compelling, a causal link between the ethanol enhancement of mito-chondrial HNE and these three pro-apoptotic responses remains to be established. An ominous finding is that prenatal ethanol exposure may elicit oxidative stress in the hypothalamus that persists, in the absence of further ethanol exposure, into adulthood (Dembele *et al.*, 2006).

Ethanol-related oxidative stress responses have also been reported in postnatal rats. The preceding studies illustrated that ethanol exposure during in utero devel-opment in the rat (first and second trimester equivalents in the human) can elicit oxidative stress in the fetus. The approximate human third trimester equivalent of brain development occurs postnatally in rodent models and ethanol can likewise elicit oxidative stress in the postnatal rat brain, albeit with windows of vulnerability. Rats acutely exposed to ethanol on postnatal day 7 expressed increased ROS in cerebral cortices, a developmental point that correlated with peak ethanol-medi-ated cell death (Heaton *et al.*, 2003). However, by postnatal day 21 the "older CNS" was resistant to these ethanol-related responses, especially those directly connected to apoptotic pathways. In the cerebellum there is more definitive evidence supporting a developmentally-related sensitivity to ethanol-related oxidative stress. Ethanol exposure on days P4, P7, P14 can increase ROS in cere-bella only on P4, a stage at which ethanol exposure generates a loss of Purkinje and granule cells (Heaton *et al.*, 2002; Light *et al.*, 2002). In short there is a grow-ing resistance to ethanol-mediated oxidative stress in the brain with development. One compelling explanation for this would be that fetal tissues, including the brain, have less active antioxidant defense systems than the adult and these increase with development (Del Maestro *et al.*, 1989; Mariucci *et al.*, 1990; Munim *et al.*, 1992; Henderson *et al.*, 1999). This holds for most of the antioxidant enzyme systems e.g., CuZn superoxide dismutase, glutathione peroxidase, glutathione S-transferase as well as for nonenzymatic antioxidant systems. Glutathione peroxidase and glutathione S-transferase activities in day 19 of the fetal brain are only 41% and 11%of adult values while glutathione and vitamin E levels are 51% and 20% of adult values, respectively (Henderson *et al.*, 1999). Clearly, such low defenses could predispose the immature brain to oxidative stress.

The setting is however, a complex one in which not only are baseline activities/ levels of these antioxidants relevant, but also the ability of the cell to up-regulate these factors in response to oxidant challenge e.g., via the antioxidant response elements, is a key factor (Rushmore et al., 1991). The latter regulation is impacted by a plethora of factors such as neurotrophic factors whose expression are age-dependent (Maisonpierre et al., 1990). A new concept that could explain both the developmental differences in vulnerability to ethanol-mediated oxidative stress as well as variations in brain area sensitivity is the neuroprotective contribution of astrocytes (Watts et al., 2005; Rathinam et al., 2006). This will be addressed in the following sections.

Apoptotic Death of Neurons

Our current understanding of the mechanisms underlying apoptosis has evolved from the coinage of its name in the early 1970's (Kerr et al., 1972) and subsequent studies using C. elegans (Horvitz et al., 1983; Ellis and Horvitz 1986). In the last ten years, elucidation of fundamental and causal mechanisms underlying apoptotic machinery has progressed exponentially, much of this following the seminal reports connecting mitochondrial release of cytochrome c to ultimate activation of effector caspases and related cellular changes associated with apoptotic death (Liu et al., 1996).

Mitochondria and Apoptosis

We are now at a stage where apoptotic responses to stressors, such as ethanol, are typically defined by anomalies in the presence or localization of these cellular events as well as end point measures typical of apoptotic death e.g., increased surface expression of phosphatidyl serine, chromosomal condensation/damage. Those most relevant to the ethanol-related research will be briefly presented as follows. A central control point for apoptotic death is mitochondrial release of a variety of pro-apoptotic compounds into the cytoplasm (Spierlings et al., 2005; Delivani and Martin, 2006; Garrido et al., 2007). The latter are proteins that are normally sequestered in the mitochondrial intermembrane space (between inner and outer membranes). These include cytochrome c, Smac/Diablo, AIF, and endo G. Their release into the cyto-plasm is elicited by permeabilization of the outer mitochondrial membrane in response to numerous stressors. Generally, these pro-apoptotic proteins can be divided into two categories, those which ultimately activate "effector" caspases (see below) and those which contribute to apoptotic death by other means e.g. DNA/ chromosomal fragmentation. The best studied is cytochrome c, the release of which ultimately elicits caspase-3 activation via an elegant, complex, multi-component, and highly regulated system.

Caspases

Caspases (**C**ysteine-dependent **ASP**artate specific prote**ASES**) are an evolutionarily conserved family of aspartate cysteine-dependent proteases which are central players in both external, membrane-mediated receptor activated apoptosis (extrinsic) and internal (intrinsic) apoptosis that is activated by many stressors within the cell (Kumar, 2007). This protease activation entails a sequence of events in which "initiator" caspases (caspase-10, caspase-9, caspase-8, and caspase-2) which exist as inactive zymogens, are proteolytically cleaved to active enzymes (Kumar, 2007). The proteolytic activation of the initiator complexes is a highly complex process relying on the formation of an adaptor protein complex. Activation of caspase-2, caspase-8, and caspase-9 rely on the formation of the PIDDosome, the "death-inducing signaling complex", and the apoptosome, respectively (Bao and Shi, 2007). The best characterized of these processes is the apoptosome-dependent activation of caspase-9. The consequence of this remarkable process is that apoptosome-bound activated caspase-9 then cleaves and activates the downstream effector caspase, caspase-3, which then performs proteolytic cleavages associated with apoptotic death (Kumar, 2007; Timmer and Salvesen, 2007). In the human, there appear to be as many as several hundred caspase-3 substrates and while many of these substrates may be of little physiological significance, many are clearly pathologically relevant (Timmer and Salvesen, 2007). Importantly, elimination of caspase-9 which is either embryo lethal or postnatally lethal, depending on the mouse strain, generates brain hyperplasia due to decreased apoptosis (Hakem *et al.*, 1998; Kuida *et al.*, 1998), responses comparable to those observed in caspase-3 knockout mice (Kumar, 2007). These untoward responses to decreased caspase-3 or caspase-9 and abnormally low cell death are in the developing brain in which neurons must be culled to enable normal development. They illustrate the sensitivity of the developing brain to properly regulated caspase activity and provide compelling justification for the use of caspase activity as a marker for stressor/ethanol-induced alterations of this enzyme in the developing brain. There are now numerous reports of ethanol-related activation of caspase-3 in the developing brain and in cultured neurons from the immature brain (see below).

Bcl-2 Family Proteins

A second group of apoptosis "markers" that connect to activation of neuron apoptosis are cellular events that mediate mitochondrial release of the pro-apoptotic proteins. Permeabilization of the outer mitochondrial membrane (MOMP) is central to the apoptosis process and it is likely that we are close to an understanding of the specific mechanisms involved. Basically, a plethora of pro-apoptotic stimuli set into play increased expression, protein-protein interactions, and subcellular relocalizations of members of Bcl-2 family of proteins (Gross *et al.*, 1999; Ryter *et al.*, 2007). The "founder" of this group of proteins is the Bcl-2 oncogene, with the group containing both pro- and anti-apoptotic proteins (Gross *et al.*, 1999). The pro-apoptotic components (Bax, Bak etc) and anti-apoptotic proteins (Bcl-2, Bcl-xL, etc) contain as many as four Bcl-2 homology domains (BH1-4) and these proteins have the ability to form homo- and heterodimers. A third grouping of Bcl-2 proteins is those possessing only BH3 domains (Bad, Bim, Bid etc). The interplays of these proteins dictate MOMP which is largely dependent on Bax and/or Bak. How Bax is activated to generate the requisite large mitochondrial membrane channels is unclear, but this process requires translocation from cytosol to mitochondria and over expression of Bcl-2 can block this relocation and subsequent apoptosis. This is only one example of the many interplays between these proteins, the bottom line being that apoptosis is finely controlled by a balance of activities and expression of the anti- and pro-apoptotic Bcl family proteins. Perturbations of these settings by stressors such as ethanol can reflect important pro-apoptotic responses to the drug.

Ethanol Induction of Apoptotic Death of Neurons in the Developing Brain

Bcl-2 Family Proteins in Ethanol-Mediated Apoptosis

There is an extensive and steadily increasing literature documenting the pro-apoptotic effects of ethanol, much of it in the central nervous system (summarized in Fig. 13.1). Most often, these responses are defined as enhanced expression of cellular components of neuron apoptotic machinery outlined above and it is now clear that ethanol can activate components of both extrinsic and intrinsic pathways in neurons (de la Monte and Wands, 2002; Ramachandran *et al.*, 2001; Mooney and Miller, 2001; Climent *et al.*, 2002). In the in vivo setting, ethanol impacts on the expression of Bcl-2 family proteins, the majority of studies focusing on the cerebral cortex and cerebellum. In the cerebral cortex exposed prenatally to ethanol, Bcl-2 (anti apoptotic) expression was lowered while BAX (pro apoptotic) remained unchanged. The resulting reduction in the Bcl-2/Bax ratio illustrates a setting in which apoptosis mediated at the mitochondrial level is likely to occur and this persisted into the postnatal period. In the murine cerebellum following early postnatal exposure, ethanol can decrease Bcl-2 expression concomitant with increases in Bax and Bcl-xs (pro apoptotic) (Heaton *et al.*, 2002; Heaton *et al.*, 2003). Temporally, the shift towards this pro-apoptotic state coincides with a period of Purkinje cell vulnerability, in postnatal day 4. At later stages of development when neurons are less vulnerable to ethanol, there is an early up-regulation of the survival promoting proteins, Bcl-2 and Bcl-xl, and a down-regulation of the pro-apoptotic Bcl-xs. Apparently, these ethanol-related protein expressions follow a rapid and earlier reduction in expression of transcripts in both the cerebellum and cerebral cortex (Moore *et al.*, 1999; Inoue *et al.*, 2002). Both Bax and Bcl-xs genes are up-regulated in the cerebellum by ethanol exposure; Bax knockout mice do present key apoptotic

Schematic of Apoptosis Components Affected by Ethanol

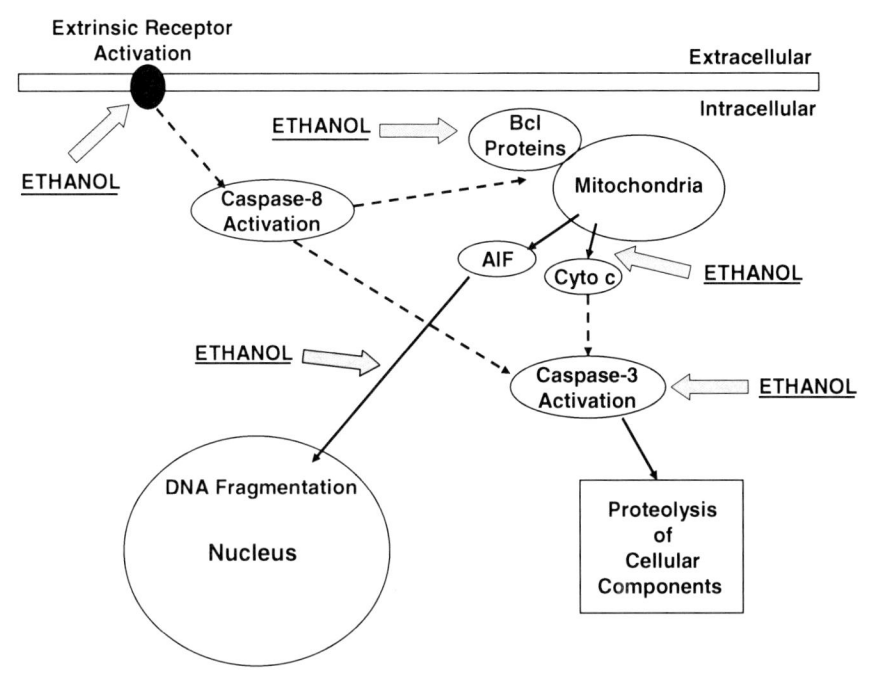

Fig. 13.1 Schematic of Apoptosis Components Affected by Ethanol Five general elements of apoptotic pathways in neurons have been shown to be impacted by ethanol. They are illustrated by the block arrows. There are two basic pathways that have been reported to be activated by ethanol, extrinsic receptor activation and intrinsic mitochondrially mediated pathways. In general, developmental exposure to ethanol shifts expression of Bcl-2 family proteins in a pro apoptotic direction which may be mechanistically connected to the observed enhanced mitochondrial release of cytochrome c (cyto c) and apoptosis inducing factor (AIF). These factors, in turn, likely play roles in the subsequent activation of the effector caspase-3 (cytochrome c) and DNA fragmentation (AIF) (*See also* Color Insert)

responses to ethanol; and over expression of Bcl-2 confers protection against ethanol-related death of cerebellar Purkinje cells (Heaton *et al.*, 1999; Young *et al.*, 2003). These observations suggest central roles for Bcl-2 and Bax in regulation of ethanol-mediated neuron death; however, this is no more than a brief glimpse into specific mechanisms underlying the release of pro-apoptotic proteins from neuron mitochondria following ethanol exposure.

Caspases in Ethanol-Mediated Apoptosis

Given the effects of ethanol on neuron Bcl protein expressions, it is not surprising that the agent elicits mitochondrial release of pro-apoptotic proteins and, ultimately, caspase activation. In vivo, binge-type maternal ethanol consumption enhances fetal brain mitochondrial release of cytochrome c and AIF, and increased caspase-3 activity, these associated with altered mitochondrial permeability transition (Ramachandran et al., 2001). Similarly, postnatal ethanol treatment can yield these responses (Light et al., 2002; Carloni et al., 2004; Young et al., 2003). In primary cultures of fetal rat cortical neurons, ethanol generates enhanced release of Cytochrome C within two hours of exposure followed by activation of caspase-3 within 3 hours of treatment (Ramachandran et al., 2003). Primary focus using cultured neurons has been on activation of caspases-3 and 8 (extrinsic pathway). Evidence is extensive that ethanol can activate caspase-3 in in vitro neuron models (Saito et al., 1999; Jacobs and Miller, 2001; Ramachandran et al., 2003) as well as in vivo models (Mooney and Miller, 2001; Ramachandran et al., 2001; Climent et al., 2002; Light et al., 2002; Olney et al., 2002; Mitchell and Snyder-Keller, 2003; Ragjopal et al., 2003; Young et al., 2003; Carloni et al., 2004). In addition to this executioner caspase, ethanol may activate the initiator caspases 2, 8, and 9 in cerebellar granule neurons although evidence suggests that ethanol-mediated apoptosis can occur in the absence of these initiators (Vaudry et al., 2002). The activation of caspase-8 and increases in Fas receptor (de la Monte and Wands, 2002) suggest that ethanol at least has the capacity to activate extrinsic pathways, however there is scarce compelling evidence that this actually occurs.

Summary

In summary, there is abundant evidence both in vivo and in cultured neurons that ethanol can induce apoptotic death, primarily by intrinsic pathways of the sort that are activated by oxidative stress. While it is tempting to flag this as the key player in this death process, it must be recognized that there may be layers of mechanisms at play. In the in vivo setting, neurons exist in an immensely complex milieu, their survival dependent on the presence and paracrine functions of multiple cell types, notably astrocytes. Ethanol impacts on these other cells as well as on receptor-mediated events on the neuron. Dissection and identification of these pathways is in progress.

Ethanol-Mediated Apoptosis and Antioxidant Interventions

Ethanol and Oxidative Stress-Mediated Apoptosis

While it is clear that multiple factors may underlie ethanol-mediated apoptotic death of neurons in the developing brain, there is abundant evidence that one important causal factor is the production of oxidative stress. While the consistent correlations between the generation of ethanol-mediated oxidative stress and apoptotic death are telling, two other experimental approaches support a causal relationship. The first approach has been to establish time lines related to the onset of oxidative stress and apoptotic events. In cultured fetal rat cortical neurons, ethanol can stimulate ROS production within minutes of treatment and this was followed within an hour by increased HNE, increased caspase-3 activity and ultimately increased measures of DNA fragmentation and Annexin V binding. Time-courses such as this with increased ROS expression occurring initially and far upstream of apoptotic responses, suggest oxidative stress as a causal factor (Ramachandran *et al.*, 2003). A second approach has been to prevent oxidative stress with antioxidants and determine if this prevents the subsequent apoptotic cascade. In the previous studies, normalization of neuron GSH content with pretreatment with N-acetylcysteine prevented the ethanol-mediated oxidative stress and blocked apoptosis (Ramachandran *et al.*, 2003). The preceding were in vitro studies and support the concept of oxidative stress as a mediator of ethanol-related neuronal apoptosis. They also suggest that antioxidant interventions could be effective approaches to mitigating an important untoward ethanol effect. To date most antioxidant interventions aimed at ethanol have utilized vitamin E (α tocopherol) which is a highly effective peroxyl radical trap (Liebler and Burr, 1992; Halliwell and Gutteridge, 1989). It is a late stage defense against the formation of toxic lipid peroxidation products such as HNE and this, combined with its relative low toxicity, make its use an alluring antioxidant intervention strategy. Treatment of cultured neonatal rat cerebellar granule cells with ethanol (87 mM and above) may impair cell viability (MTT assay) and this can be mitigated by treatment with vitamin E or β-carotene (Mitchell *et al.*, 1999). Later studies by this laboratory addressed vitamin E effects on ethanol-mediated apoptosis machinery. Ethanol enhanced activation of caspase-3 and increased at least one marker of apoptosis (Annexin V binding) and both vitamin E and the antioxidant pycnogenol (a bioflavonoid mixture) blocked the apoptotic responses (Siler-Marsiglio *et al.*, 2004). A similar protection by vitamin E was reported on cultured cerebellar granule cells along with increased levels of Bcl-2, Bcl-xl, and activated Akt kinase (Heaton *et al.*, 2002). The mechanism(s) underlying these protective effects of vitamin E are likely to be antioxidant in nature and could reflect prevention of membrane lipid peroxidation. However, exogenous antioxidant treatments can elicit other confounding events and vitamin E can normalize cellular GSH content as well as prevent reduction of neurotrophin secretion associated with ethanol exposure (Heaton *et al.*, 2004).

Neuron Glutathione Homeostasis, a Key Role for Astrocytes

There is persuasive evidence that neuronal glutathione can provide protection against ethanol-related oxidative stress and apoptotic death (Ramachandran *et al.*, 2003; Lamarche *et al.*, 2004; Watts *et al.*, 2005). Recent studies have illustrated a native system by which neuron glutathione homeostasis can be maintained in response to pro-oxidant stressors such as ethanol. Glutathione (L-γ-glutamyl-L-cysteinylglycine) (GSH) is a ubiquitous intracellular thiol that is vital to cell survival and which scavenges reactive oxygen species, serving multiple functions critical to

Schematic of the Neuroprotective Gamma Glutamyl Cycle

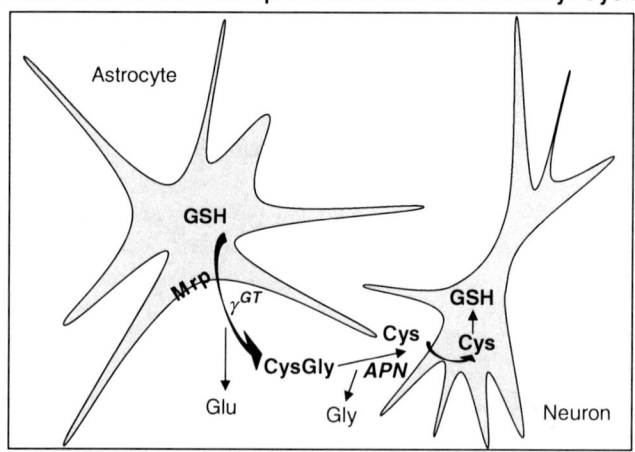

Fig. 13.2 Schematic of the Neuroprotective Gamma Glutamyl Cycle One of the neuroprotective roles of cortical astrocytes is maintenance of neuron glutathione (GSH) homeostasis. Astrocytes contain a highly effective GSH synthesis capacity and readily export the intact tripeptide into the extracellular milieu. The extrusion process is likely mediated by a multidrug resistance associated protein (Mrp). Once GSH is in the external compartment, it is hydrolyzed gamma glutamyl transpeptidase (γGT) to the dipeptide CysGly and glutamate. The dipeptide hydrolysis product is cleaved further by aminopeptidase N (APN) on the surface of the neuron to Cys and Gly. Cysteine is then transported into the neuron where it is a key determinant of neuron GSH synthesis (*See also* Color Insert)

cell survival, including detoxifying electrophiles, maintaining thiol status of proteins, and providing a reservoir for cysteine (Deleve and Kaplowitz, 1990; Meister and Anderson, 1983; Suthanthiran *et al.*, 1990). In the mature brain, the majority of CNS cells are astrocytes which exist in close and intertwined proximity to neurons. Among the many neuroprotective roles of cortical astrocytes is the maintenance of neuron GSH homeostasis (Sagara *et al.*, 1993; Dringen *et al.*, 1997; Wang and Cynader, 2000). This is by the gamma glutamyl pathway which is summarized in Fig. 13.2. Astrocytes are not immune to ethanol effects (Druse *et al.*, 2006), but they are often far more resistant than are neurons. One possible reason for this is that these cells are veritable GSH factories with GSH content as high as 20 mM, typically much more than neurons (Cooper, 1997; Dringen *et al.*, 1999; Schulz *et al.*, 2000). Also, astrocytes are much more numerous than are neurons and establish enumerable contacts with neurons (Bignami, 1991; Rohlmann and Wolff, 1996). The means by which astrocytes augment neuron glutathione is via a key contribution to what is commonly termed the "gamma glutamyl cycle". The vital first step in the process is astrocyte extrusion of GSH into the extra cellular milieu, a process that may be mediated by the multi drug resistance associated protein 1 (Wang and Cynader, 2000; Hirrlinger *et al.*, 2002). It is likely the exported GSH plays an extra cellular role as an antioxidant, but it is also hydrolyzed to the CysGly dipeptide by γ-glutamyl transpeptidase (γGT) (Drukarch *et al.*, 1989; Dringin *et al.*, 1997). The

CysGly dipeptide diffuses to contiguous neurons where it is cleaved to its two constituent amino acids, cysteine and glycine. This is mediated by an aminopeptidase, aminopeptidase N (ApN) on the neuron surface (Dringen *et al.*, 2001; Wang and Chandler, 2000). The newly generated cysteine is then transported into the neuron, probably mostly by the sodium-dependent alanine-serine-cysteine (ASC) system (Sagara *et al.*, 1993). The enhanced cellular cysteine levels are an essential determinant of neuronal GSH synthesis (Wang and Cynader, 2000).

The net effect of the astrocyte-mediated enhancement of neuron GSH synthesis is augmented protection of neurons from reactive oxygen and nitrogen species (Tanaka *et al.*, 1999, Gegg *et al.*, 2003). Since augmentation of neuron GSH can mitigate ethanol-related oxidative stress and subsequent apoptotic death in cultured fetal cortical neurons (Ramachandran *et al.*, 2003), a series of studies addressed astrocyte neuroprotection from ethanol (Watts *et al.*, 2005; Rathinam *et al.*, 2006). Co-culturing fetal rat cortical neurons with neonatal rat cortical astrocytes prevented the rapid declines in neuron GSH content following ethanol exposure (2.5 and 4.0 mg/ml) and provided protection from oxidative stress and apoptotic death associated with ethanol exposure (Watts *et al.*, 2005). In related studies, it was found that the presence of astrocytes completely prevented ethanol-related oxidative stress in neurons and that this protection was dependent on functional γGT as well as neuronal ApN (Rathinam *et al.*, 2006). Interestingly, three components of this GSH homeostasis pathway were up-regulated by ethanol exposure. Treatment of astrocytes with ethanol increased both the expression and activity of γGT as well as efflux of GSH, yet astrocyte GSH content was unaffected by the enhanced efflux. Exposure of neurons to ethanol likewise increased activity and expression of ApN.

These studies illustrate the presence of a highly effective system by which astrocytes in consort with a neuronal ectopeptidase can provide protection for neurons from the pro-oxidant responses to ethanol. Central to this neuroprotection is that this is accomplished by the maintenance of neuronal GSH homeostasis. The rapid up-regulation of the system in response to ethanol, and probably other pro-oxidants, attests to its functionality. Also, the increasing resistance to ethanol-related apoptosis with brain development (presented above) strikingly correlates with the increased brain content of astrocytes. It is tempting to speculate that the high sensitivity of the developing brain at its early stages when it is largely devoid of astrocytes is due to the absence of the astrocyte component of the gamma glutamyl cycle.

Summary and Conclusions

There is an extensive body of evidence that ethanol exposure can elicit the enhanced death of neurons in the developing brain and that this can be connected to at least some of the adverse effects of in utero ethanol on the developing brain. Additionally, there is abundant support that this neuronal death appears to be largely apoptotic. However, necrosis may also be a player (Obernier *et al.*, 2002). Interestingly, ethanol elicits neuron damage to different degrees in varying brain structures. Additionally, there are windows of vulnerability to ethanol, the more immature brain being the most sensitive to this stressor, and at a given time point ethanol only causes the death of a small portion of neurons in a given area. The mechanism(s) underlying the increased apoptotic death of neurons in the developing brain is likely multifactorial, but there is increasing evidence that oxidative stress is an important initiator. Supporting this is the presence of astrocyte-mediated neuroprotection against ethanol, which is low to nonexistent during the developmental periods of highest brain sensitivity to ethanol and it is most effective in the more mature brain which is far less sensitive to ethanol-mediated oxidative stress and apoptotic death.

References

Bailey, S.M. and Cunningham, C.C. (2002). Contribution of mitochondria to oxidative stress associated with alcoholic liver disease. *Free Rad. Biol. Med.* **32**:11–16.

Bao, Q. and Shi, Y. (2007). Apoptosome: a platform for the activation of initiator caspases. *Cell Death Diff.* **14**:56–65.

Bignami, A. (1991). Glial cells in the central nervous system. In: Magistretti, P.J. (ed.), *Discussions in Neuroscience*, vol. 8, Elsevier, Amsterdam, pp. 1–45.

Boveris, A., Llesuy, S., Azzalis, L.A., Giavarotti, L., Simon, K.A., Junqueira, V.B., Porta, E.Z., Vivelda, E.A., and Lissi, E.A. (1997). *In situ* rat brain and liver spontaneous chemiluminescence after acute ethanol intake. *Toxicol. Lett.* **93**:23–28.

Calabrese, V., Scapagnini, G., Latteri, S., Colombrita, C., Ravagna, A., Catalano, C., Pennisi, G., Calvani, M., and Butterfield, D.A. (2002). Long-term ethanol administration enhances age-dependent modulation of redox state in different brain regions in the rat: protection by acetyl carnitine. *Int. J. Tissue React.* **24**:97–104.

Carloni, S., Mazzoni, E., and Balduini, W. (2004). Caspase-3 and calpain activities after acute and repeated ethanol administration during the rat brain growth spurt. *J. Neurochem.* **89**:197–203.

Cederbaum, A.I. (1989). Introduction: role of lipid peroxidation and oxidative stress in alcohol toxicity. *Free Rad. Biol. Med.* **7**:537–539.

Chen, J., Peterson, D., Schenker, S., and Henderson, G.I. (2000). Formation of malondialdehyde adducts in livers of rats exposed to ethanol: Role in ethanol-mediated inhibition of cytochrome c oxidase. *Alcoholism: Clin. Exp. Res.* **24**(4):544–552.

Chen, J.J., Robinson, N.C., Schenker, S., Frosto, T.A., and Henderson, G.I. (1999). Formation of 4-hydroxynonenal adducts with cytochrome c oxidase in rats following short-term ethanol intake. *Hepatology* **26**:1792–1798.

Climent, E., Pascual, M., Renau-Piqueras, J., and Guerri, C. (2002). Ethanol exposure enhances cell death in the developing cerebral cortex: role of brain-derived neurotrophic factor and its signaling pathways. *J. Neurosci. Res.* **68**:213–225.

Cooper, A.J.L. (1997). Glutathione in the brain: disorders of glutathione metabolism. In: (Rosenberg, R.N., Prusiner, S.B., DiMauro, S., Barchi, R.L. and Kunk, L.M., eds). *The Molecular and Genetic Basis of Neurological Disease*, Butterworth-Heinemann, Boston, pp. 1195–1230.

Curtin, J.F., Donovan, M., and Cotter, T.G. (2002). Regulation and measurement of oxidative stress in apoptosis. *J. Immunol. Metab.* **265**:49–72.

Dawson, T.L., Gores, G.J., Nieminen, A.L., Herman, B., and Lemasters, J.J. (1993). Mitochondria as a source of reactive oxygen species during reductive stress in rat hepatocytes. *Am. J. Physiol.* **264**:C961–C967.

de la Monte, S.M. and Wands, J.R. (2001). Mitochondrial DNA damage and impaired mitochondrial function contribute to apoptosis of insulin-stimulated ethanol-exposed neuronal cells. *Alcohoism: Clin. Exp. Res.* **25**:898–906.

de la Monte, S.M. and Wands, J.R. (2002). Chronic gestational exposure to ethanol impairs insulin-stimulated survival and mitochondrial function in cerebellar neurons. *Cell Mol. Life Sci.* **59**:882–893.

DeLeve, L. and Kaplowitz, N. (1990). Importance and regulation of hepatic GSH. *Semin. Liver Dis.* **10**:251–266.

Delivani, P. and Martin, S.J. (2006). Mitochondrial membrane remodeling in apoptosis: and inside story. *Cell Death Diff.* **13**:2007–2010.Del Maestro, R. and McDonald, W. (1989). Subcellular localization of superoxide dismutases, glutathione peroxidase and catalase in developing rat cerebral cortex. *Mech. Age Dev.* **48**:15–21.

Del Maestro, R. and McDonald, W. (1989). Subcellular localization of superoxide dismutases, glutathione peroxidase and catalase in developing rat cerebral cortex. *Mech. Age Dev.* **48**:15–21.

Dembele, K., Yao, X.H., Chen, L., and Nyomba, B.L.G. (2006). Intrauterine ethanol exposure results in hypothalamic oxidative stress and neuroendocrine alterations in adult rat offspring. *Am. J. Physiol. Regul. Integr. Comp. Physiol.* **291**:R796–R802.

Devi, B.G., Henderson, G.I., Frosto, T.A., and Schenker, S. (1993). Effect of ethanol on rat fetal hepatocytes: studies on cell replication, lipid peroxidation and glutathione. *Hepatology* **18**:648–659.

Devi, B.G., Henderson, G.I., Frosto, T.A., and Schenker, S. (1994). Effects of acute ethanol exposure on cultured fetal rat hepatocytes: Relation to mitochondrial functions. *Alcoholism: Clin. Exp. Res.* **18**:1436–1442.

Devi, B.G., Schenker, S., Mazloum, B., and Henderson, G.I. (1996). Ethanol-induced oxidative stress and enzymatic defenses in cultured fetal rat hepatocytes. *Alcohol.* **13**:1–6.

Di Luzio, N.R. (1963). Prevention of acute ethanol-induced fatty liver by antioxidants. *Physiologist.* **6**:169–173.

Dreosti, I.E. (1987). Micronutrients, superoxide and the fetus. *Neuro. Toxicol.* **8**:445–450.

Dreosti, I.E., and Partick, E.J. (1987). Zinc, ethanol, and lipid peroxidation in adult and fetal rats. *Biol. Trace Element Res.* **14**:179–191.

Dringen, R., Gutterer, J.M., Gros, G., and Hirrlinger, J. (2001). Aminopeptidase N mediates the utilization of the GSH precursor CysGly by cultured neurons. *J. NeuroSci. Res.* **66**:1003–1008.

Dringen, R., Kranich, O., and Hamprecht, B. (1997). The γ-glutamyl transpeptidase inhibitor acivicin preserves glutathione released by astroglial cells in culture. *Neurochem. Res.* **22**:727–733.

Dringen, R., Kussmaul, L., Gutterer, J., Hirrlinger, J., and Hamprecht, B. (1999). The glutathione system of peroxide detoxification is less efficient in neurons than in astroglial cells. *J. Neurochem.* **72**:2523–2530.

Drukarch, B., Schepens, E., Stoof, J., Langeveld, C.H., and Van Muiswinkel, F.L. (1989). Astrocyte-enhanced neuronal survival is mediated by scavenging of extracellular reactive oxygen species. *Free Rad. Biol. Med.* **25**:217–220.

Druse, M.J., Tajuddin, N.F., Gillespie, R.A., and Le, P. (2006). The effects of ethanol and the serotonin (1A) agonist ipsapirone on the expression of the serotonin (1A) receptor and several antiapoptotic proteins in fetal rhombencephalic neurons. *Brain Res.* **1092**:79–86.

Dwivedi S. Sharma A. Patrick B. Sharma R. Awasthi YC. (2007) Role of 4-hydroxynonenal and its metabolites in signaling. Redox Report. 12(1):4–10.

Ellis, H.M. and Horvitz, H.R. (1986). Genetic control of programmed cell death in the nematode C. elegans. *Cell* **44**:817–829.

Esterbauer, H., Zollner, H., and Schaur, R.J. (1990). Aldehydes formed by lipid peroxidation: mechanisms of formation, occurrence, and determination. In: (Virgo-Peifrey, C., ed) *Membrane Lipid Oxidation*, CRC Press, Boca Raton, FL, pp 239–268.

Falkowsky, P.G. (2006). Tracing oxygen's imprint on earth metabolic evolution. *Science* **311**:1724–1725.

Fleury, C., Mignotte, B., and Vayssiere, J.L. (2002). Mitochondrial reactive oxygen species in cell death signaling. *Biochemie* **84**:131–141.

Garcia-Ruiz, C., Colell, A., Morales, A., Kaplowitz, N., and Fernandez-Checa, J.C. (1995). Role of oxidative stress generated from mitochondrial electron transport chain and mitochondrial glutathione status in loss of mitochondrial function and activation of transcription factor nuclear factor-nfk B: studies with isolated mitochondria and rat hepatocytes. *Mol. Pharmacol.* **48**:825–834.

Garrido, C., Galluzzi, L., Brunet, M., Puig, P.E., Didelot, C. and Kroemer, G. (2007). Mechanisms of cytochrome c release from mitochondria. *Cell Death Differ.* **13**:1423–1433.

Gegg, M.E., Beltran, B., Salas-Pino, S., Bolanos, J.P., Clark, J.B., Moncada, S. and Heales, S.J.R. (2003). Differential effect of nitric oxide on glutathione metabolism and mitochondrial function in astrocytes and neurons: implications for neuroprotection/ neurodegeneration?. *J. Neurochem.* **186**:228–237.

Gross, A., McDonnell, J.M., and Korsmyer, J. (1999). BCL-2 family members and the mitochondria in apoptosis. *Genes Dev.* **13**:1899–1911.

Hakem, R., Hakem, A., Duncam, G.S., Henderson, J.T., Woo, M., Soengas, M.S., Elia, A., de la Pompa, J. L., Kagi, D., Khoo, W., Potter, J., Yoshida, R., Kaufman, S. A., Lowe, S. W., Penninger, J. M., and Mak, T. W. (1998). Differential requirement for caspase-9 in apoptotic pathways in vivo. *Cell* **94**:339–352.

Halliwell, B. and Gutteridge, J.M.C. (eds). (1989). Lipid peroxidation: a radical chain reaction. *In*: Free Radicals in Biology and Medicine. 2nd edn. Clarendon Press, Oxford, pp. 188–276.

Heaton, M.B., Madoesky, I., Paiva, M., and Siler-Marsiglio, K.I. (2004). Vitamin E amelioration of ethanol neurotoxicity involves modulation of apoptosis-related protein levels in neonatal rat cerebellar granule cells. *Dev. Brain Res.* **150**:117–124.

Heaton, M.B., Moore, D.B., Paiva, M., Gibbs, T., and Bernard, O. (1999). Bcl-2 overexpression protects the neonatal cerebellum from ethanol neurotoxicity. *Brain Res.* **817**:13–18.

Heaton, M.B., Paiva, M., Madosky, I., Mayer, J., and Moore, D.B. (2003). Effects of ethanol on neurotrophic factors, apoptosis-related proteins, endogenous antioxidants, and reactive oxygen species in neonatal striatum: relationship to periods of vulnerability. *Dev. Brain Res.* **140**:237–252.

Heaton, M.B., Paiva, M., Madosky, I., and Shaw, G. (2003). Ethanol effects on neonatal rat cortex: comparative analysis of neurotrophic factors, apoptosis-related proteins, and oxidative processes during vulnerable and resistant periods. *Dev. Brain Res.* **145**:249–262.

Heaton, M.B., Paiva, M., Mayer, J., and Miller, R. (2002). Ethanol-mediated generation of reactive oxygen species in developing rat cerebellum. *Neurosci. Lett.* **334**:83–86.

Henderson, G.I., Chen, J.J., and Schenker, S. (1999). Ethanol, Oxidative stress, reactive aldehydes, and the fetus. *Front. Biosci.* **4**:541–550.

Henderson, G.I., Devi, B.G., Perez, A., and Schenker, S. (1995). In utero ethanol exposure elicits oxidative stress in the rat fetus. *Alcoholism: Clin. Exp. Res.* **19**:714–720.

Hirrlinger, J., Schulz, J.B., and Dringen, R. (2002). Glutathione release from cultured brain cells: multidrug resistance protein 1 mediates the release of GSH from rat astroglial cells. *J. Neurosci. Res.* **69**:318–326.

Horvitz, H.R., Sternberg, P.W., Greenwald, I.S., Fixsen, W., and Ellis, H.M. (1983). Mutations that affect neural cell lineages and cell fates during the development of the nematode Caenorhabditis elegans. *Cold Springs Harbor Symp. Quant. Biol.* **48**:453–463.

Inoue, M., Nakamura, K., Iwahashi, K., Ameno, K., Itoh, M., and Suwaki, H. (2002). Changes of bcl-2 and bax mRNA expressions in the ethanol-treated mouse brain. *Nihon Arukoru Yakubutsu Igakkai Zasshi* **37**:120–129.

Jacobs, J.S. and Miller, M.W. (2001). Proliferation and death of cultured fetal neocortical neurons: effects of ethanol on the dynamics of cell growth. *J. Neurocytol.* **30**:391–401.

Jones, K.L., Smith, D.L., Ulleland, C.W., and Streissguth, A.P. (1973). Pattern of malformations in offspring of chronic alcoholics. *Lancet* **2**:999–1000.

Kerr, J.F., Wyllie, A.H., and Currie, A.R. (1972). Apoptosis: a basic biological phenomenon with —ranging implications in tissue kinetics. *Brit. J. Cancer* **26**:239–257.

Kuan, C.Y., Roth, K.A., Flavell, R.A., and Rakic, P. (2000). Mechanisms of programmed cell death in the developing brain. *Trends Neurosci.* **23**:291–297.

Kuida, K., Haydar, A., Kuan, C.Y., Gu, Y., Taya, C., Karasuyama, H., Su, M.S., Rakic, P., and Flavell, R.A. (1998). Reduced apoptosis and cytochrome c-mediated caspase activation in mice lacking caspase 9. *Cell* **94**:325–337.

Kukielka, E., Dicker, E., and Cederbaum, A. (1994). Increased production of reactive oxygen species by rat liver mitochondria after chronic ethanol treatment. *Arch. Biochem. Biophys.* **309**:377–386.

Kumar, S. (2007). Caspases and their many biological functions. *Cell Death Diff.* **14**:66–72

Lamarche, F., Signorini-Allibe, N., Gonthier, B., and Barret, L. (2004). Influence of vitamin E, sodium selenite, and astrocyte-conditioned medium on neuronal survival after chronic exposure to ethanol. *Alcohol* **33**:127–138.

Lemoine, P., Harrousseau, H., Borteyro, J.P., and Menuer, J.C. (1968). Les enfants de parents alcoholiques; Anomalies observees a propos de 127 cas. *Quest Med.* **21**:467.

Liebler, D.C. and Burr, J.A. (1992). Oxidation of vitamin E during iron-catalyzed lipid peroxidation: Evidence for electron-transfer reactions of the tocoperoxyl radical. *Biochemistry* **31**:8278–8284.

Light, K.E., Belcher, S.M., and Pierce, D.R. (2002). Time course and manner of Purkinje neuron death following a single ethanol exposure on postnatal day 4 in the developing rat. *Neuroscience* **114**:327–337.

Liu, X., Kim, C.N., Yang, J., Jemmerson, R., and Wang, X. (1996). Induction of apoptotic program in cell-free extracts: requirement for dATP and cytochrome c. *Cell* **86**:47–157.

Lossi, L. and Merighi, A. (2003). In vivo cellular and molecular mechanisms of neuronal apoptosis in the mammalian CNS. *Prog. Neurobiol.* **69**:287–312.

Lovell MA., Markesbery WR. (2006) Amyloid beta peptide, 4-hydroxynonenal and apoptosis. Current Alzheimer Research. 3(4):359–64.

Maisonpierre, P.C., Belluscio, L.F., Alderson, R.F., Wiegand, S.J., Furth, M.E., Landsay, R.M., and Yancopolos, G.D. (1990). NT-3, BDNF and NGF in the developing rat nervous system: parallel as well as reciprocal patterns of expression. *Neuron* **5**:501–509.

Mariucci, G., Ambrosini, M.V., Colarieti, L., and Bruschelli, G. (1990). Differential changes in Cu, Zn and Mn superoxide dismutase activity in developing rat brain and liver. *Experientia* **46**:753–755.

Martin, L.J. (2001). Neuronal cell death in nervous system development, disease, and injury. *Intl. J. Mol. Med.* **7**:455–478.

Meister, A. and Anderson, M.E. (1983). Glutathione. *Ann. Rev. Biochem.* **52**:711–760.

Miller MW (1992) Effects of prenatal exposure to ethanol on cell proliferation and neuronal migration. In: Miller MW (ed) *Development of the Central Nervous System: Effects of Alcohol and Opiates*, Wiley-Liss, New York. pp 47–69.

Miller MW (1995a) Effect of pre- or postnatal exposure to ethanol on the total number of neurons in the principal sensory nucleus of the trigeminal nerve: cell proliferation versus neuronal death. Alcohol Clin Exp Res 19:1359–1364.

Miller, M.W. (1995b). Relationship of time of origin and death of neurons in rat somatosensory cortex: barrel versus septal cortex and projection versus local circuit neurons. *J. Comp. Neurol.* **355**:6–14.

Mitchell, E.S. and Snyder-Keller, A. (2003). c-fos and cleaved caspase-3 expression after perinatal exposure to ethanol, cocaine, or the combination of both drugs. *Dev. Brain Res.* **147**:107–117.

Mitchell, J.J., Paiva, M., and Heaton, M.B. (1999). The antioxidants vitamin E and β-carotene protect against ethanol-induced neurotoxicity in embryonic rat hippocampal cultures. *Alcohol* **17**:163–168.

Montoliu C, Sancho-Tello M, Azorin I, Burgal M, Valles S, Renau-Piqueras J, Guerri C. (1995) Ethanol increases cytochrome P4502E1 and induces oxidative stress in astrocytes. J Neurochem 65:2561–2570.

Mooney, S.M. and Miller, M.W. (2001). Effects of prenatal exposure to ethanol on the expression of bcl-2, bax and caspase 3 in the developing rat cerebral cortex and thalamus. *Brain Res.* **911**:71–81.

Moore, D.B., Walker, D.W., and Heaton, M.B. (1999). Neonatal ethanol exposure alters bcl-2 family mRNA levels in the rat cerebellar vermis. *Alcoholism: Clin. Exp. Res.* **23**:1251–1261.

Munim, A., Asayama, K., Dobashi, K., Suzuki, K., Kawaoi, A., and Kato, K. (1992). Immunohistochemical localization of superoxide dismutases in fetal and neonatal rat tissues. *J. Histochem. Cytochem.* **40**:1705–1713.

Nordmann, R., Riviere, C., and Rouach, H. (1990). Ethanol-induced lipid peroxidation and oxidative stress in extrahepatic tissues. *Alcohol Alcohol.* **25**:231–237.

Obernier, J.A., Bouldin, T.W., and Crews, F.T. (2002). Binge ethanol exposure in adult rats causes necrotic cell death. *Alcoholism: Clin. Exp. Res.* **26**:547–557.

Olney, J.W., Tenkova, T., Dikranian, K., Muglia, L.J., Jermakowicz, W.J., D'Sa, C., and Roth, K.A. (2002). Ethanol-induced caspase-3 activation in the in vivo developing mouse brain. *Neurobiol. Dis.* **9**:205–219.

Polster, B.M., and Fiskum, G. (2004). Mitochondrial mechanisms of neural cell apoptosis. *J. Neurochem.* **90**:1281–1289.

Rajgopal, Y., Chetty, C.S., and Vemuri, M.C. (2003). Differential modulation of apoptosis-associated proteins by ethanol in rat cerebral cortex and cerebellum. *Eur. J. Pharmacol.* **470**:117–124.

Ramachandran, V., Perez, A., Chen, J., Senthil, D., Schenker, S., and Henderson, G.I. (2001). *In utero* ethanol exposure causes mitochondrial dysfunction which can result in apoptotic cell death in fetal brain: A potential role for 4-hydroxynonenal. *Alcoholism: Clin. Exp. Res.* **25**:862–871.

Ramachandran, V., Watts, L.T., Maffi, S.K., Chen, J., Schenker, S., and Henderson, G. (2003). Ethanol-induced oxidative stress precedes mitochondrially mediated apoptotic death of cultured fetal cortical neurons. *J. Neurosci. Res.* **74**:577–588.

Rathinam, M.L., Watts, L.T., Stark, A.A., Mahimaimathan, L., Stewart, J., Schenker, S., and Henderson, G.I. (2006). Astrocyte control of fetal cortical neuron glutathione homeostasis: Up-regulation by ethanol. *J. Neurochem.* **96**:1289–1300.

Renis, M., Calabrese, V., Russo, A., Calderone, A., Barcellona, M.L., and Rizza, V. (1996). Nuclear DNA strand breaks during ethanol-induced oxidative stress in rat brain. *F.E.B.S. Lett.* **390**:153–156.

Reyes E, Ott S, Robinson B. (1993) Effects of in utero administration of alcohol on glutathione levels in brain and liver. Alcohol Clin Exp Res 17:877–881.

Ribiere, C., Hininger, I., Saffar-Boccara, C., Sabourault, D., and Nordmann, R. (1994). Mitochondrial respiratory activity and superoxide radical generation in the liver, brain, and heart after chronic ethanol intake. *Biochem. Pharmacol.* **47**:1827–1833.

Rohlmann, A. and Wolff, J.R. (1996). Subcellular topography and plasticity of gap junction distribution on astrocytes. In: Spray, D.C., Dermietzel, R. (eds.), *Gap Junctions in the Nervous System*, Landes, Austin, TX, pp.175–192.

Rouach, H., Park, M.K., Orfanelli, M.T., Janvier, B., and Nordmann, R. (1987). Ethanol-induced oxidative stress in the rat cerebellum. *Alcohol Alcohol.* **1**:207–211.

Rushmore, T.H., Morton, M.R., and Pickett, C.P. (1991). The antioxidant responsive element. Activation by oxidative stress and identification of the DNA consensus sequence required for functional activity. *J. Biol. Chem.* **266**:11632–11639.

Ryter, S.W., Kim, H.P., Hoetzel, A., Park, J.W., Nakahira, K., Wang, X., and Choi, A.M.K. (2007). Mechanisms of cell death in oxidative stress. *Antioxid. Redox Signal.* **9**:49–89.

Sagara J, Miura K, Bannai S. Maintenance of neuronal glutathione by glial cells. (1993) J Neurochem 61:1672–1676.

Saito, M., Saito, M., Berg, M.J., Guidotti, A., and Marks, N. (1999). Gangliosides attenuate ethanol-induced apoptosis in rat cerebellar granule neurons. *Neurochem. Res.* **24**:1107–1115.

Sastry, P.S. and Rao, K.S. (2000). Apoptosis and the nervous system. *J. Neurochem.* **74**:1–20.

Schenker, S., Becker, H.C., Randall, C., Phillips, D.K., Baskin, G.S., and Henderson, G.H. (1990). Fetal Alcohol Syndrome: Current status of pathogenesis. *Alcoholism: Clin. Exp. Res.* **14**:635–647.

Schulz, J.B., Lindenau, J., Seyfried, J. and Dichgans, J. (2000). Glutathione, oxidative stress, and neurodegeneration. *Eur. J. Biochem.* **267**:4904–4911.

Shaw, S., Rubin, K.P., and Lieber, C.S. (1983). Depressed hepatic glutathione and increased diene conjugates in alcoholic liver disease. Evidence of lipid peroxidation. *Digest. Dis. Sci.* **28**:585–589.

Siler-Marsiglio, K.I., Shaw, G., and Heaton, M.B. (2004). Pycnogenol and vitamin E inhibit ethanol-induced apoptosis in rat cerebellar granule cells. *J. Neurobiol.* **59**:261–271.

Spierlings, D., McStay, G., Saleh, M., Bender, C., Chipuk, J., Maurer, U., and Green, D.R. (2005). Connected to death: The (unexpurgated) mitochondrial pathway of apoptosis. *Science* **310**:66–67.

Streissguth, A.P., Barr, H.M., and Sampson, P.D. (1990). Moderate prenatal alcohol exposure: effects on child IQ and learning problems at age 72 years. *Alcoholism: Clin. Exp. Res.* **14**:662–669.

Suthanthiran, M., Anderson, M.E., Sharma, V.K., and Meister, A. (1990). Glutathione regulates activation-dependent DNA synthesis in highly purified normal human T lymphocytes stimulated via the CD2 and CD3 antigens. *Proc. Natl. Acad. Sci. USA.* **87**:3343–3347.

Tanaka, J., Toku, K., Zhang, B., Isihara, K., Sakanaka, M., and Maeda, N. (1999). Astrocytes prevent neuronal death induced by reactive oxygen and nitrogen species. *Glia* **28**:85–96.

Timmer, J.C. and Salvesen, G.S. (2007). Caspase substrates. *Cell Death Diff.* **14**:66–72.

Tsukamoto, H. (1993). Oxidative stress, antioxidants, and alcoholic liver fibrogenesis. *Alcohol* **10**:465–467.

Uchida, K., Szweda, L.I., Chae, H.-Z., and Stadtman, E.R. (1993). Immunochemical detection of 4-hydroxynonenal protein adducts in oxidized hepatocytes. *Proc. Natl. Acad. Sci. USA* **90**:8742–8746.

Vaudry, D., Rousselle, C., Basille, M., Falluel-Morel, A., Pamantung, T.F., Fontaine, M., Fournier, A., Vaudry, H., and Gonzalez, B.J. (2002). Pituitary adenylate cyclase-activating polypeptide protects rat cerebellar granule neurons against ethanol-induced apoptotic cell death. *Proc. Natl. Acad. Sci. USA* **99**:6398–6403.

Wang, X.F. and Cynader, M.S. (2000). Astrocytes provide cysteine to neurons by releasing glutathione. *J. Neurochem.* **74**:1434–1442.

Wass, T.S., Simmons, R.W., Thomas, J.D., and Riley, E.P. (2002). Timing accuracy and variability in children with prenatal exposure to alcohol. *Alcoholism: Clin. Exp. Res.* **26**:1887–1896.

Watts, L.T., Rathinam, M.L., Schenker, S., and Henderson, G.I. (2005). Astrocytes protect neurons from ethanol-induced oxidative stress and apoptotic death. *J. Neurosci. Res.* **80**:655–666.

Wozniak, D.F., Hartman, R.E., Boyle, M.P., Vogt, S.K., Brooks, A.R., Tenkova, T., Young, C., Olney, J.W., and Muglia, L.J. (2004). Apoptotic neurodegeneration induced by ethanol in neonatal mice is associated with profound learning/memory deficits in juveniles followed by progressive functional recovery in adults. *Neurobiol. Dis.* **17**:403–414.

Young, C., Klocke, B.J., Tenkova, T., Choi, J., Labruyere, J., Qin, Y.Q., Holtzman, D.M., Roth, K.A., and Olney, J.W. (2003). Ethanol-induced neuronal apoptosis in vivo requires BAX in the developing mouse brain. *Cell Death Differ.* **10**:1148–1155.

Chapter 14
Wernicke's Encephalopathy

Maryam R. Kashi, George I. Henderson, and Steven Schenker

Introduction

Wernicke's encephalopathy (WE) is a potentially reversible metabolic brain dysfunction resulting from thiamine deficiency. It is generally characterized by ataxia, ophthalmoplegia and global confusion. Described in Berlin in 1881 by Carl Wernicke, it was initially known as *polioencephalitis hemorrhagica superioris* and considered a fatal syndrome. The first reported cases were three patients, two with alcoholism and one with persistent vomiting after the ingestion of sulfuric acid in a suicide attempt. The common feature shared by these cases upon post-mortem exam consisted of punctate hemorrhages in the grey matter of the walls of the third and fourth ventricles and mammillary bodies (Cirignotta *et al.*, 2000; Truswell, 2000). In 1935, Strauss discovered that the cause of Wernicke's findings was vitamin B1 (thiamine) deficiency (Chiossi *et al.*, 2006). Bonhoeffer posited that Wernicke's encephalopathy and the psychosis described by Korsakoff actually represented two phases of the same pathological process (Cirignotta *et al.*, 2000). The observation that Wernicke's encephalopathy and Korsakoff's psychosis have identical neuropathology supported this belief (Charness, 1999). Thus Wernicke's encephalopathy (WE) and Korsakoff's psychosis (KP) are often used interchangeably as the Wernicke-Korsakoff Syndrome (WKS).

Korsakoff psychosis is a chronic amnestic state, characterized by retrograde amnesia (loss of memory), anterograde amnesia (defective learning) and confabulation. Loss of short-term memory is a predominant feature, while immediate and long-term memories are usually intact. (Harrison *et al.*, 2006; Chiossi *et al.*, 2005; Sivolap, 2005). Patients with KP have difficulty remembering events and facts that occur after the onset of disease (Hochhalter and Joseph, 2001). It occurs most commonly in patients with WE, usually secondary to alcoholism, but may be seen in patients without a previous diagnosis of WE (Truswell, 2000). It has been suggested that some of these patients may have had subacute cases of WE, where signs and symptoms were either mild or absent. Nevertheless, on autopsy, patients with KP have the same brain pathology as patients with WE, as mentioned earlier (Charness, 2006).

D.W. McCandless (ed.) *Metabolic Encephalopathy*,
doi: 10.1007/978-0-387-79112-8_14, © Springer Science + Business Media, LLC 2009

Pathologically, WE may be classified into acute (17%), subacute (17%) and chronic (66%) states. It has been postulated that WE may be a progressive disorder, where multiple acute and/or subclinical episodes of thiamine deficiency cause cumulative damage. These subclinical events may be devoid of the traditional symptoms associated with WE (Gui *et al.*, 2006; Harper, 1983). This may explain the findings of brain lesions at postmortem without antecedent symptoms in some patients. Analogously, Korsakoff's psychosis may occur in patients who have experienced multiple acute and/or subclinical events of WE. This may explain the discrepancy between the relatively high numbers of alcoholic patients who develop Korsakoff's psychosis as compared to nonalcoholic patients. There are at least two hypotheses to explain this observation. The first hypothesis presumes that the latter group's exposure to a thiamine-deficient state was a single event, while the former group is likely to have had long-standing thiamine deficiency and/or multiple events of WE, increasing their chances of developing KP (Homewood and Bond, 1999). The second hypothesis is that perhaps ethanol neurotoxicity and thiamine deficiency work in concert in the development of KP (Charness, 2006).

Victor *et al.* (1989), in classic studies, followed 186 alcoholic patients with Wernicke's encephalopathy for up to ten years and documented that 84% developed Korsakoff's syndrome. A subsequent study of 32 alcoholic patients followed for a period of 33 months showed the rate of progression to KP to be 56% (Wood *et al.*, 1984). Full recovery from Korsakoff's psychosis occured in only 20%; the majority of KP patients required some level of supervision and social support (Reuler *et al.*, 1985; Charness, 2006).

Wernicke's encephalopathy is associated with a mortality rate of 10–20%, predominantly as a result of sepsis, respiratory infection and decompensated liver disease; Korsakoff's psychosis is associated with a mortality rate of approximately 17% (Harrison *et al.*, 2006; Ogershok *et al.*, 2002; Merkin-Zaborsky *et al.*, 2001).

Prevalence

The prevalence rates of 0.8% –2.8% for WE come mainly from four autopsy-based studies: Norway (0.8%), New York (1.7%), Cleveland (2.2%) and Australia (2.8%) (Harper, 1983; Ogershok *et al.*, 2002). Australia's higher prevalence of WE was puzzling, as the country is not ranked high on the world league table of alcohol consumption. After conducting interviews with patients in alcohol rehabilitation units, Price concluded that Australian alcoholics were more likely to lack female social support, which may otherwise provide them with food containing thiamine (Truswell, 2000). Seventy-five percent of patients are male and the peak age incidence is in the sixth decade (41%) (Harper, 1983). Wernicke's encephalopathy remains a profoundly under-diagnosed disease, which if left untreated can progress to Korsakoff syndrome or death, as a result of irreversible cytotoxic effects (Loh *et al.*, 2004; Weidauer *et al.*, 2004). Harper found that only

20% of cases reviewed in a necropsy study had been diagnosed with WE or Wernicke-Korsakoff syndrome prior to death (Harper, 1983). Thus a high index of suspicion is the key to diagnosis.

Clinical Features

The classic triad of symptoms in WE includes ophthalmoplegia, ataxia and global confusion (see Table 14.1). However, the clinical presentation is often incomplete (Foster *et al.*, 2005). In a retrospective study, it was found that only 16.5% of patients exhibited all three signs and 19% exhibited none of these (Harper *et al.*, 1986). One study found nystagmus to be present in 85%, bilateral paralysis of the lateral rectus muscles in 54% and conjugate gaze palsies in 45% of cases (Ogershok *et al.*, 2002)). Other reported symptoms include apathy, lightheadedness, disorientation, poor memory, diplopia, inability to stand, nausea, vomiting and coma (Liu *et al.*, 2006; Giglioli *et al.*, 2004; Morcos *et al.*, 2004; Harper *et al.*, 1986; Ogershok *et al.*, 2002). In addition, WE can affect the sympathetic system, resulting in postural hypotension and syncope, and the temperature-regulating center, resulting in mild hypothermia (Worden, 1984; Reuler *et al.*, 1985). The lag time from onset of thiamine-deficiency to the start of symptoms is approximately 4–6 weeks (Harrison et al., 2006). Symptoms may range from a period of 2 days to 2 weeks before presentation for evaluation (Giglioli *et al.*, 2004; Lacasse and Lum, 2004).

Wernicke's encephalopathy is most commonly associated with alcoholism. It has been suggested that there may be a synergistic effect of alcoholism and thiamine deficiency, where a brain affected by alcoholism may be more susceptible to injury caused by thiamine deficiency (Homewood and Bond, 1999). However, WE may be found in any clinical state associated with malnutrition or thiamine deficiency

Table 14.1 Clinical features of wernicke's encephalopathy

Classical Symptoms	Other Associated Symptoms
Ataxia	Apathy
Global Confusion	lightheadedness
Ophthalmoplegia	disorientation
	poor memory
	unsteady gait
	diplopia
	vision impairment
	nystagmus, inability to stand
	nausea
	vomiting
	coma
	hypothermia

Table 14.2 Conditions associated with wernicke's encephalopathy

Anorexia Nervosa (Harrison *et al.*, 2006; Morcos *et al.*, 2004; Ogershok *et al.*, 2002)	Malignancy (Weidauer *et al.*, 2004; Ogershok *et al.*, 2002)
Chronic alcoholism	Prolonged parenteral feeding without Thiamine supplementation (Morcos *et al.*, 2004; D'Aprile *et al.*, 2000)
Diarrheal Disorders (Celik and Kaya, 2004)	
Gastric/Bariatric surgery (Attard *et al.*, 2006; Loh *et al.*, 2004; Cirignotta *et al.*, 2000)	Prolonged starvation/Malnutrition (Attard *et al.*, 2006; Ogershok *et al.*, 2002)
Hemodialysis/Peritoneal dialysis (Ogershok *et al.*, 2002; Merkin-Zaborsky *et al.*, 2001)	Regional enteritis (Ogershok *et al.*, 2002)
HIV/AIDS (Ogershok *et al.*, 2002)	Refeeding after starvation (Ogershok *et al.*, 2002)
HIV Encephalopathy (Morcos *et al.*, 2004)	Thyrotoxicosis (Ogershok *et al.*, 2002)
Hyperemesis gravidarum (Harrison *et al.*, 2006)	Uremia (Ogershok *et al.*, 2002)
Malabsorption syndromes (Ogershok et al., 2002)	

Table 14.3 Possible causes of thiamine deficiency in chronic alcoholism

1. Inadequate thiamine intake
2. Decreased activation of thiamine to thiamine pyrophosphate
3. Reduced hepatic storage of thiamine
4. Inhibition of intestinal thiamine transport
5. Impairment of thiamine absorption due to ethanol-related nutritional deficiency states

Borrowed from Hoyumpa (1980) with permission

(see Table 14.2), including hyperemesis gravidarum (Chiossi *et al.*, 2006), anorexia nervosa (Morcos et al., 2004), prolonged parenteral feeding without micronutrient supplementation (Attard *et al.*, 2006), renal disease with hemodialysis or peritoneal dialysis and gastric or bariatric surgery (Attard *et al.*, 2006; Worden and Allen, 2006; Loh et al., 2004; Cirignotta et al., 2000; Ogershok et al., 2002). Although studies have found that 23–50% of cases are actually not associated with alcohol abuse, the index of suspicion for thiamine deficiency is still low in non-alcoholic patients (Ogershok *et al.*, 2002).

The incidence of thiamine deficiency in alcoholics is 30–80% (Homewood and Bond, 1999). Factors that promote thiamine deficiency in alcoholics include poor thiamine intake, decreased activation of thiamine to thiamine pyrophosphate(TPP), decreased hepatic storage, decreased intestinal thiamine transport and impairment of thiamine absorption (see Table 14.3) (Breen *et al*, 1985; Hoyumpa, 1980). Although thiamine is stored in various sites, including skeletal muscles, heart, kidneys and brain, the liver remains the main storage site. Due to the reasons cited above, hepatic thiamine content may be reduced by 73% in patients with severe, chronic alcoholic liver disease. In addition, ethanol has been shown to promote thiamine release from the liver (Hoyumpa, 1980).

Chiossi *et al.* (2006) reviewed 49 case reports of WE due to hyperemesis gravidarum. The duration of vomiting and/or poor intake was 7.7 +/ − 2.8 weeks. The mean gestational age was 14.3 +/ − 3.4 weeks and the amount of weight loss ranged from 6–25 kg. Thirty-two percent of these patients were primigravida (Chiossi *et al.*, 2006). In laboratory rats, thiamine deficiency is a known cause of intrauterine growth retardation. In a German study, lower erythrocyte thiamine concentrations were found in patients whose pregnancies were complicated by intrauterine growth retardation than in patients with normal pregnancies (Heinze and Weber, 1990), although a causal relationship is uncertain.

Chronic renal failure patients on hemodialysis and peritoneal dialysis are at risk for thiamine deficiency due to inadequate nutrition in part and possible thiamine loss during the dialysis process. Renal failure patients are often on a diet restricted in protein and potassium, which increases the risk of thiamine deficiency (Masud, 2002; Piccoli *et al.*, 2006). Studies with detailed dietary surveys have shown poor oral intake of thiamine in chronic renal failure patients (Hung et al., 2001). There is no convincing evidence that thiamine levels are significantly altered by either hemodialysis or peritoneal dialysis (Reuler *et al.*, 1985). DeBari *et al.* (1984) measured thiamine levels of granulocytes, erythrocytes and plasma. They found no significant differences in thiamine levels in dialysis patients compared to controls. Further research in this area would benefit chronic renal failure patients and help determine possible need for supplementation of water-soluble vitamins.

Due to the obesity epidemic in the United States, bariatric surgery is becoming increasingly more common. As a result, nutrient malabsorption is becoming more common as well. A common procedure, Roux-en-y gastric bypass (RYGBP), causes both food restriction and malabsorption. Malabsorption is caused by the bypass of the distal stomach, duodenum and the first part of the jejunum. Vomiting is a common side effect of RYGBP and vitamin B12 and iron deficiency are frequently seen. Thiamine deficiency may occur in this setting as a result of bypass of the duodenum, as this is where thiamine is predominantly absorbed (Worden and Allen, 2006). Some authors routinely advocate starting parenteral thiamine administration six weeks postoperatively in malnourished patients (Loh *et al.*, 2004). Prolonged parenteral feeding without thiamine supplementation is a well-documented cause of Wernicke's encephalopathy.

Administration of intravenous glucose activates glycolysis, a process which utilizes thiamine and may enhance thiamine deficiency (Koguchi *et al.*, 2004). Thus common emergency room practice includes the administration of intravenous thiamine before intravenous glucose in order to prevent the precipitation of WE. Whether or not this practice is warranted is somewhat controversial. In a review of 49 published cases of *hyperemesis gravidarum* as the leading cause of WE, Chiossi et al. felt that approximately 30% of the cases were provoked by intravenous glucose administration without thiamine (Chiossi *et al.*, 2006). Many other case reports have also noted such a phenomenon but good evidence-based studies are lacking. Harrison notes this to be a "theoretical concern. " (Harrison *et al.*, 2006). Pazirandeh et al. (2006) point out that cellular thiamine uptake is actually slower than glucose uptake. As such, cellular thiamine repletion may not occur prior to cellular glucose exposure.

Wernicke's encephalopathy may appear inconspicuously in psychiatric patients, as it may be obscured by mental illness. Patients with schizophrenia, for example, may be particularly at risk due to poor dietary intake, high rates of homelessness and high prevalence of alcoholism (Harrison *et al.*, 2006).

Infantile WE may be found in developing countries, primarily among breast-fed infants, usually in the second to fifth months of development. Wernicke's encephalopathy is very rare in developed nations. However, in 2003, Israel was faced with an epidemic of WE due to a batch of defective soy-based vegetarian infant formula. WE was documented in 20 out of an estimated 3500 infants who were fed the formula, later found to be deficient in thiamine (Kesler *et al.*, 2005).

Diagnosis

Although the diagnosis of WE is generally considered to be a clinical one, supporting laboratory tests and neuroimaging data may be important. Generally, routine laboratory tests, such as liver profile and renal function, urinalysis, chest x-rays, electrocardiograms and echocardiograms are normal, as are cerebrospinal fluid tests. Serum lactic acid, however, has been shown to be elevated in the setting of thiamine deficiency, particularly in children (Liu *et al.*, 2006; Attard *et al.*, 2006; Weidauer *et al.*, 2004). In one case study, an electromyelogram showed diffuse sensorimotor neuropathy (Cirignotta *et al.*, 2000); in another case, an electroencephalography revealed diffuse slow activity or dysrhythmia (Chiossi *et al.*, 2006).

Serum thiamine levels may be misleading and thus should not be employed for the diagnosis of Wernicke's encephalopathy. There are two laboratory tests which are used as surrogates for body thiamine stores. The erythrocyte transketolase test (ETKA) is a reflection of thiamine reserves at a cellular level. The thiamine pyrophosphate effect (TPPE) test, expressed as a coefficient, is a measure of transketolase activity before and after the addition of thiamine. Values before added thiamine reflect the amount of coenzyme present in the cell; the values measured after thiamine addition is a reflection of the amount of apoenzyme present that lacks a coenzyme. Diagnosis is made by either a low ETKA and/or a high TPPE. Normal TPPE values range from 0% –14%. TPPE values between 15% and 24% signify marginal thiamine deficiency and values greater than 25% signify severe deficiency (Kesler *et al.*, 2005; Chiossi *et al.*, 2006). Unfortunately, ETKA and TPPE are not readily commercially available in the United States. Thus patients should be treated once suspected clinically of WE. Clinical response to treatment is the ultimate persuasive diagnostic test.

By performing thorough clinical histories, neurological and psychological exams, as well as pathological evaluations, Caine *et al.* (1997) developed a set of diagnostic criteria for WE in alcoholic patients. The diagnostic criteria, published in 1997, require two of the following four signs for the diagnosis of WE: oculomotor abnormalities, malnutrition, cerebellar dysfunction and either mild memory impairment or altered mental status (see Table 14.4). Validity testing of this approach

Table 14.4 Caine's diagnostic criteria for wernicke's encephalopathy

Two of the following criteria must be met:
1. Dietary deficiency
2. Oculomotor abnormality
3. Cerebellar dysfunction
4. Altered mental status or mild memory impairment

demonstrated improved diagnostic sensitivity from 31% (using the classic triad) to about 100%. They report that sensitivity decreased to 50% in patients with concurrent hepatic encephalopathy. This is not unexpected, as two of the elements in the above criteria, malnutrition and altered mental status, are commonly seen in patients with hepatic encephalopathy. Proper management of these conditions (i.e. serum ammonia levels and response to appropriate therapy) should elucidate the proper diagnosis. The Caine criteria were developed based on studies of patients with alcoholism and should not be applied, at present, to a general population. Future studies should assess these criteria in a nonalcoholic population.

Neuroimaging

Computed tomography (CT) and magnetic resonance imaging (MRI), and single-photon emission computed tomography (SPECT) have been studied with regard to their evaluation of WE. CT appears to be helpful only in cases with hemorrhagic lesions, which include only approximately 5% of cases (Chiossi *et al.*, 2006; Mascalchi *et al.*, 1999). Computed tomography was reported by Antunez *et al.* (1998) to have quite a low sensitivity (13%) in the diagnosis of WE.

MRI sequences typically include, pre- and post-contrast (gadolinium) T1-weighted images with gadolinium, T2-weighted images, fluid-attenuated inversion recovery (FLAIR), diffusion-weighted imaging (DWI), and apparent diffusion coefficient (ADC) mapping. MRI has a sensitivity and specificity of 53% and 93%, respectively, for WE (Ogershok *et al.*, 2002). The sensitivity may actually be higher, but MRI is often performed after the patient has been suspected of WE and empirically treated, resulting in a non-diagnostic study (Celik and Kaya, 2004). Good correlation has been found between contrast mediated magnetic resonance imaging (MRI) and neuropathological findings (Liu *et al.*, 2006). Classically, T2-weighted and FLAIR MRI images reveal symmetrical increased signal intensity of areas including the third paraventricular regions of the thalamus and hypothalamus, periaqueductal regions of the midbrain and mammillary bodies. These lesions may sometimes be enhanced with gadolinium on T1-weighted images during an acute event and may dissipate with treatment (Chiossi *et al*, 2006; Mascalchi *et al.*, 1999; Doherty *et al.*, 2002; Halavaara *et al.*, 2003). In one case, mammillary body enhancement was the only sign of acute WE (Shogry and Curnes, 1994). FLAIR sequences are reviewed to ensure that the cerebrospinal fluid has not masked high signal lesions on T2-weighted imaging (Chung *et al.*, 2003). The chronic stage of WE may be depicted by brain atrophy and diffuse signal-intensity changes in the cerebral white matter (White *et al.*, 2005).

Some of the lesion identified on MRI may be seen in other conditions, such as inferolateral and anterolateral thalamic infarcts, multiple sclerosis, Cytomegalovirus encephalitis, Behcet's disease, primary cerebral lymphoma, central pontine myelinosis, Lyme disease, Leigh's disease and variant Creutzfeldt-Jokob disease. These conditions are usually excluded from the differential diagnosis based both on their asymmetric distribution and the clinical setting (Weidauer *et al.*, 2004; Chung *et al.*, 2003). Neuroimaging may be a very useful tool in the diagnosis of WE; however, it is important to note that the absence of signs on neuroimaging does not exclude the diagnosis (Antunez *et al.*, 1998; Celik and Kaya, 2004).

Although there is insufficient evidence to suggest the presence of cytotoxic edema in acute WE, there is good evidence for vasogenic edema (Liu *et al.*, 2006). The advantage of DWI over T2-weighted and FLAIR imaging is its ability to better distinguish between cytotoxic and vasogenic edema (Chiossi *et al.*, 2006; Chung *et al.*, 2003). DWI, in conjunction with ADC mapping, is particularly useful, as it is the most sensitive method for diagnosing early injury, i.e. vasogenic edema, before the onset of necrosis, thus facilitating early diagnosis of WE (Doherty *et al.*, 2002; Halavaara *et al.*, 2003). Cytotoxic edema is represented by high intensity signal on DWI with corresponding low signal on ADC mapping. Vasogenic edema, however, will show high signal intensity on both DWI and ADC mapping (Weidauer *et al.*, 2004).

Single-photon emission computed tomography (SPECT) imaging has also been evaluated and at least one study has found it to be useful in cases where conventional MRI may be non-diagnostic (Celik and Kaya, 2004).

Pathology

Wernicke's encephalopathy, which affects the brainstem, white matter and cortex, has a characteristic appearance on autopsy (Celik and Kaya, 2004). The acute stage of WE is characterized by the inability to maintain proper osmotic gradients of cell membranes, promoting intracellular swelling and red blood cell extravasation into the perivascular space. This stage is distinguished by marked vascular dilatation, endothelial swelling and neuronal demyelinization. The chronic stage is marked by mammillary body atrophy, as well as, loss of neuropil with fibrillary astrocytosis (Weidauer *et al.*, 2004; D'Aprile *et al.*, 2000; Homewood and Bond, 1999). Neurons, however, are generally spared (Liu *et al.*, 2006; Gui *et al.*, 2006; Halavaara *et al.*, 2003). Classic neuropathological findings include petechial hemorrhages of blood vessels and small, symmetric necrotic lesions in the paraventricular areas of the thalamus, hypothalamus, mammillary bodies, periaqueductal area of the midbrain and cerebellum (McEntee, 1997; Chung *et al.*, 2003; Caine *et al.*, 1997).

Table 14.5 Macroscopic and microscopic findings of wernicke's encephalopathy on autopsy

Macroscopic lesions	Microscopic lesions
Ventricular dilatation (34%)	Mammillary bodies (99%)
Mammillary body atrophy (75%)	Third ventricular wall (61%)
Cerebellar vermal atrophy (34%)	Thalamus (61%)
Cerebellar atrophy (21%)	Midbrain (50%)
Paraventricular atrophy (5%)	Pons (50%)
	Medulla (33%)

(Harper, 1983)

It has been proposed that perhaps the paraventricular regions are more susceptible to thiamine deficiency because they have a higher rate of glucose and oxidative metabolism which require thiamine (Lacasse and Lum, 2004). The most consistent pathological findings, found in 75% of the cases, are mammillary body atrophy and brownish discoloration (Liu *et al.*, 2006; Harper, 1983). Although a small proportion of patients may have normal sized mammillary bodies, almost all have microscopic mammillary body lesions (Charness, 1999). Macrohemorrhage is found in approximately 5% of cases (Mascalchi *et al.*, 1999). Table 14.5 shows Harper's macroscopic and microscopic autopsy findings in WE (Harper, 1983).

Role of Thiamine

Thiamine (B1) is an essential coenzyme for enzymes involved in Kreb's cycle (including pyruvate dehydrogenase and alpha ketoglutarate dehydrogenase), lipid metabolism/amino acid production (transketolase) and neurotransmitter synthesis— acetylcholine and GABA (2-oxo-glutarate dehydrogenase) (Chiossi *et al.*, 2006). A commonly found water-soluble vitamin, it is found in lean pork, poultry, fish, eggs, liver, wheat germ, whole grains, beans, peas and nuts. Fruits, vegetable and dairy products are not good sources of thiamine (Table 14.6) Alcoholic beverages have virtually no thiamine (Table 14.7) (Lonsdale, 2006). However, it has been suggested that small amounts of thiamine exist in German and Australian beer (Price and Kerr, 1985). The human daily requirement of thiamine is 1.0–1.5 mg per day but this requirement is increased in states of pregnancy, lactation, thyrotoxicosis and fever. Body stores are approximately 25–30 mg and are found predominantly in skeletal muscles, heart, liver, kidneys and brain (Lacasse and Lum, 2004). These reserves are sufficient for only 2–3 weeks without continued intake (Gui *et al.*, 2006; Kesler *et al.*, 2005). Thiamine, which is excreted in the urine, has a half-life of approximately 10–20 days (Pazirandeh *et al.*, 2006). Loss of thiamine is accelerated by diuretic therapy and may be inactivated by polyphenol containing compounds found in coffee and tea (Lonsdale, 2006). Malabsorption may occur with alcoholism, with gastric surgery and with folate deficiency (Lacasse and Lum,

Table 14.6 High thiamine foods: thiamine content/100g food product

Lean Pork (avg of trimmed retail cuts, loin, and shoulder blade, lean only, cooked)	0.873 mg
Poultry (cooked):	
Turkey	0.06 mg
Chicken	0.069 mg
Fish (cooked):	
Tuna	0.278 mg
Catfish	0.42 mg
Tilapia	0.093 mg
Whitefish (mixed species)	0.171 mg
Salmon	0.34 mg
Eggs (raw)	0.069 mg
Beef Liver (cooked)	0.194 mg
Chicken Liver (cooked)	0.291 mg
Wheat germ	1.882 mg
Beans (pinto, cooked)	0.193 mg
Peas (cooked)	0.259 mg
Soybeans (cooked)	0.155 mg
Kidney (beef)	0.16 mg
Nuts	0.2 mg
Whole grains (dry):	
Whole wheat flour	0.447 mg
Bulgur	0.232 mg
Oats	0.763 mg
Whole cornmeal	0.385 mg
Buckwheat	0.101 mg
Brown rice	0.413 mg

(The Food Processor SQL, 2006)

Table 14.7 Thiamine content in alcoholic beverages

Standard beer (12 oz) - 0.00 mg
Red wine (4 oz) - 0.01 mg
White wine (4 oz) - 0.01 mg
Tequila/Gin/Bourbon/Whiskey/Vodka (80 proof/1.5 oz) - 0.00 mg

(The Food Processor SQL, 2006)

2004; Price and Kerr, 1985). Alcohol has been found to interfere with the active transport of thiamine in the gastrointestinal system, at least in rodents (Kumar *et al.*, 2000). Thiamine absorption may be significantly decreased in the setting of folate depletion but may return to normal with 4–6 weeks of folate repletion therapy. A deficiency in magnesium, required for the conversion of thiamine to thiamine pyrophosphate, may also cause thiamine deficiency (Bishai and Bozzetti, 1986; Lonsdale, 2006).

Medications postulated to affect body stores of thiamine include: 5-fluorouracil (Heier and Dornish, 1989), loop diuretics (Brady et al., 1995; Seligmann et al., 1991) and dilantin (Patrini *et al.*, 1993; Botez *et al.*, 1993).

Thiamine Absorption

Dietary thiamine exists primarily in the form of thiamine pyrophosphate (TPP), which must be hydrolyzed to free thiamine, before absorption in the small bowel (Dudeja et al., 2001). In the small bowel, thiamine absorption occurs by two processes: passive and active transport. Passive transport occurs only in the presence of high thiamine concentrations, and actually blocks the active transport process. Low doses of thiamine are absorbed by active transport (Rindi and Ventura, 1972). The details of this absorption mechanism are still not completely clear. Dudeja et al. have performed multiple studies evaluating jejunal thiamine absorption at both the brush border membrane (BBM) and the basolateral membrane (BLM). In one study, they found that human intestinal BBM absorption of thiamine is a carrier-mediated process, which is sodium-independent, pH-dependent and amiloride-sensitive. They have also proposed the possibility of a thiamine—/H+ exchange mechanism (Dudeja *et al.*, 2001). In a study of thiamine absorption in jejunal BLM, Dudeja et al. found the transport mechanism to be a pH-dependent and amiloride-sensitive carrier-mediated process (Dudeja et al., 2003). SCL19A2, believed to be a human thiamine transporter, has been shown to be expressed in all gastrointestinal tissues, with the greatest level of expression found in the liver. Reidling et al. have discovered that the minimal promoter region needed for basal activity of SLC19A2 gene is encoded between −356 and −36 (Reidling *et al.*, 2002).

Breen *et al.* evaluated the influence of acute alcohol perfusion on small bowel absorption of thiamine. They found that alcohol did not significantly decrease thiamine uptake in the jejunum, although there was a trend to lower absorption with alcohol perfusion (Breen et al., 1985). Holzbach evaluated thiamine absorption in patients after 3 days and after 4 weeks of resolution of acute delirium tremens (DT). He found no significant difference in thiamine absorption between normal patients and those with recent delirium tremens. There was, however, a significant increase in thiamine absorption 4 weeks after DT as compared to values obtained shortly (3 days) after DT. They propose the possibility of abnormal thiamine absorption in DT. It is noteworthy that the patients with visual hallucinations had lower thiamine absorption levels than those who did not have this symptom (Holzbach, 1996). Tomasulo studied thiamine deficiency in severely alcoholic patients admitted to the hospital. Forty-three percent of these patients had DT. He measured radioactive thiamine in both urine and stool and found significant differences between controls and alcoholics. The labeled thiamine excreted in 24-hour urine collections of controls and alcoholics were 45.8% and 25.3%, respectively. The reciprocal findings of stool in controls and alcoholics were 4.0% and 21.0%, respectively (Tomasulo *et al.*, 1968). Studies by Thomson *et al.* (1968) also provide evidence that chronic alcohol abuse may decrease thiamine absorption. These various studies differ, not only in their methodologies, but also in their subject populations. Breen's study assessed the effects of acute alcohol while the latter three studied chronic alcoholics. This area clearly deserves further research.

Pathogenesis of Wernicke's Encephalopathy

Serious attempts to determine the mechanism(s) of this disorder have been ongoing for at least 70 years (Peters, 1969). There are many aspects of WE which should make this task relatively easy. First, the clinical picture of this disorder is well characterized, can be readily diagnosed and is relatively specific. Second, the pathology of this entity is elegantly described and imaging by MRI, when present, is characteristic. Third, there are animal models readily available which should permit a biochemical/pathologic dissection of the problem. Fourth, the specific deficiency, a decrease of vitamin B_1, responsible for the experimental and clinical findings of WE is well-known. Most importantly, the experimentally induced and clinical syndromes are often readily reversible (if seen early) by the administration of thiamine.

With such an extensive knowledge base, what is the present state of our understanding of the mechanisms of this disorder? Not unexpectedly, initial studies, primarily in experimental animal models, focused on the known metabolic pathways which involve thiamine. Indeed, the classical studies of Peters in 1930 (Peters, 1969) showed lactate accumulation in the brainstem of thiamine deficient birds with normalization of this in vitro when thiamine was added to the tissue. This led to the concept of "the biochemical lesion" of the brain in thiamine deficiency. The enzymes which depend on thiamine are shown in Fig. 14.1. They are transketolase, pyruvate and α-ketoglutarate dehydrogenase. Transketolase is involved in the pentose phosphate pathway needed to form nucleic acids and membrane lipids, including myelin. The ketoacid dehydrogenases are key enzymes of the Krebs cycle needed for energy (ATP) synthesis and also to form acetylcholine via Acetyl CoA synthesis. Decrease in activity of this cycle would result in anaerobic metabolism and lead to lactate formation (i.e., tissue acidosis) (Fig. 14.1).

Indeed, studies in animal models of thiamine deficiency and a small number of postmortem human brain specimens have shown that transketolase and α-ketoglutarate dehydrogenase (but not consistently pyruvate dehydrogenase) were depressed. Perhaps the largest and earliest fall was seen in brain transketolase; however, when the neurological signs were reversed with thiamine there was no concomitant improvement in transketolase which rose only slightly (McCandless and Schenker, 1968). Moreover, glucose flux through the pentose phosphate pathway (dependent on transketolase) did not decrease in severe thiamine deficiency and ribose-5-phosphate (a key intermediate in the pentose cycle) did not fall (McCandless et al., 1976; McCandless, 1982). Thus, the current view is that a low transketolase is a marker of thiamine deficiency, but is likely not to be causal in the acute neurological deficits seen in thiamine deficiency (McCandless, 1982; Hazell et al., 1998). The possible

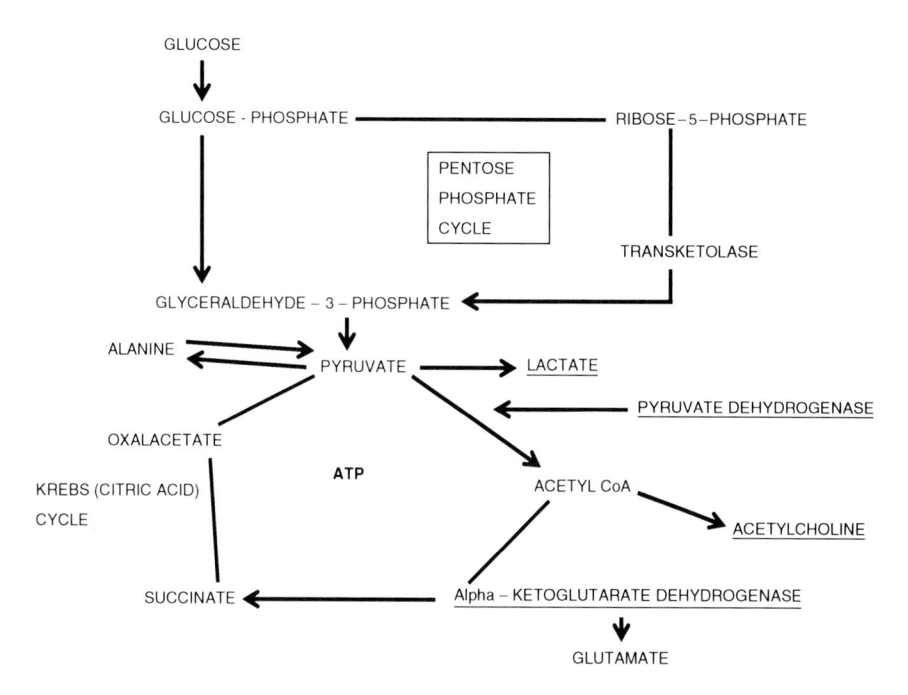

Fig. 14.1 Enzymes which depend on thiamine: there are transketolase, pyruvate and alpha-ketoglutarate dehydrogenase (see text)

effects of prolonged transketolase depression, as with chronic thiamine deficit, are uncertain (Hazell *et al.*, 1998).

Data on the role of acetylcholine deficit in thiamine deficiency are conflicting, but most recent studies do not favor a significant decrease in the synthesis of this neurotransmitter (Hazell *et al.*, 1998; Vorhees *et al.*, 1977). This would be consistent with normal pyruvate dehydrogenase activity in experimental thiamine deficiency, which should not, therefore, result in a lower Acetyl CoA level as a precursor to acetylcholine.

Perhaps the most likely mechanism of low thiamine-induced brain injury has revolved around impairment of the Krebs' cycle and deficit in available ATP (Desjardins and Butterworth, 2005). This could readily lead to apoptosis and necrosis of neurons, as has been described in such patients (Vorhees *et al.*, 1977). In this context, the data on pyruvate dehydrogenase are somewhat difficult to interpret. Postmortem brain from patients with Wernicke's encephalopathy did show a major decrease in pyruvate dehydrogenase, albeit in only a few specimens (Butterworth *et al.*, 1993). However, this was not corroborated in experimental models of this syndrome (Desjardins and Butterworth, 2005; Butterworth et al., 1993). By contrast a major decrease in brain α-ketoglutarate dehydrogenase was seen in every type of thiamine deficiency (Desjardins and Butterworth, 2005; Butterworth et al., 1993). Moreover, an impairment in this enzyme could readily explain an increase in brain lactate, due to anaerobic metabolism, and this has been observed uniformly, even as far back as 1930 (Peters, 1969; McCandless and Schenker, 1968; Desjardins and Butterworth, 2005). The major concern about the Krebs cycle deficit concept has been a difficulty in documenting consistently an impairment in brain energy stores, both ATP and phosphocreatine (PCr). Multiple studies in the brain of animals with thiamine deficiency have not shown a decrease in ATP or PCr, even when assayed in brain areas (brainstem and lateral vestibular nucleus) felt to be most affected (McCandless and Schenker, 1968; McCandless, 1982; McEntee, 1997; Holowach et al., 1968). The only exception was a study by Aikawa et al. (1984), who showed a small decrease in ATP and PCr in some parts of rat brain after exposure to pyrithiamine (thiamine antagonist). The functional significance of this small (~10%) drop in ATP is not known, but in our view is unlikely to be important. McCandless has shown *increased* levels of both ATP and PCr in the lateral vestibular nucleus of such animals, as well as in rats rendered thiamine deficient by dietary means (McCandless, 1982; McCandless and Schwartzenburg, 1981). The higher energy stores reverted to normal on restoration of thiamine. These latter data suggest that energy utilization was impaired during the symptomatic stage. Formal energy turnover studies in critically affected brain areas have not been done, to our knowledge. Clearly, the role of an impaired Krebs cycle in the pathogenesis of Wernicke's encephalopathy is unresolved.

A number of other mechanisms have been suggested recently for the brain damage caused by thiamine lack. One stipulates that an excess of extracellular glutamate induces increased neurotoxicity (McEntee, 1997). The evidence for this are increased concentrations of glutamate in the extracellular fluid (dialysate) in brains of pyrithiamine-treated rats and decrease in glutamate transporters in astrocytes of these animals (Desjardins and Butterworth, 2005; McEntee, 1997). Another concept is that of oxidative stress via the production of reactive oxygen species and/or increased expression of endothelial nitric oxide synthase (Desjardins and Butterworth, 2005). Finally, neuropathological studies in both animal models and postmortem brain sections in patients with Wernicke's have shown proliferation of astroglial cells, especially in the early stages of thiamine deficiency (Desjardins and Butterworth, 2005). Based on the known protective effects of astrocytes for neurons, this rather suggests that these cells may be activated in that setting, perhaps to provide GSH as an antioxidant (Rathinam et al., 2006). Overall, it appears that the precise mechanism by which thiamine deficiency causes brain injury is unknown. Conceivably multiple factors may be operative.

Another important question relates to the selective sensitivity of specific brain areas to thiamine deficiency. The basis for this has been discussed in terms of differences in regional metabolism, antioxidant status, or differences in thiamine turnover, but without actual data (Desjardins and Butterworth, 2005). Similar regional sensitivity has been seen with bilirubin and copper deposition/damage without explanation. Much remains to be learned.

It has been suggested that there may be a predisposition to WE in some patients, presumably on a genetic basis. Indeed, a variant of transketolase has been reported in fibroblasts of patients with WE (Blass and Gibson, 1977; Nixon *et al.*, 1984). It was proposed that this may increase the requirements of thiamine in such patients, and thus possibly make them more susceptible to thiamine deficits (Martin *et al.*, 1995). However, this concept has not been further verified (Kaufmann *et al.*, 1987; Blansjaar *et al.*, 1991). The authors are unaware of any reported familial clustering of WE, and transketolase is not felt now to be an enzyme primarily causally involved in the pathogenesis of this cerebral disorder. Studies of possible variants in other enzymes involved in thiamine metabolism have not been reported.

Treatment

Wernicke's encephalopathy is a potentially fatal but also reversible medical emergency if diagnosed and treated in the acute stage. Treatment includes supportive measures, as well as thiamine replacement; however, the basic questions of thiamine dose, frequency, route and length of treatment remain unclear (Morcos *et al.*, 2004). The Cochrane Collaboration (2007) sought to evaluate the evidence available for the use of thiamine in the treatment of Wernicke-Korsakoff Syndrome due to alcohol abuse. There were actually no studies that addressed this specifically but the Cochrane group identified one published randomized controlled study, by Ambrose et al., that compared the effects of various doses of thiamine therapy in alcoholic patients without overt clinical signs of WKS (Day *et al.*, 2004). In 2001, Ambrose evaluated the effects of differing doses of thiamine hydrochloride (5, 20, 50, 100 and 200 milligrams) given intramuscularly for 2 consecutive days in a group of alcoholic patients, none of whom had any clinical signs of WKS, in an alcohol detoxification center. Post-treatment, patients were compared based on their performance on a delayed alternation task test, established to be sensitive to the cognitive impairments of Wernicke-Korsakoff Syndrome. Patients who received the highest dose of thiamine showed superior performance (Ambrose *et al.*, 2001). However, the initial thiamine status of this patient group was not known. As a result of the paucity of data from randomized clinical trials, the Cochrane Collaboration concluded that currently, there is insufficient data available to provide clinical guidelines regarding the dose, frequency, route or duration of thiamine for the treatment of WKS due to alcoholism (Day *et al.*, 2004).

In various studies/clinical cases, thiamine 100 mg has been given intravenously for several days to two weeks, followed by maintenance doses of 50–100 mg orally per day until the patient is able to eat a well-balanced diet regularly (Lacasse and Lum, 2004; Chiossi *et al.*, 2006). Long-term treatment and prevention should include continued oral thiamine supplementation, alcohol abstinence and a balanced diet (Ogershok *et al.*, 2002), but this program is based on logic and overall good medical care, not data.

Following thiamine therapy in the acute state, one may obtain a dramatic response, which essentially confirms the diagnosis (Squirrell, 2004). Improvement in ophthalmoplegia is often the first sign of treatment benefit and may occur within hours (Koontz *et al.*, 2004; Doherty *et al.*, 2002). Ataxia may take days to weeks but 25% of cases may not improve at all. Residual peripheral neuropathy is also not uncommon (Worden and Allen, 2006; Weidauer *et al.*, 2004; Chiossi *et al.*, 2006). Chiossi, in his review of WE cases due to hyperemesis gravidarum, found that only 29% of patients obtained complete resolution of symptoms, while 53% showed resolution of most signs and symptoms within three months (Chiossi *et al.*, 2006). Improvement in mental status is variable and up to 84% of patients may develop Korsakoff's psychosis (Harrison *et al.*, 2006; Morcos *et al.*, 2004; Homewood and Bond, 1999). At a two-year follow-up visit, one patient, whose oculomotor and imaging studies had shown improvement, had persistent symptoms of severe cognitive deficits, vertigo and loss of sphincter control (Attard *et al.*, 2006). Improvement of imaging studies may be seen up to four months of treatment with thiamine (Loh *et al.*, 2004). Delay in treatment may result in irreversible neuronal death and possibly death of the patient (Gui *et al.*, 2006).

Prevention

Clearly, one of the most important prevention strategies is physician and patient education in this area. In addition, there has been much debate over thiamine fortification of alcoholic beverages in order to prevent Wernicke's encephalopathy in alcoholics, the most susceptible population,. In 1987, Australia's Mental Health Committee recommended fortification of all Australian beer and flagon wine but this was never implemented. In most developed countries, bread (white flour) is enriched with thiamine to restore what is lost from the whole wheat in the process of milling. Australia adopted this plan in 1991, using the same level of enrichment as the United States (6.4 mg thiamine hydrochloride/Kg flour). The incidence of WE in the five years after the above implementation in Australia was 40% lower (perhaps fortuitously) than in the five year period prior to bread fortification. In addition, the post-mortem diagnosis of WE in Sydney, Australia has declined from 2.1% to 1.1% (Truswell, 2000).

Conclusion

There exists a great disparity between the number of patients diagnosed with WE while alive and the number of patients diagnosed post-mortem. This issue can be improved when a high index of suspicion for WE is employed for not only alcoholic patients but any patient with malnutrition or possible thiamine deficiency. Wernicke's encephalopathy should be considered in the evaluation of any patient found to have one or more of the classic complaints, including confusion, ophthalmoplegia and ataxia, especially in the setting of malnutrition. In patients with acute altered mental status or coma, it is essential to treat empirically with thiamine, which is safe and inexpensive, even prior to the availability of neuroimaging results. We cannot overemphasize that imaging and laboratory data should not delay treatment with thiamine, which should be based initially on clinical assessment (i.e., symptoms/signs).

Since the preparation of this manuscript, a recent paper has reported total tau protein levels in cerebrospinal fluid were elevated in acute WE, but declined at follow up. This may suggest that neuronal cell death occurs transiently in acute WE.

Acknowledgement We thank Amy L. Wyatt, R.D., L.D., affiliated with the University Hospital System in San Antonio, TX, for information provided on the content of thiamine in foods, as listed in Table 14.6. We gratefully acknowledge the excellent reviews by Drs. Roger Butterworth and colleagues (Hazell, 1998; Desjardins and Butterworth, 2005) and W.J. McEntee (1997), which served as valuable source material for this review.

References

Aikawa, H., Watanabe, I.S., Furuse, T., Iwasaki, Y., Satoyoshi, E., Sumi, T., and Moroji, T. (1984). Low energy level in thiamine deficient encephalopathy. *J. Neuropathol. Exp. Neurol.* **43**:276–287.

Ambrose, M.L., Bowden, S.C., and Whelan, G. (2001). Thiamin treatment and working memory function of alcohol-dependent people: preliminary findings. *Alcohol. Clin. Exp. Res.* **25**:112–116.

Antunez, E., Estruch, R., Cardenal, C., Nicholas, J.M., Fernandez-Sola, J., and Urbano-Marquez, A. (1998). Usefulness of CT and MR Imaging in the diagnosis of acute Wernicke's Encephalopathy. *Am. J. Roentgenol.* **171**:1131–1137.

Attard, O., Dietemann, J.L., Diemunsch, P., Pottecher, T., Meyer, A., and Calon, B.L. (2006). Wernicke encephalopathy: a complication of parenteral nutrition diagnosed by magnetic resonance imaging. *Anesthesiology* **105**:847–848.

Bishai, D.M. and Bozzetti, L.P. (1986). Current progress toward the prevention of the Wernicke-Korsakoff syndrome. *Alcohol Alcohol.* **21**:315–323.

Blansjaar, B.A., Zwang, R., and Blijenberg, B.G. (1991). No transketolase abnormalities in Wernicke-Korsakoff patients. *J. Neurol. Sci.* **106**:88–90.

Blass, J.P. and Gibson, G.E. (1977). Abnormality of a thiamine-requiring enzyme in patients with Wernicke-Korsakoff syndrome. *N. Engl. J. Med.* **297**:1367–1370.

Botez, M.I., Botez, T., Ross-Chouinard, A., and Lalonde, R. (1993). Thiamine and folate treatment of chronic epileptic patients: a controlled study with the Wechsler IQ scale. *Epilepsy Res.* **16**:157–163.

Brady, J.A., Rock, C.L., and Horneffer, M.R. (1995). Thiamin status, diruretic medications, and the management of congestive heart failure. *J. Am. Diet. Assoc.* **95**:541–544.

Breen, K.J., Buttigieg, R., Iossifidis, S., Lourensz, C., and Wood, B. (1985). Jejunal uptake of thiamin hydrocholoride in man: influence of alcoholism and alcohol. *Am. J. Clin. Nutr.* **42**:121–126.

Butterworth, R.F., Kril, J.J., and Harper, C.G. (1993). Thiamine-dependent enzyme changes in the brains of alcoholics: relationship to the Wernicke-Korsakoff syndrome. *Alcohol. Clin. Exp. Res.* **17**:1084–1088.

Caine, D., Halliday, G.M., Kril, J.J., and Harper, C.G. (1997). Operational criteria for the classification of chronic alcoholics identification of Wernicke's encephalopathy. *J. Neurol. Neurosurg. Psychiatr.* **62**:51–60.

Celik, Y. and Kaya, M. (2004). Brian SPECT findings in Wernicke's encephalopathy. *Neurol. Sci.* **25**:23–26.

Charness, M.E. (1999). Intracranial voyeurism: revealing the mammillary bodies in alcoholism. *Alcohol. Clin. Exp. Res.* **23**:1941–1944.

Charness, M.E. (2006). Overview of the chronic neurologic complications of alcohol. www.upto-date.com, version 14.3; August 2006.

Chiossi, G., Neri, I., Cavazzuti, M., Basso, G., and Facchinetti, F. (2006). Hyperemesis gravi-darum complicated by Wernicke encephalopathy: background, case report, and review of the literature. *Obstet. Gynecol. Surv.* **61**:255–268.

Chung, T.I., Kim, J.S., Park, S.K., Kim, B.S., Ahn, K.J., and Yang, D.W. (2003). Diffusion weighted MR imaging of acute Wernicke's encephalopathy. *Eur. J. Radiol.* **45**:256–258.

Cirignotta, F., Manconi, M., Mondini, S., Buzzi, G., and Ambrosetto, P. (2000). Wernicke-Korsakoff encephalopathy and polyneuropathy after gastroplasty for morbid obesity. *Arch. Neurol.* **57**:1356–1359.

D'Aprile, P., Tarantino, A., Santoro, N., and Carella, A. (2000). Wernicke's encephalopathy induced by total parenteral nutrition in patient with acute leukaemia: unusual involvement of caudate nuclei and cerebral cortex on MRI. *Neuroradiology* **42**:781–783.

Day, E., Bentham, P., Callaghan, R., Kuruvilla, T., and George, S. (2004). Thiamine for Wernicke-Korsakoff Syndrome in people at risk from alcohol abuse. *Cochrane Database Systematic Review*, Issue 1, Art. No. CD004033.

DeBari, V.A., Frank, O., Baker, H., and Needle, M.A. (1984). Water soluble vitamins in granulo-cytes, erythrocytes, and plasma obtained from chronic hemodialysis patients. *Am. J. Clin. Nutr.* **39**:410–415.

Desjardins, P. and Butterworth, R.F. (2005). Role of mitochondrial dysfunction and oxidative stress in the pathogenesis of selective neuronal loss in Wernicke's encephalopathy. *Mol. Neurobiol.* **31**:17–25.

Doherty, M.J., Watson, N.F., Uchino, K., Hallam, D.K., and Cramer, S.C. (2002). Diffusion abnor-malities in patients with Wernicke encephalopathy. *Neurology* **58**:655–657.

Dudeja, P.K., Tyagi, S., Gill, R., and Said, H.M. (2003). Evidence of a carrier-mediated mecha-nism for thiamine transport to human jejunal basolateral membrane vesicles. *Dig. Dis. Sci.* **48**:109–115.

Dudeja, P.K., Tyagi, S., Kavilaveettil, R.J., Gill, R., and Said, H.M. (2001). Mechanism of thia-mine uptake by human jejunal brush-border membrane vesicles. *Am. J. Physiol. Cell Physiol.* **281**:C786–C792.

Foster, D., Falah, M., Kadom, N., and Mandler, R. (2005). Wernicke encephalopathy after bariat-ric surgery: losing more than just weight. *Neurology* **65**:1987.

Giglioli, L., Salani, B., and Mannucci, M. (2004). Case Report: Mental confusion, diplopia, and inability to stand in an 82 year-old-man. (Ltr. to Editor). *J. Am. Geriatr. Soc.* **52**:1218.

Gui, Q.P., Zhao, W.Q., and Wang, L.N. (2006). Wernicek's encephalopathy in nonalcoholic patients: clinical and pathologic features of three cases and literature reviewed. *Neuropathology* **26**:231–235.

Halavaara, J., Brander, A., Lyytinen, J., Setälä, K., and Kallela, M. (2003). *Neuroradiology* **45**:519–523.

Harper, C. (1983). The incidence of Wernicke's encephalopathy in Australia — a neuropathologi-cal study of 131 cases. *J. Neurol. Neurosurg. Psychiatr.* **46**:593–598.

Harper, C.G., Giles, M., and Finlay-Jones, R. (1986). Clinical signs in the Wernicke-Korsakoff complex: a retrospective analysis of 131 cases diagnosed at necropsy. *J. Neurol. Neurosurg. Psychiatr.* **49**:341–345.

Harrison, R.A., Vu, T., and Hunter, A.J. (2006). Wernicke's encephalopathy in a patient with schizophrenia. *J. Gen. Intern. Med.* **21**:C8–C11.

Hazell, A.S., Todd, K.G., and Butterworth, R.F. (1998). Mechanisms of neuronal cell death in Wernicke's encephalopathy. *Metab. Brain Dis.* **13**:97–122.

Heier, M.S. and Dornish, J.M. (1989). Effect of the fluoropyrimidines 5-fluorourcil and Doxifluridine on cellular uptake of thiamin. *Anticancer Res.* **9**:1073–1077.

Heinze, T. and Weber, W. (1990). Determination of thiamine (vitamin B1) in maternal blood dur-ing normal pregnancies and pregnancies with intrauterine growth retardation. *Zeit. Ernahrungswissensch.* **29**:39–46.

Hochhalter, A.K. and Joseph, B. (2001). Differential outcomes training facilitates memory in people with Korsakoff and Prader-Willi syndromes. *Integr. Physiol. Behav. Sci.* **36**:196–204.

Holowach, J., Kaufmann, F., Ikossi, M.G., Thomas, C. and McDougal, D.B. (1968). The effects of a thiamine antagonist, pyrithiamine, on levels of selected metabolic intermediates and activities of thiamine dependent enzymes in brain and liver. *J. Neurochem.* **15**:621–631.

Holzbach, E. (1996). Thiamine absorption in alcoholic delirium patients. *J. Stud. Alcohol* **57**:581–584.

Homewood, J., and Bond, N.W. (1999). Thiamin deficiency and Korsakoff's syndrome: failure to find memory impairments following nonalcoholic Wernicke's encephalopathy. *Alcohol* **19**:75–84.

Hoyumpa, A.M., Jr. (1980). Mechanisms of thiamin deficiency in chronic alcoholism. *Am. J. Clin. Nutr.* **33**:2750–2761.

Hung, S.C., Hung, S.H., Tarng, D.C., Yang, W.C., Chen, T.W., and Huang, T.P. (2001). Thiamine deficiency and unexplained encephalopathy in hemodialysis and peritoneal dialysis patients. *Am. J. Kidney Dis.* **38**:941–947.

Kaufmann, A., Uhlhaas, S., Friedl, W., and Propping, P. (1987). Human erythrocyte transketolase: no evidence for variants. *Clin. Chim. Acta.* **162**:215–219.

Kesler, A., Stolovitch, C., Hoffmann, C., Avni, I., and Morad, Y. (2005). Acute ophthalmoplegia and nystagmus in infants fed a thiamine-deficient formula: an epidemic of Wernicke encephalopathy. *J. Neuroophthalmol.* **25**:169–172.

Koguchi, K., Nakatsuji, Y., Abe, K., and Sakoda, S. (2004). Wernicke's encephalopathy after glucose infusion. *Neurology* **62**:512.

Koontz, D.W., Fernandes-Filho, J.A.F., Sagar, S.M., and Rucker, J.C. (2004). Wernicke encephalopathy. *Neurology* **63**:394.

Kumar, D., Nartsupha, C., and West, B.C. (2000). Unilateral internuclear ophthalmoplegia and recovery with thiamine in Wernicke syndrome. *Am. J. Med. Sci.* **320**:278–280.

Kumar, D., Nartsupha, C., and West, B.C. (2008). Unilateral internuclear ophthalmoplegia and recovery with thiamine in Wernicke syndrome. *Alcohol Clin Exp Res* **32**:1091–1095.

Lacasse, L. and Lun, C. (2004). Wernicke encephalopathy in a patient with T-Cell leukemia and severe malnutrition. *Can. J. Neurol. Sci.* **31**:97–98.

Liu, Y.T., Fuh, J.L., Lirng, J.F., Li, A.F.Y., and Ho, D.M.T. (2006). Correlation of magnetic resonance images with neuropathology in acute Wernicke's encephalopathy. *Clin. Neurol. Neurosurg.* **108**:682–687.

Loh, Y., Watson, W.D., Verma, A., Chang, S.T., Stocker, D.J., and Labutta, R.J. (2004). Acute Wernicke's encephalopathy following bariatric surgery: clinical course and MRI correlation. *Obesity Surg.* **14**:129–132.

Lonsdale, D. (2006). A review of the biochemistry, metabolism and clinical benefits of thiamin(e) and its derivatives. *Evid. Based Complement. Alternat. Med.* **3**:49–59.

Martin, P.R., McCool, B.A., and Singleton, C.K. (1995). Molecular genetics of transketolase in the pathogenesis of the Wernicke-Korsakoff syndrome. *Metab. Brain Dis.* **10**:45–55.

Mascalchi, M., Simonelli, P., Tessa, C., Giangaspero, F., Petruzzi, P., Bosincu, L., Conti, M., Sechi, G., and Salvi, F. (1999). Do acute lesions of Wernicke's encephalopathy show contrast enhancement? Report of three cases and review of the literature. *Neuroradiology* **41**:249–254.

Masud, T. (2002). Trace elements and vitamins in renal disease. In (W.E. Mitch, and Klahr, S. eds.), *Handbook of Nutrition and the Kidney*, 4th edn, Lippincott, Philadelphia, pp. 233–252.

McCandless, D.W. (1982). Energy metabolism in the lateral vestibular nucleus in pyrithiamin-induced thiamin deficiency. *Ann. N.Y. Acad. Sci.* **378**:355–364.

McCandless, D.W., Curley, A.D. and Cassidy, C.E. (1976). Thiamine deficiency and the pentose phosphate cycle in rats: intracerebral mechanisms. *J. Nutr.* **106**:1144–1151.

McCandless, D.W. and Schenker, S. (1968). Encephalopathy of thiamine deficiency: studies of intracerebral mechanisms. *J. Clin. Invest.* **47**:2268–2280.

McCandless, D.W. and Schwartzenburg, F.C., Jr. (1981). The effect of thiamine deficiency on energy metabolism in cells of the lateral vestibular nucleus. Res. Comm. *Psychol. Psychiat. Behav.* **6**:183–190.

McEntee, W.J. (1997). Wernicke's encephalopathy: an excitotoxicity hypothesis. *Metab. Brain Dis.* **12**:183–192.

Merkin-Zaborsky, H., Ifergane, G., Frisher, S., Valdman, S., Herishanu, Y.L., and Wirguin, I. (2001). Thiamine-responsive acute neurological disorders in nonalcoholic patients. *Eur. Neurol.* **45**:34–37.

Morcos, Z., Kerns, S.C., and Shapiro, B.E. (2004). Wernicke encephalopathy. *Arch. Neurol.* **61**:775–776.

Nixon, P.F., Kaczmarek, M.J., Tate, J., Kerr, R.A., and Price, J. (1984). An erythrocyte transketolase isoenzyme pattern associated with the Wernicke-Korsakoff Syndrome. *Eur. J. Clin. Invest.* **14**:278–281.

Ogershok, P.R., Rahman, A., Nestor, S., and Brick, J. (2002). Wernicke encephalopathy in nonalcoholic patients. *Am. J. Med. Sci.* **323**:107–111.

Patrini, C., Perucca, E., Reggiani, C., and Rindi, G. (1993). Effects of phenytoin on the in vivo kinetics of thiamine and its phosphoesters in rat nervous tissues. *Brain Res.* **628**:179–186.

Pazirandeh, S., Lo, C.W., and Burns, D.L. (2006). Overview of water-soluble vitamins. www.uptodate.com, version 14.3.

Peters, R.A. (1969). The biochemical lesion and its historical development. *Br. Med. Bull.* **25**:223–226.

Piccoli, G.B., Burdese, M., Mezza, E., Soragna, G., Tattoli, F., Consiglio, V., Maddalena, E., Bergui, M., Scarzella, G., and Segoloni, G.P. (2006). The suddenly speechless florist on chronic dialysis: the unexpected threats of a flower shop? *Nephrol. Dial. Transplant.* **21**:223–225.

Price, J. and Kerr, R. (1985). Some observations on the Wernicke Korsakoff syndrome in Australia. *Br. J. Addict.* **80**:69–76.

Rathinam, M.L., Watts, L.T., Start, A.A., Mihimainathan, L., Stewart, J., Schenker, S., and Henderson, G.I. (2006). Astrocyte control of fetal cortical neuron glutathione homeostasis: up-regulation by ethanol. *J. Neurochem.* **96**:1289–1300.

Reidling, J.C., Subramanian, V.S., Dudeja, P.K., and Siad, H.M. (2002). Expression and promoter analysis of SLC19A2 in the human intestine. *Biochim. Biophys. Acta* **1561**:180–187.

Reuler, J.B., Girard, D.E., and Cooney, T.G. (1985). Current Concepts: Wernicke's encephalopathy. *N. Eng. J. Med.* **312**:1035–1039.

Rindi, G. and Ventura, U. (1972). Thiamine intestinal transport. *Physiol. Rev.* **52**:821–827.

Seligmann, H., Halkin, H., Rauchfleisch, S., Kaufmann, N., Motro, M., Vered, Z., and Ezra, D. (1991). Thiamine deficiency in patients with congestive heart failure receiving long-term furosemide therapy: a pilot study. *Am. J. Med.* **91**:151–155.

Shogry, M.E. and Curnes, J.T. (1994). Mammillary body enhancement on MR as the only sign of acute Wernicke encephalopathy. *Am. J. Neuroradiol.* **15**:172–174.

Sivolap, Y.P. (2005). The current state of S.S. Korsakov's concept of alcoholic polyneuritic psychosis. *Neurosci. Behav. Physiol.* **35**:977–982.

Squirrell, D. (2004). Wernicke encephalopathy. *Arch. Ophthalmol.* **122**:418–419.

The Food Processor SQL. (2006). www.esha.com/foodprocessorsql; version 9.9.0

Thomson, A., Baker, H., and Leevy, C.M. (1968). Thiamine absorption in alcoholism. *Am. J. Clin. Nutr.* **21**:537.

Tomasulo, P.A., Kater, R.M.H., and Iber, F.L. (1968). Impairment of thiamine absorption in alcoholism. *Am. J. Clin. Nutr.* **21**:1340–1344.

Truswell, A.S. (2000). Australian experience with the Wernicke-Korsakoff syndrome. *Addiction* **95**:829–832.

Victor, M., Adams, R.D., and Collins, G.H., eds. (1989). Course in the illness. In: *The Wernicke-Korsakoff Syndrome and Related Neurologic Disorders Due to Alcoholism and Malnutrition*, Davis, Philadelphia, pp. 31–38.

Vorhees, C.V., Schmidt, D.E., Barrett, R.J., and Schenker, S. (1977). Effect of thiamin deficiency on acetylcholine levels and utilization in vivo in rat brain. *J. Nutr.* **107**:1902–1908.

Weidauer, S., Rosler, A., Zanella, F.E., and Lanfermann, H. (2004). Diffusion-weighted imaging in Wernicke encephalopathy associated with stomach cancer: case report and review of the literature. *Eur. Neurol.* **51**:55–57.

White, M.L., Zhang, Y., Andrew, L.G., and Hadley, W.L. (2005). MR imaging with diffusion-weighted imaging in acute and chronic Wernicke encephalopathy. *Am. J. Neuroradiol.* **26**:2306–2310.

Wood, B., Currie, J., and Breen, K. (1984). Wernicke's encephalopathy in a metropolitan hospital. A prospective study of incidence, characteristics and outcome. *Med. J. Aust.* **144**:12–16.

Worden, R.W. and Allen, H.M. (2006). Wernicke's encephalopathy after gastric bypass that masqueraded as acute psychosis: a case report. *Curr. Surg.* **63**:114–116.

Chapter 15
The Genetics of Myelination in Metabolic Brain Disease: The Leukodystrophies

John W. Rumsey

Introduction

The formation of brain myelin is of paramount importance to the proper development, survival and functioning of neurons. Myelination, consequently, is a highly regulated process dependent on the expression and temporal regulation of many genes, proteins and metabolites, as well as an adequate nutritional environment to support the post-natal brain growth spurt. Consequently, myelination is vulnerable to a wide range of perturbations. Reviews of the mechanisms and effects of myelin disorders are numerous. Here, the focus is specifically on genetic mutations affecting various cellular compartments in neuronal cells that result in either demyelination or dysmyelination. Collectively, these diseases, categorized as leukodystrophies, represent an ongoing public health problem in both developed and developing nations.

Leukodystrophies

Leukodystrophies are a broad class of diseases encompassing genetic metabolic abnormalities of the oligodendrocyte that result in either dysmyelination or demyelination (Lyon et al., 1996). The major leukodystrophies involve abnormal transport or catabolism of myelin/oligodendrocyte sphingolipids, or of the myelin proteins themselves (Morell et al., 1994). The abnormal metabolite processing can occur in the lysosome such as is the case in globoid cell leukodystrophy (Krabbe disease), at the peroxisome as in X-linked adrenoleukodystrophy (X-ALD), or affect mitochondrial functioning as in Canavan disease (CD) (Kumar et al., 2006; Moser, 1997; Moser et al., 2005; Wenger et al., 1997).

Globoid Cell Leukodystrophy (Krabbe Disease)

An autosomal recessive affliction resulting from deficient galactocerebrosidase (GALC, EC 3.2.1.46) activity in the lysosome, Krabbe disease, characterized by

D.W. McCandless (ed.) *Metabolic Encephalopathy*,
doi: 10.1007/978-0-387-79112-8_15, © Springer Science+Business Media, LLC 2009

oligodendrocyte cell death and demyelination, presents with white matter lesions on MRI scans and the presence of macrophage-like globoid cells. The disease has an estimated incidence of 1:100,000 (Wenger et al., 2001). The faulty GALC activity results in the build up of certain glactolipids, including galactosylceramide, monogalactosyldiglycerie (MGD) and galactosylsphingosine (psychosine) (Wenger et al., 2000). Most importantly, the build up of psychosine has been shown to be the toxic process responsible for the loss of myelin and the death of oligodendrocytes in Krabbe disease (Igisu and Suzuki, 1984; Suzuki and Taniike, 1998; Svennerholm et al., 1980). Additional characteristics of the disease, demyelination in the central nervous system (CNS) and the concurrent appearance of phagocytic globoid cells are also attributed to toxic levels of psychosine. Psychosine is also responsible for toxic effects in mitochondria including inhibition of the electron transport chain, alternations of mitochondrial membrane potential, cytochrome c release and induction of the transcription factor activator protein 1 (AP-1), resulting in oligodendrocyte apoptosis (Haq et al., 2003; Tapasi et al., 1998). Globoid cell leukodystrophy also affects the peripheral nervous system by causing Schwann cell death, as toxic levels of psychosine accumulate during myelination (Hogan et al., 1969).

The genetics of Krabbe disease and the GALC enzyme have been extensively studied. The cDNA for human GALC, first cloned and sequenced in 1993, facilitated studies into the structure and post-translational modifications of the protein (Chen et al., 1993). The gene is located on chromosome 14 (14q31). The protein was found to have an estimated molecular mass of 80 kilodalton (kDa) and to exist as a two subunit heterodimer (Chen and Wenger, 1993; Sakai et al., 1994) Currently, over 75 mutations in the human GALC gene have been identified, and are present in all exons excluding number 12 (Fig. 15.1) (Fu et al., 1999; Wenger et al., 1997).

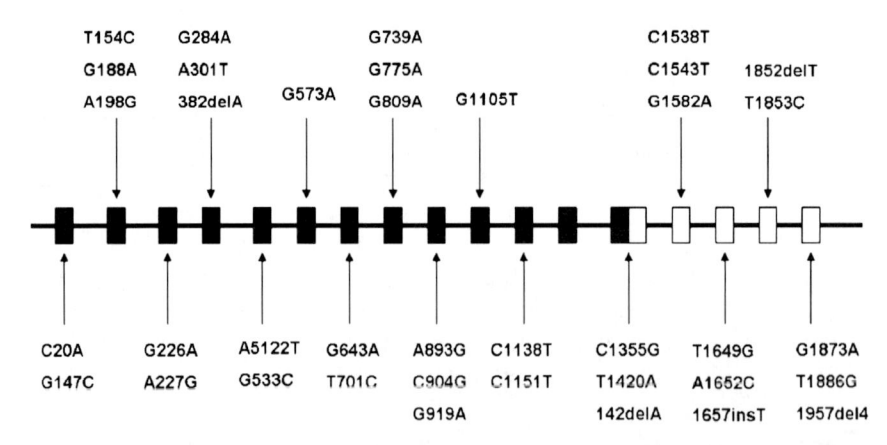

Fig. 15.1 Mutation diagram for the GALC gene. The gene contains 18 exons (boxes), 13 for the 50kDa subunit (black boxes) and 5 for the 30kDa subunit (white boxes). Homozygous mutations shown above are T154C, A198G, T701C, C1138T, C1538T, G1582A and T1886G. The G809A mutation results exclusively in juvenile or adult onset regardless of other mutations. Also indicated are some insertions and deletions. The blank line between the exons represents the introns of the gene. Adapted from Wenger et al. (2000)

Missense mutations are the most frequent, followed by deletions and insertions resulting in premature stop codons (Wenger et al., 1997). The most common mutation in individuals, present in almost 40% of all chromosomes evaluated, is a 30kb deletion resulting in a truncated large subunit and an absent small subunit (Luzi et al., 1995; Rafi et al., 1995). Expression studies using this mutant gene show no activity for the peptide produced (Rafi et al., 1995). Additional mutations result in a wide range of normal genetic polymorphisms and account for the wide variety of measured GALC activity in the general population (Wenger et al., 2001).

The clinical symptoms of the infantile form of Krabbe disease, believed to affect 90% of all patients, include developmental delays, neurological deterioration, limb stiffness, seizure, deafness, blindness and extreme irritability (Hagberg et al., 1970; Krabbe, 1916). Most of these patients regress to a decerebrate condition and the majority die before two years of age. Late onset patients (adolescent and adult) present with a variety of symptoms including loss of manual dexterity, general weakness, burning paresthesia in the extremities and loss of vision (Crome et al., 1973; Lyon et al., 1991). The clinical course of late onset Krabbe disease is unpredictable, and while the severity and time course of the disease vary, all adolescents eventually regress and all die within 2–7 years after diagnosis (Verdu et al., 1991). In the case of adult onset of the disease, progression varies, with many progressing slowly, exhibiting no or mild mental deterioration, while others progress quickly to significant physical and mental impairment. Still others deteriorate rapidly to death (De Gasperi et al., 1996; Furuya et al., 1997; Kolodny et al., 1991; Sabatelli et al., 2002).

Krabbe disease diagnosis usually occurs within six months of birth, although adolescent and adult onset cases do occur (Crome et al., 1973). Diagnosis has traditionally been done using magnetic resonance imaging (MRI) to visualize brain lesions and more recently using proton magnetic resonance spectroscopy (MRS) to image metabolic abnormalities (Brockmann et al., 2003; Kingsley et al., 2006; Sasaki et al., 1991; Zarifi et al., 2001). MRS is especially useful in diagnosing the disease before tissue alterations can be seen clearly using MRI (Brockmann et al., 2003). Conclusive diagnosis is biochemical, based on an enzymatic assay measuring galactocerebrosidase activity in leukocytes or cultured cutaneous fibroblasts (Wenger et al., 2000).

While no cure is available for globoid cell leukodystrophy, the use of an animal model, the twitcher mouse, has enabled the development and evaluation of several promising treatment options (Eto et al., 2004; Shen et al., 2001; Suzuki and Suzuki, 1995). One treatment, brain directed gene therapy using a recombinant adenovirus (AxCAGALC) encoding the β-galactocerebrosidase gene (β-GALC), showed a 15% increase in brain activity of β-GALC and a 55% decrease in psychosine concentrations. This resulted in reduced pathological observations in the brain and a prolonged lifespan. In this study, timing of the adenoviral therapy played a significant role (Shen et al., 2001). Another therapeutic technique developed in the twitcher mouse, stem cell transplantation, has also proved effective at increasing mouse lifespan. In this study, bone marrow stem cells from normal donors were effective at restoring some GALC functions and tripling the lifespan of the twitcher mutants (Suzuki and Suzuki, 1995). These findings have encouraged scientists who have used umbilical cord and hematopoietic stem cell transplantation as therapies for metabolic diseases (Boelens, 2006; Sauer et al., 2004). Specifically, reversal of the symptoms of Krabbe disease in patients receiving allogeneic hematopoietic stem cell transplantation was observed (Krivit et al., 1998). This therapy has also been effectively demonstrated in patients suffering from peripheral symptoms of the disease based on nerve conduction velocity tests (Siddiqi et al., 2006). In a similar study, umbilical cord stem cells transplanted from unrelated donors into newborns and infants with varying degrees of disease progression showed a favorable prognosis when patients received treatment prior to the onset of symptoms (Escolar et al., 2005).

Although the pathology and subsequent treatment of globoid cell leukodystrophy is seemingly simple, the disease has clearly proved to be anything but straightforward. Fortunately, the availability of an authentic mouse model, as well as cases in dogs, sheep and cats, for studying the disease provide encouraging results and directions for future therapeutic developments (Suzuki and Suzuki, 1985). This disease, like other leukodystrophies, exemplifies the scientists' limited understanding of the relationship between neurons and supporting glial cells and promises to provide insights into the homeostatic functioning of the CNS.

X-linked Adrenoleukodystrophy (X-ALD)

X-ALD is an X-linked disease arising from mutations in the ABCD1 gene. A member of subfamily D of the ATP-binding cassette (ABC) family of membrane transport proteins, the ABCD1 gene, containing 10 exons and 9 introns, encodes a peroxisomal transmembrane protein called adrenoleukodystrophy protein (ALDP) (Mosser et al., 1993). The protein is involved in the transport of substrates from the cytoplasm into the peroxisomal lumen, with the disease falling into group III peroxisomal disorders. Biochemically, this defective transport results in the accumulation of a saturated, unbranched very long chain of fatty acids (VLCFA), which are associated with progressive cerebral demyelination, adrenocortical atrophy, lymphocytic infiltration and abnormalities in testicular function (Siemerling and Creutzfeldt, 1923; van Geel et al., 1997). The most commonly accumulated VLCFAs, tetracosanoic acid (C24:0) and hexacosanoic acid (C26:0), can be found in all tissues and bodily fluids. The abnormalities in adrenal function, presumably caused by toxic VLCFA levels, result from apoptotic cell death in the adrenocortex and cause the development of Addison's disease (Powers et al., 1980).

First described by Siemerling and Creutzfeldt in 1923, the phenotypic presentation of X-ALD varies widely with age of onset, organs involved, neurological involvement and rate of progression of neurological symptoms (Siemerling and Creutzfeldt, 1923; van Geel et al., 1997). In adult onset cases, the disease mainly involves the spinal cord; consequently, it is referred to as adrenomyeloneuropathy (AMN) (Budka et al., 1976). Finally, asymptomatic carriers have been identified, adding to the heterogeneity of the disease. Overall, the incidence of disease,

1:16,800, including both hemizygous and heterozygous individuals, makes X-ALD the most frequently inherited monogenetic demyelinating disorder, as well as the most commonly occurring peroxisomal disorder (Bezman et al., 2001).

The ABCD1 gene, located on the long arm of the X chromosome at position Xq28, is 21 kilobase pairs (kb) in length, containing 10 exons and 9 introns and codes for a 745 amino acid protein (75kDa) (Migeon et al., 1981; Mosser et al., 1993). Although the protein's exact function is unknown, its location has been localized to the peroxisomal membrane, where it plays a role in either the activation of VLCFAs for import into the peroxisome or is directly involved in their import (Mosser et al., 1994; Valle and Gartner, 1993). Currently, 461 unique mutations in the ABCD1 gene have been identified and have been cataloged in an X-ALD mutation database (http://www.x-ald.nl) (Kemp et al., 2001). The most common mutations are missense (52%), while the least common is the whole exon deletion (4%). Interestingly, based on extensive genetic studies and familial case clusters, there is no correlation between the type of ABCD1 gene mutation and the clinical phenotype exhibited (Kok et al., 1995; Krasemann et al., 1996; Ligtenberg et al., 1995). This genotype/phenotype disconnect is due to the possibility of residual ALDP activity, as is the case for other ABC transport proteins, such as the multi-drug resistance transporter (MDR) p-glycoprotein. Additionally, scientists postulate the presence of a yet unidentified autosomal modifier gene based on segregation analysis (Maestri and Beaty, 1992; Moser et al., 1992). Lastly, normalization of VLCFA levels was observed in the livers of ABCD1-deficient mice and in differentiated CG4 rat oligodendrocytes after induction of the adrenoleukodystrophy—related gene (ABCD2) (Fourcade et al., 2003; Maurice and Michaela, 2001).

Mutations of the ABCD1 gene can also manifest as adrenomyeloneuropathy (AMN) with onset usually in the 30s or 40s. In these patients, spinal cord demyelination and axonal loss in the corticospinal tract are the major pathologies (Moser et al., 1992). In approximately 50% of the cases, associated cerebral involvement is also present (Schaumburg et al., 1977). The life expectancy for these patients is normal unless the cerebral demyelination progresses or the adrenocortical insufficiency remains untreated. Lastly, mutant alleles of the ABCD1 gene can cause a primary adrenocortical dysfunction (Addison's disease) (Aubourg and Chaussain, 1991). Commonly manifested Addison's disease symptoms include fatigue, hypotension and bronzing of the skin, with a high risk of developing AMN (van Geel et al., 1999).

The clinical manifestations of X-ALD vary widely but can be categorized into three distinct phenotypes based on age of symptom onset (Table 15.1). First, the most severe form, cerebral X-ALD (CCALD) appears most commonly during early childhood (3–10 years old). The characteristic rapid demyelination is associated with a large inflammatory response in the cerebral white matter (Powers, 1985; Schaumburg et al., 1975). The demyelination and inflammation are associated with oligodendrocyte cytolysis, rather than apoptosis. This raises the question as to whether the inflammatory reaction is the primary cause of demyelination or a secondary immune system response to an initial dysmyelination of neurons (Bezaire et al., 2001). These patients suffer from behavioral disturbances as well as auditory

Table 15.1 Clinical Characteristics of X-linked Adrenoleukodystrophy Phenotypes

	CCALD	AdolCALD	ACALD	AMN
Age of onset	3–10	10–21	>21	25–46
Behavioral disturbances	+	+	+	–
Impaired cognition	+	+	Frequent	–
Pyramidal tract involvement	Eventually	Eventually	Frequent	+
Cerebral MRI abnormalities	Extensive in frontal myelin	Extensive in frontal myelin	Extensive in frontal myelin	Internal capsule basal ganglia, pons
Polyneuropathy	–	Rare	Possible	Sensorimotor, mostly axonal
Impaired endocrine function	AD in most	AD in most	AD in most	AD and hypogonadism in most
Abnormal neuro-psychological exam	+	+	Frequent	–
Progression	Rapid	Rapid	Rapid	Slow

(–) = absent, (+) = present, CCALD = childhood cerebral ALD, AdolCALD = adolescent cerebral ALD, ACALD = adult cerebral ALD, AMN = adrenomyeloneuropathy, AD = adrenocortical insufficiency (Addison's disease). Adapted from van Geel et al. (1997).

and visual deterioration. As the disease progresses, seizures, dementia and spastic tetraplegia become manifest and most patients die within 3 years of neurological symptoms (Moser et al., 2001). The last two categories, adolescent (adolCALD) (11–21 years old) and adult onset (ACALD) cerebral X-ALD can occur, although less frequently, with symptoms and progression similar to the childhood form (Esiri et al., 1984; Farrell et al., 1993). AMN, a milder variant of disease, also with adult onset, can manifest with behavioral abnormalities as well as dementia or schizophrenia (Table 15.1) (van Geel et al., 1999).

The diagnosis of X-ALD employs interpretation of clinical symptoms, brain imaging and biochemical analysis (Moser et al., 2001). Clinically, the most common symptoms include intellectual deterioration, impaired vision and hearing and speech, handwriting changes, limb weakness and seizures (Theil et al., 1992). Evaluation of MRI scans from clinically symptomatic patients reveals characteristic white matter lesions in the corpus callosum and periventricular parietooccipital lobes. PET scans can also be employed to detect gray matter lesions with increased sensitivity (Volkow et al., 1987). More recently, the increased sensitivity of MR spectroscopic imaging has been used to detect white matter abnormalities undetected using conventional imaging techniques (Eichler et al., 2002). Biochemically, the impaired β-oxidation and accumulation of VLCFAs lead to increased plasma concentrations of hexacosanoic acid. This assay is 99% sensitive for males, regardless of symptoms, and 85% sensitive for female carriers. Additionally, the ratios of both tetracosanoic acid to docosanoic acid (C22:0) and hexacosanoic acid to docosanoic acid are increased in the plasma of the diseased patients (Moser et al., 2001; Wanders et al., 1992). In patients where these concentrations and ratios are not abnormal, positive diagnosis can be established by evaluating the concentrations of VLCFAs in cultured skin fibroblasts (Kennedy et al., 1994; Moser et al., 1980; Tonshoff et al., 1982). Lastly, most male patients suffer from adrenal insufficiency and, therefore, its function is assessed in all cases (Libber et al., 1986; Vanhole et al., 1994).

Over the years, several therapies have been developed for the treatment of X-ALD, including dietary treatments, immunosuppression, drug therapies and bone marrow transplantation (Korenke et al., 1997; Moser et al., 1984; Naidu et al., 1988; Rizzo et al., 1986). Unfortunately, dietary changes and the oral administration of glyceryl trioleate (GTO) and glyceryl trierucate (GTE) (Lorenzo's oil) failed to inhibit neurological progression of the disease (Aubourg et al., 1993). Immunosuppression and drug treatment regimes aimed at altering the inflammatory response have not shown any clinical benefit (Naidu et al., 1988). Bone marrow transplantation, performed on over 200 patients since 1990, has been shown to reverse or stabilize neurological progression using MRI, but this treatment shows a lag in improvement and requires extensive donor matching (Aubourg et al., 1990; Loes et al., 1994; Shapiro et al., 2000). Fortunately, advances in genetic research technologies have provided new approaches for X-ALD treatment. Using retroviral mediated transfer of the ALDP gene to both defective human fibroblasts and CD 34 + hematopoietic cells, it has been shown that VLCFA processing can be recovered (Benhamida et al., 2003; Cartier et al., 1995; Doerflinger et al., 1998). Additionally, it has been shown that the human CD 34 + hematopoietic cells will continue to express the ALDP after engraftment into SCID mice (Benhamida et al., 2003). Overall, these studies indicate promise for genetic transfer of the ALDP gene to a patient's own hematopoietic stem cells, followed by autologous transplantation.

Since the first described case of X-ALD by Haberfeld and Spieler in 1910, significant scientific progress has been achieved in elucidating the pathological, biochemical and genetic characteristics of the disease. However, an encompassing treatment regime for all phases of the disease remains unresolved, and several key questions regarding the variety of clinical manifestations and the exact role of the ALDP remain unanswered. Consequently, future research will be needed to address the in vivo role of or substrate for ALDP, as well as provide a clearer picture of the relationship between cerebral demyelination and the inflammatory process in this disease.

Canavan Disease

Canavan disease (CD), caused by mutations in the enzyme aspartoacylase (ASPA) (EC 3.5.1.15), is an autosomal recessive disorder resulting in spongy degeneration of the cerebral and cerebellar regions of the brain, as first reported by Myrtelle Canavan in 1931 (Banker et al., 1964; Canavan, 1931; Pearce, 2004). ASPA mutations result in enzyme deficiency leading to accumulation of its substrate N-acetylaspartic acid (NAA) in the white matter of the brain (Matalon et al., 1988). NAA, a highly abundant amino acid involved in brain osmoregulation, is primarily synthesized in neurons and is transaxonally transported to oligodendrocytes, where it can be catabolized to aspartic acid and acetate (Baslow, 2000; Matalon et al., 2000; Ory-Lavollee et al., 1987; Sager et al., 1999). Consequently, accumulation of NAA in oligodendrocytes leads to hydrostatic pressure buildup in the brain, resulting in cell death, spongy degeneration of the brain, and loss of myelination (Baslow, 2003).

The aspartoacylase gene, localized to a 30kb region of the short arm of chromosome 17 (17p13) by cloning, contains 5 introns and 6 exons (Fig. 15.2), which code for a 313 amino acid protein with a molecular weight of 36kDa (Kaul et al., 1993, 1994). The protein appears to be highly conserved in eukaryotes based on Southern blotting done with yeast, mouse, dog, chicken and cow genomic DNA against human cDNA, suggesting an important role for the gene and its protein in many species (Kaul et al., 1994).

In Canavan disease, the genotype strongly correlates to the phenotype and the most common mutations in the ASPA gene result in a severe disease state. The sporadic mutations, which exist in low frequency, result in milder symptoms (Kaul et al., 1994; Surendran et al., 2003). There are more than 40 identified mutations in the ASPA gene, two of which represent founder mutations among the Ashkenazi Jewish community (Fig. 15.2). These two mutations, E285A on exon 6 and Y231X on exon 5, represent severe phenotype mutations and account for 96% of all mutations in the Ashkenazi Jewish patients (Kaul et al., 1993). Additionally, the extensive screening of 4,000 Ashkenazi Jewish individuals revealed a carrier frequency for the two mutations to be 1 in 40 (Kronn et al., 1995; Matalon et al., 1995). Among non-Jewish patients, the mutations vary widely, with the most common being A305E (Kaul et al., 1996). Mutation expression profiles need to be developed for individuals because some mutations are polymorphic in nature (Surendran et al., 2003).

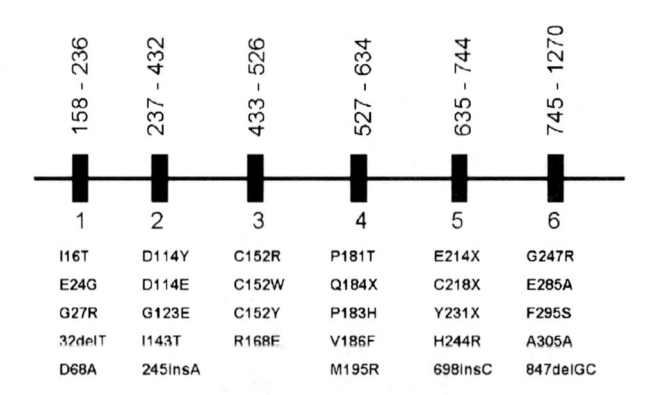

Fig. 15.2 Mutation diagram for the aspartoacylase gene. The gene contains six exons (black boxes) and 5 introns localized to chromosome 17 (17p-ter). The common Jewish mutations, E285A and Y231X are located in exon 6 and exon 5 respectively. The most common non-Jewish mutation, A305A is located in exon 6. The mutations listed below refer to the amino acid substitutions, nonsense mutations in the cDNA sequence or point mutations in the base pair sequence leading to insertions or deletions. Adapted from Surendran et al. (2003)

The clinical manifestations of severe Canavan disease, irritability, hypotonia and inadequate visual tracking, begin to appear during the first three to six months of life and typically progress rapidly as the child grows. By six months of age, developmental delay, megalencephaly and neurological deterioration, common features of severe CD, become obvious and typically progresses to death by five years of age (Traeger and Rapin, 1998; Zu Rhein et al., 1960). Less severe forms of CD have been identified that exhibit similar clinical symptoms but do not result in adolescent death, suggesting a role for modifying genes or environmental factors (Zelnik and Elpeleg, 1994; Zelnik et al., 1993). As patients with the less severe form of Canavan disease age, they exhibit progressive mental retardation and a shift from hypotonia to spasticity (Traeger and Rapin, 1998). All patients suffering from the disease have elevated levels of NAA in their brain tissue and urine (Matalon et al., 1988).

Microscopic examination of brain tissue from affected patients has revealed demyelination/dysmyelination and severe vacuolation in the cerebral cortex and the subcortical white matter (Adachi et al., 1966, 1967). Additionally, vacuoles were observed within the myelin sheaths, between the wraps at the major dense lines. There was also an increased number of astrocytes containing large nuclei and abnormal mitochondria in the cerebral cortex and cerebellum (Adachi et al., 1973; Adachi and Volk, 1968).

The easiest and most common diagnosis for Canavan disease relies on analyzing the patient's urine for high levels of NAA. The NAA levels in diseased patients' urine are greater than 50 times of normal levels. In CD patients (n = 95) NAA levels were 1440.5 ± 873.3 µmol/mmol creatinine and in normal patients (n = 53) levels were 23.5 ± 16.1 µmol/mmol creatinine (Matalon and Michals-Matalon, 1999). Additionally, diffuse white matter degeneration of the brain can be observed using CT and MRI techniques (Brismar et al., 1990; Rushton et al., 1981). Prenatal diagnosis using a biochemical assay for NAA levels in the amniotic fluid provides early diagnostic information for carrier parents, and genetic testing to identify common mutations can be used for carrier detection aiding in prenatal diagnosis (Elpeleg et al., 1994; Kelley, 1993).

Currently, effective therapies for Canavan disease are still under development. Two strategies, one aimed at developing a dietary supplement to replenish low acetate levels and the other aimed at reducing NAA using lithium, have proved unsuccessful thus far, but offer directions for further investigation (Janson et al., 2005; Mathew et al., 2005). Gene transfer of ASPA cDNA via the adeno-associated viral vector (AAV)-ASPA was approved and administered to a cohort of affected children with promising, but transient clinical results (Janson et al., 2001; Leone et al., 2000). In order to enhance efficiency of genetic transfer and stability of transfection, different AVV serotypes have been tested in rat and mouse models of Canavan disease with varying successful results, including reduction of NNA levels and seizure rescue (Davidson et al., 2000; McPhee et al., 2005; Wang et al., 2003).

The current research efforts towards understanding Canavan disease pathology and development of an effective treatment have provided much insight into the disease, and the use of an ASPA knockout mouse has provided future direction towards the goal of a cure. Specifically, understanding the molecular effects of the

ASPA mutation in the different CNS cell types will provide observations useful in finding alternative treatment options. Furthermore, the use of neural stem cells to treat Canavan disease could lead to alternative treatment therapies and enhance our understanding of the control of differentiation of stem cells in the brain.

Impacts of Leukodystrophies

Economic Impacts of Genetic Metabolic Brain Diseases

The development of genetic testing, while having obvious benefits both economically and socially, raises questions concerning the financial utility and burden of testing large, not-at-risk populations. In some diseases with available genetic testing, notably familial breast cancer (BRCA1/2), testing and population surveillance was estimated to reduce associated costs by $1681–1795 per woman and delay onset of the cancer by up to 6 months (Breheny et al., 2005). In contrast, despite the availability of genetic testing for many metabolic brain diseases, administering a test for diseases with no cures and limited therapies could have a negative financial impact on society.

Social Impacts

While the science of demyelinating metabolic brain diseases, like other genetic diseases, remains a subject primarily left to members of the scientific and medical communities; these diseases raise other concerns that must be addressed by society at large. For example, consenting to a genetic test could provide the obvious benefits of early intervention and life long therapies, but the potential quality of life disadvantages are widespread. These include the burgeoning social concerns over genetic discrimination in the workplace and by the individuals' insurance companies. Also, familial issues associated with decisions regarding whether to test, what to do with test results and when to provide access to genetic counseling remain unresolved. Finally, legal matters relating to the patentability of DNA, access to genetic tests and availability of treatments developed by individuals and companies because of donated patients' samples will be dealt with in the near future.

Conclusion

Since the initial discovery of each of the 18 distinct leukodystrophies, significant progress has been made in both the basic and clinical sciences, leading to better understanding and treatment of this class of genetic disorders. These advances began with describing each condition's unique symptoms and clinical course and culminated with identifying the specific gene responsible for demyelination or dysmyelination. Currently, research aims focus on utilizing stem cell and viral therapies to cure these abnormalities, but many of these studies have raised more questions than given answers. Specifically, determining the duration, timing and intensity of potential treatments all remain as answers for the future. Furthermore, the apparent gap between results in animal models and applicability in human disease remains to be bridged. As science moves towards those ends, these disease treatments promise to teach us more about the development of the central nervous system and the complexities of the homeostatic mechanisms regulating brain function.

References

Adachi, M., L. Schneck, J. Cara, and B. W. Volk. 1973. Spongy degeneration of the central nervous system (van Bogaert and Bertrand type; Canavan's disease). A review. Hum Pathol **4**:331–347.

Adachi, M., and B. W. Volk. 1968. Protracted form of spongy degeneration of the central nervous system (van Bogaert and Bertrand type). Neurology **18**:1084–1092.

Adachi, M., B. J. Wallace, L. Schneck, and B. W. Volk. 1966. Fine structure of spongy degeneration of the central nervous system (van Bogaert and Bertrand type). J Neuropathol Exp Neurol **25**:598–616.

Adachi, M., B. J. Wallace, and B. W. Volk. 1967. Ultramicroscopic and histochemical studies of spongy degeneration (van Bogaert and Bertrand type). J Neuropathol Exp Neurol **26**: 164–165.

Aubourg, P., C. Adamsbaum, M.-C. Lavallard-Rousseau, F. Rocchiccioli, N. Cartier, I. Jambaque, C. Jakobezak, A. Lemaitre, F. Boureau, C. Wolf, and P.-F. Bougneres. 1993. A two-year trial of oleic and erucic acids ("Lorenzo's Oil") as treatment for adrenomyeloneuropathy. N Engl J Med **329**:745–352.

Aubourg, P., S. Blanche, I. Jambaque, F. Rocchiccioli, G. Kalifa, C. Naud-Saudreau, M. O. Rolland, M. Debre, J. L. Chaussain, C. Griscelli, and et al. 1990. Reversal of early neurologic and neuroradiologic manifestations of X-linked adrenoleukodystrophy by bone marrow transplantation. N Engl J Med **322**:1860–1866.

Aubourg, P., and J. L. Chaussain. 1991. Adrenoleukodystrophy presenting as Addison's disease in children and adults. Trends Endocrinol Metab **2**:49–52.

Banker, B. Q., J. T. Roberson, and M. Victor. 1964. Spongy degeneration of the CNS in infancy. Neurology **14**:981–1001.

Baslow, M. 2003. Brain N-Acetylaspartate as a molecular water pump and its role in the etiology of canavan disease: A mechanistic explanation. J Mol Neurosci **21**:185–190.

Baslow, M. 2000. Canavan's spongiform leukodystrophy: A clinical anatomy of a genetic metabolic CNS disease — An analytical review. J. Mol. Neurosci **15**:61–69.

Benhamida, S., F. Pflumio, A. Dubart-Kupperschmitt, J. C. Zhao-Emonet, M. Cavazzana-Calvo, F. Rocchiccioli, S. Fichelson, P. Aubourg, P. Charneau, and N. Cartier. 2003. Transduced CD34 + cells from adrenoleukodystrophy patients with HIV-derived vector mediate long-term engraftment of NOD/SCID mice. Mol Ther **7**:317–324.

Bezaire, V., W. Hofmann, J. K. G. Kramer, L. P. Kozak, and M.-E. Harper. 2001. Effects of fasting on muscle mitochondrial energetics and fatty acid metabolism in Ucp3(-/-) and wild-type mice. Am J Physiol Endocrinol Metab **281**:E975–982.

Bezman, L., A. B. Moser, G. V. Raymond, P. Rinaldo, P. A. Watkins, K. D. Smith, N. E. Kass, and H. W. Moser. 2001. Adrenoleukodystrophy: Incidence, new mutation rate, and results of extended family screening. Ann Neurol **49**:512–517.

Boelens, J. 2006. Trends in haematopoietic cell transplantation for inborn errors of metabolism. J Inherit Metab Dis **29**:413–420.

Breheny, N., E. Geelhoed, J. Goldblatt, and P. O'Leary. 2005. Cost effectiveness of predictive genetic tests for familial breast and ovarian cancer. Genom Soc Pol **1**:67–79.

Brismar, J., G. Brismar, G. Gascon, and P. Ozand. 1990. Canavan disease: CT and MR imaging of the brain. AJNR Am J Neuroradiol **11**:805–810.

Brockmann, K., P. Dechent, B. Wilken, O. Rusch, J. Frahm, and F. Hanefeld. 2003. Proton MRS profile of cerebral metabolic abnormalities in Krabbe disease. Neurology **60**:819–825.

Budka, H., E. Sluga, and W. D. Heiss. 1976. Spastic paraplegia associated with Addison's disease: Adult variant of adreno-leukodystrophy. J Neurol. **213**:237–250.

Canavan, M. 1931. Schilder's encephalitis periaxialis diffusa. Report of a case in a child aged sixteen and one-half months. Arch Neurol Psychiatry **25**:299–308.

Cartier, N., J. Lopez, P. Moullier, F. Rocchiccioli, M. O. Rolland, P. Jorge, J. Mosser, J. L. Mandel, P. F. Bougneres, O. Danos, and et al. 1995. Retroviral-mediated gene transfer corrects very-long-chain fatty acid metabolism in adrenoleukodystrophy fibroblasts. Proc Natl Acad Sci **92**:1674–1678.

Chen, Y. Q., M. A. Rafi, G. de Gala, and D. A. Wenger. 1993. Cloning and expression of cDNA encoding human galactocerebrosidase, the enzyme deficient in globoid cell leukodystrophy. Hum Mol Genet **2**:1841–1845.

Chen, Y. Q., and D. A. Wenger. 1993. Galactocerebrosidase from human urine: Purification and partial characterization. Biochimica et Biophysica Acta (BBA) — Lipids and Lipid Metabolism **1170**:53–61.

Crome, L., F. Hanefeld, D. Patrick, and J. Wilson. 1973. Late onset globoid cell leukodystrophy. Brain **96**:841–848.

Davidson, B. L., C. S. Stein, J. A. Heth, I. Martins, R. M. Kotin, T. A. Derksen, J. Zabner, A. Ghodsi, and J. A. Chiorini. 2000. From the cover: Recombinant adeno-associated virus type 2, 4, and 5 vectors: Transduction of variant cell types and regions in the mammalian central nervous system. PNAS **97**:3428–3432.

De Gasperi, R., M. A. Gama Sosa, E. L. Sartorato, S. Battistini, H. MacFarlane, J. F. Gusella, W. Krivit, and E. H. Kolodny. 1996. Molecular heterogeneity of late-onset forms of globoid-cell leukodystrophy. Am J Hum Genet **59**:1233–1242.

Doerflinger, N., J. M. Miclea, J. Lopez, C. Chomienne, P. Bougneres, P. Aubourg, and N. Cartier. 1998. Retroviral transfer and long-term expression of the adrenoleukodystrophy gene in human CD34 + cells. Hum Gene Ther **9**:1025–1036.

Eichler, F. S., R. Itoh, P. B. Barker, S. Mori, E. S. Garrett, P. C. M. van Zijl, H. W. Moser, G. V. Raymond, and E. R. Melhem. 2002. Proton MR spectroscopic and diffusion tensor brain MR imaging in X-linked adrenoleukodystrophy: Initial experience. Radiology **225**:245–2452.

Elpeleg, O. N., A. Shaag, Y. Anikster, and C. Jakobs. 1994. Prenatal detection of Canavan disease (aspartoacylase deficiency) by DNA analysis. J Inherit Metab Dis **17**:664–666.

Escolar, M. L., M. D. Poe, J. M. Provenzale, K. C. Richards, J. Allison, S. Wood, D. A. Wenger, D. Pietryga, D. Wall, M. Champagne, R. Morse, W. Krivit, and J. Kurtzberg. 2005. Transplantation of umbilical-cord blood in babies with infantile Krabbe's disease. N Engl J Med **352**:2069–2081.

Esiri, M. M., N. M. Hyman, W. L. Horton, and R. H. Lindenbaum. 1984. Adrenoleukodystrophy: Clinical, pathological and biochemical findings in two brothers with the onset of cerebral disease in adult life. Neuropatol Appl Neurobiol **10**:429–445.

Eto, Y., J. S. Shen, X. L. Meng, and T. Ohashi. 2004. Treatment of lysosomal storage disorders: Cell therapy and gene therapy J Inherit Metab Dis. **27**:411–415.

Farrell, D. F., S. R. Hamilton, T. A. Knauss, E. Sanocki, and S. S. Deeb. 1993. X-linked adreno-leukodystrophy: Adult cerebral variant. Neurology **43**:1518–1522.

Fourcade, S., S. Savary, C. Gondcaille, J. Berger, A. Netik, F. Cadepond, M. El Etr, B. Molzer, and M. Bugaut. 2003. Thyroid hormone induction of the adrenoleukodystrophy-related gene (ABCD2). Mol Pharmacol **63**:1296–1303.

Fu, L., K. Inui, T. Nishigaki, N. Tatsumi, H. Tsukamoto, C. Kokubu, T. Muramatsu, and S. Okada. 1999. Molecular heterogeneity of Krabbe disease. J Inherit Metab Dis **22**:155–162.

Furuya, H., Y. Kukita, S. Nagano, Y. Sakai, Y. Yamashita, H. Fukuyama, Y. Inatomi, Y. Saito, R. Koike, S. Tsuji, Y. Fukumaki, K. Hayashi, and T. Kobayashi. 1997. Adult onset globoid cell leukodystrophy (Krabbe disease): Analysis of galactosylceramidase cDNA from four Japanese patients. Hum Genet **100**:450–456.

Hagberg, B., H. Kollberg, P. Sourander, and H. O. Akesson. 1970. Infantile globoid cell leucodystrophy (Krabbe's disease): A clinical and genetic study of 32 Swedish cases 1953–1967. Neuropaediatrie **1**:74–88.

Haq, E., S. Giri, I. Singh, and A. K. Singh. 2003. Molecular mechanism of psychosine-induced cell death in human oligodendrocyte cell line. J Neurochem. **86**:1428–1440.

Hogan, G. R., L. Gutmann, and S. M. Chou.1969. The peripheral neuropathy of Krabbe's (globoid) leukodystrophy. Neurology **19**:1094–1100

Igisu, H., and K. Suzuki. 1984. Progressive accumulation of toxic metabolite in a genetic leukodystrophy. Science **224**:753–755.

Janson, C. G., M. Assadi, J. Francis, L. Bilaniuk, D. Shera, and P. Leone. 2005. Lithium citrate for Canavan disease. Pediatr Neurol **33**:235–243.

Janson, C. G., S. W. McPhee, P. Leone, A. Freese, and M. J. During. 2001. Viral-based gene transfer to the mammalian CNS for functional genomic studies. Trends Neurosci **24**:706–712.

Kaul, R., K. Balamurugan, G. P. Gao, and R. Matalon. 1994. Canavan disease: Genomic organization and localization of human ASPA to 17p13-ter and conservation of the ASPA gene during evolution. Genomics **15**:364–370.

Kaul, R., G. P. Gao, M. Aloya, K. Balamurugan, A. Petrosky, K. Michals, and R. Matalon. 1994. Canavan disease: Mutations among Jewish and non-jewish patients. Am J Hum Genet **55**:34–41.

Kaul, R., G. P. Gao, R. Matalon, M. Aloya, Q. Su, M. Jin, A. B. Johnson, R. B. Schutgens, and J. T. Clarke. 1996. Identification and expression of eight novel mutations among non-Jewish patients with Canavan disease. Am J Hum Genet **59**:95–102.

Kaul, R., G. Ping Gao, K. Balamurugan, and R. Matalon. 1993. Cloning of the human aspartoacylase cDNA and a common missense mutation in Canavan disease. Nat Genet **5**:118–123.

Kelley, R. 1993. Prenatal detection of Canavan disease by measurement of N-acetyl-L-aspartate in amniotic fluid. J Inherit Metab Dis. **16**:918–919.

Kemp, S., A. Pujol, H. R. Waterham, B. M. van Geel, C. D. Boehm, G. Raymond, V., G. R. Cutting, R. J. Wanders, and H. W. Moser. 2001. ABCD1 mutations and the X-linked adrenoleukodystrophy mutation database: Role in diagnosis and clinical correlations. Hum Mutat **18**:499–515.

Kennedy, C. R., J. T. Allen, A. H. Fensom, S. J. Steinberg, and R. Wilson. 1994. X-linked adrenoleukodystrophy with non-diagnostic plasma very long chain fatty acids. J Neurol Neurosurg Psychiatry **57**:759–761.

Kingsley, P. B., T. C. Shah, and R. Woldenberg. 2006. Identification of diffuse and focal brain lesions by clinical magnetic resonance spectroscopy. NMR Biomed **19**:435–462.

Kok, F., S. Neumann, C. O. Sarde, S. Zheng, K. H. Wu, H. M. Wei, J. Bergin, P. A. Watkins, S. Gould, G. Sack, and et al. 1995. Mutational analysis of patients with X-linked adrenoleukodystrophy. Hum Mutat **6**:104–115.

Kolodny, E. H., S. Raghavan, and W. Krivit. 1991. Late-onset Krabbe disease (globoid cell leukodystrophy): Clinical and biochemical features of 15 cases. Dev Neurosci **13**:232–239.

Korenke, G. C., H. J. Christen, B. Kruse, D. H. Hunneman, and F. Hanefeld. 1997. Progression of X-linked adrenoleukodystrophy under interferon-$\hat{I}^2$ therapy. J Inherit Metab Dis **V20**:59–66.

Krabbe, K. 1916. A new familial, infamtile form of diffuse brain sclerosis. Brain **39**:74–114.

Krasemann, E. W., V. Meier, G. C. Korenke, D. H. Hunneman, and F. Hanefeld. 1996. Identification of mutations in the ALD-gene of 20 families with adrenoleukodystrophy/adrenomyeloneuropathy. Hum Genet 97:194–197.

Krivit, W., E. G. Shapiro, C. Peters, J. E. Wagner, G. Cornu, J. Kurtzberg, D. A. Wenger, E. H. Kolodny, M. T. Vanier, D. J. Loes, K. Dusenbery, and L. A. Lockman. 1998. Hematopoietic stem-cell transplantation in globoid-cell leukodystrophy. N Engl J Med 338:1119–1127.

Kronn, D., C. Oddoux, J. Phillips, and H. Ostrer. 1995. Prevalence of Canavan disease heterozygotes in the New York metropolitan Ashkenazi Jewish population. Am J Hum Genet 57:1250–1252.

Kumar, S., N. S. Mattan, and J. de Vellis. 2006. Canavan disease: A white matter disorder. Ment Retard Dev Disabil Res Rev 12:157–165.

Leone, P., C. G. Janson, L. Bilaniuk, Z. Wang, F. Sorgi, L. Huang, R. Matalon, R. Kaul, Z. Zeng, A. Freese, S. W. McPhee, E. Mee, and M. J. During. 2000. Aspartoacylase gene transfer to the mammalian central nervous system with therapeutic implications for Canavan disease. Ann Neurol 48:27–38.

Libber, S. M., C. J. Migeon, F. R. Brown, and H. W. Moser. 1986. Adrenal and testicular function in 14 patients with adrenoleukodystrophy or adrenomyeloneuropathy. Horm Res 24:1–8.

Ligtenberg, M. J., S. Kemp, C. O. Sarde, B. M. van Geel, W. J. Kleijer, P. G. Barth, J. L. Mandel, B. A. van Oost, and P. A. Bolhuis. 1995. Spectrum of mutations in the gene encoding the adrenoleukodystrophy protein. Am J Hum Genet 56:44–50.

Loes, D. J., A. E. Stillman, S. Hite, E. Shapiro, L. Lockman, R. E. Latchaw, H. Moser, and W. Krivit. 1994. Childhood cerebral form of adrenoleukodystrophy: short-term effect of bone marrow transplantation on brain MR observations. AJNR Am J Neuroradiol 15:1767–1771.

Luzi, P., M. A. Rafi, and D. A. Wenger. 1995. Characterization of the large deletion in the GALC gene found in patients with Krabbe disease. Hum Mol Genet 4:2335–2338.

Lyon, G., R. D. Adams, and E. H. Kolodny. 1996. Neurology of Hereditary Metabolic Diseases of Children, 2nd ed. McGraw-Hill, New York.

Lyon, G., B. Hagberg, P. Evrard, C. Allaire, L. Pavone, and V. M. 1991. Symptomatology of late onset Krabbe s leukodystrophy: The European experience. Dev Neurosci 13:240–244.

Maestri, N. E., and T. H. Beaty. 1992. Predictions of a 2-locus model for disease heterogeneity: Application to adrenoleukodystrophy. Am J Hum Genet 44:576–582.

Matalon, R., and K. Michals-Matalon. 1999. Biochemistry and molecular biology of Canavan disease. Neurochem Res 24:507–513.

Matalon, R., K. Michals, and R. Kaul. 1995. Canavan disease: From spongy degeneration to molecular analysis. J Pediatr 127:511–517.

Matalon, R., K. Michals, D. Sebesta, M. Deanching, P. Gashkoff, and J. Casanova. 1988. Aspartoacylase deficiency and N-acetylaspartic aciduria in patients with Canavan disease. Am J Med Genet 29:463–471.

Matalon, R., P. Rady, K. Platt, H. Skinner, M. Quast, G. Campbell, K. Matalon, J. Ceci, S. Tyring, M. Nehls, S. Surendran, J. Wei, E. Ezell, and S. Szucs. 2000. Knock-out mouse for Canavan disease: A model for gene transfer to the central nervous system. J Gene Med 2:165–175.

Mathew, R., P. Arun, C. N. Madhavarao, J. R. Moffett, and M. A. A. Namboodiri. 2005. Progress toward acetate supplementation therapy for Canavan disease: Glyceryl triacetate administration increases acetate, but not N-Acetylaspartate, levels in brain. J Pharmacol Exp Ther 315:297–303.

Maurice, R. E., and E. Michaela. 2001. The neurobiology and evolution of cannabinoid signalling. Phil Trans R Soc B Biol Sci 356:381–408.

McPhee, S. W., J. Francis, C. G. Janson, T. Serikawa, K. Hyland, E. O. Ong, S. S. Raghavan, A. Freese, and P. Leone. 2005. Effects of AAV-2-mediated aspartoacylase gene transfer in the tremor rat model of Canavan disease. Brain Res Mol Brain Res 135:112–121.

Migeon, B. R., H. W. Moser, A. B. Moser, J. Axelman, D. Sillence, and R. A. Norum. 1981. Adrenoleukodystrophy: Evidence for X linkage, inactivation, and selection favoring the mutant allele in heterozygous cells. Proc Natl Acad Sci 78:5066–5070.

Morell, P., R. Quarles, and W. Norton. 1994. Basic Neurochemistry, Raven, NY.

Moser, H. W. 1997. Adrenoleukodystrophy: Phenotype, genetics, pathogenesis and therapy. Brain **120**:1485–1508.

Moser, H. W., A. B. Moser, N. Kawamura, J. Murphy, K. Suzuki, H. Schaumburg, and Y. Kishimoto. 1980. Adrenoleukodystrophy: Elevated C26 fatty acid in cultured skin fibroblasts. Ann Neurol **7**:542–549.

Moser, H. W., A. B. Moser, K. D. Smith, A. Bergin, J. Borel, J. Shankroff, O. C. Stine, C. Merette, J. Ott, W. Krivit, and et al. 1992. Adrenoleukodystrophy: Phenotypic variability and implications for therapy. J Inherit Metab Dis **15**:645–664.

Moser, H. W., G. V. Raymond, and P. Dubey. 2005. Adrenoleukodystrophy: New approaches to a neurodegenerative disease. JAMA **294**:3131–3134.

Moser, H. W., K. D. Smith, P. A. Watkins, J. Powers, and A. B. Moser. 2001. X-linked adrenoleukodystrophy, p. 3257–3301. *In* AL. Beaudet (ed.), The Metabolic and Molecular Bases of Inherited Disease, 8th ed. McGraw-Hill, New York.

Moser, H. W., P. J. Tutschka, F. R. Brown, A. E. Moser, A. M. Yeager, I. Singh, S. A. Mark, A. A. Kumar, J. M. McDonnell, C. L. White, and et al. 1984. Bone marrow transplant in adrenoleukodystrophy. Neurology **34**:1410–1417.

Mosser, J., A.-M. Douar, C.-O. Sarde, P. Kioschis, R. Feil, H. Moser, A.-M. Poustka, J.-L. Mandel, and P. Aubourg. 1993. Putative X-linked adrenoleukodystrophy gene shares unexpected homology with ABC transporters. Nature **361**:726–730.

Mosser, J., Y. Lutz, M. E. Stoeckel, C. O. Sarde, C. Kretz, A. M. Douar, J. Lopez, P. Aubourg, and J. L. Mandel. 1994. The gene responsible for adrenoleukodystrophy encodes a peroxisomal membrane protein. Hum Mol Genet **3**:265–271.

Naidu, S., M. J. Bresnan, D. Griffin, S. O Toole, and H. W. Moser. 1988. Childhood adrenoleukodystrophy. Failure of intensive immunosuppression to arrest neurologic progression. Arch Neurol **45**:846–848.

Ory-Lavollee, L., R. D. Blakely, and J. T. Coyle. 1987. Neurochemical and immunocytochemical studies on the distribution of N-acetyl-aspartylglutamate and N-acetyl-aspartate in rat spinal cord and some peripheral nervous tissues. J Neurochem **48**:895–899.

Pearce, J. M. 2004. Canavan's disease. J Neurol Neurosurg Psychiatry **75**:1410.

Powers, J. 1985. Adreno-leukodystrophy (adreno-testiculo-leukomyelo-neuropathic-complex). Clin Neuropathol **4**:181–199.

Powers, J. M., H. H. Schaumburg, A. B. Johnson, and C. S. Raine. 1980. A correlative study of the adrenal cortex in adreno-leukodystrophy—evidence for a fatal intoxication with very long chain saturated fatty acids. Invest. Cell Pathol. **3**:353–376.

Rafi, M. A., P. Luzi, Y. Q. Chen, and D. A. Wenger. 1995. A large deletion together with a point mutation in the GALC gene is a common mutant allele in patients with infantile Krabbe disease. Hum Mol Genet **4**:1285–1289.

Rizzo, W. B., P. A. Watkins, M. W. Phillips, D. Cranin, B. Campbell, and J. Avigan. 1986. Adrenoleukodystrophy: Oleic acid lowers fibroblast saturated C22–26 fatty acids. Neurology **36**:357–361.

Rushton, A. R., B. A. Shaywitz, C. C. Duncan, R. B. Geehr, and E. E. Manuelidis. 1981. Computed tomography in the diagnosis of Canavan's disease. Ann Neurol **10**:57–60.

Sabatelli, M., L. Quaranta, F. Madia, G. Lippi, A. Conte, M. Lo Monaco, G. Di Trapani, M. A. Rafi, D. A. Wenger, A. M. Vaccaro, and P. Tonali. 2002. Peripheral neuropathy with hypomyelinating features in adult-onset Krabbe's disease. Neuromuscul Disord **12**:386–391.

Sager, T. N., C. Thomsen, J. S. Valsborg, H. Laursen, and A. J. Hansen. 1999. Astroglia contain a specific transport mechanism for N-acetyl-L-aspartate. J Neurochem **73**:807–811.

Sakai, N., K. Inui, N. Midorikawa, Y. Okuno, S. Ueda, A. Iwamatsu, and S. Okada. 1994. Purification and characterization of galactocerebrosidase from human lymphocytes. J Biochem (Tokyo) **116**:615–620.

Sasaki, M., N. Sakuragawa, S. Takashima, S. Hanaoka, and M. Arima. 1991. MRI and CT findings in Krabbe disease. Pediatr Neurol **7**:283–288.

Sauer, M., S. Grewal, and C. Peters. 2004. Hematopoietic Stem Cell Transplantation for Mucopolysaccharidoses and Leukodystrophies. Hämatopoetische Stammzelltransplantation bei Kindern mit Mukopolysaccharidosen und Leukodystrophien,163–168.

Schaumburg, H. H., J. M. Powers, C. S. Raine, P. S. Spencer, J. W. Griffin, J. W. Prineas, and D. M. Boehme. 1977. Adrenomyeloneuropathy: A probable variant of adrenoleukodystrophy. II. General pathologic, neuropathologic, and biochemical aspects. Neurology 27:1114–1119.

Schaumburg, H. H., J. M. Powers, C. S. Raine, K. Suzuki, and E. P. Richardson. 1975. Adrenoleukodystrophy. A clinical and pathological study of 17 cases. Arch Neurol 32:577–591.

Shapiro, E., W. Krivit, L. Lockman, I. Jambaque, C. Peters, M. Cowan, R. Harris, S. Blanche, P. Bordigoni, D. Loes, R. Ziegler, M. Crittenden, D. Ris, B. Berg, C. Cox, H. Moser, A. Fischer, and P. Aubourg. 2000. Long-term effect of bone-marrow transplantation for childhood-onset cerebral X-linked adrenoleukodystrophy. Lancet 356:713–718.

Shen, J. S., K. Watabe, T. Ohashi, and Y. Eto. 2001. Intraventricular administration of recombinant adenovirus to neonatal twitcher mouse leads to clinicopathological improvements. Gene Ther 8:1081–1087.

Siddiqi, Z. A., D. B. Sanders, and J. M. Massey. 2006. Peripheral neuropathy in Krabbe disease: Effect of hematopoietic stem cell transplantation. Neurology 67:268–272.

Siemerling, E., and H. Creutzfeldt. 1923. Bronzekrankheit und sclerosierende encephalomyelitis (Diffuse Sklerose). Archiv fur Psychiatrie 68:217–244.

Surendran, S., F. J. Bamforth, A. Chan, S. K. Tyring, S. I. Goodman, and R. Matalon. 2003. Mild elevation of N-acetylaspartic acid and macrocephaly: Diagnostic problem. J Child Neurol 18:809–812.

Suzuki, K., and K. Suzuki. 1985. Genetic galactosylceramidase deficiency (globoid cell leukodystrophy, Krabbe disease) in different mammalian species. Neurochem Pathol 3:53–68.

Suzuki, K., and K. Suzuki. 1995. The twitcher mouse: A model for Krabbe disease and for experimental therapies. Brain Pathol. 5:249–258.

Suzuki, K., and M. Taniike. 1998. Twenty five years of the "psychosine hypothesis": A personal perspective of its history and present status. Neurochem Res 23:251–259.

Svennerholm, L., M. T. Vanier, and J. E. Mansson. 1980. Krabbe disease: A galactosylsphingosine (psychosine) lipidosis. J Lipid Res 21:53–64.

Tapasi, S., P. Padma, and O. H. Setty. 1998. Effect of psychosine on mitochondrial function. Indian J Biochem Biophys 35:161–165.

Theil, A. C., R. B. Schutgens, R. J. Wanders, and H. S. Heymans. 1992. Clinical recognition of patients affected by a peroxisomal disorder: A retrospective study in 40 patients. Eur J Pediatr 151:117–120.

Tonshoff, B., W. Lehnert, and H. H. Ropers. 1982. Adrenoleukodystrophy: Diagnosis and carrier detection by determination of long-chain fatty acids in cultured fibroblasts. Clin Genet 22:25–29.

Traeger, E. C., and I. Rapin. 1998. The clinical course of Canavan disease. Pediatr Neurol 18:207–212.

Valle, D., and J. Gartner. 1993. Penetrating the peroxisome. Nature 361:682–683.

van Geel, B. M., J. Assies, E. B. Haverkort, J. H. T. M. Koelman, B. Verbeeten, Jr., R. J. A. Wanders, and P. G. Barth. 1999. Progression of abnormalities in adrenomyeloneuropathy and neurologically asymptomatic X-linked adrenoleukodystrophy despite treatment with "Lorenzo's oil". J Neurol Neurosurg Psychiatry 67:290–299.

van Geel, B. M., J. Assies, R. J. A. Wanders, and P. G. Barth. 1997. X linked adrenoleukodystrophy: Clinical presentation, diagnosis, and therapy. J Neurol Neurosurg Psychiatry 63:4–14.

Vanhole, C., F. de Zegher, P. Casaer, H. Devlieger, R. J. Wanders, G. Vanhove, and J. Jaeken. 1994. A new peroxisomal disorder with fetal and neonatal adrenal insufficiency. Arch Dis Child Fetal Neonatal Ed 71:F55–56.

Verdu, P., M. Lammens, R. Dom, A. Van Elsen, and H. Carton. 1991. Globoid cel leukodystrphy: A family with both late-infantile and adult type. Neurology 41:1382–1384.

Volkow, N. D., L. Patchell, M. V. Kulkarni, K. Reed, and M. Simmons. 1987. Adrenoleukodystrophy: Imaging with CT, MRI, and PET. J Nucl Med **28**:524–527.

Wanders, R. J., C. W. van Roermund, W. Lageweg, B. S. Jakobs, R. B. Schutgens, A. A. Nijenhuis, and J. M. Tager. 1992. X-linked adrenoleukodystrophy: Biochemical diagnosis and enzyme defect. J Inherit Metab Dis. **15**:634–644.

Wang, C., C. M. Wang, K. R. Clark, and T. J. Sferra. 2003. Recombinant AAV serotype 1 transduction efficiency and tropism in the murine brain. Gene Ther **10**:1528–1534.

Wenger, D. A., M. A. Rafi, and P. Luzi. 1997. Molecular genetics of Krabbe disease (globoid cell leukodystrophy): Diagnostic and clinical implications. Hum Mutat **10**:268–279.

Wenger, D. A., M. A. Rafi, P. Luzi, J. Datto, and E. Costantino-Ceccarini. 2000. Krabbe disease: Genetic aspects and progress toward therapy. Mol Genet Metab **70**:1–9.

Wenger, D. A., K. Suzuki, Y. Suzuki, and K. Suzuki. 2001. Globoid cell leukodystrophy (Krabbe disease), p. 3669–3694. *In* AL, Beaudet, Scriver, CR, Sly WAS, Valle D, Childs B, Vogelstein B (eds.), The Metabolic and Molecular Basis of Inherited Disease, 8th ed. McGraw-Hill, New York.

Zarifi, M. K., A. A. Tzika, L. G. Astrakas, T. Y. Poussaint, D. C. Anthony, and B. T. Darras. 2001. Magnetic resonance spectroscopy and magnetic resonance imaging findings in Krabbe's disease. J Child Neurol **16**:522–526.

Zelnik, N., and O. N. Elpeleg. 1994. Canavan disease: Clinical-genetic correlation. Dev Med Child Neurol **36**:845.

Zelnik, N., A. S. Luder, O. N. Elpeleg, V. Gross-Tsur, N. Amir, J. A. Hemli, A. Fattal, and S. Harel. 1993. Protracted clinical course for patients with Canavan disease. Dev Med Child Neurol **35**:355–358.

Zu Rhein, G. M., P. L. Eichman, and F. Puletti. 1960. Familial idiocy with spongy degeneration of the central nervous system of van Bogaert—Bertrand type. Neurology **10**:998–1006.

Chapter 16
Bilirubin Encephalopathy

Jeffrey W. McCandless and David W. McCandless

Bilirubin encephalopathy is a syndrome in newborns, in which increased plasma levels of unconjugated bilirubin outstrip the binding capacity of albumin, and gain access to the brain, producing neurological symptoms. This is a somewhat complex process in which factors such as prematurity, half-life of red blood cells, albumin binding capacity, and liver enzyme capability, contribute to the potential severity of the process. Treatment modalities exist, which can lessen, or eliminate risk for neurological effects. On the other hand, jaundice of the newborn frequently does not peak until several days after birth, when newborns have been discharged from the hospital. As with all metabolic encephalopathies, early recognition, and early treatment offer the maximum hope for a satisfactory outcome.

This chapter examines various aspects of bilirubin encephalopathy including bilirubin metabolism, bilirubin toxicity, in-vivo studies and studies in Gunn rats, human studies, and treatment modalities.

Bilirubin Metabolism

When red blood cells break down, the resulting components include heme and globin, and bilirubin is contained in the heme moiety (Tenhunen et al., 1968). Upon release, free bilirubin is lipid soluble and must be transported in plasma bound to the protein albumin (Ostrow et al., 1963). The life span of red blood cells in adults is about 120 days, and in newborns is reduced to around 90 days (Borun et al., 1957). This results in a significant amount of bilirubin to be removed, and the newborn liver must develop functional enzymatic capability quickly in order to prevent jaundice. In fact, 10–20% of newborns develop a transient "physiological jaundice" as the enzymatic ability of the liver develops to conjugate and eliminate bilirubin from the serum.

Plasma bilirubin, bound to albumin, is taken up by the liver cells, and conjugated with glucuronic acid to form bilirubin glucuronide (Schmid et al., 1957). The enzyme which catalyses the first step in this reaction is glucuronyl transferase (Brown, 1957). Subsequently, the conjugated bilirubin is excreted by the liver into the biliary system, transported to the duodenum, where further metabolism occurs,

D.W. McCandless (ed.) *Metabolic Encephalopathy*,
doi: 10.1007/978-0-387-79112-8_16, © Springer Science+Business Media, LLC 2009

and is eliminated in feces. Problems or blockage in any of these steps can result in elevated plasma bilirubin levels.

In utero, bilirubin is transported across the placenta in the unconjugated form, and metabolized by the maternal liver (Schmid et.al., 1959). Because of this, the enzymes associated with the conjugation of bilirubin are not "challenged" until birth when the placental circulation is removed. At this time, there is a short period during which the conjugating capacity is "ratcheted-up", resulting in a transient so called physiological jaundice period. The levels of plasma bilirubin during this period rarely exceed 10 mg%, and are more usually around 5 mg%. This peaks about postnatal day 4–5. The postnatal development of glucuronyl transferase function results in the disappearance of physiological jaundice by the end of the first postnatal week. Problems with this normal sequence such as delay in enzyme function, or increased bilirubin production (erythroblastosis foetalis), may result in greatly increased plasma bilirubin levels. Plasma concentration of bilirubin well in excess of 20 mg% may occur several days after birth, at a time when the newborn has been released from the hospital. Recognition and early treatment of hyperbilirubinemia are essential.

When plasma bilirubin levels reach higher than 20 mg%, the binding capacity of albumen may be exceeded (Broderson, 1980). The actual binding capacity is influenced by several variables, including the presence in plasma of fatty acids, salicylates, or sulfonamides among other substances (Broderson, 1978). The binding affinity of bilirubin and albumin is quite high, such that albumin can pull bilirubin out of the tissues, so that when administered to a jaundiced newborn, it may be an effective treatment (Diamond and Schmid, 1966).

Bilirubin Toxicity

Early studies, in vitro, were able to show that unconjugated bilirubin was a potent uncoupler of oxidative phosphorylation (Day, 1954; Brown and Waters, 1958; Zetterstrom and Ernster, 1956). These studies and others were performed in mitochondria from a variety of tissues including beef heart, rat liver, and rat brain. The results were consistent regardless of the tissue used. Biliverdin did not show these toxic effects (Brown and Waters, 1958). A variety of further studies examined the effects of bilirubin on other enzyme systems in vitro, and the results were largely consistent with the concept that bilirubin affects mitochondrial function Karp, 1979; Schenker et al., 1986).

These early in vitro studies were important, showing an effect on mitochondrial function which would later be reconfirmed in-vivo. In-vitro studies such as these frequently used bilirubin concentrations many times and measured that in brains of kernicteric animals. This was a source of criticism, but it should be remembered that pathologically, kernicteric human brain shows a highly selective localization and concentration of pigment staining. Later studies showed effects of bilirubin on many other cellular functions, such as membrane permeability. However, these results could be secondary to changes in mitochondrial function, and resultant alteration in energy metabolism.

In terms of bilirubin toxicity, it is important to note that only free unconjugated bilirubin can cross the blood—brain barrier into the brain parenchyma. Initially, the concept developed was that in the newborn, the blood—brain barrier was "immature", and that this increased permeability facilitated the development of kernicterus in hyperbilirubinemic newborns. Later studies especially on the blood brain barrier showed that it was a highly complex "transport" system, influenced by a variety of factors. Other deleterious conditions such as acidosis and hypoxia can lower the brain's threshold for the development of kernicterus. Even bilirubin itself may influence energy metabolism in endothelial cells comprising the blood—brain barrier, thus lowering its functional capacity. The finding of rare instances of adult onset of bilirubin encephalopathy argues in favor of this (Blaschke et al., 1974; Ho et al., 1980). For a further review of the blood—brain barrier in newborn bilirubin encephalopathy see Schenker et al. (1986).

The binding of bilirubin to albumin renders it water-soluble. Unbound bilirubin is lipid-soluble, and therefore has a predilection for tissue high in lipids, such as adipose tissue, and brain, and when the bilirubin-albumin bond is broken, bilirubin can enter cerebral tissue, even in adults (Diamond and Schmid, 1966). It is important to remember that the binding affinity of bilirubin to albumin is tight enough to allow albumin to pull bilirubin out of tissue. Other neurotoxic effects, such as on long-term potentiation of synaptic transmission (Zhang et al., 2003), and over stimulation of NMDA receptors (Grojean et al., 2000; Grojean et al., 2001), can be secondary to the initial toxic effect on mitochondrial function, with resultant diminution of energy metabolism.

One other feature of bilirubin worth mentioning regarding effects on cells is the apparent neuroprotective effect of mild hyperbilirubinemia (Dore et al., 1999; Dore and Snyder, 1999). The mild hyperbilirubinemia appears to have an antioxidative protective effect on cerebral neurons, and possibly on cells involved in the blood—brain barrier.

Finally, it has been reported (Mustafa et al., 1969) that albumin added to homogenates exposed to bilirubin lowered or eliminated the toxic effect. The toxic effects of bilirubin in vitro are more pronounced in brain as compared to liver, and this is attributed to the higher level of lipid in the brain mitochondria (Menken et al., 1966). This differential sensitivity of cells to the cytotoxic effects of bilirubin was also demonstrated by Cowger (Cowger, 1971). They studied the cell culture system L-929 and found bilirubin in a concentration of 2.5×10^{-6} m lowered viability, whereas ten times that amount of bilirubin was needed to produce similar toxicity in Hela cells.

In-Vivo Animal Studies

Several different animal models have been utilized to examine the effects of hyperbilirubinemia in vivo. The primary focus was to elevate bilirubin levels to the point of kernicterus, then examine changes in cerebral metabolism.

In a study involving rabbits (Morphis et al., 1982), newborn animals were infused with unconjugated bilirubin. When the newborns became symptomatic, and the brains were harvested, it was found that there was a dose-dependent protein kinase deficiency which affected protein phosphorylation. This in turn inhibited other cyclic AMP independent phosphorylation reactions. These studies were performed in whole brain homogenates.

In an early in vivo study (Diamond and Schmid, 1967), unconjugated bilirubin was infused into newborn guinea pigs, and oxidative phosphorylation was measured in whole brain extract. It was found that the neurotoxic guinea pigs had oxidative phosphorylation similar to that of control littermates. Problems with this study were at least two-fold. First, guinea pigs are born much more mature than mice, rats, or humans. Generally speaking, their eyes are open, they are ambulatory, and able to survive alone just a few days after birth. This suggests a mature brain at birth as compared to rats. Second, the studies were performed on whole brain. It is well known that the neurotoxicity of bilirubin encephalopathy is specific to comparatively small brain regions. Using whole brain for analysis could "mask" changes in more discrete brain regions.

In several papers (Rozdilsky, 1966; Rozdilsky and Olszewski, 1961, Rozdilsky, 1961), Rozdilsky and associates produced animal models of bilirubin encephalopathy in newborn kittens, dogs, and rabbits. The general experimental paradigm was to inject bilirubin sufficient to yield an initial plasma concentration of about 50 mg%, then to monitor levels at 20 mg% or more. This produced overt clinical symptoms including opisthatonis, running, twitching, etc. Nearly all animals died within 36 h, showing stupor and coma. Brains were prepared for gross and histological examination.

Results in newborn kittens showed a selective nuclear staining by bilirubin in thalamic nuclei, subthalamic nuclei, inferior colliculi, cuneate nuclei, and cochlear nuclei. The intensity of staining correlated with the time of exposure. Histologically, changes were seen in some neurons, and consisted of vacuolation and pyknotic nuclei.

In newborn dogs and rabbits, the staining of nuclear areas was much less. For example, only 8% of newborn rabbits showed nuclear staining using the above injection regimen. By contrast, pre-treating animals (dogs) with insulin to produce severe hypoglycemia resulted in over two-third demonstrating typical nuclear staining at death. The authors noted that the lesions produced in both kittens with bilirubin alone, and in dogs with a pre-insult such as hypoglycemia, produced lesions with a striking similarity to those seen in newborn children who died with severe hyperbilirubinemia. This suggests that these animal models are appropriate for studying intracerebral toxicity mechanisms.

The Gunn Rat Model

In 1938, a Wistar rat mutant was described (Gunn, 1938) which had a genetic deficiency of the liver bilirubin conjugating enzyme glucuronyl transferase. The defect in glucuronyl transferase corresponds to the defect in the human genetic disorder in bilirubin metabolism, the Crigler—Najjar syndrome. This was an autosomal recessive defect in an otherwise normal rat, and breeding and maintaining these rats was not difficult. The significance of this rat model was that here was a naturally occurring hyperbilirubinemic animal, born normal, but quickly becoming jaundiced, as do humans, and further, the rat is born quite immature, as are humans. This provided as perfect a model of human disease as could be expected. Shortly after being described, and as breeding pairs became available, a variety of studies on the pathogenesis of kernicterus in the Gunn rat took place.

A homozygous Gunn rat is not jaundiced at birth due to placental transfer of bilirubin, and its subsequent metabolism by the maternal liver. Jaundice usually occurs by day 2–3, as in humans, when the newborn liver is challenged, and the enzymatic machinery for conjugation and subsequent excretion of bilirubin is absent (Carbone and Grodsky, 1957). Bilirubin levels rise, and may peak at around the second to third week of life at about 15–25 mg%. Clinically newborn rats may appear to be lagging in growth (weight) as compared to littermates, and jaundice can clearly be seen. Frequently, homozygotes seem not to be able to gain access to nursing sites. By 4–5 weeks, survivors may only be half the weight of littermates. Neurological symptoms such as opisthotonus, wild running, and high pitch squeaking noises are noticed. Overt kernicterus can easily be produced in 100% of homozygotes by administration of albumin binding compounds such as sulfonamides.

A variety of studies have examined the neurochemical effects of cerebral bilirubin staining. In one such study (Hanefeld and Natzschka, 1971), kernicterus was produced in newborn Gunn rats, and unstained brain sections were examined by phase contrast microscopy for several key enzymes. Additional brains were prepared for routine histology. Results showed bilirubin staining in the basal ganglia, brain stem, and cerebellum. Histochemical studies showed no effect of kernicterus on lysomal enzymes, but oxidative enzymes (LDH, SDH, NAD, NADPH) all showed greatly reduced or no enzyme activity as compared to heterozygote littermate controls.

When the brains of homozygous newborn Gunn rats are examined by electron microscopy (Schutta and Johnson, 1971), several features are noted. Cerebellar neurons show cytoplasmic whorls, and significant mitochondrial changes. These changes include enlargement, and vacuoles containing glycogen. Accumulation of glycogen might not be surprising given the effect of bilirubin on oxidative phosphorylation. Also noted in this study was a diminution of the actual mass of the cerebellum.

A natural extension of these morphologic studies was to examine energy metabolism in brains of symptomatic homozygous Gunn rat newborns. In one such study (Menken and Weidenback, 1967), newborn homozygous Gunn rats were sacrificed when symptomatic, and mitochondria isolated from whole brain. Subsequently, oxidative phosphorylation and respiratory control were measured using polarographic methods. Results showed no significant differences in symptomatic pups as compared to nonsymptomatic littermate controls. As pointed out by the authors, using whole brain mitochondria in a situation in which the actual neurophathological lesion is highly circumscribed and focal, would, by necessity, include brain areas not affected by bilirubin. This would act to dilute those mitochondria from stained nuclei, rendering the data inconclusive.

In another study (Katoh et al., 1975), ATP and TCA cycle compounds were measured in the whole brains of symptomatic newborn Gunn rats. Results showed no change in energy metabolites or intermediates in symptomatic as compared to normal littermate controls. Again, however, inclusion in the sample for analysis of non-affected brain tissue diluted the sample.

This problem was addressed by refining analysis techniques to permit analysis of energy metabolites in small 20–80 mg brain samples (Schenker et al., 1966). This approach permitted analysis in cortex, subcortex, and cerebellum of ATP in symptomatic animals without inclusion in the kernicteric samples of "normal" tissue. Results showed a 13% depletion in the cerebellum of mildly symptomatic Gunn rats, and a 27% depletion of ATP in animals with advanced symptoms. Cortex and subcortex (nonstained) samples were unchanged. This supported the concept, based on symptomatology, and neuropathologic findings, that the toxicity of bilirubin in vivo is limited to the cerebral areas affected.

Even in the case of the cerebellum, it is possible that areas other than Purkinje cells (which show the greatest cellular damage) might be less affected. The micro analytical methodology developed by Passonneau and Lowry (1993), which enable the measurement of metabolites in samples as small as single cells could answer these questions. Accordingly, McCandless and Abel, (1980) applied these techniques to the cerebella of homozygous newborn Gunn rats, which were severely symptomatic. Normal littermates served as controls. ATP and phosphocreatine were measured using previously described techniques (Passonneau and Lowry, 1993). Analysis was performed in three layers of the cerebellum: molecular, Purkinje cell rich, and granular. Results showed a significant and selective decrease in both ATP and PCR (see Figs. 16.1 and 16.2). These results further verify the concept that, as shown earlier, bilirubin has an effect on cellular metabolism (oxidative phosphorylation) such that the energy source for nearly all cellular activity (ATP) is depleted. This depletion in the areas, which correlate with both site and severity of symptoms, could lead to other cellular dysfunction.

Human Bilirubin Encephalopathy-Pathology

The neuropathological findings in humans were first described by Schmorl (Schmorl, 1910), who also advanced the concept that there was a selective staining of certain cerebral areas and neurons. He speculated that somehow there was special binding attraction of bilirubin for certain neurons. In his writings he coined the term kernicterus to describe the yellow staining of cerebral nuclei. The term icterus gravis also arose because of the serious (fatal) prognosis of this disorder.

The underlying cause of human bilirubin encephalopathy is almost always erythroblastosis foetalis or a deficiency of the liver bilirubin conjugating enzyme

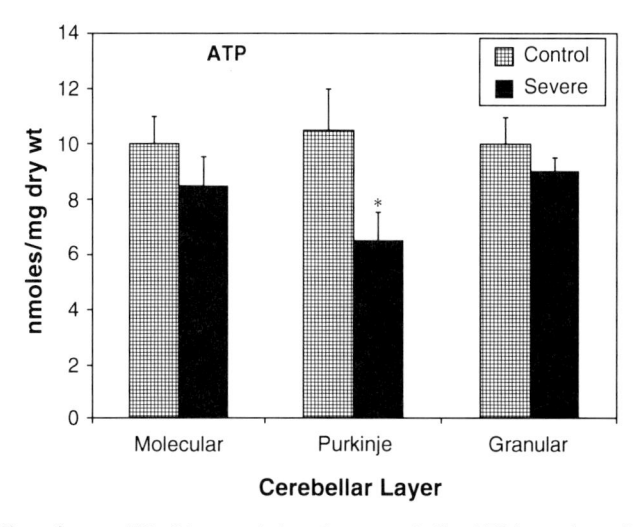

Fig. 16.1 Effect of severe bilirubin encephalopathy on cerebellar ATP in newborn Gunn rats. The mean ±SEM is expressed as nmoles mg^{-1} dry weight. * = p<0.05

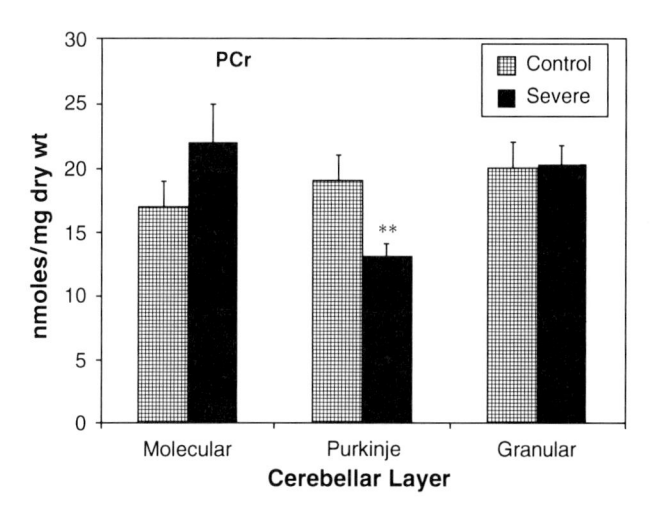

Fig. 16.2 Effect of severe bilirubin encephalopathy on cerebellar PCr in newborn Gunn rats. The mean ±SEM is expressed as nmoles mg^{-1} dry weight ** = p<0.01

system. The actual neuropathologic picture in any given case depends in part on the underlying cause, and when in the course of the disease, death occurs. Areas which are most affected include basal ganglia, cerebellum, hippocampus, medulla, thalamus and sub thalamus, etc. (Claireaux, 1961). Bilirubin crystals have been isolated from the brains of infants dying with hemolytic disease and hyperbilirubinemia (Dunn, 1951).

The focal staining of specific cerebral nuclei is noted if the patient dies during the acute phase; if the patient dies later, the staining has largely disappeared (Jervis, 1959). Microscopically, affected neurons display swollen neurons. These neurons may show vacuoles, and swelling or loss of nissl substance, and other early signs of cell death (Odell and Schutta, 1985). It is noted that these changes in kernicterus are not dissimilar to those seen in infants dying of asphyxia and anoxia.

Bilirubin Encephalopathy-Treatment

One long-standing treatment mechanism for jaundice is light therapy. The benefits of light therapy as a treatment for jaundice were first discovered in 1956 when a pediatric nurse noted that jaundiced babies who were exposed to sunlight experienced a fading of the yellow pigmentation of their skin (Cremer et al., 1958). A follow-up demonstration found that light therapy can cause the photodestruction of serum bilirubin in some infants (Cremer et al., 1958). Further controlled studies showed that phototherapy treatment effectively reduced the average serum bilirubin levels in premature infants (Lucey et al., 1968; Lucey, 1972). Although phototherapy was widely used in Europe and South America, it was not until the late 1960s that the method was used in the United States (Lucey, 1970).

Most of the pigment elimination during phototherapy is actually caused by structural isomerization as opposed to photodestruction (Ennever, 1990). Light energy causes isomerization of the bilirubin into compounds that can be excreted through urine or stools. Because bilirubin is similar to the pigment phycobilin (used to capture light energy), bilirubin changes its conformation when exposed to light.

The wavelength of light used during phototherapy impacts the extent of bilirubin reduction. In general, turquoise (blue—green) light is considered the most effective (e.g., Donzelli et al., 1995; Maisels, 1996). Several studies have compared the effects of distinct wavelengths. For example, Ebbesen et al. (2003) compared the impact of blue light (452 nm peak with 55 nm spectral width) with turquoise lights (490 nm peak with 65 nm spectral width) on total serum bilirubin concentrations in infants. Although the overall plasma bilirubin concentration was reduced equally in the two conditions, the light irradiance with turquoise light (measured in mW cm^{-2}) was about 75% of that with blue light. That finding led the authors to recommend turquoise light as a means of phototherapy. A related study by Roll and Christensen (2005) examined turquoise (490 nm) vs. blue (450 nm) irradiation in terms of cell damage. The results showed a significantly higher fraction of necrotic cells were found after exposure to blue light compared with turquoise light. Onishi, Ito and Isobe (1986) recommend an even higher wavelength (green light with a wavelength of 510 nm) because it keeps the bilirubin photoisomers at low levels. Green light also has the potential advantage of penetrating deeper in the skin than blue light (Vecchi et al., 1982). However, light of this wavelength probably has an overall limited effectiveness and is higher than the mostly commonly used wavelengths of about 425–475 nm (Madan, 2005).

Conventional fluorescent light is not the only means of providing phototherapy. Another option is a light emitting diode (LED) array, which does not have the portability limitations of fluorescent and halogen light sources. Rosen, Rosen, Rosen, Onaral and Hiatt (2005) found that light emitting diodes (LEDs) produce 71% higher intensity than a fluorescent system. LED phototherapy produced the fastest bilirubin concentration reduction and the largest overall concentration drop. Alternatively, sunlight can be used as a light source, particularly in areas where artificial light units are not available (Salih, 2001).

Numerous studies have examined the differences between intermittent vs. continuous light therapy, with sometimes inconclusive results. From a clinical standpoint, intermittent therapy has the benefit of allowing the evaluation of the infant's skin color during the dark periods (Roll, 2005). In addition, the total light energy required to decrease serum bilirubin by a certain concentration is less with intermittent therapy compared with continuous therapy (Vogl et al., 1977). However, intermittent phototherapy induces DNA repair enzymes which are error-prone, meaning the enzymes make mistakes in base pairing during the repair process (Santella et al., 1978). In addition, necrosis is more pronounced after intermittent irradiation (Roll, 2005).

One drawback to phototherapy is that blue light during the treatment causes headaches and vertigo in the hospital staff (Maisels, 1982; Maisels, 1996). In addition, phototherapy has the risk of ultraviolet light burn, which in some cases has caused death (Siegfied et al., 1992). Plexiglas shields should be used to provide the blockage of ultraviolet A (UVA) exposure. Otherwise, side effects attributed to phototherapy (such as diarrhea and rashes) are minor and resolve when phototherapy is discontinued (Ennever, 1990). Potential vision problems do not seem evident. For example, the rods and cones of infants do not appear to suffer functional damage as a result of continuous light therapy (Hamer et al., 1984). Because of the limited side effects and strong benefits, phototherapy has continued as an effective means of treatment for hyperbilirubinemia.

References

Blaschke, T.F., Berk, P.D., and Scharschmidt, B.F. (1974). Crigler-Najjar Syndrome: an unusual course with development of neurological damage at age of eighteen. Pediatr. Res. 8: 573–590

Borun, E.R., Figueroa, W.G., and Perry, S.M. (1957). The distribution of Fe-59 tagged human erythrocytes in centrifuged specimens as a function of cell age. J. Clin. Invest. 36: 676—684

Broderson, R. (1978). Intensive Care in the Newborn, Vol. 2. Ed., by Stern, L., Ol, W., and Friis-Hansen, B., Masson, NY, pp. 331–345

Broderson, R. (1980). Bilirubin transport in the newborn infant reviewed with relation to kernicterus. J. Pediatr. 96: 349–356

Brown, A. (1957). Studies on the neonatal development of the glucuronide conjugating system. Am. J. Dis. Child. 94: 510–520

Brown, W.R. and Waters, W.J. (1958). The possible mechanism and site of bilirubin inhibition of election transport. Am. J. Dis. Child. 96: 507–514

Carbone, J.V. and Grodsky, G.M. (1957). Constitutional non-hemolytic hyperbilirubinemia in the rat: Defect of bilirubin conjugation. Proc. Soc. Exp. Biol. Med. 94: 461–463

Claireaux, A.E. (1961). *Pathology of Human Kernicterus*. Ed., by Kernicterus, S.K., University of Toronto, Toronto

Cowger, M. (1971). Mechanism of bilirubin toxicity on tissue culture cells: factors that affect toxicity, reversibility by albumin, and comparison with other respiratory poisons and surfactants. Biochem. Med. 5: 1–12

Cremer, R.J., Perryman, P.W., and Richards, D.W. (1958). Influence of light on the hyperbilirubinemia of infants. Lancet 1: 1093–1097

Day, R.L. (1954). Inhibition of brain respiration in-vitro by bilirubin. Am. J. Dis. Child. 88: 504–511

Diamond, I. and Schmid, R. (1966). Experimental bilirubin encephalopathy. The mode of entry of bilirubin-14C into the central nervous system. J. Clin. Invest. 45: 678–689

Diamond, I. and Schmid, R. (1967). Oxidative phosphorylation in experimental bilirubin encephalopathy. Science 155: 1288–1289

Donzelli, G.P., Pratesi, S., Rapisardi, G., Agati, G., Fusi, F., and Pretesi, R. (1995). 1-day phototherapy of neonatal jaundice with blue-green lamp. Lancet 346: 184–185

Dore, S., and Snyder, S. (1999). Neuroprotective action of bilirubin against oxidative stress in primary hippocampal cultures. Ann. N.Y. Acad. Sci. 890: 167–172

Dore, S., et al. (1999). Bilirubin, formed by activation of heme oxygenase-2, protects neurons against oxidative stress. Proc. Natl. Acad. Sci. USA. 96: 2445–2450

Dunn, T.B. (1951). Hematoidin crystals in reticulum cell sarcoma of the mouse and newborn human tissues. Milit. Surg. 109: 352–356

Ebbesen, F., Agati, G., and Pratesi, R. (2003). Phototherapy with turquoise versus blue light. Arch. Dis. Child. Fetal Neonatal Ed. 88: F430–F431

Ennever, J.F. (1990). Blue light, green light, white light, more light: treatment of neonatal jaundice. In Maisels MJ (Ed) Neonatal jaundice. Clin. Perinatol. 17: 467–481

Grojean, S., Koziel, V., Vert, P., and Daval, J. (2000). Bilirubin induces apoptosis via activation of NMDA receptors in developing rat brain neurons. Exp. Neurol. 166: 334–341

Grojean, S., et al. (2001). Bilirubin exerts additional toxic effects in hypoxic cultured neurons from the developing rat brain by the recruitment of glutamate neurotoxicity. Pediatr. Res. 49: 507–513

Gunn, C.K. (1938). Hereditary acholuric jaundice. J. Hered. 20: 137–145

Hamer, R.D., Dobson, V., and Mayer, M.J. (1984). Absolute thresholds in human infants exposed to continuous illumination. Invest. Ophthalmol. Vis. Sci. 25: 381–388

Hanefeld, F. and Natzschka, J. (1971). Histochemical studies in infant Gunn rats with kernicterus. Neuropediatrie 2: 428–438

Ho, K.C., Hodach, R., Varma, R., Thorsteinson, V., Hess, T., and Dale, D. (1980). Kernicterus and central pontine myelinolysis in a 14-year-old boy with fulminating hepatitis. Ann. Neurol. 8: 633–641

Jervis, G.A. (1959). Constitutional non-hemolytic hyperbilirubinemia with findings resembling kernicterus. Arch. Neural. Psychiat. 81: 55–67

Karp, W. (1979). Biochemical alterations in neonatal hyperbilirubinemia and bilirubin encephalopathy: a review. Pediatrics 64: 361–368

Katoh, R., Kashiwamata, S., and Niwa, F. (1975). Studies on cellular toxicity of bilirubin. Brain Res. 83: 81–92

Lucey, J.F. (1970). Phototherapy of jaundice 1969 (1970). Birth Defects Orig. Artic. Ser. 6 (2): 63–70

Lucey, J. (1972). Neonatal phototherapy: uses, problems and questions. Semin. Hematol. 9 (2): 127–135

Lucey, J., Ferriero, M., and Hewett, J. (1968). Prevention of hyperbilirubinemia of prematurity by phototherapy. Pediatrics 41: 1046–1054

Madan, A. (2005). Phototherapy: old question, new answers. Acta Paediatr. 94 (10): 1360–1362

Maisels, J.M. (1982). Jaundice in the newborn. Pediatr. Rev. 3 (10): 305–319

Maisels, J.M. (1996). Why use homeopathic doses of phototherapy? Pediatrics 98 (2): 283–287

McCandless, D.W. and Abel, M. (1980). The effect unconjugated bilirubin on regional cerebellar energy metabolism. Neurobehav. Toxicol. 2: 81–84

Menken, M., and Weidenback, E.C. (1967). Oxidative phosphorylation and respiratory control of brain mitochondria isolated from kernicteric rats. J. Neurochem. 14: 189–193

Menken, M., Waggoner, J.G., and Berlin, N. (1966). The influence of bilirubin on oxidative phosphorylation and related reactions in brain and liver mitochondria. J. Neurochem. 13: 1241–1248

Morphis, L., Constantopoulos, A., Matsaniotis, N., and Papaphilis, A. (1982). Bilirubin induced modulation of cerebral protein phosphorylation of cerebral protein phophorylation in neonatal rabbits in-vivo. Science 218: 156–159

Mustafa, M.G., Cowger, M.L., and King, T.E. (1969). Effects of bilirubin on mitochondrial reactions. J. Biol. Chem. 244: 6403–6412

Odell, G.B.K. and Schutta, H.S. (1985). Bilirubin encephalopathy. In *Cerebral Energy Metabolism and Metabolic Encephalopathy*. Ed., by McCandless, D.W., Plenum, NY, USA

Onishi, S., Itoh, S., and Isobe, K. (1986). Wavelength dependence of the relative rate constants for the main geometric and structural photoisomerization of bilirubin IXa bound to human serum albumin. Biochem. J. 236: 23–29

Ostrow, J.D., Schmid, R., and Samuelson, D. (1963). The protein binding of 14C-bilirubin in human and murine serum. J. Clin. Invest. 42: 1286–1299

Passonneau, J., and Lowry, O. (1993). Enzymatic analysis: a practical guide. Humana Press, Totowa, NJ

Roll, E.B. (2005). Bilirubin-induced cell death during continuous and intermittent phototherapy and in the dark. Acta Paediatr. 94: 1437–1442

Roll, E.B. and Christensen, T. (2005). Formation of photoproducts and cytotoxicity of bilirubin irradiated with turquoise and blue phototherapy light. Acta Paediatr. 94: 1448–1454

Rosen, H., Rosen, A., Rosen, D., Onaral, B., and Hiatt, M. (2005). Use of a light emitting diode (LED) array for bilirubin transformation. Proc 2005 IEEE Engineering in Medicine and Biology 27th Annual Conference, Shanghai, China, September 1–4, pp. 7266–7268

Rozdilsky, B. (1961). *Experimental Studies on the Toxicity of Bilirubin*. Ed., by Kernicterus, S.K., University of Toronto, Toronto

Rozdilsky, B. (1966). Kittens as experimental model for study of kernicterus. Am. J. Dis. Child. 11: 161–176

Rozdilsky, B. and Olszewski, J. (1961). Experimental study of the toxicity of bilirubin in newborn animals. J. Neuropathol. Exp. Neurol. 20: 193–208

Salih, F.M. (2001). Can sunlight replace phototherapy units in the treatment of neonatal jaundice? An in vitro study. Photodermatol. Photoimmunol. Photomed. 17 (6): 272–277

Santella, R.M., Rosenkranz, H.S., and Speck, W.T. (1978). Intracellular deoxyribonucleic acid — modifying activity of intermittent phototherapy, J. Pediatr. 93: 106–109

Schenker, S., McCandless, D.W., and Zollman, P. (1966). Studies of cellular toxicity of unconjugated bilirubin in kernicteric brain. J. Clin. Invest. 45: 1213–1220

Schenker, S., Hoyumpa, A.M., and McCandless, D.W. (1986). Bilirubin toxicity to the brain (kernicterus) and other tissues. In *Bile Pigments and Jaundice*. Ed., by Ostrow, J.D., Marcel Dekkker, NY

Schmid, R., Hammaker, K.L., and Axelrod, J. (1957). The enzymatic formation of bilirubin glycuronide. Arch. Biochem. 70: 285–291

Schmid, R., Bukingham, S., Mendilla, G.A., and Hammaker (1959). Bilirubin metabolism in the fetus. Nature 183: 1823–1827

Schmorl, C.G. (1910) Liquor cerebrospinalis und bentrikelflussigkeit. Centrabl. F. Allg. Path. Anat. 21: 459–470

Schutta, H.S. and Johnson, L. (1971). Fine structure observations on acute bilirubin encephalopathy in Gunn rats induced by sulfadimethoxine. Lab. Invest. 24: 82–96

Siegfied, E.C., Stone, M.S., Madison, K.C. (1992). Ultraviolet light burn: a cutaneous complication of visible light phototherapy of neonatal jaundice. Pediatr. Dermatol. 9 (3): 278–282

Tenhunen, R., Marver, H.S., and Schmid, R. (1968). The enzymatic conversion of heme to bilirubin by microsomal heme oxygenase. Proc. Natl. Acad. Sci. USA. 61: 748–755

Vecchi, C, Donzelli, G.P, Migliorini, M.G, Sbrana, G., and Pratesi, R. (1982). New light in phototherapy. Lancet, 14: 390

Vogl, T.P, Cheskin, H., Blumenfeld, T.A., Speck, W.T., and Koenigsberger, M.R. (1977). Effect of intermittent phototherapy on bilirubin dynamics in Gunn rats. Pediatr Res, 11: 1021–1026

Zetterstrom, R. and Ernster, L. (1956). Bilirubin, an uncoupler of oxidative phosphorylation in isolated mitochondria. Nature 178: 1335–1338

Zhang, L., Liu, W., Tanswell, A.K., and Luo, X. (2003). The effects of bilirubin on evoked potentials and long-term potentiation in rat hippocampus in-vivo. Pediatr. Res. 53: 939–944

Chapter 17
Infectious and Inflammatory Metabolic Encephalopathies

Concepts in Pathogenesis

Kottil W. Rammohan

The normal health and metabolism of the brain can be impaired by systemic or central nervous system infections and non-infectious inflammatory disorders of the brain. The ensuing encephalopathy can manifest by symptoms as mild as listlessness and irritability or as severe as frank coma. Whether the cause is actual infection of the brain or systemic sepsis or inflammation, the presence of encephalopathy always carries a grave prognosis. Mortality and morbidity can be as high as 30–100% depending on the clinical scenario, and early recognition and appropriate supportive care can sometimes provide a favorable outcome. The cause of the encephalopathy in systemic or CNS infection remains unknown. A specific therapy other than treating the infection and providing supportive care is not available. There is concern that the encephalopathy may be neuroprotective as activation of the brain during periods of deranged metabolism may lead to irreversible injury or death. When treatment of infection is successful, an uneventful recovery can occur even in patients who endured prolonged periods of coma.

The intent of this chapter is not to discuss the various pathogens and their varied manifestations of brain injury, but specific patterns of pathogenesis to the extent of our understanding and the mechanisms that lead to brain dysfunction during infection or inflammation. Needless to say, the specific mechanisms of brain injury in these disorders are poorly understood in most instances, and some hypotheses are just that, hypotheses without evidence to support the theory. The focus of discussion will be on the pathogenesis and mechanisms of brain dysfunction. The reader is referred to in-depth discussions on symptoms and treatment to appropriate articles that focus on the clinical management of these disorders.

Normal Brain Homeostasis

All components of the brain serve specific functions to maintain brain homeostasis. Only 10% of the normal adult brain is made up of neurons which are approximately 10^{12} in number. The bulk of the brain is comprised of glial cells which are considered as nothing but the glue (glia = "glue") that holds the matrices together. However, it has become increasingly clear that glia serve more than a supportive

D.W. McCandless (ed.) *Metabolic Encephalopathy*,
doi: 10.1007/978-0-387-79112-8_17, © Springer Science + Business Media, LLC 2009

function. These cells maintain the milieu for optimal function of the neurons. Lactate, which is the main substrate of energy for neurons, is produced by the astrocytes from glycogen even under aerobic conditions and transported to neurons by metabolic coupling (Pellerin, 2003; Tekkok et al., 2005). The foot processes of astrocytes maintain the blood-brain barrier which transports glucose as well as all other metabolites into the brain compartment. Disruption of the transporters or disruption of the blood-brain barrier can significantly impair brain functions, as is often the case during CNS or systemic infections. Astrocytes also regulate reuptake of neurotransmitters, particularly glutamate and the inhibitory neurotransmitter gamma amino butyric acid (GABA) (Tilleux and Hermans, 2007). Impaired astrocyte function can therefore lead to excitotoxicity of the neurons mediated by excessive glutamate. Astrocytes and microglia serve to participate as antigen presenting cells and actively promote local brain inflammation and immune-related activities including secretion of lymphokines, chemokines and leukotrines. Oligodendrocytes produce central myelin and maintain a close relationship with neurons and axons. These cells produce nerve growth factor (NGF), brain derived neurotrophic factor (BNDF) and neurotrophin 3 (Dai et al., 2003). Ependymal cells and cells of the choroids plexus maintain the blood-CSF barrier often affected by bacterial and some viral infections. Transport of glucose into the CSF is often impaired during and sometimes after bacterial meningitis, causing persistent hypoglycorrachia. Although the return of glucose to normal levels in the CSF may never occur, patients make a complete recovery since the transport of glucose to the brain itself is not adversely affected. This will serve to remind us that the blood-CSF barrier and transport to the CSF is different from the blood-brain barrier, which is more directly related to neuronal function. Virus with tropism in the ependyma can impair CSF flow and cause hydrocephalus.

Microglia plays a prominent role in inflammation and immune function within the CNS (Dheen et al., 2007). The origin of these cells is controversial. Resident microglia is thought to be produced from bone marrow-derived macrophage precursor cells that migrated to the brain during brain development. Approximately 10% of all glia are made up of microglia. During inflammation and injury these cells proliferate and become phagocytic. They are the antigen presenting cells of the CNS and modulate intrathecal immune function.

A variety of mechanisms can lead to neuronal injury. Ischemia at the macro or microcirculation level can impair functions, leading to micro or macro infarcts in the brain. Impaired mitochondrial function can lead to "virtual hypoxia" and anaerobic metabolism with all its deleterious consequences of energy crisis and acidosis. Presence of free intracellular calcium, often seen in energy crisis states, can lead to activation of enzymes that mediate cell death. Amidst all the metabolic derangements, neurons can enter into a state of protective inactivity through generation of receptors for endogenous ligands that activate GABA receptors (Rothstein,1994; Olasmaa et al., 1990; Rothstein and Olasmaa, 1990). Such a state of hibernation can be associated with complete return of function even when prolonged coma was endured during an acute event of infection or inflammation, when successful treatment of the infection and inflammation is accomplished.

CNS Viral Infections

Direct infections of the nervous system by neurotropic viruses can lead to CNS dysfunction in many ways. Six well-defined mechanisms of encephalopathy are discussed, but many other mechanisms probably exist. In many instances the mechanism of brain dysfunction remains an enigma since by pathology, abnormalities can be minimal.

Acute Lytic Infections of the Brain with Focal or Regional Necrosis, Edema and Brain Herniation

Most neurotropic viruses that cause encephalitis belong in this category, the prototype infection being human herpes encephalitis. For discussions on clinical presentation and management, please refer to recent reviews on this topic (Tyler, 2004; Schmutzhard, 2001). Tropism is generally to the neurons although all neurons are not equally susceptible. Herpes generally produces limbic encephalitis with predominant involvement of the orbito-frontal cortex and the temporal lobes. The severe cytolytic nature of the infection leads to liquefaction hemorrhagic necrosis of the brain with severe cytotoxic edema that leads to massive herniation and death. The exact mechanism of cytolysis produced by the virus is not known. Most cytolytic viruses selectively block host protein synthesis at the expense of virus replication resulting in the death of the cell and release of progeny viruses. As discussed below, neuronal apoptosis is also a mechanism of neuronal injury by a number of viruses, including herpes simplex virus.

Acute Viral Encephalopathy Without Significant Necrosis or Inflammation

Intracerebral inoculation of hamster neurotropic measles virus in the susceptible BALB/C mouse strain leads to encephalitis and death within 2 weeks of injection. For the most part, histological examination identified minimal brain injury characterized by mild inflammation but no edema or necrosis. Viral antigen can be readily demonstrated using immunohistochemistry in selective parts of the brain including the orbito-frontal cortex, hippocampus caudate nucleus and parts of the frontal and temporal lobes (Rammohan et al., 1981; van Pottelsberghe et al., 1979). The mechanism of brain dysfunction is not readily apparent. Similarly, intracerebral inoculation of a neuroadapted strain of herpes simplex virus type 1 to the Sprague-Dawley rat resulted in clinical acute encephalitis within 3–5 days of infection, but examination of the brain did not identify significant inflammation or necrosis. In an attempt to define the nature of brain dysfunction, the investigators examined levels of enzymes known to synthesize neurotransmitters. Levels of tyrosine hydroxylase, glutamate decarboxylase, and choline acetyltransferace were measured in substantia nigra, caudate/putamen, and the frontal cortex. No changes could be demonstrated in these brains as compared to saline injected controls (Elizan et al., 1983). The mechanism of brain dysfunction in these animals remains unknown. Similar lack of inflammation and necrosis has also been demonstrated with rhabdovirus infection namely, fatal rabies infection in man, where brain inflammation is minimal as also edema and necrosis.

Oxidative Stress from Reactive Oxygen Species/Intermediates

When the production of reactive oxygen intermediates/species exceed the antioxidant capacity of the cell, oxidative stress injury occurs, resulting in peroxidation of lipids, proteins and nucleic acids. Viruses have been shown to cause oxidative injury in vitro as well as in vivo (Liao et al., 2002; Raung et al., 2001). Depending on the severity, this can lead to activation of proteases leading to apoptosis; a well-recognized mechanism of virus-mediated cellular injury. The lists of viruses that induce oxidative stress in the host are legion. One such example is HIV, which causes oxidative stress and brain injury. Retroviruses induce oxidative stress in cells that lead to neuronal apoptosis. In this regard, Tat protein of HIV is known to cause oxidative stress leading to neuronal apoptosis and resultant HIV associated dementia (Steiner et al., 2006; Pocernich et al., 2005). Of what survival advantage would there be for a virus to promote oxidative injury? Hepatitis C virus which causes cirrhosis and hepatocellular carcinoma mediates this injury in part by activation of cyclooxygenase 2 and producing prostaglandin E2. Reactive oxygen species produced in the mitochondria activate NFkB which stimulates the production of cyclooxygenase 2 (COX2) with resultant increase in prostaglandin E2 (PGE2). Both COX 2 and PGE2 modulate replication of viral RNA, and further down-stream effects promote cell survival by induction of anti apoptotic agents, although eventually these interactions lead to hepatocellular injury (Waris and Siddiqui, 2005; Tardif et al., 2005; Waris et al., 2005).

Acute Excitotoxicity and Coma

Glutamate excitotoxicity has been implicated in the pathogenesis of a number of CNS degenerative disorders, and in some instances not with good evidence. Excitotoxicity has been implicated in rabies infection as well. This is in part due to the fact that an NMDA receptor may be a receptor for the rabies virus as well (Gosztonyi, and Ludwig, 2001). In vitro as well as in vivo studies in the rat brain identified a favorable effect for ketamine, an NMDA receptor antagonist, in reducing viral replication (Lockhart et al., 1992). Subsequently neuronal degeneration was shown to be reduced in the presence of NMDA antagonists when rabies infection of neurons was carried out in vitro (Tsiang et al., 1991). On the basis of these observations a 17 year old girl who developed rabies after a bat-bite was treated using a combination of agents to block excitotoxicity (Hu et al., 2007; Willoughby et al., 2005). Therapeutic coma was induced by GABA receptor agonism by benzodiazepines and barbiturates along with NMDA receptor antagonism using ketamine and amantadine. Antiviral therapy was instituted with ribavarine but antiviral therapies alone have never been successfully used previously to treat rabies. The patient survived with residual morbidity in a disorder where survival is unprecedented and mortality the rule. The authors concluded that blocking excitotoxicity and giving additional time to allow her immune system to mount a necessary antiviral response was the reason for her recovery. Although this was an "N of one", the report of this case is a landmark event since mortality with rabies approaches 100% and now one can start thinking about possible treatment strategies based on the approach used in this case.

Viral Infection and Neuronal Apoptosis

A novel mechanism of brain injury with neurovirulent viruses is the induction of neuronal apoptosis. A variety of viruses have been shown to induce cellular apoptosis in neuronal as well as non-neuronal tissues, but a prototype virus in this category is the Sindbis virus which has been shown to induce apoptosis in susceptible cells in vitro as well as neurons of the brain and spinal cord of infected mice (Griffin, 2005; Levine et al., 1993). Neurovirulence of this virus is best correlated with its ability to induce apoptosis of the neurons and induction of apoptosis was identified to be the best correlate of mortality with this virus in young mice (Lewis et al., 1996). Sindbis virus mediates apoptosis by activation of a specific subset of caspases which set in motion activation of caspase-induced caspase activation, leading to the phenotype of the typical apoptotic cell with cell shrinkage, membrane blebbing, condensation of chromatin and DNA fragmentation (Nava et al., 1998). What are the mechanisms of activation of caspases by Sindbis virus? In vitro studies would suggest that virus replication of the infected cell is not a prerequisite for induction of apoptosis. In a series of elegant experiments, investigators were able to induce apoptosis of Chinese Hamster Ovary cell lines with Sindbis virus using UV inactivated virus (Jan and Griffin, 1999). Inhibition of protein synthesis using cyclohexamide had no effect as also the presence or absence of heparin sulfate, the initial binding molecule for Sindbis virus. The studies identified that the fusion of the virus to the host cell membrane even in the absence of virus replication, activated the appropriate caspases to set in motion the process of programmed cell death. The virulence of the strain of virus correlated best with the ability of the virus to induce apoptosis.

Chronic Virus Persistence in the Brain

Although most central nervous system viral infections result in acute encephalitis, some viruses have the ability to persistently infect the brain through the lifetime of the host. Persistence of the virus in the brain can be asymptomatic, or contribute to an encephalopathy with associated consequences. Mechanisms of virus persistence in the brain are not entirely understood but it is evident that there are factors that are host-mediated or virus-mediated that contribute to this persistent infection. Mutations in the virus can render viruses less virulent, permitting a symbiotic existence in the host. Alternatively, the virus can become "defective" with one or more structural proteins modified or missing, resulting in an inability to produce infectious virions. Abnormalities in the matrix protein of measles virus have been implicated as mechanisms of persistence in the stains of measles virus that caused sub acute sclerosis pan encephalitis (SSPE) (Choppin et al., 1981; Hall and Choppin, 1979, 1981). Host factors that contribute to viral persistence are generally related to a failing viral clearance in an immunocompromised host. Viruses that are generally non-pathogenic can under these conditions become pathogenic as in the case of papova virus induced progressive multifocal leucoencephalopahy (see separate discussion).

What are the mechanisms of brain dysfunction in persistent virus infection in the brain? Probably the most common mechanism of brain injury is a consequence of the immunopathology mediated by the persistent virus. While some of this immunopathology is antiviral, it may also be in part autoimmune in nature. The inflammation causes the release of leucotrines, prostaglandins, thromboxanes and lymphokines, which interfere with normal brain activity and cause brain degeneration. Gradually, patients succumb to progressive decline in mentation and other neurological dysfunctions.

Role of Astrocytes in Viral Encephalitis

Astrocytes are crucial for maintaining the health of neurons during normal and disease states. They also serve with the microglia as antigen presenting and phagocytic cells. Reactive astrocytosis occurs regularly during viral encephalitis. Is this protective or detrimental to the final outcome? What is the role of astrocytes in brain dysfunction during acute viral encephalitis? In a study using suckling mice infected with Japanese Encephalitis Virus, the investigators identified up-regulation of glial fibrillary acidic protein (GFAP), glutamate aspartate transporter protein (GLAST), glutamate transporter 1 (GLT-1), and the copper transporter protein ceruloplasmin. Collectively these changes would suggest activation of astrocyte functions including reducing excitotoxicity by better uptake by astrocytes of glutamate. Nevertheless, the mice succumbed to the virus in spite of these apparent favorable changes (Mishra et al., 2007).

Astrocytes are crucial in maintaining the blood-brain barrier. Foot processes of the astrocytes abut tight junctions of the endothelium and together constitute the blood —brain barrier. Disruption of the blood-brain barrier can occur during viral encephalitis. In studies using vesicular stomatitis virus administered, disruption of the blood-brain barrier occurred in wild type and NOS-1 knock out mice but not in NOS-3 KO mice or mice treated with IL12. It would appear that both IL12 and NOS-3 are essential for the integrity of maintaining the blood—brain barrier during viral encephalitis (Reiss et al., 1996; Barna et al., 1996; Bi et al., 1995). Much of these effects are mediated through activation of these agents at the astrocyte level.

Hiv Related Encephalopathy

After the advent of highly active anti-retroviral therapy (HAART) the incidence of HIV dementia has been reduced by more than 50%. While prior to HAART the incidence was 21 cases for every 100 patient years, the incidence since HAART is reduced to10.5 per 1,000 patient years. Interestingly, the *prevalence* of HIV dementia has *increased*; with HAART patients with dementia live longer. Even with advanced AIDS of CD4 counts of less than 100 mm^{-3} patients have longevities almost eightfold higher than previously.

Several syndromes need to be defined.

(a) *HIV Encephalopathy.* This is an early disorder seen prior to the occurrence of AIDS and is probably due to initial invasion of the CNS by HIV. HIV is carried to the brain predominantly by CD4 expressing cells of the monocyte macrophage lineage as well as CD4 positive T cells. In the brain, the resident microglia is the main reservoir of infection. Since fatalities are rare at this stage of the disease, there is limited pathology available regarding this phase of the disease. Microglial nodules and vacuolar degeneration of the white matter probably represent some of the earliest changes. The virus can be readily demonstrated in the microglia.

(b) *HIV Associated Demntia (HAD) (formerly AIDS Dementia Complex).* This is seen in patients with AIDS and CD4 counts of less than 400 mm^{-3}. Pathological abnormalities can sometimes be minimal. Cortical neuronal loss, particularly in a fronto-temporal distribution, can be seen in some patients as also myelin pallor and vacuolar degeneration. There is often marked microgliosis and large multinucleated giant cells. Clinically patients can be classified into five stages of progressive decline (Price and Brew, 1988):

Stage 0 Normal mentation
Stage 0.5. Sub clinical or equivocal impairment of mentation
Stage 1. Unequivocal, but mild evidence of intellectual, functional or motor impairment.
Stage 2. Moderate mental and physical impairment. May need a cane for ambulation

Stage 3. Severe mental and physical impairment. Need assistance for all activities of daily living

Stage 4. Vegetative existence. Mute, paraplegic, and incontinent

(c) *Minimal cognitive motor dysfunction.* A subcortical pattern of cognitive impairment occurs in the patients with minimal to mild motor impairment. Patients are self sufficient for all activities of daily living. A frontotemporal pattern of cognitive impairment can be demonstrated by formal neuropsychological testing. After the advent of HAART, most patients with HIV-related cognitive impairment fall into this category.

(d) *HIV Associated Progressive Encephalopathy (HPE).* This is a term restricted to children with progressive complex neuro psychiatric, cognitive and motor impairment related to HIV infection. Unlike the adult population, behavioral abnormalities occur commonly.

How does encephalopathy occur in patients with HIV infection? It is controversial if direct infection of the neuron can occur with HIV infection since most neurons lack the receptor for the virus, the CD4 molecule. Yet, significant losses of neurons occur by neuronal apoptosis in the brains of individuals with HIV infection and cognitive impairment. Accordingly several hypotheses have been postulated as the basis of the neuronal loss with HIV infection. The microglia is the main reservoir for HIV in the brain. Activated microglia secretes a number of proinflammatory cytokines, leucotrines and prostaglandins, and creates an adverse milieu for optimal neuronal function. At this stage, if treatment is instituted, reversibility can occur as damage to the neurons has not yet occurred. On the other hand, the lymphokines, particularly TNF α and IL 1β can trigger a cascade of activation of caspases that mediate apoptosis. Specifically, TRAIL (TNF related apoptosis inducing ligand) has been implicated in neuronal apoptosis induced by HIV (Miura et al., 2003a,b; and Miura et al., 2001). A number of peptides of HIV are also directly toxic to the neurons and these include gp120, gp41, Tat, Nef, Vpr, and Rev (Singh et al., personal communication 2007).

West Nile Virus Encephalitis

This disorder deserves a separate discussion only because of its novelty as a recent cause of viral encephalopathy. The virus is an arthropod-borne agent, the major transmitter in the United State being the *Culex Fatgans* mosquito. Birds are a reservoir of this virus, and it is the mortality of birds of the new-world at the Bronx zoo that brought attention to this disorder. Although the epidemic started in the North Eastern part of this country it has since then spread to almost all parts of the United States and Canada.

The disorder is seasonal and occurs during late summer and fall, the peak season for mosquitoes. Mild, often transient meningo encephalitis occurs in healthy adults and children. In the elderly it can cause coma and death (see section on the discussion of the immune compromised host). In rare instances it can cause poliomyelitis like syndrome with injury to the anterior horn cells. Recovery is variable and can be poor or incomplete. Clinically, patients can have a presentation like Guillain Barre Syndrome, but the pathology would appear to be in the spinal cord rather than the peripheral nerve roots. The mechanisms of injury in this disorder are poorly worked out.

CNS Bacterial Infections

Bacterial Meningitis

After introduction of vaccines for *Streptococcus pneumoniae* and *Haemophylus influenzae*, the incidence of meningitis in children and adults has decreased over 50%. Nevertheless, when meningitis occurs, it is truly a medical emergency. The three most common pathogens that cause bacterial meningitis are *S pneumoniae, H. influenzae type b, and meningococcus.* The exact mechanism of violation of the intrathecal space is not known but in most instances the spread is thought to be hematogenous. Spread by contiguity from otitis media, mastoiditis or the nasopharynx has also been implicated. In any event, when the bacteria enter the cerebrospinal fluid which is normally non-cellular, without protein or complement, it renders a perfect medium of culture for the bacteria with an adequate supply of glucose. Doubling of bacteria occurs every 2 h and before long, large numbers of bacteria populate the cerebrospinal compartment. Degenerating bacteria release proinflammatory proteins and lipids from the cell wall (lipoteichoic acid) which incite a powerful host response and inflammation. Large amounts of polymorphonuclear leukocytes and macrophages are attracted to the site with secretion of proinflammatory leucotrines, prostaglandins, lymphokines and nitric oxide. In particular IL 1 and TNF α mediate much of the inflammation. Severe edema and brain herniation can occur. Severe obtundation and coma occur early. If untreated, mortality and severe morbidity approach 100%. It is important to initiate multiple antibiotic therapies until cultures and sensitivities become available, especially as antibiotic resistant strains of common organisms have begun to emerge.

Metabolic Encephalopathy with Systemic Infections/ Inflammation

Criteria for Systemic Inflammatory Response Syndrome (SIRS) are fulfilled if two of the following four criteria are met (American College of Chest Physicians, 1992)

1. Heart rate > 90 per minute
2. Body temperature (core body) of < 36°C or > 38°C.
3. Respiration > 20 per minute or pCO_2 < 32 mm Hg
4. Total white count of < 4,000 mm^{-3} or > 12,000 per mm^3 with 10% immature leucocytes.

SIRS is thought to be a consequence of circulating proinflammatory cytokines, a form of "cytokine storm". When persistent, this can lead to multi organ failure, including adult respiratory distress syndrome. When SIRS is associated with a systemic infection, the condition is known as systemic sepsis.

Alteration in mental status occurring in patients with systemic sepsis always carries a serious prognosis. The mechanisms of impaired brain function are poorly understood and are probably multifactorial. Considerations include hypoxia, ischemia, mitochondrial dysfunction and anaerobic cerebral energy metabolism, blood-brain barrier dysfunction or impaired transporter function, cerebral edema, toxins like ammonia or endotoxins, and last but not least, clinical use of cerebral depressants and sedatives in severely ill patients. In patients with multi-organ failure, clearance of common short-acting sedatives can become prolonged, resulting in severe and protracted alteration of mentation.

In patients with acute liver failure during systemic infection and inflammation, there is good evidence to suggest that the liver is a major source of proinflammatory cytokines. Increased levels of TNFα, IL-6 and IL-1b are seen especially in patients with cerebral edema. Evidence that these agents are secreted by the liver was validated by the observation that after a "temporizing hepatectomy" in a patient with uncontrolled intracranial hypertension the levels of TNFα, IL-6 and IL-1b improved and the patients clinical status recovered as well (Jalan et al., 2002; Jalan and Williams, 2001). Although the role of hyperammonemia in the generation of hepatic encephalopathy is controversial, the single best correlate of increased intracranial pressure and cerebral edema in sepsis with acute liver failure is blood ammonia levels (Clemmesen et al., 1999). The molecular mechanisms that link systemic inflammation and encephalopathy remain unknown. However, cyclo-oxygenase inhibitor indomethacin has been shown to be of value in reducing increased intracranial pressure in patients with fulminant hepatic failure (Tofteng and Larsen, 2004).

Post Infectious Encephalomyelitis

Altered mental status is a pre-requisite for the diagnosis of acute disseminated encephalomyelitis (ADEM), often a consequence of non-CNS infection or immunization (Tenembaum et al., 2007) The disorder is the subject of a complete chapter in this book, so this discussion will be limited to the mechanisms of brain injury as it pertains to encephalopathies mediated by brain inflammation. There is excellent evidence to suggest that the brain injury is primarily immune mediated and therefore inflammatory in nature. Antigenic similarities between the viral antigens (or vaccine) and the brain result in injury to the brain in the course of a normal immune response of the host to the pathogen. This phenomenon, known as molecular mimicry, constitutes the basis of this disorder (Menge et al., 2007; Quaranta et al., 2006; Gout, 2001). For the most part the injury appears to be a leukoencephalitis affecting the periventricular as well as the subcortical white matter of the brain.

Axonal rather than myelin injury appears to be the basis of morbidity following inflammatory demyelination. The mechanism of axonal injury is unknown but there is some evidence to suggest that it is in part caused by the phenomenon of "virtual hypoxia" (Stys, 2004, 2005). According to this concept, injury to the mitochondria occurs probably from nitric oxide produced by inflammatory cells. Experimental studies that exposed axons to concentrations of nitric oxide seen at sites of inflammation (~ 4 μM) were capable of causing axonal conduction block (Smith et al., 2001; Kapoor et al., 1999; Redford et al., 1997) When conduction was carried out at 50–100 Hz, the conduction block outlasted the period of exposure to nitric oxide, indicating a permanent rather than transient injury. Examination of these axons identified marked edema with disorganization of myelin indicative of depolarization injury. It appeared that the axons depolarized to the stimuli but were incapable of repolarization, which is an energy dependant state. The Na^+ ions that entered the axons during depolarization could not exit the axon because the Na^+ K^+ ATPase pump requires ATP to function and in the absence of mitochondrial function there was depletion of the limited ATP available through glycolysis. The accumulation of Na^+ leads to accumulation of water and edema of the axons. This situation leads to exchange of one Ca^{2+} for two NA^+, through the Na^+–Ca^+ exchanger and this leads to intra axonal accumulation of free Ca^{2+}. Transport of Ca^{2+} to the axoplasmic reticulum is also an energy-dependent function and therefore does not occur in the state of limited ATP supply. The accumulation of Ca^{2+} then leads to activation of a variety of proteases with eventual irreversible injury to the axon. These observations suggested that blockade of Na^+ channels in a demyelinated animal could preserve axonal function. Indeed this was the case when mice with experimental autoimmune encephalomyelitis were treated with flecainide, a sodium channel blocker; preservation of axons was demonstrated in a dose-dependent manner (Bechtold et al., 2004; Kapoor et al., 2003).

Reye's syndrome is believed to be a post-infectious complication of varcella infection in children, but it will not be discussed here as it is discussed elsewhere in this book.

Encephalopathy in the Transplant Recipient

Solid organ or bone marrow transplant recipients are a vulnerable group of patients prone to infectious and non-infectious metabolic encephalopathy (Todd, 2006). Depending on the time of the occurrence of the encephalopathy, several scenarios need to be considered. During the early phase of engraftment (usually the first 30 days) patients are vulnerable to infections with candida, aspergillus, and the herpes encephalitis from herpes simplex type 1 and 6 and cytomegalovirus. It is important to recognize that HHV type 6 is a pathogen during this phase as HHV 6 has no viral thymidine kinase and cannot therefore be treated with acyclovir. Instead, patients will require treatment with intravenous gancyclovir and foscarnate. During this time, patients are also at risk for bacterial meningitis, particularly *Listeria* meningitis. From 1 to 6 months, in addition to the agents listed earlier, patients are at risk for infection with toxoplasmosis and Cryptococcus as also meningitis with tuberculosis in susceptible previously exposed population. Beyond 6 months infections are directly correlated to graft rejection and increased immune suppression.

Of recent interest is the occurrence of encephalopathy from West Nile virus infection which can be fatal in the immune compromised host. Whereas this infection is transient encephalitis in the healthy, it can result in coma and death in the elderly and the immune-compromised patient. There is nothing specific about the disorder; diagnosis is based on a high index of suspicion as the disorder is seasonal and occurs in late summer. The MRI scan may show very little if any by way of abnormalities. The cerebrospinal fluid is always abnormal with prominent pleocytosis and increased protein. Cerebrospinal fluid is usually positive for IgM antibody to the virus. Care is only supportive therapy.

At all times one should be cognizant of the adverse effects of agents used for immune suppression in the transplant recipient as a number of them can cause encephalopathy. In particular, the use of tacrolimus (FK 506) or cyclosporine should alert the physician for possible toxicity being the basis of encephalopathy (The U.S. Multicenter FK506 Liver study Group, 1994) This can be complicated as "normal" levels of cyclosporine can be toxic when the total LDL and cholesterol levels are low following organ transplants, in particular liver transplants. In patients who are recipients of orthotopic liver, who have serum total cholesterol of less than 100 mg dL^{-1} normal level of serum cyclosporine can be toxic as it is lipid transported and free levels of cyclosporine can be high (Calne, 1994).

Progressive Multifocal Leukoencephalopathy

This disorder is caused by papova virus infection of the brain in an immune-compromised host. Chronic immune suppression with AIDS, cancer, autoimmune disorders or transplantation, all predispose to the occurrence of this disorder. Especially with potent immune suppression for autoimmune disorders, this is becoming increasingly recognized as a complication that we need to contend with. There is some recent evidence that the disorder may not always be fatal if immune reconstitution can be achieved by the withdrawal of the offending agent and restoration of normal immune function. Of all the agents implicated to cause this disorder, the chronic use of steroids appears to take prominence over any other agent. A discussion of the many nuances of this disorder is beyond the scope of this chapter and the reader is referred to a treatise on this subject, which has been the topic of many recent reviews (Hou and Major, 2005; Eash et al., 2006).

Summary

Encephalopathy in the context of CNS infection and inflammation is discussed mainly from the standpoint of mechanisms of pathogenesis. It would appear that irrespective of the pathogen, mechanisms of brain injury are similar. In some instances the immune response of the host proves detrimental to the host. In other instances a lack of immune surveillance leads to detrimental infections of the CNS. During infection, pathogens utilize similar mechanisms to cause injury. Oxidative stress, neuronal apoptosis, and excitotoxicity appear to be common mechanisms for injury to the CNS by a variety of agents. Recognizing these mechanisms is crucial as it allows us to develop strategies of neuro protection in addition to treatment of the pathogen. Understanding the delicate balance of pathogen—host interactions is crucial as this information will form the foundation of our future treatment of infections and inflammations of the brain that cause encephalopathy, which will not only utilize antibiotics for treatment of infections but also neuro protective strategies.

References

American College of Chest Physicians/Society of Critical Care Medicine Consensus Conference: definitions for sepsis and organ failure and guidelines for the use of innovative therapies in sepsis. Critical Care Medicine 1992; 20(6):864–874

Barna M, Komatsu T, Reiss CS. Activation of type III nitric oxide synthase in astrocytes following a neurotropic viral infection. Virology 1996; 223(2):331–343

Bechtold DA, Kapoor R, Smith KJ. Axonal protection using flecainide in experimental autoimmune encephalomyelitis. Annals of Neurology 2004; 55(5):607–616

Bi Z, Quandt P, Komatsu T, Barna M, Reiss CS. IL-12 promotes enhanced recovery from vesicular stomatitis virus infection of the central nervous system. Journal of Immunology 1995; 155(12):5684–5689

Calne RY. Immunosuppression in liver transplantation. The New England Journal of Medicine 1994; 331(17):1154–1155

Clemmesen JO, Larsen FS, Kondrup J, Hansen BA, Ott P. Cerebral herniation in patients with acute liver failure is correlated with arterial ammonia concentration. Hepatology 1999; 29(3):648–653

Choppin PW, Richardson CD, Merz DC, Hall WW, Scheid A. The functions and inhibition of the membrane glycoproteins of paramyxoviruses and myxoviruses and the role of the measles virus M protein in subacute sclerosing panencephalitis. The Journal of Infectious Diseases 1981; 143(3):352–363

Dai X, Lercher LD, Clinton PM, Du Y, Livingston DL, Vieira C et al. The trophic role of oligodendrocytes in the basal forebrain. The Journal of Neuroscience: the Official Journal of the Society for Neuroscience 2003; 23(13):5846–5853

Dheen ST, Kaur C, Ling EA. Microglial activation and its implications in the brain diseases. Current Medicinal Chemistry 2007; 14(11):1189–1197

Eash S, Manley K, Gasparovic M, Querbes W, Atwood WJ. The human polyomaviruses. Cellular and Molecular Life Sciences 2006; 63(7–8):865–876

Elizan TS, Maker H, Yahr MD. Neurotransmitter synthesizing enzymes in experimental viral encephalitis. Journal of Neural Transmission 1983; 57(3):139–147

Gosztonyi G, Ludwig H. Interactions of viral proteins with neurotransmitter receptors may protect or destroy neurons. Current Topics in Microbiology and Immunology 2001; 253:121–144

Gout O. Vaccinations and multiple sclerosis. Neurological Sciences 2001; 22(2):151–154

Griffin DE. Neuronal cell death in alphavirus encephalomyelitis. Current Topics in Microbiology and Immunology 2005; 289:57–77

Hall WW, Choppin PW. Evidence for lack of synthesis of the M polypeptide of measles virus in brain cells in subacute sclerosing panencephalitis. Virology 1979; 99(2):443–447

Hall WW, Choppin PW. Measles-virus proteins in the brain tissue of patients with subacute sclerosing panencephalitis: absence of the M protein. The New England Journal of Medicine 1981; 304(19):1152–1155

Hou J, Major E. Management of infections by the human polyomavirus JC: past, present and future. Expert Review of Anti-Infective Therapy 2005; 3(4):629–640

Hu WT, Willoughby REJ, Dhonau H, Mack KJ. Long-term follow-up after treatment of rabies by induction of coma. The New England Journal of Medicine 2007; 357(9):945–946

Jalan R, Williams R. The inflammatory basis of intracranial hypertension in acute liver failure. Journal of Hepatology 2001; 34(6):940–942

Jalan R, Pollok A, Shah SHA, Madhavan K, Simpson KJ. Liver derived pro-inflammatory cytokines may be important in producing intracranial hypertension in acute liver failure. Journal of Hepatology 2002; 37(4):536–538

Jan JT, Griffin DE. Induction of apoptosis by Sindbis virus occurs at cell entry and does not require virus replication. Journal of Virology 1999; 73(12):10296–10302

Kapoor R, Davies M, Smith KJ. Temporary axonal conduction block and axonal loss in inflammatory neurological disease. A potential role for nitric oxide? Annals of the New York Academy of Sciences 1999; 893:304–308

Kapoor R, Davies M, Blaker PA, Hall SM, Smith KJ. Blockers of sodium and calcium entry protect axons from nitric oxide-mediated degeneration. Annals of Neurology 2003; 53(2):174–180

Liao SL, Raung SL, Chen CJ. Japanese encephalitis virus stimulates superoxide dismutase activity in rat glial cultures. Neuroscience Letters 2002; 324(2):133–136

Lockhart BP, Tordo N, Tsiang H. Inhibition of rabies virus transcription in rat cortical neurons with the dissociative anesthetic ketamine. Antimicrobial agents and chemotherapy 1992; 36(8):1750–1755

Levine B, Huang Q, Isaacs JT, Reed JC, Griffin DE, Hardwick JM. Conversion of lytic to persistent alphavirus infection by the bcl-2 cellular oncogene. Nature 1993; 361(6414):739–742

Lewis J, Wesselingh SL, Griffin DE, Hardwick JM. Alphavirus-induced apoptosis in mouse brains correlates with neurovirulence. Journal of virology 1996; 70(3):1828–1835

Menge T, Kieseier BC, Nessler S, Hemmer B, Hartung HP, Stuve O. Acute disseminated encephalomyelitis: an acute hit against the brain. Current opinion in neurology 2007; 20(3):247–254

Mishra MK, Koli P, Bhowmick S, Basu A. Neuroprotection conferred by astrocytes is insufficient to protect animals from succumbing to Japanese encephalitis. Neurochemistry international 2007; 50(5):764–773

Miura Y, Misawa N, Maeda N, Inagaki Y, Tanaka Y, Ito M et al. Critical contribution of tumor necrosis factor-related apoptosis-inducing ligand (TRAIL) to apoptosis of human CD4+ T cells in HIV-1-infected hu-PBL-NOD-SCID mice. The journal of experimental medicine 2001; 193(5):651–660

Miura Y, Koyanagi Y, Mizusawa H. TNF-related apoptosis-inducing ligand (TRAIL) induces neuronal apoptosis in HIV-encephalopathy. Journal of medical and dental sciences 2003a; 50(1):17–25

Miura Y, Misawa N, Kawano Y, Okada H, Inagaki Y, Yamamoto N et al. Tumor necrosis factor-related apoptosis-inducing ligand induces neuronal death in a murine model of HIV central nervous system infection. Proceedings of the national academy of sciences of the United States of America 2003b; 100(5):2777–2782

Nava VE, Rosen A, Veliuona MA, Clem RJ, Levine B, Hardwick JM. Sindbis virus induces apoptosis through a caspase-dependent, CrmA-sensitive pathway. Journal of virology 1998; 72(1):452–459

Olasmaa M, Rothstein JD, Guidotti A, Weber RJ, Paul SM, Spector S et al. Endogenous benzodiazepine receptor ligands in human and animal hepatic encephalopathy. Journal of neurochemistry 1990; 55(6):2015–2023

Pellerin L. Lactate as a pivotal element in neuron-glia metabolic cooperation. Neurochemistry international 2003; 43(4–5):331–338

Pocernich CB, Sultana R, Mohmmad-Abdul H, Nath A, Butterfield DA. HIV-dementia, Tat-induced oxidative stress, and antioxidant therapeutic considerations. Brain research brain research reviews 2005; 50(1):14–26

Price RW, Brew BJ. The AIDS dementia complex. The journal of infectious diseases 1988; 158(5):1079–1083

Quaranta L, Batocchi AP, Sabatelli M, Nociti V, Tartaglione T, Cuonzo F et al. Monophasic demyelinating disease of the central nervous system associated with Hepatitis A infection. Journal of neurology 2006; 253(7):944–945

Rammohan KW, McFarland HF, McFarlin DE. Induction of subacute murine measles encephalitis by monoclonal antibody to virus haemagglutinin. Nature 1981; 290(5807):588–589

Raung SL, Kuo MD, Wang YM, Chen CJ. Role of reactive oxygen intermediates in Japanese encephalitis virus infection in murine neuroblastoma cells. Neuroscience letters 2001; 315(1–2):9–12

Redford EJ, Kapoor R, Smith KJ. Nitric oxide donors reversibly block axonal conduction: demyelinated axons are especially susceptible. Brain: a journal of neurology 1997; 120 (Pt 12):2149–2157

Reiss CS, Komatsu T, Barna M, Bi Z. Interleukin-12 promotes enhanced recovery from viral infection of neurons in the central nervous system. Annals of the New York Academy of Sciences 1996; 795:257–265

Rothstein JD. Benzodiazepine-receptor ligands and hepatic encephalopathy: a causal relationship? Hepatology 1994; 19(1):248–250

Rothstein JD, Olasmaa M. Endogenous GABAergic modulators in the pathogenesis of hepatic encephalopathy. Neurochemical research 1990; 15(2):193–197

Schmutzhard E. Viral infections of the CNS with special emphasis on herpes simplex infections. Journal of neurology 2001; 248(6):469–477

Singh NN, Yahya S, Garewal M, Thomas FP. HIV-1 Encephalopathy and AIDS Dementia Complex. 5-8-2007. Personal Communication

Smith KJ, Kapoor R, Hall SM, Davies M. Electrically active axons degenerate when exposed to nitric oxide. Annals of neurology 2001; 49(4):470–476

Steiner J, Haughey N, Li W, Venkatesan A, Anderson C, Reid R et al. Oxidative stress and therapeutic approaches in HIV dementia. Antioxidants and redox signaling 2006; 8(11–12):2089–2100

Stys PK. Axonal degeneration in multiple sclerosis: is it time for neuroprotective strategies? Annals of neurology 2004; 55(5):601–603

Stys PK. General mechanisms of axonal damage and its prevention. Journal of the neurological sciences 2005; 233(1–2):3–13

Tardif KD, Waris G, Siddiqui A. Hepatitis C virus, ER stress, and oxidative stress. Trends in microbiology 2005; 13(4):159–163

Tekkok SB, Brown AM, Westenbroek R, Pellerin L, Ransom BR. Transfer of glycogen-derived lactate from astrocytes to axons via specific monocarboxylate transporters supports mouse optic nerve activity. Journal of neuroscience research 2005; 81(5):644–652

Tenembaum S, Chitnis T, Ness J, Hahn JS, Hahn JS. Acute disseminated encephalomyelitis. Neurology 2007; 68(16 Suppl 2):S23–S36

Tilleux S, Hermans E. Neuroinflammation and regulation of glial glutamate uptake in neurological disorders. Journal of neuroscience research 2007; 85(10):2059–2070

Todd C. Central Nervous System in Transplant Recipient. Continuum: Lifelong Learning in Neurology. Lippencott, Williams & Wilkins, 2006: pp. 95–110

Tofteng F, Larsen FS. The effect of indomethacin on intracranial pressure, cerebral perfusion and extracellular lactate and glutamate concentrations in patients with fulminant hepatic failure. Journal of cerebral blood flow and metabolism 2004; 24(7):798–804

Tsiang H, Ceccaldi PE, Ermine A, Lockhart B, Guillemer S. Inhibition of rabies virus infection in cultured rat cortical neurons by an N-methyl-D-aspartate noncompetitive antagonist, MK-801. Antimicrobial agents and chemotherapy 1991; 35(3):572–574

Tyler KL. Update on herpes simplex encephalitis. Reviews in neurological diseases 2004; 1(4):169–178

The U.S. Multicenter FK506 Liver Study Group. A comparison of tacrolimus (FK 506) and cyclosporine for immunosuppression in liver transplantation. The New England journal of medicine 1994; 331(17):1110–1115

Van Pottelsberghe C, Rammohan KW, McFarland HF, Dubois-Dalcq M. Selective neuronal, dendritic, and postsynaptic localization of viral antigen in measles-infected mice. Laboratory investigation: a journal of technical methods and pathology 1979; 40(1):99–108

Waris G, Siddiqui A. Hepatitis C virus stimulates the expression of cyclooxygenase-2 via oxidative stress: role of prostaglandin E2 in RNA replication. Journal of virology 2005; 79(15):9725–9734

Waris G, Turkson J, Hassanein T, Siddiqui A. Hepatitis C virus (HCV) constitutively activates STAT-3 via oxidative stress: role of STAT-3 in HCV replication. Journal of virology 2005; 79(3):1569–1580

Willoughby REJ, Tieves KS, Hoffman GM, Ghanayem NS, Amlie-Lefond CM, Schwabe MJ et al. Survival after treatment of rabies with induction of coma. The New England journal of medicine 2005; 352(24):2508–2514

Chapter 18
Major Depression and Metabolic Encephalopathy: Syndromes More Alike Than Not?

Introduction

The seminal discovery of the antidepressant imipramine in 1958, the mood-stabilizing actions of lithium ion in 1949, and that these drugs work by increasing or modulating synaptic levels of one or more monoamine such as serotonin (5HT), noradrenaline (NA) or dopamine (DA), have set the gold standard for drug research and treatment in mania and depression. However, drug discovery has not realized its full potential and in many ways is decades behind our understanding of psychiatric illness. Our newest drugs for these disorders, such as the 5HT reuptake inhibitors (SRI, e.g., fluoxetine (Prozac®)), are not any more effective than imipramine in the treatment of depression (Geddes et al., 2000) while after more than 50 years, lithium salts still remain the bench mark treatment for manic depression (Compton and Nemeroff, 2000).

The advent of sophisticated techniques in molecular neuroscience and molecular psychiatry has uncovered a wealth of knowledge pertaining to the development, susceptibility and progression of psychiatric illness, as well as identifying new candidate target molecules for drug action. For example, preclinical research has revealed that a complex array of subcellular molecules involved in cellular resilience, such as brain-derived neurotrophic factor (BDNF), cyclic AMP response element binding protein (CREB), nitric oxide (NO), B-cell lymphoma 2 (bcl-2), glycogen synthase kinase (GSK), and extracellular signal-regulated kinase (ERK) (for a review, see Manji et al., 2001; Berton and Nestler, 2006), represent the actual subcellular mediators of antidepressant drug action and indeed, very likely represent the neurobiological substrates of depressive illness (Manji et al., 2001; Berton and Nestler, 2006). Despite these advances, current drug treatment of depression remains steadfastly resigned to addressing the disorder at the synaptic level. Clearly, integrating our molecular knowledge of depression and the processes involved in resilience and vulnerability into a single working hypothesis upon which drugs can be developed has proved the biggest challenge.

Traditionally, major depression is thought to follow chronic stressful environmental conditions, where individual vulnerability to stress has a significant role in determining the impact of the stressor on the body and mind, and the later development

D.W. McCandless (ed.) *Metabolic Encephalopathy*,
doi: 10.1007/978-0-387-79112-8_9, © Springer Science+Business Media, LLC 2009

of an anxiety or mood disorder (Kendler et al., 2001). Contrary to the thinking in the 1960s, depression is not a single neurotransmitter disorder, but represents a continuum of environmental, genetic and neurochemical determinants, all of which occupy a variable, yet distinct role in the etiology, progression and treatment response of disorders as apparently distinct as depression on one end, to psychosis on the other. The monoamine hypothesis of depression has proved to have significant construct validity in that it has formed the basis for the development of our current armamentarium of antidepressant drugs. However, their delayed onset of action and that these drugs seldom exceed an expected rate of remission of 50% (Kocsis, 2003), have in recent years prompted the realization that current antidepressants are not targeting the neurobiological underpinnings of the disorder and that a time-dependent and orchestrated manipulation of the above-mentioned subcellular neurochemicals and neuronal messengers represent the end-targets of mood elevation, with current antidepressants simply acting to switch on the cascade from outside the cell (Manji et al., 2001; Berton and Nestler, 2006).

The biggest drawback of the current drugs in treating depression is their delayed onset of action, which nominally can be placed at 3–5 weeks after initiation of therapy (Manji et al., 2001; Leonard, 2003). With suicidal ideation an ever present danger in the illness, such a delay becomes a serious flaw, often requiring hospitalization of seriously ill patients until improvement is noted. Even though synaptic monoamine levels rise almost immediately after initiating antidepressant treatment (Leonard, 2003), onset of therapeutic efficacy is slow, and has been ascribed to a gradual adaptation of neuronal processes, monoamine receptors and synaptic connectivity brought into motion by the molecular mechanisms described earlier (e.g., CREB, BDNF) but initiated and maintained by chronic administration of the drug (Manji et al., 2001; Leonard, 2003). This has propagated the so-called plasticity hypothesis which is currently regarded as representing our best and most accurate understanding of how antidepressants work (D'Sa and Duman, 2002).

Recent clinical studies, however, have provided preliminary yet convincing evidence that certain synaptic active antidepressants, such as tricyclic antidepressants (TCA), SRIs and atypical agents, may have an earlier onset of action than predicted by the plasticity hypothesis. Thus, clinical studies have suggested either a more rapid onset of antidepressant action, superior efficacy and/or use as augmentation strategy, for mirtazapine (Wheatley et al., 1998; Leinonen et al., 1999; Benkert et al., 2000; Carpenter et al., 2002; Vester-Blokland and Van Oers, 2002), venlafaxine (Guelfi et al., 1995; Benkert et al., 1996; Entsuah et al., 1998), combined 5HT/NA reuptake inhibition by fluoxetine + desipramine (Nelson et al., 2004), escitalopram (Kasper et al., 2006) and SRI augmentation with pindolol (Perez et al., 1997; Tome et al., 1997; Zanardi et al., 2001). These studies argue in favor of a simple biochemical imbalance that can be corrected by acute administration of an appropriate drug that will re-establish homeostasis of a particular neurochemical. This acute change in the neurochemistry in response to antidepressant treatment is not unlike that observed in encephalopathies. In this chapter I will argue that depression, or certain types of depressive illness, may be regarded as a progressive metabolic encephalopathy. Importantly, knowledge of the latter may possibly have distinct implications for the understanding, diagnosis and treatment of major depression and may have particular value with respect to the onset of antidepressant action.

Metabolic Encephalopathy: A Brief Overview

Hepatic encephalopathy (HE), also referred to as portal-systemic encephalopathy, refers to a complex syndrome that follows impaired hepatocellular function due to liver cirrhosis or acute liver failure (Albrecht and Jones, 1999). Over 20 different compounds occur in the circulation in increased concentration following hepatocellular impairment, but of these the most important is ammonia (Zieve, 1987). Since the brain has an incomplete urea cycle, ammonia triggers a sequence of metabolic events resulting in pronounced central nervous system effects and encephalopathy due to altered neuronal function, inflammation and neurotoxicity (Cauli et al., 2007; Albrecht and Jones, 1999) resulting in diverse neuropsychiatric manifestations. HE can be classified into three stages of severity (Albrecht and Jones, 1999). Subtle psychiatric and behavioral change, including depression, is often evident in Stage I, but which progresses over Stages II–IV with impairment in mental tasks, personality changes and inappropriate behavior (Stage II), somnolence, amnesia, fits of rage, incoherent speech and confusion (Stage III), and finally coma (Stage IV) (Albrecht and Jones, 1999). The incidence of depression in HE or other encephalopathies has not been studied. However, a recent study found that 36% of patients with alcoholic liver disease met the criteria for a lifetime DSM-IV depressive disorder (Di Martini et al., 2004), while another study raised this incidence to 53% (Norris et al., 2002). The latter study also found that 29% of patients with viral liver disease and 60% of patients with cholestatic liver disease present with clinical depression (Norris et al., 2002).

The principle means of detoxification of blood-derived ammonia is reductive amidation of glutamate to glutamine by astrocyte- and glial-specific glutamine synthase (Norenberg and Martinez-Hernandez, 1979) with excess ammonia resulting in excitatory glutamate and inhibitory γ-amino butyric acid (GABA) imbalance. Patients with HE suffer from diverse disruption of glutamatergic function. The brain glutamate system is central to our understanding of the neurochemistry of liver failure, with synthesis, intercellular transport (uptake and release) and function being affected (Vaquero and Butterworth, 2006). Glutamate, the major excitatory neurotransmitter in the CNS, may be responsible for various psychiatric manifestations in HE and seizures in fulminant hepatic failure, while decreased glutamatergic tone underlies symptoms of drowsiness, lethargy, amnesia, confusion and coma.

Heightened glutamate in HE has been demonstrated in an experimental model of hyperammonia-induced encephalopathy in rats (Vogels et al., 1997). Excessive glutamate release or failed glial/neuronal reuptake (Vaquero and Butterworth, 2006) in HE results in N-methyl-d-aspartate (NMDA) receptor activation which is further enhanced by an increase in extracellular glycine, a positive regulator of the NMDA ion channel (Michalak et al., 1996). Ammonia-induced NMDA receptor activation engenders a number of important metabolic effects that may progress over time to irreversible neuronal dysfunction and damage. These would include increased NA^+, K^+, ATPase, depletion of cellular ATP (Kosenko et al., 1994; Ratnakumari et al., 1995) and proteolysis of microtubule-associated protein-2 (MAP-2) resulting in a decrease in the polymerization of microtubules, the essential building blocks of neurons (Felipo et al., 1993). Further, increased activation of nitric oxide synthase (NOS) and the release of NO results in the production of reactive nitrogen and oxygen intermediates, oxidative stress and nerve cell damage (Kosenko et al., 1995, 1998; Schliess et al., 2006). Ammonia also increases the principle second messenger of NO, cyclic guanosine monophosphate (cGMP) (Vaquero and Butterworth, 2006) which in turn exerts important neuromodulatory effects (Prast and Philippu, 2001), particularly with respect to cognition (Domek-Łopacin ska and Strosznajder, 2005). Indeed, inhibiting NOS activity not only attenuates NO-driven neuronal damage, but also results in the restoration of antioxidant enzyme activity induced by ammonia (Kosenko et al., 1998). Interestingly, bolstering brain cGMP restores cognitive impairment induced in an animal model of HE (Erceg et al., 2005), further highlighting the role of NO-cGMP in ammonia-related toxicity. Hyperammonia also inhibits the cerebral synthesis of kynurenic acid, a notable endogenous antagonist at ionotropic glutamate receptors and a putative neuroprotectant (Saran et al., 1998).

While the aforementioned are typified by excessive glutamatergic transmission, downregulation of ionotropic glutamate receptors and impaired excitatory transmission also occur in HE (see Albrecht and Jones, 1999; Vaquero and Butterworth, 2006 for a review) and these together very likely underlie symptoms of drowsiness, lethargy, amnesia, confusion and coma. Moreover, since glutamate–NO mechanisms are involved in memory formation (Riedel et al., 2003; Domek-Łopaci ska and Strosznajder, 2005), attenuated glutamatergic tone may also underlie HE-related memory and intellectual deficits.

It is not only excitatory transmission that suffers in the brain of HE patients. Convincing clinical studies concur with the evidence of excessive GABA'ergic transmission in the brains of individuals with HE (Jones, 2002; Bansky et al., 1985; Scollo-Lavizzari and Steinmann, 1985). This rationale has been reproduced in various animal models of HE as well (Bassett et al., 1987; Gammal et al., 1990). Increased GABA'ergic tone in HE may be ascribed to increased levels of circulating natural benzodiazepines (Basile et al., 1991), increased synthesis of $GABA_A$ modulating neurosteroids (Ahboucha and Butterworth, 2007), and increased availability of GABA and activation of $GABA_A$ receptors (Olasmaa et al., 1990; Wysmyk et al., 1992), while ammonia has also been found to increase GABA binding to the $GABA_A$ receptor (Ha and Basile, 1996). Increased GABA activity may also be elicited following an increase in NO-derived toxins, such as peroxynitrite (Ohkuma et al., 1995), that follows HE.

Hyperammonia stimulates the transport of aromatic amino acids across the blood brain barrier, such as enhancing glutamine/tryptophan exchange (Cangiano et al., 1983; Hilgier et al., 1992) resulting in increased cerebral tryptophan and an increase in 5HT synthesis (Zieve, 1987). This correlates with the evidence of a deficit in 5HT transporters observed in experimental HE in animals (Michalak et al., 2001). Indeed, the accumulation of tryptophan and tyrosine, which serve as precursors for the neuronal synthesis of 5HT and DA, as well as increased density and expression of monoamine oxidase A, a principle degradative enzyme of biogenic amines (Mousseau et al., 1997), may underlie altered brain and behavioral function in patients with HE. Animal studies have concurred with this suggestion (Bergqvist et al., 1996; Yurdaydin et al., 1996; Kaneko et al., 1998). With regard to regional brain 5HT neuroreceptor binding, densities for the $5HT_{1A}$ receptor in cirrhotic patients are decreased in frontal cortex by 56% and in the hippocampus by 30%, together with a 55% increase in hippocampal $5HT_{2A}$ receptor binding (Rao and Butterworth, 1994). In Wilson's disease, depressive symptomatology is related to an alteration of presynaptic serotonin transporters (Eggers et al., 2003; Hesse et al., 2003). The increased 5HT observed, and associated neuroplastic receptor changes, may contribute to the neuropsychiatric manifestations evident in these patients.

Dopamine changes, which may be linked to motor dysfunction in HE, have been noted in the brains of animals with experimental HE and in patients with HE (Mousseau et al., 1997), while a loss of striatal DA (Mousseau et al., 1993) and increased MOA A and B (Rao et al., 1993) have been described in the brain tissue of cirrhotic patients. However, both 5HT and DA are strongly regulated by GABA as well as glutamate and NO (Wheeler et al., 1995; Borkowska et al., 1999; Tao and Auerbach, 2000; Prast and Philippu, 2001) such that dysfunction in GABA/glutamate transmission may also contribute to 5HT and DA dysfunction. Raised cerebral level of 5HT is also a powerful modulator of the glutamatergic and GABA'ergic systems, with 5HT decreasing glutamate transmission and increasing GABA transmission, particularly in the hippocampus, frontal cortex and cerebellum (Ciranna, 2006) These mechanisms may partly underlie serotonergic-mediated modulation of cognitive function, analgesia, motor control and mood (Ciranna, 2006).

Another interesting topic when considering encephalopathy-based mood disturbances is the sulfur-containing amino acid, homocysteine. This amino acid is important in methylation reactions, the levels of which are regulated by folate and vitamins B6 and B12 (House et al., 1999). Elevated homocysteine levels, as well as altered folate and vitamins B6 and B12, have been implicated in depression, schizophrenia and bipolar disorder (Bottiglieri, 2005; Levine et al., 2005; Dimopoulos et al., 2007; Folstein et al., 2007), and also in certain metabolic imbalances characterized by altered folate and/or vitamin B12 availability such as eating disorders (Frieling et al., 2006), alcoholism, malnutrition, malabsorption (Frankenburg, 2007) and some severe but reversible encephalopathies (Gutiérrez-Aguilar et al., 2005). Of particular importance is that homocysteic acid, a metabolite of homocysteine, acts as an NMDA receptor agonist and is a recognized neurotoxin (Lipton et al., 1997).

While I have focused primarily on the neuropsychiatry of HE, a number of diverse liver diseases that present with metabolic encephalopathies can be linked to mood dysregulation and depression. This would include illnesses such as hepato-lenticular degeneration (Wilson's disease) (Eggers et al., 2003; Hesse et al., 2003; Chan et al., 2005), Gaucher disease (Packman et al., 2006), Fabry disease (Sadek et al., 2004), and hepatitis C infection (Forton et al., 2004). Thus, both the presentation of HE and its associated neurochemistry predict that encephalopathy has an important mood component. However, do depressed patients present with a component of encephalopathy?

Is Depressive Illness a Metabolic Encephalopathy?

While there is little to no evidence to suggest that HE and depression are alike, there are distinct similarities that warrant discussion. The biogenic amine hypothesis infers that antidepressant action is based on the immediate synaptic increase in NA, 5HT or DA. Antidepressants evoke an almost immediate increase in the synaptic levels of these neurotransmitters (Leonard, 2003). Models have emphasized the role of monoamine systems in depression and in mediating disease pathogenesis (Stahl, 1996) and treatment response (Bell et al., 2001). However, it is now well recognized that if current antidepressants act only by virtue of their ability to evoke an immediate increase in the synaptic levels of these transmitters, all antidepressants should have an almost immediate onset of action, which they do not (Manji et al., 2001; Leonard, 2003). The neuroplasticity hypothesis thus links the delayed time of onset of antidepressant action to events subsequent to their synaptic actions, involving the sequential and obligatory activation of the cascade of subcellular events described earlier, and involving second messenger activation, gene transcription and protein synthesis (Manji et al., 2001; D'sa and Duman, 2002).

Dorland's Medical Dictionary describes encephalopathy as any degenerative disease of the brain, and metabolic encephalopathy as a neuropsychiatric disturbance following acute imbalance in critical cellular metabolites occurring primarily as a result of hypoxia, ischemia, or hypoglycemia, or secondarily to disease of other organs, such as the kidney, lung, or liver. Alterations in brain functions occur in both acute and chronic HE, while gross structural brain changes are evident only over a protracted period (Vaquero and Butterworth, 2006). Thus, acute neuropsychiatric manifestations are not only evident in disorders like HE, but more importantly, they can lead to progressive neurodegenerative changes in brain cytoarchitecture over time, particularly involving astrocytes (Albrecht and Jones, 1999). That depression may be an encephalopathy is clearly a controversial statement. However, it does serve to bring face-to-face the biogenic amine and neuroplasticity hypotheses of depression. An important consideration in this analogy would be to establish whether similar monoaminergic, especially serotonergic, receptor changes are evident in the two syndromes, and also whether acutely increasing, or alternatively acutely decreasing biogenic amines in the brain will precipitate an almost immediate worsening or improvement in mood. Further, is there evidence for metabolic disturbances in depression and does major depression, like an encephalopathy, show evidence of a progressively worsening illness with anatomical involvement? Finally, is major depression associated to any degree with biochemical imbalances that are characteristic of the most widely described encephalopathies, including changes in oxidative status, ammonia, GABA and glutamate? These are discussed presently.

HE is associated with increased tryptophan and 5HT synthesis (Zieve, 1987) as well as decreased $5HT_{1A}$ receptor density in the limbic brain regions together with elevated $5HT_{2A}$ receptor density (Rao and Butterworth, 1994). Similarly, major depressive disorder is also associated with a widespread reduction in $5HT_{1A}$ receptor binding (Sargent et al., 2000; Drevets et al., 2007), as well as elevated $5HT_{2A}$ receptor density (Leonard, 2003), although a reduction in cerebral serotonin is more generally accepted as the basis for the neuropathology of depression. The widespread reduction in $5HT_{1A}$ receptors, which presynaptically are associated with the negative control of 5HT release, may explain the accumulation of 5HT in patients with HE, as well as elevated 5-hydroxyindoleacetic acid and MAOA (Rao and Butterworth, 1994). However, disagreement exists within the literature regarding the presence and direction of $5HT_{1A}$ receptor binding abnormalities in depression, and also considering the broadly accepted notion that depression is associated with a reduction in 5HT which, as pointed out, appears opposite to that observed in HE. Nevertheless, it does concur that a similar imbalance in serotonergic receptor function occurs in both HE and depression that may exert diverse direct and indirect neuromodulatory actions, and in this way contribute to the neuropsychiatric manifestations and depressed mood evident in these patients (Albrecht and Jones, 1999).

Experimental lowering of 5HT neurotransmission by acute tryptophan depletion (ATD) does indeed induce a rapid but transient depressed mood in 50–60% of patients treated with a SRI who are in remission from depression (Delgado et al., 1999; Booij et al., 2003). Clinical studies in depressed patients have also highlighted distinct metabolic alterations in medication-free patients with bipolar disorder (Dager et al., 2004), with elevated gray matter lactate and GABA noted, suggesting mitochondrial alterations and a shift in energy redox state from oxidative phosphorylation toward glycolysis. Other clinical studies have also described altered mitochondrial function in affective illness, showing that mitochondrial dysfunction is associated with vulnerability to psychopathology in selected patients (Gardner et al., 2003). Moreover, affective disorders may also present with altered glucose metabolism, with evidence for a possible redirection of glucose metabolism away from the preferred glycolytic pathway to the polyol (sorbitol) pathway (Regenold et al., 2004, 2005), a pathway that is linked to nervous tissue damage in diabetes mellitus.

The multifactorial nature of depression resembles that of other complex disorders such as diabetes mellitus or coronary heart disease. Indeed, research has increasingly discovered the relevance of depressive symptoms for the development and course of diabetes, particularly in diabetes type 2, with up to one fourth of all patients with diabetes mellitus suffering from depressive symptoms up to and including states of depressive disorders (Kruse et al., 2006). Similarly, an independent association has been found between major depression and coronary heart disease (Perlmutter et al., 2000). Indeed, the 1-month prevalence of depression in coronary heart disease patients is approximately 15%, threefold higher than that observed in the community (Rozanski et al., 1999). In patients with coronary heart disease, depression has been associated with a 2.5–4-fold increase in the occurrence of cardiovascular events (Rozanski et al., 1999). Recent evidence has brought to notice the important role of NO in this connection (Chrapko et al., 2004), which not only is an important neuromodulator in the brain, but has important regulatory functions in the cardiovascular system. Finally, both mental illness and vascular disease display high plasma homocysteine levels that may be a contributory risk factor for both illnesses (Bertsch et al., 2001; Nilsson et al., 2007) as well as their apparent association with one another.

Raised glucocorticoid levels are typical of depression (Sapolsky, 2000), such that cortisol-directed changes in intermediary metabolism, particularly glucose, can be expected. Moreover, recent evidence also suggests that stress/cortisol-induced neuronal remodeling, as well as damage, is expedited by exposure to elevated levels of glucocorticoids (Starkman et al., 1999). In fact, a more recent finding in depression is the evidence for hippocampal shrinkage, together with associated decrements in hippocampal-dependent cognitive processing, ascribed to chronically elevated cortisol levels. Hippocampal damage may lead to further inadequate hypothalamic-adrenal feedback regulation leading to a sustained elevation in cortisol (see Harvey et al., 2003 for a review). Indeed, Cushing's disease is associated with reduced hippocampal volume as well as prominent cognitive decrements and depression (Starkman et al., 1992, 2001). Glucocorticoids increase susceptibility of cells to neurotoxic events, particularly that mediated by elevated glutamate release (Sapolsky, 2000). Depression also shows a progressive worsening of symptoms over time and successive episodes, as well as a progressive worsening of hippocampal atrophy (see Harvey et al., 2003 for a review). Importantly, as in HE, postmortem studies in recurrently depressed patients show atrophic changes of the glia with associated changes in glutamate reuptake (Harvey et al., 2003), while both depression (D'sa and Duman, 2002) and transient states of hyperammonia (Montoliu et al., 2007) can evoke changes in cAMP–CREB signaling, a primary pathway involved in the neuroplastic changes associated with mood regulation and antidepressant response.

The significance of these findings with respect to the pathogenesis of depression suggest that affective disorders present with acute neurochemical changes and altered metabolic states similar to encephalopathy, possibly associated with multiple metabolic disturbances linked to liver, cardiovascular and hormonal disorders. This leads to progressive "hard wiring" over time and to intractable structural brain changes. Of particular note is the common involvement of glutamate and GABA.

Glutamate and GABA in Disorders of Mood and Their Association with Encephalopathy

The role of glutamate, GABA (Shiah and Yatham, 1998; Krystal et al., 2002; Stewart and Reid, 2002) and glutamate-mediated activation of subcellular calcium-dependent pathways, especially NO and cGMP (Harvey, 1996; McLeod et al., 2001), has in recent years become increasingly recognized as being involved in the neuropathology and treatment of affective illnesses, particularly in mediating rapid neuronal regulation throughout the CNS of especially monoaminergic pathways (Harvey, 1996). These changes may essentially originate from different etiologies but thereafter follow a common sequence of events leading to dysregulation of mood.

The Role of GABA

Although GABA is raised in HE, its concentrations are reduced in the brain, cerebrospinal fluid and plasma of depressed patients (Krystal et al., 2002; Shiah and Yatham, 1998; Sanacora et al., 2003). Postmortem studies also indicate reduced $GABA_A$ receptors in the frontal cortex of suicide victims (Pandey et al., 1997). Moreover, clinically effective antidepressants and electroconvulsive therapy (ECT) increase CSF or brain GABA levels (Shiah and Yatham, 1998; Devanand et al., 1995; Sanacora et al., 2003). This apparent contradiction between depression and HE may be due to the strong association between GABA and glutamate in cellular metabolism, with GABA originating from glutamate synthesis via glutamate decarboxylase. Thus, any excessive release of glutamate, as is evident in HE, may in turn promote GABA synthesis (Leonard, 2003). Importantly, as noted earlier, stress-evoked cortisol release increases glutamate release, while stress also increases GABA release (Engelmann et al., 2002). GABA in turn mediates inhibition of glutamatergic transmission via presynaptic $GABA_B$ heteroreceptors (Yamada et al., 1999). However, prolonged stress and reminders on the other hand, are associated with an increase in NOS activity and a profound decrease in GABA in the rat hippocampus (Harvey et al., 2004), indicating that prolonged stress leads to loss of the homeostatic function of GABA and an increase in potentially cell damaging events as a result. Furthermore, stress-induced GABA release is potentiated by peroxynitrite (Ohkuma et al., 1995) suggesting that potentially harmful metabolites arising from ischemia, hypoxia, free radicals, hypoglycemia and oxidative stress, all situations presenting themselves during HE, will enhance GABA release (Saransaari and Oja, 1997). This occurs as a protective mechanism that will attenuate further NO release (Ishizuka et al., 2000). Increased GABA thus confers resilience to chemical and psychosocial stressors. However, overt release of GABA may also be detrimental, as is typically described in certain encephalopathies (Jones, 2002). Indeed, the use of ECT to treat severe intractable depression is associated with increased GABA release, but also with memory disturbances and altered glutamate function (Chamberlin and Tsai, 1998; McDaniel et al., 2006).

The Role of Glutamate

The NMDA glutamate ionotropic receptor is a highly regulated excitatory receptor, being positively regulated by recognition sites for glutamate, glycine and various polyamines, and negatively regulated by recognition sites for magnesium, zinc, redox state, NO as well as sites for binding phencyclidine and other similar compounds (Cooper et al., 1996). Major depression is accompanied by alterations in the serine/glycine ratio as well as glutamate (Altamura et al., 1995), while chronic antidepressant treatment of depressed patients significantly reduces serum levels of aspartate and glutamate (Maes et al., 1998). Changes in the NMDA receptor complex has been noted in the frontal cortex of suicide victims (Nowak et al., 1995), while a reduction in the hippocampus of the NMDA receptor subunit, NMDAR1, has also been described (Law and Deakin, 2001). Further, zinc ion, which functions as a negative regulator of the NMDA ion channel, has been found to be lower in depressed patients than in healthy controls (Maes et al., 1994). Platelet glutamate receptor supersensitivity is evident in patients with depression (Berk et al., 2001), while elevated levels of nitrogen oxide metabolites of NO have also been observed in patients with depression (Suzuki et al., 2001) and following suicide (Kim et al., 2006).

Interestingly, depression associated with interferon alpha therapy is also associated with elevated NO levels (Suzuki et al., 2003), which concurs with evidence indicating a relationship between stress, depression and immunological influence on emotions (Dantzer, 2001; Miller et al., 2005). Various stressors induce a central production of cytokines (Dunn, 2001) that may have a negative effect on brain 5HT levels (Capuron and Dantzer, 2003; Wichers and Maes, 2004). Cytokine-mediated induction of indoleamine 2,3-dioxygenase (IDO), which is the rate limiting step in the conversion of l-tryptophan, the precursor of 5HT, to N-formylkynurenine, results in diminished synthesis of 5HT (Wichers and Maes, 2004). However, several other important metabolites are formed along the kynurenine pathway following the upregulation of IDO, one in particular being quinolinic acid, a direct agonist of the glutamate NMDA receptor (Wichers and Maes, 2004).

Inflammation contributes to the cognitive impairment in HE, while targeting the cyclooxygenase (COX) pathway has been found to normalize glutamate–NO–cGMP signaling and to restore cognitive function (Cauli et al., 2007). Certain forms of stress are associated with activation of COX and immunological NOS (iNOS) (Madrigal et al., 2006), while major depression is associated with elevated antioxidant enzyme activity and lipid peroxidation (Bilici et al., 2001). The COX II inhibitor, celecoxib, has recently been found to be effective in treating depression (Muller et al., 2006). Inflammation is associated with altered cellular redox state. Astrocyte swelling in HE triggered by ammonia, in synergism with different precipitating factors such as hyponatremia and glutamate and NO release, plays a major role in altering cellular redox state, leading to osmotic and oxidative stress (Schliess et al., 2006). The NMDA receptor presents with a redox regulatory site for the purpose of controlling oxidative stress following excessive glutamatergic activity (Cooper et al., 1996).

Preclinical studies have demonstrated the antidepressant-like properties of NMDA antagonists (Skolnick, 1999) as well as zinc ion (Kroczka et al., 2001). The mechanism whereby antidepressants are purported to achieve this is by reducing the proportion of high affinity glycine sites on the NMDA receptor (Skolnick et al., 1996) or by reducing NMDA subunit mRNA in various limbic and subcortical structures (Boyer et al., 1998), leading to an attenuation of glutamatergic transmission. The downstream events of NMDA receptor activation, particularly on NOS, have also been found to be targeted by traditional antidepressant treatment, over and above their actions on monoamine transporters (Wegener et al., 2003; Harvey et al., 2006). Consequently, NOS inhibitors (Harkin et al., 1999) and guanylyl cyclase–cGMP inhibitors (Heiberg et al., 2002) have all demonstrated distinct antidepressant-like effects in animals. Finally, it is also noteworthy that NMDA antagonists (Rogoz et al., 2002) and NOS inhibitors (Harkin et al., 2004) exert synergistic antidepressant-like effects with classical antidepressants in the forced swimming test. In agreement with these preclinical findings, robust and rapid antidepressant effects have been described following a single intravenous dose of the NMDA receptor antagonist, ketamine, with onset of action occurring within 2-h postinfusion and maintaining efficacy for 1 week (Zarate et al., 2006). Supportive evidence of the use of NMDA receptor antagonism as augmentation strategy is increasing (Barbosa et al., 2003; Stryjer et al., 2003; Rogoz et al., 2004). It is also noteworthy that ECT has also been found to involve suppression of NMDA receptor activity (Skolnick, 1999; Stewart and Reid, 2002). The above discussion thus conveys evidence that depression presents with metabolic changes akin to HE, including altered glutamate: GABA balance and changes in the immune–inflammatory response, and that a rapid onset of antidepressant action may be realized through blockade of excessive glutamate–NMDA receptor activity.

Depression as an Encephalopathy: Implications for Onset of Antidepressant Action

How does viewing major depressive illness as an encephalopathy say anything new about the pathology and treatment of the disorder? As has been alluded to earlier, a number of small clinical studies have proposed a faster onset of action for certain antidepressants, e.g., escitalopram, mirtazapine, venlafaxine, fluoxetine + desipramine and augmentation with pindolol. What is interesting is that all these agents exert a primary action on the synaptic levels of monoamines which, according to the plasticity hypothesis of depression, serves to activate a cascade of events leading to antidepressant response but which by itself is not able to "switch on" a rapid onset of action. While the overwhelming evidence supports antidepressant onset of action to be weeks rather than days, recent data using weekly or daily mood ratings demonstrate that maximum improvement can occur during the first 2 weeks, with some improvement within the first 3 days (Mitchell, 2006). In fact, differences between mirtazapine and the SRIs in favor of the former have been noted after 1 week of treatment, and appear to be due to a specific antidepressant effect rather than sedation (Thompson, 2002). Methodological differences in assessment between clinical studies make the definitive measurement of time to onset of action a controversial issue (Stahl et al., 2001; Mitchell, 2006). Nevertheless, while a faster than nominal time to onset of action in the majority of clinical cases is unlikely, it may be plausible that a faster onset of action may be possible in certain subgroups of depressive illness.

The possible mechanisms whereby an antidepressant may engender a faster onset of action has been hotly debated, with all the options described above having different sites of action in the synapse. Pindolol's action is linked to its action at the $5HT_{1A}$ autoreceptor, thereby attenuating this receptor's inhibitory actions on synaptic firing (Blier, 2003), while mirtazepine's actions at multiple 5HT'ergic and NA'ergic sites bolsters 5HT'ergic and NA'ergic firing in the limbic brain regions (Blier, 2003). Escitalopram, on the other hand, demonstrates binding to the human 5HT transporter via an affinity-modulating allosteric site in addition to its 5HT reuptake inhibitory properties (Chen et al., 2005; Sanchez, 2006). This action is hypothesized to be responsible for a longer binding to, and therefore a greater inhibition of, the serotonin transporter. This may confer improved efficacy and a faster onset of action (Kasper et al., 2006).

Faster onset may also represent a group where raised glutamatergic function represents a core neuropathology, one where its underlying psychopathology is not unlike that seen in acute metabolic disturbances found in certain encephalopathies. The use of antiglutamate drugs has not been tested in HE, although a recent study in experimental hepatic encephalopathy provides evidence that hyperammonemia-associated learning dysfunction can be normalized by the selective cGMP-phosphodiesterase inhibitor, sildenafil (Erceg et al., 2005). Clearly, the NO–cGMP cascade plays a central role in the CNS effects associated with HE. However, SRIs or similar antidepressants may also modulate NMDA-related activity, as for example NOS (Wegener et al., 2003), which will induce or drive a more rapid correction of glutamatergic synapses, contributing to an initial rapid antidepressant efficacy (Skolnick, 1999). This may be similar to the rapid onset noted with NMDA antagonists in depression described earlier (Zarate et al., 2006), but which may be followed some weeks later by full antidepressant activity mediated by the more time-dependent neuroplastic changes.

D-Cycloserine, a positive modulator of the NMDA receptor, and which has been found to inhibit 5HT function (Dall'Olio et al., 2000), has met with some, albeit small, success in exploration studies as a possible augmentation strategy in depression (Heresco-Levy et al., 2006). This illustrates the importance of an obligatory interaction between glutamate–monoamine systems. Indeed, in animals a functional noradrenergic system is necessary for changes to occur at the NMDA receptor (Harkin et al., 2000). There are other examples of monoamine–glutamate interactions in mood. Serotonin depletion, which is the basis for the monoamine hypothesis of depression, produces long-lasting increases in glutamatergic transmission (Di Cara et al., 2001) and is also known to increase downstream activation of NOS (Tagliaferro et al., 2001). Further, antidepressant discontinuation is associated with disinhibition of NMDA receptor activity and overt activation of NOS in rat hippocampus (Harvey et al., 2002, 2006).

Conclusion

Major depression is, in many ways, a multifactorial illness, resembling other complex disorders such as diabetes mellitus, coronary heart disease, liver disease and obesity, while also presenting with elements of immune–inflammatory dysfunction. Comorbid disorders of the heart, liver and endocrine systems, with their associated metabolic dysfunction, may essentially drive the neurochemical imbalance that will precipitate a depressive episode. Like hepatic-derived metabolic disorders, e.g., HE, depression may present after an acute perturbation of mood-regulating monoamines as well as GABA and glutamate. It then follows an inevitable progressive worsening over time, characterized by neuroplastic changes, inflammation and/or neurodegeneration. Glutamatergic and GABA'ergic function are not only involved in regulating cellular excitability and resilience to potentially cell damaging events and challengers, but are also responsible for rapid neurotransmission throughout the brain, leading to fast modulation of especially monoaminergic transmission. Moreover, both amino acids are intimately involved in the stress response and are also important nonspecific targets for currently used antidepressants. The correction of disturbances in GABA–glutamate activity therefore represents a putative mechanism for rapidly reverting depressed mood. While not definitive for all major depressive illnesses across the spectrum of mood disorders, depressive symptoms in some cases may be similar to an encephalopathy, presenting with neurochemical and metabolic disturbances that can be rapidly corrected by suitable and targeted pharmacotherapy. At this point, targeting pathways intimately involved in encephalopathy, e.g., glutamate, GABA and associated downstream messengers, may expedite an antidepressant response. However, the tendency of patients with depression to delay adequate treatment, and often to initiate premature discontinuation of treatment, will lead to long-standing, chronic metabolic dysfunction and a more intractable form of the illness, requiring longer and more complex treatment (Harvey et al., 2003).

Acknowledgments The author's work in the area of depression and stress is supported by the South African Medical Research Council (MRC) and the National Research Foundation (grant number 2073038).

References

Ahboucha, S. and Butterworth, R.F. 2007. The neurosteroid system: an emerging therapeutic target for hepatic encephalopathy. Metab. Brain Dis. 22:291–308.

Albrecht, J. and Jones, E.A. 1999. Hepatic encephalopathy: molecular mechanisms underlying the clinical syndrome. J. Neurol. Sci. 170:138–146.

Altamura, C., Maes, M., Dai, J., and Meltzer, H.Y. 1995. Plasma concentrations of excitatory amino acids, serine, glycine, taurine and histidine in major depression. Eur. Neuropsychopharmacol. 5:71–75.

Bansky, G., Meier, P.J., Ziegler, W.H., Walser, H., Schmid, M., and Huber, M. 1985. Reversal of hepatic coma by benzodiazepine antagonist (Ro 15-1788). Lancet 1:1324–1325.

Barbosa, L., Berk, M., and Vorster, M. 2003. A double-blind, randomized, placebo-controlled trial of augmentation with lamotrigine or placebo in patients concomitantly treated with fluoxetine for resistant major depressive episodes. J. Clin. Psychiatry 64:403–407.

Basile, A.S., Hughes, R.D., Harrison, P.M., Murata, Y., Pannell, L., Jones, E.A., Williams, R., and Skolnick, P. 1991. Elevated brain concentrations of 1,4-benzodiazepines in fulminant hepatic failure. N. Engl. J. Med. 325:473–478.

Bassett, M.L., Mullen, K.D., Skolnick, P., and Jones, E.A. 1987. Amelioration of hepatic encephalopathy by pharmacologic antagonism of the GABAA-benzodiazepine receptor complex in a rabbit model of fulminant hepatic failure. Gastroenterology 93:1069–1077.

Bell, C., Abrams, J., and Nutt, D. 2001. Tryptophan depletion and its implications for psychiatry. Br. J. Psychiatry 178:399–405.

Benkert, O., Grunder, G., Wetzel, H., and Hackett, D. 1996. A randomized, double-blind comparison of a rapidly escalating dose of venlafaxine and imipramine in inpatients with major depression and melancholia. J. Psychiatr. Res. 30:441–451.

Benkert, O., Szegedi, A., and Kohnen, R. 2000. Mirtazapine compared with paroxetine in major depression. J. Clin. Psychiatry 61:656–663.

Bergqvist, P.B., Hjorth, S., Audet, R.M., Apelqvist, G., Bengtsson, F., and Butterworth, R.F. 1996. Ammonium acetate challenge in experimental chronic hepatic encephalopathy induces a transient increase of brain 5-HT release in vivo. Eur. Neuropsychopharmacol. 6:317–322.

Berk, M., Plein, H., and Ferreira, D. 2001. Platelet glutamate receptor supersensitivity in major depressive disorder. Clin. Neuropharmacol. 24:129–132.

Berton, O. and Nestler, E.J. 2006. New approaches to antidepressant drug discovery: beyond monoamines. Nat. Rev. Neurosci. 7:137–151.

Bertsch, T., Mielke, O., Höly, S., Zimmer, W., Casarin, W., Aufenanger, J., Walter, S., Muehlhauser, F., Kuehl, S., Ragoschke, A., and Fassbender, K. 2001. Homocysteine in cerebrovascular disease: an independent risk factor for subcortical vascular encephalopathy. Clin Chem Lab Med 39:721–724.

Bilici, M., Efe, H., Koroglu, M.A., Uydu, H.A., Bekaroglu, M., and Deger, O. 2001. Antioxidative enzyme activities and lipid peroxidation in major depression: alterations by antidepressant treatments. J. Affect. Disord. 64:43–51.

Blier, P. 2003. The pharmacology of putative early-onset antidepressant strategies. Eur. Neuropsychopharmacol. 13:57–66.

Booij, L., Van der Does, A.J., and Riedel, W.J. 2003. Monoamine depletion in psychiatric and healthy populations: review. Mol. Psychiatry 8:951–973.

Borkowska, H.D., Oja, S.S., Oja, O.S., Saransaari, P., Hilgier, W., and Albrecht, J. 1999. N-methyl-d-aspartate-evoked changes in the striatal extracellular levels of dopamine and its metabolites in vivo in rats with acute hepatic encephalopathy. Neurosci. Lett. 268:151–154.

Bottiglieri, T. 2005. Homocysteine and folate metabolism in depression. Prog. Neuropsychopharmacol. Biol. Psychiatry 29:1103–1112.

Boyer, P.A., Skolnick, P., and Fossom, L.H. 1998. Chronic administration of imipramine and citalopram alters the expression of NMDA receptor subunit mRNAs in mouse brain. A quantitative in situ hybridization study. J. Mol. Neurosci. 10:219–233.

Cangiano, C., Cardelli-Cangiano, P., James, J.H., Rossi-Fanelli, F., Patrizi, M.A., Brackett, K.A., Strom, R., and Fischer, J.E. 1983. Brain microvessels take up large neutral amino acids in exchange for glutamine. Cooperative role of Na+-dependent and Na+-independent systems. J. Biol. Chem. 258:8949–8954.

Capuron, L. and Dantzer, R. 2003. Cytokines and depression: the need for a new paradigm. Brain Behav. Immun. 17(Suppl. 1):S119–S124.

Carpenter, L.L., Yasmin, S., and Price, L.H. 2002. A double-blind, placebo-controlled study of antidepressant augmentation with mirtazapine. Biol. Psychiatry 51:183–188.

Cauli, O., Rodrigo, R., Piedrafita, B., Boix, J., and Felipo, V. 2007. Inflammation and hepatic encephalopathy: ibuprofen restores learning ability in rats with portacaval shunts. Hepatology 46:514–519.

Chamberlin, E. and Tsai, G.E. 1998. A glutamatergic model of ECT-induced memory dysfunction. Harv Rev Psychiatry 5:307–317.

Chan, K.H., Cheung, R.T., Au-Yeung, K.M., Mak, W., Cheng, T.S., and Ho, S.L. 2005. Wilson's disease with depression and parkinsonism. J. Clin. Neurosci. 12:303–305.

Chen, F., Larsen, M.B., Sanchez, C., and Wiborg, O. 2005. The S-enantiomer of R,S-citalopram, increases inhibitor binding to the human serotonin transporter by an allosteric mechanism. Comparison with other serotonin transporter inhibitors. Eur. Neuropsychopharmacol. 15:193–198.

Chrapko, W.E., Jurasz, P., Radomski, M.W., Lara, N., Archer, S.L., and Le Melledo, J.M. 2004. Decreased platelet nitric oxide synthase activity and plasma nitric oxide metabolites in major depressive disorder. Biol. Psychiatry 56:129–134.

Ciranna, L. 2006. Serotonin as a modulator of glutamate- and GABA-mediated neurotransmission: implications in physiological function in pathology. Curr. Neuropharmacol. 4:101–114.

Compton, M.T. and Nemeroff, C.B. 2000. The treatment of bipolar depression. J. Clin. Psychiatry 61(Suppl. 9):57–67.

Cooper, J.R., Bloom, F.E., and Roth, R.H. 1996. The Biochemical Basis of Neuropharmacology. New York: Oxford University Press.

D'Sa, C. and Duman, R.S. 2002. Antidepressants and neuroplasticity. Bipolar Disord. 4:183–194.

Dager, S.R., Friedman, S.D., Parow, A., Demopulos, C., Stoll, A.L., Lyoo, I.K., Dunner, D.L., and Renshaw, P.F. 2004. Brain metabolic alterations in medication-free patients with bipolar disorder. Arch. Gen. Psychiatry 61:450–458.

Dall'Olio, R., Gandolfi, O., and Gaggi, R. 2000. d-Cycloserine, a positive modulator of NMDA receptors, inhibits serotonergic function. Behav. Pharmacol. 11:631–637.

Dantzer, R. 2001. Cytokine-induced sickness behavior: where do we stand? Brain Behav. Immun. 15:7–24.

Delgado, P.L., Miller, H.L., Salomon, R.M., Licinio, J., Krystal, J.H., Moreno, F.A., Heninger, G.R., and Charney, D.S. 1999. Tryptophan-depletion challenge in depressed patients treated with desipramine or fluoxetine: implications for the role of serotonin in the mechanism of antidepressant action. Biol. Psychiatry 46:212–220.

Devanand, D.P., Shapira, B., Petty, F., Kramer, G., Fitzsimons, L., Lerer, B., and Sackeim, H.A. 1995. Effects of electroconvulsive therapy on plasma GABA. Convulsive Ther. 11:3–13.

Di Cara, B., Dusticier, N., Forni, C., Lievens, J.C., and Daszuta, A. 2001. Serotonin depletion produces long lasting increase in striatal glutamatergic transmission. J. Neurochem. 78:240–248.

Di Martini, A., Dew, M.A., Javed, L., Fitzgerald, M.G., Jain, A., and Day, N. 2004. Pretransplant psychiatric and medical comorbidity of alcoholic liver disease patients who received liver transplant. Psychosomatics 45:517–523.

Dimopoulos, N., Piperi, C., Salonicioti, A., Psarra, V., Gazi, F., Papadimitriou, A., Lea, R.W., and Kalofoutis, A. 2007. Correlation of folate, vitamin B12 and homocysteine plasma levels with depression in an elderly Greek population. Clin Biochem 40:604–608.

Domek-Łopaci ska, K. and Strosznajder, J.B. 2005. Cyclic GMP metabolism and its role in brain physiology. J. Physiol. Pharmacol. 56(Suppl. 2):15–34.

Drevets, W.C., Thase, M.E., Moses-Kolko, E.L., Price, J., Frank, E., Kupfer, D.J., and Mathis, C. 2007. Serotonin-1A receptor imaging in recurrent depression: replication and literature review. Nucl Med Biol 34:865–877.

Dunn, A. 2001. Effects of cytokines and infections on brain neurochemistry. In Psychoneuroimmunology, ed. R. Ader, D. Felten, and N. Cohen, pp. 649–666. San Diego: Academic.

Eggers, B., Hermann, W., Barthel, H., Sabri, O., Wagner, A., and Hesse, S. 2003. The degree of depression in Hamilton rating scale is correlated with the density of presynaptic serotonin transporters in 23 patients with Wilson's disease. J. Neurol. 250:576–580.

Engelmann, M., Wolf, G., and Horn, T.F. 2002. Release patterns of excitatory and inhibitory amino acids within the hypothalamic supraoptic nucleus in response to direct nitric oxide administration during forced swimming in rats. Neurosci. Lett. 324:252–254.

Entsuah, R., Derivan, A., and Kikta, D. 1998. Early onset of antidepressant action of venlafaxine: pattern analysis in intent-to-treat patients. Clin. Ther. 20:517–526.

Erceg, S., Monfort, P., Hernandez-Viadel, M., Rodrigo, R., Montoliu, C., and Felipo, V. 2005. Oral administration of sildenafil restores learning ability in rats with hyperammonemia and with portacaval shunts. Hepatology 41:299–306.

Felipo, V., Grau, E., Minana, M.D., and Grisolia, S. 1993. Hyperammonemia decreases proteinkinase-C-dependent phosphorylation of microtubule-associated protein 2 and increases its binding to tubulin. Eur. J. Biochem. 214:243–249.

Folstein, M., Liu, T., Peter, I., Buell, J., Arsenault, L., Scott, T., and Qiu, W.W. 2007. The homocysteine hypothesis of depression. Am J Psychiatry 164:861–867.

Forton, D.M., Thomas, H.C., and Taylor-Robinson, S.D. 2004. Central nervous system involvement in hepatitis C virus infection. Metab. Brain Dis. 19:383–391.

Frankenburg, F.R. 2007. The role of one-carbon metabolism in schizophrenia and depression. Harv Rev Psychiatry 15:146–160.

Frieling, H., Romer, K.D., Beyer, S., Hillemacher, T., Wilhelm, J., Jacoby, G.E., de Zwaan, M., Kornhuber, J., and Bleich, S. 2006. Depressive symptoms may explain elevated plasma levels of homocysteine in females with eating disorders. J Psychiatr Res, Dec 18; doi:10.1016/j. jpsychires.2006.10.007 [Epub ahead of print].

Gammal, S.H., Basile, A.S., Geller, D., Skolnick, P., and Jones, E.A. 1990. Reversal of the behavioral and electrophysiological abnormalities of an animal model of hepatic encephalopathy by benzodiazepine receptor ligands. Hepatology 11:371–378.

Gardner, A., Johansson, A., Wibom, R., Nennesmo, I., von Dobeln, U., Hagenfeldt, L., and Hallstrom, T. 2003. Alterations of mitochondrial function and correlations with personality traits in selected major depressive disorder patients. J. Affect. Disord. 76:55–68.

Geddes, J.R., Freemantle, N., Mason, J., Eccles, M.P., and Boynton, J. 2000. SSRIs versus other antidepressants for depressive disorder. Cochrane Database Syst. Rev. CD001851.

Guelfi, J.D., White, C., Hackett, D., Guichoux, J.Y., and Magni, G. 1995. Effectiveness of venlafaxine in patients hospitalized for major depression and melancholia. J. Clin. Psychiatry 56:450–458.

Gutiérrez-Aguilar, G., Abenia-Usón, P., García-Cazorla, A., Vilaseca, M.A., and Campistol, J. 2005. Encephalopathy with methylmalonic aciduria and homocystinuria secondary to a deficient exogenous supply of vitamin B12. Rev. Neurol. 40:605–608.

Ha, J.H. and Basile, A.S. 1996. Modulation of ligand binding to components of the GABAA receptor complex by ammonia: implications for the pathogenesis of hyperammonemic syndromes. Brain Res. 720:35–44.

Harkin, A.J., Bruce, K.H., Craft, B., and Paul, I.A. 1999. Nitric oxide synthase inhibitors have antidepressant-like properties in mice. 1. Acute treatments are active in the forced swim test. Eur. J. Pharmacol. 372:207–213.

Harkin, A., Connor, T.J., Burns, M.P., and Kelly, J.P. 2004. Nitric oxide synthase inhibitors augment the effects of serotonin re-uptake inhibitors in the forced swimming test. Eur. Neuropsychopharmacol. 14:274–281.

Harkin, A., Nowak, G., and Paul, I.A. 2000. Noradrenergic lesion antagonizes desipramine-induced adaptation of NMDA receptors. Eur. J. Pharmacol. 389:187–192.

Harvey, B.H. 1996. Affective disorders and nitric oxide: a role in pathways to relapse and refractoriness. Hum. Psychopharmacol. 11:309–319.

Harvey, B.H., Jonker, L.P., Brand, L., Heenop, M., and Stein, D.J. 2002. NMDA receptor involvement in imipramine withdrawal-associated effects on swim stress, GABA levels and NMDA receptor binding in rat hippocampus. Life Sci. 71:43–54.

Harvey, B.H., McEwen, B.S., and Stein, D.J. 2003. Neurobiology of antidepressant withdrawal: implications for the longitudinal outcome of depression. Biol. Psychiatry 54:1105–1117.

Harvey, B.H., Oosthuizen, F., Brand, L., Wegener, G., and Stein, D.J. 2004. Stress–restress evokes sustained iNOS activity and altered GABA levels and NMDA receptors in rat hippocampus. Psychopharmacology (Berl.) 175:494–502.

Harvey, B.H., Retief, R., Korff, A., and Wegener, G. 2006. Increased hippocampal nitric oxide synthase activity and stress responsiveness after imipramine discontinuation: role of 5HT 2A/C-receptors. Metab. Brain Dis. 21:211–220.

Heiberg, I.L., Wegener, G., and Rosenberg, R. 2002. Reduction of cGMP and nitric oxide has antidepressant-like effects in the forced swimming test in rats. Behav. Brain Res. 134:479–484.

Heresco-Levy, U., Javitt, D.C., Gelfin, Y., Gorelik, E., Bar, M., Blanaru, M., and Kremer, I. 2006. Controlled trial of d-cycloserine adjuvant therapy for treatment-resistant major depressive disorder. J. Affect. Disord. 93:239 243.

Hesse, S., Barthel, H., Hermann, W., Murai, T., Kluge, R., Wagner, A., Sabri, O., and Eggers, B. 2003. Regional serotonin transporter availability and depression are correlated in Wilson's disease. J. Neural Transm. 110:923–933.

Hilgier, W., Puka, M., and Albrecht, J. 1992. Characteristics of large neutral amino acid-induced release of preloaded l-glutamine from rat cerebral capillaries in vitro: effects of ammonia, hepatic encephalopathy, and gamma-glutamyl transpeptidase inhibitors. J. Neurosci. Res. 32:221–226.

House, J.D., Jacobs, R.L., Stead, L.M., Brosnan, M.E., and Brosnan, J.T. 1999. Regulation of homocysteine metabolism. Adv. Enzyme Regul. 39:69–91.

Ishizuka, Y., Ishida, Y., Jin, Q.H., Mitsuyama, Y., and Kannan, H. 2000. GABA(A) and GABA(B) receptors modulating basal and footshock-induced nitric oxide releases in rat prefrontal cortex. Brain Res. 872:266–270.

Jones, E.A. 2002. Ammonia, the GABA neurotransmitter system, and hepatic encephalopathy. Metab. Brain Dis. 17:275–281.

Kaneko, K., Kurumaji, A., Watanabe, A., Yamada, S., and Toru, M. 1998. Changes in high K+-evoked serotonin release and serotonin 2A/2C receptor binding in the frontal cortex of rats with thioacetamide-induced hepatic encephalopathy. J. Neural Transm. 105:13–30.

Kasper, S., Spadone, C., Verpillat, P., and Angst, J. 2006. Onset of action of escitalopram compared with other antidepressants: results of a pooled analysis. Int. Clin. Psychopharmacol. 21:105–110.

Kendler, K.S., Thornton, L.M., and Gardner, C.O. 2001. Genetic risk, number of previous depressive episodes, and stressful life events in predicting onset of major depression. Am. J. Psychiatry 158:582–586.

Kim, Y.K., Paik, J.W., Lee, S.W., Yoon, D., Han, C., and Lee, B.H. 2006. Increased plasma nitric oxide level associated with suicide attempt in depressive patients. Prog. Neuropsychopharmacol. Biol. Psychiatry 30:1091–1096.

Kocsis, J.H. 2003. Pharmacotherapy for chronic depression. J. Clin. Psychol. 59:885–892.

Kosenko, E., Kaminsky, Y., Grau, E., Minana, M.D., Grisolia, S., and Felipo, V. 1995. Nitroarginine, an inhibitor of nitric oxide synthetase, attenuates ammonia toxicity and ammonia-induced alterations in brain metabolism. Neurochem. Res. 20:451–456.

Kosenko, E., Kaminsky, Y., Grau, E., Minana, M.D., Marcaida, G., Grisolia, S., and Felipo, V. 1994. Brain ATP depletion induced by acute ammonia intoxication in rats is mediated by activation of the NMDA receptor and Na+,K(+)-ATPase. J. Neurochem. 63:2172–2178.

Kosenko, E., Kaminsky, Y., Lopata, O., Muravyov, N., Kaminsky, A., Hermenegildo, C., and Felipo, V. 1998. Nitroarginine, an inhibitor of nitric oxide synthase, prevents changes in superoxide radical and antioxidant enzymes induced by ammonia intoxication. Metab. Brain Dis. 13:29–41.

Kroczka, B., Branski, P., Palucha, A., Pilc, A., and Nowak, G. 2001. Antidepressant-like properties of zinc in rodent forced swim test. Brain Res. Bull. 55:297–300.

Kruse, J., Petrak, F., Herpertz, S., Albus, C., Lange, K., and Kulzer, B. 2006. Diabetes and depression – a life-endangering interaction. Z. Psychosom. Med. Psychother. 52:289–309.

Krystal, J.H., Sanacora, G., Blumberg, H., Anand, A., Charney, D.S., Marek, G., Epperson, C.N., Goddard, A., and Mason, G.F. 2002. Glutamate and GABA systems as targets for novel antidepressant and mood-stabilizing treatments. Mol. Psychiatry 7(Suppl. 1):S71–S80.

Law, A.J. and Deakin, J.F. 2001. Asymmetrical reductions of hippocampal NMDAR1 glutamate receptor mRNA in the psychoses. Neuroreport 12:2971–2974.

Leinonen, E., Skarstein, J., Behnke, K., Agren, H., and Helsdingen, J.T. 1999. Efficacy and tolerability of mirtazapine versus citalopram: a double-blind, randomized study in patients with major depressive disorder. Nordic Antidepressant Study Group. Int. Clin. Psychopharmacol. 14:329–337.

Leonard, B.E. 2003. Fundamentals of Psychopharmacology. Chichester: Wiley.

Levine, J., Sela, B.A., Osher, Y., and Belmaker, R.H. 2005. High homocysteine serum levels in young male schizophrenia and bipolar patients and in an animal model. Prog. Neuropsychopharmacol. Biol. Psychiatry 29:1181–1191.

Lipton, S.A., Kim, W.K., Choi, Y.B., Kumar, S., D'Emilia, D.M., Rayudu, P.V., Arnelle, D.R., and Stamler, J.S. 1997. Neurotoxicity associated with dual actions of homocysteine at the N-methyl-d-aspartate receptor. Proc. Natl Acad. Sci. USA 94:5923–5928.

Madrigal, J.L., Garcia-Bueno, B., Caso, J.R., Perez-Nievas, B.G., and Leza, J.C. 2006. Stress-induced oxidative changes in brain. CNS Neurol. Disord. Drug Targets 5:561–568.

Maes, M., D'Haese, P.C., Scharpe, S., D'Hondt, P., Cosyns, P., and De Broe, M.E. 1994. Hypozincemia in depression. J. Affect. Disord. 31:135–140.

Maes, M., Verkerk, R., Vandoolaeghe, E., Lin, A., and Scharpe, S. 1998. Serum levels of excitatory amino acids, serine, glycine, histidine, threonine, taurine, alanine and arginine in treatment-resistant depression: modulation by treatment with antidepressants and prediction of clinical responsivity. Acta Psychiatr. Scand. 97:302–308.

Manji, H.K., Drevets, W.C., and Charney, D.S. 2001. The cellular neurobiology of depression. Nat. Med. 7:541–547.

McDaniel, W.W., Sahota, A.K., Vyas, B.V., Laguerta, N., Hategan, L., and Oswald, J. 2006. Ketamine appears associated with better word recall than etomidate after a course of 6 electroconvulsive therapies. J. ECT 22:103–106.

McLeod, T.M., Lopez-Figueroa, A.L., and Lopez-Figueroa, M.O. 2001. Nitric oxide, stress, and depression. Psychopharmacol. Bull. 35:24–41.

Michalak, A., Chatauret, N., and Butterworth, R.F. 2001. Evidence for a serotonin transporter deficit in experimental acute liver failure. Neurochem. Int. 38:163–68.

Michalak, A., Rose, C., Butterworth, J., and Butterworth, R.F. 1996. Neuroactive amino acids and glutamate (NMDA) receptors in frontal cortex of rats with experimental acute liver failure. Hepatology 24:908–913.

Miller, A.H., Capuron, L., and Raison, C.L. 2005. Immunologic influences on emotion regulation. Clin. Neurosci. Res. 4:325–333.

Mitchell, A.J. 2006. Two-week delay in onset of action of antidepressants: new evidence. Br. J. Psychiatry 188:105–106.

Montoliu, C., Piedrafita, B., Serra, M.A., del Olmo, J.A., Rodrigo, J.M., and Felipo, V. 2007. A single transient episode of hyperammonemia induces long-lasting alterations in protein kinase A. Am. J. Physiol Gastrointest Liver Physiol 292: G305–14.

Mousseau, D.D., Baker, G.B., and Butterworth, R.F. 1997. Increased density of catalytic sites and expression of brain monoamine oxidase A in humans with hepatic encephalopathy. J. Neurochem. 68:1200–1208.

Mousseau, D.D., Perney, P., Layrargues, G.P., and Butterworth, R.F. 1993. Selective loss of pallidal dopamine D2 receptor density in hepatic encephalopathy. Neurosci. Lett. 162:192–196.

Muller, N., Schwarz, M.J., Dehning, S., Douhe, A., Cerovecki, A., Goldstein-Muller, B., Spellmann, I., Hetzel, G., Maino, K., Kleindienst, N., Moller, H.J., Arolt, V., and Riedel, M. 2006. The cyclooxygenase-2 inhibitor celecoxib has therapeutic effects in major depression: results of a double-blind, randomized, placebo controlled, add-on pilot study to reboxetine. Mol. Psychiatry 11:680–684.

Nelson, J.C., Mazure, C.M., Jatlow, P.I., Bowers, M.B., Jr., and Price, L.H. 2004. Combining norepinephrine and serotonin reuptake inhibition mechanisms for treatment of depression: a double-blind, randomized study. Biol. Psychiatry 55:296–300.

Nilsson, K., Gustafson, L., and Hultberg, B. 2007. Elevated plasma homocysteine concentration in elderly patients with mental illness is mainly related to the presence of vascular disease and not the diagnosis. Dement. Geriatr. Cogn. Disord. 24:162–168.

Norenberg, M.D. and Martinez-Hernandez, A. 1979. Fine structural localization of glutamine synthetase in astrocytes of rat brain. Brain Res. 161:303–310.

Norris, E.R., Smallwood, G.A., Connor, K., McDonell, K., Martinez, E., Stieber, A.C., and Heffron, T.G. 2002. Prevalence of depressive symptoms in patients being evaluated for liver transplantation. Transplant. Proc. 34:3285–3286.

Nowak, G., Ordway, G.A., and Paul, I.A. 1995. Alterations in the N-methyl-d-aspartate (NMDA) receptor complex in the frontal cortex of suicide victims. Brain Res. 675.157–164.

Ohkuma, S., Narihara, H., Katsura, M., Hasegawa, T., and Kuriyama, K. 1995. Nitric oxide-induced [3H] GABA release from cerebral cortical neurons is mediated by peroxynitrite. J. Neurochem. 65:1109–1114.

Olasmaa, M., Rothstein, J.D., Guidotti, A., Weber, R.J., Paul, S.M., Spector, S., Zeneroli, M.L., Baraldi, M., and Costa, E. 1990. Endogenous benzodiazepine receptor ligands in human and animal hepatic encephalopathy. J. Neurochem. 55:2015–2023.

Packman, W., Wilson, C.T., Riesner, A., Fairley, C., and Packman, S. 2006. Psychological complications of patients with Gaucher disease. J. Inherit. Metab. Dis. 29:99–105.

Pandey, G.N., Conley, R.R., Pandey, S.C., Goel, S., Roberts, R.C., Tamminga, C.A., Chute, D., and Smialek, J. 1997. Benzodiazepine receptors in the post-mortem brain of suicide victims and schizophrenic subjects. Psychiatry Res. 71:137–149.

Perez, V., Gilaberte, I., Faries, D., Alvarez, E., and Artigas, F. 1997. Randomised, double-blind, placebo-controlled trial of pindolol in combination with fluoxetine antidepressant treatment. Lancet 349:1594–1597.

Perlmutter, J.B., Frishman, W.H., and Feinstein, R.E. 2000. Major depression as a risk factor for cardiovascular disease: therapeutic implications. Heart Dis. 2:75–82.

Prast, H. and Philippu, A. 2001. Nitric oxide as modulator of neuronal function. Prog. Neurobiol. 64:51–68.

Rao, V.L. and Butterworth, R.F. 1994. Alterations of [3H]8-OH-DPAT and [3H]ketanserin binding sites in autopsied brain tissue from cirrhotic patients with hepatic encephalopathy. Neurosci. Lett. 182:69–72.

Rao, V.L., Giguere, J.F., Layrargues, G.P., and Butterworth, R.F. 1993. Increased activities of MAOA and MAOB in autopsied brain tissue from cirrhotic patients with hepatic encephalopathy. Brain Res. 621:349–352.

Ratnakumari, L., Audet, R., Qureshi, I.A., and Butterworth, R.F. 1995. Na+,K(+)-ATPase activities are increased in brain in both congenital and acquired hyperammonemic syndromes. Neurosci. Lett. 197:89–92.

Regenold, W.T., Hisley, K.C., Obuchowski, A., Lefkowitz, D.M., Marano, C., and Hauser, P. 2005. Relationship of white matter hyperintensities to cerebrospinal fluid glucose polyol pathway metabolites – a pilot study in treatment-resistant affective disorder patients. J. Affect. Disord. 85:341–350.

Regenold, W.T., Phatak, P., Kling, M.A., and Hauser, P. 2004. Post-mortem evidence from human brain tissue of disturbed glucose metabolism in mood and psychotic disorders. Mol. Psychiatry 9:731–733.

Riedel, G., Platt, B., and Micheau, J. 2003. Glutamate receptor function in learning and memory. Behav. Brain Res. 140:1–47.

Rogoz, Z., Dziedzicka-Wasylewska, M., Daniel, W.A., Wojcikowski, J., Dudek, D., Wrobel, A., and Zieba, A. 2004. Effects of joint administration of imipramine and amantadine in patients with drug-resistant unipolar depression. Pol. J. Pharmacol. 56:735–742.

Rogoz, Z., Skuza, G., Maj, J., and Danysz, W. 2002. Synergistic effect of uncompetitive NMDA receptor antagonists and antidepressant drugs in the forced swimming test in rats. Neuropharmacology 42:1024–1030.

Rozanski, A., Blumenthal, J.A., and Kaplan, J. 1999. Impact of psychological factors on the pathogenesis of cardiovascular disease and implications for therapy. Circulation 99:2192–2217.

Sadek, J., Shellhaas, R., Camfield, C.S., Camfield, P.R., and Burley, J. 2004. Psychiatric findings in four female carriers of Fabry disease. Psychiatr. Genet. 14:199–201.

Sanacora, G., Mason, G.F., Rothman, D.L., Hyder, F., Ciarcia, J.J., Ostroff, R.B., Berman, R.M., and Krystal, J.H. 2003. Increased cortical GABA concentrations in depressed patients receiving ECT. Am. J. Psychiatry 160:577–579.

Sanchez, C. 2006. The pharmacology of citalopram enantiomers: the antagonism by R-citalopram on the effect of S-citalopram. Basic Clin. Pharmacol. Toxicol. 99:91–95.

Sapolsky, R.M. 2000. Glucocorticoids and hippocampal atrophy in neuropsychiatric disorders. Arch. Gen. Psychiatry 57:925–935.

Saran, T., Hilgier, W., Kocki, T., Urbanska, E.M., Turski, W.A., and Albrecht, J. 1998. Acute ammonia treatment in vitro and in vivo inhibits the synthesis of a neuroprotectant kynurenic acid in rat cerebral cortical slices. Brain Res. 787:348–350.

Saransaari, P. and Oja, S.S. 1997. Enhanced GABA release in cell-damaging conditions in the adult and developing mouse hippocampus. Int. J. Dev. Neurosci. 15:163–174.

Sargent, P.A., Kjaer, K.H., Bench, C.J., Rabiner, E.A., Messa, C., Meyer, J., Gunn, R.N., Grasby, P.M., and Cowen, P.J. 2000. Brain serotonin1A receptor binding measured by positron emission tomography with [11C]WAY-100635: effects of depression and antidepressant treatment. Arch. Gen. Psychiatry 57:174–180.

Schliess, F., Gorg, B., and Haussinger, D. 2006. Pathogenetic interplay between osmotic and oxidative stress: the hepatic encephalopathy paradigm. Biol. Chem. 387:1363–1370.

Scollo-Lavizzari, G. and Steinmann, E. 1985. Reversal of hepatic coma by benzodiazepine antagonist (Ro 15-1788). Lancet 1:1324–1325.

Shiah, I.S. and Yatham, L.N. 1998. GABA function in mood disorders: an update and critical review. Life Sci. 63:1289–1303.

Skolnick, P. 1999. Antidepressants for the new millennium. Eur. J. Pharmacol. 375:31–40.

Skolnick, P., Layer, R.T., Popik, P., Nowak, G., Paul, I.A., and Trullas, R. 1996. Adaptation of N-methyl-d-aspartate (NMDA) receptors following antidepressant treatment: implications for the pharmacotherapy of depression. Pharmacopsychiatry 29:23–26.

Stahl, S. 1996. Essential Psychopharmacology: Neuroscientific Basis and Practical Applications. Cambridge: Cambridge University Press.

Stahl, S.M., Nierenberg, A.A., and Gorman, J.M. 2001. Evidence of early onset of antidepressant effect in randomized controlled trials. J. Clin. Psychiatry 62(Suppl. 4):17–23.

Starkman, M.N., Gebarski, S.S., Berent, S., and Schteingart, D.E. 1992. Hippocampal formation volume, memory dysfunction, and cortisol levels in patients with Cushing's syndrome. Biol. Psychiatry 32:756–765.

Starkman, M.N., Giordani, B., Berent, S., Schork, M.A., and Schteingart, D.E. 2001. Elevated cortisol levels in Cushing's disease are associated with cognitive decrements. Psychosom. Med. 63:985–993.

Starkman, M.N., Giordani, B., Gebarski, S.S., Berent, S., Schork, M.A., and Schteingart, D.E. 1999. Decrease in cortisol reverses human hippocampal atrophy following treatment of Cushing's disease. Biol. Psychiatry 46:1595–1602.

Stewart, C.A. and Reid, I.C. 2002. Antidepressant mechanisms: functional and molecular correlates of excitatory amino acid neurotransmission. Mol. Psychiatry 7(Suppl. 1):S15–S22.

Stryjer, R., Strous, R.D., Shaked, G., Bar, F., Feldman, B., Kotler, M., Polak, L., Rosenzcwaig, S., and Weizman, A. 2003. Amantadine as augmentation therapy in the management of treatment-resistant depression. Int. Clin. Psychopharmacol. 18:93–96.

Suzuki, E., Yagi, G., Nakaki, T., Kanba, S., and Asai, M. 2001. Elevated plasma nitrate levels in depressive states. J. Affect. Disord. 63:221–224.

Suzuki, E., Yoshida, Y., Shibuya, A., and Miyaoka, H. 2003. Nitric oxide involvement in depression during interferon-alpha therapy. Int. J. Neuropsychopharmacol. 6:415–419.

Tagliaferro, P., Ramos, A.J., Lopez-Costa, J.J., Lopez, E.M., Saavedra, J.P., and Brusco, A. 2001. Increased nitric oxide synthase activity in a model of serotonin depletion. Brain Res. Bull. 54:199–205.

Tao, R. and Auerbach, S.B. 2000. Regulation of serotonin release by GABA and excitatory amino acids. J. Psychopharmacol. 14:100–113.

Thompson, C. 2002. Onset of action of antidepressants: results of different analyses. Hum. Psychopharmacol. 17(Suppl. 1):S27–S32.

Tome, M.B., Isaac, M.T., Harte, R., and Holland, C. 1997. Paroxetine and pindolol: a randomized trial of serotonergic autoreceptor blockade in the reduction of antidepressant latency. Int. Clin. Psychopharmacol. 12:81–89.

Vaquero, J. and Butterworth, R.F. 2006. The brain glutamate system in liver failure. J. Neurochem. 98:661–669.

Vester-Blokland, E. and Van Oers, H. 2002. Mirtazapine orally disintegrating tablets versus sertraline: response and remission in a prospective onset-of-action trial. Eur. Neuropsychopharmacol. 12:S186.

Vogels, B.A., Maas, M.A., Daalhuisen, J., Quack, G., and Chamuleau, R.A. 1997. Memantine, a noncompetitive NMDA receptor antagonist improves hyperammonemia-induced encephalopathy and acute hepatic encephalopathy in rats. Hepatology 25:820–827.

Wegener, G., Volke, V., Harvey, B.H., and Rosenberg, R. 2003. Local, but not systemic, administration of serotonergic antidepressants decreases hippocampal nitric oxide synthase activity. Brain Res. 959:128–134.

Wheatley, D.P., van Moffaert, M., Timmerman, L., and Kremer, C.M. 1998. Mirtazapine: efficacy and tolerability in comparison with fluoxetine in patients with moderate to severe major depressive disorder. Mirtazapine–Fluoxetine Study Group. J. Clin. Psychiatry 59:306–312.

Wheeler, D., Boutelle, M.G., and Fillenz, M. 1995. The role of N-methyl-d-aspartate receptors in the regulation of physiologically released dopamine. Neuroscience 65:767–774.

Wichers, M.C. and Maes, M. 2004. The role of indoleamine 2,3-dioxygenase (IDO) in the pathophysiology of interferon-alpha-induced depression. J. Psychiatry Neurosci. 29:11–17.

Wysmyk, U., Oja, S.S., Saransaari, P., and Albrecht, J. 1992. Enhanced GABA release in cerebral cortical slices derived from rats with thioacetamide-induced hepatic encephalopathy. Neurochem. Res. 17:1187–1190.

Yamada, J., Saitow, F., Satake, S., Kiyohara, T., and Konishi, S. 1999. GABA(B) receptor-mediated presynaptic inhibition of glutamatergic and GABAergic transmission in the basolateral amygdala. Neuropharmacology 38:1743–1753.

Yurdaydin, C., Herneth, A.M., Puspok, A., Steindl, P., Singer, E., and Ferenci, P. 1996. Modulation of hepatic encephalopathy in rats with thioacetamide-induced acute liver failure by serotonin antagonists. Eur. J. Gastroenterol. Hepatol. 8:667–671.

Zanardi, R., Serretti, A., Rossini, D., Franchini, L., Cusin, C., Lattuada, E., Dotoli, D., and Smeraldi, E. 2001. Factors affecting fluvoxamine antidepressant activity: influence of pindolol and 5-HTTLPR in delusional and nondelusional depression. Biol. Psychiatry 50:323–330.

Zarate, C.A., Jr., Singh, J.B., Carlson, P.J., Brutsche, N.E., Ameli, R., Luckenbaugh, D.A., Charney, D.S., and Manji, H.K. 2006. A randomized trial of an N-methyl-d-aspartate antagonist in treatment-resistant major depression. Arch. Gen. Psychiatry 63:856–864.

Zieve, L. 1987. Pathogenesis of hepatic encephalopathy. Metab. Brain Dis. 2:147–165.

Chapter 19
Attention-Deficit/Hyperactivity Disorder as a Metabolic Encephalopathy

Vivienne Ann Russell

Attention-Deficit Hyperactivity Disorder

Attention-deficit/hyperactivity disorder (ADHD) is the most commonly diagnosed psychiatric disorder of childhood (American Academy of Pediatrics, 2000; Smalley, 1997). It affects 5–10% of children worldwide and persists through adolescence into adulthood in about half of the affected individuals (Faraone et al., 2003; Meyer et al., 2004; Adewuya and Famuyiwa, 2006). Children with ADHD are characterized by severe, developmentally inappropriate, motor hyperactivity, impulsivity and inattention that results in impairment (failure at school) due to an inability to sit still, difficulty in organizing tasks, not remembering instructions, being easily distracted, fidgeting, difficulty with tasks that require sustained attention and risk-taking (American Psychiatric Association, 1994; Sagvolden et al., 2005a; Thapar et al., 2007; Abikoff, et al., 2002). There is increasing recognition that ADHD is associated with later drug and alcohol misuse and problems both socially and in the work environment (Thapar et al., 2007). In some cases the disorder is associated with antisocial behaviour and criminality (Thapar et al., 2007). Three subtypes of ADHD have been recognized, the predominantly inattentive subtype, predominantly hyperactive-impulsive subtype and the combined subtype (American Psychiatric Association, 1994).

Developmental Aspects

Development of the human brain follows a precise genetically determined programme that is subject to modification by the environment (Toga et al., 2006). During the first 3–4 years of life, sensory stimulation and experience produce an initial increase in dendritic branching and synaptic contacts on neurons (Toga et al., 2006). This is followed by dendritic pruning and synapse elimination, which occur over several years into late adolescence to produce more efficient neural circuits that continue to be remodeled throughout life (Toga et al., 2006). Any disruption of this process can result in impaired brain function.

D.W. McCandless (ed.) *Metabolic Encephalopathy*,
doi: 10.1007/978-0-387-79112-8_19, © Springer Science+Business Media, LLC 2009

There is compelling evidence that ADHD symptoms result from impaired dopamine function in the brain, specifically dopamine-mediated reward-related memory formation (Sagvolden et al., 2005a). Deficient dopamine release during development would impair strengthening of appropriate synaptic connections which would lead to impaired association of cues that predict reward with the relevant behaviour, and an inability to form long or complicated sequences of behaviour in response to specific temporal patterns of presentation of reward-predicting stimuli (Sagvolden et al., 2005a; Johanssen et al., unpublished). More recently, astrocyte function has been suggested to be impaired in ADHD, particularly in terms of formation and supply of lactate to rapidly firing neurons (Russell et al., 2006). It was proposed that this insufficiency leads to highly localized deficiencies in ATP production with resultant inability to restore ionic gradients across neuronal membranes required to maintain sustained neuronal firing. This has the immediate effect of causing the brain to switch to alternate neuronal circuits, possible consequences being an inability to hold information in the mind during a delay, impaired learning giving rise to variable behaviour and unpredictable responses to stimuli (Russell et al., 2006). This transient energy deficiency is suggested to be sufficient to impair development and delay maturation of neural circuits that control behaviour (Russell et al., 2006).

Genetics

ADHD runs in families, with first-degree relatives of affected individuals showing higher rates of the disorder (Thapar et al., 2007). Twin and adoption studies have provided consistent evidence that genetic factors contribute to the etiology of ADHD, with estimates of heritability of 60–91% (Thapar et al., 2007). The high prevalence and heritability of ADHD agrees with ADHD being caused by multiple genes with small effect size (Smalley, 1997; Faraone, 2004). The most robust associations have been found between polymorphisms in genes that encode the D4 and D5 subtypes of the dopamine receptor (DRD4 and DRD5), the dopamine transporter (DAT) gene and SNAP-25 (a protein required for neurotransmitter release as well as trafficking of glutamate NMDA receptor subunits to the plasma membrane) (Cook et al., 1995; LaHoste et al., 1996; Faraone et al., 2001; Maher et al., 2002; El-Faddagh et al., 2004; Manor et al., 2004; Gornick et al., 2006; Brookes et al., 2006; Genro et al., 2006). The 7-repeat allele of DRD4 was associated with better cognitive performance and a better long-term outcome, suggesting that this allele may be associated with a more benign form of the disorder (Gornick et al., 2006). Other gene variants have been suggested to be associated with ADHD but these need further investigation; they include genes that encode monoamine oxidase A, dopamine β-hydroxylase, the norepinephrine transporter, serotonin transporter, α_2-adrenoceptor, and serotonin1B receptor (Park et al., 2005; Thapar et al., 2005; Bobb et al., 2005; Bobb et al., 2005; Thapar et al., 2007; Kim et al., 2006; Brookes et al., 2006;

Faraone and Khan 2006). Contradictory negative findings have also been reported, suggesting that different combinations of genetic factors may be required to produce individual clusters of behavioural symptoms of ADHD (Bakker et al., 2004; Barr et al., 2002; Xu et al., 2005; Purper-Ouakil et al., 2005). The different alleles of genes encoding proteins related to dopamine function differentially affect cognitive function (Durston et al., 2005). Children possessing two copies of dopamine β-hydroxylase, the enzyme that converts dopamine to norepinephrine, had significantly poorer sustained attention than children who did not possess this allele (Bellgrove et al., 2006). It appears that the effect of a single gene on behaviour can be small, causing a slight bias towards one end of a continuum (Sagvolden et al., 2005a; Durston et al., 2005). ADHD is a heterogeneous but nevertheless highly heritable disorder resulting from complex gene–gene and gene—environment interactions (Faraone, 2004; Thapar et al., 2005).

Environmental Risk Factors

Environmental risk factors include prenatal exposure to drugs such as alcohol and nicotine, obstetric complications, head injury, and psychosocial adversity (Biederman and Faraone, 2005; Romano et al., 2006). Prenatal exposure to ethanol affects mainly dopaminergic transmission and causes hyperactivity (Gibson et al., 2000). Rats exposed to ethanol prenatally show attention deficits that are similar to those of children with fetal alcohol syndrome and ADHD (Hausknecht et al., 2005).

Epidemiological evidence reveals that ADHD is associated with prenatal exposure to nicotine (Milberger et al., 1998; Mick et al., 2002; Thapar et al., 2003). In a case-control study, children whose mothers smoked more than ten cigarettes per day during pregnancy presented a significantly higher odds ratio for the inattentive subtype of ADHD than sex and age matched controls (Schmitz et al., 2006). The odds of a diagnosis of the combined subtype of ADHD was 2.9 times greater in twins who had inherited the DAT1 polymorphism associated with ADHD and who were exposed to nicotine prenatally compared to unexposed twins without the risk allele (Neuman et al., 2006). Animal studies contributed further information; prenatal nicotine increased spontaneous locomotion in mice (Paz et al., 2006). Deletion of the gene encoding the $\beta 2$-subunit of the nicotinic acetylcholine receptor caused mice to display the defining ADHD symptoms of inattention, lack of inhibitory control and hyperactivity (Granon and Changeux, 2006). Agonists of the $\alpha 4\beta 2$-nicotinic receptor reduced the ADHD-like behavior in the mouse model (Granon and Changeux, 2006). In support of a role in the cognitive dysfunction in ADHD, nicotinic agonists also reduced spontaneous alternation deficits in young stroke-prone SHR, an effect that was prevented by an $\alpha 4\beta 2$-nicotinic receptor antagonist suggesting that $\alpha 4\beta 2$-nicotinic agonists may be useful for the treatment of attention deficits in ADHD (Ueno et al., 2002).

Ubiquitous Nature of ADHD Symptoms

ADHD symptoms are not unique to ADHD but are found in other disorders such as phenylketonuria (Sullivan and Chang, 1999; Realmuto et al., 1986), narcolepsy (Rieger et al., 2003; Naumann et al., 2006a) and fetal alcohol syndrome (Nash et al., 2006; Riikonen et al., 2005). Phenylketonuria results from high concentrations of phenylalanine that arise from an inability to convert it into tyrosine, and which inhibit the transport of neutral amino acids such as tyrosine and tryptophan across the blood—brain barrier, thereby limiting the synthesis of the three principle monoamine transmitters, norepinephrine, dopamine and serotonin. Individuals with narcolepsy have slower reaction times and more within-task variability of performance than control subjects on a variety of attentional tasks ranging from those sensitive to arousal and sustained attention, to the executive control of attention (Rieger et al., 2003). Recent studies report narcolepsy-related deficits in attentional and executive function which place high demands on inhibition and task management, but not on simple tasks of memory and attention (Rieger et al., 2003; Naumann et al., 2006b). The pattern of findings was thought to be indicative of a depletion of available cognitive processing resources because of the need for continuous allocation of resources to monitoring.

Structural Abnormalities

Numerous studies have found reduced brain volume in patients with ADHD, particularly the cerebellum, corpus callosum, prefrontal cortex and basal ganglia, especially in the right hemisphere (Castellanos et al., 1996, 2002; Durston et al., 2004; Filipek et al., 1997; Hill et al., 2003; Valera et al., 2006). Patients with lesions to the right frontal cortex displayed ADHD-like behaviour, consistent with right frontal cortex pathology in ADHD (Clark et al., 2006). Dopamine alters brain structure and function (Durston et al., 2005). The DAT1 genotype preferentially influenced caudate volume. Individuals homozygous for the 10-repeat allele which is associated with ADHD had smaller caudate volumes than individuals carrying the 9-repeat allele (Durston et al., 2005). The DRD4 genotype influenced prefrontal gray matter. Individuals homozygous for the 4-repeat allele had smaller volumes than individuals carrying other variants of the gene (Durston et al., 2005).

Functional Abnormalities

Neuroimaging studies have demonstrated functional abnormalities in the striatum, frontal cortex and cerebellum of patients with ADHD (Kim et al., 2002; Rubia et al., 1999; Moll et al., 2000; Tannock 1998; Vaidya et al., 1998; Scheres et al.,

2006). The most consistent findings in the neuroimaging literature of ADHD are deficits in neural activity within fronto-striatal and fronto-parietal circuits (Dickstein et al., 2006). Significant patterns of frontal hypoactivity were detected in patients with ADHD; affected regions included anterior cingulate, dorsolateral prefrontal and inferior prefrontal cortices, as well as basal ganglia, thalamus and portions of the parietal cortex (Dickstein et al., 2006). Functional magnetic resonance imaging (fMRI) revealed reduced ventral striatal activation in adolescents with ADHD during a reward anticipation task, suggesting impaired reward-related neuronal circuits in addition to the commonly observed prefrontal executive dysfunction (Scheres et al., 2006). Ventral striatal activation was negatively correlated with parent-rated hyperactive or impulsive symptoms (Scheres et al., 2006). Lower L-DOPA utilization especially in subcortical regions correlated specifically with symptoms of inattention (Forssberg et al., 2006). Robust increases in striatal DAT of up to 70% were found in children and adults with ADHD (Cheon et al., 2003; Dougherty et al., 1999; Krause et al., 2000) which suggests that the DAT1 gene may be overexpressed in the striatum of ADHD subjects, and that this results in reduced synaptic dopamine. However not every study found increased DAT (Jucaite et al., 2005; van Dyck et al., 2002) and more recent findings suggest that in some drug-naïve adults with ADHD, DAT levels in the left caudate and nucleus accumbens are reduced (Volkow et al., 2007).

Magnetic resonance spectroscopy (MRS) revealed lower membrane phospholipid precursor levels in prefrontal cortex and basal ganglia of children with ADHD compared to healthy children, suggesting underdevelopment of neuronal processes and synapses in ADHD (Stanley et al., 2006). Decreased white matter density and impaired integrity of myelinated neuronal pathways as revealed by diffusion tensor imaging (DTI) and MRS is consistent with delayed or impaired myelination of neuronal axons in children and adolescents with ADHD (Castellanos et al., 2002; Durston et al., 2004; Filipek et al., 1997; Mostofsky et al., 2002; Overmeyer et al., 2001; Semrud-Clikeman et al., 2000). Children with ADHD displayed a 10% reduction in white matter volume compared to children with ADHD who had been treated with stimulant medication or controls (Castellanos et al., 2002). Smaller white matter volumes were linked with slower processing speed in a colour-naming task (Semrud-Clikeman et al., 2000). These results suggest that deficits in cognitive function may be the result of delayed myelination in children with ADHD (Russell et al., 2006).

DTI provides a measure (fractional anisotropy) of the coherence and integrity of myelinated pathways. Children with ADHD displayed a reduced fractional anisotropy in the right neostriatum and premotor cortex, and in the left cerebellum and parieto-occipital cortex compared to matched controls (Ashtari et al., 2005). The lower the cerebellar fractional anisotropy, the more severe were the ratings of symptoms of inattention (Ashtari et al., 2005). These findings point to a link between white matter abnormalities and the symptomatology of ADHD.

Neurophysiology

The functional consequences of impaired/delayed developmental laying down of the myelin sheath in ADHD are seen in three types of EEG measure: (a) evoked potential latencies, (b) the topographic distribution of the power spectrum in the quantitative EEG and (c) in the coherence of the EEG waveforms between brain regions (Russell et al., 2006). Evoked potentials representing sensory information ascending in the auditory nerve (Sohmer and Student 1978) and the brain stem appear at longer than normal latencies (e.g., components III and V). The transmission times from components I–III and I–V are reported to be increased in subjects with ADHD (Lahat et al., 1995) and the latency of the steady state visual evoked potential in the frontal cortex of ADHD children is markedly delayed (Silberstein et al., 1998). Abnormal myelination of corticospinal neurons was suggested to be responsible for the delayed velocity of evoked potentials in patients with ADHD (Ashtari et al., 2005; Ucles et al., 1996).

A large proportion of patients with ADHD demonstrate an increased ratio of relative theta to alpha or beta power in the EEG, especially over anterior regions of the brain (Clarke et al., 2002b; Saletu et al., 2005; Hobbs et al., 2007; Snyder and Hall 2006). One explanation of the dominant lower firing frequencies could lie with reduced lactate availability required to sustain rapidly firing neurons (Russell et al., 2006). There is usually a marked normalization of this balance between oscillation frequencies after methylphenidate treatment (Clarke et al., 2002a). In the unmedicated sample there is a positive correlation between P2, N2 and P3 event-related potential (ERP) latencies, widely reported to be delayed (Karayanidis et al., 2000) and increased theta power (Barry et al., 2003; Lazzaro et al., 2001). A plausible reason for this shift in balance between oscillation frequencies lies in a decreased representation of the faster frequencies owing to deficient neuronal energy supply and/or reduced myelination of axons originating in brain stem reticular sources active in generating some of these rhythms (Ucles et al., 1996; Russell et al., 2006).

A more direct measure of the coupling of activity between brain regions can be estimated by EEG coherence of waveform between recording sites. Coherence can be conceptualized as the correlation in the time domain between two signals in a given frequency band. Boys with ADHD show elevated slow-wave coherences and reduced fast-wave coherences between hemispheres, although within hemispheres the coherence in the theta band is reduced, especially over frontal regions (Barry et al., 2005a,b; Chabot and Serfontein, 1996). This is most easily explained by unusual if not delayed development of the large white matter tracts connecting brain regions (Russell et al., 2006). At short distances between signals the increased coherence at slow wave frequencies in children with ADHD is viewed as consistent with a delay in the pruning back of over-produced synapses and local connections (Thatcher et al., 1986). As would be expected such long-term alterations remain unaffected by short-term methylphenidate treatment (Clarke et al., 2005).

Treatment

Psychostimulants are highly effective in ameliorating the three major clusters of behavioural symptoms of ADHD (Biederman et al., 2004). Methylphenidate produced improvements in spatial working memory, attentional set-shifting, reading performance (Mehta et al., 2004; Solanto, 1998; Keulers et al., 2006), and inhibition of previously acquired behavioural responses to non-relevant stimuli (Bedard et al., 2004; Tannock et al., 1989, 1995; Vaidya et al., 1998). Methylphenidate is short-acting, it reaches peak plasma concentrations within 2 h and its effects wear off after 4 h, suggesting that its acute pharmacological action is responsible for the therapeutic effect (Swanson and Volkow, 2003). Methylphenidate blocks DAT and this indirect dopamine agonist effect has been suggested to be critical for its action (Volkow et al., 1998; Greenhill, 2001). Like many other drugs with psychostimulant properties, methylphenidate increases extracellular concentrations of dopamine thereby, amplifying weak dopamine signals within key areas of reward-related behavioural circuits (Kuczenski and Segal, 1997, 2001; Volkow et al., 2001, 2005 Swanson and Volkow, 2003). Although considerable concern has been expressed regarding its abuse potential, there is no evidence to suggest that stimulant treatment of children with ADHD leads to substance abuse in later life (Mannuzza et al., 2003). In fact, methylphenidate administered to a widely accepted rat model for ADHD, the spontaneously hypertensive rat (SHR), at a young age, diminishes the incentive value of drugs of abuse in adulthood (Augustyniak et al., 2006; Russell et al., unpublished). The dose and route of administration of methylphenidate are important factors to consider when interpreting published data. When administered intravenously, methylphenidate like cocaine has reinforcing effects (euphoria) at doses that exceed a DAT blockade threshold of 60% (Swanson and Volkow, 2003). When administered orally at clinical doses (0.2–0.8 mg kg—¹), the pharmacological effects of methylphenidate also exceed this threshold, but reinforcing effects rarely occur (Swanson and Volkow, 2003). The pharmokinetic properties of methylphenidate in serum (and brain) differ for oral and intravenous routes of administration. Oral administration of methylphenidate (used to treat ADHD) produces a gradual increase in extracellular dopamine allowing for adaptation to occur over time which mimics the tonic firing of dopamine neurons, while intravenous administration produces a rapid rise in extracellular dopamine, mimicking the phasic effects of rapid dopamine cell firing which is considered to be a critical factor in determining the reinforcement and abuse potential of the drug (Swanson and Volkow, 2003).

Psychostimulants such as methylphenidate, cocaine and amphetamine are not specific for DAT; they bind to the noradrenergic transporter with greater affinity and are more potent in blocking the norepinephrine transporter than the dopamine transporter (Richelson and Pfenning, 1984; Easton et al., 2006b;Tanda et al., 1997). At low doses that improve cognitive function without stimulating locomotor activity, methylphenidate increased norepinephrine and dopamine release within the prefrontal cortex (blockage of the norepinephrine transporter is largely responsible for uptake of dopamine in the prefrontal cortex (Tanda et al., 1997)) without affecting dopamine release in the striatum, suggesting that the therapeutic action of low-dose psychostimulants involves the preferential activation of catecholamine neurotransmission in the prefrontal cortex (Berridge et al., 2006). At therapeutic doses given orally, methylphenidate increased extracellular norepinephrine in the somatosensory cortex and hippocampus without affecting dopamine in the nucleus accumbens and without producing sensitization to the locomotor stimulant effects of methamphetamine, suggesting that the low dose of methylphenidate does not increase drug abuse liability and that the noradrenergic system may play an important role in the mechanism of therapeutic action of this drug (Kuczenski and Segal 2002; Drouin et al., 2006). Both dopamine and norepinephrine are modulatory neurotransmitters; they enhance memory formation by strengthening synaptic connections in neural circuits that control behaviour. Increased norepinephrine in the hippocampus could affect memory formation and therefore influence behaviour. Noradrenergic neurons enhance the signal-to-noise ratio in prefrontal and parietal cortices, amplify responses to attended stimuli, and reduce responses to irrelevant stimuli (Aston-Jones et al., 1994; Himelstein et al., 2000). These functions are defective in ADHD (Himelstein et al., 2000). Dopamine activation of DRD1 receptors enhances prefrontal cortex function, complementing norepinephrine's action (Arnsten, 1998).

The locus coeruleus noradrenergic neurons innervate the entire cerebral cortex, various subcortical areas, cerebellum and spinal cord. They play an important role in attention, arousal, orienting, and vigilance (Solanto, 1998). Locus coeruleus neurons respond selectively to attended (target) stimuli, tonic locus coeruleus activity corresponds to arousal state, and both very low and very high locus coeruleus activity are associated with impaired vigilance (Arnsten, 1998; Aston-Jones et al., 1994). Noradrenergic neurons that project from the locus coeruleus to the prefrontal cortex release norepinephrine which guides behavior by modulating the transfer of information through neuronal circuits that are responsible for selective and sustained attention (Solanto, 1998). Methylphenidate increases norepinephrine release and suppresses long-latency sensory responses in the primary somatosensory cortex of freely behaving rats (Drouin et al., 2006). Children with ADHD have been suggested to have impaired perceptual processing (Oades, 2000). Methylphenidate may improve sensory processing by increasing norepinephrine release in somatosensory cortex and suppressing "noise" (Drouin et al., 2006). Methylphenidate was suggested to have both direct effects in the somatosensory cortex as well as indirect effects through top-down (prefrontal cortex) influences on primary somatosensory cortex responsivity (Arnsten, 2006).

A recent microarray study revealed that the acute effects of methylphenidate on rat striatal gene expression are consistent with its promoting increased neural plasticity, namely the formation, maturation and stabilization of new neural connections within the striatum, presumably as a result of blockade of DAT increasing extracellular dopamine (Adriani et al., 2006b). More than 700 genes were upregulated. One group of genes was involved in migration of immature neural/glial cells and/or growth of novel axons (Adriani et al., 2006b). A second group of upregulated genes were suggestive of active axonal myelination, a process that has been suggested to be impaired in ADHD (Russell et al., 2006). The third group of genes indicated the appearance and/or upregulation of mature processes, including genes for K^+ channels, gap junctions, neurotransmitter receptors, and proteins responsible for their transport and/or anchoring (Adriani et al., 2006b).

Animal Models of ADHD

ADHD is a heterogeneous disorder and so it is not surprising that many different animal models with distinctly different neural defects model the behavioral characteristics of the disorder. Diagnosis of ADHD depends on the behavioral criteria of difficulty sustaining attention, hyperactivity and impulsivity and so animal models of the disorder are required to mimic these symptoms (Sagvolden, 2000; Sagvolden et al., 2005b). Consistent with ADHD being a neurodevelopmental disorder, animal models are either genetic (SHR, dopamine transporter (DAT) knock-out mice, SNAP-25 mutant mice, mice expressing a mutant thyroid receptor) or have suffered an insult to the central nervous system during the early stages of development (anoxia, 6-hydroxydopamine) (Sagvolden, 2000; Jones et al., 1998; Zhuang et al., 2001; Siesser et al., 2006; Dell'Anna et al., 1993; Dell'Anna, 1999; Bruno et al., 2007; Shaywitz et al., 1978; Luthman et al., 1989; Gainetdinov and Caron, 2000; Gainetdinov and Caron, 2001). It appears that there are several different ways in which neural transmission is impaired in animal models of ADHD and that these involve either direct disruption of dopaminergic transmission or a more general impairment of neurotransmission, such as impaired calcium signaling in SHR or SNAP-25 in the Coloboma mutant mouse, that gives rise to compensatory changes in monoaminergic systems that are not sufficient to fully restore normal function. In general, results obtained with animal studies suggest that dopamine neurons are functionally impaired (Russell et al., 2005; Russell, 2007). However, evidence also suggests that the noradrenergic and serotonergic neurotransmitter systems may be the target of drugs that ameliorate ADHD symptoms (Russell et al., 2005; Russell, 2007). Reduction of norepinephrine with DSP-4 (N-(2-chloroethyl)-N-ethyl-2-bromobenzylamine hydrochloride) restored latent inhibition and reduced the hyperactivity of coloboma mice but did not reduce their impulsivity (Bruno et al., 2007).

Stimulant medication has been suggested to increase endogenous stimulation of α_{2A}-adrenoceptors and DRD1 receptors in the prefrontal cortex, optimizing prefrontal cortical regulation of behavior and attention (Arnsten, 2006). Electrophysiological studies in non-human primates suggest that norepinephrine enhances "signals" by suppressing "noise" through postsynaptic α_{2A}-adrenoceptors in the prefrontal cortex while dopamine decreases "noise" through DRD1 activation (Arnsten, 2006). Blockade of α_2-adrenoceptors in the monkey prefrontal cortex produces the characteristic symptoms of ADHD, impaired working memory, increased impulsivity, and increased locomotor activity. Low doses of methylphenidate increased extracellular levels of both norepinephrine and dopamine in prefrontal cortex of rats performing a delayed alternation task, strengthening prefrontal cortex regulatory output to parietal association areas, thereby inhibiting responses to irrelevant sensory stimuli and improving cognitive function (Arnsten and Dudley, 2005). Guanfacine, an α_2-adrenoceptor agonist, improved sustained attention and reduced both impulsivity and hyperactivity in SHR (Sagvolden, 2006). MRI revealed a negative blood oxygenation level dependent (BOLD) response to guanfacine in the caudate-putamen and nucleus accumbens and positive BOLD effects in frontal cortex of the rat brain, suggesting that guanfacine increases neuronal activity in the frontal cortex while decreasing striatum activity (Easton et al., 2006a). This is consistent with the activation of α_2-adrenoceptors causing inhibition of dopamine and norepinephrine release in these brain areas as well as guanfacine acting directly on postsynaptic α_{2A}-adrenoceptors in the prefrontal cortex to enhance cognitive function (Nurse et al., 1984; Russell et al., 2000; Arnsten, 1998).

There appears to be an imbalance between dopaminergic and noradrenergic neurotransmission in the prefrontal cortex of SHR (Russell, 2002). While dopamine release is decreased in SHR prefrontal cortex, norepinephrine concentrations are elevated. The noradrenergic system appears to be hyperactive as a result of impaired α_2-autoreceptor function (Russell et al., 2000; Russell, 2002). Decreased α_2-autoreceptor-mediated inhibition of norepinephrine release may be particularly disruptive to the function of target structures when the firing rate of locus coeruleus neurons is high, causing excessive spill over of norepinephrine into the extracellular space and activation of α_1-adrenoceptors in the prefrontal cortex, impairing its function (Arnsten, 1998). Other noradrenergic terminal areas in the central nervous system may be similarly affected, particularly in response to stress.

The underlying defect in SHR appears to be a disturbance in calcium metabolism not only in the brain but also in other tissues including vascular smooth muscle (Horn et al., 1995; Lehohla et al., 2001; Oshima et al., 1991; Ohno et al., 1996, 1997; Tabet et al., 2004; Fellner and Arendshorst 2002). Increased intracellular calcium concentrations have been attributed to genetic abnormalities in Ca^{2+} ATPase (Horn et al., 1995; Ohno et al., 1996, 2005). Increased intracellular calcium levels can have several consequences: (a) reduced calcium influx into neurons in response to depolarization, due to a decreased calcium gradient across the cell membrane, would decrease neurotransmitter release; (b) impaired calcium signaling (e.g., decreased NMDA-stimulated calcium influx into postsynaptic cells (Lehohla et al., 2001)) with subsequent derangement of calcium-dependent protein kinase and phosphatase activity (e.g., protein kinase C activity is increased in SHR (Tsuda et al., 2003)), and (c) impaired mitochondrial function, giving rise to increased levels of reactive oxygen species, such as the superoxide anion and hydrogen peroxide (Chan et al., 2006) and impaired ATP synthesis (Doroshchuk et al., 2004).

Attempts to compensate for impaired calcium signaling due to reduced endoplasmic reticulum Ca^{2+} ATPase function, include enhanced calcium entry through L-type calcium channels and store-operated channels in vascular smooth muscle cells in SHR (Tabet et al., 2004; Fellner and Arendshorst, 2002). Impaired vascular smooth muscle contraction could influence blood flow and impair brain function at times of high energy demand.

Energetics

Recently, the ADHD symptom of increased intra-individual variability during performance of high energy-demanding cognitive tasks, was attributed to insufficient energy (lactate) supply by astrocytes to neurons at times of rapid and/or continuous firing (Russell et al., 2006). This ubiquitous finding is not unique to ADHD but occurs in several disorders and may be attributed to inefficient information processing. Energy supply is a limiting factor in brain function (Attwell and Gibb, 2005). Inefficient neural transmission due to impaired learning in ADHD would place abnormally high demands on local energy resources and lead to intra-individual variability in responses to externally paced cognitive tasks. Consistent with impaired energy production, the synthesis rate of ATP is much lower in mitochondria of SHR brains than WKY (Doroshchuk et al., 2004). Impaired mitochondrial function was attributed to calcium overload, as a result of Ca^{2+} ATPase not being able to pump Ca^{2+} efficiently into the endoplasmic reticulum and across the cell membrane into the extracellular space. Deficient endoplasmic reticular stores of Ca^{2+} would also impair the function of neurotransmitters that act on receptors to stimulate inositol triphosphate (IP_3) production and release of calcium from intracellular stores (e.g. α_1-adrenoceptors and metabotropic glutamate receptors that regulate astrocyte function (Biber et al., 1999; Glowinski et al., 1994)). Astrocytes provide lactate as a source of energy to rapidly and/or continuously firing neurons. Increased intra-individual variability in the performance of tasks that require continual responses to rapid, externally-paced stimuli observed in subjects with ADHD as well as SHR has been attributed to the inability of astrocytes to provide sufficient lactate at times of high energy demand (Russell et al., 2006).

Methylphenidate alters brain metabolic activity. It increased the previously reduced striatal activity in patients with ADHD (Vaidya et al., 1998) and reduced cerebral blood flow in frontal and parietal cortex (Lou et al., 1984; Mehta et al., 2004). It also increased mitochondrial respiratory chain enzyme activity in several brain areas of young rats, including the cerebellum, prefrontal cortex, and hippocampus (Fagundes et al., 2006), consistent with certain aspects of ADHD behaviour being due to deficient energy supply to neurons at times of high energy demand. Methylphenidate treatment during adolescence reduced impulsivity in adulthood and increased total creatine in dorsal striatum (Adriani et al., 2006a). Total creatine was decreased in nucleus accumbens, and in prefrontal cortex the phospho-creatine/creatine ratio was increased, suggesting improved cortical energetic performance (Adriani et al., 2006a).

Conclusion

Norepinephrine, acting on α_1- or β-adrenoceptors, stimulates glycogenolysis in astrocytes to produce lactate from glucose within milliseconds of neuronal activation. Lactate is released into the extracellular fluid and is taken up by rapidly firing neurons which use this as the preferred fuel to generate ATP to drive metabolic processes. It is quite likely that the acute effect of drugs used to treat ADHD is to increase extracellular norepinephrine in prefrontal cortex, hippocampus and other terminal areas of the locus coeruleus, to enhance functioning, simultaneously stimulating astrocytes to release lactate to support the increased and sometimes sustained neuronal firing. The longer-term benefits of increased norpeinephrine and dopamine release in the prefrontal cortex and related areas of the brain might be to promote learning by strengthening synaptic connections in relevant sensorimotor circuits to produce more appropriate behavioral responses to sensory stimuli in future situations.

Acknowledgments The author wishes to thank the University of Cape Town and South African Medical Research Council for support.

References

Abikoff H. B., Jensen P. S., Arnold L. L., Hoza B., Hechtman L., Pollack S., Martin D., Alvir J., March J. S., Hinshaw S., Vitiello B., Newcorn J., Greiner A., Cantwell D. P., Conners C. K., Elliott G., Greenhill L. L., Kraemer H., Pelham W. E., Jr., Severe J. B., Swanson J. M., Wells K. and Wigal T. 2002. Observed classroom behavior of children with ADHD: Relationship to gender and comorbidity. J. Abnorm. Child Psychol. 30:349–359

Adewuya A. O. and Famuyiwa O. O. 2006. Attention deficit hyperactivity disorder among Nigerian primary school children: Prevalence and co-morbid conditions. Eur. Child Adolesc. Psychiatr. 16:10–15. Epub 28 November 2006

Adriani W., Canese R., Podo F. and Laviola G. 2006a. 1H MRS-detectable metabolic brain changes and reduced impulsive behavior in adult rats exposed to methylphenidate during adolescence. Neurotoxicol. Teratol. 29:116–125. Epub 6 December 2006

Adriani W., Leo D., Guarino M., Natoli A., Di C. E., De A. G., Traina E., Testai E., Perrone-Capano C. and Laviola G. 2006b. Short-term effects of adolescent methylphenidate exposure on brain striatal gene expression and sexual/endocrine parameters in male rats. Ann. N. Y. Acad. Sci. 1074:52–73

American Academy of Pediatrics 2000. Clinical practice guideline: Diagnosis and evaluation of the child with attention-deficit/hyperactivity disorder. Pediatrics 105:1158–1170

American Psychiatric Association 1994. *Diagnostic and statistical manual of mental disorders: DSM-IV.* American Psychiatric Association, Washington, DC

Arnsten A. F. T. 1998. Catecholamine modulation of prefrontal cortical cognitive function. Trends Cogn. Sci. 2:436–447

Arnsten A. F. 2006. Fundamentals of attention-deficit/hyperactivity disorder: Circuits and pathways. J. Clin. Psychiatr. 67(Suppl. 8):7–12

Arnsten A. F. and Dudley A. G. 2005. Methylphenidate improves prefrontal cortical cognitive function through alpha2 adrenoceptor and dopamine D1 receptor actions: Relevance to therapeutic effects in Attention Deficit Hyperactivity Disorder. Behav. Brain Funct. 1:2

Ashtari M., Kumra S., Bhaskar S. L., Clarke T., Thaden E., Cervellione K. L., Rhinewine J., Kane J. M., Adesman A., Milanaik R., Maytal J., Diamond A., Szeszko P. and Ardekani B. A. 2005. Attention-deficit/hyperactivity disorder: A preliminary diffusion tensor imaging study. Biol. Psychiatr. 57:448–455

Aston-Jones G., Rajkowski J., Kubiak P. and Alexinsky T. 1994. Locus coeruleus neurons in monkey are selectively activated by attended cues in a vigilance task. J. Neurosci. 14:4467–4480

Attwell D. and Gibb A. 2005. Neuroenergetics and the kinetic design of excitatory synapses. Nat. Rev. Neurosci. 6:841–849

Augustyniak P. N., Kourrich S., Rezazadeh S. M., Stewart J. and Arvanitogiannis A. 2006. Differential behavioral and neurochemical effects of cocaine after early exposure to methylphenidate in an animal model of attention deficit hyperactivity disorder. Behav. Brain Res. 167:379–382

Bakker S. C., van der Meulen E. M., Oteman N., Schelleman H., Pearson P. L., Buitelaar J. K. and Sinke R. J. 2004. DAT1, DRD4, and DRD5 polymorphisms are not associated with ADHD in Dutch families. Am J Med Genet B Neuropsychiatr Genet. 132B:50–52

Barr C. L., Kroft J., Feng Y., Wigg K., Roberts W., Malone M., Ickowicz A., Schachar R., Tannock R. and Kennedy J. L. 2002. The norepinephrine transporter gene and attention-deficit hyperactivity disorder. Am. J. Med. Genet. 114:255–259

Barry R. J., Johnstone S. J. and Clarke A. R. 2003. A review of electrophysiology in attention-deficit/hyperactivity disorder: II. Event-related potentials. Clin. Neurophysiol. 114:184–198

Barry R. J., Clarke A. R., McCarthy R., Selikowitz M. and Johnstone S. J. 2005a. EEG coherence adjusted for inter-electrode distance in children with attention-deficit/hyperactivity disorder. Int. J. Psychophysiol. 58:12–20

Barry R. J., Clarke A. R., McCarthy R., Selikowitz M., Johnstone S. J., Hsu C. I., Bond D., Wallace M. J. and Magee C. A. 2005b. Age and gender effects in EEG coherence: II. Boys with attention deficit/hyperactivity disorder. Clin. Neurophysiol. 116:977–984

Bedard A. C., Martinussen R., Ickowicz A. and Tannock R. 2004. Methylphenidate improves visual-spatial memory in children with attention-deficit/hyperactivity disorder. J. Am. Acad. Child Adolesc. Psychiatr. 43:260–268

Bellgrove M. A., Hawi Z., Gill M. and Robertson I. H. 2006. The cognitive genetics of attention deficit hyperactivity disorder (ADHD): Sustained attention as a candidate phenotype. Cortex 42:838–845

Berridge C. W., Devilbiss D. M., Andrzejewski M. E., Arnsten A. F., Kelley A. E., Schmeichel B., Hamilton C. and Spencer R. C. 2006. Methylphenidate preferentially increases catecholamine neurotransmission within the prefrontal cortex at low doses that enhance cognitive function. Biol. Psychiatr. 60:1111–1120

Biber K., Laurie D. J., Berthele A., Sommer B., Tolle T. R., Gebicke-Harter P. J., van C. D. and Boddeke H. W. 1999. Expression and signaling of group I metabotropic glutamate receptors in astrocytes and microglia. J. Neurochem. 72:1671–1680

Biederman J. and Faraone S. V. 2005. Attention-deficit hyperactivity disorder. Lancet 366:237–248

Biederman J., Spencer T. and Wilens T. 2004. Evidence-based pharmacotherapy for attention-deficit hyperactivity disorder. Int. J. Neuropharmacol. 7:77–97

Bobb A. J., Addington A. M., Sidransky E., Gornick M. C., Lerch J. P., Greenstein D. K., Clasen L. S., Sharp W. S., Inoff-Germain G., Wavrant-De V. F., rcos-Burgos M., Straub R. E., Hardy J. A., Castellanos F. X. and Rapoport J. L. 2005. Support for association between ADHD and two candidate genes: NET1 and DRD1. Am. J. Med. Genet. B Neuropsychiatr. Genet. 134:67–72

Brookes K., Xu X., Chen W., Zhou K., Neale B., Lowe N., Anney R., Franke B., Gill M., Ebstein R., Buitelaar J., Sham P., Campbell D., Knight J., Andreou P., Altink M., Arnold R., Boer F., Buschgens C., Butler L., Christiansen H., Feldman L., Fleischman K., Fliers E., Howe-Forbes R., Goldfarb A., Heise A., Gabriels I., Korn-Lubetzki I., Johansson L., Marco R., Medad S., Minderaa R., Mulas F., Muller U., Mulligan A., Rabin K., Rommelse N., Sethna V., Sorohan

J., Uebel H., Psychogiou L., Weeks A., Barrett R., Craig I., Banaschewski T., Sonuga-Barke E., Eisenberg J., Kuntsi J., Manor I., McGuffin P., Miranda A., Oades R. D., Plomin R., Roeyers H., Rothenberger A., Sergeant J., Steinhausen H. C., Taylor E., Thompson M., Faraone S. V. and Asherson P. 2006. The analysis of 51 genes in DSM-IV combined type attention deficit hyperactivity disorder: Association signals in DRD4, DAT1 and 16 other genes. Mol. Psychiatr. 11:934–953

Bruno K. J., Freet C. S., Twining R. C., Egami K., Grigson P. S. and Hess E. J. 2007. Abnormal latent inhibition and impulsivity in coloboma mice, a model of ADHD. Neurobiol. Dis. 25:206–216

Castellanos F. X., Giedd J. N., Marsh W. L., Hamburger S. D., Vaituzis A. C., Dickstein D. P., Sarfatti S. E., Vauss Y. C., Snell J. W., Lange N., Kaysen D., Krain A. L., Ritchie G. F., Rajapakse J. C. and Rapoport J. L. 1996. Quantitative brain magnetic resonance imaging in attention-deficit hyperactivity disorder. Arch. Gen. Psychiatr. 53:607–616

Castellanos F. X., Lee P. P., Sharp W., Jeffries N. O., Greenstein D. K., Clasen L. S., Blumenthal J. D., James R. S., Ebens C. L., Walter J. M., Zijdenbos A., Evans A. C., Giedd J. N. and Rapoport J. L. 2002. Developmental trajectories of brain volume abnormalities in children and adolescents with attention-deficit/hyperactivity disorder. JAMA 288:1740–1748

Chabot R. J. and Serfontein G. 1996. Quantitative electroencephalographic profiles of children with attention deficit disorder. Biol. Psychiatr. 40:951–963

Chan S. H., Tai M. H., Li C. Y. and Chan J. Y. 2006. Reduction in molecular synthesis or enzyme activity of superoxide dismutases and catalase contributes to oxidative stress and neurogenic hypertension in spontaneously hypertensive rats. Free Radic. Biol. Med. 40:2028–2039

Cheon K. A., Ryu Y. H., Kim Y. K., Namkoong K., Kim C. H. and Lee J. D. 2003. Dopamine transporter density in the basal ganglia assessed with [123I]IPT SPET in children with attention deficit hyperactivity disorder. Eur. J. Nucl. Med. 30:306–311

Clark L., Blackwell A. D., Aron A. R., Turner D. C., Dowson J., Robbins T. W. and Sahakian B. J. 2006. Association between response inhibition and working memory in adult ADHD: A link to right frontal cortex pathology? Biol. Psychiatr. 61:1395–1401. Epub 13 October 2006

Clarke A. R., Barry R. J., Bond D., McCarthy R. and Selikowitz M. 2002a. Effects of stimulant medications on the EEG of children with attention-deficit/hyperactivity disorder. Psychopharmacology (Berl) 164:277–284

Clarke A. R., Barry R. J., McCarthy R., Selikowitz M. and Brown C. R. 2002b. EEG evidence for a new conceptualisation of attention deficit hyperactivity disorder. Clin. Neurophysiol. 113:1036–1044

Clarke A. R., Barry R. J., McCarthy R., Selikowitz M., Johnstone S. J., Abbott I., Croft R. J., Magee C. A., Hsu C. I. and Lawrence C. A. 2005. Effects of methylphenidate on EEG coherence in attention-deficit/hyperactivity disorder. Int. J. Psychophysiol. 58:4–11

Cook E. H. J., Stein M. A., Krasowski M. D., Cox N. J., Olkon D. M., Kieffer J. E. and Leventhal B. L. 1995. Association of attention-deficit disorder and the dopamine transporter gene. Am. J. Hum. Genet. 56:993–998

Dell'Anna M. E. 1999. Neonatal anoxia induces transitory hyperactivity, permanent spatial memory deficits and CA1 cell density reduction in developing rats. Behav. Brain Res. 45:125–134

Dell' Anna M. E., Luthman J., Lindqvist E. and Olson L. 1993. Development of monoamine systems after neonatal anoxia in rats. Brain Res. Bull. 32:159–170

Dickstein S. G., Bannon K., Xavier C. F. and Milham M. P. 2006. The neural correlates of attention deficit hyperactivity disorder: An ALE meta-analysis. J. Child Psychol. Psychiatr. 47:1051–1062

Doroshchuk A. D., Postnov A. I., Afanas' eva G. V., Budnikov E. I. and Postnov I. 2004. [Decreased ATP-synthesis ability of brain mitochondria in spontaneously hypertensive rats]. Kardiologiia 44:64–65

Dougherty D. D., Bonab A. A., Spencer T. J., Rauch S. L., Madras B. K. and Fischman A. J. 1999. Dopamine transporter density in patients with attention deficit hyperactivity disorder. Lancet 354:2132–2133

Drouin C., Page M. and Waterhouse B. 2006. Methylphenidate enhances noradrenergic transmission and suppresses mid- and long-latency sensory responses in the primary somatosensory cortex of awake rats. J. Neurophysiol. 96:622–632. Epub 10 May 2006

Durston S., Hulshoff Pol H. E., Schnack H. G., Buitelaar J. K., Steenhuis M. P., Minderaa R. B., Kahn R. S. and van E. H. 2004. Magnetic resonance imaging of boys with attention-deficit/hyperactivity disorder and their unaffected siblings. J. Am. Acad. Child Adolesc. Psychiatr. 43:332–340

Durston S., Fossella J. A., Casey B. J., Hulshoff Pol H. E., Galvan A., Schnack H. G., Steenhuis M. P., Minderaa R. B., Buitelaar J. K., Kahn R. S. and van E. H. 2005. Differential effects of DRD4 and DAT1 genotype on fronto-striatal gray matter volumes in a sample of subjects with attention deficit hyperactivity disorder, their unaffected siblings, and controls. Mol. Psychiatr. 10:678–685

Easton N., Shah Y. B., Marshall F. H., Fone K. C. and Marsden C. A. 2006a. Guanfacine produces differential effects in frontal cortex compared with striatum: Assessed by phMRI BOLD contrast. Psychopharmacology (Berl) 189:369–385

Easton N., Steward C., Marshall F., Fone K. and Marsden C. 2006b. Effects of amphetamine isomers, methylphenidate and atomoxetine on synaptosomal and synaptic vesicle accumulation and release of dopamine and noradrenaline in vitro in the rat brain. Neuropharmacology 52:405–414. Epub 3 October 2006

El-Faddagh M., Laucht M., Maras A., Vohringer L. and Schmidt M. H. 2004. Association of dopamine D4 receptor (DRD4) gene with attention-deficit/hyperactivity disorder (ADHD) in a high-risk community sample: A longitudinal study from birth to 11 years of age. J. Neural Transm. 111:883–889

Fagundes A. O., Rezin G. T., Zanette F., Grandi E., Assis L. C., Dal-Pizzol F., Quevedo J. and Streck E. L. 2006. Chronic administration of methylphenidate activates mitochondrial respiratory chain in brain of young rats. Int. J. Dev. Neurosci. 25:47–51. Epub 22 December 2006

Faraone S. V. 2004. Genetics of adult attention-deficit/hyperactivity disorder. Psychiatr. Clin. North Am. 27:303–321

Faraone S. V. and Khan S. A. 2006. Candidate gene studies of attention-deficit/hyperactivity disorder. J. Clin. Psychiatr. 67(Suppl. 8):13–20

Faraone S. V., Doyle A. E., Mick E. and Biederman J. 2001. Meta-analysis of the association between the 7-repeat allele of the dopamine D(4) receptor gene and attention deficit hyperactivity disorder. Am. J. Psychiatr. 158:1052–1057

Faraone S. V., Sergeant J., Gillberg C. and Biederman J. 2003. The worldwide prevalence of ADHD: Is it an American condition? World Psychiatr. 2:104–113

Fellner S. K. and Arendshorst W. J. 2002. Store-operated Ca^{2+} entry is exaggerated in fresh preglomerular vascular smooth muscle cells of SHR. Kidney Int. 61:2132–2141

Filipek P. A., Semrud-Clikeman M., Steingard R. J., Renshaw P. F., Kennedy D. N. and Biederman J. 1997. Volumetric MRI analysis comparing subjects having attention-deficit hyperactivity disorder with normal controls. Neurology 48:589–601

Forssberg H., Fernell E., Waters S., Waters N. and Tedroff J. 2006. Altered pattern of brain dopamine synthesis in male adolescents with attention deficit hyperactivity disorder. Behav. Brain Funct. 2:40

Gainetdinov R. R. and Caron M. G. 2000. An animal model of attention deficit hyperactivity disorder. Mol. Med. Today 6:43–44

Gainetdinov R. R. and Caron M. G. 2001. Genetics of childhood disorders: XXIV. ADHD, Part 8: Hyperdopaminergic mice as an animal model of ADHD. J. Am. Acad. Child Adolesc. Psychiatr. 40:380–382

Genro J. P., Zeni C., Polanczyk G. V., Roman T., Rohde L. A. and Hutz M. H. 2006. A promoter polymorphism (– C > T) at the dopamine transporter gene is associated with attention deficit/hyperactivity disorder in Brazilian children. Am. J Med. Genet. B Neuropsychiatr. Genet. 144:215–219

Gibson M. A., Butters N. S., Reynolds J. N. and Brien J. F. 2000. Effects of chronic prenatal etha-nol exposure on locomotor activity, and hippocampal weight, neurons, and nitric oxide syn-thase activity of the young postnatal guinea pig. Neurotoxicol. Teratol. 22:183–192

Glowinski J., Marin P., Tence M., Stella N., Giaume C. and Premont J. 1994. Glial receptors and their intervention in astrocyto-astrocytic and astrocyto-neuronal interactions. Glia 11:201–208

Gornick M. C., Addington A., Shaw P., Bobb A. J., Sharp W., Greenstein D., Arepalli S., Castellanos F. X. and Rapoport J. L. 2006. Association of the dopamine receptor D4 (DRD4) gene 7-repeat allele with children with attention-deficit/hyperactivity disorder (ADHD): An update. Am. J Med. Genet. B Neuropsychiatr. Genet.144B:379–382

Granon S. and Changeux J. P. 2006. Attention-deficit/hyperactivity disorder: A plausible mouse model? Acta Paediatr. 95:645–649

Greenhill L. L. 2001. Clinical effects of stimulant medication in ADHD, in *Stimulant drugs and ADHD: Basic and clinical neuroscience,* (Solanto M. V., Arnsten A. F. T. and Castellanos F. X., eds), pp. 31–71. Oxford University Press, New York

Hausknecht K. A., Acheson A., Farrar A. M., Kieres A. K., Shen R. Y., Richards J. B. and Sabol K. E. 2005. Prenatal alcohol exposure causes attention deficits in male rats. Behav. Neurosci. 119:302–310

Hill D. E., Yeo R. A., Campbell R. A., Hart B., Vigil J. and Brooks W. 2003. Magnetic resonance imaging correlates of attention-deficit/hyperactivity disorder in children. Neuropsychology 17:496–506

Himelstein J., Newcorn J. H. and Halperin J. M. 2000. The neurobiology of attention-deficit hyperactivity disorder. Front. Biosci. 5:D461–D478

Hobbs M. J., Clarke A. R., Barry R. J., McCarthy R. and Selikowitz M. 2007. EEG abnormalities in adolescent males with AD/HD. Clin. Neurophysiol. 118:363–371

Horn J. L., Janicki P. K. and Franks J. J. 1995. Diminished brain synaptic plasma membrane Ca(2 +)-ATPase activity in spontaneously hypertensive rats: Association with reduced anesthetic requirements. Life Sci. 56:L427–L432

Jones S. R., Gainetdinov R. R., Jaber M., Giros B., Wightman R. M. and Caron M. G. 1998. Profound neuronal plasticity in response to inactivation of the dopamine transporter. Proc. Natl. Acad. Sci. USA 95:4029–4034

Jucaite A., Fernell E., Halldin C., Forssberg H. and Farde L. 2005. Reduced midbrain dopamine transporter binding in male adolescents with attention-deficit/hyperactivity disorder: Association between striatal dopamine markers and motor hyperactivity. Biol. Psychiatr. 57:229–238

Karayanidis F., Robaey P., Bourassa M., De K. D., Geoffroy G. and Pelletier G. 2000. ERP differ-ences in visual attention processing between attention-deficit hyperactivity disorder and con-trol boys in the absence of performance differences. Psychophysiology 37:319–333

Keulers E. H., Hendriksen J. G., Feron F. J., Wassenberg R., Wuisman-Frerker M. G., Jolles J. and Vles J. S. 2006. Methylphenidate improves reading performance in children with attention deficit hyperactivity disorder and comorbid dyslexia: An unblinded clinical trial. Eur. J. Paediatr. Neurol. 11:21–28. Epub 13 December 2006

Kim B. N., Lee J. S., Shin M. S., Cho S. C. and Lee D. S. 2002. Regional cerebral perfusion abnormalities in attention deficit/hyperactivity disorder. Statistical parametric mapping analy-sis. Eur. Arch. Psychiatr. Clin. Neurosci. 252:219–225

Kim C. H., Hahn M. K., Joung Y., Anderson S. L., Steele A. H., Mazei-Robinson M. S., Gizer I., Teicher M. H., Cohen B. M., Robertson D., Waldman I. D., Blakely R. D. and Kim K. S. 2006. A polymorphism in the norepinephrine transporter gene alters promoter activity and is associ-ated with attention-deficit hyperactivity disorder. Proc. Natl. Acad. Sci. USA 103:19164–19169

Krause K. H., Dresel S. H., Krause J., Kung H. F. and Tatsch K. 2000. Increased striatal dopamine transporter in adult patients with attention deficit hyperactivity disorder: Effects of methylphe-nidate as measured by single photon emission computed tomography. Neurosci. Lett. 285:107–110

Kuczenski R. and Segal D. S. 1997. Effects of methylphenidate on extracellular dopamine, serotonin, and norepinephrine: Comparison with amphetamine. J. Neurochem. 68:2032–2037

Kuczenski R. and Segal D. S. 2001. Locomotor effects of acute and repeated threshold doses of amphetamine and methylphenidate: Relative roles of dopamine and norepinephrine. J. Pharmacol. Exp. Ther. 296:876–883

Kuczenski R. and Segal D. S. 2002. Exposure of adolescent rats to oral methylphenidate: Preferential effects on extracellular norepinephrine and absence of sensitization and cross-sensitization to methamphetamine. J. Neurosci. 22:7264–7271

Lahat E., Avital E., Barr J., Berkovitch M., Arlazoroff A. and Aladjem M. 1995. BAEP studies in children with attention deficit disorder. Dev. Med. Child Neurol. 37:119–123

LaHoste G. J., Swanson J. M., Wigal S. B., Glabe C., Wigal T., King N. and Kennedy J. L. 1996. Dopamine D4 receptor gene polymorphism is associated with attention deficit hyperactivity disorder. Mol. Psychiatr. 1:121–124

Lazzaro I., Gordon E., Whitmont S., Meares R. and Clarke S. 2001. The modulation of late component event related potentials by pre-stimulus EEG theta activity in ADHD. Int. J. Neurosci. 107:247–264

Lehohla M., Russell V. and Kellaway L. 2001. NMDA-stimulated Ca^{2+} uptake into barrel cortex slices of spontaneously hypertensive rats. Metab. Brain Dis. 16:133–141

Lou H. C., Henriksen L. and Bruhn P. 1984. Focal cerebral hypoperfusion in children with dysphasia and/or attention deficit disorder. Arch. Neurol. 41:825–829

Luthman J., Fredriksson A., Lewander T., Jonsson G. and Archer T. 1989. Effects of d-amphetamine and methylphentdate on hyperactivity produced by neonatal 6-hydroxydopamine treatment. Psychopharmacology (Berl). 99:550–557

Maher B. S., Marazita M. L., Ferrell R. E. and Vanyukov M. M. 2002. Dopamine system genes and attention deficit hyperactivity disorder: A meta-analysis. Psychiatr. Genet. 12:207–215

Mannuzza S., Klein R. G. and Moulton J. L., III. 2003. Does stimulant treatment place children at risk for adult substance abuse? A controlled, prospective follow-up study. J. Child Adolesc. Psychopharmacol. 13:273–282

Manor I., Corbex M., Eisenberg J., Gritsenkso I., Bachner-Melman R., Tyano S. and Ebstein R. P. 2004. Association of the dopamine D5 receptor with attention deficit hyperactivity disorder (ADHD) and scores on a continuous performance test (TOVA). Am. J. Med. Genet. B Neuropsychiatr. Genet. 127:73–77

Mehta M. A., Goodyer I. M. and Sahakian B. J. 2004. Methylphenidate improves working memory and set-shifting in AD/HD: Relationships to baseline memory capacity. J. Child Psychol. Psychiatr. 45:293–305

Meyer A., Eilertsen D. E., Sundet J. M., Tshifularo J. G. and Sagvolden T. 2004. Cross-cultural similarities in ADHD-like behaviour amongst South African primary school children. S. Afr. J. Psychol. 34:123–139

Mick E., Biederman J., Faraone S. V., Sayer J. and Kleinman S. 2002. Case-control study of attention-deficit hyperactivity disorder and maternal smoking, alcohol use, and drug use during pregnancy. J. Am. Acad. Child Adolesc. Psychiatr. 41:378–385

Milberger S., Biederman J., Faraone S. V. and Jones J. 1998. Further evidence of an association between maternal smoking during pregnancy and attention deficit hyperactivity disorder: Findings from a high-risk sample of siblings. J. Clin. Child Psychol. 27:352–358

Moll G. H., Heinrich H., Trott G., Wirth S. and Rothenberger A. 2000. Deficient intracortical inhibition in drug-naive children with attention-deficit hyperactivity disorder is enhanced by methylphenidate. Neurosci. Lett. 284:121–125

Mostofsky S. H., Cooper K. L., Kates W. R., Denckla M. B. and Kaufmann W. E. 2002. Smaller prefrontal and premotor volumes in boys with attention-deficit/hyperactivity disorder. Biol. Psychiatr. 52:785–794

Nash K., Rovet J., Greenbaum R., Fantus E., Nulman I. and Koren G. 2006. Identifying the behavioural phenotype in fetal alcohol spectrum disorder: Sensitivity, specificity and screening potential. Arch. Women's Ment. Health 9:181–186

Naumann A., Bellebaum C. and Daum I. 2006a. Cognitive deficits in narcolepsy. J. Sleep Res. 15:329–338

Naumann A., Bellebaum C. and Daum I. 2006b. Cognitive deficits in narcolepsy. J. Sleep Res. 15:329–338

Neuman R. J., Lobos E., Reich W., Henderson C. A., Sun L. W. and Todd R. D. 2006. Prenatal smoking exposure and dopaminergic genotypes interact to cause a severe ADHD subtype. Biol. Psychiatr. 61:1320–1328.Epub 6 December 2006

Nurse B., Russell V. A. and Taljaard J. J. 1984. Alpha- and beta-adrenoceptor agonists modulate [3H]dopamine release from rat nucleus accumbens slices: Implications for research into depression. Neurochem. Res. 9:1231–1238

Oades R. D. 2000. Differential measures of 'sustained attention' in children with attention-deficit/hyperactivity or tic disorders: Relations to monoamine metabolism. Psychiatr. Res. 93:165–178

Ohno Y., Matsuo K., Suzuki H., Tanase H., Serikawa T., Takano T. and Saruta T. 1996. Genetic linkage of the sarco(endo)plasmic reticulum Ca^{2+})-dependent ATPase II gene to intracellular Ca^{2+} concentration in the spontaneously hypertensive rat. Biochem. Biophys. Res. Commun. 227:789–793

Ohno Y., Matsuo K., Suzuki H., Tanase H., Takano T. and Saruta T. 1997. Increased intracellular Ca^{2+} is not coinherited with an inferred major gene locus for hypertension (ht) in the spontaneously hypertensive rat. Am. J. Hypertens. 10:282–288

Ohno Y., Suzuki H., Tanase H., Otsuka K., Sasaki T., Suzawa T., Morii T., Ando Y., Maruyama T. and Saruta T. 2005. Quantitative trait loci mapping for intracellular calcium in spontaneously hypertensive rats. Am. J. Hypertens. 18:666–671

Oshima T., Young E. W. and McCarron D. A. 1991. Abnormal platelet and lymphocyte calcium handling in prehypertensive rats. Hypertension 18:111–115

Overmeyer S., Bullmore E. T., Suckling J., Simmons A., Williams S. C., Santosh P. J. and Taylor E. 2001. Distributed grey and white matter deficits in hyperkinetic disorder: MRI evidence for anatomical abnormality in an attentional network. Psychol. Med. 31:1425–1435

Park L., Nigg J. T., Waldman I. D., Nummy K. A., Huang-Pollock C., Rappley M. and Friderici K. H. 2005. Association and linkage of alpha-2A adrenergic receptor gene polymorphisms with childhood ADHD. Mol. Psychiatr. 10:572–580

Paz R., Barsness B., Martenson T., Tanner D. and Allan A. M. 2006. Behavioral teratogenicity induced by nonforced maternal nicotine consumption. Neuropsychopharmacology. 32(3):693–699. Epub 22 March 2006

Purper-Ouakil D., Wohl M., Mouren M. C., Verpillat P., Ades J. and Gorwood P. 2005. Meta-analysis of family-based association studies between the dopamine transporter gene and attention deficit hyperactivity disorder. Psychiatr. Genet. 15:53–59

Realmuto G. M., Garfinkel B. D., Tuchman M., Tsai M. Y., Chang P. N., Fisch R. O. and Shapiro S. 1986. Psychiatric diagnosis and behavioral characteristics of phenylketonuric children. J. Nerv. Ment. Dis. 174:536–540

Richelson E. and Pfenning M. 1984. Blockade by antidepressants and related compounds of biogenic amine uptake into rat brain synaptosomes: most antidepressants selectively block norepinephrine uptake. Eur. J. Pharmacol. 104:277–286

Rieger M., Mayer G. and Gauggel S. 2003. Attention deficits in patients with narcolepsy. Sleep 26:36–43

Riikonen R. S., Nokelainen P., Valkonen K., Kolehmainen A. I., Kumpulainen K. I., Kononen M., Vanninen R. L. and Kuikka J. T. 2005. Deep serotonergic and dopaminergic structures in fetal alcoholic syndrome: A study with nor-beta-CIT-single-photon emission computed tomography and magnetic resonance imaging volumetry. Biol. Psychiatr. 57:1565–1572

Romano E., Tremblay R. E., Farhat A. and Cote S. 2006. Development and prediction of hyperactive symptoms from 2 to 7 years in a population-based sample. Pediatrics 117:2101–2110

Rubia K., Overmeyer S., Taylor E., Brammer M., Williams S. C., Simmons A. and Bullmore E. T. 1999. Hypofrontality in attention deficit hyperactivity disorder during higher-order motor control: A study with functional MRI. Am. J. Psychiatr. 156:891–896

Russell V. A. 2002. Hypodopaminergic and hypernoradrenergic activity in prefrontal cortex slices of an animal model for attention-deficit hyperactivity disorder — the spontaneously hypertensive rat. Behav. Brain Res. 130:191–196

Russell V. A. 2007. Neurobiology of animal models of attention-deficit hyperactivity disorder. J. Neurosci. Meth. 161(2):185–198

Russell V., Allie S. and Wiggins T. 2000. Increased noradrenergic activity in prefrontal cortex slices of an animal model for attention-deficit hyperactivity disorder — the spontaneously hypertensive rat. Behav. Brain Res. 117:69–74

Russell V. A., Sagvolden T. and Johansen E. B. 2005. Animal models of attention-deficit hyperactivity disorder. Behav. Brain Funct. 1:9

Russell V. A., Oades R. D., Tannock R., Killeen P. R., Auerbach J. G., Johansen E. B. and Sagvolden T. 2006. Response variability in attention-deficit hyperactivity disorder: A neuronal and glial energetics hypothesis. Behav. Brain Funct. 2:30

Sagvolden T. 2000. Behavioral validation of the spontaneously hypertensive rat (SHR) as an animal model of attention-deficit/hyperactivity disorder (AD/HD). Neurosci. Biobehav. Rev. 24:31–39

Sagvolden T. 2006. The alpha-2A adrenoceptor agonist guanfacine improves sustained attention and reduces overactivity and impulsiveness in an animal model of attention-deficit/hyperactivity disorder (ADHD). Behav. Brain Funct. 2:41

Sagvolden T., Johansen E. B., Aase H. and Russell V. A. 2005a. A dynamic developmental theory of attention-deficit/hyperactivity disorder (ADHD) predominantly hyperactive/impulsive and combined subtypes. Behav. Brain Sci. 28:397–419

Sagvolden T., Russell V. A., Aase H., Johansen E. B. and Farshbaf M. 2005b. Rodent models of attention-deficit/hyperactivity disorder. Biol. Psychiatr. 57:1239–1247

Saletu M. T., Anderer P., Saletu-Zyhlarz G. M., Mandl M., Arnold O., Nosiska D., Zeitlhofer J. and Saletu B. 2005. EEG-mapping differences between narcolepsy patients and controls and subsequent double-blind, placebo-controlled studies with modafinil. Eur. Arch.Psychiatr. Clin. Neurosci. 255:20–32

Scheres A., Milham M. P., Knutson B. and Castellanos F. X. 2006. Ventral striatal hyporesponsiveness during reward anticipation in attention-deficit/hyperactivity disorder. Biol. Psychiatr. 61(5):720–724. Epub 1 September 2006

Schmitz M., Denardin D., Laufer S. T., Pianca T., Hutz M. H., Faraone S. and Rohde L. A. 2006. Smoking during pregnancy and attention-deficit/hyperactivity disorder, predominantly inattentive type: A case-control study. J. Am. Acad. Child Adolesc. Psychiatr. 45:1338–1345

Semrud-Clikeman M., Steingard R. J., Filipek P., Biederman J., Bekken K. and Renshaw P. F. 2000. Using MRI to examine brain-behavior relationships in males with attention deficit disorder with hyperactivity. J. Am. Acad. Child Adolesc. Psychiatr. 39:477–484

Shaywitz B. A., Klopper J. H. and Gordon J. W. 1978. Methylphenidate in 6-hydroxydopamine-treated developing rat pups. Effects on activity and maze performance. Arch. Neurol. 35:463–469

Siesser W. B., Zhao J., Miller L. R., Cheng S. Y. and McDonald M. P. 2006. Transgenic mice expressing a human mutant beta1 thyroid receptor are hyperactive, impulsive, and inattentive. Genes Brain Behav. 5:282–297

Silberstein R. B., Farrow M., Levy F., Pipingas A., Hay D. A. and Jarman F. C. 1998. Functional brain electrical activity mapping in boys with attention- deficit/hyperactivity disorder. Arch. Gen. Psychiatr. 55:1105–1112

Smalley S. L. 1997. Genetic influences in childhood-onset psychiatric disorders: Autism and attention-deficit/hyperactivity disorder. Am. J. Hum. Genet. 60:1276–1282

Snyder S. M. and Hall J. R. 2006. A meta-analysis of quantitative EEG power associated with attention-deficit hyperactivity disorder. J. Clin. Neurophysiol. 23:440–455

Sohmer H. and Student M. 1978. Auditory nerve and brain-stem evoked responses in normal, autistic, minimal brain dysfunction and psychomotor retarded children. Electroencephalogr. Clin. Neurophysiol. 44:380–388

Solanto M. V. 1998. Neuropsychopharmacological mechanisms of stimulant drug action in attention-deficit hyperactivity disorder: A review and integration. Behav. Brain Res. 94:127–152

Stanley J. A., Kipp H., Greisenegger E., MacMaster F. P., Panchalingam K., Pettegrew J. W., Keshavan M. S. and Bukstein O. G. 2006. Regionally specific alterations in membrane phospholipids in children with ADHD: An in vivo 31P spectroscopy study. Psychiatr. Res. 148:217–221

Sullivan J. E. and Chang P. 1999. Review: Emotional and behavioral functioning in phenylketonuria. J. Pediatr. Psychol. 24:281–299

Swanson J. M. and Volkow N. D. 2003. Serum and brain concentrations of methylphenidate: Implications for use and abuse. Neurosci. Biobehav. Rev. 27:615–621

Tabet F., Savoia C., Schiffrin E. L. and Touyz R. M. 2004. Differential calcium regulation by hydrogen peroxide and superoxide in vascular smooth muscle cells from spontaneously hypertensive rats. J. Cardiovasc. Pharmacol. 44:200–208

Tanda G., Pontieri F. E., Frau R. and Di C. G. 1997. Contribution of blockade of the noradrenaline carrier to the increase of extracellular dopamine in the rat prefrontal cortex by amphetamine and cocaine. Eur. J. Neurosci. 9:2077–2085

Tannock R. 1998. Attention deficit hyperactivity disorder: Advances in cognitive, neurobiological, and genetic research. J. Child Psychol. Psychiatr. 39:65–99

Tannock R., Schachar R. J., Carr R. P., Chajczyk D. and Logan G. D. 1989. Effects of methylphenidate on inhibitory control in hyperactive children. J. Abnorm. Child Psychol. 17:473–491

Tannock R., Schachar R. and Logan G. 1995. Methylphenidate and cognitive flexibility: Dissociated dose effects in hyperactive children. J. Abnorm. Child Psychol. 23:235–266

Thapar A., Fowler T., Rice F., Scourfield J., van den B. M., Thomas H., Harold G. and Hay D. 2003. Maternal smoking during pregnancy and attention deficit hyperactivity disorder symptoms in offspring. Am. J. Psychiatr. 160:1985–1989

Thapar A., O'Donovan M. and Owen M. J. 2005. The genetics of attention deficit hyperactivity disorder. Hum. Mol. Genet. 14 Spec No. 2:R275–R282

Thapar A., Langley K., Asherson P. and Gill M. 2007. Gene–environment interplay in attention-deficit hyperactivity disorder and the importance of a developmental perspective. Br. J. Psychiatr. 190:1–3

Thatcher R. W., Krause P. J. and Hrybyk M. 1986. Cortico-cortical associations and EEG coherence: A two-compartmental model. Electroencephalogr. Clin. Neurophysiol. 64:123–143

Toga A. W., Thompson P. M. and Sowell E. R. 2006. Mapping brain maturation. Trends Neurosci. 29:148–159

Tsuda K., Tsuda S. and Nishio I. 2003. Role of protein kinase C in the regulation of acetylcholine release in the central nervous system of spontaneously hypertensive rats. J. Cardiovasc. Pharmacol. 41(Suppl. 1):S57–S60

Ucles P., Lorente S. and Rosa F. 1996. Neurophysiological methods testing the psychoneural basis of attention deficit hyperactivity disorder. Childs Nerv. Syst. 12:215–217

Ueno K., Togashi H., Matsumoto M., Ohashi S., Saito H. and Yoshioka M. 2002. Alpha4beta2 nicotinic acetylcholine receptor activation ameliorates impairment of spontaneous alternation behavior in stroke-prone spontaneously hypertensive rats, an animal model of attention deficit hyperactivity disorder. J. Pharmacol. Exp. Ther. 302:95–100

Vaidya C. J., Austin G., Kirkorian G., Ridlehuber H. W., Desmond J. E., Glover G. H. and Gabrieli J. D. 1998. Selective effects of methylphenidate in attention deficit hyperactivity disorder: A functional magnetic resonance study. Proc. Natl. Acad. Sci. USA 95:14494–14499

Valera E. M., Faraone S. V., Murray K. E. and Seidman L. J. 2006. Meta-analysis of structural imaging findings in attention-deficit/hyperactivity disorder. Biol. Psychiatr. 61:1361–1369

van Dyck C. H., Quinlan D. M., Cretella L. M., Staley J. K., Malison R. T., Baldwin R. M., Seibyl J. P. and Innis R. B. 2002. Unaltered dopamine transporter availability in adult attention deficit hyperactivity disorder. Am. J. Psychiatr. 159:309–312

Volkow N. D., Wang G. J., Fowler J. S., Gatley S. J., Logan J., Ding Y. S., Hitzemann R. and Pappas N. 1998. Dopamine transporter occupancies in the human brain induced by therapeutic doses of oral methylphenidate. Am. J. Psychiatr. 155:1325–1331

Volkow N. D., Wang G., Fowler J. S., Logan J., Gerasimov M., Maynard L., Ding Y., Gatley S. J., Gifford A. and Franceschi D. 2001. Therapeutic doses of oral methylphenidate significantly increase extracellular dopamine in the human brain. J. Neurosci. 21:RC121

Volkow N. D., Wang G. J., Fowler J. S. and Ding Y. S. 2005. Imaging the effects of methylphenidate on brain dopamine: New model on its therapeutic actions for attention-deficit/hyperactivity disorder. Biol. Psychiatr. 57:1410–1415

Volkow N. D., Wang G. J., Newcorn J., Fowler J. S., Telang F., Solanto M. V., Logan J., Wong C., Ma Y., Swanson J. M., Schulz K. and Pradhan K. 2007. Brain dopamine transporter levels in treatment and drug naive adults with ADHD. Neuroimage 34:1182–1190

Xu X., Knight J., Brookes K., Mill J., Sham P., Craig I., Taylor E. and Asherson P. 2005. DNA pooling analysis of 21 norepinephrine transporter gene SNPs with attention deficit hyperactivity disorder: No evidence for association. Am. J. Med. Genet. B Neuropsychiatr. Genet. 134:115–118

Zhuang X., Oosting R. S., Jones S. R., Gainetdinov R. R., Miller G. W., Caron M. G. and Hen R. 2001. Hyperactivity and impaired response habituation in hyperdopaminergic mice. Proc. Natl. Acad. Sci. USA 98:1982–1987

Chapter 20
Retracted: Brain Damage in Phenylalanine, Homocysteine and Galactose Metabolic Disorders

Kleopatra H. Schulpis and Stylianos Tsakiris

Introduction

Inherited metabolic diseases have become a major cause of neonatal pathology, as the classical causes of neonatal distress have been markedly diminished by advances in obstetrical, prenatal, and perinatal management. Their incidence may well be underestimated, since diagnostic errors are frequent. Nevertheless, accurate diagnosis is essential to provide genetic counseling and prenatal diagnosis of subsequent pregnancies, particularly because some of these conditions have an excellent response to therapy.

Inborn errors of metabolism are individually rare but are collectively numerous. Many of them present early in the neonatal period, have a rapid fatal course and, as a whole, cannot be recognized through systematic screening tests which are too slow, too expensive, and unreliable. This makes it an absolute necessity to teach primary care physicians a simple method of clinical screening before making decisions about sophisticated biochemical investigations. Clinical diagnosis of inborn errors of metabolism in the newborn infant may occasionally be difficult. This is at least partly due to four reasons:

1. Many physicians believe that, because individual inborn errors are rare, they should be considered only after more common conditions, such as sepsis, have been excluded.
2. In view of the large number of inborn errors, it might appear that their diagnosis requires precise knowledge of a large number of biochemical pathways and their interrelationships. Actually, an adequate diagnostic approach can be based on the proper use of only a few tests.
3. The neonate has an apparently limited repertoire of responses to severe overwhelming illnesses, and the predominant clinical signs and symptoms are nonspecific: poor feeding, lethargy, failure to thrive, etc. It is certain that many patients with such defects succumb in the newborn period without having received a specific diagnosis, death often having been attributed to sepsis of some other common causes.

D.W. McCandless (ed.) *Metabolic Encephalopathy*,
doi: 10.1007/978-0-387-79112-8_20, © Springer Science+Business Media, LLC 2009

4. Classical autopsy findings in such cases are often nonspecific and unrevealing. Infection is often suspected as the cause of death, whereas sepsis is the most common accompaniment of metabolic disorders.

We describe three main treatable inborn errors of metabolism, phenylketonuria, homocystinuria, and galactosemia, giving emphasis to the neuropsychological characteristics.

Disorders of Phenylalanine Metabolism

Clinical Features

In 1934, Folling investigated two mentally retarded siblings who possessed an unusual odor. Initial studies revealed a compound in the urine that gave a green color with ferric chloride. Further studies showed this compound to be phenylpyruvic acid (the product of oxidative transamination of phenylalanine). This story, which represents medical detective work at its best, provided the first direct support of Garrods predictions regarding inborn errors of metabolism. While classical phenylketonuria (PKU) is not one of the amino acid disorders that presents in the neonatal period, one may occasionally encounter vomiting and feeding difficulties in the first few weeks of life, often in association with pyloric stenosis, a relationship still unexplained. Usually, early symptoms include excessive irritability and over-activity associated with a musty odor to the urine and sweat of the patients. Indeed, it is often the parents who are first to note the unusual odor so characteristic of these children. It is vital that the physician inquire of parents whether they have ever noted an unusual odor about their children. Eczematoid rashes plague in these infants and youngsters prior to the institution of a low phenylalanine diet were observed. Intellectual development appears normal until the period between 3 and 5 months of age when the infant begins to demonstrate apathy and listlessness, often alternating with sporadic episodes of irritability.

As these children grow older, seizures occur in about 25%. Hair and eyes are often more despigmented than in other members of the family. The behavior of these children seems to revolve around incessant activity. Uncontrollable temper tantrums are often activated by any stimulus. Unusual behavior has often been the reason why these children have been admitted to institutions (Cohn and Roth, 1983).

Biochemical Defects

The biochemical defect in classical PKU is the inability to carry out the normal hydroxylation of L-phenylalanine to tyrosine. The enzyme catalyzing this reaction is a so-called mixed function oxidase, phenylalanine hydroxylase, which is localized only to the liver, kidney, and pancreas. Dietary phenylalanine and phenylalanine produced by the catabolism of tissue protein cannot be converted further into tyrosine: as a result, phenylalanine accumulates in body fluids. Minor catabolic pathways for phenylalanine come into play because of the accumulation of phenylalanine, which makes it possible for these pathways to be activated (Fig. 20.1). Such activation occurs through the metabolic regulation of enzyme activity. Consequently, in blood and urine phenylpyruvate, phenyllactate, phenylacetate, and minor amounts of phenylethylamine, mandelic acid, and hyppuric acid are detected. An essential biochemical feature is the depression in plasma of tyrosine levels due to the enzyme defect. The pigmentary defects in PKU appear to arise as a consequence of phenylalanine acting as a competitive inhibitor on tyrosinase, the enzyme responsible for production of melanin. It is curious that with the multiplicity of metabolites produced, the only system severely deranged in phenylketonuria is the central nervous system (CNS). Considering the import of the brain to human conduct, it is an important target organ indeed.

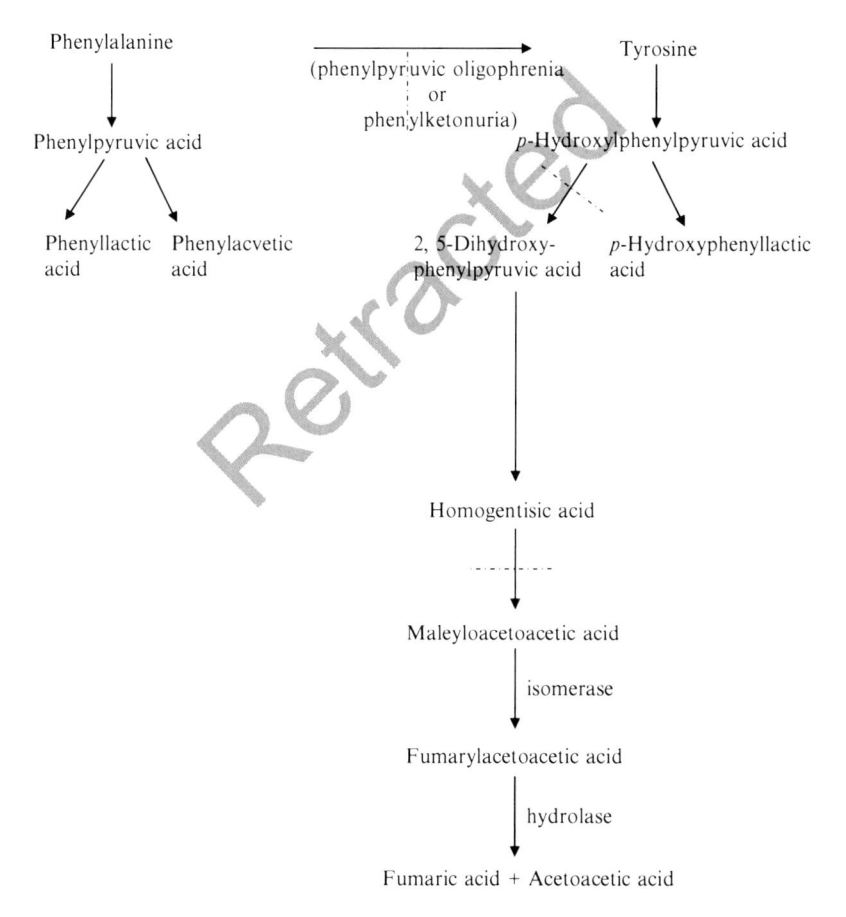

Fig. 20.1 Major catabolic pathway for phenylalanine and tyrosine. The loci of known enzymatic defects are indicated by dashed lines. Note that hereditary tyrosinemia is now believed to be due to fumarylacetoacetate hydrolase, the final step in the pathway. (Redrawn with modifications from Mazur A, Harrow B: Textbook of Biochemistry. WB Saunders, Philadelphia, 1971)

Damage to the developing CNS may be profound. Several studies indicate that the average loss in the development quotient during the first year of life in over 95% of untreated patients is approximately 50 points. A consensus has thus developed that the earlier the diet is instituted, the milder will be the effects on the brain (Mazur and Harrow, 1971).

Pathogenesis of Mental Retardation in Phenylketonuria

The noxious effects of PKU on the developing brain have been matters of great practical and experimental interest. Despite the enormous amount of research conducted on the pathogenesis of mental retardation in this disorder, it is not possible to point to any single factor as being causative. Research in this disease, and indeed in the general area of the inborn errors of amino acid metabolism, is severely hampered by the absence of a naturally occurring animal model that suffers the enzyme defect. Feeding high levels of phenylalanine results in elevated rather than depressed levels of tyrosine; use of the inhibitor *p*-chlorophenylalanine affects other enzymes as well, including pyruvate kinase. Additionally, it causes the formation of cataracts, a finding not encountered in PKU (Sershen et al., 1987; Sanchez et al., 1996; Shedlovsky et al., 1993; Brenton and Gardiner 1988).

Minor metabolites that accumulate because of the alternate pathways utilized in phenylketonuria may normally serve either physiological or pharmacological roles in the nervous system. To aggravate the deficiency of tyrosine created by shunting into these minor pathways in PKU, phenylalanine inhibits tyrosine transport across biological membranes. In turn, this curtails the source of neuroactive tyrosine derivatives that can be synthesized, including tyramine, octopamine, and the catecholamines. One can speculate that such deficiencies could interfere with neurotransmitter action.

Elevated levels of phenylalanine have been shown to inhibit transport of other amino acids besides tyrosine. This might result in an imbalance of amino acids in the brain that could disrupt protein synthesis or control of the synthesis of neurotransmitters. Several enzymes, including tyrosine hydroxylase, tryptophan, and pyruvate kinase, are inhibited in vitro by phenylalanine. Irrespective of such phenomena in a patient who died with PKU, catecholamine concentration and serotonin levels were much lower than those in control brains from patients suffering mental retardation from other causes, a finding consistent with such a possible role in vivo (Ikeda et al., 1967).

Laboratory Diagnosis

The ferric chloride test was used to suggest the diagnosis of PKU in an older child. It is positive in the presence of phenylpyruvic acid. However, during the first year of life, the test is often negative. This demonstrates the case when phenylpyruvic acid levels are less than 0.2 mg ml^{-1}, as may occur when protein intake has been very low for a period of time.

Several other metabolites, in particular from histidine and certain drugs such as chlorpromazine, may also produce a green color with the ferric chloride reagent. This test is then not an adequate screening test for PKU in infants. The method developed by Guthrie is now employed in all states in the United States and even globally as part of mandatory screening for this disorder. As performed in most screening laboratories, the test distinguishes phenylalanine levels in excess of approximately 4 mg dl^{-1} in whole blood. Moreover, tandem mass spectrometry is now widely used (Shefer et al., 2000; Rosenblatt et al., 1992). Unfortunately, approximately 8% of infants with classical PKU will give a negative Guthrie test within the first 3 days of life and it may not become positive until after the first week. Since there are occasional false negatives, the cautious physician is urged to consider the possibility of PKU in a child presenting with failure to thrive or any of the other symptoms discussed above. In such a situation a repeat Guthrie test is in order. The authors have encountered several such instances in which the initial testing by the state was negative but repeated testing, because of physician concern, was positive and led to the unequivocal diagnosis of phenylketonuria (McDonald 2000).

Variant Forms of Hyperphenylalaninemia

The most common cause of an elevation of the serum phenylalanine level about 3 mg dl^{-1} is the transient neonatal delay in the development of the tyrosine oxidizing system (transient neonatal tyrosinemia). This condition, while far more common in low-birth-weight and premature infants, does occur occasionally in full-term infants as well. The elevation of phenylalanine is secondary to a block in the metabolism of tyrosine. Administration of 100 mg of ascorbic acid will reduce the tyrosine levels, but whether this is necessary is an area of some controversy (Cohn and Roth 1983).

In an infant with a serum phenylalanine concentration greater than 4 mg dl^{-1}, the following have been recommended: administration of 100 mg of ascorbic acid 24 h prior to obtaining a repeat blood specimen, measurement of phenylalanine and tyrosine in blood, and evaluation of urine for the presence of phenylalanine, O-hydroxyphenylacetic acid, and phenylpyruvic acid, as well as tyrosine and tyrosine derivatives (p-hydroxyphenyllactic acid and p-hydroxyphenylpyruvic acid) by chromatography and spot tests. Most infants will be found to have a transient defect that demands a new testing.

In a statewide screening program, few infants will manifest persistent elevations of phenylalanine in the range of 4–20 mg dl^{-1}. Serum tyrosine will be low or normal, and the urine will be negative for phenylalanine metabolites. In all likelihood, this represents a hyperphenylalaninemia variant or a delay in the maturation of the hydroxylase system. Repetition of blood screening while on an unrestricted diet should permit distinction. With classical phenylketonuria, serum phenylalanine will rise and urinary metabolites should appear. With delayed maturation of the hydroxylase system, as activity of the system increases, serum phenylalanine concentrations will fall. By the age of 3 weeks, if blood phenylalanine levels are still high, regardless of the presence of metabolites in urine, restriction of phenylalanine intake is indicated. Multiple variant forms of PKU have been described (Shefer et al., 2000; Dyer et al., 1996).

Unfortunately, distinguishing among the various forms of PKU may not be possible by routine laboratory testing, and a number of scientists have recommended the use of a phenylalanine-loading test to separate the variant forms. The test should only be performed in centers that are equipped to quantitate the specimens obtained and to embark upon the definitive care of affected infants through the restriction of phenylalanine in the diet.

Regarding the dihydropteridine reductase deficiency (DHPR) and defect in biosynthesis of biopterin, further studies may be in order in most hyperphenylalaninemic infants.

Treatment

Treatment of PKU with a diet restricted in phenylalanine content is unquestioned in children whose phenylalanine levels are greater than 20 mg dl^{-1} and for whom the diagnosis of classical PKU appears to be unequivocal. Children whose diagnosis cannot be clearly delineated or who may have some variant of PKU pose great diagnostic difficulty for the expert in metabolic diseases. Diagnosis in such cases may require the use of either loading studies or DNA analysis. The use of diet therapy requires phenylalanine levels to be maintained above 1.5–2 mg dl^{-1}, in order not to have any restriction of growth as a consequence of amino acid deficiency. This requires that children are treated in centers with a team including a pediatrician and nutritionist and laboratory facilities to monitor amino acid levels. The diet for these children is severely restricted, requiring the creativity of nutritionist and parents to make it as palatable as possible to prolong treatment as long as necessary. At present, recommendations for duration of therapy vary from cessation in the fifth to cessation in the tenth year of life; current opinion leans towards keeping the patient on the diet lifelong because of possible deleterious effects on their CNS. Successfully treated female patients with PKU must resume the diet prior to conception because of the risk of the heightened level of phenylalanine and metabolites (maternal PKU, see below) in the fetus, even if it is genetically normal

(Baumeister and Baumeister 1998; Burgard et al., 1997; Legido et al., 1993).

Dihydropteridine Reductase Deficiency

Clinical Features

In contrast to classical PKU, this variant is unresponsive to prompt institution of dietary phenylalanine restriction. Clinical recognition comes about owing to the development of seizures, hypotonia, choreiform movements, and psychomotor retardation in a patient demonstrating elevated serum phenylalanine and depressed tyrosine levels. This is observed despite normalization of phenylalanine levels on a restricted diet (MacDonald et al., 1994).

Biochemical Defects

Assay of phenylalanine hydroxylase in a liver biopsy from one patient showed 20% of normal adult control values, but dihydropteridine reductase activity (Fig. 20.2) was less than 1% of normal in the liver, brain, and other tissues. This latter deficiency presents regeneration of tetrahydrobiopterin, the cofactor for the hydroxylase reaction. Since the reductase enzyme reaction regenerates the cofactor for tyrosine and tryptophan hydroxylase, catecholamine and serotonin synthesis are compromised as well. Patient studies are scanty, but in one patient dopamine and serotonin were decreased in the cerebrospinal fluid, brain, and various other tissues, while norepinephrine metabolites were normal. While phenylalanine hydroxylase activity was lower than that of adult controls, it was not determined whether this value represented significantly decreased activity in children.

Using high-pressure liquid chromatography (HPLC), it is possible to show that patients with this variant of hyperphenylalaninemia excrete only oxidized forms of biopterin in their urine; normal children and those with phenylalanine hydroxylase deficiency excrete predominantly tetrahydrobiopterin.

Replacement therapy with L-dopa and 5-hydroxytryptophan has been suggested as the means to circumvent the block in synthesis of dopamine and serotonin (Steinfeld et al., 2002).

Biopterin Deficiency: Hyperphenylalaninemia

Several patients have now been described with this variant of hyperphenylalaninemia. Neurological symptoms, particularly hypotonia and delay in motor development, are recognized earlier than in classical PKU. In spite of adequate control of serum phenylalanine levels, deterioration continues unabated. Seizures have not been encountered in these patients, as there is a coexistence of dihydropteridine reductase deficiency. Since both tetrahydrobiopterin and dihydropteridine reductase are essential to the hydroxylases that synthesize serotonin and norepinephrine, it is not surprising that neurological symptoms are prominent in

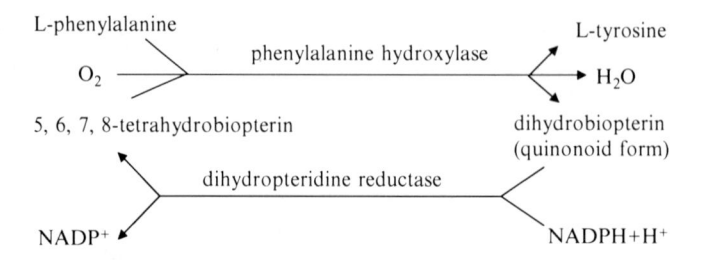

Fig. 20.2 Pterin-cofactor-dependence of phenylalanine hydroxylase. Tetrahydrobiopterin, synthesized from guanosine triphosphate, donates the electron required to convert molecular oxygen to H_2O and hydroxylate the ring of phenylalanine to produce tyrosine. The resulting dihydrobiopterin is recycled by conversion to tetrahydrobiopterin using NADPH. (From McGlivery RW. Biochemistry: A Functional Approach, 2nd edn. WB Saunders, Philadelphia, 1979)

both defects. Why seizures occur in one and not in the other defect is at present unanswered (Fig. 20.2).

These patients can be diagnosed by administration of tetrahydrobiopterin, which will cause a fall in the plasma phenylalanine level. Evidence thus far marshaled seems to support a defect in biosynthesis of biopterin, the precursor to tetrahydrobiopterin and the active cofactor for phenylalanine hydroxylase (Kure et al., 1999).

Unfortunately, tetrahydrobiopterin does not traverse the blood–brain barrier readily; therefore, exogenous administration is not likely to be effective. At present L-dopa and 5-hydroxytryptophan administration are recommended to circumvent the block in neurotransmitter synthesis (Steinfeld et al., 2002; Muntau et al., 2002).

Treated PKU

Treatment of PKU consists of a phenylalanine-restricted diet, including a special formula and foods low in phenylalanine. The formula contains all the necessary nutrients (amino acids) in protein, apart from phenylalanine. Unfortunately, amino acids in this form have a distinctive and strong taste and odor. Almost all infants accept the formula without difficulties; however, some children find it distasteful as they grow older, and many adults returning to the diet regard it as unpalatable. The diet permits sugars, fats, measured amounts of fruits and vegetables, special low-protein pastas, grains, and breads. Meat, fish, eggs, dairy products, nuts, soy products, regular grains, and corn are not allowed. When children or adults "cheat" or consume more than the allocated amount of protein, they do not immediately feel ill, although a few individuals report feeling tired or distracted. Most experience no side effects. It is only the cumulative effect of increased phenylalanine intake that is noticeable. Dietary control is monitored through frequent sampling of blood phenylalanine levels.

Until the 1980s, most clinics in North America and Europe recommended diet discontinuation during middle childhood (Schuett and Brown, 1984). At about age 5 or 6 years, most children with PKU were suddenly allowed to eat as much protein (phenylalanine) as they desired. Even though it was known that their blood phenylalanine levels would rise, it was thought that their cognitive abilities would be unaffected. The fact that high phenylalanine levels are known to affect myelin in the brain, and that myelination is essentially complete after infancy, provided the rationale for this approach to treatment for PKU. Moreover, children who did not adhere to the diet despite medical recommendations did not become mentally retarded. Therefore, the policy of diet discontinuation was adopted.

Despite the early enthusiasm for considering PKU a disease of early childhood, evidence gradually mounted demonstrating that diet discontinuation resulted in diminished IQ in a sizable proportion of these children. A North American PKU Collaborative Study was established to determine the effects of diet discontinuation in children treated early with PKU. The initial random assignment to diet continuation or discontinuation was abandoned, since some clinics came to consider it unethical to discontinue the diet for any child. At the same time, some children assigned to the diet continuation group failed to maintain metabolic control. The final sample included almost an equal number of children considered on the "off" diet. The results of the follow-up study indicated that the age at which blood phenylalanine levels consistently exceeded 15 mg dl^{-1} was the best predictor of IQ and school achievement at ages 8 and 10 (Walter et al., 2002). A retrospective study of 46 patients in Pennsylvania followed beyond age 12 reported similar results. On the contrary, a policy of diet discontinuation at age 10 was instituted in Scotland, and no decline in cognitive and motor functioning was noted after diet discontinuation in adolescents and young adults at a median age of 20. However, the individuals with PKU performed less well on all tests than the age-matched subjects without PKU did (Schweitzer-Krantz et al., 2000).

Such a therapeutic diet can be achieved with vegetable protein and can be considered nonatherogenic because of the reduction of animal protein products. We found low hemostatic variables in PKU patients on strict diet, as a result of reduced fat intake, resulting in an impaired absorption of vitamin K (Schulpis et al., 1996). In addition, low antioxidant status was determined in the plasma of PKU patients who did not adhere to their therapeutic diet. In contrast, in patients on strict diet, high antioxidant capacity was determined, as a result of their antioxidants-rich diet (Schulpis et al., 2002).

Neuropsychological Effects Despite Treatment

Early diagnosis and treatment for PKU unquestionably prevent mental retardation. Diet continuation, however, does not prevent all adverse effects of PKU. Despite the prevention of the severe neurological complications from PKU, more severe psychological consequences have been exposed.

Visual-motor deficits are prevalent. Even children treated early tend to have awkward pencil grips and poor handwriting. Fine motor speed is diminished, copying letters or figures is a laborious process, and work takes longer to complete. When asked to copy geometric designs, many children with PKU have notable difficulties, particularly when they are required to integrate figures. Visual demonstrations, diagrams, and models are less effective than verbal explanations. The children have difficulties remembering the location of objects in space. The "number line" may be incomprehensible for years after it has been taught in arithmetic class (Burgard et al., 1997).

Moreover, consistently noted in individuals treated for PKU are problems in mental processing (de Sonneville et al.,1990; Levy and Waisbren, 1994). This finding has led to increasing interest in the executive functioning of children with early treated PKU (Pardridge, 1998). "Executive functioning" is the ability to retain information and use it for problem solving. Planning, integrative processing skills (reasoning, comprehension, concept formation), and sustained attention depend on executive functioning. When information is presented slowly and simply, children with PKU learn and retain it as well as peers do. However, when the cognitive "load" is increased, or a faster processing speed (reaction time) is required, the children often become confused and overwhelmed. Children without difficulties in executive functioning are able to increase their focus and adjust to the greater processing demands. When impaired in executive functioning, individuals often react more slowly and make more errors on tests related to mental processing. Results of neuropsychological testing in early treated children with PKU demonstrate clear deficits in executive functioning and in reaction time (Welsh et al., 1990; de Sonneville et al., 1990; Schmidt et al., 1992; Schmidt et al., 1994).

Factors Related to Neuropsychological Performance in Treated PKU

Many studies have focused on the blood phenylalanine level in treated PKU. Factors such as timing of treatment initiation, lifetime level of metabolic control, and current dietary status, all have an impact. In most studies, if treatment is initially delayed past the first 3 months of life, a child performs less well than siblings with PKU who are treated earlier. If metabolic control is variable throughout childhood, the individual tends to have poorer mental processing skills, slower reaction time, diminished achievement, and lower IQ scores. One study has documented IQ loss in early treated adolescents with elevated phenylalanine levels (Beasley et al., 1994). By 18 years of age, 27% have an IQ less than 70. IQ is significantly related to the average phenylalanine control between birth and 14 years of age. The current blood phenylalanine level in an individual with PKU is also correlated with reaction time (Clarke et al., 1987; Schmidt et al., 1994) and is thought to reflect the level of brain phenylalanine. If the child is currently off diet or in poor metabolic control, he or she demonstrates compromised neuropsychological functioning. For children and adults with normal intellectual abilities, resumption of diet and maintenance of good metabolic control result in improved reaction time and concentration. Although IQ remains essentially unchanged, functioning improves when blood phenylalanine levels are reduced.

Recommendations from the various PKU clinics vary with regard to what constitutes a metabolic control in children over 6 years of age. The target for most clinics is now 2–6 mg dl^{-1}. However, owing to the restrictiveness of the diet, fewer than half of teenagers are able to maintain levels within this range. In a follow-up study of children in the United Kingdom, only 12% were following a strict diet by age 14 , and only 4% were following a strict diet by age 18 (Beasley et al., 1994). One research group suggests that levels as high as 15 mg dl^{-1} may be benign in teenagers and young adults (Griffiths et al., 1995). Other investigators reported that any elevation above 6 mg dl^{-1} might cause adverse effects (Diamond, 1994).

The possibility exists that neuropsychological performance in treated PKU is related to the extent of the enzyme block. Individuals with natural blood phenylalanine levels in the range of non-PKU mild hyperphenylalaninemia perform at a higher level than early treated individuals with mild PKU (also called "atypical PKU"), who in turn attain higher scores than those with classic PKU; therefore, the greater the activity of the phenylalanine hydroxylase enzyme, the better a person's functioning.

Neuropsychological Tests

Performances on neuropsychological tests suggest a possible localization of brain effects in treated PKU, with a suspicion that the prefrontal cortex is involved (Welsh et al., 1990). In that context, the possibility that dopamine plays an important role in the cognitive limitations in PKU needs to be considered. Projections of dopaminergic neurons in the neocortex are found primarily in the frontal lobes (Porrino and Goldman-Rakic, 1982), and the prefrontal cortex has one of the highest levels of dopamine turnover in the brain (Diamond et al., 1994; Tam et al., 1990). Investigations of individuals with PKU demonstrate that reductions in cerebrospinal fluid concentrations of dopamine and serotonin metabolites are associated with lower scores on reaction time tests (Lykkelund et al., 1988). In that perspective, Diamond et al. (1994) hypothesize that the prefrontal lobe is the most affected brain region in PKU, because it is most sensitive to mild reductions in dopamine.

Neuropsychological tests considered to be particularly sensitive to prefrontal cortical functioning – such as the tests of motor planning, visual search, and verbal fluency; the Tower of Hanoi; the Stroop Color World Test; and the Wisconsin Card Sorting Test – have been used to test this hypothesis. Diamond et al. (1994) reported that early and continuously treated children (with blood phenylalanine levels between 6 and 10 mg dl^{-1}) showed impaired performance on six tests that require memory and inhibitory abilities dependent on the dorsolateral prefrontal cortex. Other investigations have supported these findings (Welsh et al., 1990; Weglage et al., 1996). Stemerdink (1996) found that in 36 older patients (aged 8–19 years) treated early and continuously, neuropsychological performance on three out of four prefrontal tasks was impaired. The same pattern has been observed in adults with early treated PKU. Moreover, individuals with PKU attain lower standard scores on these tests than on control functions.

Researchers have also found evidence for impairment in visual contrast sensitivity, when blood phenylalanine levels are elevated. This is relevant as it is hypothesized that the retina is also highly sensitive to moderate reduction in brain dopamine (Stemerdink, 1996; Diamond, 1994; Guttler and Lou, 1986).

In vivo studies on the effects of high Phe blood concentrations on the erythrocyte membrane AChE and Na$^+$, K$^+$-ATPase activities in relation to biogenic amine blood concentrations in PKU patients (Schulpis et al., 2002), found that high Phe and/or low biogenic amine concentrations may indirectly inhibit the above-mentioned membrane enzymes activities. The observed enzyme inhibitions could be a very informative peripheral marker as regards the neurotoxic brain effects of the high blood levels of the aromatic amino acid (Schulpis et al., 2002; Tsakiris et al., 2002).

Phenylketonuria vs Dopamine

Not all studies, however, support the dopamine–prefrontal dysfunction hypothesis. Mazzocco et al. (1994), using the Tower of Hanol and visual search tests, found that children aged 6–13 who were treated early and continuously showed no deficits on the neuropsychological tests, despite a range of blood phenylalanine levels.

Variations on the dopamine hypothesis have also been proposed. Krause et al. (1985) reported a correlation between increased reaction time and decreased urinary dopamine in patients with PKU. As brain dopamine is concentrated in the corpus striatum, and the choice reaction time test requires a motor response as well as integration of stimuli, they speculate that the nigrocostriatal pathways are affected. Faust (1986–1987) obtained similar results; they suspect that the deficits are associated with complex areas of the brain, such as the anterior frontal regions, and in motor areas that represent less advanced functions.

Some studies have used magnetic resonance imaging (MRI) to investigate the relevance of myelin abnormalities in PKU. Reports conclude that the severity of the MRI changes is significantly and independently associated with the phenylalanine concentrations at the time of investigation and the timing of diet discontinuation. When metabolic control improves, the MRI picture also improves. The area of the brain in which white matter abnormalities are most commonly noted is the parieto-occipital region. Despite the provocative nature of these results, MRI findings have not been found to correlate with IQ, neuropsychological functioning, or neurological symptoms (Thompson et al., 1993). One study suggests, nevertheless, that abnormal myelination neonatally disrupts the development of inter-hemispheric connections in early treated PKU. Gourovitch et al. (1994) reported that children with early treated PKU demonstrated slowed interhemispheric transfer from the left to the right hemisphere, compared to normal controls and to children with attention-deficit/hyperactivity disorder (ADHD). Age of treatment initiation and blood phenylalanine levels at birth were correlated with reaction time on tests of interhemispheric transfer in that study.

In the CNS, oligodendrocytes extend numerous processes, and from the distal tip of each process a membrane sheet is assembled and wrapped around a segment of axon as an internode of myelin. Myelin is a highly metabolically active membrane that, under normal conditions, remains connected to and is supported by the oligodendrocyte cell body for the life span of the oligodendrocyte. Myelin is essential for the rapid conduction of action potentials and, therefore, there may be devastating consequences if oligodendrocytes fail to produce myelin (hypomyelination) or lose their myelin (demyelination) as a result of disease.

A disease in which hypomyelination occurs in specific forebrain tracts, but neurons and their axons are spared, is the autosomal recessive disorder PKU (Malamud, 1996; Dyer et al., 1996). As already mentioned, PKU is caused by a rise in blood phenylalanine (Phe) levels, due to a deficiency in the enzyme phenylalanine hydroxylase (PHA) (Scriver et al., 1995), which is expressed primarily in liver and not in brain, and catalyzes the conversion of Phe to tyrosine (Lee et al. 2003a and Lee et al. 2005). Blood Phe levels normally are about 121 mmol; however, in untreated individuals (and mice) with PKU, levels may increase to 1,200 μ or more. For the past several decades, newborns diagnosed with PKU are placed on a low Phe diet for life. The low Phe diet decreases Phe levels in blood and brain, thereby allowing myelination to proceed (Thompson et al., 1993; Pietz et al., 1995). Individuals with PKU that are continuously treated from birth avoid the severe mental retardation that occurs in untreated individuals (Levy et al., 1994).

It is well documented that although mental retardation is avoided with dietary treatment, significant problems still exist. Individuals treated for PKU from birth may continue to have elevated blood Phe levels, predominantly because the low Phe diet is distasteful and therefore difficult to maintain (MacDonald et al., 1994). If the diet is discontinued or liberalized, (a) blood Phe levels rise, (b) white matter lesions appear and increase in size, and (c) intelligence decreases (Smith and Kang, 2000; Thompson et al., 1990, 1993). In summary, high levels of circulating Phe are reported to correlate with white matter pathology in specific forebrain tracts and neurological deterioration at any age despite dietary treatment during childhood.

The molecular mechanisms underlying the neurological deficits observed in individuals with PKU are unknown. However, decreased levels of neurotransmitters, including dopamine, are likely to play a major role in the observed cognitive disabilities (Diamond et al., 1994; Puglisi-Allegra et al., 2000; Pascucci et al., 2002). To date, two distinct theories have been proposed to explain the phenomenon (for review, see Dyer et al., 1996 and 2000). The "tyrosine/dopamine theory" predicts that cognitive difficulties stem from decreased levels of tyrosine, the precursor of dopamine. Brain tyrosine levels in an experimental rat model for PKU were speculated to be low as a consequence of high blood Phe levels outcompeting tyrosine for transport across the blood–brain barrier (Diamond et al., 1994). As evidence suggests that dopamine synthesis in prefrontal cortex dopaminergic neurons is directly related to tyrosine levels (for review, see Tam and Roth, 1997), Diamond and colleagues postulated that low brain tyrosine levels lead to decreased dopamine levels and thereby to cognitive disabilities in individuals with PKU.

The second hypothesis, the "myelin/dopamine theory," takes into account the primary pathologic findings in treated PKU brain, i.e., decreased myelination within specific tracts in the brain (Malamud, 1996; Dyer et al., 1996). Myelination induces axonal maturation, i.e., myelin/axonal interactions trigger heavy phosphorylation of neurofilaments, rearrangements of the cytoskeleton, and swelling in the axon beneath the compact myelin lamellae (Colello et al., 1994; Kirkpatarick and Brady, 1994; Sanchez et al., 1996). In the myelin/dopamine theory, myelin/axonal interactions are postulated to transduce signals that upregulate the production of enzymes involved in the dopamine biosynthetic pathway, i.e., tyrosine hydroxylase (TH), which is the key regulatory enzyme in the dopamine synthetic pathway. Alternatively, myelin/axonal contact may trigger signaling pathways that result in the phosphorylation of TH, thereby activating the enzyme and upregulating dopamine synthesis. Either or both mechanisms may increase neurotransmitter production.

Animal Studies

To the best of our knowledge, the study presented below explores the relationship between Phe, tyrosine, dopamine synthesis, and myelination in frontal cortex and striatum. The genetic mouse model for PKU, which contains the PAH^{enu2} gene mutation, was considered the most experimentally appropriate animal model for the following reasons. The PAH^{enu2} gene mutation results in inactivity of the PAH gene (McDonald and Charlton, 1997), which, in turn, leads to elevated blood Phe levels; PKU mouse blood Phe levels are similar to those in individuals with PKU (Shedlovsky et al., 1993). Moreover, the neuropathology in the PKU mouse is strikingly similar to the human PKU brain, i.e., forebrain structures are hypomyelinated, including subcortical white matter and the corpus callosum in the frontal cortex, and white matter tracts within the striatum (Dyer et al., 2000). Thus, this PKU mouse model is an excellent system in which to determine whether the tyrosine/dopamine and/or the myelin/dopamine hypotheses are correct.

To determine whether a relationship between dopamine synthesis and either tyrosine levels or myelination exists, 6–8 week-old male PKU mice were placed on a low Phe diet, and levels of dopamine, tyrosine, Phe, and myelin were measured during a 4-week time course study. In the course of performing this study, several surprising findings were made concerning the differential regulation of Phe, tyrosine, and dopamine levels in frontal cortex and striatum of PKU mouse brain. The differential regulation of the amino acids Phe and tyrosine in brain was not initially a focus of the study. However, the aberrant levels of these amino acids in PKU brain made a significant contribution to the development of our conclusions that (a) tyrosine levels are not a key regulator of dopamine synthesis, and (b) a relationship appears to exist between myelination and dopamine production. Therefore, possible mechanisms involving blood–brain barrier transporter proteins that control the flux of Phe and tyrosine into and out of the brain also are discussed herein.

To determine if a relationship exists between tyrosine levels and dopamine synthesis in the treated PKU mouse brain, it was necessary to quantify tyrosine in blood and brain tissues in control heterozygous, untreated PKU, and treated PKU mice. On the basis of previous studies showing that high levels of Phe out-compete tyrosine for transport across the blood–brain barrier (Choi and Pardrige, 1986; Brenton and Gardiner, 1988; Pardridge, 1998), it was anticipated that brain tyrosine levels would be less than blood tyrosine levels in the untreated PKU mouse. However, it was unexpectedly found that tyrosine levels in frontal cortex and striatum were approximately 1.2 and 1.4 times, respectively, the level of tyrosine in the blood of untreated PKU mice. Thus, the eightfold elevation in blood Phe levels in the untreated PKU mouse did not appear to interfere with the movement of tyrosine into the brain.

In vitro studies (Tsakiris et al., 1998a,b, 2002), showed that preincubation of rat homogenates or pure eel, *E. Electricus*, with high Phe levels resulted in a decrease of acetylcholinesterase (AChE) activities, which reached about –18%. In addition, pure Na$^+$, K$^+$-ATPase activity showed an increase by 20–30% after incubation with Phe (0.24–0.9 mµ). A rat brain homogenate Na$^+$, K$^+$-ATPase activity increase by 60–65% appeared with 0.9–12.1 mM Phe incubation. These in vitro results may explain the observed muscarinic dysfunctions presented in untreated PKU patients.

Furthermore, another in vitro study (Tsakiris et al., 1998b) illustrated that preincubation of rat diaphragm homogenate with the above-mentioned Phe concentrations resulted in a decrease of AChE and Na$^+$, K$^+$-ATPase activities. These findings may explain, in some degree, the influence of high concentration of the aromatic amino acid on diaphragm synaptic acetylcholine (ACh). Moreover, a following in vitro study showed that the addition of the amino acid alanine in the preincubation mixture resulted in a complete restoration of AChE rat brain inhibition induced by high Phe concentrations in a competitive way (Tsakiris and Schulpis, 2002).

Phenylalanine vs. Blood–Brain Barrier

The basis for the discrepancy between results and previous studies, with respect to the ability of tyrosine to be transported through the blood–brain barrier in the presence of elevated Phe, may arise from the different lengths of time the blood–brain barrier was exposed to elevated Phe levels. Previous studies administered a bolus of Phe, and shortly thereafter Phe and tyrosine flux was measured in vivo into the brain or in vitro through isolated brain capillary vessels (Choi and Pardridge, 1986; Brenton and Gardiner, 1988; Pardridge, 1998). In contrast, the untreated PKU mouse blood–brain barrier was continuously exposed to abnormal Phe and tyrosine levels form shortly after birth into adulthood. In response to the abnormal long-term exposure of the blood–brain barrier to aberrant concentrations of blood Phe and tyrosine, the blood–brain barrier may have altered its expression pattern of amino acid transporters trying to achieve normal amino acid concentrations in the brain. An adaptation process within the blood–brain barrier occurs during normal development, e.g., concentrations of amino acids in brain normally change during the development of the newborn into the adult (Sershen et al., 1987; Gardiner, 1990). Therefore, it may be that the expression pattern of blood–brain barrier trans-porter proteins in untreated PKU mice altered in response to aberrant blood Phe aid tyrosine levels.

Amino acid data in untreated PKU mouse brain vs. control heterozygote brain support the possibility that blood–brain barrier transporter protein expression was abnormal in the untreated PKU mice. For example, in control heterozygote mice, striatum and blood Phe levels were approximately equal. In contrast, in untreated PKU mice, Phe levels in striatum were about 80% of blood Phe levels. Phe levels in heterozygote mouse frontal cortex were about 15% of blood levels, but in untreated PKU mice this ratio was approximately 72%. When a comparison is made between amino acid levels in PKU mouse brain and those in control hetero-zygote brain structures, two pieces of evidence suggest that the blood–brain barrier transporter proteins were abnormal in PKU mice. First, while blood Phe levels were restored to normal in treated PKU mice, Phe levels remained elevated twofold in frontal cortex and 1.5-fold in striatum. Second, tyrosine levels did not increase above 70% of normal levels in either brain structure in treated PKU mice despite the fact that tyrosine levels were near normal in blood. These data suggest that the blood–brain barrier attempted to maintain elevated levels of Phe and low levels of tyrosine in the treated PKU mouse brain, i.e., levels that had been "normal" in the untreated PKU brain. It has not been clarified whether Phe and tyrosine levels would eventually have been restored to normal in treated PKU mouse brain, since the study did not continue beyond 4 weeks. On the basis of these data, however, future studies are necessary to obtain a more profound understanding of what happens to blood–brain barrier transporter proteins in individuals with PKU follow-ing long-term changes in Phe and tyrosine levels subsequent to relaxation or resumption of a low-Phe diet.

A cellular system that appears to impact the flux of amino acids through the blood–brain barrier is glial in origin. Evidence suggests that tissue-specific gene expression in brain capillary endothelium is regulated by cells such as astrocytes; glial foot processes cover more than 95% of the endothelium (Pardridge, 1991). The activity of L-system-mediated transporters (the transporter system that moves Phe and tyrosine through the blood–brain barrier) can be modulated by several dis-tinct pathways, including factors derived from glia (Chishty et al., 2002). Thus, abnormalities in glia may affect not only the expression but also the activity of amino acid transporter systems in the blood–brain barrier of PKU mice. In fact, the primary pathology in the PKU brain is glial in origin, i.e., decreased amounts of myelin produced by oligodendroglia and increased numbers of mixed phenotype glia (Dyer and Philibotte, 1995; Dyer etal., 1996). Since mixed phenotype glia are located within the white matter tracts and along the blood vessels (Dyer et al., 2000), it is possible that they influence both the gene expression and the activity of amino acid transporters in the blood–brain barrier.

Brain tyrosine levels remained significantly reduced in the treated PKU mouse brain, and even under these conditions, dopamine still increased to near normal concentrations in both frontal cortex and striatum. The "tyrosine/dopamine" theory, however, predicted that low brain tyrosine levels would prevent increases in dopamine synthesis. Since the above augmentation did not occur, we should investigate the reasons why tyrosine levels failed to regulate dopamine synthesis in the treated PKU brain. The answer may lie in the fact that the studies that examined the sensitivity of prefrontal cortex dopaminergic neurons to various tyrosine levels were performed in otherwise normal brain conditions, i.e., Phe levels were normal (for review, see Tam and Roth, 1997). If Phe levels had been elevated, the results of these studies may have been different. Indeed, elevated levels of radiolabeled Phe added to isolated enzyme preparations (Fukami et al., 1990; Katz et al., 1976) and synaptosomal preparations (Katz et al., 1976), PC12 cells (DePietro and Fernstrom, 1999), and bovine chromaffin cells (Fukami et al., 1990) resulted in the synthesis of DOPA from the radioactive Phe. Importantly, the elevated levels of Phe used in these studies were well within the range found in individuals with PKU. TH and PAH have a very high degree of amino acid homology (Ledley et al., 1987; Dyer et al., 2000; Fitzpatrick, 2000), further strengthening the likelihood of Phe being a substrate for TH in dopaminergic neurons. Thus, in PKU brain, when tyrosine levels are low and Phe levels are elevated, an unlimited supply of substrate for dopamine synthesis appears to exist, i.e., Phe.

Significant increases in myelin-bound protein (MBP) and dopamine levels were observed in the frontal cortex and striatum during the first week the PKU mice were placed on diet. The approximate levels of Phe and tyrosine at the 1-week time point, then, should be good estimates for the concentrations of Phe and tyrosine that were conducive to the observed recovery in these brain structures. In striatum at the 1-week time point, levels of Phe fell to about 240 mμ (about 1.8-fold above control), and in frontal cortex Phe decreased to about 225 μ (about a threefold elevation above control). Tyrosine only rose to 53 and 58% of control in frontal cortex and striatum, respectively, during the first week of the study. These data re-emphasize the point that low brain tyrosine levels do not impede MBP/myelination and dopamine synthesis in the treated PKU brain.

Evidence regarding the molecular mechanism by which Phe includes pathology in the PKU brain has been previously reported. Elevated levels of Phe, in the range detected in untreated PKU mice and humans (about 1200 μ), act as a moderate noncompetitive inhibitor of HMG-CoA reductase, the key regulatory enzyme in the cholesterol biosynthetic pathway (Shefer et al., 2000). Since cholesterol is a major lipid in the myelin membrane and has been implicated in signaling pathways regulating cytoskeleton in the myelin sheath (Dyer et al., 2000), a moderate inhibition of HMG-CoA reductase is likely to have a major impact on the ability of oligodendrocytes to elaborate myelin. Indeed, exposure of cultured immature oligodendrocytes to about 1200 μ Phe resulted in the development of large numbers of mixed phenotype glia (Dyer et al., 1996). Thus, on the basis of these and additional studies (Dyer et al., 2000), it has been postulated that the mixed phenotype glia detected in the hypomyelinated brain structures in the PKU brain are oligodendrocytes that have switched to a non-myelinating phenotype. In fact, mixed phenotype glia are process bearing, do not produce membrane sheaths, and express both myelin proteins (including MBP) and glial fibrillary acid protein (Dyer et al., 2000). These data offer a plausible explanation for what on the surface appears to be conflicting histological and western blot analysis data. Histological analysis showed a paucity of myelin in the frontal cortex (e.g., subcortical white matter and myelin in corpus callosum), whereas western blot data indicated that frontal cortex and striatum of untreated PKU mice contained about 7% of the normal amount of MBP.

MBP and dopamine increases occurred at similar times in the treated PKU mouse frontal cortex and striatum. The fact that the recovery patterns were different in both brain structures suggests that a relationship exists between the two events. In frontal cortex, MBP and dopamine showed a diphasic pattern of increases in week 1 and weeks 3–4, whereas in striatum they both rose to near normal levels in a single leap during week 1 of the study. The reason for the diphasic recovery in frontal cortex is unknown and may be due to different maturation times of (a) oligodendrocytes subpopulations, and/or (b) subpopulations of non-myelinating oligodendrocytes (mixed phenotype glia) switching to myelination oligodendrocytes (Dyer and Philibotte, 1995). Evidence for adult progenitor oligodendrocytes playing a role in myelination of the PKU brain was not previously reported (Dyer et al., 2000). In any case, the data suggest that increases in MBP are associated with myelination since newly formed myelin has been verified by histological examination of treated PKU mouse brain sections. In addition, initial increases in MBP were associated with upregulation of phosphorylated neurofilaments, further confirming the presence of newly formed myelin (Colello et al., 1994; Kirkpatarick and Brady, 1994; Sanchez et al., 1996).

An approximate 20% reduction in the local rate of leucine incorporation into cerebral protein ($ICPS_{leu}$) in the PKU mouse brain vs. control has been reported (Smith and Kang, 2000). Thus, another potential mechanism regulating dopamine synthesis may be that decreased protein synthesis leads to decreased levels of enzymes that produce neurotransmitters, thereby leading to decreased neurotransmitters (Pascucci et al., 2002). Interestingly, brain regions rich in neuronal cell bodies (e.g., pyramidal cell layer of the CA2 region of the hippocampus) had the highest rates of $ICPS_{leu}$ (about 90% of control), and brain regions rich in myelin or synapses showed lower $ICPS_{leu}$ (about 80% of control) (Smith and Kang, 2000). These observations may provide a key insight into the mechanism underlying the $ICPS_{leu}$ decrease in PKU mouse brain. Myelin is a highly metabolically active and extremely large membrane (for review, see Holtzman et al., 1996; Madison et al., 1996; Dyer, 2000). Therefore, in the untreated PKU mouse brain, a significantly less protein synthesis may be expected to occur compared to the control brain, simply because a very large metabolically active membrane is not present.

Moreover, since myelin induces neuronal maturation, the decrease in protein synthesis may also impact neurons/ axons. Therefore, the observed global decrease in protein synthesis detected in PKU mouse brain (Smith and Kang, 2000) may not be surprising.

In summary, the data suggest that tyrosine/dopamine theory is not applicable to the recovering treated PKU brain. In addition to failing to take into account the above-discussed role of Phe as substrate for tyrosine hydroxylase (TH) and the possible aberrant expression pattern of blood–brain barrier amino acid transporters, the tyrosine/dopamine theory ignores the primary neuropathologic findings in untreated PKU brain. The data, instead, support the hypothesis that myelination influences dopamine synthesis. It is reasonable to expect that the observed pathology, i.e., hypomyelination with increased numbers of mixed phenotype glia, has an effect on brain function since pathologic findings generally can be correlated with clinical symptoms. Although the influence of myelin on axonal maturation and action potential conduction has been previously established, the actual molecular mechanisms by which axonal function is enhanced by signals emanating from myelin are unknown. Since (a) dopamine is not the only neurotransmitter that is significantly reduced in the untreated PKU brain, i.e., norepinephrine and serotonin are also reduced (Diamond et al., 1994; Puglisi-Allegra et al., 2000; Pascucci et al., 2002), and (b) myelin induces maturation of axons regardless of the neurotransmitters that they produce, it may be reasonable to speculate that myelin/axonal interactions elicit signals that upregulate all neurotransmitters. Moreover, diseases in which loss of myelin occurs, (such as PKU).

Brain Magnetic Resonance Spectrometry

The clinical significance of white matter alterations in patients with PKU remains obscure, as no correlation to neurologic deficits was found on the basis of results of different neuropsychological and electrophysiologic studies, as well as IQ values. The last finding is not surprising, as the MRI changes seem to depend on "most recent" Phe levels, and the IQ of the patients is predominantly influenced by Phe levels up to the age of 8–10 years. Cortical atrophy was a rare finding, especially in early treated patients, but it was frequent in series with late-treated patients (Thompson et al., 1990, 1993).

MRI results principally collaborate with pathomorphologic examinations revealing "pallor of myelin", frequently present within the parieto-occipital region, the corpus callosum, and the area of the association fibers. They do not confirm the high rate of histochemical changes within the optical tract and therefore do not explain the prolonged latencies of visual evolved potentials described in many patients with PKU (Crearly et al., 1994).

Furthermore, serum S100B protein level, evaluated in PKU patients "off diet," positively correlated with demyelinated foci in their MRI. It was suggested that determination of S100B plasma levels could be a useful peripheral marker of CNS lesions in those patients with demyelinated diseases such as PKU (Schulpis et al., 2004). Additionally, the low plasma antioxidant capacity levels in PKU patients on a "loose diet" may induce DNA oxidation as evidenced by the measured high 8-hydroxy-2-deoxyguanosine concentrations in their sera (Schulpis et al., 2005). Evaluation of the DNA oxidative marker damage in the sera may show an indirect risk for a neurodegenerative process (Schulpis et al., 2005).

It remains unclear on which morphologic alterations the white matter changes are based, as measured by MRI. Furthermore, it is unknown whether Phe is primarily toxic to oligodendrocytes or to neurons/axons, leading to secondary changes in myelin formation, as discussed in hyperphenylalaninemic rats (McDonald et al., 1990).

During the course of a Maternal PKU Study, three untreated women with classic PKU were found, but with normal or nearly normal intelligence. Their brain Phe levels were below the detection limit of proton spectroscopy (< 0.15 mmol L^{-1}). Similar exceptional patients have additionally been reported. Therefore, a series of dynamic magnetic resonance spectroscopy experiments was initialized, to examine the Phe transport kinetics at the blood–brain barrier (Rosenblat et al., 1992).

After a Phe loading test (100 mg kg^{-1} body weight), wide interindividual variations of both the apparent K_t (0.1–1.0 mmol L^{-1}) and the ratio T_{max}/V_{met} (14.0–4.3) were found. The atypical, untreated patients with PKU presented a high $K_{t, app}$ leading to a low brain uptake during the loading test, as well as a low ration of T_{max}/V_{met} indicating a high intracerebral Phe consumption rate. The results show that interindividual differences in brain Phe uptake and consumption are not limited to a few exceptional patients but seem to be "more common". Mutations that affect the function of different amino acid transport systems will influence the $K_{t, app}$ value. The consumption rate will be influenced by different rates of protein synthesis and Phe degradation, for example, by the activity of different hydroxylases, including tyrosine hydroxylase (McDonald et al., 1990).

In addition, most severe white matter abnormalities were observed in patients with small values for $K_{t, app}$ and high ratios of T_{max}/V_{met}. The brain concentrations of the classical metabolites measured by proton spectroscopy, namely N-acetylaspartate, inositol, lactate, and creatinine, were found to be normal. Even the concentration of choline, described to be elevated in acute demyelinating disorders with enhanced membrane lipid turnover, was in the normal range.

Conclusion

The aforementioned data provide evidence that interindividual differences in brain Phe uptake and consumption affect the neurologic/cognitive outcome of patients with PKU. They confirm the low correlation between mutations at the Phe hydroxylase gene and clinical outcome of untreated patients with PKU, as well as the surprisingly high rate of untreated patients with PKU with normal intellectual development.

It is still obscure whether measurements of kinetic parameters in adulthood provide an appropriate characterization of the situation during childhood. Including the limitation of the method so far, no critical brain concentration that could justify a relaxation of diet has been found.

Results of T_2 relaxometry indicate a "dysmyelination", as originally described by Hommes and Matsuo (1987) in the hyperphenylalaninemic rat, namely that the decreased synthesis of sulfatides and other myelin compartments is associated with an increased myelin turnover, leading to disruption splaying of myelin lamellae associated with increased water content.

Studies of the PAH[enu2] PKU mouse indicate a regional Phe sensitivity of oligodendrocytes, as a result of a variable inhibition of the key regulatory enzyme in cholesterol biosynthesis, 3-hydroxy-3-methyl-glutaryl-CoA-reductase. The regional distribution pattern of white matter changes might be additionally explained by different Phe-uptake constants of different brain areas as suggested by positron emission tomography studies in men.

In summary:

- The MRI studies indicate that Phe is a lifelong "toxin" for myelogenesis;
- The white matter changes are reversible (Bick et al., 1993), independent of the patient's age, and affect brain areas of short as well as prolonged cycles of myelination;
- The clinical significance of white matter alterations in patients with PKU remains obscure. The published data do not provide unequivocal information for the question of whether adolescent and adult patients should stay on a strict diet.

Efficiency of Long-Term Tetrahydrobiopterin Monotherapy in Phenylketonuria

Phenylketonuria, an inborn error of phenylalanine metabolism, occurs with a frequency of about 1 in 10,000 births and is treated with a strict dietary regimen. Recently, some patients with PKU have been found to show increased tolerance towards phenylalanine intake, while receiving tetrahydrobiopterin (BH_4) supplementation. We have treated two infants with BH_4-responsive PKU with BH_4 for more than 2 years. No additional dietary control was required to maintain blood phenylalanine concentrations in the desired range. Both children have shown normal development. Generally, these results suggest that BH_4 treatment might be an option for some patients with mild PKU, as it frees them from dietary restriction and thus improves their quality of life.

Brain Abnormalities in Maternal PKU

Dent (1957), Allan et al. (1963) and Mabry et al. (1966) described the maternal phenylketonuria (MPKU) syndrome nearly 50 years ago, but Lenke and Levy (1980) forcefully brought the syndrome to medical attention only in 1980. They reported on the offspring of more than 500 pregnancies and documented the rate of microcephaly in those offspring at 75%, mental retardation at over 90%, and congenital heart disease (CHD) at 12–13%. In various studies done in later years, MPKU syndrome has also been observed to be associated with other teratogenic effects such as facial dystrophic features mimicking fetal alcohol syndrome, intrauterine growth retardation, and malformation of other organs. Severity of syndrome has been roughly linked to maternal blood levels of phenylalanine (Lee, 2003; 2005). Collaborative studies have shown that this syndrome is preventable, if pregnancies are managed with strict dietary control of blood phenylalanine and good nutrition. The authors have recommended levels of blood phenylalanine during pregnancy range between 120–360 µmol L^{-1} (Koch et al., 1994).

The abnormalities of brain in untreated PKU patients are largely abnormalities of white matter (Huttenlocher, 2000), among other less prominent findings. One untreated adult had extensive neuronal losses in the lateral geniculate nucleus visual cortex and hippocampus (Kornguth et al., 1992), but this seems to be an anomaly, as magnetic resonance spectroscopy revealed no evidence of severe neuronal damage. MRI studies of individuals with PKU and MPKU, treated or untreated, found white matter abnormalities and hypoplasia of corpus callosum (Cleary et al., 1995), and some of these findings may be reversible with appropriate diet (Cleary, 1995) but may occur in the fetus despite an appropriate diet during gestation (Smith et al., 1979). Cerebroside and sulfatide were decreased on biopsy of one untreated three-year-old child's forebrain.

Nontheless, despite these clinical studies, the specific features of brain from human fetuses exposed to MPKU have not been extensively studied, and we have been unable to locate neuropathologic descriptions of the offspring of mothers with untreated PKU during pregnancy (Brody et al., 1987).

Disorders of Homocysteine Metabolism

Clinical Features

Patients with this most common form of homocystinuria show evidence of involvement of the eye, the skeletal system, the vascular system, and the brain. It is important to note that individuals with cystathionine β-synthase deficiency do not manifest any abnormalities at birth and that the affected pregnancies are uneventful. Thus, this disorder, as opposed to the more rare remethylation defect variants of homocystinuria (described below), is not usually part of the differential diagnosis of the catastrophically ill newborn. Ectopia lentis does not usually appear before the age of 3 years, but most patients have some manifestations by the age of 10. The initial recognition of ocular abnormalities may be an observation by parent or physician that the iris shakes, when the head is moved rapidly. While a predilection for downward dislocation of the lens seems to exist, this is not invariant. In homocystinuria, the defect is the result of thickening and fragmentation of the zonular fibers that attach the lens to the ciliary body, while in Marfan's syndrome these fibers are thin and elongated.

Osteoporosis is the most common abnormality of the skeleton, presented usually after the first decade. Because of the disordered bone tissue formation, the physician can expect, and indeed will find, scoliosis, genu valgum, and pes cavus, among other abnormalities.

Mental retardation is reported in about half of the patients, and it is infrequent for this problem to be the reason that medical assistance is first sought. Psychomotor delay may be perceived as early as the first year of life, but it may not be appreciated until later, since retardation is usually slowly progressive. Nevertheless, mental dysfunction is not the hallmark of this disease, and many patients are college graduates. Seizures occur in about to 10–15% of patients.

The complication of cystathionine β-synthase deficiency that is of most concern is the prodensity to thromembolism. This involves vessels of all diameters and is unpredictable as to when, where, and if it occurs. The malar flush and erythemous mottling of the extremities are also vascular manifestations of homocystinuria.

There is a wide range of clinical manifestations in these individuals, and with heightened awareness of the disorder patients are being diagnosed who have lens abnormalities only. Still others are apparently normal siblings of known patients who on screening and subsequent quantitation manifest increased levels of homocystine in serum and urine (Valle et al., 1980; Schulman et al., 1980) (Table 20.1).

Table 20.1 Comparison of clinical and biochemical features in three forms of Homocystinuria

Feature	Cystathionine β-Synthase Deficiency	Defective Cobalamin (B_{12}) Coenzyme synthesis
Mental retardation	Common	Common
Growth retardation	No	Common
Dislocated optic lenses	Almost always	No
Thromboembolic disease	Common	No
Megaloblastic anemia	No	Rare
Homocysteine in blood and urine	Increased	Increased
Methionine in blood and urine	Increased	Normal or decreased
Cystathionine in blood and urine	Decreased	Normal or increased
Methylmalonate in blood and urine	Normal	Increased
Serum cobalamin	Normal	Normal
Serum folate	Normal or decreased	Normal or increased
Response to vitamin	Pyridoxine	Cobalamin (B_{12})
Response to dietary methionine restriction	helpful	Harmful

(Re-created with major modifications from Bondy PK, Rosenberg LE, Metabolic Control and Disease, 8th edn, Saunders, Philadelphia, 1980)

Biochemical Defects

Methionine is an essential amino acid with a unique role in the initiation of protein synthesis. In addition, by conversion to S-adenosylmethionine, it serves as the major methyl group donor involved in the formation of creatinine and choline, in the methylation of bases in RNA, and as the source of the aminopropyl group in the formation of polyamines. Finally, in relationship to classical homocystinuria, it is converted by way of homocysteine and cystathionine in a series of reactions termed as the *transsulfuration pathway* (Fig. 20.3).

The major steps in this pathway are shown in the figure. In the fist step, S-adenosylmethionine is formed in a reaction catalyzed by methionine adenosyltransferase. This reaction involves transfer of the adenosyl partition of adenosine triphosphate (ATP) to methionine, forming a sulfonium bond, which has a high group transfer potential, i.e., it is a so-called high-energy compound. Hence, each of the groups attached to this bond can participate in a transfer reaction, much as ATP does in so many reactions within the cell.

Homocysteine lies at a metabolic crossroad; it may condense with serine to form cystathionine, or it may undergo remethylation, thereby conserving methionine. There are two pathways for remethylation in humans. In one, betaine provides the methyl groups, while in the other 5-methyltetrahydrofolate is the methyl donor. This latter reaction is catalyzed by a B_{12}-containing enzyme, 5-methyltetrahydrofolate homocysteine methyltransferase. Two defects in this latter mechanism may account for the inability to carry out remethylation. In one of them, patients are unable to synthesize or accumulate methylcobalamin, while others cannot produce the second cofactor, 5-methyltetrahydrofolate, because of a defect in 5,10-methylenetrahydrofolate reductase.

As noted above, cystathionine formation is the other major fate of methionine. The condensation of homocysteine with serine is catalyzed by the vitamin B_6-requiring enzyme cystathionine β-synthase. In the last step of the transsulfuration sequence, cystathionine undergoes cleavage to cysteine and α-ketobutyrate in yet another enzyme reaction that requires pyridoxal phosphate.

Since methionine has several pathways open to it, it is essential to know what factors control the direction that its metabolism takes. Studies in young adults have shown that the utilization of methyl groups is normally accounted for chiefly by creatinine formation. This reaction consumes more S-adenosylmethionine than all other transmethylations together. However, examination of enzyme activities from these two pathways in fetal animals leads to the conclusion that remethylation preponderates over transsulfuration. Indeed, since γ-cystathionase activity is immeasurable in human fetal liver and brain, not only is the remethylation sequence favored, but also cysteine then becomes an essential amino acid for the fetus and infant.

The predominant cause of homocystinuria is, as already mentioned, the absence of cystathionine β-synthase, a dimeric enzyme possessing two identical subunits in humans. Several lines of evidence indicate that multiple mutations may affect this enzyme. There appear to be at least three distinct types of abnormal enzyme

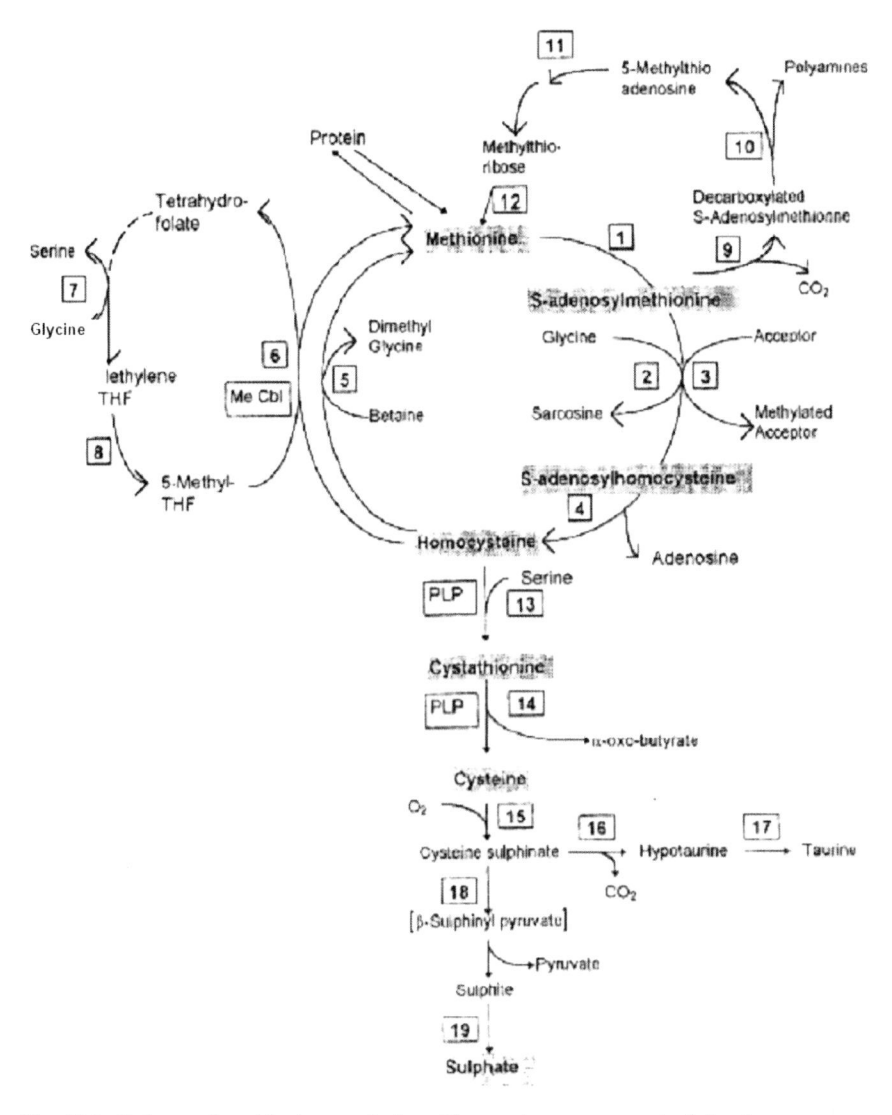

Fig. 20.3 Pathway of methionine metabolism. The numbers represent the following enzymes or sequences: (1) methionine adenosyltransferase; (2) S-adenosylmethionine-dependent transmethylation reactions; (3) glycine methyltransferase; (4) S-adenosylhomocysteine hydrolase: (5) betaine-homocysteine methyltransferase; (6) 5-methyltetrahydrofolate:homocysteine methyltransferase; (7) serine hydroxymethyltransferase; (8) 5,10-methylenetetrahydrofolate reductase; (9) S-adenosylmethionine decarboxylase; (10) spermidine and spermine synthases; (11) methylthioadenosine phosphorylase; (12) conversion of methylthioribose to methionine; (13) cystathionine β-synthase; (14) cystathionine γ-lyase; (15) cysteine dioxygenase; (16) cysteine suplhinate decarboxylase; (17) hypotaurine:NAD+oxidoreductase; (18) cysteine sulphinate:α-oxoglutarate aminotransferase; (19) sulfine oxidase. MeCbl = methylcobalamin; PLP = pyridoxal phosphate

associated with homocystinuria: absent enzyme activity, reduced enzyme activity with normal ability to bind to cofactor pyridoxal phosphate, and reduced activity with diminished ability to bind the coenzyme.

As the name implies, renal clearance of abnormal levels of homocystine in the plasma causes excessive excretion of the amino acid in the urine. In cystathionine β-synthase deficiency, plasma methionine concentrations are elevated as well – this serves as a point of distinction from the remethylation defects. At present, it appears that the pyridoxal phosphate response may be explained by the fact that this vitamin increases the steady-state concentration of the active enzymes by decreasing the rate of apoenzyme degradation and possibly by increasing the rate of holoenzyme formation. The explanation is not entirely satisfactory, however, since in vitro studies have shown detectable levels of enzyme activity in mutant fibroblasts that have no B_6 response, while in other mutant lines without detectable enzyme activity, B_6 response has occurred. Once again, a distressing lack of correspondence between in vivo observations and in vitro experiments forces investigators to probe the secrets of these diseases more deeply.

Pathophysiology

Research into the toxicity of methionine and homocysteine have uncovered a wide range of toxic effects on various organs. Nonetheless, a unifying hypothesis as to the deleterious mechanism of action of these amino acids has not yet emerged.

Evidence demonstrating that homocysteine interferes with formation of the normal cross-links of collagen has accumulated. Most collagen molecules are composed of three polypeptide chains bound together by intramolecular cross-links. Cross-link formation requires the generation of aldehyde groups by oxidation of ε-amino groups of the multiple lysyl or hydroxylysyl residues in the collagen monomers. Chemical interactions of these aldehyde groups, both by Schiff-base formation with amino groups of lysine or hydroxylysine on other chains and by Aldon condensation between two aldehydes, account for the cross-linking. How cross-link formation is disturbed in homocystinuria is not yet clear, but in vitro homocysteine forms stable thiazene ring compounds with aldehydes, thereby blocking the possibility of forming cross-links.

Hankey and Eikelboom (1999) demonstrated a direct in vivo toxic effect of homo-cyteine on the intima of large and medium-sized arteries in baboons subjected to a constant homocysteine infusion. Platelets then adhered to the disrupted intimal surface, eventually resulting in thrombosis of the vessel. As measured by platelet turnover techniques, dipyridamole (an agent effective in prevention of platelet aggregation) was capable of preventing thrombosis, but not the intimal damage. Platelet survival studies in homocystinuric patients gave comparable results. However, these data require further supporting documentation before dipyridamole can be used as an acceptable form of treatment in vitamin B_6-unresponsive homocystinurics (Mudd et al., 1989).

Laboratory Diagnosis

Elevated homocysteine (Hcy) blood levels but also the presence of Homocystine in urine is the biochemical hallmark of these disorders and can be detected by a positive urinary cyanide nitroprusside reaction. As other disulfides, including cystine and β-mercaptolactate cystine, also react, amino acid paper thin layer chromatography will be required to distinguish these compounds from homocystine. Since there are two other forms of homocystinuria (discussed later) in which plasma methionine is decreased, plasma amino acid evaluation is also in order.

Deficiency of cystathionine β-synthase may be demonstrated in cultured fibroblasts, but it is not essential to the clinical diagnosis (Stabler et al., 1988).

Treatment

Treatment in those patients who respond to pyridoxal phosphate is based on provision of approximately 250–500 mg per day. Folate deficiency may be avoided by the addition of 5 mg per day to this regimen. In patients who do not respond to the coenzyme, dietary manipulation with a low methionine, cystine-supplemented diet may be helpful. Dipyridamole, an agent effective in decreasing platelet aggregation, has been used to prevent the thromboses, but there are no reports yet as to its effectiveness.

The therapy for this disorder is often unsatisfactory and requires management and follow-up in a center equipped to deal with inborn errors of metabolism. Therefore, having made the diagnosis in a particular individual, it is essential that such children be followed periodically at a center. Since anesthesia increases the risk of intravascular coagulation, elective surgery under general anesthesia is absolutely contraindicated, as is angiography.

Defects in Homocystine Remethylation

Defective Activity of N[5]-Methyltetrahydrofolate: Homocysteine Methyl-Transferase and Cobalamin Activation

Patients have been described with this defect in remethylation of homocysteine. The initial patient presented a catastrophically ill newborn who died when 7 weeks old. While homocystine levels were elevated in blood and urine, the level of methionine in blood was quite low. Of great interest was the presence of large amounts of methylmalonic acid in urine. Three other patients presented later in childhood, two of whom were retarded mentally. One of these patients also had severe megaloblastic anemia (Doscherholmen and Hagen, 1957).

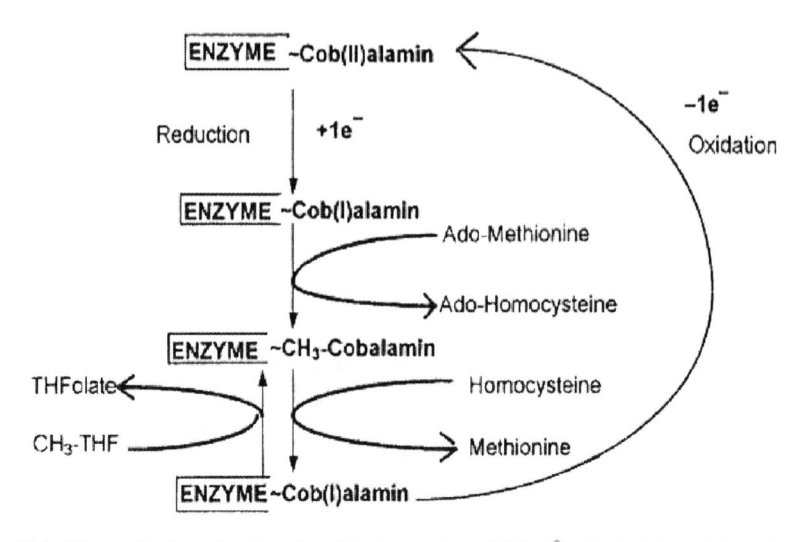

Fig. 20.4 The mechanism of action of methionine synthase. THF = tetrahydrofolate; Ado = adenosyl

Since the coenzyme from vitamin B_{12} is required in two distinct enzyme reactions, i.e., remethylation of homocystine and catabolism of methylmalonic acid, the fundamental defect must involve a step in converting B_{12} to its coenzymes. Formation of both deoxyadenosyl B_{12} and methyl B_{12} requires a prior reductive step catalyzed by cobalamin reductase, which appears to be the defective enzyme in this variant (Hogervorst et al., 2002) (Fig. 20.4).

Decreased $N^{5,10}$ Methylenetetrahydrofolate Reductase Activity

Variability of presentation characterizes this homocysteine remethylation defect as well. Several patients have been reported from one family: one with seizures and muscle weakness, a second with schizophrenia and retardation, and a third who was asymptomatic. Subsequently, patients with a more malignant neonatal presentation were reported, and hence both variants should be part of the differential diagnosis of "sepsis neonatorum".

Fibroblasts from each patient had markedly reduced levels of the above enzyme. The defect results in an inability to synthesize 5-methyltetrahydrofolate in amounts sufficient for the remethylation of homocystine to methionine. Homocystine accumulates in plasma, and the plasma methionine is decreased. As in the other remethylation defect, there is accumulation of cystathionine. Treatment with high doses of folic acid has been beneficial in several patients (Mudd et al., 1989).

Animal Studies

Evidence from the literature indicates that Hcy is an important risk factor for cognitive dysfunction (Miller 2003a,b). It has recently been suggested that chronic administration of Hcy to the rats affected both long- and short-term memory (Streck et al., 2004) and it was proposed that hyperhomocysteinemia causes memory impairment by the enhancing oxidative stress.

Apart from enhanced oxidative stress, many factors can impair cognitive functions, one of them being the altered expression pattern of neural cell adhesion molecules (NCAM) in hippocampus. NCAM are likely candidate molecules that participate in synaptogenesis in neuronal plasticity. Furthermore, recent evidence strongly suggests that they also participate in synaptic changes underlying memory formation in adult individuals (Schachner, 1977). Three major forms of NCAM generated by alternative splicing are found: NCAM 120, 140, and 180 kDa. These molecules are, in addition, involved in cellular migration, axonal growth, and regeneration of peripheral axons (Walsh and Doherty, 1997).

In that study, it was suggested that chronically giving methionine in drinking water significantly increased plasma Hcy levels, while daily administration of melatonin through the experiment partly prevented the increase of Hcy induced by methionine. This is in accordance with previous findings indicating that melatonin has the ability to reduce plasma Hcy levels (Baydas et al., 2002a–c; 2003). These results are in aggrement with previous studies demonstrating that Hcy can increase oxidative stress, possibly through its autoxidation and inhibition of GSH-Px and other antioxidant enzymes and glutathione (GSH) (Hankey and Eikeboom, 1999; Heinecke 1988; Stamler et al., 1993; Upchurch et al., 1997). Moreover, Hcy is able to prevent direct inactivation of NO by superoxide radicals generated during its autooxidation, which leads to the formation of peroxynitrite radicals; this may cause further oxidative injury (Upchurch et al., 1997).

Melatonin significantly reversed the Hcy-induced oxidative stress in the hippocampus. An established set of literature suggests that melatonin plays a significant neuromodulatory and neuroprotective role, which includes the modulation of synaptogenesis, protection against apoptosis, and antioxidant properties (Baydas et al., 2002c; Reiter, 1998). In accordance with these results, it was demonstrated by Baydas et al. (2003) and others (Osuna et al., 2002) that melatonin has the ability to reduce oxidative-stress-induced by hyperhomocysteinemia in different ways. First, it scavenges a number of reactants including the hydroxyl radical and hydrogen peroxide generated during Hcy autooxidation. Second, it may be capable of reducing peroxynitrite generation owing to its ability to scavenge NO and peroxynitrite (Zhou et al., 2004). Third, melatonin increases the activation of many antioxidant enzymes including GSH-Px, which can be inhibited by hyperhomocysteinemia (Baydas et al., 2003; Osuna et al., 2002).

In addition, it was found that chronic hyperhomocysteinemia impaired performance of rats in a water maze task and the retention of long memory in a passive avoidance task. Furthermore, Hcy in high levels may impair spatial navigation.

Consistent with these previous behavioral findings (Fukui et al., 2002), it was shown that decline in learning and memory is associated with a very significant increase in the levels of lipid peroxidant on lipid peroxidation (LPO). Further analysis revealed a correlation between time to reach the hidden platform on day 5, which indicates the learning performance, and the levels of Hcy and the activity of GSH-Px in hyperhomocysteinemic rats. Levels of plasma Hcy and concentration of hippocampal LPO are positively correlated with learning deficits, while negatively correlated with the activity of GSH-Px. Furthermore, both oxidative stress and deficits in learning and memory were prevented by treatment with melatonin, suggesting that oxidative stress was probably involved in the Hcy-induced cognitive deficits. Accumulating evidence (Fukui et al., 2002; Parle and Dhingra, 2003) has indicated that treatment with a variety of antioxidants partially reversed increase in markers of oxidative stress and decline in learning and memory. These results are in agreement with the findings of Streck et al. (2004), who reported that experimental hyperhomocysteinemia causes cognitive dysfunction because of its free-radical-generating actions.

The above study in animals demonstrated that melatonin has potential effects to prevent Hcy-induced learning and memory deficits and to modulate neural plasticity. The exact mechanism of melatonin in preventing learning and memory deficits is still in debate. Shen et al. (2002) have postulated that melatonin's ability in improving cognitive functions is related to its antioxidant action. In addition, El-Sherif et al. (2003) have suggested that melatonin may modulate specific forms of plasticity in hippocampal neurons. NCAM play a role in neural plasticity in the adult vertebrate brain and are strongly expressed in the hippocampal cells (Ronn et al., 1995), which are included in the neural circuit for processing cognitive information (Squire, 1992). Synaptic plasticity is very sensitive to environmental factors and, accordingly, synaptic dysfunction and degeneration occur early in the pathogenesis of several different neurological disorders (Mattson, 2000). It is not known if and how Hcy changes synaptic plasticity, although findings suggest that elevated Hcy levels might alter synaptic function (Christie et al., 2005).

An emerging body of literature supports the concept that a high serum homocysteine level may be a risk factor for Alzheimer's disease (AD) (Clarke et al., 1998; Seshadri et al., 2002), but the mechanisms involved have not been discovered. It has been proposed that high serum homocysteine levels may increase the risk of dementia by additive neurotoxicity (Ho et al., 2001; Kruman et al., 2002), dysregulation of gene expression (Fuso et al., 2005), or vascular damage (McCully, 1996). In this study, it was investigated whether high serum homocysteine level could modulate specific markers of AD pathology, such as increased brain levels of amyloid β-peptides (Aβ), considered one of the main abnormalities in AD.

New evidence suggests that hyperhomocysteinemia may specifically contribute to AD pathology. Correlation exists between serum homocysteine levels and brain Aβ levels, but only in female mice. The increases in Aβ10 and Aβ42 levels in APP*/PS1*/CBS* female mice as compared to APP*/PS1* mice were both significant. The mechanism leading to these gender differences is at this time unknown, even though the effects of sex hormones are obviously an important topic to be investigated (Takeuchi et al., 2000; Sai et al., 2002).

Furthermore, in vitro classical homocystinuria showed an elevation of rat hippocampal acetylcholinesterase (AChE) activity, whereas the other types of homocystinuria had no effect on the enzyme (Schulpis et al., 2006). These preliminary results may lead to the suggestion that homocystinuria due to cystathionine-6-synthase deficiency may partly implicate high Hcy levels with neurodegeneration.

Homocysteine is thought to cause oxidative-stress-related toxicity, apparently mediated by its highly reactive thiol group (Lipton et al., 1997). Several studies have shown an increase in the Girp78 ER stress protein, a marker for cell stress, in hyperhomocysteinemic models (Kokame et al., 1996). They examined the level of GRP78 in the brains of the mice used in that study, but found no differences in GRP78 expression between the female mice. It should be noted that it was a preliminary study and that additional markers of stress should be investigated in the future. Unfortunately, as that study was in progress, the laboratory infrastructure was severely affected by the hurricane Katrina with loss of the tissue bank and animals, which hindered the assessment of additional markers of stress. GFAP immunohistochemistry and unbiased stereological studies showed no evidence of gliosis or cell loss, respectively, in male or female hyperhomocysteinemic mice.

In that model, there were no changes in β-secretase activity or β-secretase expression. It only examined the whole tissue homogenates, and therefore effects that could result from redistribution of the secretase within subcellular compartments such as lipid rafts, may not have been appreciated. In addition, several important aspects of APP and Aβ metabolism remain to be investigated, including the degradation rate of Aβ, clearance, and alpha and gamma secretase activities.

It has been observed that hyperhomocysteinemia can upregulate HMG-CoA reductase gene expression in the liver (Woo et al., 2000) and stimulate cholesterol biosynthesis leading to high cholesterol concentration in the serum. Therefore, it was also important to measure serum cholesterol in APP*/PS1* and APP*/PS1*/CBS* mice since high cholesterolemia results in increased Aβ levels in the brain (Refolo et al., 2000). Of interest, no changes in cholesterol serum levels between these two groups were found. Levels of apoE expression in the brain of APP*/PS1*/CBS* versus APP*/PS1* mice were examined, because increased apoE expression may be mechanistically related to increased Aβ levels in hypercholesterolemic mice (Refolo et al., 2000). We found no significant differences in apoE expression among the experimental groups. This suggests that neither serum cholesterol nor apoE levels mediate homocysteine-induced increased in Aβ levels.

Another aspect to be examined in future studies is the impact of homocyteine levels on amyloid load. As this was a preliminary investigation, mice were sacrificed at 3.5 months because of the higher attrition rate occurring after 4 months of age. At this age, the animals showed only very sparse amyloid deposits, which were insufficient for evaluation of amyloid load. Another potential limitation of this study is that we evaluated mice resulting from early crossings. However, it is highly unlikely that this factor affected our results, as there were clear correlations (i.e., dose response effect) between homocysteine levels and Aβ levels. Likewise, it would be highly unlikely that early crossing or background strain effects could explain the observed gender differences.

In summary, individuals with elevated serum homocysteine are at increased risk of sporadic AD, but the mechanisms are controversial. This study shows that high serum homocysteine levels in female mice correlate with higher Aβ peptide levels in the brain. Since Aβ plays a central role in the pathogenesis of AD, the results advance a plausible mechanism underlying the higher risk for AD in hyperhomocysteinemia. On the basis of these results, future epidemiological studies should place emphasis on the gender differences noted.

Neuropsychologic Disturbances

Homocysteine is derived from the demethylation of the amino acid methionine (Fig. 20.4). It is metabolized by remethylation, using folate and Vitamin B_{12} as cofactors, or by trans-sulfuration, using Vitamin B_6 as a cofactor (Finkelstein, 1998). A substitution at nucleotide 677 (C-to-T transition) in the gene encoding for the enzyme methylenetetrahydrofolate reductase (MTHFR) has been identified (Kang et al., 1991) and seems to be a major genetic cause contributing to reduced enzyme activity and mildly elevated plasma homocysteine levels (Engbersen et al., 1995). Several lines of evidence suggest that increased plasma homocysteine levels or MTHFR C677T may be a risk factor for cardiovascular and cerebrovascular diseases (Clarke et al., 1991; Selhub et al., 1992 and 1995), cognitive decline, dementia, and depression (McCaddon et al., 2001; Seshadri et al., 2002; Dufouil et al., 2003; Religa et al., 2003; Bjelland et al., 2003).

Late-onset major depressive disorder (MDD) is associated with a high prevalence of brain MRI hyperintensities or infracts (Krishnan et al., 1997). Vascular mechanisms or vascular pathology have been suggested as predisposing or precipitating factors in some forms of late-life MDD (Alexopoulos et al., 1977). Theoretically, the association between higher plasma homocysteine or MTHFR C677T polymorphism and depression in older age groups may be related to cerebrovascular pathology, although this has not been examined in detail.

Elevated homocysteine levels in late-onset MDD are consistent with the results of a recent community cohort of 5948 subjects comprising two subgroups based on age: 46–49 years and 70–74 years (Bjelland et al., 2003), and a clinical study (Bottiglieri et al., 1994 and 2000). There are plausible biological mechanisms whereby higher homocysteine levels might cause depression. Since higher plasma homocysteine levels have been associated with silent brain infracts or MRI hyperintensities, which are themselves associated with an increased risk of late-onset MDD, (Hogervorst et al., 2002; Vermeer et al., 2002), vascular mechanisms may mediate the association. There was found to be a significant correlation between homocysteine levels and MRI hyperintensities, and MRI hyperintensities were more common in late-onset depression than in comparison subjects. However, the logistic-regression analysis failed to establish that cerebrovascular disease, as identified by MRI hyperintensities, plays a mediating role. This suggests that elevated homocysteine levels contribute to the presence of late-onset MDD through non-vascular mechanisms.

Possible non-vascular mechanisms underlying the association between homocysteine levels and late-onset depression include dysregulation of one-carbon metabolism and N-methyl-D-aspartate (NMDA) receptor-related neurotoxicity. The one-carbon cycle is responsible for the synthesis of methyl groups, which are ultimately utilized by S-adenosylmethionine (SAMe) in several transmethylation reactions involving the production of phospholipids, nucleotides, and monoamine neurotransmitters including serotonin, norepinephrine, and dopamine (Bottiglievie et al., 2000), which may be involved in the pathogenesis of depressive disorders. Homocysteine is involved in the one-carbon cycle. Elevated homocysteine might indicate disruption in the synthesis of methionine, an immediate precursor of SAMe. The other possible non-vascular mechanism is the excessive production of two endogenous agonists (i.e., homocysteic acid and cysteine sulfonic acid) of NMDA receptors, which results in elevated homocysteine-induced NMDA-mediated excitotoxicity (Kim and Pae, 1996). Homocysteine also markedly increases the vulnerability of hippocampal neurons to excitotoxic and oxidative injury (Kruman et al., 2000). Neuronal death through apoptosis may lead to neuropsychiatric diseases, including depression.

Patients with late-onset MDD reportedly have increased risk of developing dementia (Jorm, 2000). Three recent population-based studies examined the relationships between homocysteine levels and cognitive functioning, although depression was not systematically studied. These studies consistently indicated that elevated homocysteine levels were associated with decreased cognitive performance and suggested that vascular mechanisms do not mediate the association between homocysteine and cognition, leading to an indirect support to the possible link between elevated homocysteine levels and AD (Dufouil et al., 2003; Prins et al., 2002; Wright et al., 2004). Restriction of MMSE scores for study inclusion criteria limited the ability to examine this possible association. Nonetheless, the findings from studies suggest that elevated homocysteine may be a possible common etiologic factor for both late-onset MDD and dementia. A subgroup of patients with late-onset MDD and high homocysteine levels may be more likely to develop subsequent dementia, primarily of the Alzheimer type. Longitudinal studies that measure homocysteine levels, depression, and cognitive performance over time are needed to address this issue.

The MTHFR genotype did not differ between the patient and comparison groups. Lack of association between the MTHFR C677T genotype and depression is consistent with the results of one study (Kunugi et al., 1998), but not the other two (Bjelland et al., 2003). MTHFR polymorphism has been linked to the metabolism of homocysteine and appears to be associated with ischemic brain infarction (Kelly et al., 2002) and coronary heart disease (Klerk et al., 2002). Lack of association of this vascular-related MTHFR genotype with late-onset MDD is not in accordance with the hypothesis that the same genetic loci are responsible for depression and cerebrovascular disease. One study examining the other three vascular-related genes (APOE, VLDL-R, and DCP-1) also found no association

(Cervilla et al., 2004). Nevertheless, the absence of association does not exclude the possibility that late-onset MDD and alternative forms of the gene are associated. In addition, given the small anticipated effect size associated with any single candidate polymorphism, the sample sizes required to robustly demonstrate an association will be large. The possibility that this negative association between MTHFR C677T polymorphism and late-onset MDD in this study was due to Type II error (false negative) cannot be excluded.

Furthermore, the criterion for age at onset in late-onset depression ($\geq$ age 50) was somewhat arbitrary, but is consistent with the age cutoff used in other reports (Steffens and Krishnan, 1998). Approximately 70% of the patient group had brain MRI hyperintensities, which is consistent with most, but not all, studies in the literature (Greenwald et al., 1996; Lenze et al., 1999). There were other methodological limitations. Although there was no correlation between homocysteine levels and clinical depressive features, given the heterogeneous nature of late-life depression, further investigation of a more homogenous study sample might provide additional information about this issue. Moreover, since higher homocysteine levels may occur in elderly persons and in those with cognitive impairment (McCaddon et al., 2001), it was attempted to minimize the confounding effect of these factors by excluding subjects with cognitive impairment and by statistically controlling for the age effect. Therefore, not surprisingly, the association between elevated homocysteine and late-onset MDD was unaffected by cognitive impairment and age. Requiring high MMSE scores for study inclusion decreased the likelihood of observing significant associations between MMSE scores and homocysteine levels. Hence, even though the correlations between MMSE scores and homocysteine levels did not reach significance, it remains possible that the patients with high homocysteine levels are at greater risk of developing cognitive decline or AD on follow-up. Further studies are needed to replicate these findings, and longitudinal ones to address the possibility that a subset of patients with late-onset depression have high homocysteine levels that increase their risk of developing AD (Rabins et al., 1991).

Moreover, we very recently reported that erythrocyte membrane AChE activity in MTHFR C677T carriers was found significantly higher when Hcy levels in their blood were increased. In contrast, the erythrocyte membrane enzyme activity was restored nearly to normal when the individual Hcy levels decreased. The AChE activation observed may be implicated in cholinergic mechanisms (e.g., vasodilatation, neurological dysfunction) through modulation of acetylcholine production (Schulpis et al., 2005).

Adults with elevated total serum homocysteine levels (tHcy) are at increased risk of cardiovascular disease, stroke, and dementia (Seshadri et al., 2002). Subjects with elevated tHcy are more cognitively impaired and have more severe white matter disease on neuroimaging studies than subjects with normal tHcy (Hogervorst et al., 2002). Even after controlling for cerebrovascular-disease-promoting effects of tHcy, elevated tHcy is associated with an increased risk of cognitive impairment and dementia (Dufouil et al., 2003).

The mechanisms by which elevated tHcy increases the risk of cognitive impairment and dementia are not clear. One possibility is that homocysteine acts synergistically with β-amyloid or other cellular toxins to promote AD and/or microvascular pathology. Alternatively, it may have independent effects on cognitive function, which add to the severity of other cognitive disorders.

The biochemically confirmed case of adult-onset hyperhomocystinemia due to a vitamin B_{12} metabolic defect, cobalamin (ChlB) disease (Roze et al., 2003), has been reported. This case is unique in its presentation with primarily cognitive deficits, without other progressive neurologic signs. The primary cognitive deficit associated with combined hyperhomocystinemia and methylmalonic academia in this patient is executive dysfunction. This is supported by the patient's improvement in cognitive and functional status with tHcy-lowering medications, in parallel with a decrease in cortical white matter hyperintensity on his MRI scan. Executive function encompasses aspects of attentional control, goal setting, and cognitive flexibility. The observed improvement on all three components of the DKEFS Trails Test suggests that the patient's attention and cognitive flexibility (possibly secondary to the improved attention) were affected by the tHcy-lowering therapy (Miller et al., 2002).

The mechanisms by which elevated tHcy increases the risk for development of dementia are not clear (Seshadri et al., 2002). Although CblC disease is a rare genetic cause of combined hyperhomocystinemia and methylmalonic academia (MMA), this case demonstrates a distinct relationship between tHcy and MMA levels and cognitive function. Although a contribution from the patient's elevated MMA level to his cognitive deficit cannot be omitted, population studies suggest that tHcy is more relevant for determining cognitive status than MMA (Nilsson et al., 2000). The earlier onset and more severe neurologic deficits in the patient's sister are consistent with earlier reports of variable expressivity of deficits in CblC disease (Roze et al., 2003; Powers et al., 2001) and suggest that there are likely to be other genetic (or epigenetic) factors that can mitigate the neurologic deficits associated with this disorder.

Even in nondemented elderly subjects, tHcy levels are strongly correlated with frontal/executive function (Garcia et al., 2004), impairment of which increases the risk of progression to dementia (DeCarli et al., 2004). As frontal/executive function is often related to the degree of periventricular white matter damage (Tullberg et al., 2004), and as the latter is correlated with tHcy levels, (Hogervorst et al., 2002), tHcy may be an important mediator of executive dysfunction in subjects with cognitive impairment. After controlling for possible effects of low B-vitamin status, one recent study suggested that the white matter damage that is correlated with elevated tHcy in AD is caused by microvascular damage (Miller et al., 2002). Such microvascular damage is a known consequence of elevated tHcy due to CblC disease and is not likely to be reversible (Powers et al., 2001). This case suggests that reversible pathologic processes also contribute to homocysteine-induced cognitive dysfunction. As S-adenosyl methionine is an important cofactor for myelin synthesis, one possible reversible mechanism by which elevated tHcy may cause executive dysfunction is through demyelination (Roze et al., 2003). This case provides further support for trials of tHcy-lowering agents in adults with cognitive impairment, as tHcy-related executive dysfunction may be at least partially reversible.

Increasing evidence suggests that white matter hyperintensity lesion burden detected on MRI represents small-vessel disease (Dufouil et al., 2003), increases the risk of stroke, and is associated with cognitive impairment and dementia (Morris et al., 2001). Vascular risk and factors such as hypertension (Miller et al., 2003a–c) and, to a lesser extent, diabetes (McCaddon, 2001, 2003) are associated with a greater lesion burden and there has been increasing interest in identifying potential modifiable risk factors. One of these is the sulfur-containing amino acid homocysteine. Elevated tHcy has been associated with atherosclerotic disease and increased risk of stroke and dementia (Seshadri et al., 2002; Miller et al., 2002). Few studies have examined the effect of elevated tHcy in those with small-vessel disease (Luchsinger et al., 2004). However, white matter hyperintensities may be a marker of small-vessel disease and several studies have documented an association with elevated tHcy (Yasui et al., 2000; Rogers et al., 2003). The presented data come from mostly white populations, and therefore there is limited understanding of the effect of elevated tHcy on white matter hyperintensities in blacks and Hispanics, who are at greater risk for hypertension, diabetes, and small-vessel disease (O'suilleabhain et al., 2004; Postuma et al., 2004).

Few studies have used quantitative methods for measuring white matter hyperintensity volumes (WMHV), depending rather on semiquantitative scales that are subject to limitations in inter-rater reliability (Wright et al., 2005). Quantitative methods have been used, but the populations studied have been limited to the elderly, whites, or men only (Greenberg et al., 2000). Evidence from Framingham supports an association between various vascular risk factors and quantitative measures of white matter hyperintensities, but tHcy was not included (Van Dijk et al., 2004). The Rotterdam study found an association between elevated tHcy and the presence of silent infarcts and white matter lesions using qualitative measures (Rogers et al., 2003).

The MRI sample is healthier than the overall cohort because of a survivor effect and the functional capacity required to come in for the study, but this would tend to bias the findings towards the null. Fasting tHcy was measured at baseline, raising concerns that values were not representative of later levels. However, the intra-standard deviation (SD) was smaller than the overall SD in those with two measurements (Miller et al., 2002). Moreover, MRI scans were performed after folic acid fortification began in the United States in 1998. Although tHcy levels were lower in those enrolled after 1998, the results remained significant after controlling for the year of collection (Miller et al., 2002). Total homocysteine levels were lower in the study sample than the overall cohort, but this would likely minimize any association with measures of WMHV. Regarding potential confounders of the relationship between tHcy and WMHV, study subjects were younger and were less likely to have B_{12} deficiency than the overall cohort, but this again would tend to minimize any association.

That white matter hyperintensities on MRI represent small vessel damage has been shown by observational and pathological studies (Fazekas et al., 1993) and is supported by the association of tHcy with WMHV in this sample. Recent data suggest it may do so by contributing to endothelial dysfunction (Hassan et al., 2004). The cross-sectional nature of this analysis does not allow a conclusion. Longitudinal imaging studies will be needed to clarify whether elevated tHcy causes progression of white matter damage and whether it is on the causal pathway between elevated tHcy and outcomes such as stroke and cognitive decline.

Plasma Aβ was examined as a potential risk factor for AD and the related process of cerebral amyloid angiopathy (CAA), but was not consistently elevated in these conditions (de Leeuw et. al., 1999). Recent data from the population-based Rotterdam study, however, demonstrated an association between plasma Aβ and microvascular disease in the brain in APOE ε4 carriers (Carmelli et al., 1998; Jeerakathil et al., 2004), suggesting that Aβ might be a cause or marker of cerebrovascular dysfunction.

The hypothesis that tHcy levels were elevated in mild cognitive impairment (MCI), AD, Parkinson's disease (PD), or CAA relative to normal healthy volunteers in an outpatient clinic population is presented: unexpectedly, tHcy was independently correlated with plasma Aβ40 and Aβ42 levels. This observation raises the possibility that these factors interact to potentiate neurodegeneration.

The clinical and biochemical correlates of tHcy concentrations in plasma from AD, PD, MCI, CAA, hICH, and nondemented control individuals have been investigated. The principal clinical factors influencing tHcy levels were age, levodopa use, and multivitamin use. Biochemically, in addition to the expected correlations between tHcy levels with folate, vitamin B_{12} and Cr, it was found that tHcy levels moderately and independently correlated with plasma Aβ levels.

Several groups reported elevated tHcy in AD cases (Longstreth et al., 1996; Schmidt et al., 2004), attributed to relative deficiencies of folate and vitamin B_{12}. While lower folate levels in our AS cohort, tHcy and vitamin B_{12} were not influenced by, and AD diagnosis were also found. tHcy was not associated with cognitive status in AD. Data are consistent with those of Luchsinger et al. (2004), who found that cross-sectional and longitudinal analyses of tHcy with prevalent and incident AD were significantly confounded by age, sex, and education. Since the study is based on a single measurement of tHcy, subtle disease associations may be underestimated owing to regression dilution or prevalence bias (Clarke et al., 1998b). AD cases of short duration (< 5 years) had similar mean tHcy levels (8.5 μmol L^{-1}) relative to cases with longer duration (8.8 mmol/L) and control cases (8.7 μmol L^{-1}) (p < 0.90), arguing against significant prevalence bias by early vascular death of AD cases with higher tHcy. Other potential reasons which indicate that the results differ from previous cross-sectional studies include sample size, distribution of tHcy values, and the patient population. This study was sufficiently powered to detect a 0.4 SD difference among the AD and PD cases, and indeed detected the significant increase in PD cases. The prevalence of hyperhomocysteinemia (tHcy > 14 μmol L^{-1}) was relatively low in this population, perhaps reflecting the effects of folate supplementation of the U.S. dietary grain supply beginning in 1998 (Clarke et al., 1998a). There are, surprisingly, few North American studies examining tHcy levels in AD after 1998; the improved folate status and reduced prevalence of hyperhomocysteinemia appears to have modified the cross-sectional association of tHcy with AD in the United States. Northeastern United States- based clinic population may not be generalizable to other populations; nonetheless, other genetic and epidemiological measures in these groups are consistent with known risks for AD.

Similar to previous studies (Blandini et al., 2001; Yasui et al., 2000; Miller et al., 2003b), levodopa use was strongly associated with hyperhomocysteinemia. Patients with PD taking levodopa had almost 50% higher average tHcy levels than those not taking levodopa. tHcy results from the metabolism of levodopa and dopamine by catechol-O-methyltransferase (COMT), wherein S-adenosylmethionine serves as a methyl donor. The resulting S-adenosylhomocysteine is rapidly catabolized to homocysteine.

High levels of tHcy are implicated in nigral oxidative damage and cognitive deficits in PD. Homocysteine potentiates 1-methyl-4-phenyl-1,2,3,6-tetrahydropyridine toxicity in mice, and rotenone toxicity in cultured dopaminergic neurons. Within PD group, which excluded coexistent dementia, tHcy correlated with cognitive status, even after controlling for levodopa dosage. The association of elevated tHcy with lower cognitive status by a global screening test in nondemented PD subjects supports the results of O'suilleabhain et al. (2004), where patients with PD with elevated tHcy performed poorly on neuropsychological testing. Thus, tHcy, which was greatly increased in levodopa users relative to the other diagnostic groups, may contribute to the cognitive decline that can develop in PD. Multivitamin use was associated with lower tHcy, even in levodopa users. Therefore, the particularly high levels of tHcy associated with levodopa use, and the potential risk of cognitive impairment, may be amenable to folate and vitamin B_6 and B_{12} supplementation.

A main finding was the robust correlation of tHcy with plasma Aβ levels in most diagnostic groups. These findings confirm and extend a previous of a positive correlation between plasma Aβ40 and tHcy levels in a cross-sectional survey of community-dwelling men (Petersen et al., 2001). Possible mechanisms for the association may be that tHcy elevates Aβ levels, Aβ levels increase tHcy, or both are increased by an unknown third factor. In that sample and in other studies, plasma Aβ40, plasma Aβ42, and tHcy were correlated with age and with serum Cr levels (Mayeux et al., 2003). It cannot be excluded that another factor may result in the accumulation of both tHcy and Aβ in the plasma; however, on the basis of careful covariate analysis of data, the association is not explained by diagnosis, age, Cr, folate, vitamin B_{12}, or *APOE* polymorphisms. Cell culture data support the possibility that tHcy may increase Aβ levels: homocysteine enhances Aβ generation by upregulating a presenilin-interacting endoplasmic reticulum stress protein (Sai et al., 2002); deficient methylation upregulates presenilin gene function and Aβ generation (Scarpa et al., 2003).

Although the correlation of tHcy and Aβ was independent of diagnosis, age-related correlation increases in tHcy and Aβ may contribute to neurotoxicity and AD risk. Homocysteine potentiates Aβ oxidative toxicity in cultured neurons and smooth muscle cells and in APP transgenic mice (Scarpa et al., 2003, White et al., 2001). Elevated tHcy and elevated plasma Aβ levels are both implicated as premorbid risk factors for the development of AD (Seshadri et al., 2002) and the associated microangiopathic changes on MRI (Verneer et al., 2002). Both tHcy and Aβ40 could be markers of vascular damage in the brain; tHcy and Aβ40 were associated with white matter ischemic changes in APOE 4 carriers in the Rotterdam study (Van Dijk et al., 2004). The more general association of tHcy with plasma Aβ is independent of diagnosis (both neurodegenerative and cerebrovascular) and *APOE* genotype.

Translating the plasma measures to the tissue levels in the brain is a challenge that will require further investigation. tHcy levels in plasma and CSF Aβ levels are not correlated in AD. The relationship of plasma Aβ levels to CSF and brain levels is complicated by the kinetics of the blood–brain barrier, differential synthesis within the brain and the periphery, and the effects of Aβ deposition as amyloid deposits in the brain (Arvanitakis et al., 2002; Vanderstichele et al., 2000). Plasma tHcy and Aβ directly contact elements of the blood vessel wall, providing a potential mechanism for resulting vascular toxicity and leukoencephalopathy. Further longitudinal epidemiological studies as well as basic research investigations can address whether neurotoxicity in AD and PD is potentiated by the dual elevation of both tHcy and Aβ or whether tHcy and Aβ are markers of pathogenic processes such as vasculopathy or oxidative damage (Matsuoka et al., 2003).

The Effect of Vitamin Supplementations on Plasma tHcy Concentrations

Folic acid: Several studies on various population groups have been performed to demonstrate the effect of vitamin supplementation on plasma tHcy concentrations. It is apparent that folate is the most powerful tHcy-lowering agent. Folate has been used in daily doses ranging from 0.65 to 10 mg day^{-1}, and it seems that in apparently healthy volunteers a low daily dose of 0.65 mg or less may be sufficient to maintain plasma tHcy concentrations within the normal reference range (Ubbink et al., 1994). This low folate dose may, however, be insufficient in various pathological conditions predisposing towards coronary heart disease. In patients with severe chronic kidney failure, 10 mg of folate per day administered for 3 months failed to reduce plasma tHcy concentrations to normal in all the participants (Chauveau et al., 1996), while 5 mg of folate per day was insufficient to normalize hyperhomocyst(e)inemia observed in dialysis patients (Arnadottir et al., 1993). Low daily doses of folic acid have not yet been tested in patients with premature vascular disease and it is possible that this patient group will also require higher daily folic acid doses to maintain plasma tHcy concentrations in the normal range.

Vitamin B_{12}: Although folic acid is the most powerful tHcy-lowering agent, this does not imply that vitamin B_{12} and vitamin B_6 may be omitted in the treatment of moderate hyperhomocyst(e)inemia. Vitamin B_{12} supplementation has a small, but significant, effect on circulating tHcy concentrations (Ubbink et al., 1994; Rasmussen et al., 1996). Moreover, it has been shown that folic acid supplementation is ineffective in reducing tHcy concentrations in patients with vitamin B_{12} deficiency (Allen et al., 1990). In a general opinion, the optimum vitamin supplement to treat hyperhomocyst(e)anemia will contain at least 400 μg of vitamin B_{12} per day. At this high daily dose, even patients with intrinsic factor deficiency will absorb a sufficient amount of vitamin B_{12} by passive diffusion (Doscherholmen and Hagen, 1957). Vitamin B_{12} supplementation at high doses is innocuous (Ellenbogen and Cooper, 1991) and will eliminate the risk that folic acid supplementation may mask an underlying vitamin B_{12} deficiency.

Vitamin B_6: Owing to its dramatic effect in cystathionine β-synthase deficiency (Mudd et al., 1989), pyridoxine may be regarded as the obvious choice in the treatment of hyperhomocyst(e)inemia. However, even high-dose pyridoxine supplementation did not lower fasting plasma tHcy concentrations significantly. Selhub and Miller (1992) recently addressed the intriguing question why folic acid and vitamin B_{12} supplementation, but not pyridoxine supplementation, will reduce elevated, fasting circulating tHcy concentrations. It has been postulated that a low folate and/or vitamin B_{12} status results in low S-adenosylmethionine concentrations. S-Adenosylmethionine, however, is required to activate the enzyme cystathionine β-synthase. The supplementation of only vitamin B_6 (without folate and vitamin B_{12}) does not appear to activate the transsulfuration pathway, as the essential activator S-adenosylmethionine will remain low owing to the inadequate folate and vitamin B_{12} status. Only when the latter two vitamins are present in abundance, will remethylation proceed unimpeded with a subsequent increase in S-adenosylmethionine concentrations and activation of the transsulfuration pathway (Selhub and Miller, 1992).

Although pyridoxine supplementation has no effect on fasting plasma tHcy concentrations, it lowers the post-methionine-load tHcy peak (Dudman et al., 1993; Franken et al., 1994b; Ubbink et al., 1996). This phenomenon is also explained by Sellhub and Miller's (1992) hypothesis. The high post-methionine-load S-adenosylmethionine concentrations will inhibit remethylation and stimulate transsulfuration, but activation of transsulfuration cannot proceed during a vitamin B_6 deficiency. When vitamin B_6 is supplemented, transsulfuration can proceed unimpeded and therefore methionine loading will result in lower peak plasma tHcy concentrations (Selhub and Miller, 1992). This hypothesis assumes that the enzyme cystathionine β-synthase is sensitive to vitamin B_6 depletion, which may not be true for all population groups. For example, African Blacks have a genetically determined lower vitamin B_6 status (Vermaak et al., 1987) owing to a low pyridoxal kinase activity (Chern and Beufler, 1975). Pyridoxal kinase is required to phosphorylate pyridoxal to its physiologically active form, pyridoxal phosphate (PLP) (McCormick et al., 1961). However, Blacks do not present with high plasma tHcy concentrations when challenged with methionine. In fact, the post-methionine-load tHcy peak is significantly lower in Blacks compared to Whites, despite the poorer vitamin B_6 status of Blacks (Ubbink et al., 1995). These observations in Blacks are still unexplained, but may relate to a cystathionine β-synthase polymorphism due to which they may commonly possess a form of the enzyme that has a high affinity for PLP.

In Caucasian populations, it has been shown that a normal fasting plasma tHcy concentration is not synonymous with a normal methionine load test (Bostom et al., 1995). Elevated post-methionine-load tHcy concentrations are independently associated with diverse pathological conditions, for example premature vascular disease (Clarke et al., 1991) and neural tube defects. It is therefore prudent that vitamin therapy should normalize both fasting and post-methionine-load tHcy concentrations. Since pyridoxine supplementation attenuates the post-methionine-load tHcy peak (Dudman et al., 1993; Franken et al., 1994; Ubbink et al., 1996), this vitamin should be included in the treatment of hyperhomocyst(e)inemia. Initial studies used very high daily doses of pyridoxine (70–300 mg) to lower the post- methionine-load tHcy peak (Dudman et al., 1993; Franken et al., 1994). However, these high doses are undesirable, as patients may develop pyridoxine-induced sensory neuropathy (Schaumburg et al., 1983). The minimum daily pyridoxine dose required for the optimal reduction in post-methionine-load plasma tHcy concentrations remains unclear. A pyridoxine dose of only 20 mg day^{-1} administered for a 6-week period significantly reduced the post-methionine-load tHcy peak in asthma patients with a theophylline-induced vitamin B$_6$ deficiency (Ubbink et al., 1993 and 1996).

Despite pyridoxine supplementation, the post-methionine-load plasma tHcy concentrations in these formerly vitamin B$_6$-deficient patients remained significantly higher compared with healthy controls. This may indicate that 20 mg pyridoxine per day is not adequate for certain patient populations and higher doses may be required (Ubbink, 1998).

Galactosemia

Genetics and Epidemiology

Galactosemia is a relatively rare inherited enzyme deficiency with variable worldwide incidence reported between 1:30–40,000 in Europe (Murphy et al., 1999) and 1:1,000,000 in Japan (Hirokawa et al., 1999). The incidence in the United States, with most cases ascertained by newborn screening, is currently estimated at 1:53,000 (National Newborn Screening and Genetics Resource Center; 2002 Newborn Screening and Genetic Testing Symposium). In Greece, classical galactosemia is discovered by screening and it is estimated at 1:25,000 live births (Schulpis et al., 1993).

The galactose-1-phosphate uridyl transferase (GALT) gene is localized on 9p13, and it is a relatively compact gene with 11 exons spanning 4 kilobases (Leslie et al., 1992). One hundred sixty-seven mutations have been identified. The most prevalent mutation in Western populations is Q188R, in which an A to G transition in exon 6 converts a glutamine near the catalytic site to an arginine. Expression of this allele in yeast systems produces no detectable catalytic activity (Fridovich-Keil et al., 1995, Ross et al., 2004). Several other relatively common mutations have been characterized, the most intriguing of which is S135L, the mutation present in both African-American and African blacks (Ross et al., 2004; Lai et al., 1996). This

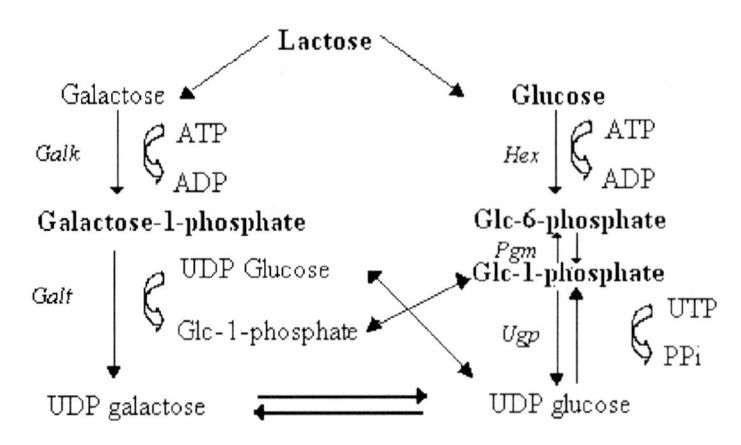

Fig. 20.5 The Leloir pathway of galactose metabolism and the uridine diphosphate (UDP) glucose pyrophosphorylase pathway. Galk, galactokinase; galt, galactose 1-phosphate uridyltransferase; gale, UDP-galactose 4-epimerase; hex, hexokinase; pgm, phosphoglucomutase; Ugp, UDP-glucose pyrophosphorylase

mutation has greater residual activity in hepatocytes and is associated with a different pattern of ability to oxidize galactose (Lai et al., 1996; Yager et al., 2001) as well as a milder phenotype (Elsas and Lai, 1998). A much more common variant, the so-called Duarte galactosemia, is also identified primarily by newborn screening. The Duarte variant denotes a compound heterozygote for a classical allele and another allele with partial GALT activity, which results in biochemical changes during early infancy, but no evidence of neonatal or long-term morbidity (Kelley and Segal, 1989).

Metabolic Derangement and Intoxication

In the presence of sufficient GALT activity, galactose is transported into cells, then phosphorylated by galactokinase, which effectively traps it as galactose-1-phosphate (gal-1-P) (Fig. 20.5). With near total absence of GALT, gal-1-P accumulates and results in product inhibition of kinase. Accumulating free galactose is diverted into secondary pathways that produce either galactitol (Van Heyningen, 1959) or galactonate (Cuatrecasas and Segal, 1966). These substances have been found in affected organs (Wang et al., 2001; Petry et al., 1991; Quan-Ma et al., 1966). Therefore, one possibility is that these substances cause direct damage to vulnerable subpopulations of neurons, for example, Purkinje cells, or to white matter.

A chronic intoxication theory has been proposed with either exogenous galactose intake from poor dietary compliance or hidden environmental sources (Acosta and Gross, 1995), or from endogenous synthesis of galactose (Berry et al., 1995) resulting in increased long-term exposure. This exposure could theoretically cause progressive neurologic symptoms in some or most patients, but mechanisms by which this impairment occurs remain under investigation (Table 20.2).

In utero toxicity has also been suggested (Holton, 1995), possibly due to elevated galactitol levels. Evidence for this theory comes from observations of fetal cataracts and early liver dysfunction. In utero exposure to galactose was demonstrated by high levels of gal-1-P in the fetus and galactitol in the amniotic fluid in mothers placed on strict galactose-restricted diets (Irons et al., 1985; Jakobs et al., 1988 and 1995). Magnetic resonance spectroscopy has documented markedly elevated brain galactitol levels in four newborn infants aged 6–15 days compared with age-matched control subjects (Wang et al., 2001). However, in the eight galactosemia patients aged 1.3–47 years, no elevation was identified in six, and slight elevation in just two. A prior magnetic resonance spectroscopy study of six adults disclosed no elevation (Moller et al., 1995). This finding suggests that although galactitol could cause early, static CNS injury, it would be unlikely to cause progressive neurologic impairment.

Dietary control and levels of gal-1-p do not appear to account for variance in cognitive outcomes (Shield et al., 2000). Nevertheless, a recent study assessing total body galactose oxidation with a breath test after a ^{13}C-galactose bolus found three related biochemical abnormalities that were more common in galactosemia patients affected neurologically with speech dyspraxia. In 24 patients with galactosemia who underwent speech evaluations, the 15 with speech dyspraxia had a significantly lower cumulative percentage dose $^{13}CO_2$ in breath, increased erythrocyte gal-1-P, and increased urinary galactitol. These abnormalities were also influenced by GALT genotype (Webb et al., 2003).

Table 20.2 Clinical and laboratory findings in classical galactosemia

Clinical findings	Anorexia
	Vomiting
	Jaundice
	Cataracts
	Hepatomegaly
	Seizures
	Mental retardation
	Ovarian failure
	Sepsis
Laboratory findings	Reducing substances (urine)
	↑ Bilirubin
	↑ AST/ALP
	↑ Prothrombin time
Special laboratory studies	Galactose-1-P (RBC)
	GALT (RBC)

Magnetic resonance spectroscopy studies have found no evidence of abnormal cerebral energy metabolism in patients with diet-restriction-treated galactosemia (Wang et al., 2001; Moller et al., 1995).

Evidence From Animal Models

Two types of animal models have been employed to study galactosemia. In the first, animals with intact GALT are exposed to elevated levels of galactose or other metabolites. The second is a genetic knockout of the GALT enzyme.

Feeding excess galactose to rats with intact GALT reduces levels of inositol and increases galactitol in synaptosomes (Warfield and Segal, 1978). Dogs fed excess galactose for 44 months displayed multiple peripheral nervous system abnormalities (Sugimoto et al., 1999). The relevance of these findings has been called into question given that these animal models had no enzyme defect, excess exogenous exposure may not mimic the treated human disease, and changes resulting from malnourishment cannot be excluded (Segal, 1992).

A galactosemia model, the GALT deficient knockout mouse, lacks measurable GALT activity and develops biochemical features similar to those in humans including high levels of galacatose-1-phosphate. However, these mice do not develop cataracts, they can reproduce normally, and they appear to have no neurologic phenotype. These mice have minimal levels of tissue galactitol. This finding suggests that GALT deficiency is necessary, but not sufficient, to cause human disease and that both elevated galactose-1-phosphate and galactitol may be necessary for the human phenotype (Leslie et al., 1996; Ning et al., 2000).

In vitro studies in galactosemia on whole rat brain AChE activity showed a direct action of galactose-1-phosphate on the enzyme, resulting in a remarkable reduction of its activity (Tsakiris and Schulpis, 2002). This finding might explain the brain cholinergic dysfunction in untreated patients with classical galactosemia. In addition, galactose and its derivatives in brain may produce free radicals, inhibit Na^+, K^+–ATPase activity and activate Mg^{2+}–ATPase, whereas the supplementation of l-cysteine or glutathione reversed their activities to normal in galactokinase in vitro galactosemia only (Tsakiris et al., 2005a). Another study on the same enzymes in rat frontal cortex, hippocampus, and hypothalamus showed the same results (Marinou et al., 2005). These in vitro findings may be valuable, as Na^+, K^+, and Mg^{2+} pump dysfunctions are closely related to most enzyme activities.

Autopsy Studies

These are published postmortem examination data on only two cases. Although these are biased towards more severe phenotypes, their importance arises from direct observations of gray and white matter damage, yielding insights into selective neuronal vulnerability in galactosemia. Future postmortem examination studies may be invaluable, and in this perspective, basic research should focus on identifying mechanisms of regional and neuronal injury, which have been initially described in the studies in question.

In 1962, Crome provided a detailed pathologic report of a severely impaired male with galactosemia and partial dietary restriction, who was institutionalized and died of pneumonia at age 8 (Crome, 1962). This individual was able to walk with a stiff gait, to feed himself with hands but not utensils, was incontinent, was socially withdrawn, and had no spoken language. Postmortem findings included diffuse white matter gliosis, particularly around the ventricles and in the brainstem, and diffuse pallor, particularly in the cerebrum. There were many areas of chronic, focal necrosis in the white matter. Cerebellar Purkinje cells were markedly depleted, largely sparing the granular layer. There was neuronal loss in the dentate nuclei and inferior olives. Extracellular fat globules were observed in the globus pallidus. This apparent neuronal selectivity is intriguing, given findings in rat brain of somewhat more abundant GALT gene expression in cerebellum (Rogers et al., 1992) as well as the prominent ataxia and tremor in some galactosemia patients.

Similar findings were reported by Haberland et al. (1971) in another severely impaired patient, untreated until diagnosis at age 13 , who died of pneumonia at age 25. This patient had severe mental retardation, extrapyramidal motor symptoms, and epilepsy. Microcephaly, cerebral cortical neuronal degeneration, cerebral white matter atrophy and sclerosis, and cerebellar findings similar to Crome's case were described. In addition, these authors reported abnormal levels of glycoproteins and glycolipids.

Central Nervous System White Matter

Both the postmortem examination reports and multiple neuroimaging studies demonstrate that white matter is abnormal in galactosemia (Wang et al., 2001; Kaufman et al., 1995a,b; Nelson et al., 1991). The mechanism for this remains obscure. It is possible that GALT deficiency leads to defective synthesis of glycoproteins and galactolipids critical for normal myelin, possibly as a result of defective transfer of galactose from uridine disphosphate-galactose (UDP-gal) (Segal, 1995 and 2004).

The main current limitation of neuroimaging research is that the diffuse CNS white matter change has not been quantified and therefore is not available to be analyzed as a continuous variable in regression analyses with an outcome like IQ. Because diffuse white matter abnormalities are almost universally present (Wang et al., 2001; Waggoner et al., 1990; Nelson et al., 1991), this also cannot be used as a class variable for group comparisons of various measures of impairment. For example, Kaufman et al. (1995a,b) reported that MRI abnormalities in 40 subjects "not correlate[d] with cognitive outcomes". White matter was diffusely abnormal in 37 of the 40 patients. Using the overall cognitive function as an outcome and "abnormal white matter" as a class predictor, this study would only have about 7% power to detect a meaningful 10-point difference. They also used the presence or absence of enlarged ventricles, focal lesions, and mild cerebral atrophy and found "no differences" in neurologic outcomes. A study using magnetic resonance spectroscopy did not identify changes in choline, *N*-acetyl aspartate, or myoinositol. However, only two spectra were presented on white matter regions (Wang et al., 2001). These comparisons would also have low statistical power.

That the abnormal-appearing white matter in galactosemic patients also functions abnormally is supported by findings of prolonged latencies of somatosensory-evoked potentials (Kaufman et al., 1995a,b). In 60 galactosemia patients, somatosensory-evoked potentials manifested one or more abnormalities (absence of expected peaks, prolonged latencies) in 34 (57%). The proportion of patients with abnormalities did not differ among the 12 patients with ataxia tremor vs. the less neurologically impaired patients. Cognitive function also did not differ between patients with normal potentials and those with abnormal potentials.

Future neuroimaging or neurophysiology research into galactosemia's pathophysiology may benefit from the identification of newer quantitative variables for use in more rigorous statistical analyses.

Influence of Genotype on Cognitive Outcome

Although galactosemia is an autosomal recessive disorder caused by mutations at a single major locus, the *GALT* gene, there appears to be considerable biochemical and phenotypic heterogeneity among patients who share common studied mutations. The basis for this variation is not well understood. There is controversy in the literature over whether cognitive outcome, e.g., IQ scores, is correlated with genotype (Shield et al., 2000; Kaufman et al., 1994; Elsas et al., 1993 and 1995). The most common mutation, the Q188R mutation, has been the most studied.

Two studies provide evidence that homozygosity for Q188R is associated with poorer cognitive outcome. Elsas et al. (1995) found a poorer outcome in 22 patients homozygous for the Q188R mutation, compared with 20 patients who were compound heterozygous for Q188R plus another disease-causing allele. Shield et al. (2000) found a significant, 20 point lower IQ in patients homozygous for Q188R, vs. those with Q188R plus an alternate mutation. There was no correlation between gal-1-P levels and IQ.

In contrast, two other studies found higher average cognitive function in patients homozygous for Q188R, although these differences fell short of statistical significance. Cleary et al. (1995) found no significant difference in IQ, comparing nine homozygous for Q188R, median IQ 81, and six heterozygous, median IQ 72. Kaufman et al. (1994), using the Woodcock Johnson Broad Cognitive score, found a score of 75 (SD 16) in 38 patients homozygous for Q188R and 67 (SD 25) in 21 patients heterozygous (Kaufman et al., 1994). The power for these two negative studies to detect a statistically significant difference ($P < 0.05$) in those samples was less than 25%. In addition, the non-Q188R/Q188R comparison groups in these studies are different. Thus, the influence of the most common genotype on phenotype remains unclear. What has been clarified is that, while the mean IQs in these studies are all low, the standard deviations are comparable to those in the general population.

This finding indicates that there is substantial heterogeneity among subjects with the same *GALT* genotype, which is similar to what has recently been described in other autosomal recessive major gene disorders (Scriver and Waters 1999).

Acute Neonatal Presentation and Diagnosis Presentation

The natural history of classical ganactosemia is that lethargy, poor feeding, jaundice, and hepatomegaly appear within days of the initiation of milk feedings. Progression of this acute neonatal toxicity syndrome may include the development of *E. coli* septicemia in the second week of life, coagulopathy, hyperchloremic metabolic acidosis with aminoaciduria, and vitreous hemorrhage (Levy et al., 1996). Cataracts may be present, but may be difficult to identify. Neurologically, these patients may develop encephalopathy and signs of increased intracranial pressure with cerebral edema, usually after several days of more nonspecific signs. In the United States, many cases are ascertained because of newborn screening, but nonspecific signs may precede diagnosis by several days (Segal, 2004) (Table 20.2).

Diagnosis

The gold standard for diagnosis is the demonstration of near total absence of GALT activity in red cells. Transfusion of red cells from a normal donor can interfere with this determination. Measurement of accumulated gal-1-P in red cells has diagnostic utility, even if dietary galactose has been withdrawn, although benign variants can also have increased gal-1-P levels, particularly if the infant is still consuming galactose. Examination of the urine for reducing substances lacks both sensitivity and specificity (Walter et al., 1999 and 2002), but can certainly raise sufficient clinical suspicion in a symptomatic infant to trigger further evaluation and empiric treatment. DNA analysis is rapidly available for several common mutations, including Q188R (Flanagan et al., 2004), and has been used in some newborn screening programs to refine the screening process. As with other mutation-selective DNA-based assays, owing to the large number of distinct mutations, a positive result has predictive value, but a negative result does not exclude disease resulting from other mutations. Prenatal diagnosis is available for couples with a known family history of galactosemia (Jakobs et al., 1995), and carrier screening for a known GALT mutation is highly effective in defining risk.

Non-Neurologic Sequelae

Growth is generally delayed in most patients with galactosemia, but final height-for-age is not significantly less than the general population, although females may be more likely to have short stature (Yager et al., 2001).

Cataracts have been reported in up to 30% of patients with galactosemia, and they usually resolve with galactose restriction (Waggoner et al., 1990). Cataracts are one of the few complications of galactosemia with a known pathophysiology. Galactitol, produced by reduction of free galactose by aldose reduction, induces swelling of lens fibers. Adolescents and adults who include amounts of galactose in their diet may rarely develop cataracts that interfere with the ability to function and may benefit from lensectomy and intraocular lens placement. A less well-known, but visually more devastating ophthalmologic complication is vitreous hemorrhage during the acute neonatal syndrome (Levy et al., 1996).

Hypergonadotropic hypogonadism is common in females with galactosemia, but has not been reported in males (Waggoner et al., 1990; Kaufman et al., 1995a). The spectrum varies from severe primary ovarian failure requiring hormone support in order to achieve secondary sexual characteristics to premature menopause. Pregnancies without hormonal intervention have been reported (Waggoner et al., 1990; de Jongh et al., 1999). The mechanism of ovarian dysfunction is unknown, but it has been hypothesized that chronic exposure of galactose-1-phosphate and galactitol may directly damage the ovaries (Gibson, 1995).

Neurologic Sequelae

Acute Elevated Intracranial Pressure in Infants

Galactose must be considered in the differential diagnosis of an infant with diffuse cerebral edema, presenting with a bulging anterior fontanel, in the setting of poor feeding, jaundice, and hepatomegaly (Huttenlocher et al., 1970). Animal models have suggested that the mechanism of cerebral edema may be that elevations in brain galactitol concentrations and alterations in glucose, ATP, and phosphocreatine levels increase osmolality (Anonym, 1972). Magnetic resonance spectroscopy has revealed elevated cerebral galactitol levels in infant brain in vivo (Wang et al., 2001). Cerebral edema in galactosemic patients responds well to dietary galactose restriction (Belman et al., 1986). The mechanism may be similar to that reported in a rat model with intact GALT but excessive dietary loading, in which galactitol accumulates in peripheral nerves and produces edema with increased pressure and eventual demyelination (Myers and Powell, 1984).

Cognitive Impairment: Gender Effects and Possible Regression with Age

Multiple studies have reported that mean IQ scores are reduced in galactosemic children and adults (Wang et al., 2001; Bosch et al., 2004a,b; Kaufman et al. 1995a; Shield et al., 2000; Antshel et al., 2004). Differences between studies result from a variety of factors. For example, differences in screening and diagnostic practices mean that the initiation of dietary treatment may vary systematically among countries. Other important differences include the test batteries used and study design.

Evidence related to a gender effect has been inconsistent. Waggoner et al. (1990) obtained developmental and IQ scores on 298 patients. Mean IQ scores of females aged 10–16 and > 16 were significantly lower than males. Kaufmann et al. (1995a) failed to identify sex differences in the broad cognitive ability scores assessed in 40 children and adults by the Woodcock–Johnson revised tests.

A number of cross-sectional studies have demonstrated lower IQ scores in adults than pediatric patients (Fishler et al., 1980), but because cross-sectional data often contain unmeasured confounders, this constitutes weak evidence of neurodegeneration. Waggoner's study, which obtained data from structured questionnaires, included responses from 88 patients tested on more than one occasion. These data revealed that mean IQ scores of galactosemic patients declined by 6.2 points from age 3–5 to 6–9 years, and by 4.4 points from 6–9 to 10–16 years (Waggoner et al., 1990). Another recent cross-sectional study, restricted to patients with the Q188R GALT mutation, revealed reduction of IQ, in particular difficulties with word retrieval and executive function, but did not support the hypothesis that IQ progressively declines. These authors suggested that older patients might diverge more from their peers in performance-IQ-related tasks as a result of slow processing speed (Antshel et al., 2004).

Speech Apraxia Commonly Occurs

Speech difficulties are common in galactosemia. One study reported that 56% of patients with galactosemia over 3 years of age have difficulties with speech, with 92% of these patients described as having delayed vocabulary (Waggoner et al., 1990). Alternatively, speech and motor difficulties may interfere with accurate testing. As in other adverse outcomes in galactosemia, there is not good evidence that development of speech difficulties correlates with age of initiation of dietary therapy, severity of presenting symptoms, or compliance with dietary therapy (Waggoner et al., 1990; Nelson et al., 1991).

Motor Function: A Subgroup Develops More Severe Ataxia and Tremor

Despite early treatment initiation and optimal dietary compliance, approximately 10–20% of galactosemia patients develop more severe and progressive ataxia and tremor (Waggoner et al., 1990; Kaufman et al., 1995a; Friedman et al., 1989).

Seizures are Uncommon

Seizures have rarely been described in galactosemia. There is a case with two galactosemic siblings who developed seizures in adulthood, described by Friedman et al. (1989). It is possible that these seizures are unrelated to galactosemia, given the low prevalence of reported seizures.

Pathophysiology of Neurology Dysfunction in Galactosemia

The chronic manifestations of galactosemia most likely result from outcomes of multiple fronts, including possibly intoxication and cell dysfunction or death, due to specific metabolic derangements and secondary disturbances in myelin production. The heterogeneity of presentation may be influenced by the inherited *GALT* alleles and amount of GALT activity, by the individual's endogenous production of galactose, by the individual's ability to oxidize galactose, and by the function of alternative enzymes in the metabolic pathways (Segal, 2004). The issues of metabolic intoxication as well as the research from animal models, postmortem tissue, neuroimaging, and genetic studies should be considered with regard to these theories.

Treatment: Standard of Care, Controversies, and Clinical Trials

Treatment of Symptomatic Infants – Standard of Care

Symptomatic infants or infants with highly suspicious newborn screening results should be evaluated promptly, and galactose-containing formulas should be withdrawn. Supportive care with intravenous fluids, phototherapy, antibiotics, and treatment of coagulopathy should be initiated as needed. Despite the nearly universal availability of newborn screening in the United States, infants with classical galactosemia still die. The use of galactose-limited formulas due to family preferences or early symptoms may completely mask the initial clinical presentation. Subsequent to diagnosis, during infancy, dietary management with galactose-free formula is straightforward (Acosta and Gross, 1995).

Long-term Nutrition Management: Standard of Care and Recent Issues

When solid foods are initiated, and when infant are weaned from formulas, nutritional issues become more complicated. An ideal, standard practice has been to reduce or attempt to eliminate galactose in the diet. However, no person, unless completely nourished with artificial formula, has ever been on a galactose-free diet. Traditional means to monitor dietary compliance have not proven clinically useful, even in some noncompliant individuals.

Prior strict dietary recommendations are tempered by data from recent technical advances, which have increased the precision of estimates of endogenous synthesis of galactose (Ning et al., 2000; Berry et al., 2004). These studies demonstrate that endogenously synthesized galactose substantially exceeds the dietary amounts acquired from fruits and vegetables (Berry et al., 1993), in galactose-containing medications, or from other sources in a strict diet. Moreover, in a controlled, 6-week study in three adolescents homozygous for the Q188R *GALT* mutation, adding much larger amounts of galactose to the diet did not significantly change any clinical or biochemical parameters (Bosch et al., 2004). Endogenous galactose production appears to be greater and much more variable in children vs. adults with galactosemia (Berry et al., 2004), but the influence of this on neurologic outcome is unknown. Taken together, these studies suggest that liberalization of the galactosemia diet with regard to fruits and vegetables is reasonable.

With dietary restriction of dairy products, adequate calcium intake is difficult to be achieved without supplements. Low bone density has been reported (Kaufman et al., 1995a; Panis et al., 2004), although the long-term complications of bone health are not well characterized.

Effects of Dietary Treatment: Does Strict Dietary Treatment After Early Childhood Improve Adult Outcomes?

Crome observed in 1962 that "it is well known that established mental retardation tends to remain refractory in cases of galactosemia, even after the withdrawal of galactose from the diet, in spite of the reversibility of other signs, e.g., cataracts, hepatic cirrhosis, enteritis, and jaundice" (Crome, 1962), and one which raises questions about the value of galactose restriction after early childhood.

Consistent with the motion that strict dietary treatment after early childhood may not be needed, a recent, to some extent controversial, case report (Lee et al., 2003b; Segal 2004) described an adult galactosemic on a normal diet since age 3 , with only mild cognitive impairment and ovarian failure. This outcome lies within the range of outcomes of patients on lifelong galactose restriction. If typical neurologic outcomes occur despite treatment with dietary restriction from childhood (Lee et al., 2003a,b), then the role of galactose restriction after infancy on neurologic outcomes may be less important.

With regard to patients with classical galactosemia "off diet," the total antioxidant capacity was found to be low, and the activities of their erythrocyte membrane AChE, (Na^+, K^+)-ATPase and Mg^{2+}-ATPase were remarkably reduced. Interestingly, these reduced enzyme activities were restored to normal when the patients followed their galactose/lactose-restricted diet and their total antioxidant capacity was increased reaching that of normal (Schulpis et al., 2005; Tsakiris et al., 2005b).

Because dietary galactose restriction has been the standard of care, and is undoubtedly lifesaving during early infancy, performing a randomized controlled trial to assess the influence of dietary treatment in early infancy on chronic neurologic outcomes is not possible. A randomized study of graded amounts of galactose in later childhood or adulthood could possibly be safe, but institutional review boards are unlikely to participate in controlled studies, in which some patients are randomized not to receive standard galactose restriction. However, several alternative quasiexperimental study designs could provide useful information (Schadewaldt et al., 2004).

As described above, phenylalanine, homocysteine, and galactose metabolic disorders affect brain function, the latter being profoundly or partially avoided by early diagnosis and proper treatment. As the ancient Greek doctor Hippocrates said, *"Prevention is better than treatment"*.

Acknowledgments The authors are highly indebted to Mrs. Anna Stamatis and Mrs. Kalliopi Tassopoulou for their carefull typing of this manuscript. Many thanks are extended to Alexios Mentis, a medical student, for his significant help in preparing this work.

References

Acosta PB, Gross KC. Hidden sources of galactose in the environment. Eur J Pediatr 1995; 154: S87–S92.

Alexopoulos GS, Meyers BS, Young RC, Campbell S, Silbersweig D, Charlson M. "Vascular depression" hypothesis. Arch Gen Psychiatr 1997; 54:915–922.

Allan JD, Holt KS, Hudson FP, Ireland JT. Phenylketonuria. Br Med J 1963; 2:498.

Allen RH, Stabler SP, Savage DG, Lindenbaum J. Diagnosis of cobalamin deficiency I Usefulness of serum methylmalonic acid and total homocysteine concentrations. Am J Hematol 1990; 34:90–98.

Anonym. Galactose toxicity in the chick: hyperosmolality or depressed brain energy reserves? Science 1972; 176:815–817.

Antshel KM, Epstein IO, Waisbren SE. Cognitive strengths and weaknesses in children and adolescents homozygous for the galactosemia Q188R mutation: a descriptive study. Neuropsychology 2004; 18:658–664.

Arnadottir M, Brattstrom L, Simonsen O, Thysell H, Hultberg B, Andersson A, Nilsson-Ehle P. The effect of high dose pyridoxine and folic acid supplementation on serum lipid and plasma homocysteine concentrations in dialysis patients. Clin Nephrol 1993; 40:236–240.

Arvanitakis Z, Lucas JA, Younkin LH, Younkin SG, Graff-Radford NR. Serum creatinine levels correlate with plasma amyloid B protein. Alzheimer Dis Assoc Disord 2002; 16:187–190.

Baumeister AA, Baumeister AA. Dietary treatment of destructive behavior associated with hyperphenylalaninemia. Clin Neuropharmacol 1998; 21:18–27.

Baydas G, Gursu MF, Cikim G, Campolat S, Yasar A, Canatan H, Kelestimur H. Effects of pinealectomy on the levels and the circadian rhythm of plasma homocysteine in rats. J Pineal Res 2002a; 33:151–155.

Baydas G, Nedzvetsky VS, Nerush PA, Kirichen SV, Demchenko HM, Reiter RJ. A novel role for melatonin: regulation of the expression of cell adhesion molecules in the rat hippocampus and cortex. Neurosci Lett 2002b; 326:109–112.

Baydas G, Yilmaz O, Celik S, Yasar A, Gursu MF. Effects of certain micronutrients and melatonin on plasma lipid, lipid peroxidation, and homocysteine levels in rats. Arch Med Res 2002c; 33:515–519.

Baydas G, Kutlu S, Naziroglu M, Canpolat S, Sandal S, Ozcan M, Kelestimur H. Inhibitory effects of melatonin on neural lipid peroxidation induced by intracerebroventricularly administered homocysteine. J Pineal Res 2003; 34:36–39.

Beasley MG, Costello PM, Smith I. Outcome of treatment in young adults with phenylketonuria detected by routine neonatal screening between 1964 and 1971. Q J Med 1994; 87:155–160.

Belman AL, Moshe SL, Zimmerman RD. Compute tomographic demonstration of cerebral edema in a child with galactosemia. Pediatrics 1986; 78:606–609.

Berry GT, Palmieri M, Gross KC, Acosta PB, Hestenburg JA, Mazur A, Reynolds R, Segal S. The effect of dietary fruits and vegetables on urinary galactitol excretion in galactose-1-phosphate uridyltransferase deficiency. J Inherit Metab Dis 1993; 16:91–100.

Berry GT, Nissim I, Lin Z, Mazur AT, Gibson JB, Segal S. Endogenous synthesis of galactose in normal men and patients with hereditary galactosemia. Lancet 1995; 346:1073–1074.

Berry GT, Moate PJ, Reynolds RA, Yager CT, Ning C, Boston RΨ, Segal S. The rate of *de novo* galactose synthesis in patients with galactose-1-phosphate uridyltransferase deficiency. Mol Genet Metab 2004; 81:22–30.

Bick U, Ullrich K, Stober U, Moller H, Schuierer G, Ludolph AC, Oberwittler C, Weglage J, Wendel U. White matter abnormalities in patients with treated hyperphenylalaninemia: magnetic resonance relaxometry and proton spectroscopy findings. Eur J Pediatr 1993; 152:1012–1020.

Bjelland I, Tell GS, Vollset SE, Refsun H, Ueland PM. Folate, vitamin B12, homocysteine and the MTHFR 677C -> T polymorphism in anxiety and depression: the Hordaland Homocysteine Study. Arch Gen Psychiatr 2003; 60:618–626.

Blandini F, Fancellu R, Martignoni E, Mangiagalli A, Pacchetti C, Samuelle A, Nappi G. Plasma homocysteine and l-dopa metabolism in patients with Parkinson disease. Clin Chem 2001; 47:1102–1104.

Bosch AM, Bakker HD, Wenniger-Prick LJ, Wanders RJ, Wijburg FA. High tolerance for oral galactose in classical galactosemia: dietary implications. Arch Dis Child 2004a; 89:1034–1036.

Bosch AM, Grootenhuis MA, Bakker HD, Heijmans HS, Wijburg FA, Last BF. Living with classical galactosemia: health-related quality of life consequences. Pediatrics 2004b; 113: e423–e428.

Bostom AG, Jacques PF, Nadeau MR, Williams RR, Ellison RC, Selhub J. Post-methionine load hyperhomocysteinemia in persons with normal fasting total plasma homocysteine: initial results from the NHLBI Family Heart Study. Atherosclerosis 1995; 116:147–151.

Bottiglieri T, Hyland K, Reynolds EH. The clinical potential of ademethionine (S-adenosylmethionine) in neurological disorders. Drugs 1994; 48:137–152.

Bottiglieri T, Laundy M, Crellin R, Toone BK, Carney MW, Reynolds EH. Homocysteine, folate, methylation, and monoamine metabolism in depression. J Neurol Neurosurg Psychiatr 2000; 69:228–232.

Brenton DP, Gardiner RM. Transport of L-phenylalanine and related amino acids at the ovine blood-brain barrier. J Physiol 1988; 402:497–514.

Brody BA, Kinney HC, Kloman AS, Gilles FH. Sequence of central nervous system myelination in human infancy I: an autopsy study of myelination. J Neuropathol Exp Neurol 1987; 46:283–301.

Burgard P, Rey F, Rupp A, Abadie V, Rey J. Neuropsychologic functions of early treated patients with phenylketonuria, on and off diet: results of a cross-national and cross-sectional study. Pediatric Res 1997; 41:368–374.

Carmelli D, DeCarli C, Swan GE, Jack LM, Reed T, Wolf PA, Miller BL. Evidence for genetic variance in white matter hyperintensity volume in normal elderly male twins. Stroke 1998; 29:1177–1181.

Cervilla J, Prince M, Joels S, Russ C, Lovestone S. Genes related to vascular disease (APOE, VLDL-R, DCP-1) and other vascular factors in late-life depression. Am J Geriatr Psychiatr 2004; 12:202–210.

Chauveau P, Chadefaux B, Coude M, Aupetit J, Kamoun P, Jungers P. Long-term folic acid (but not pyridoxine) supplementation lowers elevated plasma homocysteine level in chronic renal failure. Miner Electrolyte Metab 1996; 22:106–109.

Chern CJ, Beutler E. Pyridoxal kinase: decreased activity in red blood cells of Afro-Americans. Science 1975; 187:1084–1086.

Chishty M, Reichel A, Begley DJ, Abbott NJ. Glial induction of blood-brain barrier-like L-system amino acid transport in the EC304 cell line. Glia 2002; 39:99–104.

Choi TB, Pardidge WM. Phenylalanine transport at the human blood-brain barrier. Studies with isolated human brain capillaries. J Biol Chem 1986; 261:6536–6541.

Christie LA, Riedel G, Algaidi SA, Whalley LJ, Platt B. Enhanced hippocampal long-term potentiation in rats after chronic exposure to homocysteine. Neurosci. Lett. 2005; 373:119–124.

Clarke JT, Gates RD, Hogan SE, Barrett M, MacDonald G. Neuropsychological studies on adolescents with phenylketonuria returned to phenylalanine-restricted diets. Am J Ment Ret 1987; 92:255–262.

Clarke R, Daly L, Robinson K, Naughten E, Cahalane S, Fowler B, Graham I. Hyperhomocysteinemia: an independent risk factor for vascular disease. N Engl J Med 1991; 324:1149–1155.

Clarke R, Woodhouse P, Ulvik A, Frost C, Sherliker P, Refsum H, Ueland PM, Khaw KT. Variability and determinants of total homocysteine concentrations in plasma in an elderly population. Clin Chem 1998a; 44:102–107.

Clarke R, Smith AD, Jobst KA, Refsum H, Sutton L, Ueland PM. Folate, vitamin B12, and serum total homocysteine levels in confirmed Alzheimer disease. Arch. Neurol 1998b; 55:1449–1455.

Cleary MA, Walter JH, Wraith JE, Whiter F, Tyler K, Jenkins JP. Magnetic resonance imaging in phenylketonuria: reversal of cerebral white matter change. J Pediatr 1995; 127:251–255.

Cohn RM, Roth KS. Metabolic Diseases: a guide to early recognition. WB Saunders, Philadelphia, 1983, pp 221–226.

Colello RJ, Pott U, Schwab ME. The role of oligodendrocytes and myelin on axon maturation in the developing rat retinofugal pathway. J Neurosci 1994; 14:2594–2605.

Crome L. A case of galactosemia with the pathological and neuropathological findings. Arch Dis Child 1962; 37:415–421.

Cuatrecasas P, Segal S. Galactose conversion to D-xylulose: an alternate route of galactose metabolism. Science 1966; 153:549–551.

de Jongh S, Vreken P. IJst L, Wanders RJ, Jakobs C, Bakker HD. Spontaneous pregnancy in a patient with classical galactosemia. J Inher Metab Dis 1999; 22:754–755.

de Leeuw FE, de Groot JC, Oudkerk M, Witteman JC, Hofman A, van Gijn J, Breteler MM. A follow-up study of blood pressure and cerebral white matter lesions. Ann Neurol 1999; 46:827–833.

de Sonneville LM, Schmidt E, Michel U, Batzler U. Preliminary neuropsychological test results. Eur J Pediatr 1990; 149:S39–S44.

DeCarli C, Mungas D, Harvey D, Reed B, Weiner M, Chui H, Jagust W. Memory impairment, but not cerebrovascular disease, predicts progression of MCI to dementia. Neurology 2004; 63:220–227.

Dent C. Discussion of Armstrong MD: relation of biochemical abnormality to development of mental defect in phenylketonuria. In: Etiologic Factors in Mental Retardation: Report of Twenty-Third Ross Pediatric Research Conference November 8–9, 1956. Ross Laboratories, Columbus OH, 1957, pp. 32–33.

DePietro FR, Fernstrom JD. The relative roles of phenylalanine and tyrosine as substrates for DOPA synthesis in PC12 cells. Brain Res 1999; 831:72–84.

Diamond A. Phenylalanine levels of 6–10 mg/dl may not be as benign as once thought. Acta Paediatr. 1994; 407(Suppl.):89–91.

Diamond A, Ciaramitaro V, Donner E, Djali S, Robinson MB. An animal model of early-treated PKU. J Neurosci 1994; 14:3072–3082.

Doscherholmen A, Hagen PS. A dual mechanism of vitamin B12 plasma absorption. J Clin Invest 1957; 36:1551–1557.

Dudman NP, Wilcken DE, Wang J, Lynch JF, Macey D, Lundberg P. Disordered methionine/homocysteine metabolism in premature vascular disease. Its occurrence, cofactor therapy and enzymology. Arterioscler Thromb 1993; 13:1253–1260.

Dufouil C, Alperovitch A, Ducros V, Tzourio C. Homocysteine, white-matter hyperintensities and cognition in healthy elderly people. Ann Neurol 2003; 53:214–221.

Dyer CA. Comments on the neuropathology of phenylketonuria. Eur J Pediatr 2000; 159(Suppl. 2):S107–S108.

Dyer CA, Philibotte T. A clone of the MOCH-1 glial tumor in culture: multiple phenotypes expressed under different environmental conditions. J Neuropath Exp Neurol 1995; 54:852–863.

Dyer CA, Kendler A, Philibotte T, Gardiner P, Cruz J, Levy HL. Evidence for central nervous system glial cell plasticity in phenylketonuria. J Neuropath Exp Neurol 1996; 55:795–814.

Dyer CA, Kendler A, Jean-Guillaume D, Awatramani R, Lee A, Mason LM, Kamholz J. GFAP-positive and myelin marker-positive glia in normal and pathologic environments. J Neurosci Res 2000; 60:412–426.

Ellenbogen L, Cooper BA. Vitamin B12. In Machlin LJ, (ed) Handbook of Vitamins. Marcel Dekker, New York, 1991, pp. 491–536.

Elsas LJ 2nd, Lai K. The molecular biology of galactosemia. Genet Med 1998; 1:40–48.

Elsas L, Fridovich-Keil JL, Leslie ND. Galactosemia: a molecular approach to the enigma. Internat Pediatr 1993; 8:101–108.

Elsas LJ 2nd, Langley S, Paulk EM, Hjelm LN, Dembure PP. A molecular approach to galactosemia. Eur J Pediatr 1995; 154:S21–S27.

El-Sherif Y, Tesoriero J, Hogan MV, Wieraszko A. Melatonin regulates neuronal plasticity in the hippocampus. J Neurosci Res 2003; 72:454–460.

Engbersen AM, Franken DG, Boers GH, Stevens EM, Trijbels FJ, Blom HJ. Thermolabile 5, 10-methylenetetrahydrofolate reductase as a cause of mild hyperhomocysteinemia. Am J Hum Genet 1995; 56:142–150.

Faust D, Libon D. Pueschel S Neuropsychological functioning in treated phenylketonuria. Int J Psychiatr Med 1986–1987; 16:169–177.

Fazekas F, Kleinert R, Offenbacher H, Schmidt R, Kleinert G, Payer F, Radner H, Lechner H. Pathologic correlates of incidental MRI white matter signal hyperintensities. Neurology 1993; 43:1683–1689.

Finkelstein JD. The metabolism of homocysteine: pathways and regulation. Eur J Pediatr 1998; 157(Suppl. 2):S40–S44.

Fishler K, Koch R, Donnell GN, Wenz E. Developmental aspects of galactosemia from infancy to childhood. Clin Pediatr (Phila) 1980; 19:38–44.

Fitzpatrick PF. The aromatic amino acid hydroxylases. Adv Enzymol Relat Areas Mol Biol 2000; 74:235–294.

Flanagan JM, Tighe O, O'Neill C, Naughten E, Mayne PD, Croke DT. Identification of sequence variation in the galactose-1-phosphate uridyl transferase gene by dHPLC. Mol Genet Metab 2004; 81:133–136.

Franken DG, Boers GH, Blom HJ, Trijbels JM. Effect of various regiments of vitamin B6 and folic acid on mild hyperhomocysteinemia in vascular patients. J Inher Metab Dis 1994; 17:159–162.

Fridovich-Keil JL, Langley SD, Mazur LA, Lennon JC, Dembure PP, Elsas JL 2nd. Identification and functional analysis of three distinct mutations in the human galactose-1-phosphate uridyl-transferase gene associated with galactosemia in a single family. Am J Hum Genet 1995; 56:640–646.

Friedman JH, Levy HL, Boustany RM. Late onset of distinct neurologic syndromes in galactosemic siblings. Neurology 1989; 39:741–742.

Fukami MH, Haavik J, Flatmark T. Phenylalanine as substrate for tyrosine hydroxylase in bovine adrenal chromaffin cells. Biochem J 1990; 268:525–528.

Fukui K, Omoi NO, Hayasaka T, Shinnkai T, Suzuki S, Abe K, Urano S. Cognitive impairment of rats caused by oxidative stress and aging, and its prevention by vitamin E. Ann NY Acad Sci 2002; 959:275–284.

Fuso A, Seminara L, Cavallaro RA, D'Anselmi F, Scarpa S. S-adenosylmethionine/homocysteine cycle alterations modify DNA methylation status with consequent deregulation of PS1 and BACE and beta-amyloid production. Mol Cell Neurosci 2005; 28:195–204.

Garcia AA, Haron Y, Evans LR, Smith MG, Freedman M, Roman GC. Metabolic markers of cobalamin deficiency and cognitive function in normal older adults. J Am Geriatr Soc 2004; 52:66–71.

Gardiner RM. Transport of amino acids across the blood-brain barrier: implication for treatment of maternal phenylketonuria. J Inherit Metab Dis 1990; 13:627–633.

Gibson JB. Gonadal function in galactosemics and in galactose-intoxicated animals. Eur J Pediatr 1995; 154:S14–S20.

Gourovitch ML, Craft S, Dowton SB, Ambrose R, Sparta S. Interhemispheric transfer in children with early-treated phenylketonuria. J Clin Exp Neuropsychol 1994; 16:393–404.

Greenberg SM, Cho HS, O'Donnell HC, Rosand J, Segal AZ, Younkin LH, Younkin SG, Rebeck GW. Plasma beta-amyloid peptide, transforming growth factor-beta 1, and risk for cerebral amyloid angiopathy. Ann NY Acad Sci 2000; 903:144–149.

Greenwald BS, Kramer-Ginsberg E, Krishnan RR, Ashtari M, Aupperle PM, Patel M. MRI signal hyperintensities in geriatric depression. Am J Psychiatr 1996; 153:1212–1215.

Griffiths P, Paterson L, Harvie A. Neuropsychological effect of subsequent exposure to phenylalanine in adolescents and young adults with early-treated phenylketonuria. J Intellect Disabil Res 1995; 39:365–372.

Guttler F, Lou H. Dietary problems of phenylketonuria: effect on CNS transmitters and their possible role in behaviour and neuropsychological function. J Inherit Metab Dis 1986; 9(Suppl. 2):169–177.

Haberland C, Perou M, Brunngraber EG, Hof H. The neuropathology of galactosemia. A histopathological and biochemical study. J Neuropathol Exp Neurol 1971; 30:431–447.

Hankey GJ, Eikelboom JW. Homocysteine and vascular disease. Lancet 1999; 354:407–413.

Hassan A, Hunt BJ, O'sullivan M, Bell R, D'Souza R, Jeffery S, Bamford JM, Markus HS. Homocysteine is a risk factor for cerebral small vessel disease, acting *via* endothelial dysfunction. Brain 2004; 127:212–219.

Heinecke JW. Superoxide-mediated oxidation of low density lipoprotein by thiols. In: Cerruti PA, Fridovich I, McCord JM (eds) Oxy-Radicals in Molecular Biology and Pathology. Alan R. Liss, New York, 1988, pp. 443–457.

Hirokawa H, Okano Y, Asada M, Fujimoto A, Suyama I, Isshiki G. Molecular basis for phenotypic heterogeneity in galactosemia: prediction of clinical phenotype from genotype in Japanese patients. Eur J Hum Genet 1999; 7:757–764.

Ho PI, Collins SC, Dhitavat S, Ortiz D, Ashline D, Rogers E, Shea TB. Homocysteine potentiates beta-Amyloid neurotoxicity: role of oxidative stress. J Neurochem 2001; 78:249–253.

Hogervorst E, Ribeiro HM, Molyneux A, Budge M, Smith AD. Plasma homocysteine levels, cerebrovascular risk factors, and cerebral white matter changes (leukoaraiosis) in patients with Alzheimer disease. Arch Neurol 2002; 59:787–793.

Holton JB. Effects of galactosemia in utero. Eur J Pediatr 1995; 154:S77–S81.

Holtzman D, Mulkern R, Tsuji M, Cook C and Meyers R (1996). Phosphocreatine and creatine kinase in piglet cerebral gray and white matter in situ. Dev Neurosci. 18:535–541.

Hommes FA, Matsuo K. On a possible mechanism of abnormal brain development in experimental hyperphenylalaniemia. Neurochem Int 1987; 11:1–10

Huttenlocher PR. The neuropathology of phenylketonuria: human and animal studies. Eur J Pediatr 2000; 159(Suppl. 2):S102–S106.

Huttenlocher PR, Hillman RE, Hsia YE. Pseudotumor cerebri in galactosemia. J Pediatr 1970; 76:902–905.

Ikeda M, Levitt M, Udenfriend S. Phenylalanine as substrate and inhibitor of tyrosine hydroxylase. Arch Biochem Biophys 1967; 120:420–427.

Irons M, Levy HL, Pueschel S, Castree K. Accumulation of galactose-1-phosphate in the galactosemic fetus despite maternal milk avoidance. J Pediatr 1985; 107:261–263.

Jakobs C, Kleijer WJ, Bakker HD, Van Gennip A, Przyrembel H, Niermeijer MF. Dietary restriction of maternal lactose intake does not prevent accumulation of galactitol in the amniotic fluid of fetuses affected with galactosemia. Prenat Diagn 1988; 8:641–645.

Jakobs C, Kleijer WJ, Allen J, Holton JB. Prenatal diagnosis of galactosemia. Eur J Pediatr 1995; 154:S33–S36.

Jeerakathil T, Wolf PA, Beiser A, Massaro J, Seshadri S, D'Agostino RB, DeCarli C. Stroke risk profile predicts white matter hyperintensity volume: the Framingham Study. Stroke 2004; 35:1857–1861.

Jorm AF. Is depression a risk factor for dementia or cognitive decline? A review. Gerontology 2000; 46:219–227.

Kang SS, Wong PW, Susmano A, Sora J, Norusis M, Ruggie N. Thermolabile methylenetetrahy-drofolate reductase: an inherited risk factor for coronary artery disease. Am J Hum Genet 1991; 48:536–545.

Katz I, Lloyd T, Kaufman S. Studies on phenylalanine and tyrosine by hydroxylation by rat brain tyrosine hydroxylase. Biochim Biophys Acta 1976; 445:567–578.

Kaufman FR, Reichardt JK, Ng WG, Xu YK, Manis FR, McBride-Chang C, Wolff JA. Correlation of cognitive, neurologic, and ovarian outcome with the Q188R mutation of the galactose-1-phosphate uridyltransferase gene. J Pediatr 1994; 125:225–227.

Kaufman FR, Horton EJ, Gott P, Wolff JA, Nelson MD Jr, Azen C, Manis FR. Abnormal somato-sensory evoked potentials in patients with classic galactosemia: correlation with neurologic outcome. J Child Neurol 1995a; 10:32–36.

Kaufman FR, McBride-Chang C, Manis FR, Wolff JA, Nelson MD. Cognitive functioning, neu-rologic status and brain imaging in classical galactosemia. Eur J Pediatr 1995b; 154:S2–S5.

Kelley RI, Segal S. Evaluation of reduced activity galactose-1-phosphate uridyl transferase by combined radioisotopic assay and high-resolution isoelectric focusing. J Lab Clin Med 1989; 114:152–156.

Kelly PJ, Rosand J, Kistler JP, Shin VE, Silveira S, Plomaritoglou A, Furie KL. Homocysteine, MTHFR 677C -> T polymorphism, and risk of ischemic stroke: results of a meta-analysis. Neurology 2002; 59:529–536.

Kim WK, Pae YS. Involvement of N-methyl-d-aspartate receptor and free radical in homocysteine-mediated toxicity on rat cerebellar granule cells in culture. Neurosci Lett 1996; 216:117–120.

Kirkpatarick LL, Brady ST. Modulation of the axonal microtubule cytoskeleton by myelinating Schwann cells. J Neurosci 1994; 14:7440–7455.

Klerk M, Verhoef P, Clarke R, Blom HJ, Kok FJ, Shouten EG. MTHFR Studies Collaboration Group. MTHFR 677C -> T polymorphism and risk of coronary heart disease: a meta-analysis. JAMA 2002; 288:2023–2031.

Koch R, Levy HL, Matalon R, Rouse B, Hanley WB, Trefz F, Azen C, Friedman EG, de la Cruz F, Guttler F, Acosta PB. The international collaborative study of maternal phenylketonuria: status report 1994. Acta Paediatr Suppl 1994; 407:111–119.

Kokame K, Kato H, Miyata T. Homocysteine-respondent genes in vascular endothelial cells iden-tified by differential display analysis, GRP78/BiP and novel genes. J Biol Chem 1996; 271:29659–29665.

Kornguth S, Gilbert-Barness E, Langer E, Hegstrand L. Golgi-Kopsch silver study of the brain of a patient with untreated phenylketonuria, seizures, and cortical blindness. Am J Med Genet 1992; 44:443–448.

Krause W, Halminski M, McDonald L, Dembure P, Salvo R, Freides D, Elsas L. Biochemical and neuropsychological effects of elevated plasma phenylalanine in patients with treated phenylke-tonuria: a model for the study of phenylalanine and brain function in man. J Clin Invest 1985; 75:40–48.

Krishnan KR, Hays JC, Blazer DG. MRI-defined vascular depression. Am J Psychiatr 1997; 154:497–501.

Kruman II, Culmsee C, Chan SL, Kruman Y, Guo Z, Penix L, Mattson MP. Homocysteine elicits a DNA damage response in neurons that promotes apoptosis and hypersensitivity to excitotox-icity. J Neurosci 2000; 20:6920–6926.

Kruman II, Kumaravel TS, Lohani A, Pedersen WA, Cutler RG, Kruman Y, Haughey N, Lee J, Evans M, Mattson MP. Folic acid deficiency and homocysteine impair DNA repair in hippoc-ampal neurons and sensitize them to amyloid toxicity in experimental models of Alzheimer's disease. J Neurosci 2002; 22:1752–1762.

Kunugi H, Fukuda R, Hattori M, Kato T, Tatsumi M, Sakai T, Hirose T, Nanko S. C677T poly-morphism in methylenetetrahydrofolate reductase gene and psychoses. Mol Psychiatr 1998; 3:435–437.

450   K.H. Schulpis, S. Tsakiris

Kure S, Hou DC, Ohura T, Iwamoto H, Suzuki S, Sugiyama N, Sakamoto O, Fujii K, Matsubara Y, Narisawa K. Tetrahydrobiopterin-responsive phenylalanine hydroxylase deficiency. J Pediatr 1999; 135:375–378.

Lai K, Langley SD, Singh RH, Dembure PP, Hjelm LN, Elsas LJ 2nd. A prevalent mutation for galactosemia among black Americans. J Pediatr 1996; 128:89–95.

Ledley FD, Grennet JE, Woo LC. Structure of aromatic amino acid hydroxylases. In: Kaufman S (ed) Amino Acids in Health and Disease: New Prospectives. Alan R. Liss, New York, 1987, pp. 267–284.

Lee PJ, Ridout D, Walter JH, Cockburn F. Maternal phenylketonuria: report from the United Kingdom Registry 1978–1997. Arch Dis Child 2005; 90:143–146.

Lee PJ, Lilburn M, Baudin J. Maternal phenylketonuria: experiences from the United Kingdom. Pediatrics 2003a; 112:1553–1556.

Lee PJ, Lilburn M, Wendel U, Schadewaldt P. A woman with untreated galactosemia. Lancet 2003b; 362:446.

Legido A, Tonyes L, Carter D, Schoemaker A, Di George A, Grover WD. Treatment variables and intellectual outcome in children with classic phenylketonuria. Clin Pediatrics (Phila) 1993; 32:417–425.

Lenke RR, Levy HL. Maternal phenylketonuria and hyperphenylalaninemia. An international survey of the outcome of untreated and treated pregnancies. N Engl J Med 1980; 30::1202–1208.

Lenze E, Cross D, McKeel D, Neuman RJ, Shelina YI. White-matter hyperintensities and gray-matter lesions in physically healthy depressed subjects. Am J Psychiatr 1999; 1602–1607.

Leslie ND, Immerman EB, Flach JE, Florez M, Fridovich-Keil JL, Elsas LJ. The human galactose-1-phosphate uridyltransferase gene. Genomics 1992; 14:474–480.

Leslie ND, Yager KL, McNamara PD, Segal S. A mouse model of galactose-1-phosphate uridyl transferase deficiency. Biochem Mol Med 1996; 59:7–12.

Levy HL, Waisbren SE. PKU in adolescents: rationale and psychosocial factors in diet continuation. Acta Paediatr. 1994; 407(Suppl.):92–97.

Levy HL, Brown AE, Williams SE, de Juan E Jr. Vitreous hemorrhage as an ophthalmic complication of galactosemia. J Pediatr 1996; 129:922–925.

Lipton SA, Kim WK, Choi YB, Kumar S, D'Emilia DM, Rayudu PV, Arnelle DR, Stamler JS. Neurotoxicity associated with dual actions of homocysteine at the N-methyl-D-aspartate receptor. Proc Natl Acad Sci USA 1997; 94:5923–5928.

Longstreth WT Jr, Manolio TA, Arnold A, Burke GL, Bryan N, Jungreis CA, Enright PL, O'Leary D, Fried L. Clinical correlates of white matter findings on cranial magnetic resonance imaging of 3301 elderly people. The Cardiovascular Health Study. Stroke. 1996; 27:1274–1282.

Luchsinger JA, Tang MX, Shea S. Miller J, Green R, Mayeux R. Plasma homocysteine levels and risk of Alzheimer disease. Neurology 2004; 62:1972–1976.

Lykkelund C. Nielsen JB, Lou HC, Rasmussen V, Gerdes AM, Christensen E and Guttler F. Increased neurotransmitter biosynthesis in phenylketonuria induced by phenylalanine restriction or by supplementation of unrestricted diet with large amounts of tyrosine. Eur J Pediatr 1988; 148:238–245.

Mabry C, Denniston J, Coldwell JG. Mental retardation in children of phenylketonuric mothers. N Engl J Med 1966; 275:1331–1336.

Madison DL, Kruger WS, Kim T, Pfeiffer SE. Differential expression of rab 3 isoforms in oligodendrocytes and astrocytes. J Neurosci Res 1996; 45:258–268.

Malamud N. Neuropathology of phenylketonuria. J Neuropathol Exp Neurol 1966; 25:254–268.

Marinou K, Tsakiris S, Tsopanakis C, Schulpis KH, Behrakis P. Mg2+-ATPase activity in suckling rat brain regions in galactosaemia in vitro. L-Cysteine and glutathione effects. Toxicol Vitro. 2005; 19:167–172.

Matsuoka Y, Saito M, LaFrancois J, Saito M, Graynor K, Olm V, Wang L, Casey E, Lu Y, Shiratori C, Lemere C, Duff K. Novel therapeutic approach for the treatment of Alzheimer's disease by peripheral administration of agents with an affinity to beta-amyloid. J Neurosci 2003; 23:29–33.

Mattson MP. Apoptotic and anti-apoptotic synaptic signaling mechanisms. Brain Pathol 2000; 10:300–312.

Mayeux R, Honig LS, Tang MX, Manly J, Stern Y, Schupf N, Mehta D. Plasma A[beta]40 and A[beta]42 and Alzheimer's disease: relation to age, mortality, and risk. Neurology 2003; 61:1185–1190.

Mazur A, Harrow B. Textbook of Biochemistry. WB Saunder, Philadelphia, 1971.

Mazzocco MM, Nord AM, van Doorninck W, Greene CL, Kovar CG, Pennington BF. Cognitive development among children with early-treated phenylketonuria. Dev Neuropsychol 1994; 10:133–151.

McCaddon A, Hudson P, Davies G, Hughes A, Williams JH, Wilkinson C. Homocysteine and cognitive decline in healthy elderly. Dement Geriatr Cogn Disord 2001; 12:309–313.

McCaddon A, Hudson P, Hill D, Barber J, Lloyd A, Davies G, Regland B. Alzheimer's disease and total plasma aminothiols. Biol Psychiatr 2003; 53:254–260.

McCormick DB, Gregory ME, Snell EE. Pyridoxal phosphokinases. Assay, distribution, purification and properties. J Biol Chem 1961; 236:2076–2084.

McCully KS. Homocysteine and vascular disease. Nat Med 1996; 2:386–389.

MacDonald A, Rylance GW, Asplin DA, Hall K, Harris G, Booth IW. Feeding problems in young PKU children. Acta Paediatr 1994; 407(Suppl.):73–74.

McDonald JD. Postnatal growth rates in a mouse genetic model of classical phenylketonuria. Contemp Top Lab Amin Sci 2000; 39:54–56.

McDonald JD, Charlton CK. Characterization of mutations at the mouse phenylalanine hydroxylase locus. Genomics 1997; 39:402–405.

McDonald JD, Bode VC, Dove WF, Shedlovsky A. Pahhph-5: a mouse mutant deficiency in phenylalanine hydroxylase. Proc Natl Acad Sci USA 1990; 87:1965–1967.

Miller AL. The methionine-homocysteine cycle and its effects on cognitive diseases. Altern Med Rev 2003c; 8:7–19.

Miller JW, Green R, Mungas DM, Reed BR, Jagust WJ. Homocysteine, vitamin B6, and vascular disease in AD patients. Neurology 2002; 58:1471–1475.

Miller JW, Green R, Ramos MI, Allen LH, Mungas DM, Jagust WJ, Haan MN. Homocysteine and cognitive function in the Sacramento Area Latino Study on Aging. Am J Clin Nutr 2003a; 78:441–447.

Miller JW, Selhub J, Nadeau MR, Thomas CA, Feldman RG, Wolf PA. Effect of L-dopa on plasma homocysteine in PD patients: relationship to B-vitamin status. Neurology 2003b; 60:1125–11292.

Moller HE, Ullrich K, Vermathen P, Schuierer G, Koch H. In vivo study of brain metabolism in galactosemia by ^{1}H and ^{31}P magnetic resonance spectroscopy. Eur J Pediatr 1995; 154: S8–S131.

Morris MS, Jacques PF, Rosenberg IH, Selhub J. National Health and Nutrition Examination Survey. Hyperhomocysteinemia associated with poor recall in the third National Health and Nutrition Examination Survey. Am J Clin Nutr 2001; 73:927–933.

Mudd SH, Levy HL, Skovby F. Disorder of transsulfuration. In Scriver CR. Beaudet al., Sly WS, Valle D (eds) The Metabolic Basis of Inherited Disease, vol 1(ed 6). McGraw-Hill, New York, 1989, pp. 693–734.

Muntau AC, Roschinger E, Habich M, Demmelmair H, Hoffmann B, Sommerhoff CP, Roscher AA. Tetrahydrobiopterin as an alternative treatment for mild phenylketonuria. N Engl J Med 2002; 347:2122–2132.

Murphy M, McHugh B, Tighe O, Mayne P, O'Neill C, Naughten E, Croke DT. Genetic basis of transferase-deficient galactosemia in Ireland and the population history of the Irish Travellers. Eur J Hum Genet 1999; 7:549–554.

Myers RR, Powell HC. Galactose neuropathy: Impact of chronic endoneurial edema on nerve blood flow. Ann Neurol 1984; 16:587–594.

Nelson CD, Waggoner DD, Donnell GN, Tuerck JH, Buist NR. Verbal dyspraxia in treated galactosemia. Pediatrics 1991; 88:346–350.

Nilsson K, Gustafson L, Hultberg B. The plasma homocysteine concentration is better than that of serum methylmalonic acid as a marker for sociopsychological performance in a psychogeriatric population. Clin Chem 2000; 46:691–696.

Ning C, Reynolds R, Chen J, Yager C, Berry GT, McNamara PD, Leslie N, Segal S. Galactose metabolism by the mouse with galactose-1-phosphate uridyltransferase deficiency. Pediatr Res 2000; 48:211–217.

O'suilleabhain PE, Sung V, Hernandez C, Lacritz L, Dewey RB Jr, Bottiglieri T, Diaz-Arrastia R. Elevated plasma homocysteine level in patients with Parkinson disease: motor, affective, and cognitive associations. Arch Neurol 2004; 61:865–868.

Osuna C, Reiter RJ, Garcia JJ, Karbownik M, Tan DX, Calvo JR, Manchester LC. Inhibitory effect of melatonin on homocysteine-induced lipid peroxidation in rat brain homogenates. Pharmacol Toxicol 2002; 90:32–37.

Panis B, Forget PP, van Kroonenburgh MJ, Vermeer C, Menheere PP, Nieman FH, Rubio-Gozalbo ME. Bone metabolism in galactosemia. Bone 2004; 35:982–987.

Pardridge WM. Advances in cell biology of the blood-brain barrier transport. Semin Cell Biol 1991; 2:419–426.

Pardridge WM. Blood-brain barrier carrier-mediated transport and brain metabolism of amino acids. Neurochem Res 1998; 23:635–644.

Parle M, Dhingra D. Ascorbic Acid: a promising memory-enhancer in mice. J Pharmacol Sci 2003; 93:129–135.

Pascucci T, Ventura R, Puglisi-Allegra S, Cabib S. Deficits in brain serotonin synthesis in a genetic mouse model for phenylketonuria. Neuroreport 2002; 13:2561–2564.

Petersen RC, Stevens JC, Ganguli M, Tangalos EG, Cummings JL, DeKosky ST. Practice parameter: early detection of dementia: mild cognitive impairment (an evidence-based review). Report of the Quality Standards Subcommittee of the American Academy of Neurology. Neurology 2001; 56:1133–1142.

Petry K, Greinix HT, Nudelman E, Eisen H, Hakomori S, Levy HL, Reichardt JK. Characterization of a novel biochemical abnormality in galactosemia: deficiency of glycolipids containing galactose or N-acetylgalactosamine and accumulation of precursors in brain and lymphocytes. Biochem Med Metab Biol 1991; 46:93–104.

Pietz J, Landwehr R, Kutscha A, Schmidt H, de Sonneville L, Trefz FK. Effect of high-dose supplementation on brain function in adults with phenylketonuria. J Pediatr 1995; 127:936–943.

Porrino LJ, Goldman-Rakic PS. Brainstem innervation of prefrontal and anterior cingulate cortex in the rhesus monkey revealed by retrograde transport of HRP. J Comp Neurol 1982; 205:63–76.

Postuma RB, Lang AE. Homocysteine and levodopa: should Parkinson disease patients receive preventative therapy? Neurology 2004; 63:886–891.

Powers JM, Rosenblatt DS, Schmidt RE, Cross AH, Black JT, Moser AB, Moser HW, Morgan DJ. Neurological and neuropathologic heterogeneity in two brothers with cobalamin C deficiency. Ann Neurol 2001; 49:396–400.

Prins ND, Den Heijer T, Hofman A, Koudstaal PJ, Jolles J, Clarke R, Breteler M; Rotterdam Scan Study. Homocysteine and cognitive function in the elderly: the Rotterdam Scan Study. Neurology 2002; 59:1375–1380.

Puglisi-Allegra S, Cabib S, Pascucci T, Ventrura R, Cali F, Romano V. Dramatic brain aminergic deficit in a genetic mouse model of phenylketonuria. Neuroreport 2000; 11:1361–1364.

Quan-Ma R, Wells HJ, Wells WW, Sherman FE, Egan TJ. Galactitol in the tissues of a galactosemic child. Am J Dis Child 1966; 112:477–478.

Rabins PV, Pearlson GD, Aylward E, Kumar AJ, Dowell K. Cortical magnetic resonance imaging changes in elderly in patients with major depression. Am J Psychiatr 1991; 148:617–620.

Rasmussen K, Moller J, Lyngbak M, Pedersen M, Dybkjaer L. Age- and gender-specific intervals for total homocyteine and methylmalonic acid in plasma before and after vitamin supplementation. Clin Chem 1996; 42:630–636.

Refolo LM, Malester B, LaFrancois J, Bryant-Thomas T, Wang R, Tint GS, Sambamurti K, Duff K, Pappolla MA. Hypercholesterolemia accelerates the Alzheimer's amyloid pathology in a transgenic mouse model. Neurobiol Dis 2000; 7:321–331.

Reiter RJ. Oxidative damage in the central nervous system: protection by melatonin. Prog Neurobiol 1998; 56:359–384.

Religa D, Styczynsk M, Peplonska E, Gabryelewicz T, Pfeffer A, Chodakowska M, Luczywek E, Wasiak B, Stepien K, Golebiowski M, Winblad B, Barcikowska M. Homocysteine, apolipoprotein E, and methylenetetrahydrofolate reductase in Alzheimer's disease and mild cognitive impairment. Dement Geriatr Cogn Disord 2003; 16:64–70.

Rogers S, Heidenreich R, Mallee J, Segal S. Regional activity of galactose-1-phosphate uridyltransferase in rat brain. Pediatr Res 1992; 31:512–515.

Rogers JD, Sanchez-Saffon A, Frol AB, Diaz-Arrastia R. Elevated plasma homocysteine levels in patients treated with levodopa: association with vascular disease. Arch Neurol 2003; 60:59–64.

Ronn LC, Bock E, Linnemann D, Jahnsen H. NCAM-antibodies modulate induction of long-term potentiation in rat hippocampal CA1. Brain Res 1995; 677:145–151.

Rosenblatt J, Chinkes D, Wolfe M, Wolfer RR. Stable isotope tracer analysis by GC-MS, including quantification of isotopomer effects, Am J Physiol 1992; 263:E584–E596.

Ross KL, Davis CN, Fridovich-Keil JL. Differential roles of the Leloir pathway enzymes and metabolites in defining galactose sensitivity in yeast. Mol Genet Metab 2004; 83:103–116.

Roze E, Gervais D, Demeret S. Ogier de Baulny H, Zittoun J, Benoist JF, Said G, Pierrot-Deseilligny C, Bolgert F. Neuropsychiatric disturbances in presumed late-onset cobalamin C disease. Arch Neurol 2003; 60:1457–1462.

Sai X, Kawamura Y, Kokame K, Yamaguchi H, Shiraishi H, Suzuki R, Suzuki T, Kawaichi M, Miyata T, Kitamura T, De Strooper B, Yanagisawa K, Komano H. Endoplasmic reticulum stress-inducible protein, Herp, enhances presenilin-mediated generation of amyloid beta-protein. J Biol Chem 2002; 277:12915–12920.

Sanchez I, Hassinger L, Paskevich PA, Shine HD and Nixon RA. Oligodendroglia regulate the regional expansion of axon caliber and local accumulation of neurofilaments during development of myelin formation. J. Neurosci 1996; 16:5095–5105.

Scarpa S, Fuso A, D'Anselmi F, Cavallaro RA. Presenilin 1 gene silencing by S-adenosylmethionine: a treatment for Alzheimer disease? FEBS Lett. 2003; 541:145–148.

Schachner M. Neural recognition molecules and synaptic plasticity. Curr Opin Cell Biol 1997; 9:627–634.

Schadewaldt P, Lilburn M, Wendel U, Lee PJ. Unexpected outcome in untreated galactosemia, Reply. Molec Genet Metab 2004; 81:255–257.

Schaumburg H, Kaplan J, Windebank A, Vick N, Rasmus S, Pleasure D, Brown MJ. Sensory neuropathy from pyridoxine abuse. A new megavitamin syndrome. N Engl J Med 1983; 309:445–448.

Schmidt E, Rupp A, Burgard P and Pietz. Information processing in early treated phenylketonuria. J Clin Exp Neuropsychol 1992; 14:388.

Schmidt E, Rupp A, Burgard P, Pietz J, Weglage J, de Sonneville L. Sustained attention in adult phenylketonuria: the influence of the concurrent phenylalanine-blood-level. J Clin Exp Neuropsychol 1994; 16:681–688.

Schmidt R, Launer LJ, Nilsson LG, Pajak A, Sans S, Berger K, Breteler MM, de Ridder M, Dufouil C, Fuhrer R, Giampaoli S, Hofman A; CASCADE Consortium Study. Magnetic resonance imaging of the brain in diabetes: the Cardiovascular Determinants of Dementia (CASCADE) study. Diabetes. 2004; 53:687–692.

Schuett VE, Brown ES. Diet policies of PKU clinics in the United States. Am J Publ Health 1984; 74:501–593.

Schulman JD, Mudd SH, Shulman NR, Landvater L. Pregnancy and thrombophlebitis in homocystinuria. Blood 1980; 56:326.

Schulpis KH, Michelakakis H, Charokopos E, Papakonstantinou E, Messaritakis J, Shin Y. UDP-galactose-4-epimerase in a boy with a trisomy 21. J Inherit Metab Dis. 1993; 16:1059–1060

Schulpis KH, Platokouki H, Papakonstantinou ED, Adamtziki E, Bargeliotis A, Aronis S. Haemostatic variables in phenylketonuric patients under dietary treatment. J Inherit Metab Dis 1996; 19:603–609.

Schulpis KH, Tjamouranis J, Karikas GA, Michelakakis H, Tsakiris S. In vivo effects of high phenylalanine blood levels on Na$^+$, K$^+$-ATPase, Mg$^+$-ATPase activities and biogenic amine concentrations in phenylketonuria. Clin Biochem 2002; 35:281–285.

Schulpis KH, Tsakiris S, Karikas GA, Moukas M, Behrakis P. Effect of diet on plasma total antioxidant status in phenylketonuric patients. Eur J Clin Nutr 2003; 57:383–387.

Schulpis KH, Kariyannis C, Papassotiriou I. Serum levels of neural protein S100B in phenylketonuria. Clin Biochem 2004; 37:76–79.

Schulpis KH, Tsakiris S, Traeger-Synodinos J, Papassotiriou I. Low total antioxidant status is implicated with high 8-hydroxy-2-deoxyguanosine serum concentrations in phenylketonuria. Clin Biochem 2005; 38:239–242.

Schulpis KH, Kalimeris K, Bakogiannis C, Tsakiris T, Tsakiris S. The effect of in vitro homocystinuria on the suckkling rat hippocampal acetycholinesterase.

Schweitzer-Krantz S, Burgard P. Survey of national guidelines for the treatment of phenylketonuria. Eur J Pediatr 2000; 159(Suppl 2):S70–S73.

Scriver CR, Waters PJ. Monogenic traits are not simple: lessons from phenylketonuria. Trends Genet 1999; 15:267–272.

Scriver CR, Kaufman S, Eisensmith RC, Woo SLC. The hyperphenylalaninemias. In Scriver CR, Beauder AL, Sly WS, Valle E (eds) The metabolic and molecular bases of inherited disease. McGraw-Hill, New York, 1995, pp. 1025–1075.

Segal S. The enigma of galactosemia. Int Pediatr 1992; 7:75–82.

Segal S. Defective galactosylation in galactosemia: is low cell UDPgalactose an explanation? Eur J Pediatr 1995; 154:S65–S71.

Segal S. Another aspect of the galactosemia enigma. Mol Genet Metab 2004; 81:253–254.

Selhub J, Miller JW. The pathogenesis of homocysteinemia: interruption of the coordinate regulation of S-adenosylmethionine of the remethylation and transsulfuration of homocysteine. Am J Clin Nutr 1992; 55:131–138.

Selhub J, Jacques PE, Bostom AG, D'Agostino RB, Wilson PW, Belanger AJ, O'Leary DH, Wolf PA, Schaefer EJ, Rosenberg IH. Association between plasma homocysteine concentrations and extracranial carotid-artery stenosis. N Engl J Med 1995; 332:286–291.

Sershen H, Debler EA, Lajtha. Alterations of cerebral amino acid transport processes. In Kaufman S (ed) Amino Acids in Health and Disease: New Prospectives. Alan R. Liss, New York, 1987, pp. 87–105.

Seshadri S, Beiser A, Selhub J, Jacques PF, Rosenberg IH, D'Agostino RB, Wilson PW, Wolf PA. Plasma homocysteine as a risk factor for dementia and Alzheimer's disease. N Engl J Med 2002; 346:476–483.

Shedlovsky A, McDonald JD, Symula D, Dove WF. Mouse models for human phenylketonuria. Genetics 1993; 134:1205–1210.

Shefer S, Tint GS, Jean-Guillaume D, Daikhin E, Kendler A, Nguyen LB, Yudkoff M, Dyer CA. Is there a relationship between 3-hydroxy-3-methylglutaryl coenzyme a reductase activity and forebrain pathology in the PKU mouse? J Neurosci Res 2000; 61:549–563.

Shen YX, Xu SY, Wei W, Sun XX, Yang J, Liu LH, Dong C. Melatonin reduces memory changes and neural oxidative damage in mice treated with D-galactose. J Pineal Res. 2002; 32:173–178.

Shield J, Wadsworth E, MacDonald A, Stephenson A, Tyfield L, Holton JB, Marlow N. The relationship of genotype to cognitive in galactosemia. Arch Dis Child 2000; 83:248–250.

Smith CB, Kang J. Cerebral protein synthesis in a genetic mouse model of phenylketonuria. Proc Natl Acad Sci USA 2000; 97:11014–1109.

Smith I, Erdohazi M, Macarthney FJ, Pincott JR, Wolff OH, Brenton DP, Biddle SA, Fairweather DV, Dobbing J. Fetal damage despite low-phenylalanine diet after conception in a phenylketonuric woman. Lancet 1979; 1:17–19.

Squire LR. Memory and the hippocampus: a synthesis from findings with rats, monkeys and humans. Psychol Rev 1992; 99:195–231.

Stabler SP, Marcell PD, Podell ER, Allen RH, Savage G, Lindenbaum J. Elevation of total homocysteine in the serum of patients with cobalamin or folate deficiency detected by capillary gas chromatography-mass spectrometry. J Clin Invest 1988; 81:466–474.

Stamler US, Osborne JA, Jaraki O, Rabbani LE, Mullins M, Singel D, Loscalzo J. Adverse vascular effects of homocysteine are modulated by endothelium-derived relaxing factor and related oxides of nitrogen. J Clin Invest 1993; 91:308–318.

Steffens DC, Krishnan KR. Structural neuroimaging and mood disorders: recent findings, implications for classification, and future directions. Biol Psychiatry 1998; 43:705–712.

Steinfeld R, Kohlschutter A, Zschocke J, Lindner M, Ullrich K, Lukacs Z. Tetrahydrobiopterin monotherapy for phenylketonuria patients with common mild mutations. Eur J Pediatr 2002; 161:403–405.

Stemerdink N (1996). Early and continuously treated phenylketonuria: An experimental neuropsychological approach. Amsterdam: Academisch Proefschrift.

Streck EL, Bavaresco CS, Netto CA, Wyse AT. Chronic hyperhomocysteinemia provokes a memory deficit in rats in the Morris water maze task. Behav. Brain Res 2004; 153:377–381.

Sugimoto K, Kasahara T, Yonezawa H, Yagihashi S. Peripheral nerve structure and function in long-term galactosemic dogs: morphometric and electron microscopic analyses. Acta Neuropathol (Berl) 1999; 97:369–376.

Takeuchi A, Irizarry MC, Duff K, Saido TC, Hsiao Ashe KH, Hasegawa M, Mann DM, Hyman BT, Iwatsubo T. Age-related amyloid beta deposition in transgenic mice overexpressing both Alzheimer mutant presenilin 1 and amyloid beta precursor protein Swedish mutant is not associated with global neuronal loss. Am J Pathol 2000; 1:331–339.

Tam SY, Roth RH. Mesoprefrontal dopaminergic neurons: can tyrosine availability influence their functions? Biochem Pharmacol 1997; 53:441–453.

Tam SY, Elsworth JD, Bradberry CW, Roth RH. Mesocortical dopamine neurons: high basal firing frequency predicts tyrosine dependence of dopamine synthesis. J Neural Transm Gen Sect 1990; 81:97–110.

Thompson AJ, Smith I, Brenton D, Youl BD, Rylance G, Davidson DC, Kendall B, Lees AJ. Neurological deterioration in young adults with phenylketonuria. Lancet 1990; 336:602–605.

Thompson AJ, Tillotson S, Smith I, Kendall B, Moore SG, Brenton DP. Brain MRI changes in phenylketonuria. Associations with dietary status. Brain 1993; 116:811–821.

Tsakiris S, Kouiniotou-Krontiri P, Schulpis KH, Stavridis JC. L-phenylalanine effect on rat brain acetylcholinesterase and Na⁺, K⁺-ATPase. Z Naturforch [C] 1998a; 53:163–167.

Tsakiris S, Kouiniotou-Krontiri P, Schulpis KH. L-phenylalanine effect on rat diaphragm acetylcholinesterase and Na⁺, K⁺-ATPase. Z Naturforch [C] 1998b; 53:1055–1060.

Tsakiris S, Schulpis KH, Tzamouranis J, Michelakakis H, Karikas GA. Reduced acetylcholinesterase activity in erythrocyte membranes from patients with phenylketonuria. Clin Biochem 2002; 35:615–619.

Tsakiris S, Schulpis KH. Alanine reverses the inhibitory effect of phenylalanine on acetylcholinesterase activity. Z Naturforsch [C] 2002; 57:506–511.

Tsakiris S, Carageorgiou H, Schulpis KH. The protective effect of L-cysteine and glutathione on the adult and aged rat brain (Na + ,K+)-ATPase and Mg2+-ATPase activities in galactosemia in vitro.Metab Brain Dis. 2005a; 20:87–95.

Tsakiris S, Michelakakis H, Schulpis KH. Erythrocyte membrane acetylcholinesterase, Na+, K+-ATPase and Mg2 + -ATPase activities in patients with classical galactosaemia. Acta Paediatr 2005b; 94:1223–1226

Tullberg M, Fletcher E, DeCarli C, Mungas D, Reed BR, Harvey DJ, Weiner MW, Chui HC, Jagust WJ. White matter lesions impair frontal lobe function regardless of their location. Neurology 2004; 63:246–253.

Ubbink JB, Vermaak WJ, van der Merwe A, Becker PJ. Vitamin B-12, vitamin B-6 and folate nutritional status in men with hyperhomocysteinemia. Am J Clin Nutr 1993; 57:47–53.

Ubbink JB, Vermaak WJ, van der Merwe A, Becker PJ, Delport R, Potgieter HC Vitamin requirements for the treatment of hyperhomocysteinemia in humans. J Nutr 1994; 124:1927–1933.

Ubbink JB, Vermaak WJ, van der Merwe A, Becker PJ, Potgieter H. Effective homocysteine metabolism may protect South African Blacks against coronary heart disease. Am J Clin Nutr 1995; 62:802–808.

Ubbink JB, Van der Merwe A, Delport R, Allen RH, Stabler SP, Riezler R, Vermaak WJ. The effect of a subnormal vitamin B_6 status on homocysteine metabolism J Clin Invest 1996; 98:177–184.

Upchurch GR Jr, Welch GN, Fabian AJ, Freedman JE, Johnson JL, Keaney JF Jr, Loscalzo J. Homocyst(e)ine decreases bioavailable nitric oxide by a mechanism involving glutathione peroxidase. J Biol Chem 1997; 207:17012–17017.

Valle D, Pai GG, Thomas GH, Pyeritz RE. Homocystinuria due to a cystathionine beta-synthase deficiency: clinical manifestations and therapy. Johns Hopkins Med J 1980 146:110–117

Van Dijk EJ, Prins ND, Vermeer SE, Hofman A, van Duijn CM, Koudstaal PJ, Breteler MM. Plasma amyloid beta, apolipoprotein E, lacunar infarcts, and white matter lesions. Ann Neurol 2004; 55:570–575.

Van Heyningen R. Formation of polyols by the lens of the rat with "sugar" cataracts. Nature 1959; 184:194–195.

Vanderstichele H, Van Kerschaver E, Hesse C, Davidsson P, Buyse MA, Andreasen N, Minthon L, Wallin A, Blennow K, Vanmechelen E. Standardization of measurements of beta-amyloid (1–42) in cerebrospinal fluid and plasma. Amyloid 2000; 7:245–258.

Vermaak WJ, Barnard HC, Potgieter GM, Theron H. Vitamin B_6 and coronary artery disease. Epidemiological observations and case studies. Atherosclerosis 1987; 63:235–238.

Vermeer SE, van Dijk EJ, Koudstaal PJ, Oudkerk M, Hofman A, Clarke R, Breteler MM. Homocysteine, silent brain infarcts, and white matter lesions: The Rotterdam Scan Study. Ann Neurol 2002; 51:285–289.

Waggoner DD, Buist NR, Donnell GN. Long-term prognosis in galactosemia: Results of a survey of 350 cases. J Inherit Metab Dis 1990; 13:802–818.

Walsh FS, Doherty P. Neural cell adhesion molecules of the immunoglobulin superfamily: role in axon growth and guidance. Annu Rev Cell Dev Biol. 1997; 13:425–456.

Walter JH, Collins JE, Leonard JV. Recommendations for the management of galactosemia. UK Galactosemia Steering Group. Arch Dis Child 1999; 80:93–96.

Walter JH, White FJ, Hall SK, MacDonald A, Rylance G, Boneh A, Francis DE, Shortland GJ, Schmidt M, Vail A. How practical are recommendations for dietary control in phenylketonuria? Lancet 2002; 360:55–57.

Wang ZI, Berry GT, Dreha SF, Zhao H, Segal S, Zimmerman RA. Proton magnetic resonance spectroscopy of brain metabolites in galactosemia. Ann Neurol 2001; 50:266–269.

Warfield A, Segal S. Myoinositol and phosphatidylinositol in synaptosomes from galactose-fed rats. Proc Natl Acad Sci USA 1978; 75:4568–4572.

Webb AL, Singh RH, Kennedy MJ. Elsas LJ. Verbal dyspraxia and galactosemia. Pediatr Res 2003; 53:396–402.

Weglage J, Pietsch M, Funders B, Koch HG, Ullrich K. Deficits in selective and sustained attention processes in early treated children with phenylketonuria: result of impaired frontal lobe functions? Eur J Pediatr 1996; 155:200–204.

Welsh MC, Pennington BF, Ozonoff S, Rouse B, McCabe ERB. Neuropsychology of early-treated phenylketonuria: Specific executive function deficits. Child Dev 1990; 61:1697–1713.

White AR, Juang X, Jobling MF, Barrow CJ, Beyreuther K, Masters CL, Bush AI, Cappai R. Homocysteine potentiates copper- and amyloid beta peptide-mediated toxicity in primary neuronal cultures: possible risk factors in the Alzheimer's type neurodegenerative pathways. J Neurochem 2001; 76:1509–1520.

Woo CW, Siow YL, Pierce GN, Choy PC, Minuk GY, Mymin D, O K. Hyperhomocysteinemia induces hepatic cholesterol biosynthesis and lipid accumulation via activation of transcription factors. Am J Physiol Endocrinol Metab. 2000; 288:1002–1010.

Wright CB, Lee HS, Paik MC, Stabler SP, Allen RH, Sacco RL. Total homocysteine and cognition in a tri-ethnic cohort: The Northern Manhattan Study. Neurology 2004; 63:254–260.

Wright CB, Paik MC, Brown TR, Stabler SP, Allen RH, Sacco RL, DeCarli C. Total homocysteine is associated with white matter hyperintensity volume: the Northern Manhattan Study. Stroke 2005 Jun; 36:1207–1211.

Yager C, Gibson J, States B, Elsas LJ, Segal S. Oxidation of galactose by galactose-1-phosphate uridyltransferase-deficient lymphoblasts. J Inherit Metab Dis 2001; 24:465–476.

Yasui K, Kowa H, Nakaso K, Takeshima T, Nakashima K. Plasma homocysteine and MTHFR C677T genotype in levodopa-treated patients with PD. Neurology 2000; 55:437–440.

Zhou JL, Zhu XG, Ling YL, Li Q. Melatonin reduces peroxynitrite-induced injury in aortic smooth muscle cells. Acta Pharmacol Sin 2004; 25:186–190.

Chapter 21
Wilson Disease

Peter Ferenci

Wilson disease is an autosomal recessive inherited disorder of copper metabolism resulting in pathological accumulation of copper in many organs and tissues. The hallmarks of the disease are the presence of liver disease, neurological symptoms and Kayser-Fleischer corneal rings.

The incidence of Wilson disease was estimated to be at least 1:30,000–50,000 with a gene frequency of 1:90–1:150. Among selected groups of patients Wilson disease is certainly more frequent. About 3–6% of patients transplanted for fulminant hepatic failure and 16% of young adults with chronic active hepatitis of unknown origin have Wilson disease.

The Wilson Disease Gene

ATP7B is the gene product of the Wilson disease gene located on chromosome 13 and resides in hepatocytes in the trans-Golgi network (Bull et al., 1993; Tanzi et al., 1993). The functionally important regions of the Wilson disease gene are six copper binding domains, a transduction domain (amino acid residues 837–864; containing a Thr-Gly-Glu motif) involved in the transduction of the energy of ATP hydrolysis to cation transport, a cation channel and phosphorylation domain (amino acid residues 971-1035; containing the highly conserved Asp-Lys-Thr-Gly-Thr motif), an ATP-binding domain (amino acid residues 1240–1291) and eight hydrophobic transmembrane sequences (Bull et al., 1993; Tanzi et al., 1993; Forbes et al., 1999; Lutsenko and Petris, 2003; Kenney and Cox , 2007; de Bie et al., 2007; Lee et al., 2001; Klomp et al., 2002), in one of which (region 6) is the cys-pro-cys sequence found in all P-type ATPases (Forbes et al., 1999; Lutsenko and Petris, 2003). Alternatively spliced forms of WDP lacking transmembrane sequences three and four (exon 8) are expressed in the brain.

Molecular genetic analysis of patients reveals over 350 distinct mutations (Kenney and Cox, 2007; database maintained at the University of Alberta http://www.medgen.med.ualberta.ca). Mutations include missense and nonsense mutations, deletions, and insertions. Some mutations are associated with a severe

D.W. McCandless (ed.) *Metabolic Encephalopathy*,
doi: 10.1007/978-0-387-79112-8_21, © Springer Science + Business Media, LLC 2009

impairment of copper transport resulting in severe liver disease very early in life; other mutations appear to be less severe with the disease appearing in mid adulthood. While most reported mutations occur in single families, a few are more common. The His1069Gln missense mutation occurs in 30–60% of patients of Eastern-, Northern- and Central-European origin. It is less frequent in patients of Mediterranean descent and only rarely seen in patients of non-European origin. The 2299insC mutation can be detected in some patients of European and Japanese descent. The Arg778Leu mutation is present in up to 60% of patients from the Far-East. In Sardinia two frame shift mutations (1515insT and 2464delC) are found in about 20% of patients. These mutations have not been found in other populations.

The study of genotype–phenotype correlations is hampered by the lack of clinical data, the rarity of some mutations, and the high frequency of the presence of two different mutations in individual patients (compound heterozygotes). In an ongoing study involving 1120 patients with Wilson disease, mostly from Europe, mutations on both chromosomes were identified in 67% of the patients and at least one mutation in 21%. Sufficient information is available only for the H1069Q mutation. Homozygosity for H1069Q is associated with late onset of the neurological disease. In contrast, patients with mutations in exons 8 and 13 commonly present with liver disease.

Other genes may be involved in the pathogenesis of Wilson disease. ATP7B interacts with COMMD1, a protein that is deleted in Bedlington terriers with hereditary copper toxicosis. COMMD1 (de Bie et al., 2007) exerts its regulatory role in copper homeostasis through the regulation of ATP7B stability. COMMD1 mutations may modify ATP7B function resulting in impaired copper trafficking.

Hepatic Copper Metabolism and the Role of ATP7B

Copper is an essential nutrient needed for such diverse processes as mitochondrial respiration (cytochrome C), melanin biosynthesis (tyrosinase), dopamine metabolism (DOPA-β-monooxygenase), iron homeostasis (ceruloplasmin), antioxidant defense (superoxyde dismutase), connective tissue formation (lysyl oxidase), and peptide amidation.

Dietary copper intake is approximately 1–2 mg/day. Quoted copper contents of foods are unreliable. While some foods, such as meats and shellfish, have consistently high concentrations, others such as dairy produce are consistently low in copper. However, the copper content of cereals and fruits varies greatly with soil copper content and the method of food preparation. Estimates of copper intake should include water copper content, and the permitted upper copper concentration for drinking water is $2 \, \text{mg L}^{-1}$. Approximately 10% of dietary copper is absorbed in the upper intestine, transported in the blood loosely bound to albumin, certain amino acids and peptides. Finally, most of the ingested copper is taken up by the liver. Copper homeostasis is critically dependent on the liver because this organ provides the only physiologically relevant mechanism for excretion of this metal.

Within the hepatic parenchyma, the uptake and storage of copper occurs in hepatocytes, which regulate the excretion of this metal into the bile. Copper appears in the bile as an unabsorbable complex, and as a result, there is no enterohepatic circulation of this metal.

The hepatic uptake of diet-derived copper occurs via the copper transporter 1 (Ctr1), which transports copper with high affinity in a metal-specific, saturable fashion at the hepatocyte plasma membrane (Lee et al., 2001; Klomp et al., 2002). After uptake copper is bound to metallothionein (MT), a cytosolic, low molecular weight, cystein-rich, metal binding protein. MT I and MT II are ubiquitously expressed in all cell types including hepatocytes, and have a critical role to protect intracellular proteins from copper toxicity (Palmiter, 1998; Kelley and Palmiter, 1996). The copper stored in metallothionein can be donated to other proteins. Specific pathways allow the intracellular trafficking and compartmentalization of copper, ensuring adequate cuproprotein synthesis while avoiding cellular toxicity (Fig.21.1).

Metallochaperones (like ATOX 1) transfer copper to the site of synthesis of copper containing proteins (Rae et al., 1999; Huffman and O'Halloran, 2002). The cytoplasmic copper chaperone ATOX1 is required for copper delivery to ATP7B by direct protein–protein interaction (Hamza et al., 1999; Walker et al., 2002). ATP7B is abundantly expressed in hepatocytes and is localized in these cells to the late secretory pathway, predominantly the *trans*-Golgi network. With increasing intracellular copper concentrations, this ATPase traffics to a cytoplasmic vesicular compartment that distributes near the canalicular membrane in polarized hepatocytes and

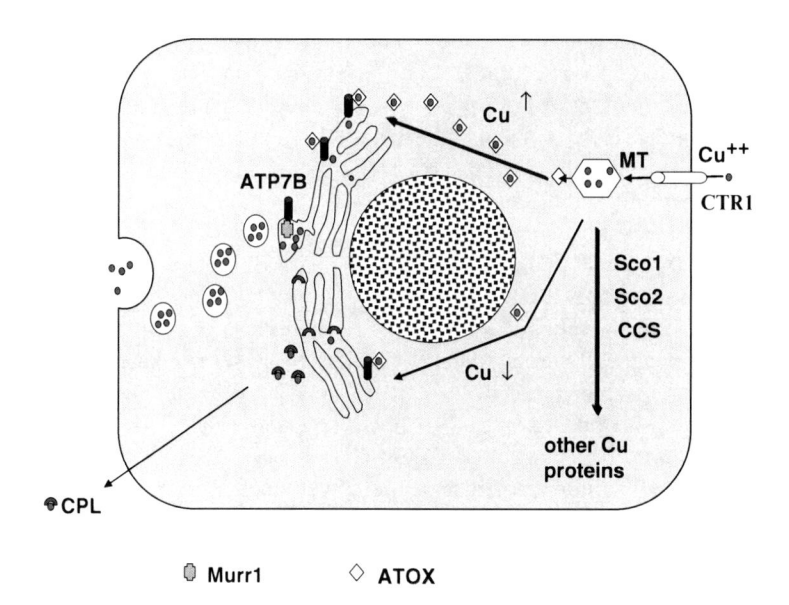

Fig. 21.1 Model of hepatobiliary copper transport CTR1= coppers transporter 1, MT= Metallothionein, CPL= ceruloplasmin ATOX, Sco1, Sco2, CCS — copper chaperones (*See also* Color Insert)

is critical for copper excretion (Schaefer et al., 1999). Copper is incorporated into the ceruloplasmin at the level of the Golgi compartment (Hellman et al., 2002). Ceruloplasmin contains six tightly bound copper atoms. Its main function is to carry copper to various tissues. Another important physiological role of ceruloplasmin is to act as ferrooxidase, converting Fe^{2+} to Fe^{3+}. Other chaperones (Sco1, Sco2, Cox17, lys7) carry copper for synthesis of the other cuproenzymes and do not require an interaction with ATP7B.

Biliary excretion is the only mechanism for copper elimination, and the amount of copper excreted in the bile is directly proportional to the size of the hepatic copper pool. Because hepatic uptake of dietary copper in not saturable, hepatic copper accumulation can easily be induced. Toxicity of copper, however, depends on its molecular association and subcellular localization rather than on its concentration in the liver. Metallothionein-bound copper is nontoxic. Several metals including zinc can induce metallothionein synthesis.

Pathogenesis

The basic defect is the impaired biliary excretion of copper resulting in the accumulation of copper in various organs including the liver, the cornea and the brain. The consequence of copper accumulation is the development of severe hepatic and neurological disease. Hepatic copper accumulation initially causes mitochondrial damage with alteration of lipid oxidation, resulting in marked hepatic steatosis. It is suggested that free-radical formation and oxidative damage, probably mediated via mitochondrial copper accumulation, are important in WD pathogenesis (Gu et al., 2000). Accumulation of prooxidant copper within hepatic mitochondria leads to premature oxidative aging of mitochondrial DNA by causing somatic mutations of the mitochondrial genome (Mansouri et al., 1997). Further damage to hepatocytes, inflammation, and fibrogenesis are caused by the copper released from necrotic hepatocytes. Recently a new mechanism for copper toxicity was described (Lang et al., 2007). Cu^{2+} triggers hepatocyte apoptosis through activation of acid sphingomyelinase and release of ceramide. Genetic deficiency or pharmacological inhibition of sphingomyelinase prevented Cu^{2+}-induced hepatocyte apoptosis and protected rats, genetically prone to develop Wilson disease, from acute hepatocyte death, liver failure and early death.

The pathogenesis of neurological Wilson disease is less clear. Copper is released into the circulation if the capacity of the liver to store copper is exhausted, and taken up by virtually all organs. Since copper is not taken up by neurons (Watt and Hooper, 2000), increased amounts of extracellular copper may explain the mechanism of neuronal damage in Wilson disease. It is conceivable that increased copper uptake into the brain is a direct result of certain mutations resulting in specific functional alterations of cerebral ATP7B. Neuronal damage is mediated by copper deposition in the brain (Watt and Hooper, 2000). Genetic variation of Apolipoprotein E (ApoE) has an important impact on the onset of neurological symptoms in H1069Q homozygotes. It is known that ApoE is able to bind metal ions with the highest affinity for copper and increases the resistance of cell cultures to oxidative stress (Miyata and Smith, 1996). However, ApoE isoforms vary in their neuroprotective properties. Patients who carry the wild type (apoE ε3) appear to be protected from copper toxicity to a certain degree (Schiefermeier et al., 2000).

Metals play an important role in neurobiology. Copper binding proteins are able to display oxidant or anti-oxidant properties, which would impact on neuronal function or in the triggering of neurodegenerative process. A major source of free radical production in the brain derives from copper. To prevent metal-mediated oxidative stress, cells have evolved complex metal transport systems. In neurodegenerative diseases, two proteins have been described as copper binding proteins: A protein related to Alzheimer's disease (AD), the amyloid precursor protein (APP), and a protein related to Creutzfeldt-Jakob disease, the Prion protein (PrP). The AD amyloid precursor protein is a major regulator of neuronal copper homeostasis which has a copper binding domain (CuBD). The surface location of this site, structural homology of CuBD to copper chaperones, and the role of APP in neuronal copper homeostasis are consistent with the CuBD acting as a neuronal metallotransporter (Barnham et al., 2003). In health the brain strictly regulates the movement of metals across the blood-brain barrier (BBB), which is relatively impermeable to fluctuations in blood levels. This barrier is relevant in AD because the disease is characterized by the accumulation in the brain of β-amyloid (Aβ) a copper-zinc-metalloprotein that aggregates and becomes redox-active in the presence of excessive amounts of these metals. We are only beginning to unravel the age-dependent failure of metal homeostatic mechanisms of the brain that contribute to abnormal Aβ biochemistry in AD (Bush et al., 2003). The ingestion of low amounts of copper in drinking water impairs trace conditioning and increases neuronal and brain parenchymal Aβ immuno-reactivity in cholesterol-supplemented rabbits (Sparks and Schreurs, 2003). This finding suggests that Aβ metabolism is extraordinarily sensitive to small changes in copper concentrations that might be transduced across the BBB. Clioquinol, an antibiotic and bioavailable Cu/Zn chelator decreased brain Aβ deposition by 49% in APP2576 transgenic mice treated orally for 9 weeks and improved their general health and body weight parameters (Beyreuther et al., 2001).

Ceruloplasmin is a ferroxidase that oxidizes toxic ferrous iron to its nontoxic ferric form (Patel et al., 2002). If ceruloplasmin is not present, iron concentration may increase.

In addition to the direct toxic effects of copper, in certain brain areas, as in the pineal gland, ATP7B is expressed and functionally active (Borjigin et al., 1999). Glucose metabolism, especially in striatal and cerebellar areas, is disturbed in patients with WD and correlates with the severity of extrapyramidal motor symptoms. The most severe cases are characterized by the lowest consumption in the striatal area. When there is marked improvement of extrapyramidal motor symptoms impaired glucose consumption reveals a persistent brain lesion (Hermann et al., 2002).

Furthermore the impaired synthesis of dopamine beta hydroxylase, a copper containing enzyme, may explain the preferential affection of basal ganglia. By single photon emission CT (SPECT) specific striatal binding ratios of two tracers ([123]2β-carbomethoxy-3β-4[123]iodophenyl tropane [123]βa-CIT and [123]iodobenzamide [123]IBZM) were reduced. The concordant bicompartmental dopaminergic deficit in neurological WD provides in vivo evidence for assigning WD to the group of secondary parkinsonian syndromes (Barthel et al., 2003). Few other neurotransmitter systems have been studied in WD. In vivo neuroimaging studies suggest that depression is associated with central serotonergic deficits (Hesse et al., 2003). [1]H spectra demonstrated a reduction of N-acetylaspartate and N-acetylaspartylglutamate in patients with neurological WD. Choline was also reduced (Page et al., 2004).

Clinical Presentations

Wilson disease may present at any age (Ferenci et al., 2007); the oldest reported case was 76 years at the time of diagnosis. The clinical symptoms are highly variable, the most common ones being liver disease and neuropsychiatric disturbances. Children usually present with liver disease, while in older patients neurological disease is more common. None of the clinical signs is typical and diagnostic. One of the most characteristic features of Wilson disease is that no two patients, even within a family, are ever quite alike. With increased awareness for Wilson disease patients are generally diagnosed earlier, thus "late" consequences of the disease like Kayser-Fleischer rings or severe neurological symptoms are less frequently seen. Early symptoms, if present at all, are uncharacteristic and nonspecific. Patients presenting with acute or chronic hepatic Wilson disease are indistinguishable from patients with liver diseases of other etiology. Early neurological symptoms are also quite untypical, and may progress slowly over many years before diagnosis is made based on "typical signs." About half of the patients are referred for psychological testing because of poor school performance or behavioral problems.

Kayser-Fleischer Rings

Characteristically, the ring starts as a small crescent of golden brown granular pigment seen at the top of the limbus. This is followed by the appearance of a lower crescent, and these two crescents gradually broaden, meet laterally and form complete rings (Fig. 21.2).The finding of a complete ring therefore suggests long-standing disease and is a useful indicator of severe copper overload. The ring is not always detected by clinical inspection. If doubt exists, the cornea should be examined under a slit lamp by experienced ophthalmologists. Kayser-Fleischer rings are present in 95% of patients with neurological symptoms, in 50-60% of patients without neurological symptoms, and only in 10% of asymptomatic siblings.

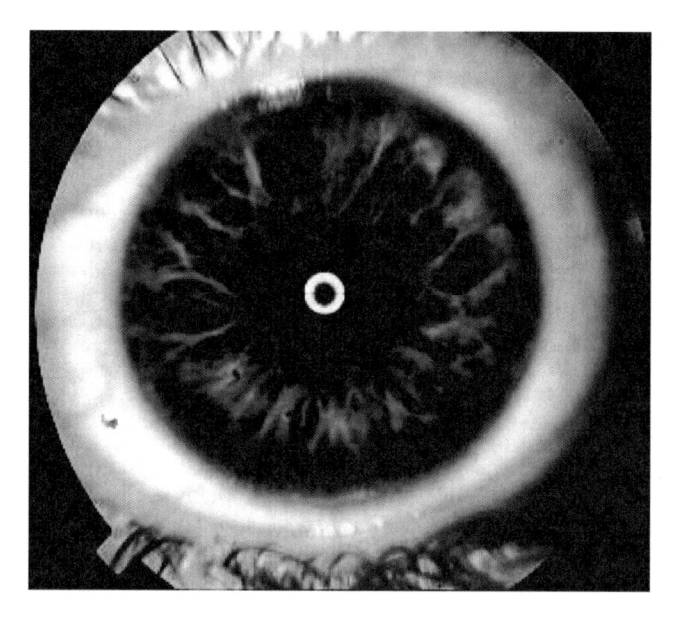

Fig. 21.2 Kayser-Fleischer ring in a 15 year old patient with neurological Wilson disease (*See also* Color Insert)

Liver Disease

Most patients with Wilson disease, whatever their clinical presentation, have some degree of liver disease. Chronic liver disease (if undiagnosed and untreated) may precede manifestation of neurological symptoms for more than 10 years. Patients can present with liver disease at any age. The most common age of hepatic manifestation is between 8 and 18 years, but cirrhosis may already be present in children below the age of 5. On the other hand, Wilson disease is diagnosed also in patients presenting with advanced chronic liver disease in their 50s or 60s, without neurological symptoms and without Kayser-Fleischer rings.

Depending on referral patterns the proportion of patients presenting with liver disease alone varies from 20 to 46%. Liver disease may mimic any forms of common liver conditions, ranging from asymptomatic transaminasemia to acute hepatitis, fulminant hepatic failure (about one out of six patients with hepatic presentation), chronic hepatitis, and cirrhosis (about one out of three patients) with all of its complications.

Acute Wilsonian Hepatitis and Fulminant Wilson Disease

Acute Wilsonian hepatitis is indistinguishable from other forms of acute (viral or toxic) liver diseases. It should be suspected in young patients with acute hepatitis nonA–E.

Liver histology often reveals the presence of cirrhosis. This initial episode of liver damage may be self-limiting and may resolve without treatment, and diagnosis is made retrospectively, when neurological symptoms occur years later.

On the other hand, the disease may rapidly deteriorate and resemble fulminant hepatic failure with massive jaundice, hypoalbuminemia, ascites, severe coagulation defects, hyperammonemia and hepatic encephalopathy. Hepatocellular necrosis results in the release of large amounts of stored copper. Hypercupriemia results in hemolysis and severe hemolytic anemia complicates acute liver disease. Although Wilson disease is a rare disease, in patients presenting with fulminant hepatic failure it is not uncommon and accounts for 6–12% of patients with fulminant hepatic failure referred for emergency liver transplantation.

Although fulminant and subfulminant liver failure due to Wilson disease has several distinctive features, rapid diagnosis may be very difficult. Serum aminotransferase activity is usually not increased above ten times normal and much lower than the values commonly recorded in fulminant hepatitis. The combination of anemia, marked jaundice and relatively low aminotransferase activities in young patients should always raise the suspicion of acute Wilson disease. The conventionally used parameters of copper metabolism are of little use. Kayser-Fleischer corneal rings and neurological abnormalities are absent in most patients presenting with acute liver disease. An alkaline phosphatase-total bilirubin ratio below 2.0 has been claimed to provide 100% sensitivity and specificity to diagnose Wilsonian fulminant liver failure, but the usefulness of this test was not confirmed in larger series. The best diagnostic test is the quantification of copper in biopsy material or in the explanted liver. One puzzling feature of fulminant Wilson disease is the preponderance of female sex (female: male ratio 3:1).

Chronic Hepatitis due to Wilson Disease

Wilson disease may present, particularly in young patients, with a clinical syndrome indistinguishable from chronic active hepatitis of other etiology (Scott et al., 1978). Symptoms include malaise, fatigue, anorexia, and vague abdominal complaints. Arthralgias, amenorrhea, delayed puberty, and low grade jaundice may be present. Frequently, Kayser Fleischer rings are absent and plasma ceruloplasmin is in the normal range. Liver biopsy shows severe chronic active hepatitis but diagnosis is missed if hepatic copper content is not measured. Suspicion for Wilson disease should be high in young persons with chronic active hepatitis of unclear etiology. In this group Wilson disease is a common diagnosis. Without treatment, patients progressively deteriorate with ascites, edema and occasionally jaundice within a few months, and eventually die of liver failure.

About half of the patients presenting with neurological symptoms may also suffer from significant liver disease. In a substantial proportion symptomatic liver disease predates the occurrence of neurological signs.

Neurological Presentation

Neurological symptoms usually develop in mid-teenage or in the twenties. However, there are well documented cases in which neurological symptoms developed much later (45–70 years). The initial symptoms may be very subtle abnormalities such as mild tremor, speech and writing problems and are frequently misdiagnosed as behavioral problems associated with puberty. The symptoms may remain constant or progress steadily. The hallmark of neurological Wilson disease is a progressive movement disorder. The most common symptoms are dysarthria, dysphagia, apraxia, and a tremor-rigidity syndrome ("juvenile Parkinsonism"). Because of increasing difficulty in controlling movement, patients become bedridden and unable to care for themselves. Ultimately, the patient becomes helpless – usually alert, but unable to talk. In patients presenting with advanced liver disease, neurological symptoms are mistaken as signs of hepatic encephalopathy.

Psychiatric Presentation

About one-third of patients initially present with psychiatric abnormalities. Symptoms can include reduced performance in school or at work, depression, very labile mood, sexual exhibitionism, and frank psychosis. Frequently, adolescents with problems in school or work are referred for psychological counseling and psychotherapy. Among our patients two were hospitalized in psychiatric institutions for psychosis, one having committed several suicide attempts and two for severe alcohol abuse before the diagnosis of Wilson disease was made. The delay in diagnosis in one case was 12 years.

Other Clinical Manifestations

Hypercalciuria and nephrocalcinosis are the presenting signs in patients with Wilson disease. Hypercalciuria is possibly the consequence of a tubular defect in calcium reabsorption. Penicillamine therapy was accompanied by a decrease in urinary calcium excretion to normal values in half of the patients studied.

Cardiac manifestations in Wilson disease include arrhythmias, cardiomyopathy, cardiac death, and autonomic dysfunction; 34% patients with Wilson disease have electrocardiographic abnormalities. Two cases of cardiac deaths were reported (one died of repeated ventricular fibrillation, the other, of dilated cardiomyopathy). In one of them copper content in the myocardium was measured and was markedly elevated.

Chondrocalcinosis and osteoarthritis may be due to copper accumulation similar to the arthropathy of hemochromatosis.

Table 21.1 Routine tests for diagnosis of Wilson disease

Test	Typical finding	False " negative "	False " positive "
Serum ceruloplasmin	Decreased	Normal levels in pts. with marked hepatic inflammation	Low levels in: – Malabsorption
		Overestimation by immunologic assay	– Aceruloplasminemia
			– Liver insufficiency – Heterozygotes
24 h urinary copper	> 100 µg dl^{-1}	Normal:	Increased:
		– Incorrect collection	–Hepatocellular necrosis
		–Children without liver disease	– Contamination
Serum "free" copper	> 10 µgdl^{-1}	Normal if ceruloplasmin overestimated by immunologic assay	
Hepatic copper	> 250 µ g^{-1}dry weight	Due to regional variation – In pts with active liver disease – In pts with regenerative nodules	Cholestatic syndromes
Kayser-Fleischer rings by slit lamp	Present	– In up to 40% of patients with Hepatic Wilson disease – In most asymptomatic siblings	Primary biliary cirrhosis

Diagnosis

The diagnosis of Wilson disease is usually made on the basis of clinical findings and laboratory abnormalities (see Table 21.1). According to Scheinberg and Sternlieb (Scheinberg and Sternlieb, 1984), diagnosis of Wilson disease can be made if two of the following symptoms are present: Kayser-Fleischer rings, typical neurological symptoms and low serum ceruloplasmin levels.

Patients with Neurological Disease

In a patient presenting with typical neurological symptoms and having Kayser-Fleischer rings the diagnosis is straightforward. Clinical neurological examination is more sensitive than any other method to detect neurological abnormalities. No further diagnostic procedures are necessary to establish the diagnosis. Kayser Fleischer rings are rarely absent in neurologically symptomatic patients. However, there are a few well documented cases of neurological Wilson disease without demonstrable Kayser-Fleischer rings. In such patients diagnosis is usually made by a low serum ceruloplasmin level.

Brain magnetic resonance imaging (MRI) is useful to document the extent of changes in the central nervous system. The most common abnormalities are changes in signal intensity of gray and white matter, and atrophy of the caudate nucleus, brain stem, cerebral, and cerebellar hemispheres. A characteristic finding in Wilson disease is the "face of the giant panda" sign, but is found only in a minority of patients. In Wilson disease, an abnormal striatum or an abnormal pontocerebellar tract correlates with pseudoparkinsonian, and an abnormal dentatothalamic tract with cerebellar signs. On treatment some of the MRI abnormalities are fully reversible.

Auditory evoked brainstem potentials are helpful to document the degree of functional impairment and the improvement by decoppering treatment (Grimm et al., 1990; Grimm et al., 1992).

Patients with Liver Disease and Hemolytic Anemia

Diagnosis is far more complex in patients presenting with liver diseases. None of the commonly used parameters alone allows for a definite diagnosis of Wilson disease. Usually a combination of various laboratory parameters is necessary to establish the diagnosis. Kayser-Fleischer rings may be absent in up to 50% of patients with Wilsonian liver disease and even in a higher proportion in fulminant Wilson disease. Kayser-Fleischer rings can occasionally be detected in patients with primary biliary cirrhosis.

Laboratory Parameters

Routine Laboratory Parameters of Liver Disease

In general, transaminases are only mildly increased, and deep jaundice combined with mild elevation of liver enzymes should raise the suspicion for fulminant Wilson disease. However, increases of transaminases may be indistinguishable from findings seen in acute hepatitis. Sometimes alkaline phosphatase activities are relatively low in patients with Wilson disease. A ratio of total serum bilirubin concentration and alkaline phosphatase activity (>2) may differentiate fulminant Wilson disease from other forms of fulminant hepatic failure. However, the usefulness of this test was not confirmed in larger series.

Serum Ceruloplasmin

Serum ceruloplasmin can be measured by an immunologic assay or by the oxidase method. Since the immunologic ceruloplasmin assay can be automated by nephelometric methods, it is widely used in clinical laboratories. The oxidase method is only performed in specialized centers. Whereas serum ceruloplasmin is decreased in most patients with neurological Wilson disease, it may be in the low normal range in up to 45% of patients with hepatic disease (Steindl et al., 1997). On the other hand, even a low ceruloplasmin level is not diagnostic for Wilson disease in the absence of Kayser Fleischer rings. It may be low in subjects with familial hypoceruloplasminemia, in celiac disease, in severely malnourished subjects, and in heterozygous carriers of the Wilson disease gene (Cauza et al., 1997). Thus, in patients with liver disease a normal ceruloplasmin level cannot exclude, nor is a low level sufficient to make the diagnosis of, Wilson disease.

An overestimation of serum ceruloplasmin can be suspected if the serum copper concentration is lower than expected by the measured ceruloplasmin (which contains 0.3% of copper) level. Finally, ceruloplasmin is an acute phase reactant and its serum concentration increases as a consequence of inflammation. Most patients with normal ceruloplasmin had marked liver disease. Similarly serum ceruloplasmin may increase in pregnancy to high normal values.

Serum Copper

In general, serum copper values parallel those of ceruloplasmin. Therefore, serum copper is frequently low in patients with Wilson disease. However, about half of patients have serum copper levels in the normal range. Patients with fulminant Wilson disease and/or hemolytic anemia may even have markedly increased levels. Most of the copper in serum is bound to ceruloplasmin, and under normal conditions, less than 5% circulates as "free copper" and does not exceed 10 µg dl^{-1} in normal subjects. The "free" copper concentration can be calculated by subtracting from the total copper concentration the ceruloplasmin bound copper (ceruloplasmin times 3.3).

Urinary Copper Excretion

Urine copper excretion is markedly increased in patients with Wilson disease; however, its usefulness in clinical practice is limited. The estimation of urinary copper excretion may be misleading due to incorrect collection of 24-h urine volume or to copper contamination. In presymptomatic patients urinary copper excretion may be normal, but increase after D-penicillamine challenge (Da Costa et al., 1992). This test is valuable in the diagnosis of Wilson disease with active liver disease, but is unreliable to exclude the diagnosis in asymptomatic siblings (Muller et al., 2007). On the other hand urinary copper excretion is also increased in any disease with extensive hepatocellular necrosis.

Hepatic Copper Content

Hepatic copper content exceeding 250 µg g^{-1} dry weight (normal: up to 50) is increased in 82°% of patients with Wilson disease. In the absence of other tests suggestive for abnormal copper metabolism, diagnosis of Wilson disease cannot be made based on an increased hepatic copper content alone. Patients with chronic cholostatic

Table 21.2 Scoring system developed at the eighth International Meeting on Wilson disease, Leipzig 2001 (Ferenci et al., 2003)

Typical clinical symptoms and signs		Other tests	
KF rings		**Liver copper** (in absence of cholestasis)	
Present	2	> 5xULN (> 250µg g⁻¹)	2
Absent	0	50-250µg g⁻¹	1
Neurological symptoms		Normal (< 50µg g⁻¹)	−1
Severe	2	Rhodanine pos. granules[a]	1
Mild	1	**Urinary copper** (in absence of acute hepatitis)	
Absent	0		
Serum Caeruloplasmin		**Normal**	0
Normal(> 0.2g l⁻¹)	0	1–2x ULN	1
0.1–0.2g l⁻¹	1	> 2x ULN	2
< 0.1g l⁻¹	2	Normal, but > 5xULN after D-pen	2
Coombs' neg. hemolytic		Mutation Analysis	
Anemia		2 chromosome mutations	4
Present	1	1 chromosome mutation	1
Absent	0	No mutations detected	0
TOTAL SCORE		Evaluation:	
4 or more		Diagnosis established	
3		Diagnosis possible, more test needed	
2 or less		Diagnosis very unlikely	

[a]if no quantitative liver copper available

diseases, neonates and young children and possibly also subjects with exogenous copper overload have increased hepatic copper concentration >250 µg g⁻¹. On the other hand, hepatic copper content may be normal or borderline in about 18 ° of patients with unquestionable Wilson disease due to sampling, given the great regional differences in hepatic copper distribution, especially in the cirrhotic liver. Thus, estimates from a single biopsy specimen may be misleading.

Hepatic copper content was measured in 106 liver biopsies obtained at diagnosis of Wilson disease, in 212 patients with a variety of noncholestatic liver diseases, and 26 without evidence of liver disease (Ferenci et al., 2005). Liver copper content was >250 µg g⁻¹ dry weight in 87 (82%) patients, between 50 and 250 µg g⁻¹ in 15, and in the normal range in four. Liver copper content did not correlate with age, the grade of fibrosis, or the presence of stainable copper. Liver copper content was >250 or between 50 and 250 µg g⁻¹ dry weight in 3 (1.4%) and 20 (9.1%) of 219 patients with noncholostatic liver diseases, respectively. By lowering the cut off from >250 to 75 µg g⁻¹ dry weight the sensitivity of liver copper content to diagnose Wilson disease increased from 81.2 to 96%, the negative predictive value from 88.2 to 97.1%but the specificity (98.6 to 90.%) and the positive predictive value (97.6 to 87.4%) decreased. Thus, although liver copper content is a useful parameter it neither proves nor excludes Wilson disease with certainty.

Diagnosis of Wilson disease requires a combination of a variety of clinical and biochemical tests. A diagnostic scoring system (Table 21.2) was developed at the

8th International Meeting on Wilson disease, Leipzig, Germany (Ferenci et al., 2003) and its validity was confirmed by a retrospective analysis of a larger cohort of pediatric cases (Koppikar and Dhawan, 2005) and in patients with "atypical" disease (Xuan et al., 2007).

Liver Biopsy

Light Microscopy

Liver biopsy findings are generally nonspecific and not directly helpful to make the diagnosis of Wilson disease. Liver pathology includes early changes like fatty intracellular accumulations, which often proceed to marked steatosis. At later stages, hepatic inflammation with portal and periportal lymphocytic infiltrates, presence of necrosis and of fibrosis may be indistinguishable from other forms of hepatitis. Some patients have cirrhosis without any inflammation. The detection of focal copper stores by the Rhodanin stain is a pathognomic feature of Wilson disease but is only present in a minority (about 10%) of patients.

Electron Microscopy

The ultrastructural abnormalities include pathological changes of mitochondria and peroxisomes. Hepatocellular mitochondria are pleomorphic, with varying combinations of abnormalities including enlargement, bizarre shapes, and increased matrix density, separation of the normally apposed inner and outer membranes, widened intercristal spaces, enlarged granules, and crystalline, vacuolated, or dense inclusions. Sometimes peroxisomes are abnormally enlarged, rounded, or misshapen, and contain a granular or flocculent matrix of varying electron density.

Mutation Analysis

Direct Mutation Analysis

Direct molecular-genetic diagnosis is difficult because of the occurrence of many mutations, each of which is rare (Ferenci, 2005). Furthermore, most patients are compound heterozygotes (i.e., carry two different mutations). Direct mutation diagnosis is only helpful if a mutation occurs with a reasonable frequency in the population. In Northern, Central and Eastern Europe (Ferenci, 2005) the most common mutations are: H1069Q mutation (allele frequency: 43.5%), mutations of exon 8 (6.8%), 3400delC (3%) and P969Q (1.6%). In other parts of the world the pattern of mutations is different (i.e. Sardinia: UTR –441/-427del, 2463delC (Loudianos et al., 1999);

Far East: R778L (Kim et al., 1998; Nanji et al., 1997)). Screening for mutations is done by denaturating HPLC analysis followed by direct sequencing of exon suspected to carry a mutation. This approach is impractical for clinical diagnosis. In contrast, using allele-specific probes, direct mutation diagnosis is rapid and clinically very helpful, if a mutation occurs with a reasonable frequency in the population (Table 21.3.) In Austria, the H1069Q mutation is present in 61% of Wilson disease patients, and a two-step PCR based test for this mutation became very useful. A multiplex PCR for the most frequent mutations makes direct mutation analysis for diagnosis feasible.

Haplotype Analysis

Because of the complexity in identifying the many mutations in Wilson disease, haplotypes can be used to screen for mutations and to examine asymptomatic siblings of index patients. A number of highly polymorphic microsatellite markers have been described that closely flank the gene and are highly variable: D13S316, D13S314, D13S301, D13S133 (Cox, 1996). Where the markers are different at each locus in a patient, testing of at least one parent/or child of the patient is necessary to obtain the haplotype. The identification of unusual haplotypes can lend support, but is not sufficient to confirm the diagnosis of Wilson disease.

Microsatellite markers are also useful to study the segregation of the Wilson disease gene in most families. By these approaches, diagnostic dilemmas in differentiating heterozygote gene carriers and affected asymptomatic siblings can be solved (Maier-Dobersberger et al., 1995; Vidaud et al., 1996). For such analysis, at least one first degree relative and the index patient are required.

Table 21.3 Common mutations of the WD gene in various populations

Area (Ref)	Most common mutation (exon)	Other common mutations (exon)
Northwestern-, Central-, Eastern-Europe (Ferenci, 2005; Czlonkowska et al., 2005; Czlonkowska et al., 2005)[*a]	H1069Q (14)	3400delC (15), exon 8 (multiple), P969Q (13)
Sardinia (Loudianos et al., 1999)	-441/-421 del (5' UTR)	2463delC, V1146M
Canary Islands (Czlonkowska et al., 2005)	L708P (8)	
Spain Czlonkowska et al., (2005)	M645R (6)	L1120X (15)
Turkey (Loudianos et al., 1999, p. 68)	P969Q, A1003T (13)	Exon 8, H1069Q,
Brasil (Czlonkowska et al., 2005)	3400delC (15)	
Saudi Arabia (Czlonkowska et al., 2005)	Q1399R (21)	
Far East (Kim et al., 1998; Nanji et al., 1997)	R778L (8)	

[*a]Austria, Hungary, Czech Republic, Slovakia, Germany, Benelux, Poland, Russia, Bjelorus, Bulgaria, former Yugoslavia, Greece

Family Screening

Once diagnosis of Wilson disease is made in an index patient, evaluation of his family is mandatory. The likelihood of finding a homozygote among siblings is 25%, among children 0.5%. Testing of second degree relatives is only useful if the gene was found in one of the immediate members of his/her family. No single test is able to identify affected siblings or heterozygote carriers of Wilson diseasegene with sufficient certainty. Today, mutation analysis is the only reliable tool for screening the family of an index case with known mutations; otherwise haplotype analysis can be used. A number of highly polymorphic microsatellite markers that closely flank the gene allow tracing Wilson diseasegene in a family.

Treatment

Treatments for Wilson disease progressed from the intramuscular administration of BAL to the more easily administered oral penicillamine. Alternative agents to penicillamine like trientine were developed and introduced for patients with adverse reactions to penicillamine. Zinc was developed separately, as was tetrathiomolybdate, which was used for copper poisoning in animals. Today, the mainstay of treatment for Wilson disease remains lifelong pharmacologic therapy, but the choice of the drug mostly depends on the opinion of the treating physician and is not based on comparative data. Based on the recent AASLD practice guideline on Wilson disease, initial treatment for symptomatic patients should include a chelating agent (penicillamine or trientine). Treatment of presymptomatic patients or maintenance therapy of successfully treated symptomatic patients can be accomplished with the chelating agent penicillamine or trientine, or with zinc (Roberts and Schilsky, 2003).

D-Penicillamine

Penicillamine was first reported to be effective in treating Wilson disease by Walshe in 1956 and is since the "gold standard" for therapy. Penicillamine acts by reductive chelation: it reduces copper bound to protein and decreases thereby the affinity of the protein for copper. Reduction of copper thus facilitates the binding of copper to the drug. The copper mobilized by penicillamine is then excreted in the urine. Within a few weeks to months, penicillamine brings the level of copper to a subtoxic threshold, and allows tissue repair to begin. The great majority of symptomatic patients, whether hepatic, neurological or psychiatric, respond within months of starting treatment. Among neurological patients, a significant number may experience an initial worsening of symptoms before they get better.

The usual dose of penicillamine is 1–1.5 g/day. Initially, this dose will cause a large cupriuresis, but copper excretion later on decreases to 0.5 mg d^{-1}. To prevent deficiency induced by penicillamine pyridoxine (vitamin B_6) should be supplemented (50 mg/week). Once the clinical benefit is established, it is possible to reduce the dosage of penicillamine to 0.5–1 g d^{-1}. A lower maintenance dose will decrease the likelihood of late side effects of the drug.

A major problem of penicillamine is its high level of toxicity. In our series 20° of patients had major side-effects and were switched to other treatments. Other series report even higher frequencies of side effects. There are two broad classes of penicillamine toxicity: direct, dose dependent side effects and immunologically induced lesions. Direct side effects are pyridoxine deficiency, and interference with collagen and elastin formation. The later results in skin lesions like cutis laxa and elastosis perforans serpingiosa. By routine skin biopsies 1 year after initiation of treatment we found signs of elastic and collagen fiber abnormalities in every patient, but none has developed symptomatic skin disease so far. These side effects can be prevented or mitigated by decreasing the dosage of penicillamine. Immunologic mediated side effects include leukopenia and thrombocytopenia, systemic lupus erythematodes, immune complex nephritis, pemphigus, buccal ulcerations, myasthenia gravis, optic neuritis, and Goodpasture syndrome. Immunologic mediated side effects occur within the first three months of treatment and require immediate cessation of penicillamine. To diagnose these side effects as soon as possible, patients should be monitored in weekly intervals during the first six weeks of therapy. If the drug is well tolerated, control intervals can be gradually prolonged.

Trientine

Trientine is a copper chelator, acting primarily by enhancing urinary copper excretion. Trientine is licensed for treatment of Wilson disease and is now generally available. Experience with trientine is not as extensive as with penicillamine. It seems to be as effective as penicillamine, with far fewer side effects. Its efficacy was evaluated in patients with intolerance to penicillamine (Scheinberg et al., 1987). Discontinuation of penicillamine resulted in death from hepatic decompensation or fulminant hepatitis in 8 of 11 patients who stopped their own treatment after an average survival of only 2.6 years. In contrast, 12 of 13 patients with intolerance to penicillamine switched to trientine (1–1.5 g/day) were alive at 2–15 years later. The remaining patient was killed accidentally. However, the efficacy of trientine was not compared with penicillamine as initial treatment of Wilson disease. Uncontrolled anecdotal reports and our own experience indicate that trientine is a satisfactory first line treatment for Wilson disease. In the early phase of treatment trientine appears to be more potent to mobilize copper than penicillamine, but cupriuresis diminishes more rapidly than with penicillamine. The cupriuretic power of trientine may be disappointing but is sufficient to keep the patient clinically well.

Ammonium Tetrathiomolybdate

This drug has two mechanisms of action. First, it complexes with copper in the intestinal tract and thereby prevents absorption of copper. Second, the absorbed drug forms a complex with copper and albumin in the blood and renders the copper unavailable for cellular uptake. There is very limited experience with this drug. Tetrathiomolybdate appears to be a useful form of initial treatment in patients presenting with neurological symptoms (Brewer et al., 1996). In contrast to penicillamine therapy, treatment with tetrathiomolybdate does not result in initial neurological deterioration. This agent is particularly effective at removing copper from the liver. Because of its effectiveness, continuous use can cause copper deficiency. Besides, bone marrow depression was observed in a few patients treated with this drug.

Zinc

Zinc interferes with the intestinal absorption of copper by two mechanisms. Both metals share the same carrier in enterocytes and pretreatment with zinc blocks this carrier for copper transport (with a half-life of about 11 days). The impact of zinc induced blockade of other copper transport by other carriers into the enterocytes was not studied. Second, zinc induces metellothionein in enterocytes (Yuzbasiyan-Gurkan et al., 1992), which acts as an intracellular ligand binding metals which are then excreted in the feces with desquamated epithelial cells. Indeed, fecal excretion of copper is increased in patients with Wilson disease on treatment with zinc. Furthermore, zinc also induces metallothionein in the liver protecting hepatocytes against copper toxicity (Lee et al., 1989; Gonzalez et al., 2005). Zinc also induces metallothionein in enterocytes, which acts as an intracellular ligand binding metals (Yuzbasiyan-Gurkan et al., 1992) which are then excreted in the feces with desquamated epithelial cells. Indeed, fecal excretion of copper is increased in patients with Wilson disease on treatment with zinc.

Data on zinc in the treatment of Wilson disease are derived from uncontrolled studies using different zinc preparations (zinc sulfate, zinc acetate) at different doses (75–250 mg d^{-1}) (Ferenci, 1997). The efficacy of zinc was assessed by four different approaches. First, patients successfully decoppered by d-penicillamine were switched to zinc and the maintenance of their asymptomatic condition was monitored. Most patients maintained a negative copper balance and no symptomatic recurrences occurred. Some patients, however, died of liver failure after treatment was switched to zinc. Stremmel observed the occurrence of severe neurological symptoms in a 25 year old asymptomatic sibling 4 months after switching from d-penicillamine to zinc (Ferenci, 1997).

The Second group of symptomatic patients switched to zinc as alternate treatment due to intolerance to D-penicillamine. Sixteen case histories have been published so far. Liver function and neurological symptoms improved in three and five patients, respectively. One patient further deteriorated neurologically and improved on re-treatment with d-penicillamine. The remaining patients remained in stable condition. Follow-up studies in 141 patients demonstrated that zinc is effective as sole therapy in the long-term maintenance treatment of Wilson disease. In a third group, zinc was used as first line therapy. About one-third of them were asymptomatic siblings, two-third presented with neurological or hepatic symptoms. Most patients remained free of symptoms or improved. In 15° neurological symptoms worsened and improved on d-penicillamine. Three patients died of progressive liver disease. Finally, in a prospective study in 67 newly diagnosed cases, the efficacy of d-penicillamine and zinc was similar. This was not a randomized study; every other patient was treated with zinc. Zinc was better tolerated than D-penicillamine. However, two zinc-treated patients died of progressive liver disease.

It is unknown whether a combination of zinc with chelation therapy is useful or not. Theoretically these drugs may have antagonistic effects. Interactions in the maintenance phase of zinc therapy with penicillamine and trientine were investigated by Cu balance studies and absorption of orally administered ^{64}Cu as endpoints. The result on Cu balance was about the same with zinc alone as it is with zinc plus one of the other agents. Thus, there appears to be no advantages to concomitant administration.

Monitoring Therapy

If a decoppering agent is used for treatment, the compliance can be tested by repeated measurements of the 24 h urinary copper excretion. This approach is not useful if patients are treated with zinc. The dose of d-penicillamine can be lowered if in a compliant patient urinary copper excretion decreases over time and stabilizes at < 500 μ g/day. Efficacy of treatment can be monitored by the determination of " free " copper in serum, and depending on the presenting symptoms, Liver disease can be assessed by routine liver function tests. Repeated liver biopsies with measurement of hepatic copper content are not helpful. Improvement of neurological symptoms can be documented by clinical examination or by auditory evoked brainstem potentials. In addition, some of the MRI abnormalities are fully reversible on treatment.

Liver Transplantation

Liver transplantation is the treatment of choice in patients with fulminant Wilson disease and in patients⁻ with decompensated cirrhosis. Besides improving survival, liver transplantation also corrects the biochemical defect underlying Wilson disease. However, the role of this procedure in the management of patients with neurological Wilson disease in the absence of hepatic insufficiency is still uncertain.

Schilsky analyzed 55 transplants performed in 33 patients with decompensated cirrhosis and 21 with Wilsonian fulminant hepatitis in the United States and Europe. The median survival after orthotopic liver transplantation was 2.5 years, the longest survival time after transplantation was 20 years. Survival at 1 year was 79%. Nonfatal complications occurred in five patients. Fifty-one orthotopic liver transplants (OLT) were performed on 39 patients (16 pediatric, 23 adults) at the University of Pittsburgh. The rate of primary graft survival was 73 % and patient survival was 79.4 %. Survival was better for those presenting with a chronic advanced liver disease (90%) than it was for those presenting with a fulminant hepatic failure (73%). In the Mayo clinic series 1-year survival ranged from 79 to 87%, with an excellent chance to survive long term. The outcome of neurological disease following OLT is uncertain. In the retrospective survey four of the seven patients with neurological or psychiatric symptoms due to Wilson disease improved after OLT. Anecdotal reports documented a dramatic improvement in neurological function within 3–4 months after OLT. In contrast, central pontine and extrapontine myelinolysis and new extrapyramidal symptoms developed in a patient 19 months after OLT. Some patients transplanted for decompensated cirrhosis have had psychiatric or neurological symptoms, which improved following OLT.

Prognosis

Untreated, symptomatic Wilson disease progresses to death in all patients. The majority of patients will die of complications of advanced liver failure, some of progressive neurological disease. The overall mortality from Wilson disease treated medically (in most cases by d-penicillamine) has not been assessed prospectively. In a German study, in 51 patients the cumulative survival was slightly reduced during the early period of follow up but was not different from an age- and sex matched control population after 15 years of observation (96%). In a recent survey 63 of 300 patients diagnosed between 1948 and 2000 had died, but in a substantial proportion the cause of death was unrelated to Wilson disease (Walshe , 2007).

Liver Disease

In general, prognosis depends on the severity of liver disease at diagnosis. In patients without cirrhosis or with compensated cirrhosis, liver disease does not progress after initiation of therapy. Liver function improves gradually and will become normal in most patients within 1–2 years. In compliant patients treated with d-penicillamine or trientine, liver functions remain stable and no progressive liver disease is observed.

Schilsky followed 20 patients with Wilsonian chronic active hepatitis. Treatment with D-penicillamine was promptly initiated in 19 patients. One refused treatment and died 4 months later. Treated patients received D-penicillamine or trientine for a total of 264 patient-years (median: 14). In 18, symptomatic improvement and virtually normal levels of serum albumin, bilirubin, aspartate aminotransferase, and alanine aminotransferase followed within 1 year. One woman died after 9 months of treatment. Two patients, who became noncompliant after 9 and 17 years of successful pharmacological treatment, required liver transplants.

In patients presenting with fulminant Wilson disease, medical treatment is rarely effective. Without emergency liver transplantation mortality is very high. The modified Nazer score (Dhawan et al., 2005) is a useful guide to assess short term mortality in the setting of liver transplantation.

If diagnosed and treated early, hemolysis subsides within a few days after initiation of d-penicillamine therapy. Spontaneous remissions may occur even without treatment but relapse usually within a few months. Hemolysis associated with active liver disease may progress to fulminant Wilson disease rapidly.

Neurological Disease

Patients presenting with neurological symptoms have a better prognosis than those presenting with liver disease. The prognosis for survival is favorable (Stremmel et al., 1991), provided that therapy is introduced early. In Brewer´s series, two out of 54 patients died due to complications which were attributed to their impaired neurological function (Brewer and Yuzbasiyan-Gurkan, 1992).

Neurological symptoms are partly reversible and improvement occurs gradually over several months. Initially, neurological symptoms may worsen, especially on treatment with D-penicillamine. In some patients neurological symptoms disappear completely, and abnormalities documented by evoked responses or MR-imaging may completely resolve within 18–24 months. Brain function was assessed by repeated recording of short latency sensory potentials, auditory brain stem potentials and cognitive P300 evoked potentials in 10 followed prospectively after diagnosis for 5 years (Grimm et al., 1990). Electrophysiological and clinical improvement was observed as early as 3 months after initiation of therapy and continued until final assessment after 5 years. Three patients became completely normal but residual symptoms were detectable in seven. Czlonkowska et al (Czlonkowska et al., 2005) studied 164 patients diagnosed over a 11-year period. Twenty died during the observation period. The relative survival rate of all patients in our group was statistically smaller than in the Polish population. The main cause of death was the diagnosis in advanced stage of disease, but in six patients presenting with mild signs, the disease progressed despite treatment. There was no difference in mortality rate in patients treated with d-penicillamine or zinc sulphate as initial therapy.

References

Barnham KJ, McKinstry WJ, Multhaup G, Galatis D, Morton CJ, Curtain CC, et al. Structure of the Alzheimer's disease amyloid precursor protein copper binding domain. A regulator of neuronal copper homeostasis. J Biol Chem. 2003;278:17401–17407

Barthel H, Hermann W, Kluge R, Hesse S, Collingridge DR, Wagner A, Sabri O. Concordant pre- and postsynaptic deficits of dopaminergic neurotransmission in neurological Wilson disease. Am J Neuroradiol. 2003;24:234–238

Beyreuther K, Zheng H, Tanzi RE, Masters CL, Bush AI. Treatment with a copper-zinc chelator markedly and rapidly inhibits beta-amyloid accumulation in Alzheimer's disease transgenic mice. Neuron. 2001;30:665–676

Borjigin J, Payne AS, Deng J, Li X, Wang MM, Ovodenko B, Gitlin J. A novel pineal night specific ATPase encoded by the Wilson disease gene. J Neurosci. 1999;19:1018–1026

Brewer GJ, Johnson V, Dick RD, Kluin KJ, Fink JK, Brunberg JA. Treatment of Wilson disease with ammonium tetrathiomolybdate. II. Initial therapy in 33 neuroligicalaly affected patients and follow-up with zinc therapy. Arch Neurol. 1996;53:1017–1025

Brewer GJ, Yuzbasiyan-Gurkan V: Wilson disease. Medicine 1992;71:139–164

Bull PC, Thomas GR, Rommens JM, Forbes JR, Cox DW. The Wilson disease gene is a putative copper transporting P-type ATPase similar to the Menkes gene. Nat Genet. 1993;5:327–337

Bush AI, Masters CL,Tanzi RE. Copper, β-amyloid, and Alzheimer's disease: Tapping a sensitive connection. PNAS 2003;100:11193–11194

Cox DW. Molecular advances in Wilson disease. in: Progress in Liver Disease, Vol. X, 1996, Chapter 10; pp. 245–263

Czlonkowska A, Tarnacka B, Litwin T, Gajda J, Rodo M. Wilson's disease-cause of mortality in 164 patients during 1992–2003 observation period. J Neurol. 2005;25:698–703

de Bie P, van de Sluis B, Burstein E, van de Berghe PVE, Muller P, Berger R, Gitlin JD, Wijmenga C, Klomp LWJ. Distinct Wilson's Disease Mutations in ATP7B Are Associated With Enhanced Binding to COMMD1 and Reduced Stability of ATP7B. Gastroenterology 2007;133,1316–1326

Dhawan A, Taylor RM, Cheeseman P, De Silva P, Katsiyiannakis L, Mieli-Vergani G. Wilson's disease in children: 37-year experience and revised King's score for liver transplantation. Liver Transpl. 2005;11:441–448

Ferenci P. Zinc treatment of Wilson's disease. In: Zinc and diseases of the digestive tract. Kruse-Jarres JD, Schölmerich J (eds.). Kluwer Academic publishers, Lancaster. 1997:117–124

Ferenci P. Wilson's disease. (Clinical Genomics). Clinical Gastroenterol Hepatol 2005;3:726–733

Ferenci P, Caca K, Loudianos G, Mieli-Vergani G, Tanner S, Sternlieb I, Schilsky M, Cox D, Berr F. Diagnosis and phenotypic classification of Wilson disease. Final report of the proceedings of the working party at the 8th International Meeting on Wilson disease and Menkes disease, Leipzig/Germany, 2001. Liver International 2003;23:139–142

Ferenci P, Członkowska A, Merle U, Szalay F, Gromadzka G, Yurdaydin C, Vogel W, Bruha R, Schmidt HT, Stremmel W. Late onset Wilson disease. Gastroenterology. 2007;132:1294–1298

Ferenci P, Steindl-Munda P, Vogel W, Jessner W, Gschwantler M, Stauber R, Datz Ch, Hackl F, Wrba F, Lorenz O. Diagnostic value of quantitative hepatic copper determination in patients with Wilson disease. Clinical Gastroenterol Hepatol 2005;3:811–818

Forbes J, Hsi G, Cox D. Role of the copper-binding domain in the copper transporter function of ATP7B, the P-type ATPase defective in Wilson disease. J Biol Chem 1999;274:12408–12413

Gonzalez BP, Nino Fong R, Gibson CJ, Fuentealba IC, Cherian MG. Zinc supplementation decreases hepatic copper accumulation in LEC rat: a model of Wilson's disease. Biol Trace Elem Res. 2005;105:117–134

Grimm G, Madl Ch, Katzenschlager R, Oder W, Ferenci P, Gangl A. Detailed evaluation of brain dysfunction in patients with Wilson's disease. EEG Clin Neurophysiol 1992;82:119–124

Grimm G, Oder W, Prayer L, Ferenci P, Madl Ch. Prospective follow-up study in Wilson's disease. Lancet 1990;336:963–964

Gu M, Cooper JM, Butler P, Walker AP, Mistry PK, Dooley JS, Schapira AH. Oxidative-phosphorylation defects in liver of patients with Wilson's disease. Lancet 2000;356:469–474

Hamza I, Schaefer M, Klomp LW, Gitlin JD. Interaction of the copper chaperone HAH1 with the Wilson disease protein is essential for copper homeostasis. Proc Natl Acad Sci U S A 1999;96:13363

Hellman NE, Kono S, Mancini GM, Hoogeboom AJ, de Jong GJ, Gitlin JD. Mechanisms of copper incorporation into human ceruloplasmin. J Biol Chem 2002;277:46632–46638

Hermann W, Barthel H, Hesse S, Grahmann F, Kuhn HJ, Wagner A, Villmann T. Comparison of clinical types of Wilson's disease and glucose metabolism in extrapyramidal motor brain regions. J Neurol. 2002;249:896–901

Hesse S, Barthel H, Hermann W, Murai T, Kluge R, Wagner A, Sabri O, Eggers B. Regional serotonin transporter availability and depression are correlated in Wilson's disease. J Neural Transm. 2003;110:923–933

Kelley EJ, Palmiter RJ. A murine model of Menkes disease reveals a physiological function of metallothionein. Nat Genet 1996;13:219

Kenney SM, Cox DW. Sequence variation database for the Wilson disease copper transporter, ATP7B. Hum Mutat. 2007;28:1171–1177

Kim EK, Yoo OJ, Song KY, Yoo HW, Choi SY, Cho SW, et al. Identification of three novel mutations and a high frequency of the Arg778Leu mutation in Korean patients with Wilson disease. Hum Mutat 1998;11:275–278.

Klomp AE, Tops BB, Van Denberg IE, Berger R, Klomp LW. Biochemical characterization and subcellular localization of human copper transporter 1 (hCTR1). Biochem J 2002;364:497

Koppikar S, Dhawan A. Evaluation of the scoring system for the diagnosis of Wilson's disease in children. Liver Int. 2005;25:680–681

Lang PA, Schenck M, Nicolay JP, Becker JU, Kempe DS, Lupescu A, Koka S, et al. Liver cell death and anemia in Wilson disease involve acid sphingomyelinase and ceramide. Nat Med. 2007;13:164–170

Lee DY, Brewer GJ, Wang Y. Treatment of Wilson's disease with zinc. VII. Protection of the liver from copper toxicity by zinc-induced metallothionein in a rat model. J Lab Clin Med 1989;114:639–645

Lee J, Pena MM, Nose Y, Thiele DJ. Biochemical characterization of the human copper transporter Ctr1. J Biol Chem 2001;277:4380–4387

Loudianos G, Dessi V, Lovicu M, Angius A, Figus AL, Lilliu F, et al. Molecular characterization of Wilson disease in the Sardinian population – evidence of a founder effect. Hum Mutat 1999;14:294–303

Lutsenko S, Petris MJ. Function and regulation of the mammalian copper-transporting ATPases: Insights from biochemical and cell biological approaches. J Membr Biol 2003;191:1–12

Maier-Dobersberger Th, Rack S, Granditsch G, Korninger L, Steindl P, Mannhalter Ch, Ferenci P. Diagnosis of Wilson's disease in an asymptomatic sibling by DNA linkage analysis. Gastroenterology 1995;109:2015–2018

Mansouri A, Gaou I, Fromenty B, Berson A, Letteron P, Degott C, Erlinger S, Pessayre D. Premature oxidative aging of hepatic mitochondrial DNA in Wilson's disease. Gastroenterology. 1997;113:599–605

Miyata M, Smith JD. Apolipoprotein E allele-specific antioxidant activity and effects on cytotoxicity by oxidative insults and betaamyloid peptides. Nature Genet 1996;14:55–61

Muller T,Koppikar S, Taylor RM, Carragher F, Schlenck B,Heinz-Erian P, Kronenberg F, Ferenci P, Tanner S, Siebert U, Staudinger R, Mieli-Vergani G, Dhawan A. Re-evaluation of the penicillamine challenge test in the diagnosis of Wilson's disease in children. J Hepatol. 2007;47:270–276

Nanji MS, Nguyen VT, Kawasoe JH, Inui K, Endo F, Nakajima T, et al. Haplotype and mutation analysis in Japanese patients with Wilson disease. Am J Hum Genet 1997;60:1423–1426

Palmiter RD. The elusive function of metallothioneins. Proc Natl Acad Sci U S A 1998;95:8428–8430

Patel BN, Dunn RJ, Jeong SY, Zhu Q, Julien JP, David S. Ceruloplasmin regulates iron levels in the CNS and prevents free radical injury. J Neurosci. 2002;22:6578–6586

Page RA, Davie CA, MacManus D, Miszkiel KA, Walshe JM, Miller DH, Lees AJ,Schapira AH. Clinical correlation of brain MRI and MRS abnormalities in patients with Wilson disease. Neurology 2004;63:638–643

Rae T, Schmidt P, Pufahl R, Culotta VC, O'Halloran TV. Undetectable intracellular free copper: the requirement of a copper chaperone for superoxide dismutase. Science 1999;284:805–808

Roberts EA, Schilsky ML. AASLD practice guidelines: A practice guideline on Wilson disease. Hepatology 2003;37:1475–1492

Schaefer M, Hopkins R, Failla M, Gitlin JD. Hepatocyte-specific localization and copper-dependent trafficking of the Wilson–G646

Schiefermeier M, Kollegger H, Madl C, Polli C, Oder W, Kuhn H, Berr F, et al. The impact of apolipoprotein E genotypes on age at onset of symptoms and phenotypic expression in Wilson's disease. Brain. 2000;123:585–590

Scheinberg IH, Jaffe ME, Sternlieb I. The use of trientine in preventing the effects of interrupting penicillamine therapy in Wilson's disease. N Engl J Med 1987;317:209–213

Scheinberg IH, Sternlieb I. Wilson's disease. Vol 23. Major Problems in Internal Medicine. Saunders, Philadelphia,1984

Scott J, Gollan JL, Samourian S, Sherlock S. Wilson's disease, presenting as chronic active hepatitis. Gastroenterology 1978;74:645–651

Sparks DL, & Schreurs BG. Trace amounts of copper in water induce beta-amyloid plaques and learning deficits in a rabbit model of Alzheimer's disease. Proc.Natl.Acad.Sci.USA. 2003;100:11065–11069

Steindl P, Ferenci P, Dienes HP, Grimm G, Pabinger I, Madl Ch et al. Wilson's disease in patients presenting with liver disease: a diagnostic challenge. Gastroenterology 1997;113:212–218

Stremmel W, Meyerrose KW, Niederau C, Hefter H, Kreuzpaintner G, Strohmeyer G: Wilson's disease: Clinical presentation, treatment, and survival. Ann Int Med 1991;115:720–726

Tanzi RE, Petrukhin K, Chernov I, Pellequer JL, Wasco W, Ross B, Romano DM, et al. The Wilson disease gene is a copper transporting ATPase with homology to the Menkes disease gene. Nat Genet. 1993;5:344–350

Vidaud D, Assouline B, Lecoz P, Cadranel JF, Chappuis P. Misdiagnosis revealed by genetic linkage analysis in a family with Wilson disease. Neurology 1996;46:1485–1486

Walker JM, Tsivkovskii R, Lutsenko S. Metallochaperone Atox1 transfers copper to the NH2-terminal domain of the Wilson's disease protein and regulates its catalytic activity. J Biol Chem 2002;277:27953–27959

Walshe JM. Cause of death in Wilson disease. Mov Disord. 2007 [Epub ahead of print]

Watt NT, Hooper NM. The response of neurones and glial cells to elevated copper. Brain Res Bull. 2000;55:219–224

Xuan A, Bookman I, Cox DW, Heathcote J. Three atypical cases of Wilson disease: Assessment of the Leipzig scoring system in making a diagnosis. J Hepatol. 2007;47:428–433

Yuzbasiyan-Gurkan V, Grider A, Nostrant T, Cousins RJ, Brewer GJ.: Treatment of Wilson's disease with zinc: X. Intestinal metallothionein induction. J Lab Clin Med 1992;120:380–386

Chapter 22
Metabolic Abnormalities in Alzheimer Disease

Florian M. Gebhardt and Peter R. Dodd

Alzheimer's Disease

History

In 1906, German neuropathologist and psychiatrist Alois Alzheimer described "eine eigenartige Erkrankung der Hirnrinde" (a peculiar disease of the cerebral cortex). Alzheimer noted two abnormalities in autopsied brain tissue from his index case: senile plaques, proteinaceous structures previously described in the brain of normal elderly people; and abnormal cells delineated with silver stain that became known as neurofibrillary tangles (NFTs). The distribution and abundance of tangle-filled neurons are now the main criteria used to diagnose Alzheimer disease (AD) at autopsy.

Epidemiology

About 25 million people are suffering from dementia worldwide, a figure expected to pass 42 million by 2020; there is a new case every seven seconds globally (Ferri et al., 2005; Dua et al., 2006). Dementia is the fourth leading cause of death for those over 65, and the second-largest source of disability burden (Katzman, 1986). AD, the commonest form of dementia, accounts for 50–70% of cases (Gearing et al., 1995). Prevalence increases with age: 2.9% of the population aged 65–74 are diagnosed with dementia, increasing to 10.9% for those aged 75–84 and 30.2% for those over 85 (Gurland et al., 1999). AD causes emotional suffering to the patient's family and places a major economic burden on society through provision of care and loss of wages. The cost of dementia is estimated at $US248 billion per annum (Jonsson et al., 2006; Wimo et al., 2006). Despite these consequences, the etiology of AD is poorly understood and there is no cure.

D.W. McCandless (ed.) *Metabolic Encephalopathy*,
doi: 10.1007/978-0-387-79112-8_22, © Springer Science + Business Media, LLC 2009

Diagnosis

Clinical

Since AD can be confirmed only by neuropathological examination, clinical diagnosis relies on ruling out other forms of dementia. A diagnosis of *probable AD* according to DSM-IV (Diagnostic and Statistical Manual of Mental Disorders) guidelines is over 80% accurate when checked against neuropathology (American Psychiatric Association, 2000; Knopman et al., 2001). A probable AD clinical diagnosis is determined from deficits in two or more areas of cognition, specifically aphasia, apraxia, agnosia, or executive functioning, resulting in impairment in an occupational or social context. These changes must represent a decline from a previous level of mental functioning. Individuals will often be unable to learn new material and/or forget material known prior to illness. Defects must not result from central nervous system (CNS) conditions that cause progressive worsening of memory or cognition (e.g., Parkinson disease), systemic conditions known to cause dementia (e.g., human immunodeficiency virus infection), trauma, substance abuse, or mental disorder (e.g., schizophrenia) (American Psychiatric Association, 2000).

Pathological

AD is characterized by plaques surrounded by dystrophic neurites, NFTs, and regional atrophy caused by neuronal loss. The pattern of damage is not distributed uniformly through the brain. Cell loss is particularly severe in pyramidal neurons in layers III and IV of the neocortex, as well as in glutamate-innervated cortical and hippocampal neurons (Albin and Greenamyre, 1992). The destruction of this neuronal population correlates with clinical severity ante mortem; more than 90% are lost by end-stage (Bussiere et al., 2003).

Presentation and Metabolism

AD patients commonly show marked weight loss and cachexia (Finch and Cohen, 1997; Poehlman and Dvorak, 2000). This has led to speculation that the regulation of body weight may be affected, and there is some evidence that patients with normal hypothalamic pituitary adrenal axis (HPA) and thyroid function may have aberrant circulating levels of leptin and tumor nrcrosis factor-α (TNFα) (Intebi et al., 2002). Moreover, a decline in body mass index (BMI) is reportedly associated with increased risk of AD (Buchman et al., 2005). However, a review of the available literature does not support the view that body wasting results from generalized hypometabolism; rather, the patients may show aberrant levels of physical activity and energy intake (Poehlman and Dvorak, 2000). Nevertheless, brain metabolism appears to be disordered in most patients compared with matched controls (Fig. 22.1). The present

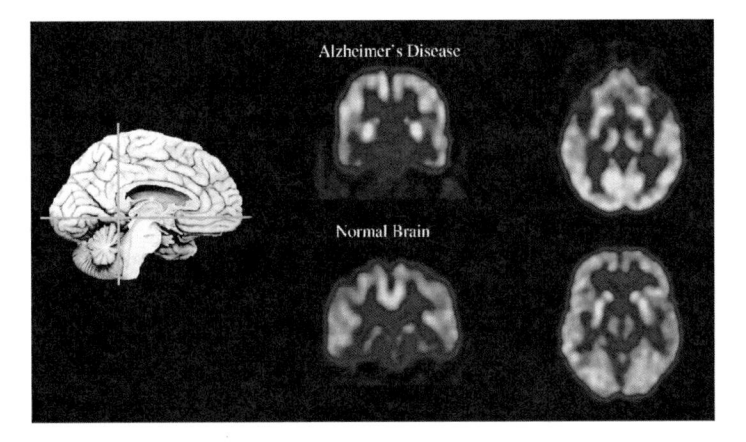

Fig. 22.1 Positron emission tomography image of Alzheimer disease and normal brain (coronal and transverse sections through the hippocampus) (*See also* Color Insert)

chapter will explore the likely origins and impact of this aspect of the disease. The AD brain is marked by neuropathological alterations that include the loss of synapses, atrophy, and the selective depletion of neurotransmitter systems (Hansen et al., 1993; Giasson et al., 2003). We will first consider the mechanisms that give rise to these key pathognomonic features, and then discuss the role that metabolic abnormalities may play in generating those features.

β-Amyloid Precursor Protein

AD plaques are extracellular deposits of aggregated amyloid-β (Aβ) peptide that is proteolytically derived from the β-amyloid precursor protein (APP). *In vitro* and *in vivo* studies have shown that APP abnormalities can affect synaptic vesicle and axonal transport. Axonal dysfunction might lead to impaired synaptic plasticity and reduced neuronal viability.

There are two major forms of plaques: neuritic and diffuse. Neuritic plaques are composed of extracellular deposits of Aβ surrounded by dystrophic neurites. Diffuse plaques are amorphous extracellular deposits of Aβ-immunoreactive granular material that lack neurites, and are thought to precede neuritic plaques.

The amyloid cascade hypothesis, which is probably the most widely held formulation of AD pathogenesis (Hardy and Selkoe, 2002), postulates that neurodegeneration and cognitive decline in AD are primarily related to a toxic effect of Aβ on cell bodies and adjacent cell processes (dendrites and synapses). It gained widespread acceptance after it was shown that fibrillar Aβ is cytotoxic *in vitro* (Pike et al., 1991, 1993). Studies of synaptosomes showed that Aβ induces membrane lipid peroxidation and impairs ion, glucose, and glutamate transporter functions. Aggregated or fibrillar forms of Aβ are toxic when injected into the brains of aged primates (Geula et al., 1998).

The Aβ sequence includes 28 amino acids of the extracellular region, and 12–15 residues of the membrane-spanning region, of APP. There are two major variants of Aβ: Aβ40 is the main form secreted from cultured cells and is found in cerebrospinal fluid, whereas Aβ42 is the major component of amyloid deposits. Aβ40 is produced in greater quantities in normal subjects, in whom Aβ42 comprises only a small fraction of total Aβ. The two Aβ forms differ in cytotoxicity, solubility, and tendency to generate fibrils. C-terminal alanine and isoleucine residues render Aβ42 more hydrophobic, and it aggregates more rapidly and forms Aβ oligomers more readily (Burdick et al., 1992; Jarrett et al., 1993; Bitan et al., 2003; Chen and Glabe, 2006). Aβ42 is more neurotoxic, and is prominent within neurons of AD brains (Gouras et al., 2000). Aβ42 can aggregate into a non-β-sheet, non-fibrillar state, or a β-sheet fibrillar state. The latter is cytotoxic and tends to deposit in plaques (Cuajungco and Faget, 2003). Aβ42 monomer oligomerizes to form protofibrils, and then fibrils, which are neurotoxic. Oligomers and protofibrils appear to be more neurotoxic than fibrils or plaques (Dahlgren et al., 2002). The amount of soluble Aβ42 isolated from brain correlates with the severity of neurodegeneration in AD (Lue et al., 1999; McLean et al., 1999). Plaques in familiar AD cases have higher Aβ42 to Aβ40 ratios than sporadic AD plaques (Iwatsubo et al., 1994). However, there is no close correlation between plaque deposition and the clinical severity of dementia, and neuronal and synaptic losses do not correlate with the brain Aβ burden (Lippa et al., 2000; Carter and Lippa, 2001).

The strongest support for the amyloid hypothesis has come from molecular-genetic studies of mutations in the *APP*, *PS1*, and *PS2* genes that are responsible for most familial AD (FAD) cases (these comprise ~1% of all AD cases (Campion et al., 1999)), and associated biochemical studies on the effects of these mutations on the properties of Aβ or its production from APP. More than 160 mutations have been found in the three genes. Twenty-seven FAD mutations have been described in the *APP* gene on chromosome 21, clustered around the Aβ sequence within APP; most increase the production of Aβ42. (Note, however, that only 3% of FAD cases are due to APP mutations (Tanzi et al., 1992)). For example, enhanced β-secretase cleavage and increased Aβ production have been found in the *APP* "Swedish" mutation (Haass et al., 1995). Aβ assembles more slowly into fibrils and protofibrils in subjects with the APP "Flemish" mutation than in wild-type normal subjects (Walsh et al., 2001). The *APP* "Dutch" mutant causes faster assembly of Aβ than wild type (Kirkitadze et al., 2001). Trisomy-21 (Down syndrome) subjects develop Aβ deposits before other pathologies such as NFTs.

A central question on the role of Aβ in AD is whether Aβ deposition initiates the disease process. The cascade hypothesis has been extended to include the idea that neuronal death is due to toxic influences of non-fibrillary Aβ oligomer production within the neuronal cytoplasm, which manifests prior to, and independent of, extracellular lesions. Alternatively, extracellular Aβ oligomers may be neurotoxic before they become fibrillary and aggregate into plaques. Transgenic mice harboring constructs that target intracellular Aβ develop neurodegeneration, whereas mice with extracellular Aβ targeting do not (La Ferla et al., 1995). The emergence of intraneuronal Aβ is correlated with cognitive dysfunction (Billings et al., 2005).

Gouras et al. (2000) showed that neurons in AD-vulnerable brain regions accumulate Aβ42 (in both the absence and presence of NTF) but not the more abundantly secreted Aβ40. Studies on autopsied Down syndrome and APP transgenic mouse brains suggest that intraneuronal Aβ42 increases with aging. Once plaque pathology develops, intraneuronal Aβ42 decreases (Haass and Steiner, 2001).

The mechanism by which Aβ induces neuronal cell loss remains elusive, but it may involve interactions with integrins. Integrins influence long-term potentiation in hippocampal neurons, regulate neurite outgrowth induced by growth factors, and respond to toxicity associated with neurodegeneration (Giancotti and Ruoslahti, 1999; Williamson et al., 2002; Grace and Busciglio, 2003). Fibrillar Aβ binds to integrin-activating focal-adhesion and downstream pathways, regulating the cell cycle through various effectors that include MAPK, PI3-K, and GSK3β kinases (see Tanzi et al., 1992 for review; Zhang et al., 1994; Berg et al., 1997; Williamson et al., 2002).

APP is part of a superfamily from which 16 homologous amyloid precursor-like proteins (APLPs) and APP homologs are derived (McLean et al., 1999). The *APP* gene comprises 18 exons and spans more than 30 kb, translated into a 110–130 kDa type-1 membrane glycoprotein, which is limited to the brain but almost ubiquitous in all cell types. APP is inserted in the membrane in such a way that only a small *C*-terminal part remains in the cytosol. The *N*-terminal part forms a large extracellular domain. APP has at least three isoforms arising from the alternative splicing of exons 7, 8, and 15 (Sandbrink et al., 1996): APP695, APP751, and APP770. Few physiological differences have been observed between the isoforms (Kang et al., 1987; Tanzi et al., 1988). The APP695 isoform predominates in human cerebral cortex; it differs from APP751, which is more abundant in non-neuronal cells, by its lack of a Kunitz-type protease inhibitor (KPI) domain.

The cytoplasmic domain of APP contains trafficking signals that localize it to synaptic regions. Kamal et al. (2000) showed that the axonal transport of APP in neurons is mediated by its binding to the kinesin I light chain subunit. APP positioned on the cell surface (Storey et al., 1999) has a neurite-promoting activity, whereas secreted APP does not (Qiu et al., 1995). It also co-localizes with plaque adhesion components (Yamazaki et al., 1997) and participates in synaptic vesicle recycling (Marquez-Sterling et al., 1997). This suggests that APP could function as a cell-surface receptor, transducing signals from the extracellular matrix to the interior of the cell.

APP undergoes post-translational processing. In the endoplasmatic reticulum (ER), APP is *N*-glycosylated, and in the Golgi apparatus it is glycosylated, phosphorylated, and tyrosine-sulfated. Its proteolytic cleavage involves three secretase enzymes (α-, β-, and γ-), defined by their cleavage sites that determine the onset and progression of neurodegeneration. The major proteolytic pathway involves cleavage of APP by α-secretase, a member of the ADAM (a disintegrin and metalloproteinase) family of proteinases that include ADAM10, ADAM7 (TNFα-converting enzyme, TACE), and ADAM9 (Lammich et al., 1999). Besides the Golgi apparatus, the plasma membrane is the other major site where α-secretase cleavage of APP occurs (Lammich et al., 1999; Skovronsky et al., 2000). α-Secretase cleaves

APP within the Aβ sequence, between Lys[16] and Leu[17], producing a 105–125 kDa soluble N-terminal fragment (sAPPα) and a membrane-anchored fragment (C83 or p10) of 10 kDa (Kojro et al., 2001).

Further processing of p10 by γ-secretase yields truncated Aβ fragments of about 3 kDa (p3) and the APP intracellular domain (AICD). p3 is generally not found in amyloid cores of classical plaques, or in amyloid deposits in the cerebral vasculature. The AICD fragment can form multiprotein complexes, which are transported to the nucleus and may have a role in nuclear signaling (Shoji et al., 1992).

sAPPα plays important roles in modulating neuronal excitability, and in developmental and synaptic plasticity. It modulates dendritic outgrowth *in vitro* (Butterfield et al., 1994). sAPPα also protects neurons against excitotoxic, metabolic, and oxidative insults in cell culture and *in vivo* (Goodman and Mattson, 1994). However, the mechanisms whereby sAPPα affects neuronal plasticity and survival have not been clearly established. sAPPα activates microglia largely through the JNK and p38-MAPK pathways (*v.i.*), which can generate proinflammatory factors and cause neurotoxicity (Bodles and Barger, 2005).

The upregulation of α-secretase activity might preclude the formation of Aβ peptides and their deposition as plaques. To investigate this concept *in vivo*, Postina et al. (2004) generated transgenic mice overexpressing ADAM10. Moderate neuronal overexpression increased the release of sAPPα, reduced the formation of Aβ, and prevented the latter's deposition in plaques. Because α-secretase-cleaved APP prevents the formation of amyloidogenic fragments (Aβ), this pathway is called *non-amyloidogenic pathway*, which is thought not to lead to AD pathology. The pathway can be enhanced by, for example, neuropeptides such as pituitary adenylate cyclase-activating polypeptide (PACAP; Kojro et al., 2006), PKC activators (Etcheberrigaray et al., 2004), lovastatin (Kojro et al., 2001), simvastatin (Hoglund et al., 2005), and retinoids (Goodman and Pardee, 2003).

The amyloidogenic pathway that generates Aβ from APP is also a two-step proteolytic process involving β-and γ-secretases (Kaether and Haass, 2004). First, the ectodomain of APP is shed N-terminally by β-secretase (BACE1 and or BACE2) into an extracellularly released soluble sAPPβ and a membrane-bound C-terminal fragment. This is followed by attack by the γ-secretase complex, which can cleave at different transmembrane-domain sites of APP: the γ-site, which is variable and can occur after amino acids 38, 40, and 42 (referring to the sequence of Aβ), the ε-site at amino acid 49, and the ζ-site at amino acid 46. It is likely that the C-terminus of Aβ is generated by a series of cleavages. The first cleavage at the ε-site, which is close to the cytoplasmic border of the transmembrane domain, releases the free AICD (intracellular part) in the cytosol (Sastre et al., 2001; Weidemann et al., 2002). The remaining membrane-anchored fragment undergoes an intermediate scission at the ζ-site; this is followed by γ-cleavage to generate peptides mainly 40 or 42 amino acids in length (Aβ42 and Aβ40; Qi-Takahara et al., 2005; Zhao et al., 2005). AICD controls p53 at the transcriptional level and increases its activity *in vitro* and *in vivo* (Alves da Costa et al., 2006).

BACE1 and BACE2 are highly homologous type 1 single-transmembrane aspartyl proteases. Both enzymes are expressed in neural tissues; BACE2 is the less abundant

(Bennett et al., 2000a). BACE 1 is synthesized as a large precursor protein that can be modified by glycosylation. Mature BACE1 is cleaved by a furin-like endoprotease (Bennett et al., 2000b; Creemers et al., 2001). Its major locations are endosomal/lysosomal compartments, Golgi-derived vesicles, endoplasmic reticulum/intermediate compartments in postmitotic neurons, and, in relative low abundance, the cell surface. It can cycle between the cell surface and endosomes (Huse et al., 2000). However, the mechanisms by which it is trafficked and activated have not been elucidated.

Harada et al. (2006) showed that BACE1 is exclusively expressed in neurons, not in glial cells. They further suggest that surviving neurons in AD brains might generate more BACE1 than neurons in control brains. This conforms to the idea that neurons are the primary source of the extracellular $A\beta$ deposited in plaques. BACE2 transcripts are expressed in the CNS and many peripheral tissues; it is much less abundant in neurons than BACE1. The proteins have similar structural organization and share 51% identity at the amino acid level. BACE2 is localized predominantly in post-Golgi structures and at the cell surface (Ehehalt et al., 2002). It can be distinguished from BACE1 by subcellular localization and on the basis of APP cleavage specificity: in addition to the BACE1 sites, BACE2 also cleaves between Phe^{19}-Phe^{20} and Phe^{20}-Ala^{21} within the $A\beta$ peptide (Farzan et al., 2000).

BACE protein expression and enzyme activity are significantly higher in susceptible brain regions of sporadic AD cases than in nondemented control tissue, whereas there are no differences in cerebellum. BACE mRNA expression is elevated in frontal cortex. A significant correlation has been demonstrated between levels of BACE and plaque numbers in AD brains. Higher BACE expression may lead to increased $A\beta$ production and plaque deposition in sporadic AD patients (Li et al., 2004a).

γ-Secretase is a transmembrane aspartyl protease and belongs to the family of intramembrane cleaving proteases (I-CLiPs) that cleave proteins within the lipid bilayer (Wolfe et al., 1999a). It interacts with and cleaves single transmembrane receptors such as Notch and APP (Chyung et al., 2005). In contrast with other I-CLiPs, γ-secretase requires the formation of a large multiprotein complex to be active. The complex contains presenilins, nicastrin, and anterior pharynx defective-1 (APH-1) and presenilin enhancer-2 (PEN-2) proteins. Zhou et al. (2005) identified an integral part of the γ-secretase complex, CD147, a transmembrane glycoprotein with two Ig-like domains. *In vitro*, $A\beta$ production is downregulated in the presence of the CD147 subunit within the γ-secretase complex.

The presenilins are the proteolytically active components and undergo endoproteolysis into an N-terminal fragment (NTF) and a C-terminal fragment (CTF) that remain associated (Thinakaran et al., 1996). This NTF—CTF heterodimer is the mature, active form of presenilin. Two conserved aspartates within adjacent transmembrane domains are essential for both presenilin endoproteolysis and γ-secretase activity (Wolfe et al., 1999b). Presenilin-1 mRNA levels are significantly higher in AD brain than in controls, but the high early-stage level decreases as the disease advances (Ikeda et al., 1998). Presenilin-2 mRNA levels are lower in the frontal cortex and hippocampus of subjects with mild AD and in those with late-onset sporadic AD (McMillan et al., 2000).

The role of the other three γ-secretase subunits is less well understood. Nicastrin and APH-1 might form a precomplex to which presenilin is first added. The ecto-domain of nicastrin interacts with the *N*-terminus of the substrate, an essential step in substrate recognition and processing (Shah et al., 2005). The addition of PEN-2 facilitates presenilin endoproteolysis, leading to the generation of the active com-plex. γ-Secretase also cleaves other substrates, notably the Notch receptor that is required for the development and maintenance of multicellular organisms.

Many questions still remain, such as whether extracellular Aβ modulates intrac-ellular Aβ, or the mechanism by which Aβ accumulation leads to synaptic dysfunc-tion. Other factors, such as oxidative stress, which is extensive in AD, may aid the early accumulation of Aβ (Butterfield et al., 2002b). Aβ peptides stimulate oxidative stress by direct and indirect mechanisms. Aβ-induced oxidative stress may result from an imbalance between reactive oxygen species (ROS) and reactive nitrogen species (RNS), which could react with a number of cellular macromolecular targets including proteins, lipids, carbohydrates, DNA, and RNA. An early marker for oxidative stress is the formation of protein carbonyls, 4-hydroxy-2-*trans*-nonenal (4-HNE) and 3-nitrotyrosine (3-NT), a marker for the nitration of proteins (Butterfield, 2002). Aβ peptide can bind to mitochondrial proteins to generate free radicals, or it can promote oxidative stress via neuroinflammation. Aβ peptides stimulate microglial cells to release a neurotoxin, quinolinic acid, which may also play a role in neurotoxicity (Guillemin et al., 2003).

The intracellular domain of APP contains several residues that are phosphor-ylated by different protein kinases. Phosphorylation of the *C*-terminal domain of APP with protein kinase-C (PKC), calcium/calmodulin-dependent kinase II (CaM kinase II), cyclin-dependent kinases 2 and 5 (CDK2, CDK5), or glycogen synthase kinase-3 (GSK3) have different effects (see below for further details on these enzymes). Phosphorylation of APP on Ser[655] can occur through PKC and CaM kinase II and on Thr[654] by PKC (numbering for the APP695 isoform). In tissue culture, treatment with phosphatase inhibitors or PKC activators increases the production of APP C-terminal fragments and reduces the level of Aβ. This leads to the idea that PKC phosphorylation may favor non-amyloidogenic processing of APP, since it prevents β-amyloid formation (Buxbaum et al., 1993). PKC can increase the activity of TACE, the TNFα converting enzyme, one of the metallopro-teases that acts as α-secretase. Activation of PKC also leads to increased sAPPα, which is neuroprotective and promotes neurite outgrowth (Stein et al., 2004). *In vitro* and *in vivo* studies identified PKCβ and PKCε as the APP-modulating kinases. PKCα is located post-synaptically, whereas PKCε is pre-synaptic, which suggest there may be a PKC-mediated protective role on both sides of the synapse (Tanaka and Nishizuka, 1994).

Phosphorylation of APP on Thr[668] by CDK2, CDK5, GSK3, and *c-jun* kinase (JNK) have been described, but reports are conflicting. Phosphorylation at Thr[668] has a dual role, regulating both neuronal development and neurodegeneration. For example, CDK5-mediated phosphorylation regulates neuronal differentiation, but also the amyloidogenic processing of APP (Iijima et al., 2000). GSK3 phosphoryla-tion of APP is linked to neurodegeneration and β-amyloid production (Lee et al.,

2003b). Phosphorylation of APP by JNK3 at Thr[688] is more problematic. On one hand, it affects AICD-mediated transcriptional activity by destabilizing the ACID-Fe65 complex. On the other, JNK3 phosphorylation relocates APP to neurites and promotes their extension.

Pastorino et al. (2006) showed that the prolyl isomerase PIN1 binds to phosphorylated Thr[668] in APP and regulates its isomerization between the cis and trans conformations. This can help prevent the amyloidogenic processing of APP, and thus affect mechanisms leading to AD pathology. These and other results suggest that APP phosphorylation is regulated by a number of kinases through different signaling pathways. The regulation of APP phosphorylation could be a target for AD treatment.

Tau

Independent of the etiology, neurofibrillary degeneration of abnormally hyperphosphorylated *tau* in neurons is a hallmark lesion of neurodegenerative tauopathies, which include AD. Paired helical filaments (PHFs) are abnormal structures generated by self-aggregation of hyperphosphorylated forms of *tau* protein that form a compact filamentous network. PHF formation from hyperphosphorylated *tau* follows several steps: *tau* phosphorylation, conformational change, and finally, polymerization. NFTs are intracellular aggregates composed of arrays of PHFs within neuronal cell bodies (Grundke-Iqbal et al., 1986a; Goedert et al., 1988).

Although AD is the most frequent form of dementia in the elderly, NFTs are, in the absence of β-amyloid plaques, also abundant in other neurodegenerative diseases, including Pick disease (Hof et al., 1994), progressive supranuclear palsy (Hof et al., 1992), corticobasal degeneration (Paulus and Selim, 1990), and amyotrophic lateral sclerosis/parkinsonism-dementia of Guam A argyrophilic grain disease, and a mutation in *tau* has been demonstrated to be a major factor in frontotemporal dementia with Parkinsonism-17 (FTDP-17; Spillantini et al., 2000).

In AD, the *tau* phosphorylation/dephosphorylation imbalance in affected neurons is generated at least in part by the activities of kinases such as CDK5, GSK-3, CaM kinase II, and PKA. Pyramidal cells of the CA 1 region of the hippocampus, layer III and IV of the association cortex, and layers II and IV of the entorhinal cortex are the main regions where PHFs can be found. The number and distribution of NTFs correlate with the severity of disease (Pearson et al., 1985) and can be used to distinguish six stages of disease progression (Braak and Braak, 1991, 1995). The transentorhinal stages I–II represent clinically silent cases; the limbic stages III–IV, incipient AD; the neocortical stages V–VI, fully developed AD. Giannakopoulos et al. (2003) showed that total NFT counts in specific brain areas such as the entorhinal and frontal cortex, as well as neuron numbers in the CA1 region of the hippocampus, are the best predictors of cognitive deficits in brain aging and AD. Although NTF pathology correlates well with AD severity, it does not correlate well with the extent of neuronal death (Gomez-Isla et al., 1997). A quantitative analysis of NFT in human brain revealed that a substantial number of pyramidal cells may persist either unaffected or in a transitional stage of NFT formation. While it is not possible to assess whether such transitional neurons are fully functional, they might respond positively to therapeutic strategies aimed at protecting cells against neurofibrillary degeneration (Bussiere et al., 2003).

Tau is mainly expressed in neurons, although non-neuronal cells usually have trace amounts. Under pathological conditions, *tau* protein can also be expressed in glial cells (Chin and Goldman, 1996). It can also be found in peripheral tissues such as heart, kidney, lung, muscle, pancreas, testis, and fibroblasts (Ingelson et al., 1996; Vanier et al., 1998).

The *tau* gene is located on chromosome 17 and contains 16 exons, although exon 4A, 6, and 8 are not present in human brain mRNA, as they are specific to peripheral tissues (Himmler et al., 1989; Nelson et al., 1996). Exons 1, 4, 5, 7, 9, 11, 12, 13 are constitutive, whereas exon 14 is found in mRNA but not translated into protein. Six *tau* isoforms result from alternative splicing of exons 2, 3, and 10 (Goedert et al., 1989a, b; Himmler et al., 1989). *Tau* isoforms range from 352 to 441 amino acids, with a molecular weight from 45 to 65 kDa. The isoforms differ by the presence of either three or four microtubule-binding domains of 31 or 32 amino acids near the C-terminal. Those encoded by exon 10 have four repeat regions (4R *tau*), whereas those lacking exon 10 have three repeat regions (3R *tau*). The different *C*-termini are combined with two, one, or no inserts of 29 amino acids each in the *N*-terminal portion that are coded by exons 2 and 3. The *N*-terminal region of *tau* binds to spectrin and actin filaments, plasma membrane components, src tyrosine kinases (fyn), and phospholipase C (Buee et al., 2000). Only a single isoform occurs in fetal brain, corresponding to the smallest of the *tau* isoforms. Goedert et al. (1992b) found all six isoforms in neurofibrilllary lesions in AD brain.

Function of Tau

The three members of microtubule-associated proteins (MAPs) appear to have the same functions, which include the nucleation and promotion of microtubule polymerization, the regulation of microtubule polarity and dynamics, and the spacing and bundling of axonal microtubules. *Tau* proteins (MAPT) bind to spectrin and actin filaments (Griffith and Pollard, 1982; Carlier et al., 1984; Correas et al., 1990) and allow microtubules to interconnect with other cytoskeletal components such as neurofilaments. Microtubules are essential for axoplasmic flow, which in turn is essential for neuronal activity. *Tau* can interact with cytoplasmic organelles, which allows connection between microtubules and mitochondria (Rendon et al., 1990) or between microtubules and the neural plasma membrane (Brandt et al., 1995). *Tau* has been assigned roles in signal transduction, intracellular vesicle transport, and anchoring phosphatases and kinases (Götz et al., 2001). For example, *tau* can associate with fyn (an src-family nonreceptor tyrosine kinases; Lee et al., 1998) or phospholipase C-γ (PLC- γ) isozymes (Hwang et al., 1996).

Phosphorylation of Tau

The biological activity of *tau* is regulated by its degree of phosphorylation and dephosphorylation. Normal *tau* contains 2–3 mol phosphate/mol of protein, whereas in AD it is three- to fourfold more hyperphosphorylated (Grundke-Iqbal et al., 1986b; Kopke et al., 1993). The abnormal hyperphosphorylation of *tau* appears to precede its accumulation in affected neurons in AD. Abnormally hyperphosphorylated *tau* is found not only in NFTs (Grundke-Iqbal et al., 1986b) but also in cytosol in AD brain cells (Iqbal et al., 1986).

Tau proteins contain more than 30 phosphorylation sites (Morishima-Kawashima et al., 1995; Lovestone and Reynolds, 1997; Hanger et al., 1998; Johnson and Hartigan, 1999) for a large number of protein kinases, including PKA, PKC, CDK5, mitogen-activated extracellular signal-regulated protein kinases (MAPK/ERK), stress-activated protein kinases (SAPK/JNK), p38 kinases, CaM kinase II, microtubule affinity-regulating kinase (MARK), and casein kinases (Buee et al., 2000; Anderton et al., 2001; Lee et al., 2001). Phosphorylation at many of these sites occurs in a significant fraction of fetal *tau*; studies of biopsy-derived material have shown that they are also phosphorylated in a small fraction of *tau* from adult human brain (Matsuo et al., 1994). However, some sites are phosphorylated in abnormal *tau*, but not in *tau* from normal adult brain (Mercken et al., 1992; Hasegawa et al., 1996; Zheng-Fischhofer et al., 1998).

Abnormal Tau

Interruptions of post-translational modifications play a major role in the hyperphosphorylation of *tau*, which lead to structural and conformational changes. Hyperphosphorylation of *tau* affects its binding to tubulin, leading to destabilization of microtubular network and impairment of microtubule-associated axonal transport.

Mitogen-Activated Protein (MAP) Kinases

The MAP kinase family belongs to the proline-directed protein kinases and has three members: stress-activated protein kinase/c-Jun amino terminal kinase (SAPK/JNK) p38 kinase, and extracellular signal-regulated protein kinases (ERKs; Roux and Blenis, 2004). These serine/threonine protein kinases are activated in response to a variety of extracellular stimuli. They mediate signal transduction from the cell surface to the nucleus (Boulton et al., 1991; Goedert et al., 1997). For activation, all three kinases have a tripeptide motif that has to be phosphorylated. For ERK this motif is $Thr^{202}Glu^{203}Tyr^{204}$; for JNK it is $Thr^{183}Pro^{184}Tyr^{185}$; and for p38, $Thr^{180}Gly^{181}Tyr^{182}$ (Raingeaud et al., 1995). They have all received attention with respect to *tau* hyperphosphorylation, because they are all overexpressed in AD (Zhu et al., 2000a, 2001b; Ferrer et al., 2001a,b).

Mitogen-Activated Protein Kinase ERK1/2 (MAPK/ERK)

ERKs include ERK1 (p44 ERK1) and ERK2 (p42 ERK2). Both are able to phosphorylate recombinant *tau* at the same sites as PHF (Drewes et al., 1992; Goedert et al., 1992a; Lu et al., 1993). ERKs are activated by MAPK/ERK kinases (MEKs), which in turn are activated by MEK kinases (MEKKs). Activation of this cascade results in transcriptional regulation through different downstream activators, such as cAMP-response-element-binding-protein (CREB). CREB is necessary for cell proliferation, differentiation, and survival (Derkinderen et al., 1999). ERK1 and ERK2 phosphorylate *tau* and neurofilament proteins *in vitro* (Drewes et al., 1992; Goedert et al., 1992a; Veeranna et al., 1998). ERK1 mRNA in human hippocampus is limited to dentate gyrus cells, whereas ERK2 mRNA is present in dentate gyrus granule cells and in pyramidal neurons of the hippocampus and temporal cortex. ERK1 and ERK2 proteins occur in neurons, dendrites, axons, and reactive astrocytes (Hyman et al., 1994).

Several observations support the role of ERK in the pathogenesis of AD (Mc Shea et al., 1999; Ferrer et al., 2001a; Zhu et al., 2002). Nonactivated ERK levels are almost the same in AD and control brains, whereas phosphorylated MAPK/ERK is increased in AD. The reason for MAPK/ERK activation in AD is unknown, but MAPK kinase 1 (MEK1), an activator of ERK, is increased in AD. Trojanowski et al. (1993) reported reduced levels of ERK2 in AD brain, which could be also due to neuronal loss. They localized ERK2 in NFTs and dystrophic plaque neurites in the AD hippocampus. They also found ERK2 immunoreactivity in neurons with neurofibrillary tangles and in dystrophic neurites of senile plaques in AD (Trojanowski and Lee, 1995). Ferrer et al. (2001a) found a close relationship between ERK phosphorylation and *tau* deposition, not only in neurons but also in glia cells in AD.

p38

p38 MAP kinase is a stress-activated enzyme responsible for transducing inflammatory signals and initiating apoptosis. Four isoforms of p38 MAPKs (p38α, β, γ, and δ) have been identified (Ono and Han, 2000). Additionally, p38α is alternatively spliced, and an isoform of p38β, p38β2, is the major isoform in human brain (Enslen et al., 1998). In contrast, p38γ and p38δ are not expressed in human brain (Li et al., 1996a; Kumar et al., 1997). Among the other functions, p38 kinase has the capability to phosphorylate *tau in vitro* at several sites that are seen in PHF-*tau* (Reynolds et al., 1997a,b).

Elevated activated-p38 immunoreactivity in AD brain was reported by Hensley et al. (1999). They found increased phospho-p38 activity in the AD hippocampus. There was also an accumulation of activated p38 kinase in neurons filled with NFTs. An association of phosphorylated p38 with neurofibrillary tangles has also been reported, together with co-precipitation of p38 and PHF-*tau* (Zhu et al., 2000b). Pei et al. (2001) correlated the appearance of activated p38 with neurofibrilllary degeneration in AD brain regions at different Braak stages. Entorhinal cortex showed an intense phospho-p38 staining in early stages. With AD progression, the phosphor-p38 immunoreactivity intensity spread to other brain regions. This accumulation of activated p38 in neurons occurred earlier than the deposition of amyloid in the extracellular space. In contrast, Sun et al. (2003) showed co-localization of phosphorylated *tau* and phosphor-p38 in AD cases that did not correlate with NTFs. This may suggest that activation of p38 decreases its association with phosphorylated *tau*.

c-Jun Kinase (JNK) also called Stress Activated Protein Kinases, SAPK

Three different genes encode JNKs: *JNK1, JNK2*, and *JNK3* (Gupta et al., 1996). *JNK1* and *JNK2* are ubiquitously expressed, whereas *JNK3* mRNA is restricted to brain, heart, and testis. All transcripts are alternatively spliced, but functional differences between the variants have not been reported. In the prenatal brain, *JNK1* mRNA is widely expressed and rapidly downregulated after birth; *JNK3* expression does not change. *In vitro*, JNK phosphorylates *tau* on threonine and serine residues and can interfere with tau induction (Goedert et al., 1997; Reynolds et al., 1997b; Sadot et al., 1998). JNK2 phosphorylates the most sites in *tau*, followed by JNK3 and JNK1. Phosphorylation of *tau* by JNK isoforms also leads to a lower ability of *tau* to promote microtubule assembly *in vitro* (Yoshida et al., 2004).

In hippocampal and cortical AD brain samples, activated JNK is localized exclusively with NFTs; it is absent from controls without pathology. A translocation of activated JNK from the nucleolus to the cytoplasm was found in controls with some pathology and in mild AD cases, indicating a correlation of disease state and activation/redistribution of JNK. Zhu et al. (2001a) showed that JNK and JNK3 co-localize with *tau* pathology in AD and that they are significantly more activated (phosphorylated) in AD. There is an increase in JNK phosphorylation from early Braak stages to the limbic stages of AD (Yoshida et al., 2004).

Phosphorylated JNK regulates the activity of caspase-3, which can cleave a small *tau* fragment that aggregates more rapidly than full-length *tau* (Berry et al., 2003; Gamblin et al., 2003). A major contributing factor for the activation of JNK, p38, and other *tau* kinases could be reduced activity of PP-2A/PP1 and tyrosine phosphatases, resulting in hyperphosphorylated *tau*.

Glycogen Synthase Kinase 3 (GSK3)

Glycogen synthase kinase-3 (GSK3) is a proline-directed serine-threonine kinase that was initially identified through its phosphorylation and inactivation of glycogen synthase (Embi et al., 1980). GSK3 is capable of phosphorylating, and thereby regulating, over 40 known substrates. GSK3 has been implicated in signal transduction pathways downstream of phosphoinositide 3-kinase and mitogen-activated protein kinases (MAPKs). GSK3 exists as two related isoforms called GSK3α and

GSK3β (Woodgett, 1990). GSK3β is involved in energy metabolism, neuronal cell development, and body pattern formation (Plyte et al., 1992). GSK3 phosphorylates *tau* at multiple sites, some of which are aberrant in the abnormally hyperphosphorylated *tau* (Mandelkow et al., 1992; Sperber et al., 1995; Lee et al., 2003a).

There is no consistent data about the concentration of GSK3 protein in AD brain. Active GSK3 localizes to dystrophic neurites and NFTs, and GSK3β activity is significantly upregulated in AD. This is thought to be initiated through β-amyloid-induced impairment of PI3/Akt signaling, leading to increased NTF formation and synaptic death (Takashima, 2006). Cells co-transfected with *tau* and GSK3β show abnormal microtubule bundles (Sang et al., 2001). The effects of GSK3α and β largely depend on their state of activation and inhibition, brought about through the action of other kinases and phosphatases. Stambolic et al. (1996) showed that Li$^+$; inhibits the GSK3-dependent phosphorylation of *tau* (Zhang et al., 2003). Increased GSK3 activity inhibits the phosphatase-1 (PP1) inhibitor 2 (I-2). Increased PP1 activity then dephosphorylates GSK3 (Sakashita et al., 2003).

Cyclin-dependent Kinases

Cyclin-dependent kinase-5 (CDK5) is a protein serine/threonine kinase with close structural homology to the mitotic CDKs. CDK5 is predominantly expressed in terminally differentiated cells such as neurons. It has a unique function in neurons unrelated to cell cycle control, unlike other CDKs. CDK5 must associate with p35 to be active in postmitotic neurons (Lew et al., 1994). This complex is involved in CNS development and maturity, including neuronal migration, corticogenesis, neurite outgrowth, regulation of the synaptic vesicle cycle, neurotransmitter release, and postsynaptic neurotransmitter receptor regulation and signaling (Tsai et al., 1994; Tang and Wang, 1996; Dhavan and Tsai, 2001; Hisanaga and Saito, 2003). Paudel et al. (1993) showed that CDK5 phosphorylates *tau* protein. There is also growing evidence that CDK5 is involved in the mechanisms of synaptic plasticity underlying learning and memory (Fischer et al., 2002).

Patrick and coworkers (1999) showed that deregulation of CDK5 contributes to the pathogenesis AD through the accumulation of a truncated fragment of p35 called p25, which constitutively activates and changes the subcellular distribution of active CDK5 from membranes to the cytosol. Cleavage of p35 to p25 occurs through the Ca^{2+}-activated protease calpain (Kusakawa et al., 2000). The p25/CDK5 complex is more effective in phosphorylating *tau* than CDK5/p35. p25/CDK5 also causes morphological degeneration and apoptosis in primary neuronal cultures (Lew et al., 1994; Poon et al., 1997). Exogenous overexpression of p25 in transgenic mice results in a neurodegenerative phenotype, including the formation of PHFs, *tau* aggregation, and neuronal loss similar to that observed in AD (Cruz et al., 2003; Noble et al., 2003). p25 may cause harm by preventing the p35/CDK5 kinase from phosphorylating its normal substrates, such as PAK1, synapsin, syntaxin, MUNC18, and DARP32, by sequestering CDK5 from the cell periphery and nerve terminals; this will contribute further to neuronal dysfunction.

cAMP-dependent Protein Kinase A (PKA)

PKA-dependent phosphorylation of substrate requires a rise in intracellular cAMP levels and the translocation of PKA, which is regulated by A kinase anchoring proteins (AKAPs; Coghlan et al., 1993). Phosphorylation of *tau* by PKA significantly promotes subsequent tau phosphorylation by GSK3β (but not MAP kinase or CDK5) *in vitro* (Singh et al., 1996; Wang et al., 1998). Liu et al. (2004b) showed that PKA-catalyzed phosphorylation facilitates *tau* phosphorylation by GSK3 in rat brain and suggested a role for PKA upstream of GSK3 in *tau* hyperphosphorylation. PKA-induced phosphorylation of *tau* can be detected in early and advanced AD, but not in normal brain tissue. PKA phosphorylation of *tau* precedes or is coincident with the initial appearance of filamentous *tau* aggregates (Jicha et al., 1999).

Calcium/Calmodium-dependent Protein Kinase II (CaM Kinase II)

CaM kinase II is a ubiquitous serine/threonine protein kinase. Four similar isoforms of CaM kinase II (α, β, γ and δ) can be found, each encoded by distinct genes (Rostas and Dunkley, 1992; Braun and Schulman, 1995; Bayer et al., 1999). The α and β isoforms are brain specific, whereas the γ and δ forms are ubiquitous, which suggests there are functional differences between the isoforms. CaM kinase II is highly abundant in brain and is a major constituent of the postsynaptic density (Kelly et al., 1984). It plays important roles in the synthesis and release of neurotransmitters, ion-channel regulation, structural modification of the cytoskeleton, axonal transport, synaptic plasticity, and gene expression (Braun and Schulman, 1995).

CaM kinase II participates in the phosphorylation of *tau* (Baudier and Cole, 1987; Steiner et al., 1990). It phosphorylates a number of other substrates *in vitro*, including MAP-2, tyrosine hydroxylase, synapsin 1, APP, and various intermediate filament proteins such as vimentin, desmin, and GFAP (Colbran et al., 1989). *In vitro*, CaM kinase II-phosphorylated *tau* inhibits microtubule assembly (Yamamoto et al., 1983; Yamamoto et al., 1985, 1988). It phosphorylates *tau in vitro* in such a way to slow its electrophoretic mobility (Baudier and Cole, 1987). Serine[416] (numbered according to the longest human *tau* isoform) is one of the major phosphorylation sites for CaM kinase II *in vitro* (Steiner et al., 1990). Immunostaining of AD brain showed serine[416] phosphorylation only in neuronal soma, but not in neuropil threads and dystrophic neurites (Yamamoto et al., 2005). *Tau* phosphorylation in glia cells is unlikely due to CaM kinase II.

Levels of CaM kinase II and *tau* mRNA differ between AD cases and controls only in layer I of the entorhinal cortex. No changes have been detected in CA regions of hippocampus or other areas prone to NFT formation. Mc Kee and coworkers (1990) reported increased immunoreactivity of CaM kinase II in the CA1 area of the hippocampus and the subiculum despite NFT formation and neuronal depletion in AD. PHF fractions from brain homogenates are enriched in CaM kinase II (Xiao et al., 1996).

Casein Kinase 1 (Ck 1)

Casein kinase I is a highly abundant serine/threonine protein kinase, widely distributed in nuclear, cytoplasmic, and membrane fractions. It occurs in multiple isoforms
encoded by distinct genes (CK1-α, -β, -γ1, -γ2, -γ3, -δ, -ε) that share an N-terminal
catalytic domain combined with various unique C-termini, which regulates the
catalytic activity (Xu et al., 1995; Knippschild et al., 2005). CK1 isoforms can
phosphorylate *tau in vitro* and in cell culture and modulate its binding to mictotu-
bules (Li et al., 2004c). All CK1 isoforms are more abundant in AD hippocampus
CA1 than in controls (Ghoshal et al., 1999). Isoform CK 1δ seems to be most
closely linked with AD. It is solely expressed in neurons and shows a more than
30-fold higher expression in AD hippocampus than in controls. Overexpression of
CK1δ is also observed at the mRNA level (Yasojima et al., 2000).

In the same brain region, CK1α is only 2.4-fold increased. In purified *tau*
filaments from AD brains, it comprises up to 0.5% of the preparation by weight
(Ghoshal et al., 1999). Kannanayakal et al. (2006) showed that CK1α is associated
with neurofibrillary lesions, while CK1δ is primarily connected with granulovascular
degeneration bodies. Taken together, CK1 protein is increased in AD tissue, it can
phosphorylate *tau*, and it co-localizes with AD lesions.

In conclusion, *tau* pathology appears to be an essential and primary cause of
neurodegeneration in AD and other tauopathies. A large amount of data implicates
several kinases in *tau* hyperphosphorylation. *Tau* phosphorylation is modulated by
the sequential and collective action of distinct kinases that are essential for PHF
formation in neurons and glial cells in AD. Inhibition of abnormal *tau* hyperphosphorylation is one of the most promising therapeutic targets for AD, but the mechanism
underlying the interaction between amyloid and tangle pathologies in AD remains
to be elucidated.

The Presenilins PS1 and PS2

The presenilin gene *PS1* resides on chromosome 14 in humans; the *PS2* gene is on
chromosome 1. The proteins share about 62% amino acid identity. They are
transcribed in many tissues, both within the CNS and peripherally. Presenilin
proteins have multiple transmembrane (TM) domains and in their mature form are
localized to the cell surface to regulate the cleavage of single-TM domain proteins
(Chyung et al., 2005). They also reside in the ER and the Golgi (De Strooper et al.,
1997). A clue to the roles of *PS1* and *PS2* in AD comes from the observation of a
significant increase in Aβ42 levels in plasma, and in media from cultured fibroblasts, from *PS1* and *PS2* mutation carriers (Scheuner et al., 1996).

The structure of *PS1*, derived from its cDNA sequence, predicts a 467 amino
acid polypeptide (~50 kDa) with between six and nine TM domains, an aqueous
N-terminus, and a large hydrophilic loop region (amino acids 267–403) between the
putative sixth and seventh TM (TM6–TM7) helices. Its active site, which contributes
to γ-secretase activity, is in TM7.

Presenilins are translated as full-length holoproteins that are not modified by glycosylation, sulfation, or phosphorylation (Walter et al., 1996). They are endo-proteolytically cleaved into a 27 kDa NTF and a 17 kDa CTF. Endoproteolysis seems to be essential for presenilin stability and activity. The NTFs and CTFs dimerise to form the active, functional presenilin heterodimer. Only a small portion of full-length presenilin protein is present in brain tissue (Podlisny et al., 1997).

Miss-sense mutations in *PS1* and *PS2* are associated with early-onset familial AD. Analyses of brains from AD patients with *PS1* mutations have demonstrated a significant increase in the amount of $A\beta42$-immunoreactive deposits in hippocampus, cerebral cortex, and other brain regions (Lemere et al., 1996). The majority of miss-sense mutations occur in highly conserved sites (Fraser et al., 2000). They are predominantly located in highly conserved TM domains, at/near putative membrane interfaces, or in the large hydrophilic loop. Dimerization models suggest that *PS1* and *PS2* mutations influence the assembly or stability of the functional presenilin complex (Cervantes et al., 2001).

Besides their role in APP γ-secretase, the functions of the presenilins have not yet been determined. Presenilins are essential for the maintenance of cortical structures and functions (Feng et al., 2004). They are involved in cell-signaling pathways, trafficking of membrane proteins, and Ca^{2+} signaling. Presenilin holo-protein accounts for 80% of Ca^{2+} leakage from the ER, which appears to be independent of its γ-secretase function. Increased levels of Ca^{2+} might initiate a signal transduction pathway in which Ca^{2+} and diacylglycerol act together to activate Ca^{2+}-sensitive protein kinases that are responsible for *tau* hyperphosphor-ylation. High intracellular Ca^{2+} levels also promote the influx of extracellular Ca^{2+}, which acts as a positive feedback mechanism to enhance excitation and release more glutamate until this cycle becomes toxic to the cell (Hynd et al., 2004).

Protein Phosphatases

Phosphorylation at serine and threonine residues is an important regulator for cell signaling and transcription. The level of phosphorylation is controlled by the opposing actions of protein kinases and phosphatases. Four major protein phos-phatases (PP1, PP2A, PP2B, and PP2C), responsible for dephosphorylation at serine and threonine residues, have been identified in the cytosol of eukaryotic cells. Although they have overlapping functions, type-1 protein phosphatases are inhibited by nanomolar concentrations of two thermostable proteins, IPP1 and IPP2. Type-2 phosphatases are unaffected by these inhibitors. To distinguish type-2 phosphatases, their activity regulation has to be taken into account. PP2A activity is independent of metal ions, whereas Ca^{2+}/calmodulin activates PP2B and Mg^{2+} activates PP2C.

All of these phosphatases except PP2C dephosphorylate *tau in vitro* and possibly *in vivo* (Gong et al., 2005). Their contribution in the regulation of *tau* dephosphor ylation depends on their affinities for *tau* at different phosphorylation sites, their specific enzymatic activities toward *tau*, and their relative abundance in the brain.

PP2A is co-localized with *tau* and microtubules in brain and is apparently the major *tau* dephosphorylating enzyme. Liu et al. (2005a) found that PP2A accounts for ~70% of total brain *tau* phosphatase activity, whereas PP1, PP2B, and PP5 each account for only ~10%. PP2A can dephosphorylate almost all phosphorylation sites of *tau* reported in AD and it also can regulate the activities of several *tau* kinases (Bennecib et al., 2001; Kins et al., 2003; Pei et al., 2003). In AD brain tissue, PP2A activity is ~20% compromised in comparison to healthy controls (Gong et al., 1993, 1995). The mRNA level is also decreased in AD (Vogelsberg-Ragaglia et al., 2001).

PP2A is inhibited by two heat-stable proteins: I1 PP2A (Li et al. 1995) and I2 PP2A (Li et al., 1996b). The level of I1 PP2A is ~20% increased in AD brains, which could contribute to the decrease in PP2A activity in AD brain. Lui et al. (2007) associated PP2A activity with GSK3β through accumulation of I2PP2A in response to GSK3β activation. *In vitro* experiments showed that inhibition of PP2A activity results in the abnormal hyperphosphorylation of *tau* and the upregulation of kinases such as PKA, MAPK, and CaM kinase II (Gong et al., 2000). Wang et al. (2007) showed that PPA2-dependent dephosphorylation of hyperphosphorylated *tau* inhibits its self-assembly into PHFs and enables *tau* to bind to microtubule again. However, the mechanism leading to the inactivation of PP2A in AD brain is currently not known.

PP2B (calcineurin) is one of the major serine/threonine phosphatases and present in nearly all mammalian cells studied. It accounts for 1% of total proteins in brain tissue, which is considerably higher than that of other protein phosphatases (Herzig and Neumann, 2000). PP2B is expressed in neurons, including the tangle-bearing neurons in AD hippocampus (Pei et al., 1994), but its role in dephosphorylating *tau* is conflicting. Upregulation of PP2B expression in pyramidal AD hippocampal neurons has been reported (Hata et al., 2001), although Ryan et al. (2001) showed the opposite. Decreased PP2B activity in AD was reported by Lian et al. (2001).

PP1 and PP2A together account for over 90% of total mammalian brain protein phosphoseryl/phosphothreonyl phosphatase activity. In comparison to PP2A, the activity of PP1 is relatively low (Gong et al., 1994a). PP1 activity is mainly regulated by inhibitor-1 (I-1), which is activated by PKA and inactivated at basal Ca^{2+} levels by PP2A. The 20% reduction of PP1 activity in AD brains could result from reduced PP2A activity allowing the up-regulation of I-1 (Gong et al., 1993; Iqbal et al., 2005).

PP5 is a novel phosphoseryl/phosphothreonyl protein phosphatase, ubiquitously expressed at high abundance in human brain tissue. It dephosphorylates *tau in vitro* and may associate with microtubules, raising the possibility that it may participate in regulation of *tau* phosphorylation in the brain. Its *C*-terminal catalytic domain is 40% homologous to PP1, PP2A, and PP2B, whereas its *N*-terminal domain consists of three tetratricopeptide repeats (TPR) which are not shared with other members of the PP1/PP2A family (Becker et al., 1994). PP5 has little phosphatase activity *in vitro* in the absence of activators. However, it effectively dephosphorylates *tau* after the removal of the TPR domain (Liu et al., 2005b). In AD the activity of PP5 is decreased but its protein level is unaltered. The mechanism leading to this is not known (Liu et al., 2005b).

Oxidative Stress

Alzheimer's disease is a disease of aging, and aging is associated with an increase in oxidative stress. The early involvement of oxidative stress in AD has been demonstrated by many studies on cell culture models, transgenic mice, and autopsy brains from subjects with mild cognitive impairment (MCI) and AD. It can be either hypothesized that senile plaques and NFTs are closely associated with oxidative stress and contribute equally to neuronal death, or that oxidative stress together with Aβ and *tau* cause neuronal death and the formation of plaques, and NTF occurs later (Butterfield et al., 2006). Nunomura et al. (2001) showed that oxidative damage is high in early stages of AD and reduces with disease progression. Areas with many amyloid plaques and NTF-positive neurons have less oxidative damage than those without. In cell culture, Aβ accumulation and *tau* phosphorylation are induced by oxidative stress (Misonou et al., 2000; Gomez-Ramos et al., 2003).

Oxidative stress is a condition in which free radicals and their products are present in excess of antioxidant defence mechanisms, leading to oxidative damage and structural and functional modifications of proteins, DNA, and RNA. Radicals can be formed through different mechanisms. ROS are formed by the reduction of molecular oxygen to water: superoxide (O_2^-), peroxyl radical ($OH^\bullet$), and hydrogen peroxide (H_2O_2). RNS (NO and peroxynitrite) can also cause oxidative stress.

High oxygen consumption and high level of oxidizable unsaturated fatty acids make the brain susceptible to oxidative damage. The brain accounts for only 2% of body weight but consumes 20% of the total O_2. The levels of antioxidant enzymes are relatively low in brain compared with other organs.

RNA or DNA exposed to oxygen radicals generates various modified bases. More than 20 different types of oxidatively altered purines and pyrimidines have been detected (Demple and Harrison, 1994). Guanine, through its low redox potential, is the base most readily oxidized by free-radical attack, leading to the formation of 8-hydroxyl-2-deoxyguanosine (8OHdG) and 8-hydroxyguanosine (8OHG). These can be directly oxidized in the RNA molecule or be incorporated during transcription (Breen and Murphy, 1995). 8OHdG and 8OHG can have an impact on translation accuracy because they can pair with adenine and cytosine (Ishibashi et al., 2005). Regardless of its origin, oxidized guanine in RNA cannot be eliminated, in contrast to DNA where oxidized bases can be removed by glycosylases. There are some proteins in mammalian cells, such as MTH1(an oxidized purine nucleoside triphosphatase) and NUDT5 (an ADP-ribose pyrophosphatase), that can degrade 8OHG. MTH1 protein, for example, is lower in the CA1 and CA3 region of AD brain than in non-neurological controls (Nakabeppu et al., 2006) Polynucleotide phosphorylase (PNP) protein can bind specifically to RNA containing 8OdHG (Hayakawa and Sekiguchi, 2006). Y-box-binding protein 1 (YB-1 protein) can bind to RNA containing 8OHG, and may separate oxidatively damaged RNA from other cellular components (Hayakawa et al., 2002).

Neurons in vulnerable regions of AD brain are positive for 8OHG. Cytoplasmic regions show strong immunoreactivity, which leads to the proposition that RNA and mitochondrial DNA are the major targets of oxidative damage. These results have been confirmed by electron microscopy, which showed that most oxidized nucleosides are associated with ribosomes (Nunomura et al., 2001). Oxidatively damaged RNA is not only present in FAD but also in sporadic AD (Nunomura et al., 2004). There is an interaction with APOE: 30–70% of mRNA from the frontal cortex of AD *APOE*-ε4 carriers was oxidized, whereas only 2% is oxidized in age-matched controls (Shan and Lin, 2006). Some of the oxidized transcripts are related to AD, such as those for MAPK, APOE, and calpains. However, oxidation of β-amyloid or *tau* transcripts has not been demonstrated.

The regional distribution of oxidized RNA correlates with the neuronal vulnerability. For example, the levels of 8OHG are higher in hippocampus and neocortex than in cerebellum in AD brain (Ding et al., 2005; Shan and Lin, 2006). Ding et al. (2005) also showed that besides mRNA, rRNA and tRNA are highly oxidized in AD, which suggests that oxidized RNA can alter phenotypic expression.

DNA Oxidation

ROS can lead to more than 20 oxidized base variations, which can cause DNA strand breaks, DNA—DNA and DNA—protein cross-links, and to sister—chromatid exchange leading to replication faults, including mutations that could ultimately alter protein synthesis (Cooke et al., 2003). In addition, DNA bases may be modified by 4-HNE and acrolein, leading to the formation of exocyclic adducts.

Mullaart et al. (1990) reported a twofold increase in DNA strand breaks in AD brain. There is a higher level of oxidative DNA damage in AD than in controls. Neocortical regions show higher levels than cerebellum, but there is no correlation with NTFs or senile plaques (Gabbita et al., 1998). Mitochondrial DNA (mtDNA) is much more susceptible to oxidation: nuclear DNA has protective histones, and owing to it position in the nucleus it is not accessible to oxidative species. mtDNA has only limited DNA repair capacity. Overall, mtDNA has approximately tenfold higher levels of oxidized bases than nuclear DNA (Wang et al., 2005). Comparison of oxidative DNA damage in samples from patients with MCI and AD suggests that it is an early event in AD pathogenesis.

There are different ways for DNA repair, such as excision of nucleotides and repair of mismatches. In nuclei, the repair of 8OHG is specifically carried out by the base excision repair enzyme 8-oxoguanine glycosylase (OGG1). OGG1 activity is lower in AD hippocampus than in age-matched controls (Lovell et al., 2000). DNA double strand breaks can be fixed via non-homologous end-joining using a DNA-dependent protein kinase (DNA—PK) complex. Levels of this complex are lower in AD cortex than in controls (Shackelford, 2006).

Proteins

Oxidative modification of proteins is indicated by the high concentrations of protein carbonyls, by nitration of tyrosine residues, and by the cross-linking of proteins

(Cras et al., 1995). Lipid peroxidation is indicated by the high concentrations of the thiobarbituric acid reactive substances malondialdehyde, 4-HNE, acrolein, isoprostanes, and altered phospholipid composition (Markesbery and Lovell, 1998; Butterfield et al., 2001). Many different proteins involved in AD are oxidatively modified (see Butterfield et al., 2006 for review; Sultana et al., 2006). All these modifications are usually linked to a functional loss of proteins leading to their degradation or aggregation, as observed in AD.

4-HNE is one of the most common products and toxic markers of oxidative stress. It reacts covalently with cysteine, lysine, and histidine residues in proteins, and disrupts their structure to alter function (Esterbauer et al., 1991; Subramaniam et al., 1997). It is relatively stable and can diffuse to different subcellular compartments. It interacts with many cell proteins, including *tau* and histones (Gomez-Ramos et al., 2003). 4-HNE concentration is increased in amyloid plaques in AD patients (Esterbauer et al., 1991; Ando et al., 1998). Aβ can induce lipid peroxidation (Butterfield et al., 1994) and HNE formation (Mark et al., 1997a). Through covalent modification of Aβ via 4-HNE, the aggregation of Aβ into oligomers is initiated, but the conversion of Aβ into fibrils is inhibited (Siegel et al., 2007). This links oxidative damage to excitotoxicity, in which high extracellular glutamate leads to excessive activation of glutamate receptors, Ca^{2+} influx, and neural cell death. Lauderback et al. (2001) showed that 4-HNE can oxidatively modify the glutamate transporter EAAT2 and reduce its ability to transport glutamate. Glutamine synthase is inhibited in cells or synaptosomes treated with Aβ or HNE (Harris et al., 1996; Keller et al., 1997a, b). 4-HNE has been linked to growth inhibition, alterations in glutathione levels, inhibition of enzymes that are critical for neuronal survival such as Na^+/K^+ ATPase, and altered glucose transport (Mark et al., 1997a; Pedersen et al., 1999).

Hemeoxygenases

Hemeoxygenase-1 (HO1) and hemeoxygenase-2 (HO2), members of the stress protein superfamily, are deployed in the ER. Three HO isoforms (HO1, HO2, and HO3) have been identified. HO1 is a 32 kDa heat-shock protein with a low expression in brain that increases after heat shock. HO2 is a constitutively synthesized 36 kDa protein that has its major activity in the brain (Trakshel et al., 1988). HO3 is related to HO2 but is the product of a different gene. They catalyze the oxidative degradation of pro-oxidant metalloporphyrins such as heme (potential pro-oxidant) to biliverdin, which is subsequently converted to bilirubin (potential antioxidant) by biliverdin reductase and carbon monoxide, finally releasing iron (pro-oxidant; Ewing and Maines, 1991).

Heme is synthesized in mitochondria and plays a role in number of metabolic processes of the cell, controlling the activity of transcription factors, specific signaling pathways, cholesterol synthesis, and iron homeostasis. Heme can also convert less reactive oxidants to highly reactive free radicals, and disturbed heme metabolism can lead to mitochondria decay, oxidative stress, accumulation of iron, and cell death.

Upregulation of hemeoxygenases (Schipper et al., 1995) and heme levels (Atamna and Frey, 2004) occur in AD and other neurodegenerative diseases (Castellani et al., 1995; Schipper et al., 1998). Spared brain regions in AD have extremely low levels of HO1, whereas temporal cortex and hippocampus show intense HO1 staining (Schipper et al., 1998). This raises the question whether HOs have a neuroprotective or neurodegenerative function. On one hand, HOs protect cells by promoting the catabolism of pro-oxidant metalloporphyrins to biliverdin, which has free-radical scavenging capabilities (Dore et al., 1999). But in the course of heme catabolism, free iron and carbon monoxide can be discharged, mediating oxidative stress and cellular injury by promoting free-radical generation within mitochondria or other cellular compartments (Zhang and Piantadosi, 1992). Decreased heme levels lead to loss of mitochondrial complex IV, oxidative stress, accumulation of iron, and cell death (Atamna et al., 2001, 2002; Killilea et al., 2003).

HO1 expression is coincident with NFTs and senile plaques (Kimpara et al., 2000). Heme inhibits PHF phosphorylation catalyzed by PHF-dependent kinase (Vincent and Davies, 1992). Takeda et al. (2000a) showed that the expression of HO1 in AD brain is closely associated with pathological changes in *tau* protein. Overexpression of HO1 reduces the expression of *tau* and the activity of ERK1 and ERK2 (Takeda et al., 2000b). Heme also inhibits the aggregation of Aβ *in vitro* (Howlett et al., 1997), and low levels of intracellular heme causes aggregation of APP (Atamna et al., 2002). Atamna and Boyle (2006) demonstrated that Aβ binds to heme, which reduces intracellular heme levels and leads to metabolic disturbances.

Glucose

The high metabolic activity in human brain is primarily fueled by glucose, and the brain accounts for 25% of the glucose metabolism. To enter the brain, glucose must be transported across the blood–brain barrier (BBB), because the brain itself is incapable of synthesizing or storing glucose. Glucose is metabolized by glycolysis to pyruvate, which is converted to acetyl-CoA, the major carbon source for the citric acid cycle. Acetylcholine and cholesterol, among many other substances, are derived from acetyl-CoA (Michikawa and Yanagisawa, 1999). The citric acid cycle produces ATP, NADH, and FADH2. Oxidative phosphorylation in mitochondria uses NADH and FADH2 to produce more ATP. ATP is important for protein synthesis, membrane transport, and synaptic transmission (Erecinska and Silver, 1989; Hoyer, 2000).

The transport of glucose into the brain is complex, involving neurons, astrocytes, and endothelial cells of small blood vessels. It is dependent on the density of glucose transporters in conjunction with capillary density and net blood flow. Thirteen members of the glucose transporter (GLUT) protein family have been described so far, including GLUT1—12 and the H+/myosinositol cotransporter. (Mc Ewen and Reagan, 2004). A large number of glucose transporters are found in brain, each with a specific cellular distribution (Mc Ewen and Reagan, 2004). GLUT1 is the primary glucose transporter found in capillary endothelial cells of the BBB and in glia. GLUT3 is the main neuronal glucose transporter, while GLUT4 and GLUT8 are also found on a few neurons. GLUT3 has a five to sevenfold higher glucose turnover number than GLUT1. GLUT expression levels are regulated through metabolic demand for glucose and regional rates of cerebral glucose utilization.

Glucose metabolism is significantly reduced in the cerebral cortex in early-stage AD (Jagust et al., 1991; Hoyer, 2000), leading to diminished ATP-dependent membrane transport. A 50% lower ATP production, through disturbed glucose metabolism, can be seen in the beginning of sporadic AD (Hoyer, 1992). This energy deficit may compromise ATP-dependent processes, such as molecular mechanisms in the ER and Golgi. ATP reduction can block protein secretion (Dorner et al., 1990), and lead to misfolding of protein complexes (Kaufman, 1999) and the degradation of membrane phospholipids (Sun et al., 1993b). This may be followed by membrane depolarization and impaired functioning of ligand- and voltage-gated ion channels, causing disruption of neuronal homeostasis.

Impairment of glucose uptake and/or metabolism in AD brain can arise in multiple ways, such as reduced cerebral blood flow or altered GLUT1 and GLUT3 activity. Glucose metabolism is reduced in temporal and parietal regions of AD cortex. The abundance of GLUT1 and GLUT3 protein is reduced in AD cortex (Simpson et al., 1994). In another study, the levels of GLUT1 mRNA did not differ, whereas GLUT1 protein expression was reduced (Mooradian et al., 1997).

Impaired cerebral glucose utilization in AD might result from the reduced capacities of key enzymes in the glycolytic chain and mitochondrial enzyme complexes. For example, the activity of the pyruvate dehydrogenase (PDHC) complex, which catalyzes the conversion of pyruvate to acetyl-CoA, is reduced in AD (Sorbi et al., 1983; Butterworth and Besnard, 1990). The activity of the α-ketoglutarate dehydrogenase complex (KGDHC) is reduced in affected and unaffected areas of AD brain (Butterworth and Besnard, 1990; Mastrogiacomo et al., 1993). KGDHC catalyzes oxidation of α-ketoglutarate to succinyl-CoA and is an important enzyme in glutamate metabolism (Blass et al., 1997). The decreased activity of KGDHC is not correlated with protein levels, i.e., lower enzyme activity cannot be explained by loss of protein.

Another consequence of impaired glucose metabolism or uptake is the reduced intracellular level of uridine 5′-diphospho-β-N-acetylglucosamine (UDP-GlcNAc), leading to the downregulation of protein O-GlcNAcylation, a type of protein O-glycosylation, which might be an alternative path for phosphorylation (Dong et al., 1993; Hart, 1997; Wells et al., 2001; Marshall et al., 2004). Many proteins are modified by O-GlcNAcylation, including transcription factors, cytoskeletal proteins, hormone receptors, nuclear-pore proteins, and *tau* (Liu et al., 2004a). Reduced glucose uptake and/or metabolism leads to a reduction of O-GlcNAcylation and a simultaneous elevation of *tau* phosphorylation (Liu et al., 2004a). Griffith et al. (1995) reported that APP can be modified by O-GlcNAc.

Aβ induces impaired glucose transport in neurons, followed by decreased ATP levels, by a mechanism involving plasma membrane lipid peroxidation and conjugation of 4-HNE to the neuronal glucose transporter GLUT3 (Mark et al., 1997b). The activity of transcription factors such as hypoxia-inducible factor one (HIF-1) can also play a role in glucose utilization. Hypoxia and/or hypoglycemia activates HIF-1, which helps restore oxygen homeostasis by inducing glycolysis (see review by Schubert, 2005).

Glutamate, the major neurotransmitter in the brain, is converted to glutamine in astrocytes by glutamine synthetase (GS). Astrocytes then recycle glutamine back to neurons (reviewed by Walton and Dodd, 2007). This cycle requires ATP, and it has been estimated that ~70% of the energy from glucose oxidation is used in the process (Shulman et al., 2004). Intracellular glutamate is nontoxic (Danbolt, 2001), but high levels of glutamate inside astrocytes due to absent or reduced GS activity could cause a reversal of glutamate transport and thus contribute to excitotoxicity (Nicholls and Attwell, 1990). In addition, glutamate transporters in rat hippocampal slices reverse glutamate transport after ATP depletion (Madl and Burgesser, 1993).

APOE

Apolipoprotein E (APOE) is the major lipoprotein within the CNS; it is synthesized by astrocytes (Pitas et al., 1987) and by neurons under physiological and pathological conditions (Harris et al., 2004). APOE is a polymorphic 299-amino acid protein. The gene is located on chromosome 19 and has three possible alleles, ε2, ε3, and ε4. *APOE*-ε3 is the most common (frequency in population 60–70%), followed by *APOE*-ε4 with a frequency of 15–20%, and *APOE*-ε2 with a frequency of 5–10%. The three isoforms differ by single amino acid substitutions (cysteine to arginine) at two positions. APOE-ε4 lacks both cysteine residues (Cys[112] and Cys[158]), while APOE-ε3 and APOE-ε2 contain 1 and 2 cysteine residues, respectively (Mahley and Huang, 2006).

The ε4 allele of *APOE* is the major genetic risk factor for AD; the age of disease onset is inversely related to its gene dosage (Esterbauer et al., 1991; Corder et al., 1993; Butterfield et al., 2002a), whereas the ε2 allele appears to have a protective effect (Corder et al., 1994). Individuals with the ε4/ε4 genotype have a 50–90% higher risk to develop AD at the age of 85; those with one ε4 allele have a 45% higher chance of developing AD (Corder et al., 1993; Farrer et al., 1997). On average the mean age at AD onset is 68 years for carriers of two ε4 alleles, whereas in carriers of one ε4 allele the onset of AD was on average 7 years later. The age of onset in subjects without an ε4 allele is 16 years later than in subjects with two ε4 alleles. The importance of these associations is underscored by the fact that *APOE*-ε4 is linked to about 50% of AD cases (Hyman et al., 1996a; Hyman et al., 1996b).

APOE has important roles in neurobiology. It is involved in lipid transport in the brain, where it is proposed to function as a ligand directing the delivery of lipids for neuronal repair and remodeling after injury. APOE is also required for neuronal plasticity (Mahley, 1988; Mahley and Rall, 2000).

APOE genotype reportedly affects all three hallmarks of Alzheimer disease: neuronal loss, neurofibrillary tangles, and amyloid plaques. APOE-ε4 either inhibits Aβ clearance or stimulates Aβ deposition, whereas APOE-ε3 promotes clearance of Aβ. The formation of insoluble complexes with Aβ is faster in *APOE*-ε4 than in *APOE*-ε3 subjects (La Du et al., 1994; Holtzman et al., 2000; Huang et al., 2004). APOE-ε4 also promotes more Aβ production than APOE -ε3. (Ye et al., 2005). APOE-ε4 dosage correlates with Aβ40 levels, although Aβ42 levels show no significant association with genotype (Ishii et al., 1997). Aβ42-induced lysosomal leakage (Ditaranto et al., 2001) and cell death are enhanced to a greater extent through *APOE*-ε4 than *APOE* -ε3 (Ji et al., 2002, 2006). APOE can bind sAPPα in serum and may regulate its clearance. *In vitro*, APOE-ε4 is less potent in binding sAPP than APOE -ε3 (Barger and Mattson, 1997).

APOE-ε3 and ε4 differ in their effects on *tau* phosphorylation and aggregation. *In vitro*, APOE-ε3, but not -ε4, binds to *tau* and protects it from hyperphosphorylation. (Strittmatter et al., 1994; Lovestone et al., 1996). Transgenic mice expressing human *APOE*-ε4 in neurons have increased *tau* phosphorylation, whereas *APOE*-ε4 expression in astrocytes did not alter *tau* phosphorylation (Tesseur et al., 2000a, b), suggesting an ε4-neurospecific effect on *tau* phosphorylation. NFTs show no association with *APOE* genotype, but are strongly associated with dementia severity (Gomez-Isla et al., 1996).

PET studies reveal that *APOE*-ε4 is associated with lowered parietal, temporal, and posterior cingulate cerebral glucose metabolism in patients with a clinical diagnosis of AD. Cognitively, normal late-middle-aged and young carriers of the *APOE*-ε4 allele have abnormally low rates of glucose metabolism (Small et al., 1995, 2000; Reiman et al., 2001, 2004). *APOE*-ε4 is associated with cognitive decline in individuals with and without dementia (Growdon et al., 1996; Jonker et al., 1998) but does not induce a change in the rate of progression of dementia (Gomez-Isla et al., 1996).

APOE-ε4 interacts with mitochondrial dysfunction. Mitochondria play a critical role in synaptogenesis, and transgenic mice expressing APOE-ε4 showed a significant loss of synapto-dendritic connections (Li et al., 2004b; see also Mahley et al., 2006). APOE isoform-specific effects have been reported on neurite extensions in tissue culture. APOE-ε3 stimulates neural outgrowth, whereas -ε4 does not (Nathan et al., 1994). It is likely that APOE-ε4 affects AD pathogenesis through interactions with different factors or pathways. However, the mechanisms of these APOE-ε4-mediated effects are still largely unknown.

Insulin and Related Factors

Insulin, insulin-like growth factor (IGF-I), and the insulin receptor (IR) are present in all tissues of the body. They are highly expressed in brain, particularly in hippocampus and cortex (Werther et al., 1987). There is strong evidence that insulin and IGF-I are important for energy homeostasis, and metabolic, neurotrophic, neuromodulatory, and neuroendocrine actions in the brain (Wozniak et al., 1993; O'Kusky et al., 2000; D'Ercole et al., 2002; Popken et al., 2004). Insulin is actively transported across the BBB into the cerebrospinal fluid (King and Johnson, 1985), although the molecular mechanism is not known (Banks, 2004). Insulin might also be produced locally in the brain (Schulingkamp et al., 2000).

Insulin resistance is the condition where cells fail to respond normally to insulin, and could be caused through reduced local CNS insulin levels. AD patients have lower cerebrospinal fluid and higher plasma insulin concentrations (Craft et al., 1998). This might indicate altered insulin metabolism in the brain. Supportive evidence for this comes from insulin-administration trials with AD patients, which have led to improved memory and performance (Craft et al., 1999). Insulin resistance could be caused by the reduced binding of insulin or IGF-I to their receptors, although the data are contradictory. Some studies show similar binding of IGF-I to its receptors in AD, whereas in others the binding is increased and suggested an upregulation of IGF-I receptors due to low IGF levels (Crews et al., 1992; Jafferali et al., 2000). Resistance might derive from altered receptor expression. IGF-I receptors are more abundant, whereas insulin receptors are scarcer, in AD brains than in controls (Frolich et al., 1998, 1999).

Binding of insulin to its receptor, and of IGF-I to IGF-I receptors, leads to activation of insulin receptor tyrosine kinase and autophosphorylation of the receptor, followed by tyrosine phosphorylation of the insulin receptor substrate (IRS) protein (Shpakov and Pertseva, 2000). This then activates downstream pathways such as phospholipase Cγ (PLC-γ; Gasparini et al., 2002), phosphoinositide 3-kinase (PI3K; Sun et al., 1993a), and MAPK (Giovannone et al., 2000; see Plum et al., 2005 for review). The activities of tyrosine kinases are lower in late-onset AD brains than in age-matched controls (Frolich et al., 1998, 1999).

Schubert et al. (2004) found no changes in the proliferation or survival of neurons, or of brain glucose metabolism, in a brain-insulin-receptor-knockout (NIRKO) mouse model. However, insulin-mediated tyrosine kinase phosphorylation and PI3K activation were inhibited, and neuronal apoptosis was reduced. PI3K activation leads to reduced phosphorylation, and hence activation, of GSK3β; this in turn can lead to *tau* hyperphosphorylation (Hong and Lee, 1997). IRS-2-deficient mice develop intracellular deposits of hyperphosphorylated *tau* as they age (Schubert et al., 2003). The increased GSK3β activity can be explained by impaired insulin signaling in neurons. In AD, increased levels of CDK5 and MAPK, which are activated through insulin and IGF-I, cannot be attributed the impaired insulin and IGF-I signaling (Patrick et al., 1999; Zhu et al., 2002). However, these increased levels could be due to oxidative stress (Patrick et al., 1999; Crossthwaite et al., 2002).

Insulin affects APP metabolism. sAPPβ secretion can be regulated by several growth factors that act at receptors with intrinsic tyrosine kinase activity (which includes IRs) (Refolo et al., 1989; Schubert et al., 1989). Gasparini et al. (2001) showed that insulin decreases intracellular levels of Aβ and increases extracellular levels of Aβ through accelerated βAPP trafficking to the plasma membrane mediated by tyrosine kinase/MAPK pathways. Extracellular Aβ can inhibit insulin signaling by competing with insulin binding or reducing the affinity of IRs and IGF-I receptors (Ling et al., 2002; Xie et al., 2002). The neurotoxic effects of Aβ can be prevented by IGF-I (Dore et al., 1997) through a decrease in GSK3β activity and an increase in PI3K signaling (Zheng et al., 2000).

Insulin signaling can also be regulated through insulin degrading enzymes (IDEs), which are lower by 50% in hippocampus of AD brains in subjects with *APOE*-ε4 alleles than in non-*APOE*-ε4 carriers (Cook et al., 2003). Mc Dermott and Gibson (1997) showed that IDE is the main soluble Aβ-degrading enzyme in human brain. Mice lacking IDE show increased levels of extracellular Aβ, whereas overexpression of IDE reduces the level of Aβ and Aβ-positive plaques (Leissring et al., 2003).

Taken together, these data suggest that metabolic abnormalities, *tau* hyperphosphorylation, and Aβ-aggregation in AD can be attributed to reduced levels of insulin, impaired insulin receptor signaling, and receptor dysfunction.

Concluding Remarks

The web of metabolic interactions in AD is complex, but there is an interplay between genetics, the external and cellular environments, a range of metabolic indices, and the key etiological and pathogenic players, which lead to the devastating cerebral consequences of this major disease of later life.

References

Albin, R.L. and Greenamyre, J.T. 1992. Alternative excitotoxic hypotheses. Neurology 42:733–738

Alves da Costa, C., Sunyach, C., Pardossi-Piquard, R., Sevalle, J., Vincent, B., Boyer, N., Kawarai, T., Girardot, N., St George-Hyslop, P. and Checler, F. 2006. Presenilin-dependent γ-secretase-mediated control of p53-associated cell death in Alzheimer's disease. J. Neurosci. 26:6377–6385

American Psychiatric Association 2000. Diagnostic and Statistical Manual of Mental Disorders, Text Revision. 4th ed. Washington, DC: American Psychiatric Association

Anderton, B.H., Betts, J., Blackstock, W.P., Brion, J.P., Chapman, S., Connell, J., Dayanandan, R., Gallo, J.M., Gibb, G., Hanger, D.P., Hutton, M., Kardalinou, E., Leroy, K., Lovestone, S., Mack, T., Reynolds, C.H. and Van Slegtenhorst, M. 2001. Sites of phosphorylation in *tau* and factors affecting their regulation. Biochem. Soc. Symp. 67:73–80

Ando, Y., Brannstrom, T., Uchida, K., Nyhlin, N., Nasman, B., Suhr, O., Yamashita, T., Olsson, T., El Salhy, M., Uchino, M. and Ando, M. 1998. Histochemical detection of 4-hydroxynonenal protein in Alzheimer amyloid. J. Neurol. Sci. 156:172–176

Atamna, H. and Boyle, K. 2006. Amyloid-β peptide binds with heme to form a peroxidase: Relationship to the cytopathologies of Alzheimer's disease. Proc. Natl. Acad. Sci. USA 103:3381–3386

Atamna, H. and Frey, W.H., II 2004. A role for heme in Alzheimer's disease: Heme binds amyloid β and has altered metabolism. Proc. Natl. Acad. Sci. USA 101:11153–11158

Atamna, H., Liu, J. and Ames, B.N. 2001. Heme deficiency selectively interrupts assembly of mitochondrial complex IV in human fibroblasts: Revelance to aging. J. Biol. Chem. 276:48410–48416

Atamna, H., Killilea, D.W., Killilea, A.N. and Ames, B.N. 2002. Heme deficiency may be a factor in the mitochondrial and neuronal decay of aging. Proc. Natl. Acad. Sci. USA 99:14807–14812

Banks, W.A. 2004. The source of cerebral insulin. Eur. J. Pharmacol. 490:5–12

Barger, S.W. and Mattson, M.P. 1997. Isoform-specific modulation by apolipoprotein E of the activities of secreted β-amyloid precursor protein. J. Neurochem. 69:60–67

Baudier, J. and Cole, R.D. 1987. Phosphorylation of *tau* proteins to a state like that in Alzheimer's brain is catalyzed by a calcium/calmodulin-dependent kinase and modulated by phospholipids. J. Biol. Chem. 262:17577–17583

Bayer, K.U., Lohler, J., Schulman, H. and Harbers, K. 1999. Developmental expression of the CaM kinase II isoforms: Ubiquitous γ- and δ-CaM kinase II are the early isoforms and most abundant in the developing nervous system. Brain Res. Mol. Brain Res. 70:147–154

Becker, W., Kentrup, H., Klumpp, S., Schultz, J.E. and Joost, H.G. 1994. Molecular cloning of a protein serine/threonine phosphatase containing a putative regulatory tetratricopeptide repeat domain. J. Biol. Chem. 269:22586–22592

Bennecib, M., Gong, C.X., Grundke-Iqbal, I. and Iqbal, K. 2001. Inhibition of PP-2A upregulates CaMKII in rat forebrain and induces hyperphosphorylation of *tau* at Ser 262/356. FEBS Lett. 490:15–22

Bennett, B.D., Babu-Khan, S., Loeloff, R., Louis, J.C., Curran, E., Citron, M. and Vassar, R. 2000a. Expression analysis of BACE2 in brain and peripheral tissues. J. Biol. Chem. 275:20647–20651

Bennett, B.D., Denis, P., Haniu, M., Teplow, D.B., Kahn, S., Louis, J.C., Citron, M. and Vassar, R. 2000b. A furin-like convertase mediates propeptide cleavage of BACE, the Alzheimer's β-secretase. J. Biol. Chem. 275:37712–37717

Berg, M.M., Krafft, G.A. and Klein, W.L. 1997. Rapid impact of β-amyloid on paxillin in a neural cell line. J. Neurosci. Res. 50:979–989

Berry, R.W., Abraha, A., Lagalwar, S., La Pointe, N., Gamblin, T.C., Cryns, V.L. and Binder, L.I. 2003. Inhibition of *tau* polymerization by its carboxy-terminal caspase cleavage fragment. Biochemistry 42:8325–8331

Billings, L.M., Oddo, S., Green, K.N., McGaugh, J.L. and La Ferla, F.M. 2005. Intraneuronal Aβ causes the onset of early Alzheimer's disease-related cognitive deficits in transgenic mice. Neuron 45:675–688

Bitan, G., Kirkitadze, M.D., Lomakin, A., Vollers, S.S., Benedek, G.B. and Teplow, D.B. 2003. Amyloid β-protein (Aβ) assembly: Aβ40 and Aβ42 oligomerize through distinct pathways. Proc. Natl. Acad. Sci. USA 100:330–335

Blass, J.P., Sheu, K.F., Piacentini, S. and Sorbi, S. 1997. Inherent abnormalities in oxidative metabolism in Alzheimer's disease: Interaction with vascular abnormalities. Ann. NY Acad. Sci. 826:382–385

Bodles, A.M. and Barger, S.W. 2005. Secreted β-amyloid precursor protein activates microglia *via* JNK and p38-MAPK. Neurobiol. Aging 26:9–16

Boulton, T.G., Nye, S.H., Robbins, D.J., Ip, N.Y., Radziejewska, E., Morgenbesser, S.D., De Pinho, R.A., Panayotatos, N., Cobb, M.H. and Yancopoulos, G.D. 1991. ERKs: A family of protein-serine/threonine kinases that are activated and tyrosine phosphorylated in response to insulin and NGF. Cell 65:663–675

Braak, H. and Braak, E. 1991. Neuropathological stageing of Alzheimer-related changes. Acta Neuropathol. (Berlin) 82:239–259

Braak, H. and Braak, E. 1995. Staging of Alzheimer's disease-related neurofibrillary changes. Neurobiol. Aging 16:271–284

Brandt, R., Leger, J. and Lee, G. 1995. Interaction of *tau* with the neural plasma membrane mediated by *tau*'s amino-terminal projection domain. J. Cell Biol. 131:1327–1340

Braun, A.P. and Schulman, H. 1995. The multifunctional calcium/calmodulin-dependent protein kinase: From form to function. Annu. Rev. Physiol. 57:417–445

Breen, A.P. and Murphy, J.A. 1995. Reactions of oxyl radicals with DNA. Free Radic. Biol. Med. 18:1033–1077

Buchman, A.S., Wilson, R.S., Bienias, J.L., Shah, R.C., Evans, D.A. and Bennett, D.A. 2005. Change in body mass index and risk of incident Alzheimer disease. Neurology 65:892–897

Buee, L., Bussiere, T., Buee-Scherrer, V., Delacourte, A. and Hof, P.R. 2000. *Tau* protein isoforms, phosphorylation and role in neurodegenerative disorders. Brain Res. Brain Res. Rev. 33:95–130

Burdick, D., Soreghan, B., Kwon, M., Kosmoski, J., Knauer, M., Henschen, A., Yates, J., Cotman, C. and Glabe, C. 1992. Assembly and aggregation properties of synthetic Alzheimer's A4/β amyloid peptide analogs. J. Biol. Chem. 267:546–554

Bussiere, T., Giannakopoulos, P., Bouras, C., Perl, D.P., Morrison, J.H. and Hof, P.R. 2003. Progressive degeneration of nonphosphorylated neurofilament protein-enriched pyramidal neurons predicts cognitive impairment in Alzheimer's disease: Stereologic analysis of prefrontal cortex area 9. J. Comp. Neurol. 463:281–382

Butterfield, D.A. 2002. Amyloid β-peptide(1–42)-induced oxidative stress and neurotoxicity: Implications for neurodegeneration in Alzheimer's disease brain. A review. Free Radic. Res. 36:1307–1313

Butterfield, D.A., Hensley, K., Harris, M., Mattson, M.P. and Carney, J.M. 1994. β-Amyloid peptide free radical fragments initiate synaptosomal lipoperoxidation in a sequence-specific fashion: Implications to Alzheimer's disease. Biochem. Biophys. Res. Commun. 200:710–715

Butterfield, D.A., Drake, J., Pocernich, C. and Castegna, A. 2001. Evidence of oxidative damage in Alzheimer's disease brain: Central role for amyloid β-peptide. Trends Mol. Med. 7:548–554

Butterfield, D.A., Castegna, A., Lauderback, C.M. and Drake, J. 2002a. Evidence that amyloid β-peptide-induced lipid peroxidation and its sequeæ in Alzheimer's disease brain contribute to neuronal death. Neurobiol. Aging 23:655–664

Butterfield, D.A., Griffin, S., Münch, G. and Pasinetti, G.M. 2002b. Amyloid β-peptide and amyloid pathology are central to the oxidative stress and inflammatory cascades under which Alzheimer's disease brain exists. J. Alzheimer's Dis. 4:193–201

Butterfield, D.A., Perluigi, M. and Sultana, R. 2006. Oxidative stress in Alzheimer's disease brain: New insights from redox proteomics. Eur. J. Pharmacol. 545:39–50

Butterworth, R.F. and Besnard, A.M. 1990. Thiamine-dependent enzyme changes in temporal cortex of patients with Alzheimer's disease. Metab. Brain Dis. 5:179–184

Buxbaum, J.D., Koo, E.H. and Greengard, P. 1993. Protein phosphorylation inhibits production of Alzheimer amyloid β/A4 peptide. Proc. Natl. Acad. Sci. USA 90:9195–9198

Campion, D., Dumanchin, C., Hannequin, D., Dubois, B., Belliard, S., Puel, M., Thomas-Anterion, C., Michon, A., Martin, C., Charbonnier, F., Raux, G., Camuzat, A., Penet, C., Mesnage, V., Martinez, M., Clerget-Darpoux, F., Brice, A. and Frebourg, T. 1999. Early-onset autosomal dominant Alzheimer disease: Prevalence, genetic heterogeneity, and mutation spectrum. Am. J. Hum. Genet. 65:664–670

Carlier, M.F., Simon, C., Cassoly, R. and Pradel, L.A. 1984. Interaction between microtubule-associated protein *tau* and spectrin. Biochimie 66:305–311

Carter, J. and Lippa, C.F. 2001. β-amyloid, neuronal death and Alzheimer's disease. Curr. Mol. Med. 1:733–737

Castellani, R., Smith, M.A., Richey, P.L., Kalaria, R., Gambetti, P. and Perry, G. 1995. Evidence for oxidative stress in Pick disease and corticobasal degeneration. Brain Res. 696:268–271

Cervantes, S., Gonzalez-Duarte, R. and Marfany, G. 2001. Homodimerization of presenilin N-terminal fragments is affected by mutations linked to Alzheimer's disease. FEBS Lett. 505:81–86

Chen, Y.R. and Glabe, C.G. 2006. Distinct early folding and aggregation properties of Alzheimer amyloid-β peptides Aβ40 and Aβ42: Stable trimer or tetramer formation by Aβ42. J. Biol. Chem. 281:24414–24422

Chin, S.S. and Goldman, J.E. 1996. Glial inclusions in CNS degenerative diseases. J. Neuropathol. Exp. Neurol. 55:499–508

Chyung, J.H., Raper, D.M. and Selkoe, D.J. 2005. γ-secretase exists on the plasma membrane as an intact complex that accepts substrates and effects intramembrane cleavage. J. Biol. Chem. 280:4383–4392

Coghlan, V.M., Bergeson, S.E., Langeberg, L., Nilaver, G. and Scott, J.D. 1993. A-kinase anchoring proteins: A key to selective activation of cAMP-responsive events? Mol. Cell. Biochem. 127–128:309–319

Colbran, R.J., Schworer, C.M., Hashimoto, Y., Fong, Y.L., Rich, D.P., Smith, M.K. and Soderling, T.R. 1989. Calcium/calmodulin-dependent protein kinase II. Biochem. J. 258:313–325.

Cook, D.G., Leverenz, J.B., McMillan, P.J., Kulstad, J.J., Ericksen, S., Roth, R.A., Schellenberg, G.D., Jin, L.W., Kovacina, K.S. and Craft, S. 2003. Reduced hippocampal insulin-degrading enzyme in late-onset Alzheimer's disease is associated with the apolipoprotein E-ε4allele. Am. J. Pathol. 162:313–319

Cooke, M.S., Evans, M.D., Dizdaroglu, M. and Lunec, J. 2003. Oxidative DNA damage: Mechanisms, mutation, and disease. FASEB J. 17:1195–1214

Corder, E.H., Saunders, A.M., Strittmatter, W.J., Schmechel, D.E., Gaskell, P.C., Small, G.W., Roses, A.D., Haines, J.L. and Pericak-Vance, M.A. 1993. Gene dose of apolipoprotein E type 4 allele and the risk of Alzheimer's disease in late onset families. Science 261:921–923

Corder, E.H., Saunders, A.M., Risch, N.J., Strittmatter, W.J., Schmechel, D.E., Gaskell, P.C., Jr., Rimmler, J.B., Locke, P.A., Conneally, P.M., Schmader, K.E., Small, G.W., Roses, A.D., Haines, J.L. and Pericak-Vance, M.A. 1994. Protective effect of apolipoprotein E type 2 allele for late onset Alzheimer disease. Nat. Genet. 7:180–184

Correas, I., Padilla, R. and Avila, J. 1990. The tubulin-binding sequence of brain microtubule-associated proteins, *tau* and MAP-2, is also involved in actin binding. Biochem. J. 269:61–64

Craft, S., Peskind, E., Schwartz, M.W., Schellenberg, G.D., Raskind, M. and Porte, D., Jr. 1998. Cerebrospinal fluid and plasma insulin levels in Alzheimer's disease: Relationship to severity of dementia and apolipoprotein E genotype. Neurology 50:164–168

Craft, S., Asthana, S., Newcomer, J.W., Wilkinson, C.W., Matos, I.T., Baker, L.D., Cherrier, M., Lofgreen, C., Latendresse, S., Petrova, A., Plymate, S., Raskind, M., Grimwood, K. and Veith, R.C. 1999. Enhancement of memory in Alzheimer disease with insulin and somatostatin, but not glucose. Arch. Gen. Psychiatry 56:1135–1140

Cras, P., Smith, M.A., Richey, P.L., Siedlak, S.L., Mulvihill, P. and Perry, G. 1995. Extracellular neurofibrillary tangles reflect neuronal loss and provide further evidence of extensive protein cross-linking in Alzheimer disease. Acta Neuropathol. (Berlin) 89:291–295

Creemers, J.W., Ines-Dominguez, D., Plets, E., Serneels, L., Taylor, N.A., Multhaup, G., Craessaerts, K., Annaert, W. and De Strooper, B. 2001. Processing of β-secretase by furin and other members of the proprotein convertase family. J. Biol. Chem. 276:4211–4217

Crews, F.T., McElhaney, R., Freund, G., Ballinger, W.E., Jr. and Raizada, M.K. 1992. Insulin-like growth factor I receptor binding in brains of Alzheimer's and alcoholic patients. J. Neurochem. 58:1205–1210

Crossthwaite, A.J., Hasan, S. and Williams, R.J. 2002. Hydrogen peroxide-mediated phosphorylation of ERK1/2, Akt/PKB and JNK in cortical neurones: Dependence on Ca^{2+} and PI3-kinase. J. Neurochem. 80:24–35

Cruz, J.C., Tseng, H.C., Goldman, J.A., Shih, H. and Tsai, L.H. 2003. Aberrant Cdk5 activation by p25 triggers pathological events leading to neurodegeneration and neurofibrillary tangles. Neuron 40:471–483

Cuajungco, M.P. and Faget, K.Y. 2003. Zinc takes the center stage: Its paradoxical role in Alzheimer's disease. Brain Res. Brain Res. Rev. 41:44–56

Dahlgren, K.N., Manelli, A.M., Stine, W.B., Jr., Baker, L.K., Krafft, G.A. and LaDu, M.J. 2002. Oligomeric and fibrillar species of amyloid-β peptides differentially affect neuronal viability. J. Biol. Chem. 277:32046–32053

Danbolt, N.C. 2001. Glutamate uptake. Prog. Neurobiol. 65:1–105

Demple, B. and Harrison, L. 1994. Repair of oxidative damage to DNA: Enzymology and biology. Annu. Rev. Biochem. 63:915–948

D'Ercole, A.J., Ye, P. and O'Kusky, J.R. 2002. Mutant mouse models of insulin-like growth factor actions in the central nervous system. Neuropeptides 36:209–220

Derkinderen, P., Enslen, H. and Girault, J.A. 1999. The ERK/MAP-kinases cascade in the nervous system. Neuroreport 10:R24–R34

De Strooper, B., Beullens, M., Contreras, B., Levesque, L., Craessaerts, K., Cordell, B., Moechars, D., Bollen, M., Fraser, P., St George-Hyslop, P. and Van Leuven, F. 1997. Phosphorylation, subcellular localization, and membrane orientation of the Alzheimer's disease-associated presenilins. J. Biol. Chem. 272:3590–3598

Dhavan, R. and Tsai, L.H. 2001. A decade of CDK5. Nat. Rev. Mol. Cell Biol. 2:749–759

Ding, Q., Markesbery, W.R., Chen, Q., Li, F. and Keller, J.N. 2005. Ribosome dysfunction is an early event in Alzheimer's disease. J. Neurosci. 25:9171–9175

Ditaranto, K., Tekirian, T.L. and Yang, A.J. 2001. Lysosomal membrane damage in soluble Aβ-mediated cell death in Alzheimer's disease. Neurobiol. Dis. 8:19–31

Dong, D.-L., Xu, Z.-S., Chevrier, M.R., Cotter, R.J., Cleveland, D.W. and Hart, G.W. 1993. Glycosylation of mammalian neurofilaments. Localization of multiple O-linked N-acetylglucosamine moieties on neurofilament polypeptides L and M. J. Biol. Chem. 268:16679–16687

Dore, S., Kar, S. and Quirion, R. 1997. Insulin-like growth factor I protects and rescues hippocampal neurons against β-amyloid- and human amylin-induced toxicity. Proc. Natl. Acad. Sci. USA 94:4772–4777

Dore, S., Takahashi, M., Ferris, C.D., Zakhary, R., Hester, L.D., Guastella, D. and Snyder, S.H. 1999. Bilirubin, formed by activation of heme oxygenase-2, protects neurons against oxidative stress injury. Proc. Natl. Acad. Sci. USA 96:2445–2450

Dorner, A.J., Wasley, L.C. and Kaufman, R.J. 1990. Protein dissociation from GRP78 and secretion are blocked by depletion of cellular ATP levels. Proc. Natl. Acad. Sci. USA 87:7429–7432

Drewes, G., Lichtenberg-Kraag, B., Doring, F., Mandelkow, E.M., Biernat, J., Goris, J., Doree, M. and Mandelkow, E. 1992. Mitogen activated protein (MAP) kinase transforms tau protein into an Alzheimer-like state. EMBO J. 11:2131–2138

Dua, T., Cumbrera, M.G., Mathers, C. and Saxena, S. 2006. Global burden of neurological disorders: Estimates and projections, in Neurological Disorders: Public Health Challenges, eds J.A. Aarli, G. Avanzini, J.M. Bertolote, H. de Boer, H. Breivik, T. Dua, N. Graham, A. Janca, J. Kesselring, C. Mathers, A. Muscetta, L. Prilipko, B. Saraceno, S. Saxena, T.J. Steiner, pp. 27–40. Geneva, Switzerland: WHO Press

Ehehalt, R., Michel, B., De Pietri-Tonelli, D., Zacchetti, D., Simons, K. and Keller, P. 2002. Splice variants of the β-site APP-cleaving enzyme BACE1 in human brain and pancreas. Biochem. Biophys. Res. Commun. 293:30–37

Embi, N., Rylatt, D.B. and Cohen, P. 1980. Glycogen synthase kinase-3 from rabbit skeletal muscle. Separation from cyclic-AMP-dependent protein kinase and phosphorylase kinase. Eur. J. Biochem. 107:519–527

Enslen, H., Raingeaud, J. and Davis, R.J. 1998. Selective activation of p38 mitogen-activated protein (MAP) kinase isoforms by the MAP kinase kinases MKK3 and MKK6. J. Biol. Chem. 273:1741–1748

Erecinska, M. and Silver, I.A. 1989. ATP and brain function. J. Cereb. Blood Flow Metab. 9:2–19

Esterbauer, H., Schaur, R.J. and Zollner, H. 1991. Chemistry and biochemistry of 4-hydroxynonenal, malonaldehyde and related aldehydes. Free Radic. Biol. Med. 11:81–128

Etcheberrigaray, R., Tan, M., Dewachter, I., Kuiperi, C., van der Auwera, I., Wera, S., Qiao, L., Bank, B., Nelson, T.J., Kozikowski, A.P., van Leuven, F. and Alkon, D.L. 2004. Therapeutic effects of PKC activators in Alzheimer's disease transgenic mice. Proc. Natl. Acad. Sci. USA 101:11141–11146

Ewing, J.F. and Maines, M.D. 1991. Rapid induction of heme oxygenase 1 mRNA and protein by hyperthermia in rat brain: Heme oxygenase 2 is not a heat shock protein. Proc. Natl. Acad. Sci. USA 88:5364–5368

Farrer, L.A., Cupples, L.A., Haines, J.L., Hyman, B., Kukull, W.A., Mayeux, R., Myers, R.H., Pericak-Vance, M.A., Risch, N. and van Duijn, C.M. 1997. Effects of age, sex, and ethnicity on the association between apolipoprotein E genotype and Alzheimer disease. A meta-analysis. APOE and Alzheimer disease meta analysis consortium. J. Am. Med. Assoc. 278:1349–1356

Farzan, M., Schnitzler, C.E., Vasilieva, N., Leung, D. and Choe, H. 2000. BACE2, a β-secretase homolog, cleaves at the β site and within the amyloid-β region of the amyloid-β precursor protein. Proc. Natl. Acad. Sci. USA 97:9712–9717

Feng, R., Wang, H., Wang, J., Shrom, D., Zeng, X. and Tsien, J.Z. 2004. Forebrain degeneration and ventricle enlargement caused by double knockout of Alzheimer's presenilin-1 and presenilin-2. Proc. Natl. Acad. Sci. USA 101:8162–8167

Ferrer, I., Blanco, R., Carmona, M. and Puig, B. 2001a. Phosphorylated mitogen-activated protein kinase (MAPK/ERK-P), protein kinase of 38 kDa (p38-P), stress-activated protein kinase (SAPK/JNK-P), and calcium/calmodulin-dependent kinase II (CaM kinase II) are differentially expressed in tau deposits in neurons and glial cells in tauopathies. J. Neural Transm. 108:1397–1415

Ferrer, I., Blanco, R., Carmona, M., Ribera, R., Goutan, E., Puig, B., Rey, M.J., Cardozo, A., Vinals, F. and Ribalta, T. 2001b. Phosphorylated map kinase (ERK1, ERK2) expression is associated with early tau deposition in neurones and glial cells, but not with increased nuclear DNA vulnerability and cell death, in Alzheimer disease, Pick's disease, progressive supranuclear palsy and corticobasal degeneration. Brain Pathol. 11:144–158

Ferri, C.P., Prince, M., Brayne, C., Brodaty, H., Fratiglioni, L., Ganguli, M., Hall, K., Hasegawa, K., Hendrie, H., Huang, Y., Jorm, A., Mathers, C., Menezes, P.R., Rimmer, E. and Scazufca, M. 2005. Global prevalence of dementia: A delphi consensus study. Lancet 366:2112–2117

Finch, C.E. and Cohen, D.M. 1997. Aging, metabolism, and Alzheimer disease: Review and hypotheses. Exp. Neurol. 143:82–102

Fischer, A., Sananbenesi, F., Schrick, C., Spiess, J. and Radulovic, J. 2002. Cyclin-dependent kinase 5 is required for associative learning. J. Neurosci. 22:3700–3707

Fraser, P.E., Yang, D.S., Yu, G., Levesque, L., Nishimura, M., Arawaka, S., Serpell, L.C., Rogaeva, E. and St George-Hyslop, P. 2000. Presenilin structure, function and role in Alzheimer disease. Biochim. Biophys. Acta 1502:1–15

Frolich, L., Blum-Degen, D., Bernstein, H.G., Engelsberger, S., Humrich, J., Laufer, S., Muschner, D., Thalheimer, A., Turk, A., Hoyer, S., Zochling, R., Boissl, K.W., Jellinger, K. and Riederer, P. 1998. Brain insulin and insulin receptors in aging and sporadic Alzheimer's disease. J. Neural Transm. 105:423–438

Frolich, L., Blum-Degen, D., Riederer, P. and Hoyer, S. 1999. A disturbance in the neuronal insulin receptor signal transduction in sporadic Alzheimer's disease. Ann. NY Acad. Sci. 893:290–293

Gabbita, S.P., Lovell, M.A. and Markesbery, W.R. 1998. Increased nuclear DNA oxidation in the brain in Alzheimer's disease. J. Neurochem. 71:2034–2040

Gamblin, T.C., Chen, F., Zambrano, A., Abraha, A., Lagalwar, S., Guillozet, A.L., Lu, M., Fu, Y., Garcia-Sierra, F., LaPointe, N., Miller, R., Berry, R.W., Binder, L.I. and Cryns, V.L. 2003. Caspase cleavage of tau: Linking amyloid and neurofibrillary tangles in Alzheimer's disease. Proc. Natl. Acad. Sci. USA 100:10032–10037

Gasparini, L., Gouras, G.K., Wang, R., Gross, R.S., Beal, M.F., Greengard, P. and Xu, H. 2001. Stimulation of β-amyloid precursor protein trafficking by insulin reduces intraneuronal β-amyloid and requires mitogen-activated protein kinase signaling. J. Neurosci. 21:2561–2570

Gasparini, L., Netzer, W.J., Greengard, P. and Xu, H. 2002. Does insulin dysfunction play a role in Alzheimer's disease? Trends Pharmacol. Sci. 23:288–293

Gearing, M., Mirra, S.S., Hedreen, J.C., Sumi, S.M., Hansen, L.A. and Heyman, A. 1995. The consortium to establish a registry for Alzheimer's disease (CERAD). Part X. Neuropathology confirmation of the clinical diagnosis of Alzheimer's disease. Neurology 45:461–466

Geula, C., Wu, C.K., Saroff, D., Lorenzo, A., Yuan, M. and Yankner, B.A. 1998. Aging renders the brain vulnerable to amyloid β-protein neurotoxicity. Nat. Med. 4:827–831

Ghoshal, N., Smiley, J.F., De Maggio, A.J., Hoekstra, M.F., Cochran, E.J., Binder, L.I. and Kuret, J. 1999. A new molecular link between the fibrillar and granulovacuolar lesions of Alzheimer's disease. Am. J. Pathol. 155:1163–1172

Giancotti, F.G. and Ruoslahti, E. 1999. Integrin signaling. Science 285:1028–1032

Giannakopoulos, P., Herrmann, F.R., Bussiere, T., Bouras, C., Kovari, E., Perl, D.P., Morrison, J.H., Gold, G. and Hof, P.R. 2003. Tangle and neuron numbers, but not amyloid load, predict cognitive status in Alzheimer's disease. Neurology 60:1495–1500

Giasson, B.I., Lee, V.M. and Trojanowski, J.Q. 2003. Interactions of amyloidogenic proteins. Neuromol. Med. 4:49–58

Giovannone, B., Scaldaferri, M.L., Federici, M., Porzio, O., Lauro, D., Fusco, A., Sbraccia, P., Borboni, P., Lauro, R. and Sesti, G. 2000. Insulin receptor substrate (IRS) transduction system: Distinct and overlapping signaling potential. Diabetes Metab. Res. Rev. 16:434–441

Goedert, M., Wischik, C.M., Crowther, R.A., Walker, J.E. and Klug, A. 1988. Cloning and sequencing of the cDNA encoding a core protein of the paired helical filament of Alzheimer disease: Identification as the microtubule-associated protein *tau*. Proc. Natl. Acad. Sci. USA 85:4051–4055

Goedert, M., Spillantini, M.G., Jakes, R., Rutherford, D. and Crowther, R.A. 1989a. Multiple isoforms of human microtubule-associated protein *tau*: Sequences and localization in neurofibrillary tangles of Alzheimer's disease. Neuron 3:519–526

Goedert, M., Spillantini, M.G., Potier, M.C., Ulrich, J. and Crowther, R.A. 1989b. Cloning and sequencing of the cDNA encoding an isoform of microtubule-associated protein *tau* containing four tandem repeats: Differential expression of *tau* protein mRNAs in human brain. EMBO J. 8:393–389

Goedert, M., Cohen, E.S., Jakes, R. and Cohen, P. 1992a. p42 MAP kinase phosphorylation sites in microtubule-associated protein *tau* are dephosphorylated by protein phosphatase 2A1. Implications for Alzheimer's disease [corrected]. FEBS Lett. 312:95–99

Goedert, M., Spillantini, M.G., Cairns, N.J. and Crowther, R.A. 1992b. *Tau* proteins of Alzheimer paired helical filaments: abnormal phosphorylation of all six brain isoforms. Neuron 8:159–168

Goedert, M., Cuenda, A., Craxton, M., Jakes, R. and Cohen, P. 1997. Activation of the novel stress-activated protein kinase SAPK4 by cytokines and cellular stresses is mediated by SKK3 (MKK6); comparison of its substrate specificity with that of other SAP kinases. EMBO J. 16:3563–3571

Gomez-Isla, T., West, H.L., Rebeck, G.W., Harr, S.D., Growdon, J.H., Locascio, J.J., Perls, T.T., Lipsitz, L.A. and Hyman, B.T. 1996. Clinical and pathological correlates of apolipoprotein E ε4 in Alzheimer's disease. Ann. Neurol. 39:62–70

Gomez-Isla, T., Hollister, R., West, H., Mui, S., Growdon, J.H., Petersen, R.C., Parisi, J.E. and Hyman, B.T. 1997. Neuronal loss correlates with but exceeds neurofibrillary tangles in Alzheimer's disease. Ann. Neurol. 41:17–24ss

Gomez-Ramos, A., Diaz-Nido, J., Smith, M.A., Perry, G. and Avila, J. 2003. Effect of the lipid peroxidation product acrolein on *tau* phosphorylation in neural cells. J. Neurosci. Res. 71:863–870

Gong, C.X., Singh, T.J., Grundke-Iqbal, I. and Iqbal, K. 1993. Phosphoprotein phosphatase activities in Alzheimer disease brain. J. Neurochem. 61:921–927

Gong, C.X., Grundke-Iqbal, I. and Iqbal, K. 1994a. Dephosphorylation of Alzheimer's disease abnormally phosphorylated *tau* by protein phosphatase-2A. Neuroscience 61:765–772

Gong, C.X., Shaikh, S., Wang, J.Z., Zaidi, T., Grundke-Iqbal, I. and Iqbal, K. 1995. Phosphatase activity toward abnormally phosphorylated *tau*: decrease in Alzheimer disease brain. J. Neurochem. 65:732–738

Gong, C.X., Lidsky, T., Wegiel, J., Zuck, L., Grundke-Iqbal, I. and Iqbal, K. 2000. Phosphorylation of microtubule-associated protein *tau* is regulated by protein phosphatase 2A in mammalian brain. Implications for neurofibrillary degeneration in Alzheimer's disease. J. Biol. Chem. 275:5535–5544

Gong, C.X., Liu, F., Grundke-Iqbal, I. and Iqbal, K. 2005. Post-translational modifications of *tau* protein in Alzheimer's disease. J. Neural Transm. 112:813–838

Goodman, Y. and Mattson, M.P. 1994. Secreted forms of β-amyloid precursor protein protect hippocampal neurons against amyloid β-peptide-induced oxidative injury. Exp. Neurol. 128:1–12

Goodman, A.B. and Pardee, A.B. 2003. Evidence for defective retinoid transport and function in late onset Alzheimer's disease. Proc. Natl. Acad. Sci. USA 100:2901–2905

Götz, J., Chen, F., van Dorpe, J. and Nitsch, R.M. 2001. Formation of neurofibrillary tangles in P301L tau transgenic mice induced by Aβ42 fibrils. Science 293:1491–1495

Gouras, G.K., Tsai, J., Naslund, J., Vincent, B., Edgar, M., Checler, F., Greenfield, J.P., Haroutunian, V., Buxbaum, J.D., Xu, H., Greengard, P. and Relkin, N.R. 2000. Intraneuronal Aβ42 accumulation in human brain. Am. J. Pathol. 156:15–20

Grace, E.A. and Busciglio, J. 2003. Aberrant activation of focal adhesion proteins mediates fibrillar amyloid β-induced neuronal dystrophy. J. Neurosci. 23:493–502

Griffith, L.M. and Pollard, T.D. 1982. The interaction of actin filaments with microtubules and microtubule-associated proteins. J. Biol. Chem. 257:9143–9151

Griffith, L.S., Mathes, M. and Schmitz, B. 1995. β-amyloid precursor protein is modified with O-linked *N*-acetylglucosamine. J. Neurosci. Res. 41:270–278

Growdon, J.H., Locascio, J.J., Corkin, S., Gomez-Isla, T. and Hyman, B.T. 1996. Apolipoprotein E genotype does not influence rates of cognitive decline in Alzheimer's disease. Neurology 47:444–448

Grundke-Iqbal, I., Iqbal, K., Quinlan, M., Tung, Y.C., Zaidi, M.S. and Wisniewski, H.M. 1986a. Microtubule-associated protein *tau*. A component of Alzheimer paired helical filaments. J. Biol. Chem. 261:6084–6089

Grundke-Iqbal, I., Iqbal, K., Tung, Y.C., Quinlan, M., Wisniewski, H.M. and Binder, L.I. 1986b. Abnormal phosphorylation of the microtubule-associated protein τ (*tau*) in Alzheimer cytoskeletal pathology. Proc. Natl. Acad. Sci. USA 83:4913–4917

Guillemin, G.J., Williams, K.R., Smith, D.G., Smythe, G.A., Croitoru-Lamoury, J. and Brew, B.J. 2003. Quinolinic acid in the pathogenesis of Alzheimer's disease. Adv. Exp. Med. Biol. 527:167–176

Gupta, S., Barrett, T., Whitmarsh, A.J., Cavanagh, J., Sluss, H.K., Derijard, B. and Davis, R.J. 1996. Selective interaction of JNK protein kinase isoforms with transcription factors. EMBO J. 15:2760–2770

Gurland, B.J., Wilder, D.E., Lantigua, R., Stern, Y., Chen, J., Killeffer, E.H. and Mayeux, R. 1999. Rates of dementia in three ethnoracial groups. Int. J. Geriat. Psychiatry 14:481–493

Haass, C. and Steiner, H. 2001. Protofibrils, the unifying toxic molecule of neurodegenerative disorders? Nat. Neurosci. 4:859–860

Haass, C., Lemere, C.A., Capell, A., Citron, M., Seubert, P., Schenk, D., Lannfelt, L. and Selkoe, D.J. 1995. The Swedish mutation causes early-onset Alzheimer's disease by β-secretase cleavage within the secretory pathway. Nat. Med. 1:1291–1296

Hanger, D.P., Betts, J.C., Loviny, T.L., Blackstock, W.P. and Anderton, B.H. 1998. New phosphorylation sites identified in hyperphosphorylated *tau* (paired helical filament-*tau*) from Alzheimer's disease brain using nanoelectrospray mass spectrometry. J. Neurochem. 71:2465–2476

Hansen, L.A., Masliah, E., Galasko, D. and Terry, R.D. 1993. Plaque-only Alzheimer disease is usually the Lewy body variant, and *vice versa*. J. Neuropathol. Exp. Neurol. 52:648–654

Harada, H., Tamaoka, A., Ishii, K., Shoji, S., Kametaka, S., Kametani, F., Saito, Y. and Murayama, S. 2006. β-Site APP cleaving enzyme 1 (BACE1) is increased in remaining neurons in Alzheimer's disease brains. Neurosci. Res. 54:24–29

Hardy, J.A. and Selkoe, D.J. 2002. The amyloid hypothesis of Alzheimer's disease: Progress and problems on the road to therapeutics. Science 297:353–356

Harris, M.E., Wang, Y., Pedigo, N.W., Jr., Hensley, K., Butterfield, D.A. and Carney, J.M. 1996. Amyloid β peptide (25–35) inhibits Na-dependent glutamate uptake in rat hippocampal astrocyte cultures. J. Neurochem. 67:277–286

Harris, F.M., Tesseur, I., Brecht, W.J., Xu, Q., Mullendorff, K., Chang, S., Wyss-Coray, T., Mahley, R.W. and Huang, Y. 2004. Astroglial regulation of apolipoprotein E expression in neuronal cells. Implications for Alzheimer's disease. J. Biol. Chem. 279:3862–3868

Hart, G.W. 1997. Dynamic O-linked glycosylation of nuclear and cytoskeletal proteins. Annu. Rev. Biochem. 66:315–335

Hasegawa, M., Jakes, R., Crowther, R.A., Lee, V.M., Ihara, Y. and Goedert, M. 1996. Characterization of mAb AP422, a novel phosphorylation-dependent monoclonal antibody against *tau* protein. FEBS Lett. 384:25–30

Hata, R., Masumura, M., Akatsu, H., Li, F., Fujita, H., Nagai, Y., Yamamoto, T., Okada, H., Kosaka, K., Sakanaka, M. and Sawada, T. 2001. Up-regulation of calcineurin Aβ mRNA in the Alzheimer's disease brain: Assessment by cDNA microarray. Biochem. Biophys. Res. Commun. 284:310–316

Hayakawa, H. and Sekiguchi, M. 2006. Human polynucleotide phosphorylase protein in response to oxidative stress. Biochemistry 45:6749–6755

Hayakawa, H., Uchiumi, T., Fukuda, T., Ashizuka, M., Kohno, K., Kuwano, M. and Sekiguchi, M. 2002. Binding capacity of human YB-1 protein for RNA containing 8-oxoguanine. Biochemistry 41:12739–12744

Hensley, K., Floyd, R.A., Zheng, N.Y., Nael, R., Robinson, K.A., Nguyen, X., Pye, Q.N., Stewart, C.A., Geddes, J., Markesbery, W.R., Patel, E., Johnson, G.V. and Bing, G. 1999. p38 kinase is activated in the Alzheimer's disease brain. J. Neurochem. 72:2053–2058

Herzig, S. and Neumann, J. 2000. Effects of serine/threonine protein phosphatases on ion channels in excitable membranes. Physiol. Rev. 80:173–210

Himmler, A., Drechsel, D., Kirschner, M.W. and Martin, D.W., Jr. 1989. *Tau* consists of a set of proteins with repeated C-terminal microtubule-binding domains and variable N-terminal domains. Mol. Cell. Biol. 9:1381–1388

Hisanaga, S. and Saito, T. 2003. The regulation of cyclin-dependent kinase 5 activity through the metabolism of p35 or p39 Cdk5 activator. Neurosignals 12:221–229

Hof, P.R., Delacourte, A. and Bouras, C. 1992. Distribution of cortical neurofibrillary tangles in progressive supranuclear palsy: A quantitative analysis of six cases. Acta Neuropathol. (Berlin) 84:45–51

Hof, P.R., Bouras, C., Perl, D.P. and Morrison, J.H. 1994. Quantitative neuropathologic analysis of Pick's disease cases: Cortical distribution of Pick bodies and coexistence with Alzheimer's disease. Acta Neuropathol. (Berlin) 87:115–124

Hoglund, K., Thelen, K.M., Syversen, S., Sjogren, M., von Bergmann, K., Wallin, A., Vanmechelen, E., Vanderstichele, H., Lutjohann, D. and Blennow, K. 2005. The effect of simvastatin treatment on the amyloid precursor protein and brain cholesterol metabolism in patients with Alzheimer's disease. Dement. Geriatr. Cogn. Disord. 19:256–265

Holtzman, D.M., Bales, K.R., Tenkova, T., Fagan, A.M., Parsadanian, M., Sartorius, L.J., Mackey, B., Olney, J., McKeel, D., Wozniak, D. and Paul, S.M. 2000. Apolipoprotein E isoform-dependent amyloid deposition and neuritic degeneration in a mouse model of Alzheimer's disease. Proc. Natl. Acad. Sci. USA 97:2892–2897

Hong, M. and Lee, V.M. 1997. Insulin and insulin-like growth factor-1 regulate *tau* phosphorylation in cultured human neurons. J. Biol. Chem. 272:19547–19553

Howlett, D., Cutler, P., Heales, S. and Camilleri, P. 1997. Hemin and related porphyrins inhibit β-amyloid aggregation. FEBS Lett. 417:249–251

Hoyer, S. 1992. Oxidative energy metabolism in Alzheimer brain. Studies in early-onset and late-onset cases. Mol. Chem. Neuropathol. 16:207–224

Hoyer, S. 2000. Brain glucose and energy metabolism abnormalities in sporadic Alzheimer disease. Causes and consequences: An update. Exp. Gerontol. 35:1363–1372

Huang, Y., Weisgraber, K.H., Mucke, L. and Mahley, R.W. 2004. Apolipoprotein E: Diversity of cellular origins, structural and biophysical properties, and effects in Alzheimer's disease. J. Mol. Neurosci. 23:189–204

Huse, J.T., Pijak, D.S., Leslie, G.J., Lee, V.M. and Doms, R.W. 2000. Maturation and endosomal targeting of β-site amyloid precursor protein-cleaving enzyme. The Alzheimer's disease β-secretase. J. Biol. Chem. 275:33729–33737

Hwang, S.C., Jhon, D.Y., Bae, Y.S., Kim, J.H. and Rhee, S.G. 1996. Activation of phospholipase Cγ by the concerted action of tau proteins and arachidonic acid. J. Biol. Chem. 271:18342–18349

Hyman, B.T., Elvhage, T.E. and Reiter, J. 1994. Extracellular signal regulated kinases. Localization of protein and mRNA in the human hippocampal formation in Alzheimer's disease. Am. J. Pathol. 144:565–572

Hyman, B.T., Gomez-Isla, T., Rebeck, G.W., Briggs, M., Chung, H., West, H.L., Greenberg, S., Mui, S., Nichols, S., Wallace, R. and Growdon, J.H. 1996a. Epidemiological, clinical, and neuropathological study of apolipoprotein E genotype in Alzheimer's disease. Ann. NY Acad. Sci. 802:1–5

Hyman, B.T., Gomez-Isla, T., West, H., Briggs, M., Chung, H., Growdon, J.H. and Rebeck, G.W. 1996b. Clinical and neuropathological correlates of apolipoprotein E genotype in Alzheimer's disease. Window on molecular epidemiology. Ann. NY Acad. Sci. 777:158–165

Hynd, M.R., Scott, H.L. and Dodd, P.R. 2004. Glutamate-mediated excitotoxicity and neurode-generation in Alzheimer's disease. Neurochem. Int. 45:583–595

Iijima, K., Ando, K., Takeda, S., Satoh, Y., Seki, T., Itohara, S., Greengard, P., Kirino, Y., Nairn, A.C. and Suzuki, T. 2000. Neuron-specific phosphorylation of Alzheimer's β-amyloid precursor protein by cyclin-dependent kinase 5. J. Neurochem. 75:1085–1091

Ikeda, K., Urakami, K., Isoe, K., Ohno, K. and Nakashima, K. 1998. The expression of presenilin-1 mRNA in skin fibroblasts from Alzheimer's disease. Dement. Geriatr. Cogn. Disord. 9:145–148

Ingelson, M., Vanmechelen, E. and Lannfelt, L. 1996. Microtubule-associated protein tau in human fibroblasts with the Swedish Alzheimer mutation. Neurosci. Lett. 220:9–12

Intebi, A.D., Garau, L., Brusco, I., Pagano, M., Gaillard, R.C. and Spinedi, E. 2002. Alzheimer's disease patients display gender dimorphism in circulating anorectic adipokines. Neuroimmunomodulation 10:351–358

Iqbal, K., Grundke-Iqbal, I., Zaidi, T., Merz, P.A., Wen, G.Y., Shaikh, S.S., Wisniewski, H.M., Alafuzoff, I. and Winblad, B. 1986. Defective brain microtubule assembly in Alzheimer's disease. Lancet 2:421–426

Iqbal, K., Alonso Adel, C., Chen, S., Chohan, M.O., El-Akkad, E., Gong, C.X., Khatoon, S., Li, B., Liu, F., Rahman, A., Tanimukai, H. and Grundke-Iqbal, I. 2005. Tau pathology in Alzheimer disease and other tauopathies. Biochim. Biophys. Acta 1739:198–210

Ishibashi, T., Hayakawa, H., Ito, R., Miyazawa, M., Yamagata, Y. and Sekiguchi, M. 2005. Mammalian enzymes for preventing transcriptional errors caused by oxidative damage. Nucleic Acids Res. 33:3779–3784

Ishii, K., Tamaoka, A., Mizusawa, H., Shoji, S., Ohtake, T., Fraser, P.E., Takahashi, H., Tsuji, S., Gearing, M., Mizutani, T., Yamada, S., Kato, M., St George-Hyslop, P.H., Mirra, S.S. and Mori, H. 1997. Aβ1–40 but not Aβ1–42 levels in cortex correlate with apolipoprotein E ε4 allele dosage in sporadic Alzheimer's disease. Brain Res. 748:250–252

Iwatsubo, T., Odaka, A., Suzuki, N., Mizusawa, H., Nukina, N. and Ihara, Y. 1994. Visualization of Aβ42(43) and Aβ40 in senile plaques with end-specific Aβ monoclonals: Evidence that an initially deposited species is Aβ42(43). Neuron 13:45–53

Jafferali, S., Dumont, Y., Sotty, F., Robitaille, Y., Quirion, R. and Kar, S. 2000. Insulin-like growth factor-I and its receptor in the frontal cortex, hippocampus, and cerebellum of normal human and Alzheimer disease brains. Synapse 38:450–459

Jagust, W.J., Seab, J.P., Huesman, R.H., Valk, P.E., Mathis, C.A., Reed, B.R., Coxson, P.G. and Budinger, T.F. 1991. Diminished glucose transport in Alzheimer's disease: Dynamic PET studies. J. Cereb. Blood Flow Metab. 11:323–330

Jarrett, J.T., Berger, E.P. and Lansbury, P.T., Jr. 1993. The carboxy terminus of the β amyloid protein is critical for the seeding of amyloid formation: Implications for the pathogenesis of Alzheimer's disease. Biochemistry 32:4693–4697

Ji, Z.-S., Miranda, R.D., Newhouse, Y.M., Weisgraber, K.H., Huang, Y. and Mahley, R.W. 2002. Apolipoprotein E4 potentiates amyloid β peptide-induced lysosomal leakage and apoptosis in neuronal cells. J. Biol. Chem. 277:21821–21828

Ji, Z.-S., Mullendorff, K., Cheng, I.-H., Miranda, R.D., Huang, Y. and Mahley, R.W. 2006. Reactivity of apolipoprotein E4 and amyloid β peptide: Lysosomal stability and neurodegeneration. J. Biol. Chem. 281:2683–2692

Jicha, G.A., Weaver, C., Lane, E., Vianna, C., Kress, Y., Rockwood, J. and Davies, P. 1999. cAMP-dependent protein kinase phosphorylations on *tau* in Alzheimer's disease. J. Neurosci. 19:7486–7494

Johnson, G.V. and Hartigan, J.A. 1999. *Tau* protein in normal and Alzheimer's disease brain: An update. J. Alzheimer's Dis. 1:329–351

Jonker, C., Schmand, B., Lindeboom, J., Havekes, L.M. and Launer, L.J. 1998. Association between apolipoprotein E ε4 and the rate of cognitive decline in community-dwelling elderly individuals with and without dementia. Arch. Neurol. 55:1065–1069

Jonsson, L., Eriksdotter-Jonhagen, M., Kilander, L., Soininen, H., Hallikainen, M., Waldemar, G., Nygaard, H., Andreasen, N., Winblad, B. and Wimo, A. 2006. Determinants of costs of care for patients with Alzheimer's disease. Int. J. Geriatr. Psychiatry 21:449–459

Kaether, C. and Haass, C. 2004. A lipid boundary separates APP and secretases and limits amyloid β-peptide generation. J. Cell Biol. 167:809–812

Kamal, A., Stokin, G.B., Yang, Z., Xia, C.H. and Goldstein, L.S. 2000. Axonal transport of amyloid precursor protein is mediated by direct binding to the kinesin light chain subunit of kinesin-I. Neuron 28:449–459

Kang, J., Lemaire, H.G., Unterbeck, A., Salbaum, J.M., Masters, C.L., Grzeschik, K.H., Multhaup, G., Beyreuther, K. and Müller-Hill, B. 1987. The precursor of Alzheimer's disease amyloid A4 protein resembles a cell-surface receptor. Nature 325:733–736

Kannanayakal, T.J., Tao, H., Vandre, D.D. and Kuret, J. 2006. Casein kinase-1 isoforms differentially associate with neurofibrillary and granulovacuolar degeneration lesions. Acta Neuropathol. (Berlin) 111:413–421

Katzman, R. 1986. Alzheimer's disease. N. Engl. J. Med. 314:964–973

Kaufman, R.J. 1999. Stress signaling from the lumen of the endoplasmic reticulum: Coordination of gene transcriptional and translational controls. Genes Dev. 13:1211–1233

Keller, J.N., Mark, R.J., Bruce, A.J., Blanc, E., Rothstein, J.D., Uchida, K., Waeg, G. and Mattson, M.P. 1997a. 4-hydroxynonenal, an aldehydic product of membrane lipid peroxidation, impairs glutamate transport and mitochondrial function in synaptosomes. Neuroscience 80:685–696

Keller, J.N., Pang, Z., Geddes, J.W., Begley, J.G., Germeyer, A., Waeg, G. and Mattson, M.P. 1997b. Impairment of glucose and glutamate transport and induction of mitochondrial oxidative stress and dysfunction in synaptosomes by amyloid β-peptide: Role of the lipid peroxidation product 4-hydroxynonenal. J. Neurochem. 69:273–284

Kelly, P.T., McGuinness, T.L. and Greengard, P. 1984. Evidence that the major postsynaptic density protein is a component of a Ca^{2+}/calmodulin-dependent protein kinase. Proc. Natl. Acad. Sci. USA. 81:945–949

Killilea, D.W., Atamna, H., Liao, C. and Ames, B.N. 2003. Iron accumulation during cellular senescence in human fibroblasts in vitro. Antioxid. Redox Signal. 5:507–516

Kimpara, T., Takeda, A., Yamaguchi, T., Arai, H., Okita, N., Takase, S., Sasaki, H. and Itoyama, Y. 2000. Increased bilirubins and their derivatives in cerebrospinal fluid in Alzheimer's disease. Neurobiol. Aging 21:551–554

King, G.L. and Johnson, S.M. 1985. Receptor-mediated transport of insulin across endothelial cells. Science 227:1583–1586

Kins, S., Kurosinski, P., Nitsch, R.M. and Gotz, J. 2003. Activation of the ERK and JNK signaling pathways caused by neuron-specific inhibition of PP2A in transgenic mice. Am. J. Pathol. 163:833–843

Kirkitadze, M.D., Condron, M.M. and Teplow, D.B. 2001. Identification and characterization of key kinetic intermediates in amyloid β-protein fibrillogenesis. J. Mol. Biol. 312:1103–1119

Knippschild, U., Gocht, A., Wolff, S., Huber, N., Lohler, J. and Stoter, M. 2005. The casein kinase 1 family: participation in multiple cellular processes in eukaryotes. Cell Signal. 17:675–689

Knopman, D.S., DeKosky, S.T., Cummings, J.L., Chui, H., Corey-Bloom, J., Relkin, N., Small, G.W., Miller, B. and Stevens, J.C. 2001. Practice parameter: Diagnosis of dementia (an evidence-based review). Report of the Quality Standards Subcommittee of the American Academy of Neurology. Neurology 56:1143–1153

Kojro, E., Gimpl, G., Lammich, S., Marz, W. and Fahrenholz, F. 2001. Low cholesterol stimulates the nonamyloidogenic pathway by its effect on the α-secretase ADAM 10. Proc. Natl. Acad. Sci. USA. 98:5815–5820

Kojro, E., Postina, R., Buro, C., Meiringer, C., Gehrig-Burger, K. and Fahrenholz, F. 2006. The neuropeptide PACAP promotes the α-secretase pathway for processing the Alzheimer amyloid precursor protein. FASEB J. 20:512–514

Kopke, E., Tung, Y.C., Shaikh, S., Alonso, A.C., Iqbal, K. and Grundke-Iqbal, I. 1993. Microtubule-associated protein *tau*. Abnormal phosphorylation of a non-paired helical filament pool in Alzheimer disease. J. Biol. Chem. 268:24374–24384

Kumar, S., McDonnell, P.C., Gum, R.J., Hand, A.T., Lee, J.C. and Young, P.R. 1997. Novel homologues of CSBP/p38 MAP kinase: Activation, substrate specificity and sensitivity to inhibition by pyridinyl imidazoles. Biochem. Biophys. Res. Commun. 235:533–538

Kusakawa, G., Saito, T., Onuki, R., Ishiguro, K., Kishimoto, T. and Hisanaga, S. 2000. Calpain-dependent proteolytic cleavage of the p35 cyclin-dependent kinase 5 activator to p25. J. Biol. Chem. 275:17166–17172

La Du, M.J., Falduto, M.T., Manelli, A.M., Reardon, C.A., Getz, G.S. and Frail, D.E. 1994. Isoform-specific binding of apolipoprotein E to β-amyloid. J. Biol. Chem. 269:23403–23406

La Ferla, F.M., Tinkle, B.T., Bieberich, C.J., Haudenschild, C.C. and Jay, G. 1995. The Alzheimer's Aβ peptide induces neurodegeneration and apoptotic cell death in transgenic mice. Nat. Genet. 9:21–30

Lammich, S., Kojro, E., Postina, R., Gilbert, S., Pfeiffer, R., Jasionowski, M., Haass, C. and Fahrenholz, F. 1999. Constitutive and regulated α-secretase cleavage of Alzheimer's amyloid precursor protein by a disintegrin metalloprotease. Proc. Natl. Acad. Sci. USA. 96:3922–3927

Lauderback, C.M., Hackett, J.M., Huang, F.F., Keller, J.N., Szweda, L.I., Markesbery, W.R. and Butterfield, D.A. 2001. The glial glutamate transporter, GLT-1, is oxidatively modified by 4-hydroxy-2-nonenal in the Alzheimer's disease brain: The role of $A\beta_{1-42}$. J. Neurochem. 78:413–416

Lee, G., Newman, S.T., Gard, D.L., Band, H. and Panchamoorthy, G. 1998. *Tau* interacts with src-family non-receptor tyrosine kinases. J. Cell Sci. 111:3167–3177

Lee, V.M., Goedert, M. and Trojanowski, J.Q. 2001. Neurodegenerative tauopathies. Annu. Rev. Neurosci. 24:1121–1159

Lee, C.W., Lau, K.F., Miller, C.C. and Shaw, P.C. 2003a. Glycogen synthase kinase-3 β-mediated *tau* phosphorylation in cultured cell lines. Neuroreport 14:257–260

Lee, M.S., Kao, S.C., Lemere, C.A., Xia, W., Tseng, H.C., Zhou, Y., Neve, R., Ahlijanian, M.K. and Tsai, L.H. 2003b. APP processing is regulated by cytoplasmic phosphorylation. J. Cell Biol. 163:83–95

Leissring, M.A., Farris, W., Chang, A.Y., Walsh, D.M., Wu, X., Sun, X., Frosch, M.P. and Selkoe, D.J. 2003. Enhanced proteolysis of β-amyloid in APP transgenic mice prevents plaque formation, secondary pathology, and premature death. Neuron 40:1087–1093

Lemere, C.A., Lopera, F., Kosik, K.S., Lendon, C.L., Ossa, J., Saido, T.C., Yamaguchi, H., Ruiz, A., Martinez, A., Madrigal, L., Hincapie, L., Arango, J.C., Anthony, D.C., Koo, E.H., Goate, A.M. and Selkoe, D.J. 1996. The E280A presenilin 1 Alzheimer mutation produces increased Aβ42 deposition and severe cerebellar pathology. Nat. Med. 2:1146–1150

Lew, J., Huang, Q.Q., Qi, Z., Winkfein, R.J., Aebersold, R., Hunt, T. and Wang, J.H. 1994. A brain-specific activator of cyclin-dependent kinase 5. Nature 371:423–426

Li, M., Guo, H. and Damuni, Z. 1995. Purification and characterization of two potent heat-stable protein inhibitors of protein phosphatase 2A from bovine kidney. Biochemistry 34:1988–1996

Li, Z., Jiang, Y., Ulevitch, R.J. and Han, J. 1996a. The primary structure of p38γ: A new member of p38 group of MAP kinases. Biochem. Biophys. Res. Commun. 228:334–340

Li, M., Makkinje, A. and Damuni, Z. 1996b. Molecular identification of I1PP2A, a novel potent heat-stable inhibitor protein of protein phosphatase 2A. Biochemistry 35:6998

Li, R., Lindholm, K., Yang, L.B., Yue, X., Citron, M., Yan, R., Beach, T., Sue, L., Sabbagh, M., Cai, H., Wong, P., Price, D. and Shen, Y. 2004a. Amyloid β peptide load is correlated with increased β-secretase activity in sporadic Alzheimer's disease patients. Proc. Natl. Acad. Sci. USA. 101:3632–3637

Li, Z., Okamoto, K., Hayashi, Y. and Sheng, M. 2004b. The importance of dendritic mitochondria in the morphogenesis and plasticity of spines and synapses. Cell 119:873–887

Li, G., Yin, H. and Kuret, J. 2004c. Casein kinase 1δ phosphorylates tau and disrupts its binding to microtubules. J. Biol. Chem. 279:15938–15945

Lian, Q., Ladner, C.J., Magnuson, D. and Lee, J.M. 2001. Selective changes of calcineurin (protein phosphatase 2B) activity in Alzheimer's disease cerebral cortex. Exp. Neurol. 167:158–165

Ling, X., Martins, R.N., Racchi, M., Craft, S. and Helmerhorst, E. 2002. Amyloid β antagonizes insulin promoted secretion of the amyloid β protein precursor. J. Alzheimer's Dis. 4:369–374

Lippa, C.F., Schmidt, M.L., Nee, L.E., Bird, T., Nochlin, D., Hulette, C., Mori, H., Lee, V.M. and Trojanowski, J.Q. 2000. AMY plaques in familial AD: Comparison with sporadic Alzheimer's disease. Neurology 54:100–104

Liu, F., Iqbal, K., Grundke-Iqbal, I., Hart, G.W. and Gong, C.-X. 2004a. O-GlcNAcylation regulates phosphorylation of tau: A mechanism involved in Alzheimer's disease. Proc. Natl. Acad. Sci. USA 101:10804–10809

Liu, S.-J., Zhang, J.-Y., Li, H.-L., Fang, Z.-Y., Wang, Q., Deng, H.-M., Gong, C.-X., Grundke-Iqbal, I., Iqbal, K. and Wang, J.-Z. 2004b. Tau becomes a more favorable substrate for GSK-3 when it is prephosphorylated by PKA in rat brain. J. Biol. Chem. 279:50078–50088

Liu, F., Grundke-Iqbal, I., Iqbal, K. and Gong, C.X. 2005a. Contributions of protein phosphatases PP1, PP2A, PP2B and PP5 to the regulation of tau phosphorylation. Eur. J. Neurosci. 22:1942–1950

Liu, F., Iqbal, K., Grundke-Iqbal, I., Rossie, S. and Gong, C.X. 2005b. Dephosphorylation of tau by protein phosphatase 5: Impairment in Alzheimer's disease. J. Biol. Chem. 280:1790–1796

Liu, G.-P., Zhang, Y., Yao, X.-Q., Zhang, C.-E., Fang, J., Wang, Q. and Wang, J.-Z. 2007. Activation of glycogen synthase kinase-3 inhibits protein phosphatase-2A and the underlying mechanisms. Neurobiol. Aging

Lovell, M.A., Xie, C. and Markesbery, W.R. 2000. Decreased base excision repair and increased helicase activity in Alzheimer's disease brain. Brain Res. 855:116–123

Lovestone, S. and Reynolds, C.H. 1997. The phosphorylation of tau: A critical stage in neurodevelopment and neurodegenerative processes. Neuroscience 78:309–324

Lovestone, S., Anderton, B.H., Hartley, C., Jensen, T.G. and Jorgensen, A.L. 1996. The intracellular fate of apolipoprotein E is tau dependent and APOE allele-specific. Neuroreport 7:1005–1008

Lu, Q., Soria, J.P. and Wood, J.G. 1993. p44mpk MAP kinase induces Alzheimer type alterations in tau function and in primary hippocampal neurons. J. Neurosci. Res. 35:439–444

Lue, L.F., Kuo, Y.M., Roher, A.E., Brachova, L., Shen, Y., Sue, L., Beach, T., Kurth, J.H., Rydel, R.E. and Rogers, J. 1999. Soluble amyloid β peptide concentration as a predictor of synaptic change in Alzheimer's disease. Am. J. Pathol. 155:853–862

Madl, J.E. and Burgesser, K. 1993. Adenosine triphosphate depletion reverses sodium-dependent, neuronal uptake of glutamate in rat hippocampal slices. J. Neurosci. 13:4429–4444

Mahley, R.W. 1988. Apolipoprotein E: Cholesterol transport protein with expanding role in cell biology. Science 240:622–630

Mahley, R.W. and Huang, Y. 2006. Apolipoprotein (apo) E4 and Alzheimer's disease: Unique conformational and biophysical properties of apoE4 can modulate neuropathology. Acta Neurol. Scand. Suppl. 185:8–14

Mahley, R.W. and Rall, S.C., Jr. 2000. Apolipoprotein E: Far more than a lipid transport protein. Annu. Rev. Genomics Hum. Genet. 1:507–537

Mahley, R.W., Weisgraber, K.H. and Huang, Y.-D. 2006. Apolipoprotein E4: A causative factor and therapeutic target in neuropathology, including Alzheimer's disease. Proc. Natl. Acad. Sci. USA 103:5644–5651

Mandelkow, E.M., Drewes, G., Biernat, J., Gustke, N., van Lint, J., Vandenheede, J.R. and Mandelkow, E. 1992. Glycogen synthase kinase-3 and the Alzheimer-like state of microtubule-associated protein *tau*. FEBS Lett. 314:315–321

Mark, R.J., Lovell, M.A., Markesbery, W.R., Uchida, K. and Mattson, M.P. 1997a. A role for 4-hydroxynonenal, an aldehydic product of lipid peroxidation, in disruption of ion homeostasis and neuronal death induced by amyloid β-peptide. J. Neurochem. 68:255–264

Mark, R.J., Pang, Z., Geddes, J.W., Uchida, K. and Mattson, M.P. 1997b. Amyloid β-peptide impairs glucose transport in hippocampal and cortical neurons: involvement of membrane lipid peroxidation. J. Neurosci. 17:1046–1054

Markesbery, W.R. and Lovell, M.A. 1998. Four-hydroxynonenal, a product of lipid peroxidation, is increased in the brain in Alzheimer's disease. Neurobiol. Aging 19:33–36

Marquez-Sterling, N.R., Lo, A.C., Sisodia, S.S. and Koo, E.H. 1997. Trafficking of cell-surface β-amyloid precursor protein: Evidence that a sorting intermediate participates in synaptic vesicle recycling. J. Neurosci. 17:140–151

Marshall, S., Nadeau, O. and Yamasaki, K. 2004. Dynamic actions of glucose and glucosamine on hexosamine biosynthesis in isolated adipocytes: Differential effects on glucosamine 6-phosphate, UDP-N-acetylglucosamine, and ATP levels. J. Biol. Chem. 279:35313–35319

Mastrogiacomo, F., Bergeron, C. and Kish, S.J. 1993. Brain α-ketoglutarate dehydrogenase complex activity in Alzheimer's disease. J. Neurochem. 61:2007–2014

Matsuo, E.S., Shin, R.W., Billingsley, M.L., Van deVoorde, A., O'Connor, M., Trojanowski, J.Q. and Lee, V.M. 1994. Biopsy-derived adult human brain *tau* is phosphorylated at many of the same sites as Alzheimer's disease paired helical filament *tau*. Neuron 13:989–1002

McDermott, J.R. and Gibson, A.M. 1997. Degradation of Alzheimer's β-amyloid protein by human and rat brain peptidases: Involvement of insulin-degrading enzyme. Neurochem. Res. 22:49–56

McEwen, B.S. and Reagan, L.P. 2004. Glucose transporter expression in the central nervous system: Relationship to synaptic function. Eur. J. Pharmacol. 490:13–24

McKee, A.C., Kosik, K.S., Kennedy, M.B. and Kowall, N.W. 1990. Hippocampal neurons predisposed to neurofibrillary tangle formation are enriched in type II calcium/calmodulin-dependent protein kinase. J. Neuropathol. Exp. Neurol. 49:49–63

McLean, C.A., Cherny, R.A., Fraser, F.W., Fuller, S.J., Smith, M.J., Beyreuther, K., Bush, A.I. and Masters, C.L. 1999. Soluble pool of Aβ amyloid as a determinant of severity of neurodegeneration in Alzheimer's disease. Ann. Neurol. 46:860–866

McMillan, P.J., Leverenz, J.B. and Dorsa, D.M. 2000. Specific downregulation of presenilin 2 gene expression is prominent during early stages of sporadic late-onset Alzheimer's disease. Brain Res. Mol. Brain Res. 78:138–145

McShea, A., Zelasko, D.A., Gerst, J.L. and Smith, M.A. 1999. Signal transduction abnormalities in Alzheimer's disease: Evidence of a pathogenic stimuli. Brain Res. 815:237–242

Mercken, M., Vandermeeren, M., Lubke, U., Six, J., Boons, J., van de Voorde, A., Martin, J.J. and Gheuens, J. 1992. Monoclonal antibodies with selective specificity for Alzheimer *tau* are directed against phosphatase-sensitive epitopes. Acta Neuropathol. (Berlin) 84:265–272

Michikawa, M. and Yanagisawa, K. 1999. Inhibition of cholesterol production but not of nonsterol isoprenoid products induces neuronal cell death. J. Neurochem. 72:2278–2285

Misonou, H., Morishima-Kawashima, M. and Ihara, Y. 2000. Oxidative stress induces intracellular accumulation of amyloid β-protein (Aβ) in human neuroblastoma cells. Biochemistry 39:6951–6959

Mooradian, A.D., Chung, H.-C. and Shah, G.N. 1997. GLUT-1 expression in the cerebra of patients with Alzheimer's disease. Neurobiol. Aging 18:469–474

Morishima-Kawashima, M., Hasegawa, M., Takio, K., Suzuki, M., Yoshida, H., Watanabe, A., Titani, K. and Ihara, Y. 1995. Hyperphosphorylation of *tau* in PHF. Neurobiol. Aging 16:365–371

Mullaart, E., Boerrigter, M.E., Ravid, R., Swaab, D.F. and Vijg, J. 1990. Increased levels of DNA breaks in cerebral cortex of Alzheimer's disease patients. Neurobiol. Aging 11:169–173

Nakabeppu, Y., Kajitani, K., Sakamoto, K., Yamaguchi, H. and Tsuchimoto, D. 2006. MTH1, an oxidized purine nucleoside triphosphatase, prevents the cytotoxicity and neurotoxicity of oxidized purine nucleotides. DNA Repair 5:761–772

Nathan, B.P., Bellosta, S., Sanan, D.A., Weisgraber, K.H., Mahley, R.W. and Pitas, R.E. 1994. Differential effects of apolipoproteins E3 and E4 on neuronal growth in vitro. Science 264:850–852

Nelson, P.T., Stefansson, K., Gulcher, J. and Saper, C.B. 1996. Molecular evolution of *tau* protein: Implications for Alzheimer's disease. J. Neurochem. 67:1622–1632

Nicholls, D. and Attwell, D. 1990. The release and uptake of excitatory amino acids. Trends Pharmacol. Sci. 11:462–468

Noble, W., Olm, V., Takata, K., Casey, E., Mary, O., Meyerson, J., Gaynor, K., LaFrancois, J., Wang, L., Kondo, T., Davies, P., Burns, M., Veeranna, Nixon, R., Dickson, D., Matsuoka, Y., Ahlijanian, M., Lau, L.F. and Duff, K. 2003. Cdk5 is a key factor in *tau* aggregation and tangle formation in vivo. Neuron 38:555–565

Nunomura, A., Perry, G., Aliev, G., Hirai, K., Takeda, A., Balraj, E.K., Jones, P.K., Ghanbari, H., Wataya, T., Shimohama, S., Chiba, S., Atwood, C.S., Petersen, R.B. and Smith, M.A. 2001. Oxidative damage is the earliest event in Alzheimer disease. J. Neuropathol. Exp. Neurol. 60:759–767

Nunomura, A., Chiba, S., Lippa, C.F., Cras, P., Kalaria, R.N., Takeda, A., Honda, K., Smith, M. A. and Perry, G. 2004. Neuronal RNA oxidation is a prominent feature of familial Alzheimer's disease. Neurobiol. Dis. 17:108–113

O'Kusky, J.R., Ye, P. and D'Ercole, A.J. 2000. Insulin-like growth factor-I promotes neurogenesis and synaptogenesis in the hippocampal dentate gyrus during postnatal development. J. Neurosci. 20:8435–8442

Ono, K. and Han, J. 2000. The p38 signal transduction pathway: Activation and function. Cell Signal. 12:1–13

Pastorino, L., Sun, A., Lu, P.-J., Zhou, X.-Z., Balastik, M., Finn, G., Wulf, G., Lim, J., Li, S.-H., Li, X., Xia, W., Nicholson, L.K. and Lu, K.-P. 2006. The prolyl isomerase Pin1 regulates amyloid precursor protein processing and amyloid-β production. Nature 440:528–534

Patrick, G.N., Zukerberg, L., Nikolic, M., de la Monte, S., Dikkes, P. and Tsai, L.H. 1999. Conversion of p35 to p25 deregulates Cdk5 activity and promotes neurodegeneration. Nature 402:615–622

Paudel, H.K., Lew, J., Ali, Z. and Wang, J.H. 1993. Brain proline-directed protein kinase phosphorylates *tau* on sites that are abnormally phosphorylated in *tau* associated with Alzheimer's paired helical filaments. J. Biol. Chem. 268:23512–23518

Paulus, W. and Selim, M. 1990. Corticonigral degeneration with neuronal achromasia and basal neurofibrillary tangles. Acta Neuropathol. (Berlin) 81:89–94

Pearson, R.C.A., Esiri, M.M., Hiorns, R.W., Wilcock, G.K. and Powell, T.P.S. 1985. Anatomical correlates of the distribution of the pathological changes in the neocortex in Alzheimer disease. Proc. Natl. Acad. Sci. USA 82:4531–4534

Pedersen, W.A., Cashman, N.R. and Mattson, M.P. 1999. The lipid peroxidation product 4-hydroxynonenal impairs glutamate and glucose transport and choline acetyltransferase activity in NSC-19 motor neuron cells. Exp. Neurol. 155:1–10

Pei, J.J., Sersen, E., Iqbal, K. and Grundke-Iqbal, I. 1994. Expression of protein phosphatases (PP-1, PP-2A, PP-2B and PTP-1B) and protein kinases (MAP kinase and P34cdc2) in the hippocampus of patients with Alzheimer disease and normal aged individuals. Brain Res. 655:70–76

Pei, J.J., Braak, E., Braak, H., Grundke-Iqbal, I., Iqbal, K., Winblad, B. and Cowburn, R.F. 2001. Localization of active forms of C-jun kinase (JNK) and p38 kinase in Alzheimer's disease brains at different stages of neurofibrillary degeneration. J. Alzheimer's Dis. 3:41–48

Pei, J.J., Gong, C.X., An, W.L., Winblad, B., Cowburn, R.F., Grundke-Iqbal, I. and Iqbal, K. 2003. Okadaic-acid-induced inhibition of protein phosphatase 2A produces activation of mitogen-activated protein kinases ERK1/2, MEK1/2, and p70 S6, similar to that in Alzheimer's disease. Am. J. Pathol. 163:845–858

Pike, C.J., Walencewicz, A.J., Glabe, C.G. and Cotman, C.W. 1991. In vitro aging of β-amyloid protein causes peptide aggregation and neurotoxicity. Brain Res. 563:311–314

Pike, C.J., Burdick, D., Walencewicz, A.J., Glabe, C.G. and Cotman, C.W. 1993. Neurodegeneration induced by β-amyloid peptides in vitro: The role of peptide assembly state. J. Neurosci. 13:1676–1687

Pitas, R.E., Boyles, J.K., Lee, S.H., Foss, D. and Mahley, R.W. 1987. Astrocytes synthesize apolipoprotein E and metabolize apolipoprotein E-containing lipoproteins. Biochim. Biophys. Acta 917:148–161

Plum, L., Schubert, M. and Bruning, J.C. 2005. The role of insulin receptor signaling in the brain. Trends Endocrinol. Metab. 16:59–65

Plyte, S.E., Hughes, K., Nikolakaki, E., Pulverer, B.J. and Woodgett, J.R. 1992. Glycogen synthase kinase-3: Functions in oncogenesis and development. Biochim. Biophys. Acta 1114:147–162

Podlisny, M.B., Citron, M., Amarante, P., Sherrington, R., Xia, W., Zhang, J., Diehl, T., Levesque, G., Fraser, P., Haass, C., Koo, E.H., Seubert, P., St George-Hyslop, P., Teplow, D.B. and Selkoe, D.J. 1997. Presenilin proteins undergo heterogeneous endoproteolysis between Thr291 and Ala299 and occur as stable N- and C-terminal fragments in normal and Alzheimer brain tissue. Neurobiol. Dis. 3:325–337

Poehlman, E.T. and Dvorak, R.V. 2000. Energy expenditure, energy intake, and weight loss in Alzheimer disease. Am. J. Clin. Nutr. 71:650S–655S

Poon, R.Y., Lew, J. and Hunter, T. 1997. Identification of functional domains in the neuronal Cdk5 activator protein. J. Biol. Chem. 272:5703–5708

Popken, G.J., Hodge, R.D., Ye, P., Zhang, J., Ng, W., O'Kusky, J.R. and D'Ercole, A.J. 2004. In vivo effects of insulin-like growth factor-I (IGF-I) on prenatal and early postnatal development of the central nervous system. Eur. J. Neurosci. 19:2056–2068

Postina, R., Schroeder, A., Dewachter, I., Bohl, J., Schmitt, U., Kojro, E., Prinzen, C., Endres, K., Hiemke, C., Blessing, M., Flamez, P., Dequenne, A., Godaux, E., van Leuven, F. and Fahrenholz, F. 2004. A disintegrin-metalloproteinase prevents amyloid plaque formation and hippocampal defects in an Alzheimer disease mouse model. J. Clin. Invest. 113:1456–1464

Qi-Takahara, Y., Morishima-Kawashima, M., Tanimura, Y., Dolios, G., Hirotani, N., Horikoshi, Y., Kametani, F., Maeda, M., Saido, T.C., Wang, R. and Ihara, Y. 2005. Longer forms of amyloid β protein: Implications for the mechanism of intramembrane cleavage by γ-secretase. J. Neurosci. 25:436–445

Qiu, W.Q., Ferreira, A., Miller, C., Koo, E.H. and Selkoe, D.J. 1995. Cell-surface β-amyloid precursor protein stimulates neurite outgrowth of hippocampal neurons in an isoform-dependent manner. J. Neurosci. 15:2157–2167

Raingeaud, J., Gupta, S., Rogers, J.S., Dickens, M., Han, J., Ulevitch, R.J. and Davis, R.J. 1995. Pro-inflammatory cytokines and environmental stress cause p38 mitogen-activated protein kinase activation by dual phosphorylation on tyrosine and threonine. J. Biol. Chem. 270:7420–7426

Refolo, L.M., Salton, S.R., Anderson, J.P., Mehta, P. and Robakis, N.K. 1989. Nerve and epidermal growth factors induce the release of the Alzheimer amyloid precursor from PC 12 cell cultures. Biochem. Biophys. Res. Commun. 164:664–670

Reiman, E.M., Caselli, R.J., Chen, K., Alexander, G.E., Bandy, D. and Frost, J. 2001. Declining brain activity in cognitively normal apolipoprotein E ε4 heterozygotes: a foundation for using positron emission tomography to efficiently test treatments to prevent Alzheimer's disease. Proc. Natl. Acad. Sci. USA 98:3334–3339

Reiman, E.M., Chen, K., Alexander, G.E., Caselli, R.J., Bandy, D., Osborne, D., Saunders, A.M. and Hardy, J. 2004. Functional brain abnormalities in young adults at genetic risk for late-onset Alzheimer's dementia. Proc. Natl. Acad. Sci. USA 101:284–289

Rendon, A., Jung, D. and Jancsik, V. 1990. Interaction of microtubules and microtubule-associated proteins (MAPs) with rat brain mitochondria. Biochem. J. 269:555–556

Reynolds, C.H., Nebreda, A.R., Gibb, G.M., Utton, M.A. and Anderton, B.H. 1997a. Reactivating kinase/p38 phosphorylates tau protein in vitro. J. Neurochem. 69:191–198

Reynolds, C.H., Utton, M.A., Gibb, G.M., Yates, A. and Anderton, B.H. 1997b. Stress-activated protein kinase/c-jun N-terminal kinase phosphorylates *tau* protein. J. Neurochem. 68:1736–1744

Rostas, J.A. and Dunkley, P.R. 1992. Multiple forms and distribution of calcium/calmodulin-stimulated protein kinase II in brain. J. Neurochem. 59:1191–1202

Roux, P.P. and Blenis, J. 2004. ERK and p38 MAPK-activated protein kinases: A family of protein kinases with diverse biological functions. Microbiol. Mol. Biol. Rev. 68:320–344

Ryan, M., Starkey, M., Faull, R., Emson, P. and Bahn, S. 2001. Indexing-based differential display studies on post-mortem Alzheimer's brains. Brain Res. Mol. Brain Res. 88:199–202

Sadot, E., Jaaro, H., Seger, R. and Ginzburg, I. 1998. Ras-signaling pathways: Positive and negative regulation of tau expression in PC12 cells. J. Neurochem. 70:428–431

Sakashita, G., Shima, H., Komatsu, M., Urano, T., Kikuchi, A. and Kikuchi, K. 2003. Regulation of type 1 protein phosphatase/inhibitor-2 complex by glycogen synthase kinas-3β in intact cells. J. Biochem. (Tokyo) 133:165–171

Sandbrink, R., Masters, C.L. and Beyreuther, K. 1996. APP gene family. Alternative splicing generates functionally related isoforms. Ann. NY Acad. Sci. 777:281–287

Sang, H., Lu, Z., Li, Y., Ru, B., Wang, W. and Chen, J. 2001. Phosphorylation of *tau* by glycogen synthase kinase 3β in intact mammalian cells influences the stability of microtubules. Neurosci. Lett. 312:141–144

Sastre, M., Steiner, H., Fuchs, K., Capell, A., Multhaup, G., Condron, M.M., Teplow, D.B. and Haass, C. 2001. Presenilin-dependent γ-secretase processing of β-amyloid precursor protein at a site corresponding to the S3 cleavage of Notch. EMBO Rep. 2:835–841

Scheuner, D., Eckman, C., Jensen, M., Song, X., Citron, M., Suzuki, N., Bird, T.D., Hardy, J., Hutton, M., Kukull, W., Larson, E., Levy-Lahad, E., Viitanen, M., Peskind, E., Poorkaj, P., Schellenberg, G., Tanzi, R., Wasco, W., Lannfelt, L., Selkoe, D. and Younkin, S. 1996. Secreted amyloid β-protein similar to that in the senile plaques of Alzheimer's disease is increased in vivo by the presenilin 1 and 2 and APP mutations linked to familial Alzheimer's disease. Nat. Med. 2:864–870

Schipper, H.M., Cisse, S. and Stopa, E.G. 1995. Expression of heme oxygenase-1 in the senescent and Alzheimer-diseased brain. Ann. Neurol. 37:758–768

Schipper, H.M., Liberman, A. and Stopa, E.G. 1998. Neural heme oxygenase-1 expression in idiopathic Parkinson's disease. Exp. Neurol. 150:60–68

Schubert, D. 2005. Glucose metabolism and Alzheimer's disease. Ageing Res. Rev. 4:240–257

Schubert, D., Jin, L.W., Saitoh, T. and Cole, G. 1989. The regulation of amyloid β protein precursor secretion and its modulatory role in cell adhesion. Neuron 3:689–694

Schubert, M., Brazil, D.P., Burks, D.J., Kushner, J.A., Ye, J., Flint, C.L., Farhang-Fallah, J., Dikkes, P., Warot, X.M., Rio, C., Corfas, G. and White, M.F. 2003. Insulin receptor substrate-2 deficiency impairs brain growth and promotes *tau* phosphorylation. J. Neurosci. 23:7084–7092

Schubert, M., Gautam, D., Surjo, D., Ueki, K., Baudler, S., Schubert, D., Kondo, T., Alber, J., Galldiks, N., Kustermann, E., Arndt, S., Jacobs, A.H., Krone, W., Kahn, C.R. and Bruning, J.C. 2004. Role for neuronal insulin resistance in neurodegenerative diseases. Proc. Natl. Acad. Sci. USA 101:3100–3105

Schulingkamp, R.J., Pagano, T.C., Hung, D. and Raffa, R.B. 2000. Insulin receptors and insulin action in the brain: Review and clinical implications. Neurosci. Biobehav. Rev. 24:855–872

Shackelford, D.A. 2006. DNA end joining activity is reduced in Alzheimer's disease. Neurobiol. Aging 27:596–605

Shah, S., Lee, S.F., Tabuchi, K., Hao, Y.H., Yu, C., La Plant, Q., Ball, H., Dann, C.E., III, Sudhof, T. and Yu, G. 2005. Nicastrin functions as a γ-secretase-substrate receptor. Cell 122:435–447

Shan, X. and Lin, C.L. 2006. Quantification of oxidized RNAs in Alzheimer's disease. Neurobiol. Aging 27:657–662

Shoji, M., Golde, T.E., Ghiso, J., Cheung, T.T., Estus, S., Shaffer, L.M., Cai, X.-D., McKay, D.M., Tintner, R., Frangione, B. and Younkin, S.G. 1992. Production of the Alzheimer amyloid β protein by normal proteolytic processing. Science 258:126–129

Shpakov, A.O. and Pertseva, M.N. 2000. Structural and functional characterization of insulin receptor substrate proteins and the molecular mechanisms of their interaction with insulin superfamily tyrosine kinase receptors and effector proteins. Membr. Cell. Biol. 13:455–484

Shulman, R.G., Rothman, D.L., Behar, K.L. and Hyder, F. 2004. Energetic basis of brain activity: Implications for neuroimaging. Trends Neurosci. 27:489–495

Siegel, S.J., Bieschke, J., Powers, E.T. and Kelly, J.W. 2007. The oxidative stress metabolite 4-hydroxynonenal promotes Alzheimer protofibril formation. Biochemistry 46:1503–1510

Simpson, I.A., Chundu, K.R., Davies-Hill, T., Honer, W.G. and Davies, P. 1994. Decreased concentrations of GLUT1 and GLUT3 glucose transporters in the brains of patients with Alzheimer's disease. Ann. Neurol. 35:546–551

Singh, T.J., Zaidi, T., Grundke-Iqbal, I. and Iqbal, K. 1996. Non-proline-dependent protein kinases phosphorylate several sites found in tau from Alzheimer disease brain. Mol. Cell. Biochem. 154:143–151

Skovronsky, D.M., Moore, D.B., Milla, M.E., Doms, R.W. and Lee, V.M. 2000. Protein kinase C-dependent α-secretase competes with β-secretase for cleavage of amyloid-β precursor protein in the trans-Golgi network. J. Biol. Chem. 275:2568–2575

Small, G.W., Mazziotta, J.C., Collins, M.T., Baxter, L.R., Phelps, M.E., Mandelkern, M.A., Kaplan, A., La Rue, A., Adamson, C.F., Chang, L., Guze, B.H., Corder, E.H., Saunders, A.M., Haines, J.L., Pericak-Vance, M.A. and Roses, A.D. 1995. Apolipoprotein E type 4 allele and cerebral glucose metabolism in relatives at risk for familial Alzheimer disease. J. Am. Med. Assoc. 273:942–947

Small, G.W., Ercoli, L.M., Silverman, D.H., Huang, S.C., Komo, S., Bookheimer, S.Y., Lavretsky, H., Miller, K., Siddarth, P., Rasgon, N.L., Mazziotta, J.C., Saxena, S., Wu, H.M., Mega, M.S., Cummings, J.L., Saunders, A.M., Pericak-Vance, M.A., Roses, A.D., Barrio, J.R. and Phelps, M.E. 2000. Cerebral metabolic and cognitive decline in persons at genetic risk for Alzheimer's disease. Proc. Natl. Acad. Sci. USA 97:6037–6042

Sorbi, S., Bird, E.D. and Blass, J.P. 1983. Decreased pyruvate dehydrogenase complex activity in Huntington and Alzheimer brain. Ann. Neurol. 13:72–78

Sperber, B.R., Leight, S., Goedert, M. and Lee, V.M. 1995. Glycogen synthase kinase-3β phosphorylates tau protein at multiple sites in intact cells. Neurosci. Lett. 197:149–153

Spillantini, M.G., van Swieten, J.C. and Goedert, M. 2000. Tau gene mutations in frontotemporal dementia and parkinsonism linked to chromosome 17 (FTDP-17). Neurogenetics 2:193–205

Stambolic, V., Ruel, L. and Woodgett, J.R. 1996. Lithium inhibits glycogen synthase kinase-3 activity and mimics wingless signalling in intact cells. Curr. Biol. 6:1664–1668

Stein, T.D., Anders, N.J., De Carli, C., Chan, S.L., Mattson, M.P. and Johnson, J.A. 2004. Neutralization of transthyretin reverses the neuroprotective effects of secreted amyloid precursor protein (APP) in APPSW mice resulting in tau phosphorylation and loss of hippocampal neurons: Support for the amyloid hypothesis. J. Neurosci. 24:7707–7717

Steiner, B., Mandelkow, E.-M., Biernat, J., Gustke, N., Meyer, H.E., Schmidt, B., Mieskes, G., Söling, H.D., Drechsel, D., Kirschner, M.W., Goedert, M. and Mandelkow, E. 1990. Phosphorylation of microtubule-associated protein tau: Identification of the site for Ca^{2+}-calmodulin dependent kinase and relationship with tau phosphorylation in Alzheimer tangles. EMBO J. 9:3539–3544

Storey, E., Katz, M., Brickman, Y., Beyreuther, K. and Masters, C.L. 1999. Amyloid precursor protein of Alzheimer's disease: Evidence for a stable, full-length, trans-membrane pool in primary neuronal cultures. Eur. J. Neurosci. 11:1779–1788

Strittmatter, W.J., Weisgraber, K.H., Goedert, M., Saunders, A.M., Huang, D., Corder, E.H., Dong, L.-M., Jakes, R., Alberts, M.J., Gilbert, J.R., Han, S.-H., Hulette, C., Einstein, G., Schmechel, D.E., Pericak-Vance, M.A. and Roses, A.D. 1994. Hypothesis: Microtubule instability and paired helical filament formation in the Alzheimer disease brain are related to apolipoprotein E genotype. Exp. Neurol. 125:163–171

Subramaniam, R., Roediger, F., Jordan, B., Mattson, M.P., Keller, J.N., Waeg, G. and Butterfield, D.A. 1997. The lipid peroxidation product, 4-hydroxy-2-trans-nonenal, alters the conformation of cortical synaptosomal membrane proteins. J. Neurochem. 69:1161–1169

Sultana, R., Poon, H.F., Cai, J., Pierce, W.M., Merchant, M., Klein, J.B., Markesbery, W.R. and Butterfield, D.A. 2006. Identification of nitrated proteins in Alzheimer's disease brain using a redox proteomics approach. Neurobiol. Dis. 22:76–87

Sun, X.J., Crimmins, D.L., Myers, M.G., Jr., Miralpeix, M. and White, M.F. 1993a. Pleiotropic insulin signals are engaged by multisite phosphorylation of IRS-1. Mol. Cell. Biol. 13:7418–7428

Sun, F.F., Fleming, W.E. and Taylor, B.M. 1993b. Degradation of membrane phospholipids in the cultured human astroglial cell line UC-11MG during ATP depletion. Biochem. Pharmacol. 45:1149–1155

Sun, A., Liu, M., Nguyen, X.V. and Bing, G. 2003. P38 MAP kinase is activated at early stages in Alzheimer's disease brain. Exp. Neurol. 183:394–405

Takashima, A. 2006. GSK-3 is essential in the pathogenesis of Alzheimer's disease. J. Alzheimer's Dis. 9:309–317

Takeda, A., Perry, G., Abraham, N.G., Dwyer, B.E., Kutty, R.K., Laitinen, J.T., Petersen, R.B. and Smith, M.A. 2000a. Overexpression of heme oxygenase in neuronal cells, the possible interaction with tau. J. Biol. Chem. 275:5395–5399

Takeda, A., Smith, M.A., Avila, J., Nunomura, A., Siedlak, S.L., Zhu, X., Perry, G. and Sayre, L.M. 2000b. In Alzheimer's disease, heme oxygenase is coincident with Alz50, an epitope of tau induced by 4-hydroxy-2-nonenal modification. J. Neurochem. 75:1234–1241

Tanaka, C. and Nishizuka, Y. 1994. The protein kinase C family for neuronal signaling. Annu. Rev. Neurosci. 17:551–567

Tang, D. and Wang, J.-H. 1996. Cyclin-dependent kinase 5 (Cdk5) and neuron-specific Cdk5 activators. Prog. Cell Cycle Res. 2:205–216

Tanzi, R.E., McClatchey, A.I., Lamperti, E.D., Villa-Komaroff, L., Gusella, J.F. and Neve, R.L. 1988. Protease inhibitor domain encoded by an amyloid protein precursor mRNA associated with Alzheimer's disease. Nature 331:528–530

Tanzi, R.E., Vaula, G., Romano, D.M., Mortilla, M., Huang, T.L., Tupler, R.G., Wasco, W., Hyman, B.T., Haines, J.L., Jenkins, B.J. and et al. 1992. Assessment of amyloid β-protein precursor gene mutations in a large set of familial and sporadic Alzheimer disease cases. Am. J. Hum. Genet. 51:273–282

Tesseur, I., van Dorpe, J., Bruynseels, K., Bronfman, F., Sciot, R., van Lommel, A. and van Leuven, F. 2000a. Prominent axonopathy and disruption of axonal transport in transgenic mice expressing human apolipoprotein E4 in neurons of brain and spinal cord. Am. J. Pathol. 157:1495–1510

Tesseur, I., van Dorpe, J., Spittaels, K., van den Haute, C., Moechars, D. and van Leuven, F. 2000b. Expression of human apolipoprotein E4 in neurons causes hyperphosphorylation of protein tau in the brains of transgenic mice. Am. J. Pathol. 156:951–964

Thinakaran, G., Borchelt, D.R., Lee, M.K., Slunt, H.H., Spitzer, L., Kim, G., Ratovitsky, T., Davenport, F., Nordstedt, C., Seeger, M., Hardy, J., Levey, A.I., Gandy, S.E., Jenkins, N.A., Copeland, N.G., Price, D.L. and Sisodia, S.S. 1996. Endoproteolysis of presenilin 1 and accumulation of processed derivatives in vivo. Neuron 17:181–190

Trakshel, G.M., Kutty, R.K. and Maines, M.D. 1988. Resolution of the rat brain heme oxygenase activity: Absence of a detectable amount of the inducible form (HO-1). Arch. Biochem. Biophys. 260:732–739

Trojanowski, J.Q. and Lee, V.M. 1995. Phosphorylation of paired helical filament tau in Alzheimer's disease neurofibrillary lesions: Focusing on phosphatases. FASEB J. 9:1570–1576

Trojanowski, J.Q., Mawal-Dewan, M., Schmidt, M.L., Martin, J. and Lee, V.M.-Y. 1993. Localization of the mitogen activated protein kinase ERK2 in Alzheimer's disease neurofibrillary tangles and senile plaque neurites. Brain Res. 618:333–337

Tsai, L.H., Delalle, I., Caviness, V.S., Jr., Chae, T. and Harlow, E. 1994. p35 is a neural-specific regulatory subunit of cyclin-dependent kinase 5. Nature 371:419–423

Vanier, M.T., Neuville, P., Michalik, L. and Launay, J.F. 1998. Expression of specific *tau* exons in normal and tumoral pancreatic acinar cells. J. Cell Sci. 111:1419–1432

Veeranna, Amin, N.D., Ahn, N.G., Jaffe, H., Winters, C.A., Grant, P. and Pant, H.C. 1998. Mitogen-activated protein kinases (Erk1,2) phosphorylate Lys-Ser-Pro (KSP) repeats in neurofilament proteins NF-H and NF-M. J. Neurosci. 18:4008–4021

Vincent, I.J. and Davies, P. 1992. A protein kinase associated with paired helical filaments in Alzheimer disease. Proc. Natl. Acad. Sci. USA 89:2878–2882

Vogelsberg-Ragaglia, V., Schuck, T., Trojanowski, J.Q. and Lee, V.M. 2001. PP2A mRNA expression is quantitatively decreased in Alzheimer's disease hippocampus. Exp. Neurol. 168:402–412

Walsh, D.M., Hartley, D.M., Condron, M.M., Selkoe, D.J. and Teplow, D.B. 2001. In vitro studies of amyloid β-protein fibril assembly and toxicity provide clues to the ætiology of Flemish variant (Ala692-- > Gly) Alzheimer's disease. Biochem. J. 355:869–877

Walter, J., Capell, A., Grunberg, J., Pesold, B., Schindzielorz, A., Prior, R., Podlisny, M.B., Fraser, P., Hyslop, P.S., Selkoe, D.J. and Haass, C. 1996. The Alzheimer's disease-associated presenilins are differentially phosphorylated proteins located predominantly within the endoplasmic reticulum. Mol. Med. 2:673–691

Walton, H.S. and Dodd, P.R. 2007. Glutamate-glutamine cycling in Alzheimer's disease. Neurochem. Int. 50(7–8):1052–1066

Wang, J., Xiong, S., Xie, C., Markesbery, W.R. and Lovell, M.A. 2005. Increased oxidative damage in nuclear and mitochondrial DNA in Alzheimer's disease. J. Neurochem. 93:953–962

Wang, J.-Z., Wu, Q., Smith, A., Grundke-Iqbal, I. and Iqbal, K. 1998. *Tau* is phosphorylated by GSK-3 at several sites found in Alzheimer disease and its biological activity markedly inhibited only after it is prephosphorylated by A-kinase. FEBS Lett. 436:28–34

Wang, J.-Z., Grundke-Iqbal, I. and Iqbal, K. 2007. Kinases and phosphatases and *tau* sites involved in Alzheimer neurofibrillary degeneration. Eur. J. Neurosci. 25:59–68

Weidemann, A., Eggert, S., Reinhard, F.B., Vogel, M., Paliga, K., Baier, G., Masters, C.L., Beyreuther, K. and Evin, G. 2002. A novel ε-cleavage within the transmembrane domain of the Alzheimer amyloid precursor protein demonstrates homology with Notch processing. Biochemistry 41:2825–2835

Wells, L., Vosseller, K. and Hart, G.W. 2001. Glycosylation of nucleocytoplasmic proteins: Signal transduction and O-GlcNAc. Science 291:2376–2378

Werther, G.A., Hogg, A., Oldfield, B.J., McKinley, M.J., Figdor, R., Allen, A.M. and Mendelsohn, F.A. 1987. Localization and characterization of insulin receptors in rat brain and pituitary gland using in vitro autoradiography and computerized densitometry. Endocrinology 121:1562–1570

Williamson, R., Scales, T., Clark, B.R., Gibb, G., Reynolds, C.H., Kellie, S., Bird, I.N., Varndell, I.M., Sheppard, P.W., Everall, I. and Anderton, B.H. 2002. Rapid tyrosine phosphorylation of neuronal proteins including *tau* and focal adhesion kinase in response to amyloid-β peptide exposure: Involvement of Src family protein kinases. J. Neurosci. 22:10–20

Wimo, A., Jonsson, L. and Winblad, B. 2006. An estimate of the worldwide prevalence and direct costs of dementia in 2003. Dement. Geriatr. Cogn. Disord. 21:175–181

Wolfe, M.S., Xia, W., Moore, C.L., Leatherwood, D.D., Ostaszewski, B., Rahmati, T., Donkor, I.O. and Selkoe, D.J. 1999a. Peptidomimetic probes and molecular modeling suggest that Alzheimer's γ-secretase is an intramembrane-cleaving aspartyl protease. Biochemistry 38:4720–4727

Wolfe, M.S., Xia, W., Ostaszewski, B.L., Diehl, T.S., Kimberly, W.T. and Selkoe, D.J. 1999b. Two transmembrane aspartates in presenilin-1 required for presenilin endoproteolysis and γ-secretase activity. Nature 398:513–517

Woodgett, J.R. 1990. Molecular cloning and expression of glycogen synthase kinase-3/factor A. EMBO J. 9:2431–2438

Wozniak, M., Rydzewski, B., Baker, S.P. and Raizada, M.K. 1993. The cellular and physiological actions of insulin in the central nervous system. Neurochem. Int. 22:1–10

Xiao, J., Perry, G., Troncoso, J. and Monteiro, M.J. 1996. α-Calcium-calmodulin-dependent kinase II is associated with paired helical filaments of Alzheimer's disease. J. Neuropathol. Exp. Neurol. 55:954–963

Xie, L., Helmerhorst, E., Taddei, K., Plewright, B., van Bronswijk, W. and Martins, R. 2002. Alzheimer's β-amyloid peptides compete for insulin binding to the insulin receptor. J. Neurosci. 22:RC221

Xu, R.-M., Carmel, G., Sweet, R.M., Kuret, J. and Cheng, X. 1995. Crystal structure of casein kinase-1, a phosphate-directed protein kinase. EMBO J. 14:1015–1023

Yamamoto, H., Fukunaga, K., Tanaka, E. and Miyamoto, E. 1983. Ca^{2+}- and calmodulin-dependent phosphorylation of microtubule-associated protein 2 and *tau* factor, and inhibition of microtubule assembly. J. Neurochem. 41:1119–1125

Yamamoto, H., Fukunaga, K., Goto, S., Tanaka, E. and Miyamoto, E. 1985. Ca^{2+}, calmodulin-dependent regulation of microtubule formation *via* phosphorylation of microtubule-associated protein 2, *tau* factor, and tubulin, and comparison with the cyclic AMP-dependent phosphorylation. J. Neurochem. 44:759–768

Yamamoto, H., Saitoh, Y., Fukunaga, K., Nishimura, H. and Miyamoto, E. 1988. Dephosphorylation of microtubule proteins by brain protein phosphatases 1 and 2A, and its effect on microtubule assembly. J. Neurochem. 50:1614–1623

Yamamoto, H., Hiragami, Y., Murayama, M., Ishizuka, K., Kawahara, M. and Takashima, A. 2005. Phosphorylation of *tau* at serine 416 by Ca^{2+}/calmodulin-dependent protein kinase II in neuronal soma in brain. J. Neurochem. 94:1438–1447

Yamazaki, T., Koo, E.H. and Selkoe, D.J. 1997. Cell surface amyloid β-protein precursor colocalizes with β1 integrins at substrate contact sites in neural cells. J. Neurosci. 17:1004–1010

Yasojima, K., Kuret, J., DeMaggio, A.J., McGeer, E. and McGeer, P.L. 2000. Casein kinase 1δ mRNA is upregulated in Alzheimer disease brain. Brain Res. 865:116–120

Ye, S., Huang, Y., Mullendorff, K., Dong, L., Giedt, G., Meng, E.C., Cohen, F.E., Kuntz, I.D., Weisgraber, K.H. and Mahley, R.W. 2005. Apolipoprotein (apo) E4 enhances amyloid β peptide production in cultured neuronal cells: ApoE structure as a potential therapeutic target. Proc. Natl. Acad. Sci. USA 102:18700–18705

Yoshida, H., Hastie, C.J., McLauchlan, H., Cohen, P. and Goedert, M. 2004. Phosphorylation of microtubule-associated protein *tau* by isoforms of c-Jun N-terminal kinase (JNK). J. Neurochem. 90:352–358

Zhang, J. and Piantadosi, C.A. 1992. Mitochondrial oxidative stress after carbon monoxide hypoxia in the rat brain. J. Clin. Invest. 90:1193–1199

Zhang, C., Lambert, M.P., Bunch, C., Barber, K., Wade, W.S., Krafft, G.A. and Klein, W.L. 1994. Focal adhesion kinase expressed by nerve cell lines shows increased tyrosine phosphorylation in response to Alzheimer's Aβ peptide. J. Biol. Chem. 269:25247–25250

Zhang, F., Phiel, C.J., Spece, L., Gurvich, N. and Klein, P.S. 2003. Inhibitory phosphorylation of glycogen synthase kinase-3 (GSK-3) in response to lithium. Evidence for autoregulation of GSK-3. J. Biol. Chem. 278:33067–33077

Zhao, G., Cui, M.Z., Mao, G., Dong, Y., Tan, J., Sun, L. and Xu, X. 2005. γ-Cleavage is dependent on ζ-cleavage during the proteolytic processing of amyloid precursor protein within its transmembrane domain. J. Biol. Chem. 280:37689–37697

Zheng, W.H., Kar, S., Dore, S. and Quirion, R. 2000. Insulin-like growth factor-1 (IGF-1): A neuroprotective trophic factor acting *via* the Akt kinase pathway. J. Neural Transm. Suppl.:261–272

Zheng-Fischhofer, Q., Biernat, J., Mandelkow, E.M., Illenberger, S., Godemann, R. and Mandelkow, E. 1998. Sequential phosphorylation of *tau* by glycogen synthase kinase-3β and protein kinase A at Thr212 and Ser214 generates the Alzheimer-specific epitope of antibody

AT100 and requires a paired-helical-filament-like conformation. Eur. J. Biochem. 252:542–552

Zhou, S., Zhou, H., Walian, P.J. and Jap, B.K. 2005. CD147 is a regulatory subunit of the γ-secretase complex in Alzheimer's disease amyloid β-peptide production. Proc. Natl. Acad. Sci. USA 102:7499–7504

Zhu, X., Rottkamp, C.A., Boux, H., Takeda, A., Perry, G. and Smith, M.A. 2000a. Activation of p38 kinase links *tau* phosphorylation, oxidative stress, and cell cycle-related events in Alzheimer disease. J. Neuropathol. Exp. Neurol. 59:880–888

Zhu, X., Rottkamp, C.A., Raina, A.K., Brewer, G.J., Ghanbari, H.A., Boux, H. and Smith, M.A. 2000b. Neuronal CDK7 in hippocampus is related to aging and Alzheimer disease. Neurobiol. Aging 21:807–813

Zhu, X., Raina, A.K., Rottkamp, C.A., Aliev, G., Perry, G., Boux, H. and Smith, M.A. 2001a. Activation and redistribution of c-jun N-terminal kinase/stress activated protein kinase in degenerating neurons in Alzheimer's disease. J. Neurochem. 76:435–441

Zhu, X., Rottkamp, C.A., Hartzler, A., Sun, Z., Takeda, A., Boux, H., Shimohama, S., Perry, G. and Smith, M.A. 2001b. Activation of MKK6, an upstream activator of p38, in Alzheimer's disease. J. Neurochem. 79:311–318

Zhu, X., Lee, H.G., Raina, A.K., Perry, G. and Smith, M.A. 2002. The role of mitogen-activated protein kinase pathways in Alzheimer's disease. Neurosignals 11:270–281

Chapter 23
Prions and the Transmissible Spongiform Encephalopathies

Richard C. Wiggins

Introduction and Overview

The purpose of this chapter is to highlight the prion protein, which is expressed in a wide range of tissues and most likely has a variety of important cellular functions, and its role in producing the neurodegenerative diseases known as the transmissible spongiform encephalopathies (TSEs). The protein is a normal cellular protein; however, it possess a unique property, so that when the normal α-helix-rich conformation is converted to a misfolded, β-sheet-rich conformation, the resultant particulate protein is infectious and produces a TSE, whose clinical features vary somewhat depending on the species (e.g., cow, sheep, human, etc.). The basic mechanism of infectivity remains mostly unknown, but it seems to be correctly thought of as a slowly progressing type of chain reaction. This model is often described as "recruitment and conversion." The resultant TSEs are akin to other neurodegenerative diseases in that the TSEs are extraordinarily slow progressing diseases in humans and animals. In humans, after presymptomatic periods of as long as several decades, they are fatal and incurable. The unique primary and higher-order structure of the prion proteins of each species imparts a unique character to the TSEs of each species. Additionally, allelic variation at key structural sites appears to impart a unique character to the resultant disease. Variances in the character of disease resultant from these structural differences appear to account for the concept of "prion strains".

While most of the TSEs are relatively rare, a recent epidemic (1986–2000) of the prion disease known as bovine spongiform encephalopathy (BSE), or *mad cow*, thrust the TSEs into the public awareness, especially since the consumption of beef contaminated with infectious central nervous system (CNS) tissue seems to have transmitted the mad cow disease to humans (Will et al., 1996; Collinge et al., 1996; Bruce et al., 1997; Will et al., 1999). The original event that triggered the BSE epidemic will not likely ever be known with certainty; however, it is generally thought to be from the introduction in 1926 in the United Kingdom (BSE Inquiry, 2000) of cattle feed containing offals (including brain and spinal cord) and mammalian meat and bone meal (MBM) animal by-products. It is thought that at some point the process was contaminated with infectious prion material, possibly

D.W. McCandless (ed.) *Metabolic Encephalopathy*,
doi: 10.1007/978-0-387-79112-8_23, © Springer Science+Business Media, LLC 2009

sheep offals infected with scrapie, which precipitated the mad cow epidemic (BSE Inquiry, 2000). Scrapie disease in sheep is also a TSE, and it has been epidemic in sheep in the United Kingdom for 200 years. The removal of offals and mammalian meat and bone meal additives appeared to end the epidemic, which supports the offal—scrapie theory for the origin of the epidemic; however, there are other theories (Chesebro, 2004). There are also indications that the epidemic may have featured multiple strains of the prion protein (Capobianco et al., 2007). The transmitted disease in humans is known as a new variant of the Creutzfledt–Jacob disease (vCJD). The mad cow outbreak peaked in 1992 and all but disappeared in the United Kingdom by 2000. Because the incubation period for the TSEs is typically extraordinarily long, years or even decades, speculation abounds as to whether vCJD in humans has already peaked or remains dormant in the form of a future epidemic waiting to emerge (Brown, 2001).

There is currently a less well publicized epidemic of the prion disease (*chronic wasting disease*, CWD) in Western deer and elk populations in the Rocky Mountain regions of the United States. Its rate of spread and appearance from the Dakotas to New York indicate that the disease may soon extend nationwide, and it may transmit to other animals, such as moose.

The TSEs are not recent in their origins. Scrapie disease in sheep has been known for hundreds of years, and the human prion disease, *kuru*, has been well known for some time in the Fore-speaking population of New Guinea, where it was attributed to the funeral ritual of cooking and eating dead relatives, especially by the women and children. The infectivity of these diseases was shown by Gajdusek, who received the Nobel Prize in 1976 for demonstrating that kuru was transmissible by inoculating chimpanzees with brain material from human kuru patients. At that time, the nature of infectious pathogen was completely unknown, although it was commonly assumed that the TSEs were caused by atypical "slow viruses. " Prusiner (Bolton et al., 1982; Prusiner, 1982; Diener et al., 1982) used the sheep scrapie model and provided evidence for the fact that the TSEs are actually a novel type of infectious disease caused by proteinaceous infectious particles, which he named *prions* (Prusiner, 1982). Although contrary to prevalent scientific wisdom, a number of investigators had thought for some time that the scrapie disease agent was a protein and the idea was known as the "protein only" hypothesis. Prusiner received the Nobel Prize (1997) for his discovery of the prion, which confirmed the protein-only hypothesis.

Transmission of Prion Diseases Between Humans and Animals

Human TSEs are rare but include the Creutzfeldt–Jacob disease (CJD), Gertstmann–Straussler–Scheinker (GSS) disease, and familial fatal insomnia. All three human diseases appear to have both familial and sporadic occurrence, and appear to be different manifestations of pathology of the same prion protein. Kuru is the human disease caused by ritual cannibalism of brain. Similarly, human vCJD appears to have originated in the mad cow BSE epidemic in the United Kingdom. The hypothesis is that BSE manifested in humans as vCJD after eating beef contaminated with infectious CNS material. Human kuru and familial fatal insomnia are also known to be transferable across species, as shown by the inoculation of laboratory animals with infectious human CNS tissue.

However, it appears that not all TSEs are transferable to all species. Scrapie disease in sheep is not known to be infectious in humans, although it seems infectious in cows. Thus, there appear to be species barriers to infectivity. The potential infectivity of CWD disease to humans is not known, although testing of game animals taken by hunters in the CWD-infected areas has been encouraged. The mechanism for the horizontal spread of CWD in deer and elk is without explanation, and it is not clear if transmission to humans is possible, or what the disease would look like if it were transmitted (Xie et al., 2006). It is intriguing to speculate that prion infectivity could leap from one species to another nonpermissible species by way of a permissible intermediate, possibly as evidenced by sheep scrapie to mad cow to human vCJD.

All evidence points to TSEs as a family of related diseases caused by the same homologous prion protein in all species (see reviews by Brown, 2005; Soto, 2006; Watts et al., 2006). Some comfort can be taken that the sporadic occurrences of TSEs are rare in all species and in human populations; however, it is equally clear from recent events that a persistent threat of epidemics of TSEs exists. Such epidemics have potentially enormous consequences, such as the economically catastrophic epidemic of mad cow disease in the United Kingdom. Thus far, there is no early diagnosis during the long period of latency, although current research is demonstrating the possible presence of infectious prion protein in tissues at some time before clinical symptoms appear.

Given the fundamental uncertainties about (1) what causes sporadic TSEs, (2) what initiated the epidemic of BSE, and (3) the unknown vector for the current epidemic of CWD, one has to be concerned about long-term health risks to humans and economically important animal species. The long latency between exposure (such as by eating contaminated meat) and the progression of the clinical disease is especially troublesome. Prudence requires us to consider two unwelcome hypotheses for the acquisition of TSEs from animals:

1. The human population has been seeded to an unknown extent with latent infectious prion protein from the mad cow epidemic (Enserink, 2005) and possibly also from the CWD epidemic now underway in deer and elk.
2. An unknown environmental or animal reservoir might exist and present a continuing threat to humans and animals (Johnson et al., 2006).

Gaps and Enigmas in the Prion Concept

Because of the economically catastrophic mad cow epidemic and the ensuing risk to humans, intense research has been brought to focus on prions and the corresponding TSEs. Likewise, basic research on the normal prion protein has accelerated.

However, even after nearly three decades of intense biomedical research, the prion story remains enigmatic:

- The prion protein is actually a normal protein that is expressed in many tissues, yet it is particularly highly expressed in neurons in the brain to perform functions that remain unknown;
- The prion protein produces disease only when it adopts a misfolded configuration, yet how misfolding occurs remains unknown;
- In addition to the transmissible form, the prion diseases include familial and sporadic (isolated) occurrences;
- When normal brain tissue is inoculated with misfolded prion protein, normal prion protein in the host is converted to the misfolded state, yet precisely how this recruitment of normal protein and its conversion take place remains unknown;
- Prion disease is potentially "infectious" by eating contaminated food, yet how the pathogenic prion protein passes from the gut to infect the brain remains uncertain;
- The time between consumption of prion-contaminated food and the onset of the clinical symptoms of spongiform encephalopathy is exceedingly long (Prusiner et al., 1982), yet what happens during the years between the consumption of contaminated food and the onset of clinical symptoms remains unknown;
- Prion diseases occur through different strains, as evidenced by different progression patterns and clinical features of the disease that seem to be associated with different genetic and/or post-translational polymorphisms in the prion protein, yet how these differences translate into different disease outcomes remains unknown.

Prion Disease Prevention

While the prion encephalopathies remain relatively rare in humans, public concern has been heightened recently by the outbreak of an economically catastrophic epidemic of mad cow disease in the United Kingdom (1986–2000) and a number of isolated, sporadic occurrences in cows in the United States and Canada. Because of the potential for enormous risk to public health, disease prevention is aimed at protecting the public from the consumption of contaminated food, or the accidental inoculation with contaminated clinical materials or surgical instruments. Another troublesome problem is that the infectious conformation of the prion protein is extraordinarily resistant to normal cleaning and degradation procedures, so that contaminated tissues and work environments are exceedingly difficult to decontaminate (Taylor, 1990; Brown and Gajdusek, 1991; Zobeley et al., 1999; Brown, et al., 2004). Given the full extent of the potential routes of exposure and transmission to humans, the risk from the mad cow epidemic remains unknown because there are so many uncertainties (Brown, 2001). While a spike of human clinical disease was observed immediately in the wake of the mad cow epidemic, the prion diseases are known to have long incubation periods. Therefore, one must consider the possibility that this rapid onset of vCJD is evidence that human population has been "seeded" to an unknown extent, which presents long-term concerns. Alternatively, the spike of vCJD in the immediate wake of the BSE epidemic may reflect a pattern of more rapid disease progression in highly susceptible individuals, attributable to age, genetic polymorphisms in the prion protein, etc., while other similarly exposed individuals may prove susceptible only after long latency.

The potential for presymptomatic detection of prions in the blood of infected individuals has profound importance for humans exposed, or potentially exposed, to contaminated food. Saá et al. (2006) show that the scrapie disease prion (PrP^{Sc}) can be detected biochemically in the blood of scrapie-infected hamsters during most of the presymptomatic phase of the disease. Infectious prions have also been found in the blood of CWD-positive deer (Mathiason et al., 2006). Because the human population may have been seeded with vCJD disease as a result of the BSE epidemic, the potential to detect prions in presymptomatic individuals would seem to have profound importance for human blood donation and organ transplantation, as well as for the screening of farm and game animals. The potential for development of immunity (Peretz et al., 2001; Heppner et al., 2001; Pankiewicz et al., 2006) or of effective vaccines (Sadowski and Wisniewski, 2004) may open the door to effective management or cure, which are at present out of reach.

The Mechanism of Protein Infectivity

More than 20 years after the beginning of work to unravel the prion mystery, the critical mechanisms of disease production still remain enigmatic. Prior to Prusiner's work, the spongiform encephalopathies now known to be caused by prions were commonly thought to be caused by undiscovered "slow viruses", since the incubation period for the appearance of clinical disease seemed to span decades. During the search for these pathogens, Prusiner (Bolton et al., 1982; Prusiner, 1982) discovered the proteinaceous infectious particles that cause disease, and he proposed the name *prion*. Most importantly, this work, for which Prusiner received the Nobel Prize (1997), lent decisive support for a contentious, earlier "protein only" hypothesis for disease transmission. While it is now clear that the normal isoform of the prion protein has important cellular functions, how the misfolded isoform produces a lethal spongiform encephalopathy remains elusive in detail. The progression of brain disease seems to involve cellular functions related to the normal protein in its neural cell-surface glycosylphosphadidylinositol (GPI) linkage (Chesebro et al., 2005), as when this linkage is broken, TSE^{Sc} mice accumulate plaques but never develop clinical disease (Aguzzi, 2005). In other words, the β-sheet-rich prion isoform is not intrinsically neurotoxic. Toxicity is imparted by disrupting the signaling and/ or metabolic functions of the normal GPI-linked protein. Furthermore, prion protein gene knockouts (PRNP$^{0/0}$) seem to be asymptomatic upon inoculation with scrapie prions (Brandner et al., 1996).

At the heart of the prion enigma and the protein-only hypothesis is the mechanism of infectivity of the prion particles. Part of the mystery of TSEs is that, in addition to infectious transmission, there are also instances of sporadic (unknown cause) as well as genetic instances of occurrence. Furthermore, human and nonhuman TSEs have strains in which the clinical course of disease features strain-specific variability. The complexity of these strain differences in disease manifestation provide a complicated accounting of mechanisms under the protein-only hypothesis (review Soto, 2006), although genetic polymorphism PrP in humans and animals almost certainly contributes to "strain" specific differences (Schoch et al., 2006). Because of the difficulty intrinsic to the idea of a single protein adopting multiple conformations, Safar et al. (2005) searched exhaustively for small polynucleotides possibly hidden within PrP^{Sc} (Fig. 23.1) particles and found that the small polynucleotides found in prion preparations are of host origin and cannot account for infectivity or strain differences.

An additional difficulty is that other neurodegenerative diseases that feature protein misfolding similar to TSE, such as Parkinson's disease, Alzheimer's disease, etc., are not thought to be infectious, as is the case with the prion proteins. However, it now seems possible that the assumption of noninfectivity in other

Protein-Only Infection

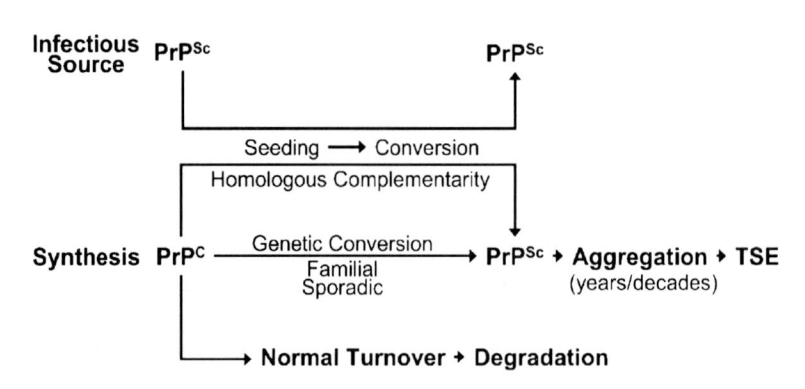

Fig. 23.1 Hypothetical mechanisms of protein-only infectivity. Exposure to infectious PrP^{Sc} is thought to seed and convert native PrP^{C} protein in normal brain. Here, PrP^{Sc} from an exogenous infectious source, such as contaminated food, seeds and converts the normal cellular pool of endogenously synthesized PrP^{C} from the native PrP^{C} conformation to the infectious PrP^{Sc} conformation. Amino acid sequence homology between the infectious source of prion and the host PrP^{C} permits crossing the species barrier, such as in the case of BSE spreading to humans in the form of vCJD. Likewise in this concept, infectious PrP^{Sc} from animal species that lacks critical features of homology with the amino acid sequence of human PrP would not be infectious to humans, such as scrapie disease in sheep. Once PrP^{C} has been converted to PrP^{Sc} through the slow process of seeding and conversion, the misfolded protein is no longer recognized by the normal processes for turnover and degradation of cellular proteins. Consequently, the brain burden of PrP^{Sc} increases, producing a gradual accumulation of plaques in the brain and eventual clinical disease

fibrillar plaque diseases may be for lack of looking. For example, experimental inflammatory amyloidosis in mice is accelerated when the mice are also administered purified amyloid fibrils (Lundmark et al., 2002). Similar seeding was observed in mice administered extracts from human Alzhemier's brain (Meyer-Luehmann et al., 2006). So, seeding and transmission (or infectivity) may be a pathogenic mechanism most robustly shown with the prion protein, but potentially and weakly applicable to other fibrillar plaque proteins as well. The idea of a basic similarity of the seeding, misfolding, and aggregation processes of Alzheimer's disease and the prion diseases is, in fact, supported by sequence similarities in the presumptive amyloidogenic domains of the β-amyloid protein in Alzheimer's disease and PrP in the TSEs (Come et al., 1993).

There is no known cure for prion disease; however, optimism is offered by experimental observation that noncytotoxic dosages of branched polyamines cure prion-infected neuroblastoma cells in culture (Supattapone et al., 1999, 2001), apparently by rendering the aggregated protein to normal proteolytic process in the cell.

Origin and Acceptance of the Prion Hypothesis

Before the discovery of the prion mechanism of infectivity, the assumption was that kuru in humans and scrapie in sheep were caused by unconventional viruses, virus-like pathogens, or "slow viruses, " because the interval between exposure and onset of clinical disease was years, or even decades (Prusiner, 1982). However, it is important to distinguish between the actual discovery of the prion particles (Bolton et al., 1982; Prusiner, 1982) and the "protein only" hypothesis for the transmission of the diseases now attributed to prions – the TSEs. Specifically, long before the discovery of the prion particles, the replication and transmission of the scrapie disease agent without nucleic acid was suspected (Alper et al., 1966, 1967), leading Griffith (1967) to propose a self-replicating, protein-only agent for disease transmission.

During the search for the actual pathogen responsible for the TSEs, Prusiner (1982) coined the term prion to distinguish between the molecular properties of proteinaceous infectious particles associated with scrapie disease (Bolton et al., 1982) and the molecular properties of known viruses, plasmids, viroids, etc. that feature a core of nucleic acid as the informational molecule. In the original description, the purified prion particles were characterized as composed of a single protein, roughly 27,000–30,000 D in molecular weight, and resistant to proteinase K digestion. In particular, enrichment of the protein correlated with increasing titer of the infectious agent (Bolton et al., 1982) and the material failed to exhibit the spectra indicative of the presence of nucleic acids (Prusiner et al., 1984).

Because the idea of an infective protein is such a radical departure from more than a century of belief that all infectious diseases are caused by microorganisms, such as bacteria and viruses, acceptance of the prion hypothesis was slow. Key links in gaining almost universal acceptance of the prion hypothesis of infectious disease include several recent studies:

- Demonstration in knock-out mice that expression of the prion gene is necessary for the establishment and progression of disease (Brandner et al., 1996);
- Demonstration that synthetic polypeptides of prion protein produced in a genetically engineered *E. coli* cell-free system produced a TSE disease in transgenic mice overexpressing prion protein (Legname et al., 2004);
- Conformation in vitro of the core idea that normal prion protein can be changed to an infectious conformation by seeding normal brain homogenates with scrapie protein and amplification of the infective protein by cyclic amplification (Castilla et al., 2005).

Suspicion that the TSEs might yet have a viral origin has not disappeared, and it is possible to point to weaknesses in the prion studies to sustain the possibility that accidental contamination by nucleic acids best explains the observed infectivity. While the use of transgenic mice overexpressing PrP (Legname et al., 2004) provides a more robust experimental model for the development of disease after inoculation with synthetic prion protein, others see transgenic mice overexpressing prion protein as predisposed to producing prion disease spontaneously. In what some might consider direct support for a viral origin of the prion diseases, Manuelidis et al. (2007) reported the presence of 25-nm virus-like particles in cell cultures that also propagate TSE infectivity. Similar particles in the range of 20–35 nm have been often reported in TSEs; however, no viruses or virus-like particles have ever been purified and shown to produce TSE in animals. Although the papers of Manuelidis are important reminders of the need for caution, the prion hypothesis appears to remain incontrovertible to most scientists by the sheer weight of positive evidence.

Through the work of more than two decades, it is now known that the prion protein is a misfolded isoform of a normal cell-membrane protein, and various terminologies have appeared as numerous researchers have joined the investigations (Liberski and Ironside, 2005). The normal isoform of the cellular protein is designated PrP^C, whereas the infectious isoform in scrapie is PrP^{Sc}. Because there are a variety of isoforms producing different transmissible spongiform encephalopathies, PrP^d has been introduced recently to connote any disease isoform. Many authors seem to use PrP^{Sc} as a generic for any of the TSEs. Also, PrP^{res} has been used to connote a purified protein or transgenic polypeptide fragment that features the classical resistance to proteinase K digestion.

Function of Normal Prion Protein

The normal PrP polypeptide is encoded by a highly conserved gene, which is variously abbreviated as PRNP, prnp, Prnp, etc. (for a review, see Sakaguchi, 2005). The prion gene is located on chromosome 20 in humans and on chromosome 2 in mice (Oesch, et al., 1985; Basler, et al., 1986), and it appears to be conserved across species from yeast to primates (Westaway and Prusiner, 1986), indicating the PrP proteins perform essential cellular functions spanning almost the whole time span of living organisms on earth. Across the vertebrates, amino acid sequence homology between frog, turtle, chicken, and mammal is conserved at a rate of about 30%, which is sufficient to preserve the same molecular architecture of mammalian PrP^C (Harris et al., 1991; Calzolai et al., 2005). In mammals PrP is expressed in many cell types and tissues; however, it is comparatively highly expressed in brain neurons, especially in the synapses. In particular, PrP^C appears to be associated with the development of neural precursor cells in the developing nervous system, as well as their differentiation into mature neurons (Steele et al., 2006). In the adult nervous system, PrP^C may normally potentiate acetylcholine release in neuromuscular junctions (Re et al., 2006) and may have an important role in regulation of receptor density at the neuormuscular junction (Harris et al., 1991). A role in memory has also been postulated (Bailey et al., 2004).

In its CNS expression, PrP^C is a GPI-anchored cell-membrane protein (for reviews, see Brown, 2005; Soto, 2006), whereas the infectious prion is a misfolded isoform of the same polypeptide (Stahl et al., 1990). The normal and infectious isoforms have identical amino acid sequences. The GPI linkage has been characterized, but the purpose of the linkage remains unknown. Mouillet-Richard et al. (2000) have postulated that PrP^C may be a signal transduction protein. The GPI linkage does not appear to have a role in the formation of the infections prion (Lewis et al., 2006), although the GPI-linked PrP^C is necessary for disease progression (Chesebro et al., 2005).

PrP^C is also a copper-binding metalloprotein (Hornshaw et al., 1995; Brown et al., 1997), and it may function as an antioxidant. Physiologically relevant copper binding now appears to be a widely accepted, normal function of PrP^C (for a review, see Soto, 2006). Furthermore, copper binding at the β-sheet conversion site my actually stabilize the normal PrP^C isoform and prevent the misfolding that produces the β-sheet-rich infectious prion (Cox et al., 2006). PrP has also been shown to confer anti-apoptotic activity in cultured neuronal cells (Kuwahara et al., 1999; Lee et al., 2006).

Because misfolded isoforms of PrP produce fatal spongiform encephalopathies, the PrP has largely been studied in the brain. However, new evidence indicates PrP^C has important functions in non-neural tissues. Zhang et al., (2006) show that the prion protein is a biomarker for hematopoetic stem cells, where it is normally expressed on the cell surface. When bone marrow is depleted of prion-protein-expressing stem cells, the capacity to regenerate is also lost. This evidence indicates normal prion protein expression is required for bone marrow stem cells to regenerate. Because prion protein is expressed in blood, especially mononuclear cells, blood transfusion has been implicated as a route of transmission of infectious prions (Halliday et al., 2005). Platelets have been variously described as either expressing (Perini, 1996; Robertson et al., 2006) or not expressing prion protein (Herrmann et al., 2001), and it is probable that prion expression in platelets is derived from precursor cells in the megakaryocyte lineage (Starke et al., 2005). Doubt has been expressed that platelets are a source of transmissible infection (Holada et al., 2002). The normal function of PrP^C in mononuclear cells remains unresolved.

PrP^C is also constitutively expressed in the human gut; however, its function there remains elusive. Infection by *Helicobacter pylori*, as is the case in 50% of the world's population, significantly upregulates PrP^C expression, and may be an important factor in human susceptibility to infectious prions in contaminated food (Konturek et al., 2005).

Mechanism of the Prion Protein Misfolding and Disease

The transmission, or "infectivity" of prion diseases appears inseparable from the mechanisms of fibrillogenesis, in which the native form of PrP, rich in α-helical domains, is converted to PrP^{Sc}, which is rich in β-sheet domains. The central concept that a protein molecule, specifically, a single polypeptide chain, is infectious requires a fundamental change in the traditional view of infectious agents. While the aggregation of prion particles is probably similar to the formation of aggregate neurotoxic proteins found in other neurodegenerative diseases (Taylor et al., 2002), the mechanism of the prion protein's infectivity remains speculative, although the fact of infectivity appears inescapable. It is clear is that neurotoxic PrP isoforms are produced by misfolding of normal PrP^C, leading to the aggregation of infectious PrP^{Sc} prion particles and their lethal accumulation in neurons.

As in the case of the prion diseases, the spectrum of amyloid diseases features the conversion of native proteins to a fibrillar isoform comprising predominantly β-sheet structures. There is recent evidence that PrP may be the most striking example of a potential that other fibrillar proteins also have to seed the conversion of the native conformation of the same, or of a similar protein, to the fibrillar conformation (Lundmark et al., 2002; Meyer-Luehmann et al., 2006). For example, whether administered orally or i.v., purified amyloid-A protein is a potent enhancing factor in the disease progression of amyloid-A protein amyloidosis in mice. As noted by Lundmark et al., (2002), the fact that amyloidosis can be accelerated by eating a homologous protein molecule indicates a seeding and propagation phenomenon remarkably akin to the transmission of prion diseases.

Events that initiate misfolding of the native PrP^C polypeptide remain mostly unknown; however, it has been shown in a model system of neuroblastoma cell cultures that normal PrP^C is precursor to PrP^{Sc} (Borchelt et al., 1990). These studies employed a 2-h pulse of [^{35}S]-methionine to label the newly synthesized PrP^C. Kinetic analysis of [^{35}S]-radioactivity in purified protein showed that labeling peaked immediately in the PrP^C pool and rapidly declined to less than 90% of peak activity within 14–22 h, whereas radioactivity was still increasing after 46 h in the PrP^{Sc} pool. Given a pulse duration of only 2 h, followed by chase, it is reasonable to assume labeled PrP^C is the direct precursor of labeled PrP^{Sc}. Furthermore, studies with recombinant PrP polypeptides show that α-helix-rich polypeptides, which are characteristic of the native PrP^C configuration, can be reversibly converted to monomeric β-sheet-rich conformations in the absence of denaturing additives (Jackson et al., 1999). The authors interpret the ease with which monomeric helical PrP polypeptides can be converted in high yield to β-sheet isoforms as evidence that the amino acid backbone of PrP may actually have a thermodynamic preference for the β-sheet isoform. If so, then native PrP[C] would have to be held in the native isoform, rich in helical structure, by protective microenvironments.

Such a concept raises possibility that the α-helical-rich and β-sheet-rich isoforms could be held in dynamic balance with an intervening high energy barrier. Such a barrier would have to be overcome as PrP[C] is converted to PrP[Sc]. Keeping in mind that these are polypeptides produced through recombinant DNA methods, this hypothesis of an energy barrier separating the α-helical PrP from an even more stable β-sheet conformation would explain transmission and infectivity of prion disease as a breakdown in this energy barrier, such as by seeding with PrP[Sc] (Come et al, 1993). In such a view, seeding would depend on amino acid sequence homology, so that dissimilar sequences would account for species barriers in the transmission of the TSEs (Krebs et al., 2004). In the absence of seeding (as in the case of consuming contaminated food or inoculating experimental animals with PrP[Sc]), a breakdown in the energy barrier would ordinarily be a rare but readily available mechanism for the occurrence of sporadic prion disease. Likewise, the hypothesis might also account for certain genotypes that would predispose conversion of helical PrP to β-sheet-rich isoforms, such as Met/Met or Val/Val homozygotes at position 129 in humans.

Several breakthroughs have been made in the several years, revealing how toxic aggregates of prion isoforms kill neurons. This work resolves the question of whether scrapie is produced by direct neurotoxicity of the PrP[Sc] or by some other mechanism. It is now known that the infectious PrP[Sc] polypeptide is not intrinsically neurotoxic to neurons. Development of prion disease requires the presence of normal PrP expression. Furthermore, uncoupling normal PrP from the cell surface also uncouples the PrP[Sc] from the progression of disease. Key steps in this discovery include the following:

- Grafting transgenic tissue overexpressing PrP into embryonic forebrain of PrP deficient mice (PRNP[0/0]) and inoculating with PrP[Sc] shows that grafted tissues accumulate PrP[Sc] and exhibit scrapie histology. Even though substantial amounts of graft-derived PrP[Sc] migrated into PRNP[0/0] tissue, no pathology appeared in Prnp[0/0] tissue after 16 months (Brandner et al., 1996).
- Arresting the conversion of PrP[C] to PrP[Sc] in neurons during infection reverses disease progression, even though PrP[Sc] continues to accumulate (Mallucci et al., 2003). Transgenic mice engineered to disable the PRNP gene in neurons at 12 weeks of age were inoculated with PrP[Sc] several weeks before PRNP inactivation. Within days after the PRNP gene was inactivated, the mice became depleted of PrP[C], and PrP[Sc] accumulation reached levels in brain tissue typical of terminally ill animals. Nonetheless, the PRNP-inactivated animals remained alive and disease free a year later, despite high PrP[Sc] levels.

    Inoculated control mice lacking the transgenic inactivation of PRNP all died during this period.

- Disruption of normal PrP function in vivo by binding with monoclonal antibodies produces rapid cell death in neurons (Sulforosi et al., 2004), which is consistent with indications that prion disruption of the normal PrP function in vivo would also produce neuronal cell death.
- Normal PrP protein is anchored to the neuronal cell membrane by a GPI linkage. Disabling the GPI linkage in transgenic mice does not prevent the accumulation of PrP^{Sc} plaques in inoculated mice, but does prevent the development of scrapie disease (Chesebro et al., 2005).

In vivo, there are genetic predispositions to PrP^{Sc} aggregate formation, which would account for the so-called familial varieties of prion diseases. For example, from studies of human kuru, it is known that homozygous Met or Val at codon 129 in the human *PRNP* gene predisposes the PrP protein to misfolding. In the case of kuru disease in the Fore-speaking population in the remote highlands of Papua New Guinea, homozygotes for Met or Val at position 129 have an earlier onset of disease following the consumption of contaminated CNS tissues during ritual cannibalism (Lee et al., 2001; Mead et al., 2003). Similarly, Wadsworth et al. (2004) show that sequence variation at position 129 drastically alters the infectivity and clinical consequences of BSE and vCJD. Given that all the vCJD deaths in the aftermath of the mad cow epidemic in the United Kingdom have been Met/Met homozygotes (Brown et al., 2001), there may be a large and undetected pool of dormant carriers of vCJD in the population (Carrell, 2004). Possibly, the infectious strain of BSE in the mad cow epidemic may not be able to replicate in other human genotypes (Brown et al., 2001). Heterozygosity at codon 129 confers some resistance to prion disease, possibly by decelerating homologous protein interactions that might lead to misfolding. Worldwide, genetic polymorphism for the *PRNP* gene reveals high allelic frequencies for TSE-favoring combinations. This high frequency of apparently unfavorable genotypes indicates a dynamic balance of strong positive and negative selection pressures for different alleles of *PRNP* during the evolution of humans.

    While there is a clear negative selection against homozygosity of Met and Val at codon 129 because of the predisposition to prion disease, the positive selection pressures that would account for the high frequencies of these alleles remain unknown – presumably hidden within the still unknown full range of normal PrP^{C} functions in a wide range of cells and tissues. As reviewed in Mead et al. (2003), Met homozygotes comprise 37% of the human population in the United Kingdom, and Val homozygotes comprise 12%. Once the chain reaction of misfolding begins, because of sporadic disease, familial predisposition, or the consumption of contaminated foods, the progression of disease from cell to cell and to new hosts appears accountable in general principles, but remains unresolved in precise detail.

    In the case of CJD, it has been shown that the wide range of patterns in the expression of the disease is associated with combinations of point mutations and insertion mutations, which have familial transmission (Goldfarb et al., 1991). Various numbers of coding repeats within an unstable region are a prominent feature of the mechanism of disease.

Progression from Exposure to Disease

While there appears to be little doubt that infectious PrP^{Sc} particles, once in the brain, initiate the slow progression of a TSE disease, the picture of how ingested PrP^{Sc} from contaminated food sources (such as mad cow or kuru) reaches the brain is not clear. In particular, the time course of the TSEs, over periods of years and even decades, is also without clear explanation.

While PrP^{C} is expressed in many tissues in the body, CNS neurons may be the only cells killed by infectious PrP^{Sc}. Whereas cells in most tissues are renewed and turn over comparatively rapidly, neurons are nonproliferating cells in most of the brain. Cells in non-neural tissues may simply turn over so rapidly that a disease process requiring years may not have time to become evident. In contrast, the nerve cells produced during the period of brain growth and development remain alive for the full life span of a mammal. So, it seems reasonable that such long-lived cells would be selectively vulnerable to a slow progressing disease that requires years, or decades, to produce clinical effects. It also seems plausible that PrP^{Sc} may not initiate a recognizable disease process in other cells and tissues because of its positive selectivity for nerve cells. The unique functional roles and corresponding metabolic signatures of nerve cells may render neurons particularly susceptible to disorders caused by the prion protein. What is known is that non-neural brain cells respond differently. For example, the accumulation of PrP^{Sc} is lethal to neurons, but astrocytes are stimulated to proliferate (DeArmond et al., 1992).

The central issue in the progression of disease is how PrP^{Sc} propagates itself from the source of exposure (food) to the brain. In the examples provided by kuru, the recent mad cow epidemic in the United Kingdom, and the resultant transmission of vCJD to humans, the common link is that infectious prions in contaminated food reached the brain by some mechanism, and once in the brain, seeded the conversion of native host PrP^{C} to an infective PrP conformation.

The lymphatic system is implicated in some studies. Lemurs experimentally inoculated with food containing BSE-infected cattle brain demonstrated PrP immunohistochemical staining in the tonsils, gut and associated lymphatic tissues, spleen, and cervical dorsal and ventral nerve roots, whereas control lemurs did not show detectable PrP staining (Bons et al., 1999). Because of the importance of resolving the mechanism of disease by oral exposure, Jeffrey et al. (2006) show that ingested prion protein in gut is quickly cleared and taken up into the lymphatic system. However, a slower process of de novo formation of infectious prion particles seems to occur in parallel, involving the Peyer's patches. These results indicate that the lymphatic system and the enteric nervous system are exposed to PrP^{Sc} simultaneously; however, the route by which PrP^{Sc} reaches the brain is not clear. Interestingly, transmission along nerve fibers would also account for how experimental intracerebral inoculation with PrP^{Sc} spreads peripherally to other tissues (DeJoia et al., 2006).

Horizontal Spread of Prion Disease

In the examples of kuru, mad cow disease, and vCJD, food contaminated with infected CNS components seems to account for the horizontal spread of disease between individuals. High titers of infectious prions are present in the CNS of diseased animals, so that infected brain tissues have been used in countless experiments to transmit TSEs to normal animals by inoculation. The potential infectivity of other tissues from infected animals appears unclear. For example, muscle tissue from deer infected with CWD successfully transferred the prion disease to mice inoculated intracerebrally (Angers et al., 2006). Whereas PrP^{Sc} has also been detected in muscle of scrapie-infected sheep, there has been no similar detection of prion in muscle in BSE-affected cows, or in deer with natural CWD. It seems likely that these results might differ because of relatively low levels of prion expression in muscle compared to brain, or because a far less aggressive disease process manifests in muscle tissue. In either case, the immunohistochemical detection of prion protein is likely to be more difficult in muscle. Angers et al. (2006) find relatively high titers of prion in muscle of innoculated mice, and high titers in muscle also correlated with the most rapid progression of brain disease. However, when infectivity of prions from infectious muscle and brain were compared, the incubation time was 30% longer for prions from muscle than with CNS prion preparations from the same animals.

It has also been observed that chronic lymphocytic inflammation from other underlying disease causes the prion protein to be expressed in otherwise prion-free tissues (Heikenwalder et al., 2005), which might have significant implications for the potential infectivity of non-neural tissues. For example, while the prion protein is not normally excreted in urine, chronic inflammatory kidney disease does result in prionuria in PrP^{Sc}-infected mice. Inoculation of noninfected mice with urinary PrP^{Sc} produced scrapie (Seeger et al., 2005), which might be a source of horizontal prion infectivity in animal populations, such as the epidemic of CWD in North American deer and elk. Infectious prions have also been detected in the saliva of CWD-positive deer and may contribute to the spread to other individuals (Mathiason et al., 2006). The stability of infectious prions in soil, possibly for years, indicates a potential for an environmental reservoir of prions (Brown and Gajdusek, 1991; Johnson et al., 2006).

Conclusions

The prion protein is a normal cellular protein. It is expressed in many tissues from a highly conserved gene, although the levels of expression in neurons are relatively high. In the brain, the protein appears to be a GPI-linked cell-surface glycoprotein, and it may also be a copper-binding metalloprotein. It is relatively rich in α-helix domains, but the protein possesses a unique property, so that under certain conditions the protein converts to a relatively β-sheet-rich conformation. In this new conformation, the protein particles are known as prions, which exhibit a unique principle of infectivity and produce the class of diseases known as TSEs. The basic mechanisms of conversion and infectivity remain mostly unknown, but it seems to be a chain reaction type of propagation, in which protein in the prion conformation seeds and converts normal protein. This model is often described as recruitment and conversion. The resultant TSEs are plaque-forming diseases in the brain and they are extraordinarily slow progressing. The accumulation of prion plaques can be uncoupled from the progression of clinical disease by disrupting the GPI linkage of the normal protein. How prions in food produce disease in the brain is unknown, but after periods of several to many years, prions in the brain produce a fatal and incurable spongiform degenerative disease. Similarities to other plaque-forming diseases, such as Alzhemier's, have often been noted, although other plaque proteins are not known to be infectious, or at least not as robustly infectious as the prion protein. The primary means of prevention is to protect food from contamination with infectious CNS material. The recent BSE epidemic in the United Kingdom appears to have subsided; however, the long-term risk to humans from consumption of contaminated beef remains unknown. In humans, certain genetic polymorphisms in the prion protein, especially homozygotes for Met and Val at position 129, are predisposed to rapid development of clinical disease. Enigmatically, the allele frequencies for either Met or Val at 129 in most human populations are high, which indicates strong positive selection pressure for these variants at some time in the antiquity of humans. In the wake of the mad cow epidemic, long-range implications of TSEs for human organ transplantation, blood donation, etc. remain unclear, although prudence would dictate extreme caution. The development of presymptomatic detection methods, or even vaccines, is an active area of research. A growing epidemic of CWD is currently underway in North American deer and elk, although neither the cause nor the implications to human are known.

Disclaimer: This manuscript has been reviewed by the National Health and Environmental Effects Research Laboratory and approved for publication. Approval does not signify that the contents reflect the views of the Environmental Protection Agency.

References

Aguzzi, A. 2005. Prion toxicity: all sail and no anchor. Science 308:1420–1421

Alper, T., Haig, D. A., and Clarke, M. C. 1966. The exceptionally small size of the scrapie agent. Biochem. Biophys. Res. Commun. 22: 278–284

Alper, T., Cramp, W. A., Haig, D. A., and Clarke, M. C. 1967. Does the agent of scrapie replicate without nucleic acid? Nature 214: 764–766

Angers, R. C., Browning, S. R., Seward, T. S., Sigurdson, C. J., Miller, M. W., Hoover, E. A., and Telling, G. C. 2006. Prions in skeletal muscles of deer with chronic wasting disease. Science 311: 1117

Bailey, C. H., Kandel, E. R., and Si, K. 2004. The persistence of long-term memory: a molecular approach to self-sustaining changes in learning-induced synaptic growth. Neuron 44: 49–57

Basler, K., Oesch, M., Scott, M., Westaway, D., Walchli, M., Groth, D. F., McKinley, M. P., Prusiner, S. B., and Weissmann, C. 1986. Scrapie and cellular PrP isoforms are encoded by the same chromosomal gene. Cell 46: 417–428

Bolton, D. C., McKinley, M. P., and Prusiner, S. B. 1982. Identification of a protein that purifies with the scrapie prion. Science 218: 1309–1311

Bons, N., Mestre-Frances, N., Belli, P., Cathala, F., Gajdusek, D. C., and Brown, P. 1999. Natural and experimental oral infection of nonhuman primates by bovine spongiform encephalopathy agents. Proc. Natl. Acad. Sci. 96: 4046–4051

Brandner, S., Isenmann, S., Raeber, A., Fischer, M., Sailer, A., Kobayashi, Y., Marino, S., Weissmann, C., and Aguzzi, A. 1996. Normal host prion protein necessary for scrapie-induced neurotoxicity. Nature 379: 339–343.

Borchelt, D. R., Scott, M., Taraboulos, A., Stahl, N., and Prusiner, S. B. 1990. Scrapie and cellular prion proteins differ in their kinetics of synthesis and topology in cultured cells. J. Cell Biol. 110: 743–752

Brown, P. 2001. Bovine spongiform encephalopathy and variant Creutzfeldt—Jacob diseae. BMJ 322: 841–844

Brown, D. R (Editor). 2005. *Neurodegeneration and Prion Disease*. Springer, Berlin Heidelberg New York

Brown, P. and Gajdusek, D. C. 1991. Survival of scrapie virus after 3 year's interment. Lancet 337: 269–270

Brown, D. R., Qin, K., Herms, J. W., Madlung, A., Manson, J., Strome, R., Fraser, P. E., Kruck, T., von Bohlen, A., Schulz-Schaeffer, W., Giese, A., Westaway, D., and Kretzschmar, H. 1997. The cellular prion protein binds copper in vivo. Nature 390: 684–687

Brown, P., Will, R. G., Bradley, R., Asher, D. M., and Detwiler, L. 2001. Bovine spongiform encephalopathy and variant Creutzfeldt—Jacob disease: background, evolution, and current concerns. Emerging Infect. Dis. 7: 6–16

Brown, P., Rau, E. H., Lemieux, P., Johnson, B. K., Bacote, A. E., and Gajdusek, D. C. 2004. Infectivity studies of both ash and air emissions from simulated incineration of scrapie-contaminated tissues. Environ. Sci. Technol. 38: 6155–6160

Bruce, M. E., Will, R. G., Ironside, J. W., McConnell, I., Drummond, D., Suttie, A., McCardle, L., Chree, A., Hope, J., Birkett, C., Cousens, S., Fraser, H., and Bostock, C. J. 1997. Transmission to mice indicate tht "new variant" CJD is caused by BSE agent. Nature 389: 498–501

BSE Inquiry Report. 2000. www.bseinquiry.gov.uk/report

Calzolai, L., Lysek, D. A., Pérez, D. A., Güntert, P., and Wüthrich, K. 2005. Prion protein NMR structures of chickens, turtles, and frogs. Proc. Nat. Acad. Sci. 102: 651–655

Capobianco, R., Casalone, C., Suardi, S., Mangieri, M., Miccolo, C., Limido, L., Catania, M., Rossi, G., Di Fede, G., Giaccone, G., Bruzzone, M. G., Minati, L., Corona, C., Acutis, P., Gelmetti, D., Lombardi, G., Groschup, M. H., Buschmann, A., Zanusso, G., Monaco, S., Caramelli, M., and Tagliavini, F. 2007. Conversion of the BASE prion strain into the BSE strain: The origin of BSE. PLoS Pathog. 3: e31

Carrell, R. W. 2004. Prion dormancy and disease. Science 306: 1692–1693

Castilla, J., Saá, P., Hetz, C., and Soto, C. 2005. In vitro generation of indectious scrapie prions. Cell 121: 195–206

Chesebro, B. 2004. A fresh look at BSE. Science 305:1918–1921

Chesebro, B., Trifilo, M., Race, R., Meade-White, K., Teng, C., LaCasse, R., Raymond, L., Favara, C., Baron, G., Priola, S., Caughey, B., Masliah, E., and Oldstone, M. 2005. Anchorless prion protein results in infectious amyloid disease without clinical scrapie. Science 308: 1435–1439

Collinge, J., Sidle, K. C. L., Meads, J., Ironside, J., and Hill, A. F. 1996. Molecular analysis of prion strain variation and the aetiology of a "new variant" CJD. Nature 383: 685–690

Come, H. H., Fraser, P. E., and Lansbury, P. T., Jr., 1993. A kinetic model for amyloid formation in the prion diseases: importance of seeding. Proc. Nat. Acad. Sci 90: 5959–5963

Cox, D. L., Pan, J., and Singh, R. R. P. 2006. A mechanism for copper inhibition of infectious prion conversion. Biophys. J. 91: L11–L13

DeArmond, S. J., Kristensson, and Bowler, R. P. 1992. PrP^{Sc} causes nerve cell death and stimulates astrocyte proliferation: a paradox. Prog. Brain Res. 94: 437–446

DeJoia, C., Moreaux, B., O'Connell, K., and Bessen, R. A. (2006). Prion infection of oral and nasal mucosa. J. Virol. 80: 4546–4556

Diener, T. O., McKinley, M. P., and Prusiner, S. B. 1982. Viroids and prions. Proc. Natl. Acad. Sci. 79: 5220–5224

Enserink, M. 2005. After the crisis: more questions about prions. Science 310: 1756–1758

Goldfarb, L. G., Brown, P., McCombie, W. R., Goldgaber, D., Swergold, G. D., Wills, P. R., Cervenakova, L., Baron, H., Gibbs, Jr. C. J., Gajdusek, C. 1991. Transmissible familial Creutzfeldt—Jacob disease associated with five, seven, and eight octapeptide coding repeats in the PRNP gene. Proc. Natl. Acad. Sci. 88: 10926–10930

Griffith, J. S. 1967. Self-replication and scrapie. Nature 215: 1043–1044

Halliday, S., Houston, F. and hunter, N. 2005. Expression of PrP^C on cellular components of sheep blood. J. Gen. Virol. 86: 1571–1579

Harris, D. A., Falls, D. L., Johnson, F. A., and Fischbach, G. D. 1991. A prion-like protein from chicken brain copurifies with an acetylcholine-inducing activity. Proc. Natl. Acad. Sci. 88: 7664–7668

Heikenwalder, M., Zeller, N., Seeger, H., Prinz, M., Klöhn, P. C., Schwarz, P., Ruddle, N. H., Weissmann, C., and Aguzzi, A. 2005. Chronic lymphocytic inflammation specifies the organ trophism of prions. Science 307: 1107–1110

Heppner, F. L., Musahl, C., Arrighi, I., Klein, M. A., Rülicke, T., Oesch, B., Zinkernagel, R. M., Kalinke, U., and Aguzzi, A. 2001. Prevention of scrapie pathogenesis by transgenic expression of anti-prion protein antibodies. Science 294: 178–182

Herrmann, L. M., Davis, W. C., Knowles, D. P., Wardrop, K. J., Sy, M. S., Gambetti, P., and O'Rourke, K. I. 2001. Cellular prion protein is expressed on peripheral blood mononuclear cells but not on platelets of normal and scrapie-infected sheep. Haematologica 86: 146–153

Holada, K., Vostal, J. G., Thiesen, P. W., MacAuley, C., Gregori, L., and Rohwer, R. G. 2002. Scrapie infectivity in hamster blood is not associated with platelets. J. Virol. 76: 4649–4650

Hornshaw, M. P., McDermott, J. R., and Candy, J. M. 1995. Copper binding to the N-terminal tandem repeat regions of mammalian and avian prion protein. Biochem. biophys. Res. Commun. 207: 621–629

Jackson, G. S., Hosszu, L. L. P., Power, A., Hill, A. F., Kenney, J., Saibil, H., Craven, C. J., Waltho, J. P., Clarke, A. R., and Collinge, J. 1999. Reversible conversion of monomeric human prion protein between native and fibrillogenic conformations. Science 283: 1935–1937

Jeffrey, M., González, L., Espenes, A., Press, C. M., Martin, S., Chaplin, M., DAvis, L., Landsverk, T., MacAldowie, C., Eaton, S., and McGovern, G. 2006. Transportation of prion protein across the intestinal mucosa of scrapie-susceptible and scrapie-resistant sheep. J. Pathol. 209: 4–14

Johnson, C. J., Phillips, K. E., Schramm, P. T., McKenzie, D., Aiken, J. M., and Pedersen, J. A. 2006. Prions adhere to soil minerals and remain infectious. PLoS Pathog. 2: e32

Konturek, P. C., Bazela, K., Kukharskyy, V., Bauer, M., Hahn, E. G., and Schuppan, D. 2005. *Helicobacter pylori* upregulates prion protein expression in gastric mucosa: A possible link to prion disease. World J. Gastroenterol. 11: 7651–7656

Krebs, M. R. H., Morozova-Roche, L. A., Daniel, K., Robinson, C. V., and Dobson, C. M. 2004. Observation of sequence specificity in the seeding of protein amyloid fibrils. Protein Sci. 13: 1933–1938

Kuwahara, C., Takeuchi, A. M., Nishimura, T., Haraguchi, K., Kubosaki, A., Matsumoto, Y., Saeki, K., Matsumoto, Y., Yokoyama, T., Itohara, S., and Onodera, T. 1999. Prions prevent neuronal cell-line death. Nature 400: 225–226

Lee, H. S., Brown, P., Cervenáková, L., Garruto, R. M., Alpers, M. P., Gajkusek, D. C., and Goldfarb, L. G. 2001. Increased susceptibility to kuru of carriers of the PRNP 129 Methionine/Methionine genotype. J. Infect. Dis. 183: 192–196

Lee, D. C., Sakudo, A., Chi-keyong, K., Nishimura, T., Saeki, K., Matsumoto, Y., Yokoyama, T., Chen, S. G., Itohara, S., and Onodera, T. 2006. Fusion of Doppel to octapeptide repeat and N-terminal half of hydophobic region of prion protein confers resistence to serum deprivation. Microbiol. Immunol. 50: 203–209

Legname, G., Baskakov, I. V., Nguyen, O. A. B., Riesner, D., Cohen, F. E., DeArmond, S. J., and Prusiner, S. B. 2004. Synthetic mammalian prions. Science 305: 673–676

Lewis, P. A., Properzi, F., Prodromidou, K., Clarke, A. R., Collinge, J., and Jackson, G. S. 2006. Removal of the glycosylphosphatidylinositol anchor from PrP^{Sc} by cathepsin D does not reduce prion infectivity. Biochem. J. 395: 443–448

Liberski, P. P. and Ironside, J. W. 2005. Neuropathology of transmissible spongiform encephalopathies (prion disease). In, Brown, D. R (Editor). 2005. *Neurodegeneration and Prion Disease*. Springer, Berlin Heidelberg New York, pp. 13–48

Lundmark, K., Westermark, G. T., Nyström, S., Murphy, C. L., Solomon, A., and Westermark, P. 2002. Transmissibility of systemic amyloidosis by a prion-like mechanism. Proc. Natl. Acad. Sci. 99: 6979–6984

Mallucci, G., Dickinson, A., Linehan, J., Klöhn, P. C., Brandner, S., and Collinge, J. 2003. Depleting neuronal PrP in Prion infection prevents disease and reverses spongiosis. Science 302: 871–874

Manuelidis, L., Yu, Z. X., Banquero, N., and Mullins, B. 2007. Cells infected with scrapie and Creutzfeldt—Jacob disease agents produce intracellular 25-nm virus-like particles. Proc. Natl. Acad. Sci. 104: 1965–1970

Mathiason, C. K., Powers, J. G., Dahmes, S. J., Osborn, D. A., Miller, K. V., Warren, R. J., Mason, G. L., Hays, S. A., Hayes-Klug, J., Seelig, D. M., Wild, M. A., Wolfe, L. L., Spraker, T. R., Miller, M. W., Sigurdson, C. J., Telling, G. C., and Hoover, E. A. 2006. Infectious prions in the saliva and blood of deer with chronic wasting disease. Science 314: 133–136

Mead, S., Stumpf, M. P. H., Whitfield, J., Beck, J. A., Poulter, M., Campbell, T., Uphill, J. B., Goldstein, D., Alpers, M., Fisher, E. M. C., and Collinge, J. 2003. Balancing selection at the prion protein gene consistent with prehistoric kurulike epidemics. Science 300: 64–643

Meyer-Luehmann, M., Coomaraswamy, J., Bolmont, T., Kaeser, S., Schaefer, C., Kilger, E., Neuenschwander, A., Abramowski, D., Frey, P., Jaton, A. L., Vigouret, J.- M., Paganetti, P., Walsh, D. M., Mathews, P. M., Ghiso, J., Staufenbeil, M., Walker, L. C., and Jucker, M. 2006. Exogenous induction of cerebral β-amyloidogenesis is governed by agent and host. Science 313: 1781–1784

Mouillet-Richard, S., Ermonval, M., Chebassier, C., Laplanche, J. L., Lehmann, S., Launay, J. M., and Kellermann, O. 2000. Signal transduction through prion protein. Science 289: 1925–1928

Oesch, B., Westaway, D., Walchli, M., McKinley, M. P., Kent, S. B. H., Aebersold, R., Barry, R. A., Tempst, P., Teplow, D. B., Hood, L. E., Prusiner, S. B., and Weissmann, C. 1985. A cellular gene encodes scrapie Prp 27–30 protein. Cell 40: 735–746

Pankiewicz, J., Prelli, F., ManSun, S. Y., Kascsak, R. J., Kascsak, R. B., Spinner, D. S., Carp, R. I., Meeker, H. C., Sadowski, M., and Wisniewski, T. 2006. Clearance and prevention of prion infection in cell culture by anti-PrP antibodies. Eur. J. Neurosci. 23: 2635–2647

Peretz, D., Williamson, R. A., Kaneko, K., Vergara, J., Leclerc, E., Schmitt-Ulms, G., Mehlhorn, I. R., Legname, G., Wormald, M. R., Rudd, P. M., Dwek, R. A., Burton, D. R., and Prusiner, S. B. 2001. Antibodies inhibit prion propagation and clear cell cultures of prion infectivity. Nature 412: 739–743

Perini, F., Frangione, B., and Prelli, F. 1996. Prion protein release by platelets. Lancet 347: 1635–1636

Prusiner, S. B. 1982. Novel proteinaceous infectious particles cause scrapie. Science 216: 136–144

Prusiner, S. B., Gajdusekk, C., and Aplers, M. P. 1982. Kuru with incubation periods exceeding two decades. Ann. Neurol. 12: 1–9

Prusiner, S. B., Groth, D. F., Bolton, D. C., Kent, S. B., and Hood, L. E., Prusiner, S. B. 1984. Purification and structural studies of a major scrapie prion protein. Cell 38: 127–134

Re, L., Rossini, F., Re, F., Bordicchia, M., Mercanti, A., Fernandez, O. S. L., and Barocci, S. 2006. Prion protein potentiates acetylcholine release at the neuromuscular junction. Pharmacol. Res. 53: 62–68

Robertson, C., Booth, S. A., Beniac, D. R., Coulthart, M. B., Booth, T. F., and McNicol, A. 2006. Cellular prion protein is released on exosomes from activated platelets. Blood 107: 3907–3911

Saá, P., Castilla, J., and Soto, C. 2006. Presymptomatic detection of prions in blood. Science 313: 92–94

Sadowski, M. and Wisniewski, T. 2004. Vaccines for conformational disorders. Expert Rev. Vaccines 3: 279–290

Safar, J. G., Kellings, K., Serban, A., Groth, D., Cleaver, J. E., Prusiner, S. B., and Riesner, D. 2005. Search for a prion-specific nucleic acid. J. Virol. 79: 10796–10806.

Sakaguchi, S. 2005. Prion protein, prion protein-like protein and neurodegeneration. In Brown D. R. *Neurodegeneration and Prion Disease*, Springer, Berlin Heidelberg New York, pp. 167–193

Schoch, G., Seeger, H., Bogousslavsky, J., Tolnay, M., Janzer, R. C., Aguzzi, A., and Glatzel, M. 2006. An anlaysis of prion strains by PrP^{Sc} profiling in sporadic Creutzfeldt—Jacob disease. PLoS Med. 3: e14 (236–244)

Seeger, H., Heikenwalder, M., Zeller, N., Kranich, J., Schwarz, P., Gaspert, A., Seifert, B., Miele, G., and Aguzzi, A. 2005. Coincident scrapie infection and nephritis lead to urinary prion excretion. Science 310: 324–326

Soto, C. 2006. Prions. *The New Biology of Proteins*. CRC Press, Boca Raton, USA

Stahl, N., Borchelt, D. R., and Prusiner, S. B. 1990. Differential release of cellular and scrapie prion protein form cellular membranes of phosphatidylinositol specific phospholipase. Biochemistry 29: 5405–5412

Starke, R., Harrison, P., Mackie, I., Wang, G., Erusalimsky, J. D., Gale, R., Massé, J. M., Cramer, E., Pizzey, A., Biggerstaff, J., and Machin, S. 2005. The experssion of prion protein (PrP^C) in the magakaryocyte lineage. J. Thromb. Haemost. 3: 1266–1273

Steele, A. D., Emsley, J. G., Özdinler, P. H., lindquist, S., and Macklis, J. D. 2006. Prion protein (PrP^C) positively regulates neuroal precursor proliferation during developmental and adult mammalian neurogenesis. Proc. Natl. Acad. Sci. 103: 3416–3421

Sulforosi, L., Criado, J. R., McGavern, D. B., Wirz, S., Sánchez-Alvarez, M., Sugama, S., DeGiorgio, L. A., Volpe, B. T., Wiseman, E., Abalos, G., Masliah, E., Gilden, D., Oldstone, M. B., Conti, B., and Williamson, R. A. 2004. Cross-linking cellular prion protein triggers neuronal apoptosis in vivo. Science 303: 1514–1516

Supattapone, S., Nguyen, H. O. B., Cohen, F. E., Prusiner, S. B., and Scott, M. R. 1999. Elimination of prions by branched polyamines and implications for therapeutics. Proc. Natl. Acad. Sci. 96: 14529–14534

Supattapone, S., Wille, H., Uyechi, L., Safar, J., Tremblay, P., Szoka, F., Cohen, F. E., Prusiner, S. B., and Scott, M. R. 2001. Branched polyamines cure prion-infected neuroblastoma cells. J. Virol. 75: 3453–3461

Taylor, D. M. 1990. Inactivation of the BSE agent. Dev. Biol. Stand. 75:97–102

Taylor, J. P., Hardy, J., and Fischbeck, K. H. 2002. Toxic proteins in neurodegenerative disease. Science 296: 1991–1995

Wadsworth, J. D. F., Asante, E. A., Desbruslais, M., Linehan, J. M., Joiner, S., Gowland, I., Welch, J., Stone, L., Lloyd, S. E., Hill, A. F., Brandner, S., and Collinge, J. 2004. Human prion protein with Valine 129 prevents expression of variant CJD phenotype. Science 306: 1793–1796

Watts, J. C., Balanchandran, A., and Westaway, D. 2006. The expanding universe of prion diseases. PLoS Pathog. 2(3): e26

Westaway, D. and Prusiner, S. B. 1986. Conservation of the cellular gene encoding scrapie prion protein. Nucleic Acids Res. 14: 2035–2044

Will, R. G., Ironside, J. W., Zeidler, M., Cousens, S. N., Estiberio, K., Alperovitch, A., Posner, S., Pocchiari, M., Hofman, A., and Smith, P. G. 1996. A new variant of Creutzfeldt—Jacob disease in the UK. Lancet 347: 921–925

Will, R. G., Cousens, S. N., Farrington, C. P., Smith, P. G., Knight, R. S. G., and Ironside, J. W. 1999. Deaths from variant Creutzfeldt—Jacob disease. Lancet 353: 979

Xie, Z., O'Rourke, K. I., Dong, Z., Jenny, A. L., Langenberg, J. A., Belay, E. D., Schonberger, L. B., Petersen, R. B., Zou, W., Kong, Q., Gambetti, P., and Chen, S. G. 2006. Chronic wasting disease of elk and deer and Creutzfeldt—Jacob disease. 2006. J. Biol. Chem. 281: 4199–4206

Zhang, C. C., Steele, A. D., Lindquist, S., and Lodish, H. F. 2006. Prion protein is expressed on long-term repopulating hematopoetic stem cells and is important to their self-renewal. Proc. Natl. Acad. Sci. 103: 2184–2189

Zobeley, E., Flechsig, E., Cozzio, A., Enari, M., and Weissman, C. 1999. Infectivity of scrapie prions bound to stainless steel surface. Mol. Med. 5: 240–243

Chapter 24
Lead Encephalopathy

Ivan J. Boyer

Introduction

Lead is arguably the most studied of the neurotoxicants (Silbergeld, 1992; Dietrich, 1995). Epidemiological and experimental studies of the toxic effects and mechanisms of action of lead have grown exponentially since the 1960s and 1970s. This work has produced an unparalleled body of scientific information on the effects of lead on the central nervous system (CNS). This chapter provides a brief review of the current knowledge and the key hypotheses about the toxicity of lead to the central nervous system (CNS).

An enormous amount of lead has been released to the environment through human activities since the 1920s, when tetraethyl lead was first sold as an antiknock agent in gasoline for automobile engines. The use of inorganic lead as an anticorrosive agent in paints and primers represents another important source of lead released to the environment. These practices are largely responsible for widespread and persistent environmental contamination often exceeding concentrations now known to have the potential to cause serious harm to human health.

The great depth and breadth of the research on the toxicity of lead has implicated numerous cellular and molecular mechanisms of action in the CNS. Lead can bind to many different cell components to affect multiple cellular processes, depending on exposure rate and duration, body burden, developmental, nutritional and health status and, ultimately, the concentrations of lead at target sites in specific brain regions. The overall effects of lead exposure on the brain depend, in turn, on perturbations of many complex functions and interrelationships among the diverse cell types and regions of the brain.

Despite the substantial research, no single mechanism or simple set of mechanisms has emerged to explain the wide spectrum of effects of lead exposure on the brain. Indeed, all of the molecular mechanisms proposed to date can still be fairly characterized as tentative hypotheses about the many ways that lead might alter brain function. On the other hand, the extensive data and hypotheses gained from this research may serve as the best model available for investigating other neurotoxicants with effects that can range from subclinical CNS impairment to overt encephalopathy. There are many such

D.W. McCandless (ed.) *Metabolic Encephalopathy*,
doi: 10.1007/978-0-387-79112-8_24, © Springer Science+Business Media, LLC 2009

toxicants, including other metals (for example, arsenic, cadmium, manganese, methylmercury, thallium, and triethyltin), ethanol, polychlorinated biphenyls (PCBs), drugs, and pesticides (Rice, 1989; Barone et al., 1995; Tsai et al., 2006; See Mendola et al., 2002, for review).

Clinical Lead Encephalopathy

High levels of lead exposure can cause overt clinical encephalopathy in both children and adults. The effects include brain edema and demyelination of the cerebral and cerebellar white matter, which are thought to be secondary to vascular injury (Reyes et al., 1986).

Clinical lead encephalopathy was once frequently observed in the developed world, especially in children. Lead exposures high enough to cause overt encephalopathy in children were often attributable to ingesting chips of deteriorating paint, dust from lead-contaminated soil, or lead leached into foods or beverages from lead solders or ceramic glazing. Children with encephalopathy often had chips of lead paint visible in abdominal radiographs (Gordon et al., 1998; AAP, 2005). Before the advent of chelation therapy, only about 50% of children hospitalized for lead encephalopathy survived, and more than 25% of the survivors suffered mental retardation, seizures, cerebral palsy, optic atrophy or other severe sequelae (Chisolm and Harrison, 1956; Christian et al., 1964; Perlstein and Attala, 1966). Lead encephalopathy was also occasionally reported in adults from, for example, exposures to inorganic or organic lead compounds in occupational settings or from drinking moonshine whiskey or intentionally inhaling leaded gasoline (Holstege et al., 2004; Cairney et al., 2005).

These sources, among others (e.g., lead-containing folk remedies, cosmetics and toys), are responsible for relatively high lead exposures that still occur in many parts of the world (for reviews see Hackley and Katz-Jacobson, 2003; Papanikolaou et al., 2005; AAP, 2005).

Young children are much more sensitive to lead exposure than adults are. Thus, overt lead encephalopathy is rarer in adults (Bressler and Goldstein, 1991). However, both adults and children exposed to lead can present with serious neurodeterioration and death (Perelman et al., 1993).

Differences in susceptibility have been explained by many developmental, physiological and metabolic differences between young children and adults. The developing brain in young children is exquisitely sensitive to the disruptive effects of lead. Young children ingest more lead, absorb a greater proportion of lead in the digestive tract, and accumulate a greater proportion of circulating lead in the brain (Ziegler et al., 1978; Clark et al., 1985; Philip and Gerson, 1994; Leggett, 1993; Cory-Slechta and Schaumburg, 2000; Markowitz, 2000; Lanphear et al., 2002). In addition, resistance in the adult may be partially explained by the greater capacity of the mature brain to sequester lead away from target sites (Holtzman et al., 1984; Tiffany-Castiglioni et al., 1989; Lindahl et al., 1999).

Lead encephalopathy is now rare in the developed world. This is attributed to eliminating leaded gasoline and lead-based paints, remediating lead contamination, and preventing or reducing lead exposures in both residential and occupational settings. For example, the mean blood lead concentration of the U.S. population fell from about 16 to 3 μg dl^{-1} between 1976 and 1991, after banning leaded gasoline and regulating smokestack emissions in the 1970s (Fischbein, 1998; AAP, 2005).

However, occasional cases of lead encephalopathy in both children and adults continue to appear in the literature. For example, Atre et al. (2006) reported the case of a 41-year-old man who, for about 6 months, ingested ayurvedic herbal medicines containing up to 37 mg g^{-1} lead (Saper et al., 2004). His blood lead concentration was 161 μg dl^{-1} when he was hospitalized. Clinical signs included mild disorientation, short-term memory loss, disinterest in his surroundings, spastic gait, and loss of balance. Magnetic resonance imaging (MRI) revealed bilateral, symmetrical lesions in the gray matter, gray-white matter junction, and subcortical white matter in the parasagital occipital lobes, edema and sulcal effacement. Subcortical white matter lesions were found in the frontal regions, parieto-occipital lobe, body of the corpus callosum and cerebellum. These lesions resolved nearly completely in this patient after ten days of treatment with a chelator.

Overt lead encephalopathy is almost always associated with blood lead concentrations >100 μg dl^{-1} in both children and adults, although concentrations as low as 50–70 μg dl^{-1} may cause encephalopathy, especially in children (Reyes et al., 1986; Adams and Victor, 1993; Mani et al., 1998; Cory-Slechta and Schaumburg, 2000).

There are differences in the presentation of encephalopathy after acute lead exposures compared with subchronic and chronic exposures in both children and adults. Thus, "acute lead encephalopathy" is distinguished from "chronic lead encephalopathy."

Acute lead encephalopathy typically presents with cerebral edema (Saryan and Zenz, 1994; Papanikolaou et al., 2005). The clinical signs and symptoms may include headache, persistent vomiting, ataxia, paralysis, stupor, convulsions, and coma (Chisolm and Kaplan, 1968; Bressler and Goldstein, 1991; Landrigan and Todd, 1995; al Khayat et al., 1997).

In contrast, chronic lead encephalopathy presents mainly with extensive tissue destruction, cavity formation and thickening and cellular disorganization of the walls of the veins, suggesting that the cerebral damage is secondary vasotoxicity (Mani et al., 1998). This can progress to cerebral or cerebellar edema, proteinaceous perivascular exudation, hemorrhage, neuronal loss, and gliosis (Teo et al., 1997). Clinical signs include loss of memory, sensory perception, and the ability to concentrate. Other signs and symptoms may include drowsiness, restlessness, irritability, disorientation, dizziness, tremor, ataxia, syncope, behavioral abnormalities, diminished libido, depression, headaches, and seizures (Reyes et al., 1986; Saryan and Zenz, 1994; Landrigan and Todd, 1995; Mani et al., 1998). Severe chronic lead encephalopathy may also present with vomiting, apathy, stupor, paralysis, and coma (Saryan and Zenz, 1994; Landrigan and Todd, 1995).

Leaded Gasoline Encephalopathy

Severe encephalopathy caused by abusing leaded gasoline presents with nystagmus, myoclonus or chorea, hyperreflexia and hallucinations, and other findings typically associated with chronic lead encephalopathy, including "clouding" of consciousness, tremor, limb and gait ataxia, and convulsions (see Cairney et al., 2002 for review). The risk of death is high without emergency hospitalization and prolonged intensive care treatment (Currie et al., 1994; Goodheart and Dunne, 1994; Burns and Currie, 1995).

In addition, these patients display considerable neurological and cognitive impairment after discharge from the hospital. Residual signs include nystagmus, postural tremor, dysdiadochokinesia, dysmetria, brisk deep reflexes, palmomental reflex, and gait ataxia. More subtle effects include impaired performance in tests of visual attention, recognition memory, and paired associate learning (Cairney et al., 2004a,b; Cairney et al., 2005).

Cairney et al. (2004b) evaluated neurological and cognitive functions in three groups of young men in a remote indigenous community in Australia. These groups included gasoline abusers with a history of encephalopathy, gasoline abusers with no history of encephalopathy, and controls with no history of substance abuse. All of the abusers exhibited impaired neurological and cognitive functions compared with controls. However, the frequency and severity of many of these effects were much greater in the abusers with a history of encephalopathy ("encephalopathic" group) than those without such a history ("nonencephalopathic" group). Further, these effects were not correlated with blood lead concentrations, number of years of gasoline abuse or volume of gasoline inhaled per week in the encephalopathic group.

Cairney et al. (2004b) hypothesized that the persistent, severe effects observed in encephalopathic abusers might be explained by lasting catastrophic damage to cortical and cerebellar functions in these patients. For example, hyperreflexia and the emergence of a palmomental reflex are consistent with cortical damage. Nystagmus, ataxia, and poor hand-to-foot coordination are signs of cerebellar damage. This hypothesis is consistent with postmortem and radiological studies that show prominent abnormalities in the cortical and cerebellar regions associated with leaded gasoline encephalopathy (Valpey et al., 1978; Kaelan et al., 1986; Roger et al., 1990).

Further, Cairney et al. (2005) reexamined encephalopathic and nonecephalopathic leaded gasoline abusers after 2 years of abstinence from leaded gasoline sniffing. The abstinence was credited to an intervention strategy that included replacing leaded gasoline with aviation gas (Avgas) and increasing social, occupational, and recreational opportunities in the community. The subjects with a history of encephalopathy continued to show substantial signs of frontocerebellar and brainstem dysfunction, including dysdiadochokinesia, palmomental reflex, hyperreflexia, postural tremor, and ataxia.

Delayed Effects in Adults

Occupational Exposures

Exposures to lead during adulthood can accelerate the aging of the brain, manifested by increased rates of accumulating white matter lesions, progressive brain atrophy, and cognitive decline (Rowland and McKinstry, 2006).

Stewart et al. (2006) showed that tibial lead concentration in a cohort of former chemical workers was directly correlated with white matter lesions and inversely correlated with total brain volume and the volumes of several specific brain regions (frontal and total gray matter, parietal white matter, cingulate gyrus, and insula). The men in this cohort were exposed to tetraethyl lead, tetramethyl lead and inorganic lead, although their exposures ended more than 15 years before the study (Rowland and McKinstry, 2006; Stewart et al., 2006). Previous investigations revealed that cumulative lead exposure, as measured by tibial lead concentrations, was correlated with declining cognitive function years after their blood and brain lead concentrations had decreased substantially (Stewart et al., 1999; Schwartz et al., 2000; Links et al., 2001).

The overall results of these studies suggest that past lead exposures can accelerate age-related decline in cognitive functions by causing persistent or progressive damage to the brain (Stewart et al., 2006). Greater burdens of white matter lesions are generally associated with greater risks of impaired cognitive and motor functions, mood disorders, and strokes (Rowland and McKinstry, 2006). This damage may begin, for example, with altered synaptic connections, other changes in cellular architecture and cell death in specific areas of the adult brain during occupational lead exposure (Stewart et al., 2006). Brain regions where myelination normally continues until late in life may be more susceptible to early injury. In turn, the initial lesions in smaller, more specific areas of the brain may trigger a progressive atrophy that encompasses larger brain regions, and eventually the whole brain, later in life. Long-term hypertension and other vascular effects of lead exposure could prove to be important causative or contributing factors in this overall process.

Amyloidogenesis

Experimental animal studies indicate that lead exposure in early life may increase amyloidogenesis in the brain later in life (Basha et al., 2005; Bolin et al., 2006). The results lend substantial support for the hypothesis that toxicant exposures during critical periods in childhood may "reprogram" brain development to yield neurological and cognitive deficits that appear only much later during adulthood.

For example, the brains of senescent rats exposed to lead during early development exhibited delayed overexpression of β-amyloid precursor protein (APP) mRNA, altered activities of several signal-dependent and development-specific transcription factors, and elevated concentrations of both APP and the amyloidogenic β-amyloid (Aβ) product of APP. The effects included a substantial elevation in one of the regulators (Sp1) of the APP gene.

The blood and tissue lead concentrations in the affected animals were indistinguishable from background, indicating that the effects were not attributable to mobilization of lead from bone during old age (Basha et al., 2005). On the other hand, APP gene expression and Sp1 activity were not affected in rats exposed to lead during old age, supporting the hypothesis that these effects are attributable to "reprogramming" triggered by lead exposure during early development.

The Aβ peptide is known to yield reactive species (free radicals) that can cause significant oxidative damage in the aging brain (Bush, 2003). One measure of oxidative damage is the concentration of 8-hydroxy-2'-deoxyguanosine (oxo^{8}dG) in DNA. Bolin et al. (2006) found elevated concentrations of oxo^{8}dG in the cerebral cortex of aging rats exposed to lead during postnatal days 1—20 (46 µg dl^{-1}). In contrast, they found no change in oxo^{8}dG in senescent rats exposed during 18—20 months of age (60 µg dl^{-1}). In addition, lead exposure did not alter the activities of the DNA repair enzyme 8-oxoguanine DNA glycosylase (Ogg1), copper/zinc-superoxide dismutase (SOD1) or manganese-SOD (SOD2), or the concentrations of reduced glutathione (GSH) in the cerebral cortex of the affected rats.

Bolin et al. (2006) observed that the APP gene and several other genes that display delayed upregulation after lead exposure have fewer CpG sequences upstream of their transcription start sites (TSSs) than genes that do not respond in this way. Thus, the delayed upregulation of the APP gene may depend on the low density of potential methylation sites (that is, CpG) upstream of the TSS. For example, the low density of CpG may render the gene hypersusceptible to a single lead-induced methylation event in this region.

In addition, Bolin et al. (2006) noted unpublished data showing that Ogg1 cannot repair oxo^{8}dG preceded by methylcytosine. This suggests that lead-induced methylation of CpGs early in life would not be amenable to repair.

Alternatively, Bolin et al. (2006) suggested that lead may interact with the trace-metal-mediated post-transcriptional action of the 5'-untranslated region (UTR) of the APP gene. This interaction during early postnatal development could cause delayed accumulation of APP and, thus, the accumulation of Aβ in the cerebral cortex.

Previous studies reported increases in oxo^{8}dG in brain regions of patients with Alzheimer's disease, Parkinson's disease, and other progressive neurodegenerative diseases associated with aging (Sanchez-Ramos et al., 1994; Gabbita et al., 1998; Lovell et al., 1999). Accumulating Aβ is conducive to the formation of Aβ aggregates and, in susceptible individuals, to the development of amyloid plaques, which are the hallmarks of Alzheimer's disease.

The findings of Basha et al. (2005) and Bolin et al. (2006) substantially support the hypothesis that neurotoxicant exposures early in life can contribute to neurodegenerative events in old age (Weiss, 1991; Lu et al., 2004). Specifically, their studies demonstrate that lead may cause delayed effects through epigenetic alterations that DNA repair enzymes cannot recognize as damage (Basha et al., 2005; Bolin et al., 2006). Such alterations during childhood may alter gene regulation and responsiveness throughout life and could, therefore, help to explain the effects observed during adulthood.

Calcification

Ide-Ektessabi et al. (2005) used X-ray fluorescence spectrosopy to measure the concentrations of lead and several trace elements in the brains of three groups of patients. The groups included patients with (1) diffuse neurofibrillary tangles with calcification (DNTC), (2) Alzheimer's disease (AD) with calcification, and (3) AD without calcification. They found high concentrations of lead, zinc, and calcium in the brain tissues of DNTC and AD patients with calcification, but no lead in the tissues of AD patients without calcification. There was a strong correlation between degree of calcification and lead concentration in the calcified areas of the brain in these patients.

Although no cause and effect relationship can be determined from this study, the results support the hypothesis that lead may be involved in the development of some progressive neurodegenerative diseases, particularly those associated with calcification in brain tissues.

Vulnerability of the Injured Brain

Brain areas undergoing neurodegeneration appear to be much more vulnerable to the toxicity of lead and other divalent metals than the normal areas of the brain.

For example, astrocytes and neurons in the normal brain contain the divalent metal transporter DMT-1 (Burdo et al., 1999; Wang et al., 2001; Huang et al., 2004). DMT-1 has a broad substrate range that includes iron, zinc, manganese, cobalt, cadmium, copper, nickel, and lead (Gunshin et al., 1997). Neuronal injury caused by the excitotoxin, kainate, is associated with elevated expression of DMT-1 in astrocytes in the hippocampus (Wang et al., 2002a). This effect is correlated with elevated iron concentrations, number of iron-positive cells, and expression of the iron storage protein, ferritin (Ong et al., 1999; Wang et al., 2002b; Huang and Ong 2005). Thus, elevated iron concentrations in the injured hippocampus appear to be mediated by elevated expression of DMT-1.

Ong et al. (2006) reported substantially elevated lead and cadmium concentrations in the hippocampus and other brain areas 1-month after intracerebroventricular injection of kainate in rats exposed simultaneously to these metals in drinking water. The lead concentrations in the hippocampus were nearly three times greater in the kainate injected rats than in the saline injected rats exposed to lead and cadmium. The concentrations of lead in the hippocampus and in the frontal cortex and striatum adjacent to the kainate injection site remained elevated 2 months after the injection. Kainate injection did not alter the concentrations of zinc in the brain, with or without exposure to lead and cadmium in the drinking water.

Ong et al. (2006) speculated that the elevated lead and cadmium uptake in areas of neuronal degeneration may be attributable, at least in part, to the action of the DMT-1 transporter.

In addition, the elevated uptake of these metals may have, in turn, contributed to the stimulation of DMT-1 expression by aggravating the neuroinflammation initiated by the kainate-induced damage (Ong and Farooqui 2005; Ong et al., 2006). The results of these studies are consistent with the hypothesis that initial damage to circumscribed areas can trigger gradual lead accumulation and the spread of the damage to larger areas of the brain (Stewart et al., 2006; Rowland and McKinstry, 2006).

Subclinical Effects in Children

Pb^{2+} exposures lower than those associated with overt effects produce "asymptomatic" cognitive impairment and biochemical and morphological changes in the brain (e.g., Needleman et al., 1990; Silbergeld, 1992; Fergusson et al., 1997; Factor-Litvak et al., 1999; Zawia, 2003; Basha et al., 2004).

Blood lead concentrations >10 μg dl^{-1} are associated with lower IQ, behavioral problems, hyperactivity, and poor school performance (Yule et al., 1981; Ernhart et al., 1989; Needleman and Gatsonis, 1990; Baghurst et al., 1992; Bellinger et al., 1992; Feldman and White, 1992; Dietrich et al., 1993; Rice, 1993; Pocock et al., 1994; Cory-Slechta, 1997; Wasserman et al., 1997, 2000). Several epidemiologic studies indicate that lead exposure impairs fine-motor coordination, balance, and social-behavioral modulation (Dietrich et al., 2000). Teachers were more likely to characterize students with elevated tooth lead concentrations as distractible, inattentive, dependent, disorganized, hyperactive, impulsive, and unable to follow directions (Needleman et al., 1979; Sciarillo et al., 1992). These students were also more likely to display reading disabilities and fail to graduate from high school than those with lower dentin lead concentrations (Needleman et al., 1979, 1990). In addition, elevated bone and blood lead concentrations have been correlated with greater displays of aggression and delinquency (Needleman et al., 1996, 2002; Dietrich et al., 2001).

However, neurobehavioral effects such as these have received much less attention than the potential effects of lead on IQ in children, at least in part because IQ is relatively easy to measure (WHO, 1995; Bellinger, 2004).

Two meta-analyses of epidemiological studies on the effects of early childhood lead exposure on IQ appeared in 1994 (Pocock et al., 1994; Schwartz, 1994). Both analyses indicated that blood lead concentrations in young children, which peaks around 2 years of age, is responsible for cognitive deficits observed at ≥ 4 years of age, when IQ measurements become stable and reliable. For example, over the range of about 0–25 µg dl^{-1}, a 10 µg dl^{-1} increase in blood lead concentration at 2 years of age was associated with a 5.8-point decline in IQ and an 8.9-point decline in a test of educational achievement measured at 10 years of age (Bellinger et al., 1992). Unlike the blood lead concentrations measured at 2 years of age, maximum blood lead concentrations observed from birth to 1, 2 or 10 years of age were not significantly associated with IQ tested at 10 years of age. Thus, these authors suggested that the timing of exposure may be more important than the amplitude of the exposure (Bellinger et al., 1992). In sum, a number of early studies supported the hypothesis that lead exposure during a critical phase of brain development early in childhood results in persistent damage that is not fully expressed as cognitive deficits until school age.

However, many other studies suggested that concurrent or average lifetime blood lead concentrations in school-age children (that is ~6–13 years of age) is a better predictor of cognitive impairment than peak blood lead concentrations measured at ~2 years of age (Baghurst et al., 1992; Dietrich et al., 1993; Dietrich, 1995; Wasserman et al., 1997; Tong et al., 1996, 1998; Lanphear et al., 2000; Chen et al., 2005). In addition, numerous studies reported an inverse correlation between concurrent or average lifetime blood lead concentrations <10 µg dl^{-1} and both cognitive development and academic achievement (Lanphear et al., 2000; Bellinger and Needleman 2003; Canfield et al., 2003).

These observations are important because the effects were detected at blood lead concentrations lower than the current 10 µg dl^{-1} level of concern for lead in children (CDC, 1991, 2002; WHO, 1995; AAP, 2005) and approaching current background concentrations. Thus, no threshold lead exposure has been established below which no effects can be expected on cognitive function and academic performance in children (CDC 1991; WHO 1995).

In addition, many studies suggest that the concentration-response curve is steeper in children at low concentrations (for example, blood lead concentrations $\leq$10 µg dl^{-1}) than at higher concentrations (Fulton et al., 1987; Pocock et al., 1994; Schwartz, 1994; Lanphear et al., 2000; Bellinger and Needleman, 2003; Canfield et al., 2003). Thus, a small increase in blood lead concentrations from a relatively low baseline concentration may have a disproportionately greater effect on cognitive and academic performance than the same increase from a higher baseline (Canfield et al., 2003).

Other studies examined whether declines in blood lead concentration from the peak at around 2 years of age are associated with improvements in cognition later in life (Rogan et al., 1991; Ruff et al., 1993, 1996; Tong et al., 1998; Liu et al., 2002; Dietrich et al., 2004; Chen et al., 2005). Overall, these studies support the view that children are vulnerable to lead exposures occurring anytime from early childhood through school age. Further, chelation therapy does not appear to reverse the effects of lead in children with blood lead concentrations below about 45 µg dl^{-1} (Rogan et al., 2001; CDC, 2002; Liu et al., 2002; Dietrich et al., 2004; AAP, 2005).

Mechanisms of Action

Lidsky and Schneider (2003) provided a comprehensive review of the numerous mechanisms of toxicity implicated to explain the wide-range of effects of lead exposure on the CNS. Tables 24.1 and 24.2, which are adapted from Lidsky and Schneider (2003), illustrate the substantial diversity of mechanisms and effects associated with lead toxicity.

Many of the molecular and biochemical effects of Pb^{2+} in the brain appear to result from its ability to mimic Ca^{2+} and alter Ca^{2+}-dependent pathways (Lidsky and Scheinder, 2003). For example, Pb^{2+} can alter neurotransmitter release by interacting with protein kinase C (PKC), Ca^{2+}-calmodulin, or Ca^{2+}-dependent potassium channels (Goldstein and Ar, 1983; Habermann et al., 1983; Minnema et al., 1986; Markovac and Goldstein, 1988a,b; Laterra et al., 1992). Pb^{2+} can pass through

Table 24.1 Mechanisms of lead toxicity[a]

- Sequestration and mobilization of Pb^{2+} from bone storage sites
- Accumulation in brain by astrocytes
- Long half-life in brain (2 years) and slow release from accumulation sites
- Substitution for Zn^{2+} in Zn^{2+}-mediated processes
- Disruption of Ca^{2+} homeostasis
- Competition with and substitution for Ca^{2+}
- Stimulation of release of Ca^{2+} from mitochondria
- Opening of mitochondrial transition pore
- Direct damage to mitochondria and mitochondrial membranes
- Inhibition of antioxidative enzymes (for example SOD)
- Alteration of lipid metabolism

[a]Adapted from Lidsky and Schneider (2003)

Table 24.2 Effects of lead toxicity[a]

- Impaired heme biosynthesis and anemia
- Disruption of thyroid hormone transport to the brain
- Disruption of the blood-brain barrier
- Decreased cellular energy metabolism
- Oxidative stress
- Lipid peroxidation
- Altered neurotransmitter release
- Excitotoxicity
- Altered neurotransmitter receptor density
- Altered activity of second messenger systems
- Abnormal myelin formation
- Impaired development and function of oligodendrocytes
- Abnormal dendritic branching patterns
- Altered regulation of gene transcription
- Abnormal expression of neurotrophic factors
- Apoptosis
- Impaired neuropsychological functioning
- Lowered IQ
- Impaired academic achievement

[a]Adapted from Lidsky and Schneider (2003)

endothelial cells of the blood-brain barrier through the action of the Ca^{2+}-ATPase pump (Bradbury and Deane, 1993; Kerper and Hinkle, 1997a,b). In addition, Pb^{2+} can pass through or inhibit voltage-sensitive Ca^{2+} channels in the cell membrane of astroglia and neurons (Audesirk and Audesirk, 1993; Bernal et al., 1997; Peng et al., 2002).

Alternatively, some of the effects of Pb^{2+} may be attributable to its ability to substitute for Zn^{2+} or to interfere with the regulation of neural cell adhesion molecules during postnatal development (Bressler and Goldstein, 1991; Regan, 1989, 1991, 1993). Lee et al. (2002) suggested that selective increase in tyrosine hydroxylase levels observed in the locus ceruleus of lead exposed neonatal rats may help explain lead-induced hyperactivity in children, because this enzyme is rate-limiting in the biosynthesis of catecholamies.

The following provides a brief summary of some of the mechanisms of lead toxicity addressed in several recent papers to illustrate the complexity of this issue. The reader should consult reviews by Silbergeld (1992), Finkelstein et al. (1998), Bressler et al. (1999), Lidsky and Schneider (2003) and Johnston (2004) for more comprehensive information on this topic.

Mechanisms Proposed for Brain Edema

Lead is vasotoxic to the developing brain (Pentschew and Garro, 1966; Press, 1977; Holtzman et al., 1982; Goldstein, 1990). Sensitivity to lead decreases substantially with the maturation of the brain microvasculature and the blood-brain barrier (Pentschew and Garro, 1966; Goldstein et al., 1974; Press, 1977; Toews et al., 1978; Holtzman et al., 1982; Goldstein, 1990). The greater susceptibility of the cerebellum compared with the cerebrum may be attributable to the relatively late maturation of the cerebellum and the cerebellar microvasculature (Hossain et al., 2004). However, the molecular and biochemical mechanisms that underlie the major manifestations of acute lead encephalopathy (including edema and microhemorrhage), the differential sensitivity of children and adults, and the comparative susceptibilities of specific brain regions remain to be explained.

Bouton et al. (2001b) used gene expression profiling in cultured astrocytes to identify lead-responsive genes that modulate neuronal and microvascular functions in the brain. They identified many potential new mediators of lead toxicity, including annexins, thrombospondin, collagens, and tRNA synthetases. Most notable, lead induced vascular endothelial growth factor (VEGF) in the astrocytes (Hossain et al., 2004).

VEGF is a potent angiogenic and vascular permeability factor that stimulates the migration, proliferation and proteolytic enzyme activity of endothelial cells, as well as vasogenic cerebral edema (Leung et al., 1989; Wang et al., 1996; Bates and Curry, 1996, 1997). Hossain et al. (2004) found increased VEGF expression in the cerebellum of lead-treated rats concurrent and co-located with acute vasogenic edema, increased vascular permeability to serum albumin, and microvascular hemorrhage. The edema, but not the increased permeability and microhemorrhaging, was associated with elevated phosphorylation of Flk-1 VEGF receptors in the cerebellum of the exposed rats.

Hossain et al. (2004) suggested that lead exposure interferes specifically with compensatory mechanisms of interstitial fluid homeostasis by stimulating VEGF expression in astrocytes and, in turn, activating Flk-1 VEGF receptors in the cerebellum. This mechanism appears to be independent of lead-induced increases in blood-brain barrier permeability and microhemorrhaging. Different susceptibilities of the cerebellum and cortex to lead-induced edema could be explained, at least in part, by the greater induction of VEGF observed in the cerebellum. These authors also suggested that lead-induced edema may be caused by changes in the activity of aquaporins mediated by VEGF and the VEGF receptor.

Aquaporins are water channels that mediate rapid osmotically driven water transport across cell membranes (Preston et al., 1992; Agre et al., 2002; Badaut et al., 2002). Brain tissues, including microvessels, contain relatively high levels of aquaporins. In particular, Aquaporin 4 (AQP4) is abundantly expressed in the brain, mainly in astrocytes, which are commonly observed to be swollen during cerebral edema (Jung et al., 1994; Kimelberg, 1995; Nielsen et al., 1997).

Gunnarson et al. (2005) found that lead specifically elevates water permeability through AQP4 in astrocytes. They observed this effect in a rat astrocyte line transfected with cDNA constructs encoding mouse AQP4 and in primary astroglial cell cultures from the hippocampi of lead-exposed and control rats. The primary cultures expressed AQP4 endogenously, unlike the astrocyte line, which did not express AQP4 without transfection.

The effect of Pb^{2+} on AQP4 appears to depend on the activity of Ca^{2+}-calmodulin-dependent protein kinase II (CaMK-II), but not PKC (Gunnarson et al., 2005). In addition, treatment with a chelator, but not simple Pb^{2+} washout, reversed the effect, suggesting that Pb^{2+} enters and acts inside the astrocytes. AQP4 was not upregulated or redistributed in the brains of the lead-treated rats.

The results of this study showed that AQP4 may play an important role in the development of edema associated with acute lead encephalopathy (Gunnarson et al., 2005). Lead is the only heavy metal demonstrated, to date, to alter the activity of AQP4. However, other heavy metals, including mercury, nickel, copper, gold and silver, can alter the activities of other aquaporins (Yasui et al., 1999; Nicchia et al., 2000; Niemietz and Tyerman, 2002; Zelenina et al., 2003, 2004). Thus, aquaporins, as a group, may represent an important target for heavy metals and other toxicants that can cause edema in the brain.

Mechanisms Proposed for Developmental Effects

Critical neurodevelopmental processes occur in the human CNS during the first three years of life, including synaptogenesis, myelination, and programmed cell death (apoptosis). Studies of the effects of low-level lead exposure in animals suggest that these processes are highly vulnerable to lead toxicity, yet no single mechanism of lead toxicity has been identified to explain these effects (Silbergeld, 1992; McCall, 1983; Dietrich, 1999).

The mechanisms by which lead damages the developing brain probably involve multiple molecular, biochemical and cellular processes, including neurotransmitter release, second messenger signaling pathways, immediate-early and delayed gene expression, and neural network plasticity (Minnema et al., 1986; Bressler et al., 1999; Kim et al., 2000; Wilson et al., 2000).

Experimental animal studies have demonstrated that lead can disrupt several features of neuronal plasticity, including the capacity of the brain to be shaped by experience, to learn and remember, and to reorganize and recover after injury (Johnston, 2003). The enhanced brain plasticity of childhood is attributable to several factors, including postnatal persistence of neurogenesis in certain parts of the brain, deletion of neurons through apoptosis, and activity-dependent refinement of synaptic connections (Raff et al., 1993; Johnston et al., 2001; Wilson et al., 2000).

In particular, synaptic connections are refined through the activity of N-methyl-D-aspartate (NMDA) and α-amino-3-hydroxy-5-methyl-4-isoxazolepropionic acid (AMPA) type glutamate receptors (McDonald and Johnston, 1990; Penn and Shatz, 1999). Trophic factors that maintain synaptic connections during development are released when Ca^{2+} enters post-synaptic neurons. Ca^{2+} enters these cells through Ca^{2+} channels opened by NMDA-type receptors during the depolarization mediated by AMPA-type receptors (McDonald and Johnston, 1990; Penn and Shatz, 1999; Kovalchuk et al., 2002). The elevated intracellular Ca^{2+}, in turn, stimulates the phosphorylation of transcription factors by activating Ca^{2+}-calmodulin kinase IV (CaMK-IV), PKC, and the Ras-mitogen activated protein (MAP) kinase cascade (Sweatt, 2001; Sheng and Kim, 2002). Elevated intracellular Ca^{2+} also stimulates the phosphorylation of AMPA-type receptors by CaMK-II. Many other neurotransmitters and several growth factors also appear to play important roles in this process (Johnston et al., 2003).

Ultimately, changes in the activities of both NMDA- and AMPA-type receptors strengthen signal transmission across synapses, a phenomenon referred to as long-term potentiation (LTP). LTP appears to be critical for the development of memory and learning in both adults and children (Malinow and Malenka, 2002). However, LTP develops much more readily in children than in adults, at least in part because of the enhanced function of NMDA-type receptors in the developing brain (McDonald and Johnston, 1990; Crair and Malenka, 1995).

Animal studies have demonstrated that lead can disrupt several elements of this process, including the synaptic release of glutamate, the activities of NMDA-type receptors, PKC and the Ras-MAP kinase cascade, and LTP (Johnston and Goldstein, 1998; Wilson et al., 2000; Bouton et al., 2001a,b; Kim et al., 2002; Lasley and Gilbert, 2002; Toscano et al., 2002).

Conclusions

Extensive research has been conducted over the last several decades to describe the toxic effects and clarify the likely mechanisms of action of lead in the CNS. The mechanisms by which lead affects the CNS have not been fully elucidated, although many potential mechanisms have been studied by numerous groups of investigators. The findings of such studies can help to advance the development of general neurotoxicological principles, including explanations for the selective vulnerability of the CNS to environmental toxicants.

Among the most important lessons learned from these studies is that exposures to toxicants can yield a wide spectrum of neurotoxic effects ranging from overt encephalopathy and severe mental retardation to subtle deficits in sensory, motor, and cognitive functions. These effects depend on the timing, magnitude, and duration of exposure and many other factors (Mendola et al., 2002). Multiple mechanisms, target molecules, biochemical pathways, and cellular processes may explain the assembly of effects attributable to exposures to these substances.

Moreover, some of the effects may remain undetectable many years after exposure. Thus, early childhood exposures may not be fully expressed until school age or much later in life, potentially as late as senescence.

References

AAP. (2005) Lead exposure in children: Prevention, detection, and management. American Academy of Pediatrics, Committee on environmental health. Pediatrics **116(4)**: 1036–1046

Adams, R.D. and M. Victor. (1993) Principles of neurology. 5th ed., McGraw-Hill, NewYork

Agre, P., L.S. King, M. Yasui, W.B. Guggino, O.P. Ottersen, Y. Fujiyoshi, A. Engel and S. Nielsen. (2002) Aquaporin water channels: From atomic structure to clinical medicine. J. Physiol. **542**: 3–16

al Khayat, A., N.S. Menon and M.R. Alidina. (1997) Acute lead encephalopathy in early infancy: clinical presentation and outcome. Ann. Trop. Paediatr. **17**: 39–44

Atre, A.L., P.R. Shinde, S.N. Shinde, R.S. Wadia, A.A. Nanivadekar, S.J. Vaid and R.S. Shinde. (2006) Pre- and post-treatment MR imaging findings in lead encephalopathy. AJNR Am. J. Neuroradiol. **27**: 902–903

Audesirk, G. and T. Audesirk. (1993) The effects of inorganic lead on voltage sensitive calcium channels differ among cell types and among channel subtypes. Neurotoxicology **14**: 259–265

Badaut, J., F. Lasbennes, P.J. Magistretti and L. Regli. (2002) Aquaporins in brain: Distribution, physiology, and pathophysiology. J. Cereb. Blood Flow Metab. **22**: 367–378

Baghurst, P.A., A.J. McMichael, N.R. Wigg, G.V. Vimpani, E.F. Robertson, R.J. Roberts and S.L. Tong. (1992) Environmental exposure to lead and children's intelligence at the age of seven years: The Port Pirie cohort study. N. Engl. J. Med. **327**: 1279–1284

Barone, S., Jr., M.E. Stanton and W.R. Mundy. (1995) Neurotoxic effects of neonatal triethyltin (TET) exposure are exacerbated with aging. Neurobiol. Aging **16**: 723–735

Basha, M.R., W. Wei, G.R. Reddy, and N.H. Zawia. (2004) Zinc finger transcription factors mediate perturbations of brain gene expression elicited by heavy metals. In: Molecular Neurotoxicology, Environmental Agents, and Transcription-Transduction Coupling. N.H. Zawia (editor), CRC Press, Boca Raton, pp 43–64

Basha, M.R., W. Wei, S.A. Bakheet, N. Benitez, H.K. Siddiqi, Y.W. Ge, D.K. Lahiri and N.H. Zawia. (2005) The fetal basis of amyloidogenesis: Exposure to lead and latent overexpression of amyloid precursor protein and beta-amyloid in the aging brain. J. Neurosci. **25**: 823–829

Bates, D.O. and F.E. Curry. (1996) Vascular endothelial growth factor increases hydraulic conductivity of isolated perfused microvessels. Am. J. Physiol. **271**: H2520–H2528

Bates, D.O. and F.E. Curry. (1997) Vascular endothelial growth factor increases microvascular permeability via a Ca(2+)-dependent pathway. Am. J. Physiol. **273**: H687–H694

Bellinger, D.C. (2004) Lead. Pediatrics **113**: 1016–1022

Bellinger, D.C. and H.L. Needleman. (2003) Intellectual impairment and blood lead levels. N. Engl. J. Med. **349**: 500–502

Bellinger, D.C., K.M. Stiles and H.L. Needleman. (1992) Low-level lead exposure, intelligence and academic achievement: A long-term follow-up study. Pediatrics **90**: 855–861

Bernal, J., J.H. Lee, L.L. Cribbs and E. Perez-Reyes. (1997) Full reversal of Pb^{2+} block of L-type Ca^{2+} channels requires treatment with heavy metal antidotes. J. Pharmacol. Exp. Ther. **282**: 172–180

Bolin, C.M., R. Basha, D. Cox, N.H. Zawia, B. Maloney, D.K. Lahiri, and F. Cardozo-Pelaez. (2006) Exposure to lead (Pb) and the developmental origin of oxidative DNA damage in the aging brain. FASEB J. **20(6)**: 788–790

Bouton, C.M., L.P. Frelin, C.E. Forde, H.A. Godwin and J. Pevsner. (2001a) Synaptotagmin I is a molecular target for lead. J. Neurochem. **76**: 1724–1735

Bouton, C.M., M.A. Hossain, L.P. Frelin, J. Laterra, and J. Pevsner. (2001b) Microarray analysis of differential gene expression in lead-exposed astrocytes. Toxicol. Appl. Pharmacol. **176**: 34–53

Bradbury, M.W. and R. Deane. (1993) Permeability of the blood-brain barrier to lead. Neurotoxicology **14**: 131–136

Bressler, J.P. and G.W. Goldstein. (1991) Mechanisms of lead neurotoxicity. Biochem. Pharmacol. **41**: 479–484

Bressler, J., K. Kim, T. Chakraborti and G. Goldstein. (1999) Molecular mechanisms of lead neurotoxicity. Neurochem. Res. **24(4)**: 595–600

Burdo, J.R., J. Martin, S.L. Menzies, K.G. Dolan, M.A. Romano, R.J. Fletcher, M.D. Garrick, L.M. Garrick and J.R. Connor. (1999) Cellular distribution of iron in the brain of the Belgrade rat. Neuroscience **93**: 1189–1196

Burns, C.B. and B. Currie. (1995) The efficacy of chelation therapy and factors influencing mortality in lead intoxicated petrol sniffers. Aust. N. Z. J. Med. **25**: 197–203

Bush, A.I. (2003) The metallobiology of Alzheimer's disease. Trends Neurosci. **26**: 207–214

Cairney, S., P. Maruff, C. Burns and B. Currie. (2002) The neurobehavioural consequences of petrol (gasoline) sniffing. Neurosci. Biobehav. Rev. **26**: 81–89

Cairney, S., P. Maruff, C.B. Burns, J. Currie and B.J. Currie. (2004a) Saccade dysfunction associated with chronic petrol sniffing and lead encephalopathy. J Neurol Neurosurg. Psychiatr. **75**: 472–476

Cairney, S., P. Maruff, C.B. Burns, J. Currie and B.J. Currie. (2004b) Neurological and cognitive impairment associated with leaded gasoline encephalopathy. Drug Alcohol Depend. **73**: 183–188

Cairney, S., P. Maruff, C.B. Burns, J. Currie and B.J. Currie. (2005) Neurological and cognitive recovery following abstinence from petrol sniffing. Neuropsychopharmacol. **30**: 1019–1027

Canfield, R.L., C.R. Henderson, D.A. Cory-Slechta, C. Cox, J.A. Jusko and B.P. Lanphear. (2003) Intellectual impairment in children with blood lead concentrations below 10 micrograms per deciliter. N. Engl. J. Med. **348**: 1517–1526

CDC. (1991) Preventing lead poisoning in young children: A statement by the Centers for Disease Control. Centers for Disease Control and Prevention, Atlanta, GA

CDC. (2002) Managing elevated blood lead levels among young children: Recommendations from the Advisory Committee on Childhood Lead Poisoning Prevention. Centers for Disease Control and Prevention, Atlanta, GA.www.cdc.gov/nceh/lead/CaseManagement/caseManage_main.htm

Chen, A., K.N. Dietrich, J.H. Ware, J. Radcliffe and W.J. Rogan. (2005) IQ and blood lead from 2 to 7 years of age: are the effects in older children the residual of high blood lead concentrations in 2-year-olds? Environ. Health Perspect. **113(5)**: 597–601

Chisolm, J.J. Jr. and H.E. Harrison. (1956) The exposure of children to lead. Pediatrics **18**: 943–957

Chisolm, J.J. Jr. and E. Kaplan. (1968) Lead poisoning in childhood comprehensive management and prevention. J. Pediatr. **73**: 942–950

Christian, J.R., B.S. Celewycz and S.H. Andelman. (1964) A three-year study of lead poisoning in Chicago. Am. J. Public Health **54**: 1241–1245

Clark, C.S., R.L. Bornschein, P. Succop, S.S. Que Hee, P.B. Hammond and B. Peace. (1985) Condition and type of housing as an indicator of potential environmental lead exposure and pediatric blood lead levels. Environ. Res. **38**: 46–53

Cory-Slechta, D.A. (1997) Relationships between Pb-induced changes in neurotransmitter system function and behavioral toxicity. Neurotoxicology **18**: 673–688

Cory-Slechta, D.A. and H.H. Schaumburg. (2000) Lead, inorganic. In: Experimental and Clinical Neurotoxicology (second edition), P.S. Spencer, H.H. Schaumburg, A.C. Ludolph (editors). Oxford University Press, New York, pp. 708–720

Crair, M.C. and R.C. Malenka. (1995) A critical period for long-term potentiation at thalamocortical synapses. Nature **375**: 325–328

Currie, B., J. Burrow, D. Fisher, D. Howard, M. McEiver and C. Burns. (1994) Petrol sniffer's encephalopathy. Med. J. Aust. **160**: 800

Dietrich, K.N. (1995) A higher level of analysis: Bellinger's, interpreting the literature on lead and child development. Neurotoxicol. Teratol. **17**: 223–225

Dietrich, K.N. (1999) Environmental neurotoxicants and psychological development. In: Pediatric Neuropsychology: Research, Theory and Practice. G. Taylor, D. Ris and K.O. Yeates (editors). Guilford Press, New York, pp. 206–34

Dietrich, K.N., O.G. Berger, P.A. Succop, P.B. Hammond and R.L. Bornschein. (1993) The developmental consequences of low to moderate prenatal and postnatal lead exposure: Intellectual attainment in the Cincinnati lead study cohort following school entry. Neurotoxicol. Teratol. **15**: 37–44

Dietrich, K.N., O.G. Berger, A. Bhattacharya. (2000) Symptomatic lead poisoning in infancy: A prospective case analysis. J. Pediatr. **137**: 568–571

Dietrich, K.N., M.D. Ris, P.A. Succop, O.G. Berger and R.L. Bornschein. (2001). Early exposure to lead and juvenile delinquency. Neurotoxicol. Teratol. **23**: 511–518

Dietrich, K.N., J.H. Ware, M. Salganik, J. Radcliffe, W.J. Rogan, G.G. Rhoads, M.E. Fay, C.T. Davoli, M.B. Denckla, R.L. Bornschein, D. Schwarz, D.W. Dockery, S. Adubato and R.L. Jones. (2004) Effect of chelation therapy on the neuropsychological and behavioral development of lead exposed children after school entry. Pediatrics **114**: 19–26

Ernhart, C.B., M. Morrow-Tlucak, A.W. Wolf, D. Super and D. Drotar. (1989) Low level lead exposure in the prenatal and early preschool periods: Intelligence prior to school entry. Neurotoxicol. Teratol. **11**: 161–170

Factor-Litvak, P., G. Wasserman, J.K. Kline and J. Graziano. (1999) The Yugoslavia prospective study of environmental lead exposure. Environ. Health Perspect. **107**: 9–15

Feldman, R.G. and R.F. White. (1992) Lead neurotoxicity and disorders of learning. J. Child Neurol. **7**: 354–359

Fergusson, D.L., Horwood and M. Lynskey. (1997) Early dentine lead levels and educational outcomes at 18 years. J. Child Psychol. Psychiatr. **38**: 471–478

Finkelstein, Y., M.E. Markowitz and J.F. Rosen. (1998) Low-level lead-induced neurotoxicity in children: An update on central nervous system effects. Brain Res. Rev. **27**: 168–176

Fischbein, A. (1998) Occupational and environmental exposure to lead. In: Environmental and Occupational Medicine. W. Rom (editor), Lippincott-Raven, Philadelphia, pp. 973–996

Fulton, M., G. Raab, G. Thomson, D. Laxen, R. Hunter, W. Hepburn. (1987) Influence of blood lead on the ability and attainment of children in Edinburgh. Lancet **1**: 1221–1226

Gabbita, S.P., M.A. Lovell and W.R. Markesbery. (1998) Increased nuclear DNA oxidation in the brain in Alzheimer's disease. J. Neurochem. **71**: 2034–2040

Goldstein, G.W. (1990) Lead poisoning and brain cell function. Environ. Health Perspect. **89**: 91–94

Goldstein, G.W. and D. Ar. (1983) Lead activates calmodulin sensitive processes. Life Sci. **33**: 1001–1006

Goldstein, G.W., A.K. Asbury and I. Diamond. (1974) Pathogenesis of lead encephalopathy: Uptake of lead and reaction of brain capillaries. Arch. Neurol. **31**: 382–389

Goodheart, R.S. and J.W. Dunne. (1994) Petrol sniffer's encephalopathy. Med. J. Aust. **160**: 178–181

Gordon, R.A., G. Roberts, Z. Amin, R.H. Williams and F.P. Paloucek. (1998) Aggressive approach in the treatment of acute lead encephalopathy with an extraordinarily high concentration of lead. Arch. Pediatr. Adolesc. Med. **152**: 1100–1104

Gunnarson, E., G. Axehult, G. Baturina, S. Zelenin, M. Zelenina and A. Aperia. (2005) Lead induces increased water permeability in astrocytes expressing Aquaporin 4. Neuroscience **136**: 105–114

Gunshin, H., B. Mackenzie, U.V. Berger, Y. Gunshin, M.F. Romero, W.F. Boron, S. Nussberger, J.L. Gollan and M.A. Hediger. (1997) Cloning and characterization of a mammalian proton-coupled metal-ion transporter. Nature **388**: 482–488

Habermann, E., K. Crowell, P. Janicki. (1983) Lead and other metals can substitute for Ca^{2+} in calmodulin. Arch. Toxicol. **54**: 61–70

Hackley, B. and A. Katz-Jacobson. (2003) Lead poisoning in pregnancy: A case study with implications for midwives. Journal of Midwifery and Women's Health **48(1)**: 30–38

Holstege, C.P., J.D. Ferguson, C.E. Wolf, A.B. Baer and A. Poklis. (2004) Analysis of moonshine for contaminants. J. Toxicol. **42(5)**: 597–601

Holtzman, D., C. DeVries, H. Nguyen, J.H. Jameson, J. Olson, M. Carrithers and K. Bensch. (1982) Development of resistance to lead encephalopathy during maturation in the rat pup. J. Neuro. Exp. Neuro. **41**: 652–663

Holtzman, D., C. DeVries, H. Nguyen, J. Olson and K. Bensch. (1984) Maturation of resistance to encephalopathy: Cellular and subcellular mechanism. Neurotoxicology **5**: 97–124

Hossain, M.A., J.C. Russell, S. Miknyoczki, B. Ruggeri, B. Lal and J. Laterra. (2004) Vascular endothelial growth factor mediates vasogenic edema in acute lead encephalopathy. Ann. Neurol. **55(5)**: 660–667

Huang, E., W.Y. Ong and J.R. Connor. (2004) Distribution of divalent metal transporter-1 in the monkey basal ganglia. Neuroscience **128**: 487–496

Huang, E. and W.Y. Ong. (2005) Distribution of ferritin in the rat hippocampus after kainate-induced neuronal injury. Exp. Brain Res. **161**: 502–511

Ide-Ektessabi, A., Y. Ota, R. Ishihara, Y. Mizuno and T. Takeuchi. (2005) Distribution of lead in the brain tissues from DNTC patients using synchrotron radiation microbeams. Nucl. Instrum. Methods Phys. Res. B **241**: 681–684

Johnston, M.V. (2003) Brain plasticity in paediatric neurology. Eur. J. Paediatr. Neurol. **7**: 105–113

Johnston, M.V. (2004) Clinical disorders of brain plasticity. Brain and Develop. **26**: 73–80

Johnston, M.V. and G.W. Goldstein. (1998) Selective vulnerability of the developing brain to lead. Curr. Opin. Neurol. **11**: 689–693

Johnston, M.V., A. Nishimura, K. Harum, J. Pekar and M.E. Blue. (2001) Sculpting the developing brain. Adv. Pediatr. **48**: 1–38

Johnston, M.V., L. Alemi and K.H. Harum. (2003) Learning, memory and transcription factors. Pediatr. Res. **53**: 369–74

Jung, J.S., R.V. Bhat, G.M. Preston, W.B. Guggino, J.M. Baraban and P. Agre. (1994) Molecular characterization of an aquaporin cDNA from brain: Candidate osmoreceptor and regulator of water balance. Proc. Natl. Acad. Sci. USA **91**: 13052–13056

Kaelan, C., C. Harper and B. Vieira. (1986) Acute encephalopathy and death due to petrol sniffing: Neuropathological findings. Aust. N. Z. J. Med. **16**: 804–807

Kerper, L.E. and P.M. Hinkle. (1997a) Lead uptake in brain capillary endothelial cells: activation by calcium store depletion. Toxicol. Appl. Pharmacol. **146**: 127–133

Kerper, L.E. and P.M. Hinkle. (1997b) Cellular uptake of lead is activated by depletion of intracellular calcium stores. J. Biol. Chem. **272**: 8346–8352

Kim, K.A., T. Chakraborti, G.W. Goldstein and J.P. Bressler. (2000) Immediate early gene expression in PC-12 cells exposed to lead: Requirement for protein kinase C. J. Neurochem. **74**: 1140–1146

Kim, S.A., T. Chakraborti, G. Goldstein, M. Johnston and J. Bressler. (2002) Exposure of lead elevates induction of Zif268 and Arc mRNA in rats after electroconvulsive shock: The involvement of PKC. J. Neurosci. Res. **69**: 268–277

Kimelberg, H.K. (1995) Current concepts of brain edema: Review of laboratory investigations. J. Neurosurg. **83**: 1051–1059

Kovalchuk, Y., E. Hanse, K.W. Kafitz and A. Konnerth. (2002) Postsynaptic induction of BDNF-mediated long-term potentiation. Science **295**: 1729–1734

Landrigan, P.J. and A.C. Todd. (1995) Lead poisoning. West. J. Med. **161**: 153–159

Lanphear, B.P., K. Dietrich, P. Auinger and C. Cox. (2000) Cognitive deficits associated with blood lead levels < 10 μg/dl in US children and adolescents. Public Health Rep. **115**: 521–529

Lanphear, B.P., R. Hornung, M. Ho, C.R. Howard, S. Eberly and K. Knauf. (2002) Environmental lead exposure during early childhood. J. Pediatr. **140**: 40–47

Lasley, S.M. and M.E. Gilbert. (2002) Rat hippocampal glutamate and GABA release exhibit biphasic effects as a function of chronic lead exposure level. Toxicol. Sci. **66**: 139–47

Laterra, J., J.P. Bressler, R.R. Indurti, L. Belloni-Olivi and G.W. Goldstein. (1992) Inhibition of astroglial-induced endothelial differentiation by inorganic lead: A role for protein kinase C. Proc. Natl. Acad. Sci. USA **89**: 10748–10752

Lee, W.T., H. Yoon, D.J. Lee, C.H. Koo and K.A. Park. (2002) Effects of postnatally administered inorganic lead on the tyrosine hydroxylase immunoreactive norepinephrinergic neurons of the locus ceruleus of the rat. Arch. Histol. Cytol. **65**: 45–53

Leggett, R.W. (1993) An age-specific kinetic model of lead metabolism in humans. Environ. Health Perspect. **101**: 598–616

Leung, D.W., G. Cachianes, W.J. Kuang, and N. Ferrara. (1989) Vascular endothelial growth factor is a secreted angiogenic mitogen. Science **246**: 1306–1309

Lidsky, T.I. and J.S. Schneider. (2003) Lead neurotoxicity in children: Basic mechanisms and clinical correlates. Brain. **126**: 5–19

Lindahl, L.S., L. Bird, M.E. Legare, G. Mikeska, G.R. Bratton and E. Tiffany-Castiglioni. (1999) Differential ability of astroglia and neuronal cells to accumulate lead: Dependence on cell type and on degree of differentiation. Toxicol. Sci. **50**: 236–243

Links, J.M., B.S. Schwartz, D. Simon, K. Bandeen-Roche and W.F. Stewart. (2001) The influence of toxicant "residence time" and bioavailability from body stores in estimation of cumulative target organ dose: Application to lead-associated neurocognitive decline. Environ. Health Perspect. **109**: 361–368

Liu, X., K.N. Dietrich, J. Radcliffe, N.B. Ragan, G.G. Rhoads and W.J. Rogan. (2002) Do children with falling blood lead levels have improved cognition? Pediatrics **110**: 787–791

Lovell, M.A., S.P. Gabbita and W.R. Markesbery. (1999) Increased DNA oxidation and decreased levels of repair products in Alzheimer's disease ventricular CSF. J. Neurochem. **72**: 771–776

Lu, T., Y. Pan, S.Y. Kao, C. Li, I. Kohane, J. Chan and B. Yankner. (2004) Gene regulation and DNA damage in the ageing human brain. Nature **429**: 883–891

Malinow, R. and R.C. Malenka. (2002) AMPA receptor trafficking and synaptic plasticity. Annu. Rev. Neurosci. **25**: 103–26

Mani, J., N. Chaudhary, M. Kanjalkar and P.U. Shah. (1998) Cerebellar ataxia due to lead encephalopathy in an adult. J. Neurol. Neurosurg. Psychiatr. **65**: 797–798

Markovac, J. and G.W. Goldstein. (1988a) Lead activates protein kinase C in immature rat brain microvessels. Toxicol. Appl. Pharmacol. **96**: 14–23

Markovac, J. and G.W. Goldstein. (1988b) Picomolar concentrations of lead stimulate brain protein kinase C. Nature **334**: 71–73

Markowitz, M. (2000) Lead Poisoning. Pediatr. Rev. **21**: 327–35

McCall, R.B. (1983) A conceptual approach to early mental development. **In**: Origins of Intelligence. M. Lewis (editor). Plenum Press, New York, pp. 107–133

McDonald, J.W. and M.V. Johnston. (1990) Physiological and pathophysiological roles of excitatory amino acids during central nervous system development. Brain Res. Rev. **15**: 41–70

Mendola, P., S.G. Selevan, S. Gutter and D. Rice. (2002) Environmental factors associated with a spectrum of neurodevelopmental deficits. Ment. Retard. Dev. Disabil. **8**: 188–197

Minnema, D.J., R.D. Greenland and I.A. Michaelson. (1986) Effect of in vitro inorganic lead on dopamine release from superfused rat striatal synaptosomes. Toxicol. Appl. Pharmacol. **84**: 400–411

Needleman, H.L. and C.A. Gatsonis. (1990) Low-level lead exposure and the IQ of children: A meta-analysis of modern studies. JAMA **263**: 673–678

Needleman, H.L., C. Gunnoe, A. Leviton, R. Reed, H. Peresie, C. Maher and P. Barrett. (1979) Deficits in psychologic and classroom performance of children with elevated dentine lead levels. N. Engl. J. Med. **300**: 689–695

Needleman, H.L., A. Schell, D. Bellinger, A. Leviton and E.N. Allred. (1990) The long-term effects of exposure to low doses of lead in childhood: An 11-year follow-up report. N. Engl. J. Med. **322**: 83–88

Needleman, H.L. J.A. Riess, M.J. Tobin, G.E. Biesecker and J.B. Greenhouse. (1996) Bone lead levels and delinquent behavior. JAMA **275**: 363–369

Needleman, H.L., C. McFarland, R.B. Ness, S.E. Fienberg and M.J. Tobin. (2002) Bone lead levels in adjudicated delinquents. A case control study. Neurotoxicol. Teratol. **24**: 711–717

Nicchia, G.P., A. Frigeri, G.M. Liuzzi, M.P. Santacroce, B. Nico, G. Procino, F. Quondamatteo, R. Herken, L. Roncali and M. Svelto. (2000) Aquaporin-4-containing astrocytes sustain a temperature- and mercury-insensitive swelling in vitro. Glia **31**: 29–38

Nielsen, S., E.A. Nagelhus, M. Amiry-Moghaddam, C. Bourque, P. Agre and O.P. Ottersen. (1997) Specialized membrane domains for water transport in glial cells: High-resolution immunogold cytochemistry of aquaporin-4 in rat brain. J. Neurosci. **17**: 171–180

Niemietz, C.M. and S.D. Tyerman. (2002) New potent inhibitors of aquaporins: Silver and gold compounds inhibit aquaporins of plant and human origin. FEBS Lett. **531**: 443–447

Ong, W.Y. and A.A. Farooqui. (2005) Iron, neuroinflammation, and Alzheimer's disease. J. Alzheimer's Dis. **8**: 183–200

Ong, W.Y., M.Q. Ren, J. Makjanic, T.M. Lim and F. Watt. (1999) A nuclear microscopic study of elemental changes in the rat hippocampus after kainate-induced neuronal injury. J. Neurochem. **72**: 1574–1579

Ong, W.Y., X. He, L.H. Chua and C.N. Ong. (2006) Increased uptake of divalent metals lead and cadmium into the brain after kainite-induced neuronal injury. Exp. Brain Res. **173**: 468–474

Papanikolaou, N.C., E.G. Hatzidaki, S. Belivanis, G.N. Tzanakakis, A.M. Tsatsakis. (2005) Lead toxicity update: A brief review. Med. Sci. Monit. **11(10)**: RA329–RA336

Peng, S., R.K. Hajela and W.D. Atchison. (2002) Characteristics of block by Pb^{2+} of function of human neuronal L-, N-, and R-type Ca^{2+} channels transiently expressed in human embryonic kidney cells. Mol. Pharmacol. **62**: 1418–1430

Penn, A.A. and C.J. Shatz. (1999) Brain waves and brain wiring: The role of endogenous and sensory driven neural activity in development. Pediatr. Res. **45**: 447–458

Pentschew, A. and F. Garro. (1966) Lead encephalomyelopathy of the suckling rat and its implications on the porphyrinopathic nervous diseases: With special reference to the permeability disorders of the nervous system capillaries. Acta Neuropathol. **6**: 266–278

Perelman, S., L. Hertz-Pannier, M. Hassan, and A. Bourrillon. (1993) Lead encephalopathy mimicking a cerebellar tumor. Acta Paediatr. **82**: 423–425

Perlstein, M.A. and R. Attala. (1966) Neurologic sequelae of plumbism in children. Clin. Pediatr. **5**: 292–298

Philip, A.T. and B. Gerson. (1994) Lead poisoning — Part I. Clin. Lab. Med. **14**: 423–44

Pocock, S.J., M. Smith and P. Baghurst. (1994) Environmental lead and children's intelligence: A systematic review of the epidemiological evidence. Br. Med. J. **309**: 1189–1197

Press, M.F. (1977) Lead encephalopathy in neonatal long-evans rats: Morphologic studies. J. Neuropathol. Exp. Neurol. **36**: 169–193

Preston, G.M., T.P. Carroll, W.P. Guggino and P. Agre. (1992) Appearance of water channels in Xenopus oocytes expressing red cell CHIP28 protein. Science **256**: 385–387

Raff, M.C., B.A. Barres, J.F. Burne, H.S. Coles, Y. Ishizaki, and M.D. Jacobson. (1993) Programmed cell death and the control of cell survival: Lessons from the nervous system. Science **262**: 695–700

Regan, C.M. (1989) Lead-impaired neurodevelopment: Mechanisms and threshold values in the rodent. Neurotoxicol. Teratol. **11**: 533–537

Regan, C.M. (1991) Neural cell adhesion molecules: Neuronal development, and lead toxicity. **In**: Proceedings, Ninth International Neurotoxicology Conference, J. Cranmer (editor), Little Rock

Regan, C.M. (1993) Neural cell adhesion molecules, neuronal development and lead toxicity. Neurotoxicology **14**: 69–76

Reyes, P.F., C.F. Gonzalez, M.K. Zalewska and A. Besarab. (1986) Intracranial calcification in adults with chronic lead exposure. AJR Am. J. Roentgenol. **146**: 267–270

Rice, D. (1989) Delayed neurotoxicity in monkeys exposed developmentally to methyl mercury. Neurotoxicology. **10**: 645–650

Rice, D.C. (1993) Lead-induced changes in learning: Evidence for behavioral mechanisms from experimental animal studies. Neurotoxicology **14**: 167–178

Rogan, W.J., K.N. Dietrich, J.H. Ware, D.W. Dockery, M. Salganik, J. Radcliffe, R.L. Jones, N.B. Ragan, J.J. Chisolm and G.G. Rhoads. (2001) The effect of chelation therapy with succimer on neuropsychological development in children exposed to lead. N. Engl. J. Med. **344**: 1421–1426

Roger, S.D., D. Crimmins, C. Yiannikas and D.C. Harris. (1990) Lead intoxication in an anuric patient: Management by intraperitoneal EDTA. Aust. N. Z. J. Med. **20**: 814–817

Rowland, A.S. and R.C. McKinstry. (2006) Lead toxicity, white matter lesions, and aging. Neurology **66**: 1464–1465

Ruff, H.A., P.E. Bijur, M. Markowitz, Y.C. Ma and J.F. Rosen. (1993) Declining blood lead levels and cognitive changes in moderately lead-poisoned children. JAMA **269**: 1641–1646

Ruff, H.A., M.E. Markowitz, P.E. Bijur and J. Rosen. (1996) Relationships among blood lead levels, iron deficiency and cognitive development in two-year-old children. Environ. Health Perspect. **104(2)**: 180–185

Sanchez-Ramos, J., E. Overvik and B. Ames. (1994) A marker of oxyradical-mediated DNA damage (8-hydroxy-2'-deoxyguanosine) is increased in nigrostriatum of Parkinson's disease brain. Neurodegen. **3**: 197–204

Saper, R.B., S.N. Kales, J. Paquin, M.J. Burns, D.M. Eisenberg, R.B. Davis and R.S. Phillips. (2004) Heavy metal content of ayurvedic herbal medicine products. JAMA **292**: 2868–2873

Saryan, L.A. and C. Zenz. (1994) Lead and its compounds. **In**: Occupational Medicine (third edition). L.A. Saryan and C. Zenz (editors), Mosby, St. Louis, pp. 506–541

Schwartz, J. (1994) Low-level lead exposure and children's IQ: A meta-analysis and search for a threshold. Environ. Res. **65**: 42–55

Schwartz, B.S., W.F. Stewart, K.I. Bolla, D. Simon, K. Bandeen-Roche, B. Gordon, J.M. Links, A.C. Todd, W. Shi, S. Bassett and P. Youssem. (2000) Past adult lead exposure is associated with longitudinal decline in cognitive function. Neurology **55**: 1144–1150

Sciarillo, W.G., G. Alexander and K.P. Farrell. (1992) Lead exposure and child behavior. Am. J. Public Health **82**: 1356–1360

Sheng, M. and M.J. Kim. (2002) Postsynaptic signaling and plasticity mechanisms. Science **298**: 776–80

Silbergeld, E.K. (1992) Mechanisms of lead neurotoxicity, or looking beyond the lamppost. FASEB J. **6**: 3201–3206

Stewart, W.F., B.S. Schwartz, D. Simon, K.I. Bolla, A.C. Todd and J. Links. (1999) The relation between neurobehavioral function and tibial and chelatable lead levels in former organolead manufacturing workers. Neurology **52**: 1610–1617

Stewart, W.F., B.S. Schwartz, C. Davatzikos, D. Shen, D. Liu, X. Wu, A.C. Todd, W. Shi, S. Bassett and D. Youssem. (2006) Past adult lead exposure is linked to neurodegeneration measured by brain MRI. Neurology **66**: 1476–1484

Sweatt, J.D. (2001) The neuronal MAP kinase cascade: A biochemical signal integration system subserving synaptic plasticity and memory. J. Neurochem. **76**: 1–10

Teo, J.G.C., K.Y.C. Goh, A. Ahuja, H.K. Ng and W.S. Poon. (1997) Intracranial vascular calcifications, glioblastoma multiforme, and lead poisoning. AJNR Am. J. Neuroradiol. **18**: 576–579

Tiffany-Castiglioni, E., E.M. Sierra, J.-N. Wu and T.K. Rowles. (1989) Lead toxicity in neuroglia. Neurotoxicology **10**: 417–443

Toews, A.D., A. Kolber, J. Hayward, M.R. Krigman and P. Morell. (1978) Experimental lead encephalopathy in the suckling rat: Concentration of lead in cellular fractions enriched in brain capillaries. Brain Res. **147**: 131–138

Tong, S., P. Baghurst, A. McMichael, M. Sawyer and J. Mudge. (1996) Lifetime exposure to environmental lead and children's intelligence at 11–13 years: The Port Pirie cohort study. Br. Med. J. **312**: 1569–1575

Tong, S., P.A. Baghurst, M.G. Sawyer, J. Burns and A.J. McMichael. (1998) Declining blood lead levels and changes in cognitive function during childhood: The Port Pirie cohort study. JAMA **280**: 1915–1919

Toscano, C.D., H. Hashemazadeh-Gargari, J.L. McGlothan and T.R. Guilarte. (2002) Developmental $Pb^{(2+)}$ exposure alters NMDA subtypes and reduces CREB phosphorylation in the rat brain. Dev. Brain Res. **139**: 217–226

Tsai, Y.-T., C.-C. Huang, H.-C. Kuo, H.-M. Wang, W.-S. Shen, T.-S. Shih and N.-S. Chu. (2006) Central nervous system effects in acute thallium poisoning. Neurotoxicol. **27**: 291–295

Valpey, R., M. Sumi, M. Copass and G. Goble. (1978) Acute and chronic progressive encephalopathy due to gasoline sniffing. Neurology **28**: 507–510

Wang, W., M.J. Merrill and R.T. Borchardt. (1996) Vascular endothelial growth factor affects permeability of brain microvessel endothelial cells in vitro. Am. J. Physiol. **271**: C1973–C1980

Wang, X.S., W.Y. Ong and J.R. Connor. (2001) A light and electron microscopic study of the iron transporter protein DMT-1 in the monkey cerebral neocortex and hippocampus. J. Neurocytol. **30**: 353–360

Wang, X.S., W.Y. Ong and J.R. Connor. (2002a) A light and electron microscopic study of divalent metal transporter-1 distribution in the rat hippocampus, after kainate-induced neuronal injury. Exp. Neurol. **177**: 193–201

Wang, X.S., W.Y. Ong and J.R. Connor. (2002b) Increase in ferric and ferrous iron in the rat hippocampus with time after kainate-induced excitotoxic injury. Exp. Brain Res. **143**: 137–148

Wasserman, G.A., X. Liu, N.J. Lolacono, P. Factor-Litvak, J.K. Kline, D. Popovac, N. Morina, A. Musabegovic, N. Vrenezi, S. Capuni-Paracka, V. Lekic, E. Preteni-Redjepi, S. Hadzialjevic, V. Slavkovich and J.H. Graziano. (1997) Lead exposure and intelligence in 7-year-old children: The Yugoslavia prospective study. Environ. Health Perspect. **105**: 956–962

Wasserman, G.A., X. Liu, D. Popovac, P. Factor-Litvak, J. Kline, C. Waternaux, N. Lolacono and J.H. Graziano. (2000) The Yugoslavia prospective lead study: Contributions of prenatal and postnatal lead exposure to early intelligence. Neurotoxicol. Teratol. **22**: 811–818

Weiss, B. (1991) Cancer and the dynamics of neurodegenerative processes. Neurotoxicol. **12**: 379–386

Wilson, M.A., M.V. Johnston, G.W. Goldstein and M.E. Blue. (2000) Neonatal lead exposure impairs development of rodent barrel field cortex. Proc. Natl. Acad. Sci. USA **97**: 5540–5545

WHO. (1995) Environmental health criteria 165, inorganic lead. World Health Organization International Programme for Chemical Safety, Geneva, Switzerland

Yasui, M., T.H. Kwon, M.A. Knepper, S. Nielsen and P. Agre. (1999) Aquaporin-6: An intracellular vesicle water channel protein in renal epithelia. Proc. Natl. Acad. Sci. USA **96**: 5808–5813

Yule, W., R. Lansdown, I.B. Millar and M.A. Urbanowicz. (1981) The relationship between blood lead concentrations, intelligence and attainment in a school population: A pilot study. Dev. Med. Child Neurol. **23**: 567–576

Zawia, N.H. (2003) Transcriptional involvement in neurotoxicity. Toxicol. Appl. Pharmacol. **190**: 177–188

Zelenina, M., A.A. Bondar, S. Zelenin and A. Aperia. (2003) Nickel and extracellular acidification inhibit the water permeability of human Aquaporin-3 in lung epithelial cells. J. Biol. Chem. **278**: 30037–30043

Zelenina, M., S. Tritto, A.A. Bondar, S. Zelenin and A. Aperia. (2004) Copper inhibits the water and glycerol permeability of Aquaporin-3. J. Biol. Chem. **279**: 51939–51943

Ziegler, E.E., B.B. Edwards, R.I. Jensen, K.R. Mahaffey and S.J. Fomon. (1978) Absorption and retention of lead by infants. Pediatr. Res. **12**: 29–34

ERRATUM

Retracted: Brain Damage in Phenylalanine, Homocysteine and Galactose Metabolic Disorders

Kleopatra H. Schulpis and Stylianos Tsakiris

D.W. McCandless (ed.) *Metabolic Encephalopathy*,
DOI: 10.1007/978-0-387-79112-8, pp. 393–458, © Springer Science+Business Media, LLC 2009

DOI 10.1007/978-0-387-79112-8_25

This chapter is being retracted as significant portions of the chapter are identical to "An Updated Review of the Long-Term Neurological Effects of Galactosemia", Ridel et al., *Pediatric Neurology*, 2005 (doi:10.1016/j.pediatrneurol.2005.02.015), which was not cited by the chapter authors.

The online version of the original chapter can be found at
http://dx.doi.org/10.1007/978-0-387-79112-8_20

Index

Color Plates

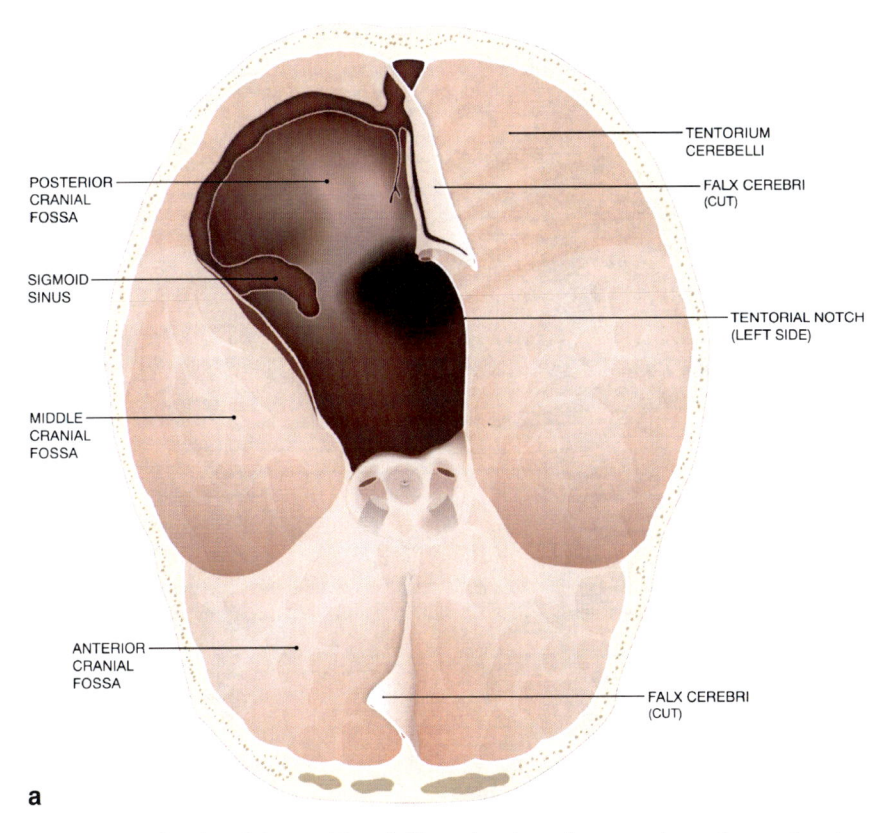

POSTERIOR CRANIAL FOSSA

SIGMOID SINUS

MIDDLE CRANIAL FOSSA

ANTERIOR CRANIAL FOSSA

TENTORIUM CEREBELLI

FALX CEREBRI (CUT)

TENTORIAL NOTCH (LEFT SIDE)

FALX CEREBRI (CUT)

a

Figure 1.1 (a) Interior of the cranial vault illustrating the major supporting and protective structures, including the cranial fossae. Reflections of dura mater off the calvaria form important structures that help support the weight and restrict the motion of the brain within the cranial vault.

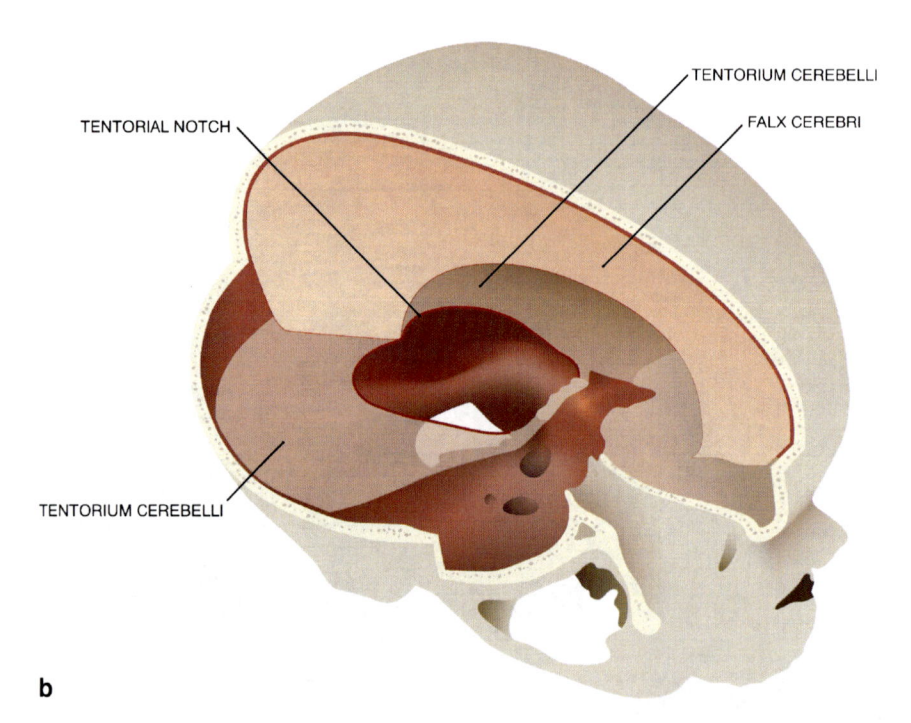

TENTORIAL NOTCH

TENTORIUM CEREBELLI

FALX CEREBRI

TENTORIUM CEREBELLI

b

Figure 1.1 (b) Vertical reflections are falces and include the large falx cerebri between the two cerebral hemispheres and the smaller falx cerebelli (not shown) that separates the cerebellar hemispheres. The most significant horizontal reflection is the tentorium cerebelli which supports the occipital lobe of the brain and prevents it from crushing the underlying cerebellum, which resides in the posterior cranial fossa (redrawn after Drake et al., 2005).

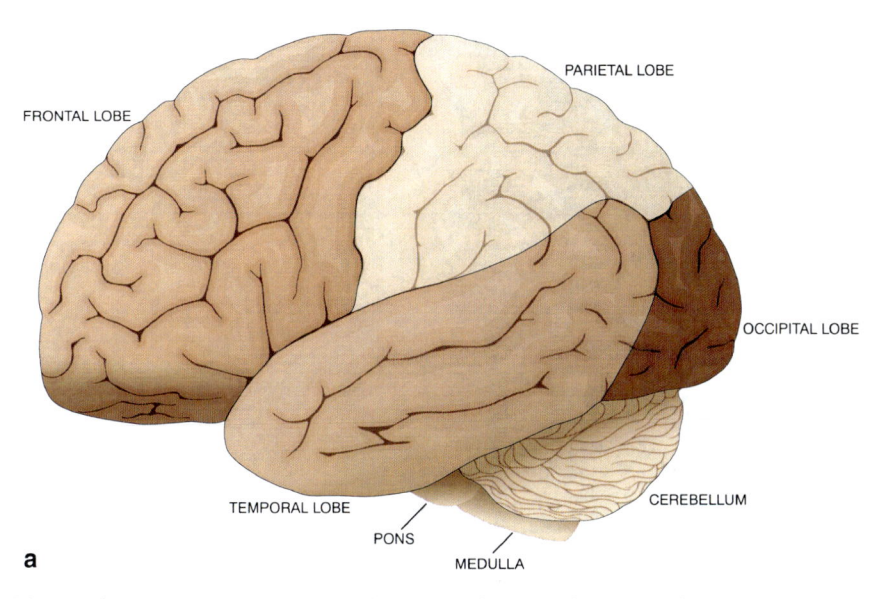

FRONTAL LOBE

PARIETAL LOBE

OCCIPITAL LOBE

CEREBELLUM

TEMPORAL LOBE

PONS

MEDULLA

a

Figure 1.2 Diagrams illustrate the location and relationships of the lobes of the cerebrum and the segments of the brainstem. (a) Lateral view of the gross brain. Note the frontal, parietal, temporal and occipital lobes. The cerebellum is the small region inferior to the caudal region of the cerebrum.

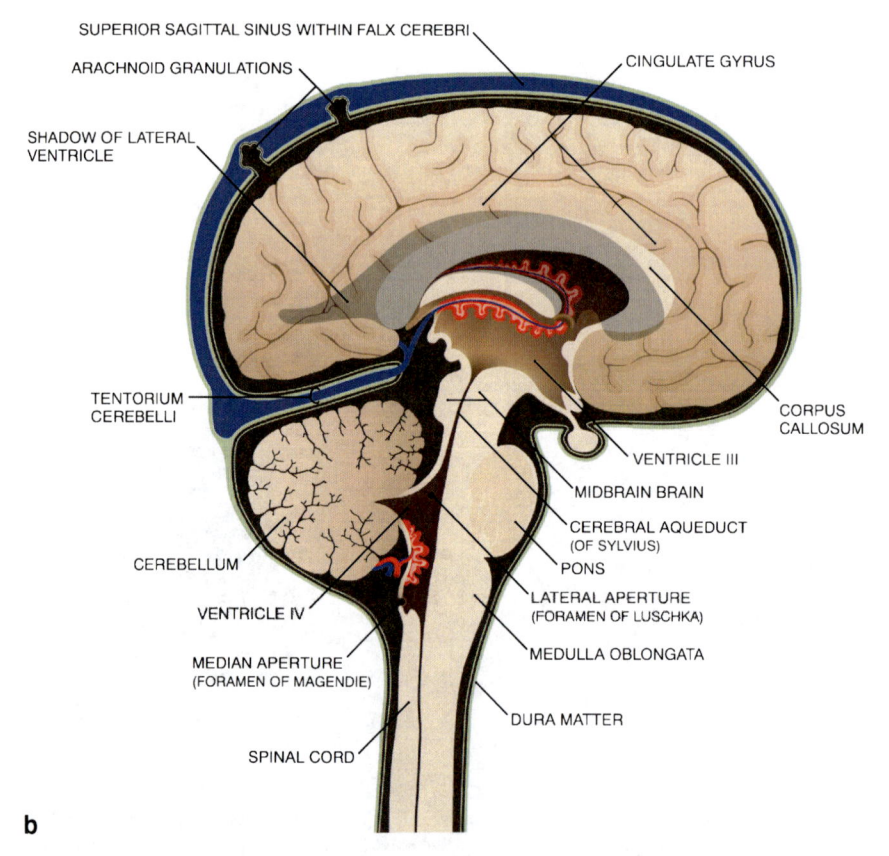

SUPERIOR SAGITTAL SINUS WITHIN FALX CEREBRI

ARACHNOID GRANULATIONS

CINGULATE GYRUS

SHADOW OF LATERAL VENTRICLE

TENTORIUM CEREBELLI

CORPUS CALLOSUM

VENTRICLE III

MIDBRAIN BRAIN

CEREBRAL AQUEDUCT (OF SYLVIUS)

CEREBELLUM

PONS

VENTRICLE IV

LATERAL APERTURE (FORAMEN OF LUSCHKA)

MEDIAN APERTURE (FORAMEN OF MAGENDIE)

MEDULLA OBLONGATA

DURA MATTER

SPINAL CORD

b

Figure 1.2 (b) Sagittal view that allows one to appreciate the position of the cingulate gyrus (a major constituent of the limbic lobe, which surrounds the corpus callosum and thalamus), as well as the segments of the brainstem: midbrain, pons and medulla. The ventricular system of the brain is also demonstrated. Each portion of the brain has a cerebrospinal fluid-filled cavity at its core, and these cavities are all connected. Cerebrospinal fluid is produced by the choroid plexus within the ventricles and it escapes the ventricular system through foramina in the roof over the fourth ventricle at the pontomedullary junction.

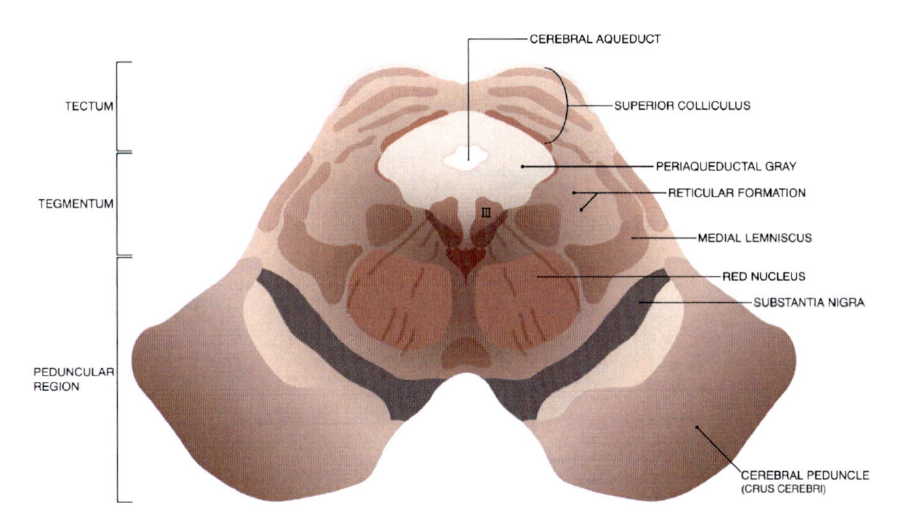

CEREBRAL AQUEDUCT

SUPERIOR COLLICULUS

PERIAQUEDUCTAL GRAY

RETICULAR FORMATION

MEDIAL LEMNISCUS

RED NUCLEUS

SUBSTANTIA NIGRA

CEREBRAL PEDUNCLE
(CRUS CEREBRI)

TECTUM

TEGMENTUM

PEDUNCULAR
REGION

III

Figure 1.3 A cross-section through the midbrain that illustrates the tectum, tegmentum and peduncular regions. Within the wall of the midbrain, note the positions of the periaqueductal gray matter, red nucleus, medial lemniscus, substantia nigra, and territory occupied by the reticular formation. (III = nucleus of cranial nerve III, the oculomotor nerve)

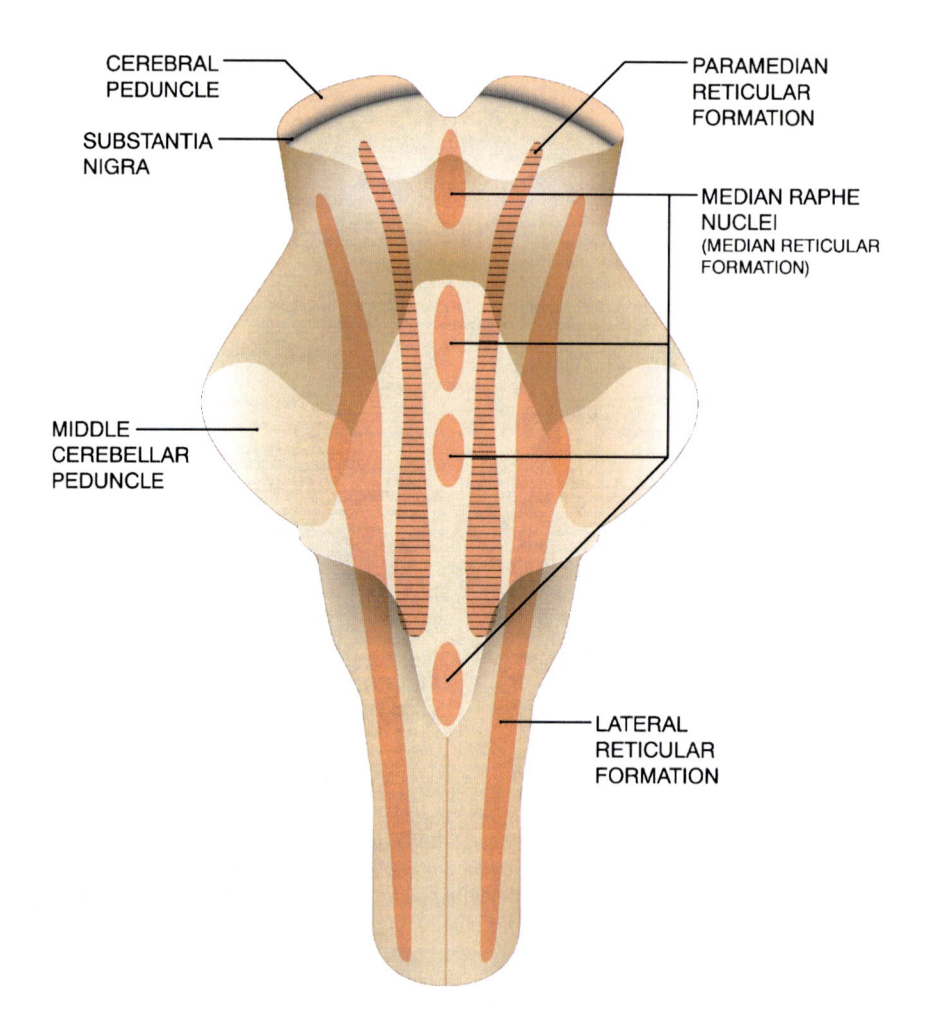

Figure 1.4 Diagram of a dorsal view of the brainstem; the cerebellum has been removed. The extent of the reticular formation within the brainstem is illustrated. The reticular formation is a polysynaptic network that consists of three regions: a series of midline raphe nuclei (the median reticular formation, which is the site of origin of the major serotonergic pathways in the nervous system); this is flanked bilaterally by the paramedian reticular formation (an efferent system of magnocellular neurons with ascending and descending projections); and farthest from the midline, the lateral reticular formation, consisting of parvocellular neurons that project transversely.

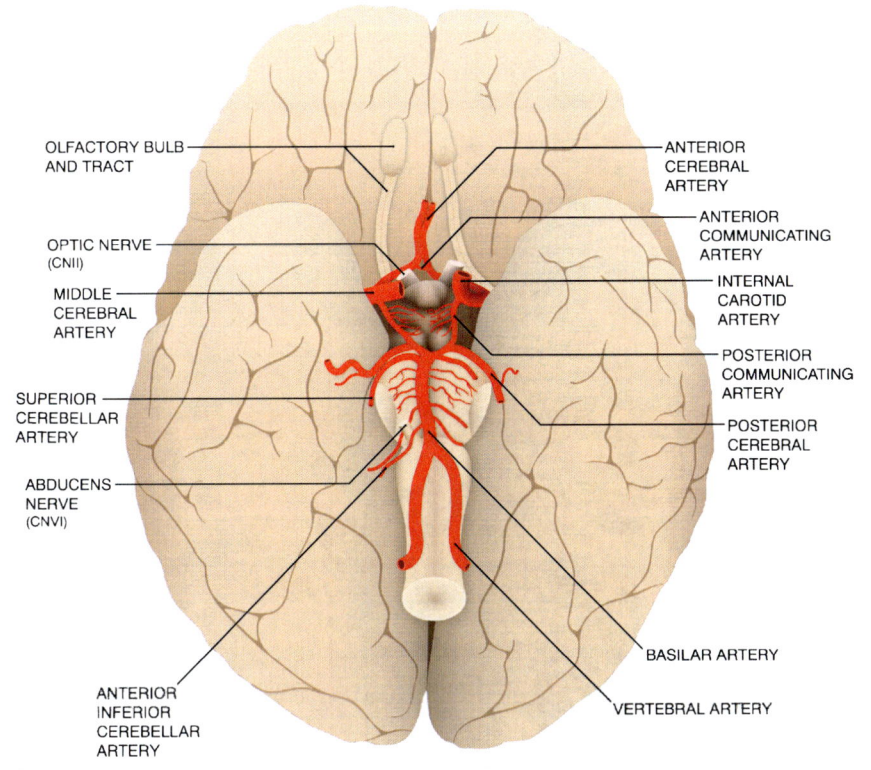

OLFACTORY BULB AND TRACT

OPTIC NERVE (CNII)

MIDDLE CEREBRAL ARTERY

SUPERIOR CEREBELLAR ARTERY

ABDUCENS NERVE (CNVI)

ANTERIOR CEREBRAL ARTERY

ANTERIOR COMMUNICATING ARTERY

INTERNAL CAROTID ARTERY

POSTERIOR COMMUNICATING ARTERY

POSTERIOR CEREBRAL ARTERY

BASILAR ARTERY

VERTEBRAL ARTERY

ANTERIOR INFERIOR CEREBELLAR ARTERY

a

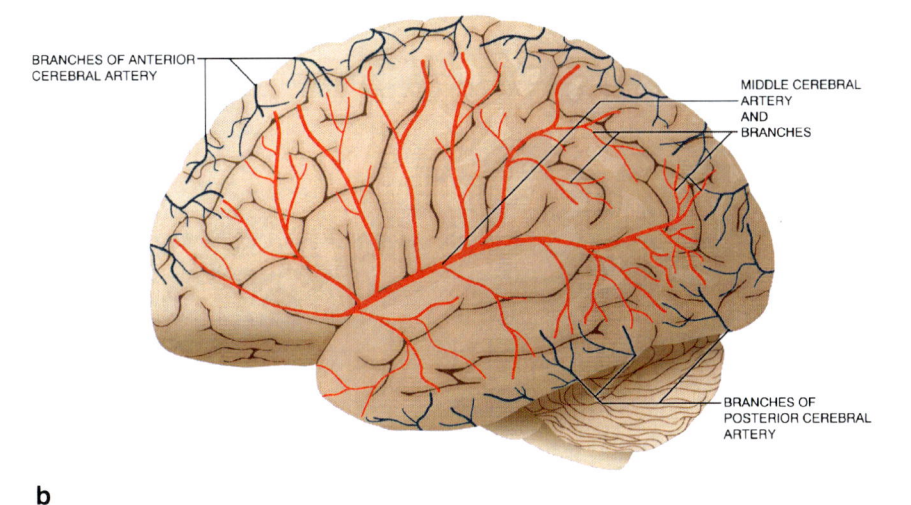

BRANCHES OF ANTERIOR CEREBRAL ARTERY

MIDDLE CEREBRAL ARTERY AND BRANCHES

BRANCHES OF POSTERIOR CEREBRAL ARTERY

b

Figure 1.5   Blood supply to the brain. (a) Vessels that contribute to the arterial circle of Willis at the base of the brain. Note contributions from the vertebral and internal carotid systems. Throughout its length, each vessel that participates in the arterial circle gives off numerous small,

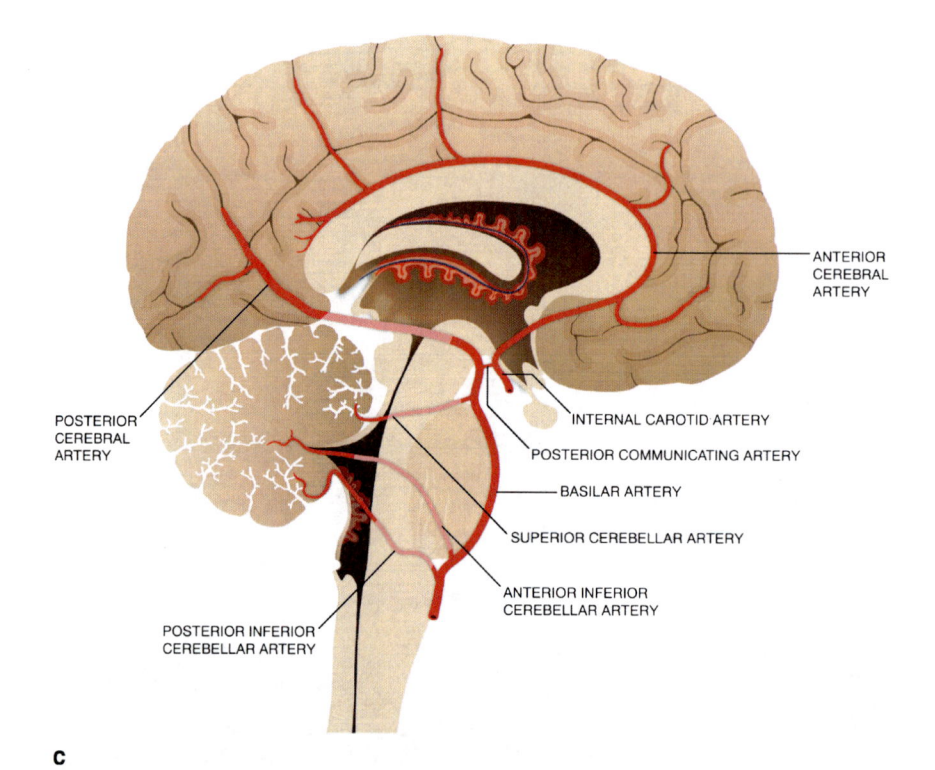

c

Figure 1.5 (continued) unnamed branches that penetrate the brainstem. (b) Distribution of blood supply to the lateral surface of the cerebrum is illustrated. The middle cerebral artery is the prominent vessel, the anterior cerebral artery vascularizes the territory on either side of the falx cerebri and a narrow strip of superior surface of the cerebrum. (c) Sagittal section that depicts the distribution of blood flow to the cerebrum. Note that the anterior and middle cerebral arteries carry blood from the internal carotid arteries, whereas the blood to the posterior cerebral arteries comes from the vertebral/basilar artery system.

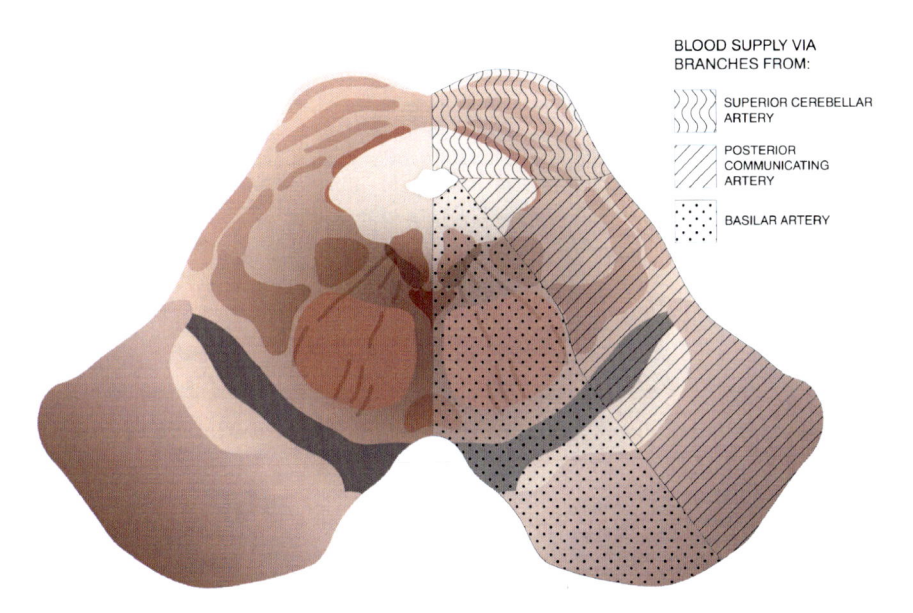

BLOOD SUPPLY VIA
BRANCHES FROM:

SUPERIOR CEREBELLAR
ARTERY

POSTERIOR
COMMUNICATING
ARTERY

BASILAR ARTERY

Figure 1.6 Cross section illustrating the distribution of blood to the walls of the midbrain. The tectum is supplied by branches of the superior cerebellar artery. The medial aspects of the peduncular and tegmental regions are vascularized by branches of the basilar artery. The lateral peduncular and tegmental regions are supplied by branches of the posterior communicating artery. Note that these small arteries, which enter the walls of the central nervous system from the periphery, are functional end arteries and do not anastomose with adjacent arteries.

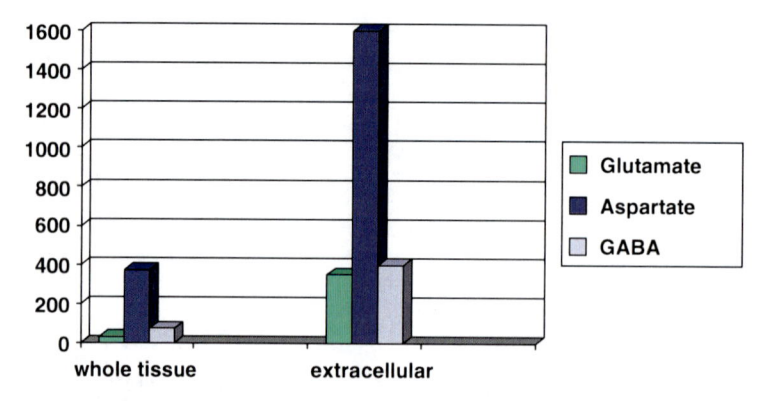

Figure 3.2 – Hypoglycemia causes an increase in tissue aspartate and decrease in glutamate, while both amino acids flood the extracellular space of the brain. GABA similarly floods the extracellular space, but its inhibitory effects are often insufficient to prevent hypoglycemic convulsions in the face of the excitatory amino acids released. Data from Norberg & Siesjö 1976 (28)for whole tissue and extracellular data from Sandberg et al 1986. (32)

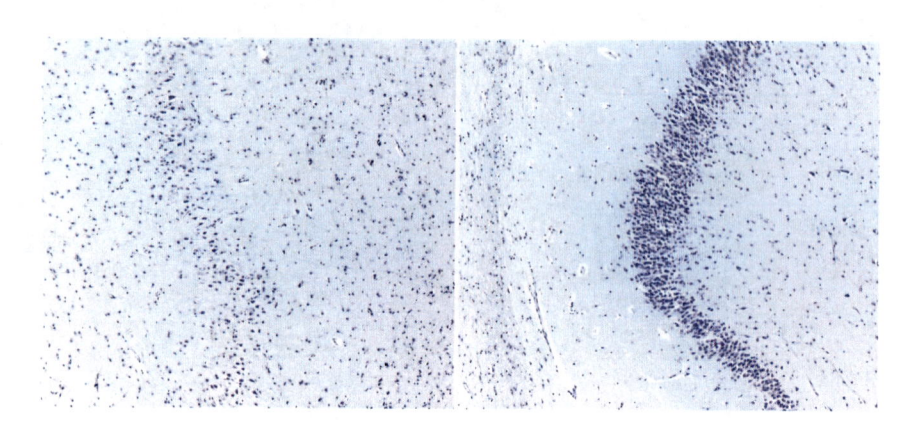

Figure 3.4 Left and right hippocampus in human hypoglycemia to show asymmetry of brain damage. The dentate granule cells are depleted on the left, but the normal band of well-populated dentate granule cells is seen on the right. Cresyl violet stain.

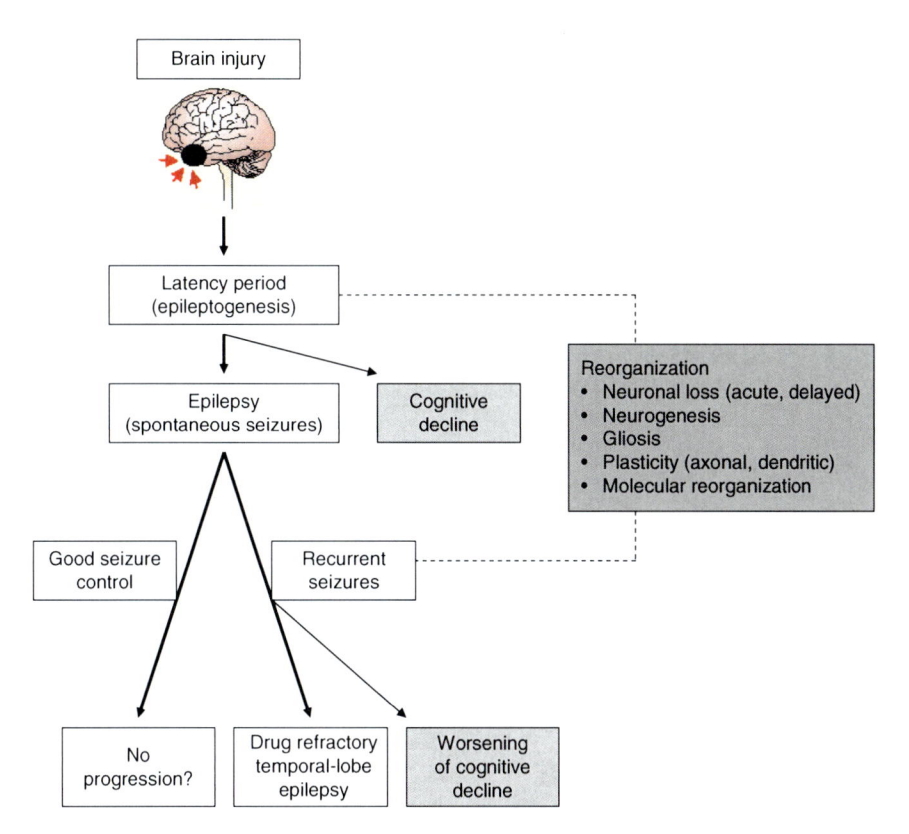

Fig. 7.1 Epileptic process in symptomatic temporal-lobe epilepsy. Adapted from Pitkanen, A., and Sutula, T.P. 2002. Is epilepsy a progressive disorder? Prospects for new therapeutic approaches in temporal-lobe epilepsy. The Lancet Neurology 1:173, with permission from Elsevier

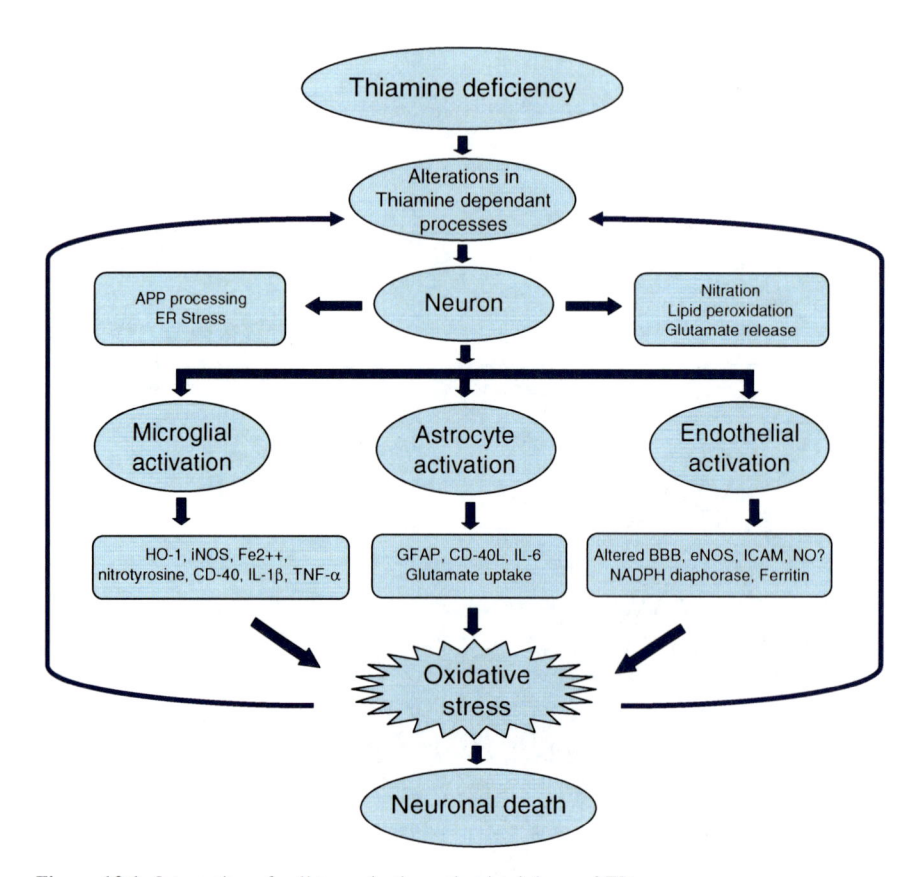

Figure 12.1 Interaction of cell types in the pathophysiology of TD

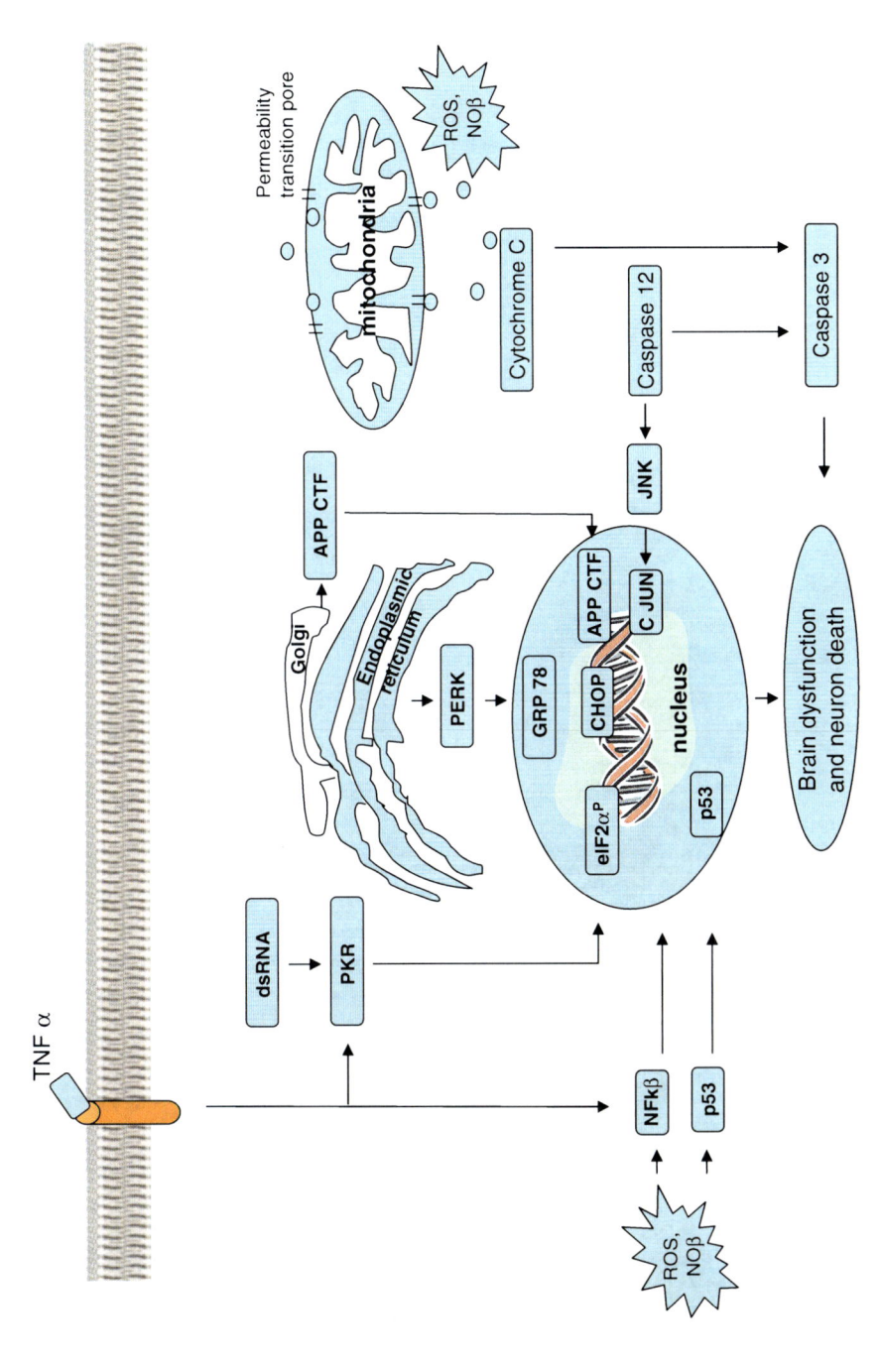

Figure 12.2 Interactions of protein processing and nuclear translocation in the pathophysiology of TD.

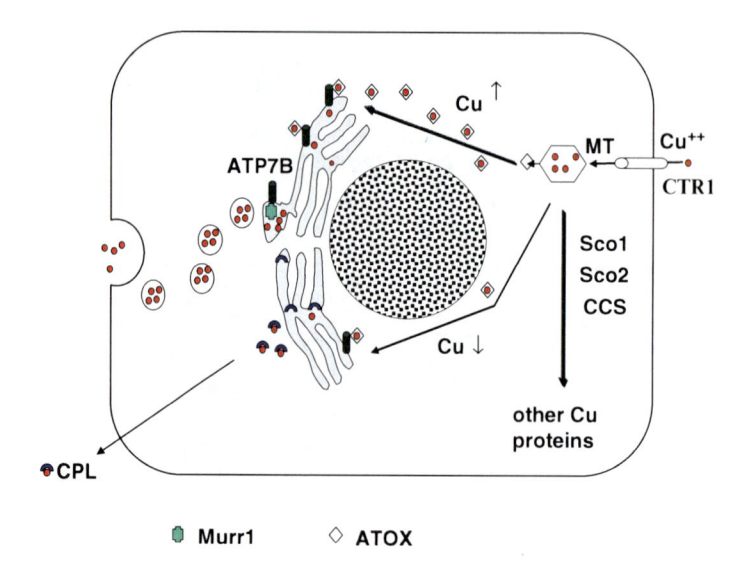

Figure 21.1 Model of hepatobiliary copper transport
CTR1= copper transporter 1, MT= Metallothionein, CPL= ceruloplasmin
ATOX, Sco1, Sco2, CCS – copper chaperones

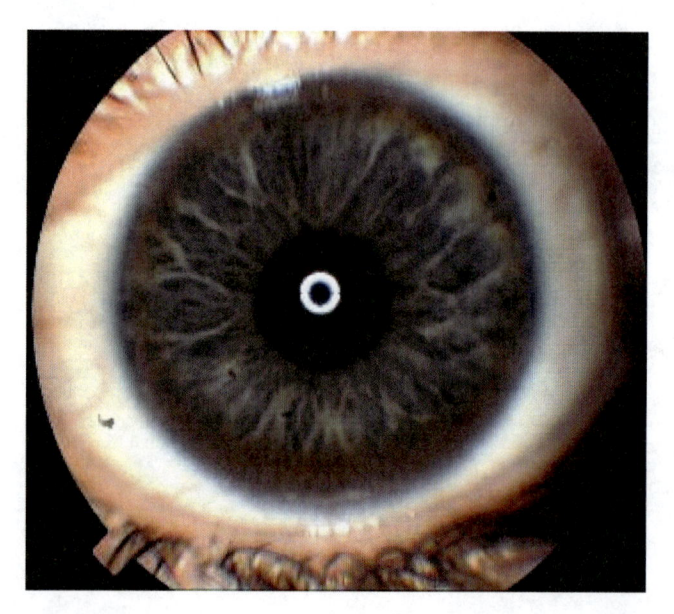

Figure 21.2 Kayser-Fleischer ring in a 15 year old patient with neurologic Wilson disease.

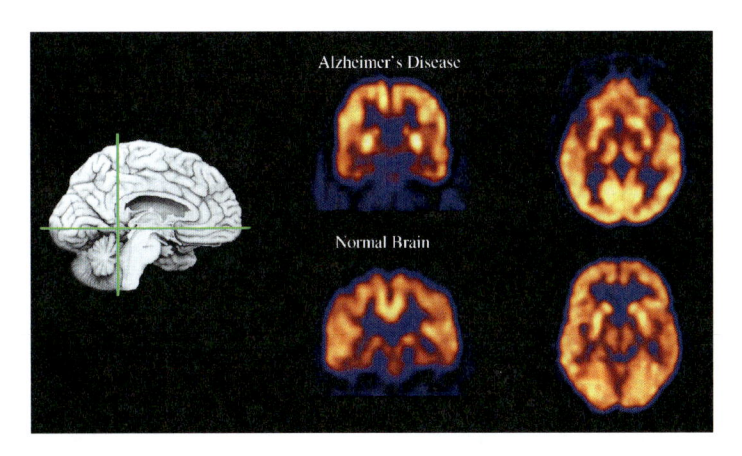

Figure 22.1 PET image of AD and normal brain (coronal and transverse sections through the hippocampus).